Unsupervised and Semi-Supervised Learning

Series Editor

M. Emre Celebi, Computer Science Department, Conway, AR, USA

Springer's Unsupervised and Semi-Supervised Learning book series covers the latest theoretical and practical developments in unsupervised and semi-supervised learning. Titles – including monographs, contributed works, professional books, and textbooks – tackle various issues surrounding the proliferation of massive amounts of unlabeled data in many application domains and how unsupervised learning algorithms can automatically discover interesting and useful patterns in such data. The books discuss how these algorithms have found numerous applications including pattern recognition, market basket analysis, web mining, social network analysis, information retrieval, recommender systems, market research, intrusion detection, and fraud detection. Books also discuss semi-supervised algorithms, which can make use of both labeled and unlabeled data and can be useful in application domains where unlabeled data is abundant, yet it is possible to obtain a small amount of labeled data.

Topics of interest in include:

- Unsupervised/Semi-Supervised Discretization
- Unsupervised/Semi-Supervised Feature Extraction
- Unsupervised/Semi-Supervised Feature Selection
- Association Rule Learning
- Semi-Supervised Classification
- Semi-Supervised Regression
- Unsupervised/Semi-Supervised Clustering
- Unsupervised/Semi-Supervised Anomaly/Novelty/Outlier Detection
- Evaluation of Unsupervised/Semi-Supervised Learning Algorithms
- Applications of Unsupervised/Semi-Supervised Learning

While the series focuses on unsupervised and semi-supervised learning, outstanding contributions in the field of supervised learning will also be considered. The intended audience includes students, researchers, and practitioners.

** Indexing: The books of this series indexed in zbMATH **

Y-h. Taguchi

Unsupervised Feature Extraction Applied to Bioinformatics

A PCA Based and TD Based Approach

Second Edition

Springer

Y-h. Taguchi
Department of Physics
Chuo University
Tokyo, Japan

ISSN 2522-848X ISSN 2522-8498 (electronic)
Unsupervised and Semi-Supervised Learning
ISBN 978-3-031-60981-7 ISBN 978-3-031-60982-4 (eBook)
https://doi.org/10.1007/978-3-031-60982-4

This Springer imprint is published by the registered company Springer Nature Switzerland AG
The registered company address is: Gewerbestrasse 11, 6330 Cham, Switzerland

To all the scientists who have ever written at least one peer-reviewed paper. . . .

Foreword

Machine learning techniques serve as powerful tools in bioinformatics, specifically for predicting the structure and function of proteins, identifying disease causing mutations, biomarkers, potential drug-like molecules, and so on. However, it is not straightforward to relate the features with performance. On the other hand, a simple statistical analysis can provide insights to understand the relationship; for example, increase in long-range contacts slows down the folding of proteins, positive charged residues tend to dominate in DNA binding domains, etc. Hence, linear algebra has the capability to reveal complicated genomic structures in a more direct manner than machine learning.

Almost 10 years ago, Prof. Taguchi and I published a paper on predicting protein folding types using principal component analysis (PCA), one of the liner algebra methods. He has continued his research to investigate the applications of PCA on various biological problems. Recently, he successfully moved to tensors. These methods provide insights to understand the concepts due to the fact that the data are easily interpreted, and "trace back" the output from input features. It is amazing that such a simple strategy can be applied to a wide range of biological problems discussed in this book.

Prof. Taguchi has elegantly designed the book to understand the concepts easily. He has provided mathematical foundations on all important aspects followed by feature extractions. At the end of the book, he shows that PCA and tensors are powerful tools, which perform similar to machine learning techniques in the study of biological problems, namely, biomarker identification, gene expression, and drug discovery, evidenced with his numerous high-quality publications in reputed international journals.

In essence, this book is a valuable resource for students, research scholars and faculty members to simultaneously grasp the fundamentals and applications of PCA and tensors. Although the applications listed in this book are limited to bioinformatics, the approach is extendable to other fields as well since they are general linear methods, which are easily understandable.

With these appreciations, I recommend this well-written book to the readers.

Chennai, India
25 March, 2019

M. Michael Gromiha

Preface to the Second Edition

Magic is most fun when you're looking for it.
Frieren, Frieren, Season 1, Episode 21

Since the publication of the first edition, we drastically improved the methods described in the first edition. The sections added are 5.7, 5.8, 5.9, 7.10, 7.11, 7.12, 7.13, 7.14, 7.15, 7.16, 7.17, 7.18, 8.1, 8.2, 8.3. In the sections added in Chaps. 5 and 7, we described newly proposed strategy that enables us to treat with more complicated integration of multiple profiles. In the sections added in Chap. 8, we have discussed why proposed method can work well. Based upon the theoretical discussion, we proposed the new strategy where we optimized the standard deviation used in Gaussian distribution by which P values are attributed to the features and the features associated with significant P values are selected. These methods are also implemented in the form of R packages as described in the Appendix C. We hope that this book can help many people who need to solve the problems described in this book.

Tokyo, Japan
March, 2024

Y-h. Taguchi

Preface to the First Edition

He stole something unexpected..., your heart.
Inspector Zenigata, Lupin III: The castle of Cagliostro, movie, Episode 1

This is a book about very classical mathematical techniques: principal component analysis and tensor decomposition. Because these two are essentially based upon linear algebra, one might think that these are no more than text book level matters. Actually, when I started to make use of them for the cutting-edge researches, many reviewers who reviewed my manuscripts complained about the usage of these old-fashioned techniques. They said, for example, "Why not using more modernized methods, e.g., kernel tricks?" or "Principal component analysis is a very old method for which no new findings can exist." In spite of these criticisms, I have continuously published numerous papers where I discussed how principal component analysis or tensor decomposition can be used for data science in a completely new way.

The principal reason why such old techniques can work pretty well is because of the topic targeted: feature selection in large p small n problem. Large p small n problem means that there are huge number of variables of which very small number of observations are available. In such situations, it is of course difficult to know what has happened in the system, because there are not enough number of points that cover the whole state space. This situation is also known as "the curse of dimensionality" which means the lack of enough number of observations compared with the number of dimensions. This problem remains unsolved over long period.

In this book, I apply principal component analysis and tensor decomposition in order to tackle this difficult problem. There are several reasons why these two can work well in this difficult problem. At first, these two are unsupervised methods. In contrast to the conventional supervised methods, unsupervised methods are more robust. Especially, it is free from overfitting that can easily occur when supervised methods are applied to small number of samples with large number of dimensions, because unsupervised methods do not learn from labeling from which supervised methods must learn. Second, unsupervised methods are more stable than supervised methods, because unsupervised methods are independent of labeling. Another advantage of principal component analysis and tensor decomposition is that they consider the interaction between variables not after the features are selected but before they are selected.

The main purpose of this book is feature selection, which means selecting small or limited number of critical variables among huge number of variables. Although there have been numerous proposals for feature selection, there are very few fitted to apply to the large p small n problems. One typical approach among those not fitted to large p small n problems is a statistical test. When we would like to find features that satisfy some required properties, statistical test can compute the probability that the desired property can appear by chance. If some features are associated with small enough probability, we can regard that the feature is truly associated with this property. In large p small n problem, this strategy often fails. Smaller number of samples can increase the probability that the desired property can happen by chance. On the other hand, if the number of features are large, small probability can happen by chance; if the number of features is as many as 10^4, features associated with the probability as small as 10^{-4} can appear with the probability of 1 (i.e., almost always). Because of the same reason, even if we try to find the features best fitted with the desired property, it might be simply accidental.

The basic idea to resolve these difficulties using principal component analysis and tensor decomposition is as follows. First, before features are selected, whole data set is embedded into lower dimensional space. Because feature selection is performed within this lower dimensional space, it is not a large p small n problem any more. Thus, it is also free from "The curse of dimensionality." Then the dimension in which feature selection is performed is selected with variety of methods fitted to desired properties. As can be seen in the later parts of this book, this simple idea works surprisingly well.

In Chap. 1, I re-introduce basic concepts including scalar, vector, matrix, and tensor, from data science point of views. Chapters 2 and 3 introduce two embedding methods by which dimensions are reduced, principal component analysis as a part of matrix factorization and tensor decomposition, respectively. The following two chapters explain how we can make use of these two for the feature selection by applying them to synthetic data sets. The last two chapters are dedicated to the applications of two methods to bioinformatics where large p small n problems are very usual.

Although the application of the proposed methods is limited to genomic science, because general workframe of the methodologies is very universal, readers are expected to apply these two to their own problems in data science. I am happy to hear from their achievements when the methods proposed in this book are applied to various problems.

Tokyo, Japan
March, 2019

Y-h. Taguchi

Acknowledgments

Unfortunately, because I have developed this method almost by my own, I have no persons to which I would acknowledge their contributions. On the other hand, I have many researchers who would like to make use of my own methods for their problems as applications. These researchers include Prof. Yoshiki Murakami who is a medical doctor and wrote many medical papers with me, Prof. Hideaki Umeyama who is a pharmacologist and a specialist about in silico drug design and has performed a university running project with me, and Prof. Mitsuo Iwadate who was once a member of Prof. Umeyama's lab. In addition to this, two Taiwanese professors, Hsiuying Wang at the Institute of Statistics, National Chiao Tung University, and Ka-Lok Ng at the Department of Bioinformatics and Medical Engineering, Asia University, have published a few papers with me. I would like to thank all of the researchers, including these above-mentioned professors, who wrote papers with me using the methods proposed in this book. At last, but not at least, I would like to thank all of my family members, my wife, Tomoko, and sons, Yuu and Koki, for their continuous mental supports for me. Without their help, I could not write this book. In addition to the acknowledgment above when the first edition was written, I would like to add two more persons to be acknowledged. Prof. Turki Turki who has written almost all papers published since the publication of the first edition with me and Prof. Sanjiban Sekhar Roy who has published a few papers with me and edited two volumes with me. Without the contributions from these two, I could not write the second edition at all.

Contents

Acronyms

AD	Alzheimer's disease
ALL	acute lymphoblastic leukemia
ALS	alternating least square
AY	amygdala
BAHSIC	backward elimination using Hilbert-Schmidt norm of the cross-covariance operator
BH	Benjamini Hochberg
BP	biological process
CC	cellular component
CHB	chronic hepatitis B
CHC	chronic hepatitis C
ChIP	chromatin immunoprecipitation
CP	canonical polyadic
DAVID	The Database for Annotation, Visualization and Integrated Discovery
DBTSS	DataBase of Transcriptional Start Sites
DEG	differentially expressed gene
DF	dengue fever
DHF	dengue hemorrhagic fever
DMS	differentially methylated site
DNA	deoxyribonucleic acid
FACS	fluorescence activated cell sorting
FDR	false discovery rate
FE	feature extraction
FN	false negative
FP	false positive
GEO	gene expression omnibus
GO	gene ontology
HC	hippocampus
HDAC	histone deacetylase
HOOI	higher orthogonal iteration of tensors
HOSVD	higher order singular value decomposition

HTS	high-throughput sequencing
KEGG	Kyoto Encyclopedia of Genes and Genomes
KO	knock out
LBDD	ligand-based drug design
LDA	linear discriminant analysis
limma	Linear Models for Microarray Data
LOOCV	leave one out cross validation
MF	matrix factorization
MF	molecular function
miRNA	microRNA
MPFC	medial prefrontal cortex
MSigDB	the molecular signatures database
NASH	nonalcoholic steatohepatitis
NP	non-deterministic polynomial-time
NSCLC	non-small cell lung cancer
OE	overexpression
PC	principal component
PCA	principal component analysis
PTSD	post-traumatic stress disorder
RFE	recursive feature elimination
RGB	red, green, and blue
RNA	ribonucleic acid
RPKM	reads per kilobase of exon per million
SAM	significance analysis of microarrays
SBDD	structure-based drug design
scRNA-seq	single cell RNA sequencing
SD	standard deviation
SE	septal nucleus
SNP	single-nucleotide polymorphism
SOTA	State-of-the-Art
ST	striatum
SVD	Singular value decomposition
TCGA	The cancer genome atlas
TD	Tensor decomposition
TF	transcription factor
TGE	transgenerational epigenetics
TN	true negative
TP	true positive
TSS	transcription start site
UDB	universal disease biomarker
UFF	unsupervised feature filtering
UPGMA	unweighted pair group method using arithmetic average
UMAP	Uniform manifold approximation and projection
UTR	untranslated region
VS	ventral striatum

Part I
Mathematical Preparations

In this part, we briefly introduce mathematical basics required for understanding the content of this book. Most of the part is usually taught in the first grade of undergraduate course of university. Thus, some reader might skip this part. It is tried to reintroduce basic mathematical concept from the data science point of views.

Chapter 1
Introduction to Linear Algebra

None can extinguish souls!
Momo Minamoto, Release the Spyce, Season 1, Episode 12

1.1 Introduction

Linear algebra is composed of simple arithmetic operations: addition, subtraction, multiplication, and division. In spite of their simpleness, it is often powerful enough to represent some complicated data set. In some sense, linear algebra is something like scissors. Although scissors can do only one thing, cutting, it can be used for various purposes if it is used by skilled persons. A piece of paper can be a beautiful art called as a cutting picture that looks like a very complicated sculpture. A skilled hairdresser can use scissors to change a female outlook so beautiful. Likewise, linear algebra can be used to understand very complicated data set that is difficult to understand otherwise, if you can make use of it so as to let it to demonstrate the maximum power. In this chapter, we prepare the knowledge that can be used in the later chapters for the application as data science technology.

1.2 Scalars

1.2.1 Scalars

Scalars are numbers that take real values. In the data science context, scalars are usually numbers that describe samples. Here samples correspond to some objects that will be targeted under the investigation. The examples of pairs of samples and associated scalars are:

Y-h. Taguchi, *Unsupervised Feature Extraction Applied to Bioinformatics*, Unsupervised and Semi-Supervised Learning,
https://doi.org/10.1007/978-3-031-60982-4_1

- Person and weight
- Food and price
- Star and brightness

Thus, in contrast to the generic algebra, scalars are not always able to be added with each other; brightness cannot be added to price, price cannot be added to weight, and so on. Not only addition, but also division, multiplication, or subtraction are not always possible, either. Arithmetic is possible only between same scalars: brightness plus brightness, weight plus weight. In this sense, data science algebra is more restricted than usual algebra.

In the data science, it is critically important to remember that all scalars analyzed have origins in the real world; no scalars are purely ideal numbers. This is primarily distinct from simple mathematical numbers that do not always have counterpart in the real world. Scalars in data science always represent something that exists in the real world.

Exercise
1.1 List ten pairs of samples and associated scalars.

1.2.2 Dummy Scalars

In contrast to scalars that describe samples, samples are often associated with features that cannot be described with real values. Such examples are color. Although it is possible to artificially attribute real values to colors, e.g., using RGB (red, green, and blue) color model, it is empirically useless. In RGB color system, colors are represented as combinations of three scalars. For example, red corresponds to (1,0,0) and blue corresponds to (0,0,1). Formal addition of distinct colors, e.g., red plus blue, results in completely distinct third colors, (1,0,1), which corresponds to pink. Thus, it does not make sense. More severely, there are generally no ways to add distinct features. What comes if American is added to Japanese (in this case, feature is nationality)? In order to avoid this difficulty, dummy scalars are usually introduced. All features that cannot be described using real values are converted into 1 or 0. If a sample has the feature, corresponding dummy scalar takes 1 otherwise 0. In the example of colors, the number of scalars is as many as the number of colors. If all samples under the investigation can take one hundred colors, we have to prepare same number of dummy scalars and add 1 or 0 to them dependent on color association with each sample. All samples with red have dummy scalar, to which red color is attributed, of 1. Introduction of dummy scalars is critically important since its introduction enables us to deal with any features that cannot be easily represented by real values.

Exercise
1.2 List ten features that must be treated as dummy scalars.

1.2.3 Generating New Features by Arithmetic

Although distinct scalars cannot be added with each other, in the real application we need to generate new features from scalars. In order to perform arithmetic between distinct scalars, multipliers are introduced. Suppose that there are three distinct scalars, x, y and z. In order to enable addition among these, multipliers α, β and γ are multiplied to scalars as αx, βy, and γz. Now, it is possible to add them as $\alpha x + \beta y + \gamma z$. Multipliers have two functions. The first function is to make scalars nondimensional. Nondimensional scalars mean those without unit. For example, if one would like to add weight, price, and brightness, the multipliers of these should have unit of inverse of weight, price, and brightness. Then products of scalars and associated multipliers are nondimensional. In order to perform arithmetic between scalars, introduction of multipliers is essential. The second function of multipliers is to equalize the amount of scalars. If weight is measured in kg, it has values between 0 and 100. If price is defined in Japanese currency, yen, it typically has values between 0 and 1,000,000. Brightness can be measured by various units. If lumen is employed as unit, brightness typically takes values as large as several thousands. Without multipliers, individual contributions of distinct scalars to newly generated feature cannot be balanced. Thus, the introduction of multipliers is required in order to control contributions of scalars to generated feature. Once scalars are multiplied with multipliers, the product of scalars and multipliers can be arguments of any arithmetic functions, e.g., sin and log. Thus, new features can be generated not only by arithmetic but also using functions, e.g., $\log(\alpha x + \beta y + \gamma z)$.

In this context, dummy scalars can also be combined with usual scalars that take real values. In this sense, any of x, y and z can also be dummy scalars. Since dummy scalars are nondimensional, multipliers associated with dummy scalars are also nondimensional.

Exercise

1.3 Generate ten new features using three scalars x, y and z as well as three associated multipliers α, β and γ.

1.3 Vectors

1.3.1 Vectors

Vectors are composed of a set of scalars. For convenience, the elements of vectors are represented by adding suffix to scalars, e.g., x_j, where x is scalar and j is suffix that spans integers. By employing these notations, we are free from introducing the numerous characters to represent a set of many scalars.

In order to be free from representing vectors as a set of many scalars with suffix, we can introduce a vector notation, $\boldsymbol{x}$,

Table 1.1 An example of vector: foods vs. prices

Foods	Prices
Bread	100 Yen
Beef	1000 Yen
Pork	300 Yen
Fish	200 Yen

Table 1.2 Another example of vector: foods vs. weights

Foods	Weights
Bread	200 g
Beef	300 g
Pork	100 g
Fish	150 g

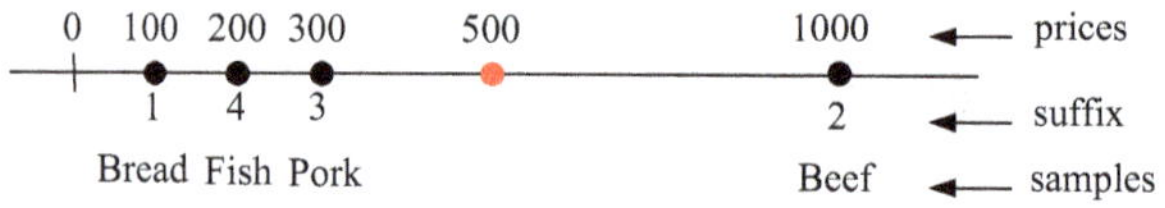

Fig. 1.1 A geometrical interpretation of vector $\boldsymbol{x} = (100, 1000, 300, 200)$. Individual components of the vector that correspond to prices of four samples are considered to be four coordinates of four points aligned along a line. Prices considered to be coordinates are displayed above the line, while suffix that corresponds to four samples is displayed below the line. A red point represents an imaginary sample with the price of 500 yen

$$\boldsymbol{x} = (x_1, x_2, \ldots, x_M), \tag{1.1}$$

where M is the number of samples. In short, it is often represented as $\boldsymbol{x} \in \mathbb{R}^M$. This says that there are M samples, each of which a scalar x_j is attributed to. A typical example of $\boldsymbol{x}$ is that there are M foods, each of which prices are attributed to, e.g., (Table 1.1) where $M = 4$ and $\boldsymbol{x} = (100, 1000, 300, 200)$.

Exercise
1.4 Generate some vectors that represent a set of samples.

It is very usual that samples are accompanied with more than one scalar. For example, we can attribute weights to foods (Table 1.2).

Then, a set of foods is accompanied with additional vector, $\boldsymbol{y} = (200, 300, 100, 150)$.

1.3.2 Geometrical Interpretation of Vectors: One Dimension

It is often very useful to interpret the vectors geometrically. For example, $\boldsymbol{x} = (100, 1000, 300, 200)$ can be considered to be coordinates of four points aligned along a line (Fig. 1.1).

There are several advantages of the geometrical representation of vectors. At first, it can give samples the order that can be easily visually recognized. By simply glancing the sequence of scalars, it is hard to recognize the rank order of scalars. Second, the distances between samples can be introduced. Then, from the prices, we can say that two pairs of samples, the pair of bread and fish and the pair of pork and fish, are equally separated. If we specifically define measure of distance, say Euclid distance, we can compute the distance between samples numerically as

$$\text{distance between bread and beef} = \sqrt{(100-1000)^2} = 900 \tag{1.2}$$

$$\text{distance between bread and fish} = \sqrt{(100-200)^2} = 100, \tag{1.3}$$

where Euclid distance between two points j and j' having coordinates of x_j and $x_{j'}$, respectively, can be defined as

$$\sqrt{\left(x_j - x_{j'}\right)^2}. \tag{1.4}$$

Using the numerical distances, we can quantitatively compare two pairs of samples on how far they are apart from each other. In this case, bread is nine times apart from beef than fish. These two points, the definition of rank order of samples and numerical distances between pairs of samples, will turn out to be critical for data science analysis.

An additional advantage of geometrical interpretation is that any points along the line automatically have prices. For example, if a point is placed on the line with the coordinate of 500 yen (a red point in Fig. 1.1), this point represents a sample with the price of 500 yen. This allows us to think about an imaginary sample with this price without specifying what it is. This is also a great advantage for data science, which must predict something unknown. With geometrical representation, we can discuss about samples with arbitrary scalars without specifying what it is. This abstraction is very important as can be seen later.

Exercise

1.5 Draw geometrical representation of Table 1.2.

1.3.3 Geometrical Interpretation of Vectors: Two Dimensions

As denoted in the Sect. 1.3.2, samples can be associated with more than one scalar (Tables 1.1 and 1.2). In this case, geometrical representation must also be altered from a line to a plane. Figure 1.2 shows geometrical representation of four foods according to the scalars shown in Tables 1.1 and 1.2.

Now, using two scalars simultaneously, the relationship among four foods becomes clearer. Beef is apart from other three, because it has the largest weight and highest price. As in the one dimension, any points in the plane are automatically associated with pairs of scalars: prices and weight. A red point in Fig. 1.2 represents an imaginary sample associated with a price of 500 yen and a weight of 250 g.

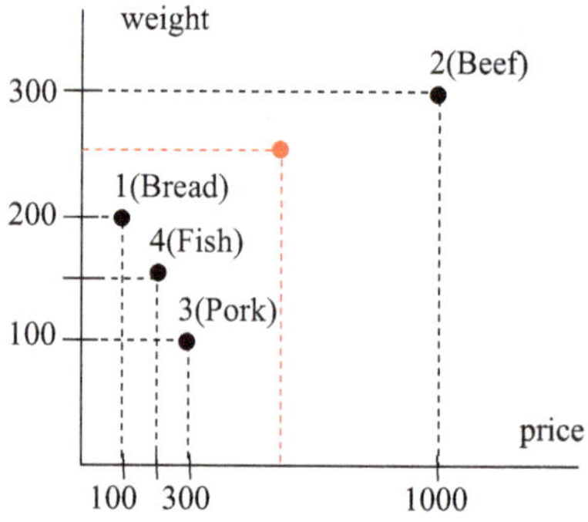

Fig. 1.2 A geometrical interpretation of Tables 1.1 and 1.2. Horizontal axis and vertical axis correspond to prices (Table 1.1) and weights (Table 1.2), respectively. A red point represents an imaginary sample with the price of 500 yen and the weight of 250 g

Table 1.3 Foods vs. prices with using dollar as price

Foods	Prices
Bread	1 dollar
Beef	10 dollars
Pork	3 dollars
Fish	2 dollars

If one thinks that there are nothing unclear, one might miss an important point: scale. In Fig. 1.2, length that corresponds to 100 yen does differ from length that corresponds to 100 g. Nevertheless, there are no reasons to make them equal to each other. When length of 100 yen is made to be equal to 100 g, the plot will be elongated toward horizontal direction. The problem is that there are no criteria to decide scale, since prices can never be related to weight.

One may wonder that it is not a problem, since numerical distance can be defined independent of scale. For example, the Euclidean distance between fish and pork in the plane shown in Fig. 1.2 can be defined as

$$\sqrt{(200-300)^2+(150-100)^2} \simeq 111 \tag{1.5}$$

that is independent of scale.

Exercise

1.6 Compute Euclidean distances of any pairs of samples (points) in Fig. 1.2.

Although it apparently seems to work, it actually does not. Suppose that we use dollar instead of yen for prices. For example, if we can assume that 1 dollar costs 100 yen, Table 1.1 now becomes Table 1.3.

Then, the Euclidean distance between fish and pork is not about 111 but

$$\sqrt{(2-3)^2+(150-100)^2} \simeq 50. \tag{1.6}$$

Now it is clear that there are many problems in two dimensional representations. At first, the distance cannot be determined independent of the unit of scalars. As soon as the foods are imported from Japan to the USA, the distances between foods

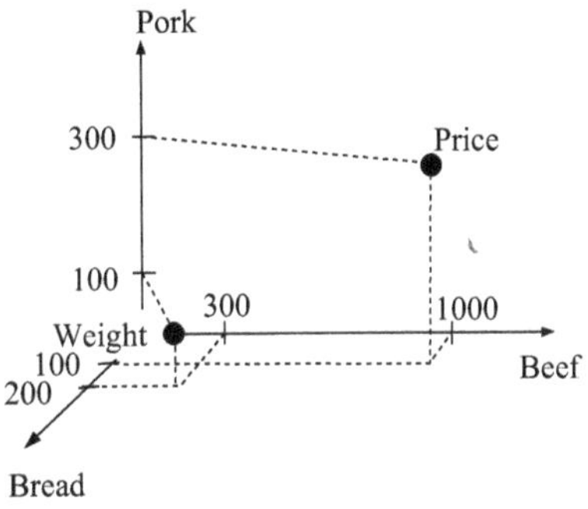

Fig. 1.3 An alternative geometrical interpretation of two vectors $\boldsymbol{x} = (100, 1000, 300, 200)$ for prices and $\boldsymbol{y} = (200, 300, 100, 150)$ for weights. Because of the limitation of the spatial dimension that we can recognize (up to three), the fourth scalars in $\boldsymbol{x}$ and $\boldsymbol{y}$ that represent Fish are omitted

might change. It does not make sense. In addition to this, in the system of dollar-gram unit, the prices are almost ignored on the computing distances. It also does not make sense.

Unfortunately, there are no definite ways to address this problem uniquely. How we should scale different scalars must be decided dependent upon what we would like to know from the data given. It is highly context dependent. Thus, we have to postpone this discussion later when we apply mathematics to real data set.

1.3.4 Geometrical Interpretation of Vectors: Features

In the previous sections, geometrical representations were applied to samples, i.e., four foods. In the Sect. 1.3.1, two vectors $\boldsymbol{x} = (100, 1000, 300, 200)$ for prices and $\boldsymbol{y} = (200, 300, 100, 150)$ for weights were defined, respectively. These two vectors can also be interpreted as a geometrical representation of two features, price and weight (Fig. 1.3). Excluding the omission of fish for easier visual recognition, Figs. 1.2 and 1.3 are mathematically equivalent. In spite of the mathematical equivalence, it is not very popular to interpret vectors as geometrical representation of not samples but features. This is primarily because we have to plot different scalars, i.e., prices and weights, on the common axes. In the Sect. 1.3.3, the ambiguity of scale was pointed out. The problem of scale is more visible in the geometrical interpretation of vectors for features (Fig. 1.3) than that for samples (Figs. 1.1 and 1.2). In the third (vertical) axis in Fig. 1.3 that corresponds to pork, 300 yen is more distant from origin than 100 g. It is apparent that this spatial relationship between price and weight of bread is not informative at all, since as soon as we use dollar (Table 1.3) instead of yen, the price (now it is "only" three dollars) becomes closer to the origin than the weight (100 g). Second, it is not recommended to plot distinct units (in this case, price and weight) along the same axis in physical sciences where this kind of coordinate representation was firstly developed (for example, energy and force can never be plotted on the same axis).

In spite of these difficulties, the emphasis of the equivalence between two geometrical representations (either that of samples or that of features) will turn out to be practically very useful for the main topics of this book.

Exercise

1.7 Draw geometrical representations of prices and weights using combinations of samples distinct from those used in Fig. 1.3, e.g., beef, pork, and fish.

1.3.5 Generating New Features by Arithmetic

As has been down in scalars (Sect. 1.2.3), new features can be generated from vectors, too, e.g., $\alpha \boldsymbol{x} + \beta \boldsymbol{y} + \gamma \boldsymbol{z}$, where α, β, and γ are multipliers similar to the cases in scalars and $\boldsymbol{x}$, $\boldsymbol{y}$, and $\boldsymbol{z}$ are vectors. One distinction from generations of new features using scalars is that function must be applied to individual new features generated from scalars. Then, generating new feature with applying a function to vector should be denoted as like $\log(\alpha x_i + \beta y_i + \gamma z_i)$, which corresponds to the ith scalar that consists of new features in the form of vectors.

Exercise

1.8 Generate new features in the vector form, using scalars shown in Tables 1.1 and 1.2 with arbitrary multipliers (and if possible, with applying functions to scalars).

1.3.6 Dummy Vectors

As features that cannot be described with real values were treated as dummy scalars, vectors can also be composed of dummy scalars. In some sense, dummy scalars themselves could be interpreted as vectors. For examples, three colors in RGB representation, $(1, 0, 0)$, $(0, 0, 1)$ and $(1, 1, 0)$, can be now geometrically interpreted in three dimensional vectors that consist of three integer scalars. They are also geometrical representations of features introduced in Sect. 1.3.4. Thus, from this point of views, i.e., unified treatment of dummy scalars with usual scalars that can be treated as real numbers, introduction of geometrical vector representation of features is critical, although it is rarely emphasized in the text books that introduce data science.

In the later part of this book, we try to select a part of features from all features for the practical reasons. Colors represented in geometrical vector representation are very useful for this purpose, since these allow us to select, for example, only the first scalars of RGB representations. Such a decomposition of colors never be possible without vector representations.[1]

On the other hand, in contrast to vector representation of scalars that can be represented as real values, dummy vectors can be placed only at grid points whose coordinates are composed of integer. Of course, as can be seen in RGB representation of colors, dummy scalars are often allowed to be extended to take real values as well ($(0.5, 0.5, 0)$ can make sense in RGB representation of colors), it is not always true. For example, if the dummy scalars represent whether sample is book, chair, or stick, although dummy scalars can be represented as $(1, 0, 0)$, $(0, 1, 0)$, $(0, 0, 1)$, $(0.5, 0.5, 0)$ does not make sense at all, since $(0.5, 0.5, 0)$ means a sample associated with a feature composed of 50% book and 50% chair.

In contrast to vectors that can be represented as real numbers, e.g., prices and weights, not all points in the geometrical representation of dummy scalars do not have anything real. For example, the dummy vector that represents if a sample is book, chair, or stick cannot take $(1, 1, 0)$ since no samples cannot be book and chair simultaneously.

Exercise
1.9 Think about dummy vectors assuming some.

1.4 Matrices

As vectors are composed of scalars, matrices, X, are composed of vectors, as

$$X = \left(\boldsymbol{x}_1^T, \boldsymbol{x}_2^T, \ldots, \boldsymbol{x}_M^T\right), \tag{1.7}$$

where M is the number of features, e.g., price, weight, and color. $\boldsymbol{x}^T$ represents transposition of a vector $\boldsymbol{x}$ where

$$\boldsymbol{x}_j = \left(x_{1j}, x_{2j}, \ldots, x_{Nj}\right) \tag{1.8}$$

corresponds to the vector of ith feature (M is the number of samples). When prices in Table 1.1 and weights in Table 1.2 are represented as matrix, it should be Table 1.4. In this case, a matrix X is

$$X = \begin{pmatrix} 100 & 1000 & 300 & 200 \\ 200 & 300 & 100 & 150 \end{pmatrix} \tag{1.9}$$

[1] Practically, employing only the first scalars in RGB representation is equivalent to the usage of red sunglass through which only red color can penetrate. Now, colors are transformed to real values that describe red color intensity of colors, although in this example only integers are allowed since colors are treated as example dummy scalars.

Table 1.4 The matrix that represents Tables 1.1 and 1.2 in the unified format

	i	1	2
j	Sample	Prices (Yen)	Weight (gram)
1	Bread	100	200
2	Beef	1000	300
3	Pork	300	100
4	Fish	200	150

and vectors are

$$\boldsymbol{x}_1 = (100, 200) \quad (1.10)$$

$$\boldsymbol{x}_2 = (1000, 300) \quad (1.11)$$

$$\boldsymbol{x}_3 = (300, 100) \quad (1.12)$$

$$\boldsymbol{x}_4 = (200, 150), \quad (1.13)$$

where $N = 2$ and $M = 4$. For example, x_{24} is 150, since x_{ij} corresponds to the ith scalar attributed to jth sample.

Exercise

1.10 Write down the matrix X that corresponds to the table generated by merging Tables 1.2 and 1.3.

1.4.1 Equivalences to Geometrical Representation

There are several advantages for matrix representation. The first advantage is coincidence with geometrical representation. Matrix representation is highly coincident with geometrical representations. When rows in Table 1.4 are considered to be vectors as in equations, from (1.10) to (1.13), it is equivalent to Fig. 1.2; bread, beef, pork, and fish correspond to $\boldsymbol{x}_1$, $\boldsymbol{x}_2$, $\boldsymbol{x}_3$, and $\boldsymbol{x}_4$.

On the other hand, when columns in Table 1.4 are considered to be vectors as,

$$\boldsymbol{x}_i = (x_{i1}, x_{i2}, \ldots, x_{iM}), \quad (1.14)$$

i.e.,

$$\boldsymbol{x}_1 = (100, 1000, 300, 200) \quad (1.15)$$

$$\boldsymbol{x}_2 = (200, 300, 100, 150), \quad (1.16)$$

they are equivalent to Fig. 1.3; price corresponds to $\boldsymbol{x}_1$ and weight corresponds to $\boldsymbol{x}_2$.

Thus, conversely $X \in \mathbb{R}^{N \times M}$ can be considered to be either M-dimensional vectors as many as N (Fig. 1.3) or N-dimensional vectors as many as M (Fig. 1.2).

Thus, matrix representation is not only convenient to represent a set of vectors attributed to samples (Table 1.4) but also useful for geometrical representations. Since two distinct geometrical representations (Figs. 1.2 and 1.3) are important for the purpose of this book as mentioned in the above, matrix that can represent two distinct geometrical representations in the unified way is very important and useful.

Exercise

1.11 Write down the two geometrical interpretation of matrix X generated in the previous exercise.

1.4.2 Matrix Manipulation and Feature Generation

Any feature generation in the form, $\alpha \boldsymbol{x} + \beta \boldsymbol{y} + \gamma \boldsymbol{z}$, can be performed with matrix manipulation; it is another advantage of matrix representation in data science. Suppose $\boldsymbol{x}, \boldsymbol{y}, \boldsymbol{z}$ are vectors attributed to three features, e.g., price, weight, and color. Define matrix X as

$$X = \begin{pmatrix} \boldsymbol{x} \\ \boldsymbol{y} \\ \boldsymbol{z} \end{pmatrix}. \tag{1.17}$$

Then, $\alpha \boldsymbol{x} + \beta \boldsymbol{y} + \gamma \boldsymbol{z}$ can be represented as

$$\alpha \boldsymbol{x} + \beta \boldsymbol{y} + \gamma \boldsymbol{z} = \boldsymbol{\alpha} X = (\alpha, \beta, \gamma) \begin{pmatrix} x_1 & x_2 & \cdots & x_M \\ y_1 & y_2 & \cdots & y_M \\ z_1 & z_2 & \cdots & z_M \end{pmatrix} \tag{1.18}$$

with defining multiplier vector, $\boldsymbol{\alpha}$ as

$$\boldsymbol{\alpha} = (\alpha, \beta, \gamma). \tag{1.19}$$

In data science, it is very important to describe samples with newly generated features. Otherwise, we cannot make use of newly generated features in order to describe the relationship between samples. In order to describe samples with newly generated features, we need generally at least new features as many as N that is the number of original features. Thus, the number of multiplier vectors must be as many as N as well. In this case, since there are three feature vectors, $\boldsymbol{x}, \boldsymbol{y}, \boldsymbol{z}$, the number of multiplier vectors must be three as well, i.e.,

$$\boldsymbol{\alpha}_1 = (\alpha_1, \beta_1, \gamma_1) \tag{1.20}$$

$$\boldsymbol{\alpha}_2 = (\alpha_2, \beta_2, \gamma_2) \tag{1.21}$$

$$\boldsymbol{\alpha}_3 = (\alpha_3, \beta_3, \gamma_3). \tag{1.22}$$

With multiplier matrix, A, being defined as

$$A = \begin{pmatrix} \boldsymbol{\alpha}_1 \\ \boldsymbol{\alpha}_2 \\ \boldsymbol{\alpha}_3 \end{pmatrix} = \begin{pmatrix} \alpha_1 & \beta_1 & \gamma_1 \\ \alpha_2 & \beta_2 & \gamma_2 \\ \alpha_3 & \beta_3 & \gamma_3 \end{pmatrix}, \quad (1.23)$$

new features, $\boldsymbol{x}', \boldsymbol{y}', \boldsymbol{z}'$, that describe M samples can be obtained as matrix form

$$X' = \begin{pmatrix} \boldsymbol{x}' \\ \boldsymbol{y}' \\ \boldsymbol{z}' \end{pmatrix} = AX = \begin{pmatrix} \alpha_1 & \beta_1 & \gamma_1 \\ \alpha_2 & \beta_2 & \gamma_2 \\ \alpha_3 & \beta_3 & \gamma_3 \end{pmatrix} \begin{pmatrix} x_1 & x_2 & \cdots & x_M \\ y_1 & y_2 & \cdots & y_M \\ z_1 & z_2 & \cdots & z_M \end{pmatrix}. \quad (1.24)$$

Then jth sample is now described with new feature

$$\begin{pmatrix} x'_j \\ y'_j \\ z'_j \end{pmatrix} = \begin{pmatrix} \alpha_1 x_j + \beta_1 y_j + \gamma_1 z_j \\ \alpha_2 x_j + \beta_2 y_j + \gamma_2 z_j \\ \alpha_3 x_j + \beta_3 y_j + \gamma_3 z_j \end{pmatrix}. \quad (1.25)$$

Now it is obvious that Euclidean distance between two samples j and j' computed using X differs from that using X';

$$\sqrt{(x_j - x_{j'})^2 + (y_j - y_{j'})^2 + (z_j - z_{j'})^2}$$
$$\neq \sqrt{(x'_j - x'_{j'})^2 + (y'_j - y'_{j'})^2 + (z'_j - z'_{j'})^2}. \quad (1.26)$$

Thus, by selecting A, we can gain more suitable features adapted for the purpose (e.g., discrimination between samples that belong to more than two distinct groups). The problem is how to tune the *best* A. This is nothing but one of the critical topics that will be discussed in the later part of this book. It is also the reason why I decided to write this book.

Table 1.5 is an example of generated new features X' from X by $X' = AX$ with

$$A = \begin{pmatrix} 1 & \frac{1}{2} & 1 \\ \frac{1}{2} & 1 & 1 \\ 1 & 1 & \frac{1}{2} \end{pmatrix}. \quad (1.27)$$

Figure 1.4 is the geometrical representation of X and $X' = AX$ shown in Table 1.5. It reveals various problems associated with the generation of new features. First, the separation of beef (red point) from other three foods is enhanced after the new feature, X', is generated. This means that generation of new feature can alter relationships among samples drastically. Thus, we have to be careful when

Table 1.5 An example of generation of new features. Upper: Original X, lower: generated $X' = XA$ where A is given in Eq. (1.27)

X	i	1	2	3
j	Sample	Prices (Yen)	Weight (gram)	Size (cm)
1	Bread	100	200	6
2	Beef	1000	300	10
3	Pork	300	100	6
4	Fish	200	150	14

X'	i	1	2	3
j	Sample	1st feature	2nd feature	3rd feature
1	Bread	200	256	303
2	Beef	1160	810	1305
3	Pork	356	256	403
4	Fish	289	264	357

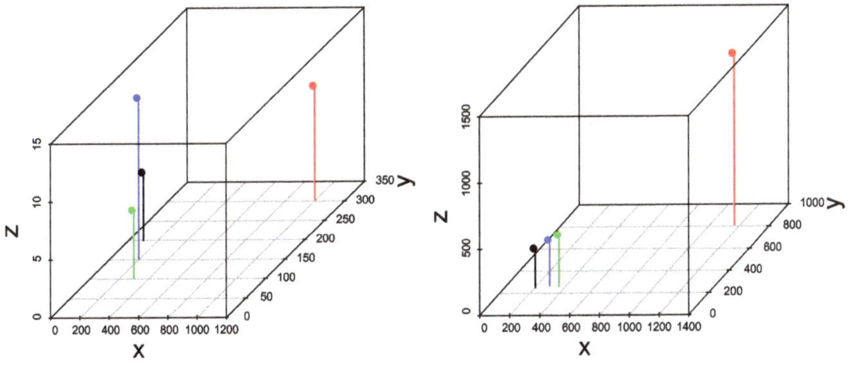

Fig. 1.4 A geometrical interpretation of X (left) and X' (right) shown in Table 1.5. Black: bread, red: beef, green: pork, blue: fish

generating new features. It might cause artifacts. Second, the dependence upon size (z-axis) in X is almost destroyed in X'. This is simply because the numerical values of sizes shown in Table 1.5 for X are much smaller than prices and weights. Basically, because A shown in Eq. (1.27) gives similar weights to three features, price, weight, and size, to generate new feature, the dependence upon size is smeared out. This is related to the problem of scale. Appreciate rescaling of individual features will recover this problem. However, we have to be also careful if rescaling is reasonable.

Exercise

1.12 Generate new features using the matrix X shown in (1.9) with arbitrary multiplier matrix A.

1.5 Tensors

1.5.1 Introduction of Tensors

As vectors are composed of scalars and matrices are composed of vectors, tensors can be composed of matrices. Suppose Table 1.6 represents foods in two shops.

Now, we can define tensor, x_{ijk}, that describes the jth feature attributed to the ith food in the kth shop. Visually, this can be represented as a cuboid (Fig. 1.5), whose three edges correspond to i (features = price and weight), j (samples = foods), and k (two shops).

We can even extend tensor further. For example, we can add days as the fourth suffix, ℓ. Then $x_{ijk\ell}$ represents the ith feature attributed to the jth sample (food) in the kth shop at the ℓth day. As you can easily suspect, this extension is unlimited. We can add as many suffix as we hope as long as data is available. Additional suffix can represent city, country, year, month, and so on.

As integration of scalars is named as vector, as that of vectors is named as matrix, and as that of matrices is named as tensor, we can name tensors with more than three suffixes alternatively. Nevertheless, we are not willing to do this, since it is unrealistic to prepare infinite sequences of names attributed to tensors with a distinct number of suffixes. Instead of that, we name the tensor with m suffix as m-mode tensor. Table 1.6 and Fig. 1.5 are the two distinct representations of three mode tensor.

Table 1.6 Two tables that describe a list of foods in two shops

$k = 1$	i	1	2
j	Sample	Prices (Yen)	Weight (gram)
1	Bread	100	200
2	Beef	1000	300
3	Pork	300	100
4	Fish	200	150

$k = 2$	i	1	2
j	Sample	Prices (Yen)	Weight (gram)
1	Bread	200	250
2	Beef	1500	200
3	Pork	200	150
4	Fish	100	1500

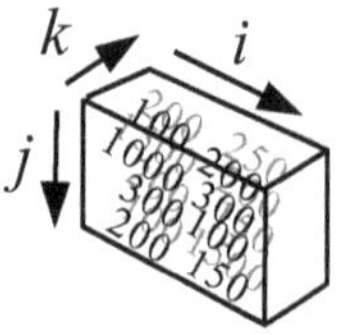

Fig. 1.5 A cuboid that represents Table 1.5. Black and gray numbers correspond to $k = 1$ and $k = 2$, respectively

As long as we follow this convention, conversely, we can name scalars, vectors, and matrices as tensors as well; i.e., scalars, vectors matrices can be considered to be zero, one, and two mode tensors, although this kind of convention is rarely employed. What I would like to emphasize is that scalars, vectors, matrices, and tensors should be treated in the unified way, not in the distinct ways at least in the data science. This is because in contrast to conventional sciences that make use of the these concepts, i.e., scalars, vectors, matrices, and tenors, distinction among these are not associated with any real distinct meaning.

In physics, potential energy is scalar, velocity is vector, and stress is tensor. This is simply because their physical realization inevitably requires them. Multiplications between distinct layers are not arbitrary, but strictly decided. Product between energy and vector does not make sense (although it may occasionally have meaning of energy flow). In data science, we can generate any kind of new features as long as they work. In this sense, in data science, scalars, vectors, matrices, and tensors should be treated samely. Thus, introduction of tenors is natural in data science.

Exercise
1.13 Generate a three mode tensor whose components are $x_{ijk} \in \mathbb{R}^{3\times3\times3}$.

1.5.2 Geometrical Representation of Tensors

In contrast to scalars, vectors, and matrices for which geometrical representations can be obtained straightly, geometrical representation of tensors is harder. This is primarily because we live in three dimensional physical space. This difficulty has partially already existed when we introduced the concept of matrices. For example, if we have to represent 4×4 matrices geometrically, we cannot avoid to deal with four-dimensional vectors which we cannot visually represent anymore.

This limitation is severer for tensors. If we hope to get geometrical representation of data shown in Table 1.6, the most easiest way is to prepare two planes on each of which four two dimensional vectors are drawn; $k = 1$ and $k = 2$ correspond to Figs. 1.2 and 1.6, respectively. One possible drawback of this geometrical representation is the difficulty of comparison between $k = 1$ and $k = 2$, since even scale of horizontal and vertical axes differs between Figs. 1.2 and 1.6. Although k takes only two values ($k = 1, 2$) in the present case, k can span over more shops. In that case, geometrical representation might become more difficult to interpret.

In the following, this kind of "vectoralization" can be named as unfolding. In the unfolding of the tensor $x_{ijk} \in \mathbb{R}^{2\times4\times2}$ shown in Table 1.6 as well as in Figs. 1.5, 1.2 and 1.6 can be expressed as a matrix $X^{i\times(jk)}$ whose elements are $x_{i(jk)} \in \mathbb{R}^{2\times8}$

$$X^{i\times(jk)} = \begin{pmatrix} 100 & 1000 & 300 & 200 & 200 & 1500 & 200 & 100 \\ 200 & 300 & 100 & 150 & 250 & 200 & 150 & 1500 \end{pmatrix} = (\boldsymbol{x}_1, \dots, \boldsymbol{x}_8) . \tag{1.28}$$

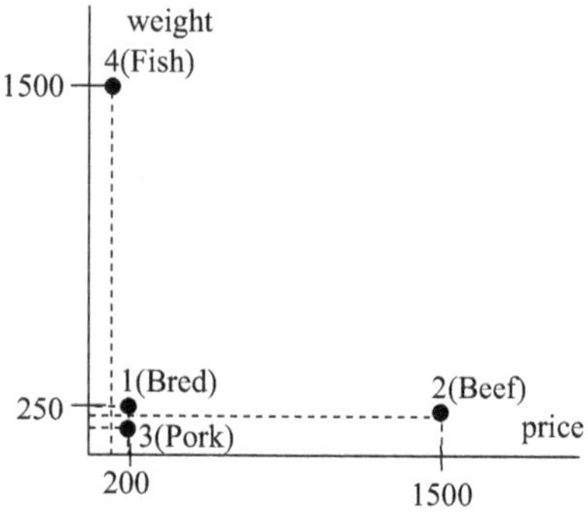

Fig. 1.6 A geometrical interpretation of $k = 2$ in Table 1.6. Horizontal axis and vertical axis correspond to prices and weights, respectively

Here, j (samples) and k (shops) are expanded as one suffix that is put in parentheses together; the first four columns, i.e., $\boldsymbol{x}_1$ to $\boldsymbol{x}_4$, and the second four columns, i.e., $\boldsymbol{x}_5$ to $\boldsymbol{x}_8$, correspond to Figs. 1.2 and 1.6, respectively. Then, the tensor can be represented as a 2(features) $\times$ 8(= 4(samples) $\times$ 2(shops)) matrix that is equivalent to eight two dimensional vectors, $\boldsymbol{x}_1$ to $\boldsymbol{x}_8$.

In this unfolding, data set shown in Table 1.6 as well as in Fig 1.5 is represented as eight points in the two dimensional space spanned by two features (prices and weight). It is obvious that there can be more unfolding. For example, x_{ijk} can also be unfolded as a matrix $X^{k\times(ij)}$ whose elements are $x_{k(ij)} \in \mathbb{R}^{2\times 8}$,

$$X^{k\times(ij)} = \begin{pmatrix} 100 & 1000 & 300 & 200 & 200 & 300 & 100 & 150 \\ 200 & 1500 & 200 & 100 & 250 & 200 & 150 & 1500 \end{pmatrix} = (\boldsymbol{x}_1, \ldots, \boldsymbol{x}_8) . \tag{1.29}$$

Here each column represents a combination of feature and sample. For example, the first column, i.e., $\boldsymbol{x}_1$, represents prices ($i = 1$) of bread ($j = 1$) at two shops, and the seven column, i.e., $\boldsymbol{x}_7$, represents weights ($i = 2$) of pork ($j = 2$) at two shops (see Table 1.6).

Because these two unfolding, Eqs. (1.28) and (1.29), are occasionally represented as two dimensional vectors, geometrical representations are possible. However, there is yet another unfolding, which is represented as a matrix $X^{j\times(ik)}$ whose elements are $x_{k(ij)} \in \mathbb{R}^{4\times 4}$,

$$X^{j\times(ik)} = \begin{pmatrix} 100 & 200 & 200 & 250 \\ 1000 & 300 & 1500 & 200 \\ 300 & 100 & 200 & 150 \\ 200 & 150 & 100 & 1500 \end{pmatrix} = (\boldsymbol{x}_1, \ldots, \boldsymbol{x}_4) . \tag{1.30}$$

Because they are represented as four four-dimensional vectors, $\boldsymbol{x}_1$ to $\boldsymbol{x}_4$, unfortunately nothing can be represented visibly anymore; here, $\boldsymbol{x}_1$ to $\boldsymbol{x}_4$ correspond to $(ik) = (11), (12), (21)$, and (22). In other words, they correspond to price at the first shop, weight and the first shop, price at the second shop, and the weight at the second shop (see Table 1.6).

Generally, m-mode tensors have m kinds of representations of unfolding. This makes it difficult to interpret geometrical representation. Moreover, because unfolding mixes more than one feature into one, the interpretation becomes more difficult. These difficulties of interpretation are possibly the reason why tensor is not employed frequently in data science. In data science, how to interpret outcomes is the central topic; if the introduction of tensor makes the interpretation more difficult, it is not hard to imagine that prople will avoid the tensor representation of data set.

Exercise

1.14 Unfold the three mode tensor generated at the last exercise as $X^{i\times(jk)}$.

1.5.3 Generating New Features

In contrast to the matrix representation where generating new features can be easily represented as linear algebra, generating new features in tensor representation of data set is much harder.

The primary reason of this is mixture of the features. For example, in Eq. (1.30), $\boldsymbol{x}_1$ and $\boldsymbol{x}_3$ represent prices, while $\boldsymbol{x}_2$ and $\boldsymbol{x}_4$ represent weights. Thus, manipulation of this matrix inevitably results in mixture of distinct features, i.e., price and weights. On the other hands, $\boldsymbol{x}_1$ and $\boldsymbol{x}_2$ are the data at the first shop, while $\boldsymbol{x}_3$ and $\boldsymbol{x}_4$ are the data at the second shop. Thus, careless manipulation also makes the interpretation of generation of new features difficult. Simple mixing of distinct columns includes both mixture of price and weight and that between two shops.

In order to avoid these difficulties, it is better to generate new feature before unfolding. This inevitably requires "tensor" algebra.

1.5.4 Tensor Algebra

As in the case of matrix, it is possible to introduce algebra to tensor. Addition and subtraction are straightforward: simply adding or subtracting two corresponding components of tensor: x_{ijk} and x'_{ijk}.

Nevertheless, multiplication is not easy. In order to extend matrix multiplication to tensor, we introduce tensor multiplication between three mode tensor $\mathcal{X}$ whose component is x_{ijk} and vector $\boldsymbol{x}$ whose length is as large as the first mode of $\mathcal{X}$

$$\{\mathcal{X} \times_i \boldsymbol{x}\}_{jk} = \sum_i x_{ijk} x_i, \tag{1.31}$$

where $\{X\}_{jk}$ is the jth row and kth column component of generated matrix X.

Scalar, vector, and matrix can be considered to be zero, one, two mode tensors. Similarly, tensor multiplication includes scalar, vector, and matrix multiplication.

For example, inner product between two vectors can be represented as tensor product between two one mode tensor,

$$\boldsymbol{x} \cdot \boldsymbol{y} = \sum_i x_i y_i = \boldsymbol{x} \times_i \boldsymbol{y}. \tag{1.32}$$

Matrix product can also be represented as multiplication between two mode tensors,

$$\{XY\}_{ij} = \sum_k x_{ik} y_{kj} = \{X \times_k Y\}_{ij}. \tag{1.33}$$

Using tensor multiplication operator, we can easily generate new features as

$$\mathcal{X}' = A \times_i \mathcal{X}. \tag{1.34}$$

Here $x_{ijk} \in \mathbb{R}^{N \times M \times K}$ is the element of $\mathcal{X}$ and A is multiplier matrix whose element is $a_{\ell i} \in \mathbb{R}^{N \times N}$ where new features are defined as

$$x'_{\ell jk} = \sum_i a_{\ell i} x_{ijk}. \tag{1.35}$$

In order to be coincident with matrix representation, Eq. (1.24), we introduce the notation $\mathcal{X}^{\cdot k} \in \mathbb{R}^{N \times M}$, which represents the matrix generated from $\mathcal{X} \in \mathbb{R}^{N \times M \times K}$. In this representation, Eq. (1.34) can also be written

$$\mathcal{X}'^{\cdot k} = A\mathcal{X}^{\cdot k}. \tag{1.36}$$

Exercise

1.15 Execute Eq. (1.34) using a three mode tensor $\mathcal{X}$ whose elements are $x_{ijk} \in \mathbb{R}^{3 \times 3 \times 3}$ and a 3×3 matrix A.

In order for later use, we further introduce additional tensor multiplication. Zero multiplication is defined as

$$\left\{\boldsymbol{x} \times^0 \boldsymbol{y}\right\}_{ij} = x_i y_j. \tag{1.37}$$

Generally, multiplication of m-mode tensor and m'-mode tensor with the operator $\times^0$ results in $(m + m')$-mode tensor.

We can further extend the operator $\times^0$ into the other way as

$$\{\mathcal{X} \times^k \mathcal{X}'\}_{i_1 i_2 \ldots i_{m-1} k j_1 j_2 \cdots j_{m'-1}} = x_{i_1 i_2 \ldots i_{m-1} k} x'_{k j_1 j_2 \cdots j_{m'-1}}, \tag{1.38}$$

where $\mathcal{X}$ and $\mathcal{X}'$ are m-mode and m'-mode tensors whose components are $x_{i_1 i_2 \ldots i_{m-1} k}$ and $x'_{k j_1 j_2 \cdots j_{m'-1}}$, respectively. Thus, the multiplication of m-mode tensor and m'-mode tensor with the operator $\times^k$ results in $(m + m' - 1)$-mode tensor.

The number of suffixes attached to the operator $\times$ does not have to be restricted to one. When the number of suffixes attached to the operator $\times$ is more than one, they are defined as

$$\{\mathcal{X} \times_{k\ell} \mathcal{X}'\}_{i_1 i_2 \ldots i_{m-2} j_1 j_2 \cdots j_{m-2}} = \sum_{k\ell} x_{i_1 i_2 \ldots i_{m-2} k\ell} x'_{k\ell j_1 j_2 \cdots j_{m-2}} \tag{1.39}$$

and

$$\{\mathcal{X} \times^{k\ell} \mathcal{X}'\}_{i_1 i_2 \ldots i_{m-2} k\ell j_1 j_2 \cdots j_{m-2}} = x_{i_1 i_2 \ldots i_{m-2} k\ell} x'_{k\ell j_1 j_2 \cdots j_{m'-2}}. \tag{1.40}$$

These are $(m + m' - 4)$-mode tensor and $(m + m' - 2)$-mode tensor, respectively. In this sense the operator to which lower suffix is added can be represented using the operator to which upper suffix is added as

$$\mathcal{X} \times_{k\ell} \mathcal{X}' = \sum_{k\ell} \mathcal{X} \times^{k\ell} \mathcal{X}'. \tag{1.41}$$

We can also add upper and lower suffixes to the operator $\times$ together as

$$\{\mathcal{X} \times_k^{\ell} \mathcal{X}'\}_{i_1 i_1 \ldots i_{m-2} j_1 \ell j_2 \cdots j_{m'-2}} = \sum_{k} x_{i_1 i_2 \ldots i_{m-2} \ell k} x'_{k\ell j_1 j_2 \cdots j_{m'-2}}. \tag{1.42}$$

Adding multiple upper and lower suffixes is straightforward.

There is one problem on adding both upper and lower suffixes to the operator $\times$ when the same suffix is added as both lower and upper suffixes. This is equivalent to adding the suffix as lower suffix only, e.g.,

$$\begin{aligned} \{\mathcal{X} \times_k^k \mathcal{X}\}_{i_1 i_2 \ldots i_{m-2} j_1 j_2 \cdots j_{m'-2}} &= \{\mathcal{X} \times_k \mathcal{X}'\}_{i_1 i_2 \ldots i_{m-2} j_1 j_2 \cdots j_{m'-2}} \\ &= \sum_{k} x_{i_1 i_2 \ldots i_{m-2} k\ell} x'_{k\ell j_1 j_2 \cdots j_{m'-2}}. \end{aligned} \tag{1.43}$$

Thus as a rule, when the same suffix appears as both upper and lower suffixes, we erase upper one since it does make any changes.

$$\times_k^k \to \times_k. \tag{1.44}$$

The usefulness of these additional tensor multiplications from equations from (1.37) to (1.44) might be unclear. For example, by applying (1.37) to Eqs. (1.15) and (1.16), although we can get

$$\boldsymbol{x}_1^T \times^0 \boldsymbol{x}_2^T = \begin{pmatrix} 100 \\ 1000 \\ 300 \\ 200 \end{pmatrix} \times^0 \begin{pmatrix} 200 \\ 300 \\ 100 \\ 150 \end{pmatrix} = \begin{pmatrix} 20{,}000 & 30{,}000 & 10{,}000 & 15{,}000 \\ 200{,}000 & 300{,}000 & 100{,}000 & 150{,}000 \\ 60{,}000 & 90{,}000 & 30{,}000 & 45{,}000 \\ 40{,}000 & 60{,}000 & 20{,}000 & 30{,}000 \end{pmatrix}, \tag{1.45}$$

it is unclear how this generated matrix can help us to interpret the data; these represent prices $\times$ weight that does not seemingly make any senses. In order to make use of the matrix representation (1.45), we need mathematical technique to be introduced later in this book.

Exercise

1.16 Generate a matrix by applying $\times^0$ to a pair of arbitrary vectors.

Appendix

Rank

In this appendix, rank of matrix is briefly introduced because the concept of rank is important in the next chapter. Suppose that matrix $X \in \mathbb{R}^{N \times M}$ is represented as M N-dimensional vectors, $\boldsymbol{x}_j \in \mathbb{R}^N$, as

$$X = \left(\boldsymbol{x}_1, \ldots, \boldsymbol{x}_j, \ldots, \boldsymbol{x}_M\right). \tag{1.46}$$

If there are vectors, $\boldsymbol{c}_j \in \mathbb{R}^{M'}$, $1 \leq j \leq M$, such that

$$\boldsymbol{x}_j = \sum_{j' \in J} c_{jj'} \boldsymbol{x}_{j'}, \tag{1.47}$$

where J is a set of M' integers taken from $[1, M]$ without repetitions, the smallest M' is called as the rank of matrix otherwise the rank of matrix is equal to M. In other words, not all $\boldsymbol{x}_j$s are independent but at most M' out of M $\boldsymbol{x}_j$s are independent. This means that $\boldsymbol{x}_j$s span not M-dimensional space but at most $M'(< M)$ dimensional space. Thus, the rank of tensor at most min(M,N) because the number of independent vectors cannot exceed the number of dimensions.

Chapter 2
Matrix Factorization

Don't tie me down!
Zero Two, Darling in the FranXX, Season 1, Episode 12

2.1 Introduction

Similar to scalars that can be represented as a product of smaller numbers, e.g., $18 = 3 \times 3 \times 2$, matrices can also be represented as a product of smaller (lower ranked) matrices. As can be seen in the following, there are no unique ways to represent a matrix as a product of smaller matrices. Presenting a matrix as a product of smaller matrices is called as matrix factorization (MF). What kind of MF should be employed highly depends upon the purpose. In this chapter, I introduce some MFs fitted for the later applications in this book.

2.2 Matrix Factorization

The aim of MF is to represent an $N \times M$ matrix $X \in \mathbb{R}^{N \times M}$ as a matrix product of an $N \times K$ matrix $Y \in \mathbb{R}^{N \times K}$ and a $K \times M$ matrix $Z \in \mathbb{R}^{K \times M}$ as

$$X = YZ. \tag{2.1}$$

Generally, whether Eq. (2.1) has at least a solution or not depends upon many factors. When

$$(N + M)K \geq NM \tag{2.2}$$

stands, an MF can exist because the number, NM, of equations that must be fulfilled is smaller than the number of variables, $(N + M)K$. Even when Eq. (2.2) is not satisfied, in some case when the rank of X is equal to or smaller than $\min(N, M)$,

Y-h. Taguchi, *Unsupervised Feature Extraction Applied to Bioinformatics*, Unsupervised and Semi-Supervised Learning,
https://doi.org/10.1007/978-3-031-60982-4_2

Eq. (2.1) can be performed. For example, in the case of Eq. (1.37), it is obvious that Eq. (2.1) is possible with $K = 1$.

On the other hand, it is also obvious that Eq. (2.1) cannot always have the unique solution. Let us consider when $X = I$ where I is a $K \times K$ unit matrix whose diagonal components are 1, while other components are zero, i.e.,

$$I = AB, \tag{2.3}$$

where A and B are $K \times L$ and $L \times K$ matrices, respectively. Using this, Eq. (2.1) can be rewritten as

$$X = YZ = YIZ = YABZ = (YA)(BZ). \tag{2.4}$$

Now, a pair of YA and BZ is also an MF of an $N \times M$ matrix, X. Thus, in order to perform MF, we need some additional restriction to Y and Z other than simple requirement that Y and Z are $N \times K$ and $K \times M$ matrices, respectively. Dependent upon the restriction applied, there are several fashions of MF.

Although there are numerous MFs, most of them are limited to be applied to square matrices, X. Because matrix representation of data set in data science is not generally restricted to square matrix, here we consider only MFs applicable to nonsquare matrices.

Exercise
2.1 For $I \in \mathbb{R}^{2\times 2}$, generate $A, B \in \mathbb{R}^{2\times 2}$ that satisfy Eq. (2.3).

2.2.1 Rank Factorization

Rank factorization is an MF applicable to any $N \times M$ nonsquare rank K matrix. Thus, in this case, Eq. (2.2) does not need to be fulfilled.

Rank factorization is directly related to geometrical representation. Without losing generality, we can assume $N \geq M$ (if not, we can consider the transposed matrix). Then $N \times M$ matrix can be interpreted as M N-dimensional vectors as

$$X = (\boldsymbol{x}_1^T, \dots \boldsymbol{x}_M^T), \tag{2.5}$$

where $\boldsymbol{x}_j \in \mathbb{R}^N$. Suppose that X is rank K matrix. Then, the number of independent vectors in $\boldsymbol{x}_j$ is K. Then each $\boldsymbol{x}_j$ can be represented by a linear combination of K independent N-dimensional vectors, $\boldsymbol{c}_k \in \mathbb{R}^N$ as

$$\boldsymbol{x}_j = \sum_{k=1}^{K} \boldsymbol{c}_k f_{kj}, \tag{2.6}$$

where f_{kj} are coefficients of linear combination. If we define matrix F such that kth row and jth column component is f_{kj} and a matrix $C \in \mathbb{R}^{N \times K}$ as

$$C = (\boldsymbol{c}_1^T, \ldots, \boldsymbol{c}_K^T), \tag{2.7}$$

then we can write

$$X = CF. \tag{2.8}$$

This is nothing but rank factorization. Computing F is nothing but solving simultaneously linear equations that correspond to Eq. (2.6), and thus it is not difficult at all.

Now the data set is not M points in N-dimensional space, but those in $K(< N)$-dimensional space spanned by K $\boldsymbol{c}_j$s. In this sense, rank factorization is directly related to geometrical representation of data set in the sense that rank factorization is a projection of N-dimensional space to K-dimensional space.

Exercise

2.2 Apply rank factorization to Eq. (1.30) with

$$C = \begin{pmatrix} -1 & 1 & 1 & 1 \\ 1 & -1 & 1 & 1 \\ 1 & 1 & -1 & 1 \\ 1 & 1 & 1 & -1 \end{pmatrix}. \tag{2.9}$$

2.2.2 *Singular Value Decomposition*

Singular value decomposition (SVD) is one of special cases of rank factorization. In SVD, an $N \times M$ matrix is represented not by a product of two matrices but by a product of three matrices,

$$X = U \Sigma V^T, \tag{2.10}$$

where U is an $N \times N$ orthogonal matrix, V is an $M \times N$ orthogonal matrix,[1] and Σ is $N \times N$ diagonal matrix if $N < M$. Oppositely, if $N > M$, U is an $N \times M$ orthogonal matrix, V is an $M \times M$ orthogonal matrix, and Σ is $M \times M$ diagonal matrix. Here orthogonal matrix is that multiplication with its transposition results in unit matrix, i.e.,

[1] The term "orthogonal matrix" can be used only for square matrix. In this sense, when U or V is not a square matrix, it is not very correct to call them "orthogonal matrix," but it is true that Eq. (2.11) is satisfied even when U or V is not a square matrix, because column vectors of them are orthogonal with each other.

$$U^T U = V^T V = I. \tag{2.11}$$

Since a matrix X is represented not by a product of two matrices but by a product of three matrices, arbitrarity of MF shown in Eq. (2.4) is removed because

$$X = U\Sigma V^T = UI\Sigma V^T = UAB\Sigma V^T \neq UA\Sigma BV^T. \tag{2.12}$$

Thus, SVD can be a unique representative MF of a matrix X.

Here, I am not willing to mathematically prove the existence of SVD for arbitrary matrix because any fundamental linear algebra textbook should have a proof. Instead of that, I briefly introduce SVD from data science point of views.

2.2.2.1 How to Compute SVD

Suppose that Eq. (2.10) is obtained. In the following we assume $N < M$. Then,

$$XX^T = U\Sigma V^T (U\Sigma V^T)^T = U\Sigma V^T V \Sigma^T U^T = U\Sigma^2 U^T \tag{2.13}$$

$$XX^T U = U\Sigma^2 U^T U = U\Sigma^2. \tag{2.14}$$

Because Σ is a diagonal matrix, Σ can be expressed as

$$\Sigma = \begin{pmatrix} \lambda_1 & & & \\ & \lambda_2 & & \\ & & \ddots & \\ & & & \lambda_N \end{pmatrix}, \tag{2.15}$$

and $N \times N$ matrix U can be expressed using $\boldsymbol{u}_i \in \mathbb{R}^N$,

$$U = (\boldsymbol{u}_1^T, \boldsymbol{u}_2^T, \ldots, \boldsymbol{u}_N^T), \tag{2.16}$$

then Eq. (2.14) can be rewritten as

$$XX^T \boldsymbol{u}_i = \lambda_i^2 \boldsymbol{u}_i, 1 \leq i \leq N. \tag{2.17}$$

Thus, computing U is equivalent to the diagonalization of XX^T if we also require $|\boldsymbol{u}_i| = 1$ such that U is an orthogonal matrix as required, because eigen vectors, $\boldsymbol{u}_i$s, are known to be orthogonal with one another.

From Eq. (2.10),

$$X^T = (U\Sigma V^T)^T = V\Sigma U^T \tag{2.18}$$

$$X^T U\Sigma^{-1} = V\Sigma U^T U\Sigma^{-1} = V\Sigma\Sigma^{-1} = V \tag{2.19}$$

$$V = X^T U \Sigma^{-1}. \tag{2.20}$$

Thus we can get

$$\boldsymbol{v}_i = \frac{1}{\lambda_i} X^T \boldsymbol{u}_i \tag{2.21}$$

if we express $M \times N$ matrix $V \in \mathbb{R}^{M \times N}$ using $\boldsymbol{v}_i \in \mathbb{R}^M$ as

$$V = \left(\boldsymbol{v}_1^T, \boldsymbol{v}_2^T, \ldots, \boldsymbol{v}_N^T\right). \tag{2.22}$$

Then from Eqs. (2.17) and (2.21)

$$X^T X X^T \boldsymbol{u}_i = \lambda_i^2 X^T \boldsymbol{u}_i \tag{2.23}$$

$$X^T X \cdot \lambda_i \boldsymbol{v}_i = \lambda_i^2 \cdot \lambda_i \boldsymbol{v}_i \tag{2.24}$$

$$X^T X \boldsymbol{v}_i = \lambda_i^2 \boldsymbol{v}_i. \tag{2.25}$$

Thus, $\boldsymbol{v}_i$ is an eigen vector of $X^T X$. This means V, defined by Eq. (2.22), is an orthogonal matrix if we also require $|\boldsymbol{v}_i| = 1$.

Thus performing diagonalization of Eq. (2.17) together with applying Eq. (2.21), or performing diagonalization of Eq. (2.25) together with applying Eq. (2.26),

$$\boldsymbol{u}_i = \frac{1}{\lambda_i} X \boldsymbol{v}_i \tag{2.26}$$

that is equivalent to

$$U = X V \Sigma^{-1}, \tag{2.27}$$

which can be derived as $V = X^T U \Sigma^{-1}$, we can perform SVD shown in Eq. (2.10).

Exercise
2.3 Apply SVD to

$$X = \begin{pmatrix} 1 & 1 \\ 1 & 1 \\ -1 & 1 \\ 1 & -1 \end{pmatrix}. \tag{2.28}$$

2.2.2.2 Applying SVD to Shop Data

Here we apply SVD to transposed matrix $X_1 = X^T \in \mathbb{R}^{4 \times 2}$ of the matrix $X \in \mathbb{R}^{2 \times 4}$ defined in Eq. (1.9).

$$X_1^T X_1 = \left(X^T\right)^T X^T = XX^T = \begin{pmatrix} 100 & 1000 & 300 & 200 \\ 200 & 300 & 100 & 150 \end{pmatrix} \begin{pmatrix} 100 & 200 \\ 1000 & 300 \\ 300 & 100 \\ 200 & 150 \end{pmatrix}$$

$$= \begin{pmatrix} 1140{,}000 & 380{,}000 \\ 380{,}000 & 162{,}500 \end{pmatrix}. \tag{2.29}$$

We diagonalize $X_1^T X_1$, Eq. (2.29). Eigen equation that eigen value λ should satisfy is

$$\begin{vmatrix} \lambda - 1140{,}000 & 380{,}000 \\ 380{,}000 & \lambda - 162{,}500 \end{vmatrix} = 0 \tag{2.30}$$

$$(\lambda - 1140{,}000)(\lambda - 162{,}500) - 380{,}000^2 = 0 \tag{2.31}$$

$$\lambda^2 - (1140{,}000 + 162{,}500)\lambda + 114{,}000 \cdot 162{,}500 - 380{,}000^2 = 0 \tag{2.32}$$

$$\lambda^2 - 1{,}302{,}500\lambda + 185{,}250{,}000{,}000 - 144{,}400{,}000{,}000 = 0 \tag{2.33}$$

$$\lambda^2 - 1{,}302{,}500\lambda + 40{,}850{,}000{,}000 = 0 \tag{2.34}$$

$$\lambda_\pm = \frac{1{,}302{,}500 \pm \sqrt{1{,}302{,}500^2 - 4 \cdot 40{,}850{,}000{,}000}}{2} \tag{2.35}$$

$$= \frac{1{,}302{,}500 \pm \sqrt{1{,}533{,}106{,}250{,}000}}{2} \tag{2.36}$$

$$= \frac{1{,}302{,}500 \pm 2500\sqrt{245{,}297}}{2} = 651{,}250 \pm 1250\sqrt{245{,}297}. \tag{2.37}$$

Eigen vector $\boldsymbol{v} = (v_1, v_2)^T$ should satisfy

$$\begin{pmatrix} \lambda_\pm - 1{,}140{,}000 & 380{,}000 \\ 380{,}000 & \lambda_\pm - 162{,}500 \end{pmatrix} \boldsymbol{v} = 0 \tag{2.38}$$

$$\begin{pmatrix} -488{,}750 \pm 1250\sqrt{245{,}297} & 380{,}000 \\ 380{,}000 & 488{,}750 \pm 1250\sqrt{245{,}297} \end{pmatrix} \boldsymbol{v} = 0. \tag{2.39}$$

Then we get

$$v_1^\pm = \frac{488{,}750 \pm 1250\sqrt{245{,}297}}{380{,}000} v_2^\pm = \frac{391 \pm \sqrt{245{,}297}}{304} v_2^\pm. \tag{2.40}$$

In order that $V = (\boldsymbol{v}^+, \boldsymbol{v}^-)$ is orthogonal matrix, $|\boldsymbol{v}^+| = |\boldsymbol{v}^-| = 1$.

$$(v_1^{\pm})^2 + (v_2^{\pm})^2 = 1 \tag{2.41}$$

$$\left(\frac{391 \pm \sqrt{245{,}297}}{304}\right)^2 (v_2^{+})^2 + (v_2^{+})^2 = 1 \tag{2.42}$$

$$\frac{391^2 \pm 2 \cdot 391\sqrt{245{,}297} + 245{,}297 + 304^2}{304^2}(v_2^{\pm})^2 = 1 \tag{2.43}$$

$$\frac{490{,}594 \pm 782\sqrt{245{,}297}}{92{,}416}(v_2^{\pm})^2 = 1 \tag{2.44}$$

$$v_2^{\pm} = \sqrt{\frac{92{,}416}{490{,}594 \pm 782\sqrt{245{,}297}}} \simeq 0.3244, 0.9459 \tag{2.45}$$

$$v_1^{\pm} = \frac{391 \pm \sqrt{245{,}297}}{304} v_2^{\pm} \simeq 0.9459, -0.3244 \quad . \tag{2.46}$$

$\boldsymbol{u}^{\pm}$ can be computed via Eq. (2.26),

$$\boldsymbol{u}^{\pm} = \frac{1}{\lambda^{\pm}} X_1 \boldsymbol{v}^{\pm} = \frac{1}{\lambda^{\pm}} \begin{pmatrix} 100 & 200 \\ 1000 & 300 \\ 300 & 100 \\ 200 & 150 \end{pmatrix} \begin{pmatrix} v_1^{\pm} \\ v_2^{\pm} \end{pmatrix} . \tag{2.47}$$

Here we consider what $\boldsymbol{u}^{\pm}$ represents. Because X_1 can be considered to be a set of two four-dimensional vectors $\boldsymbol{x}_1 \in \mathbb{R}^4$ and $\boldsymbol{x}_2 \in \mathbb{R}^4$ as $X_1 = (\boldsymbol{x}_1, \boldsymbol{x}_2)$, their relations should be represented in two dimensional space, since there can be only two independent vectors. In this sense, $\boldsymbol{u}^{\pm}$ represents how $\boldsymbol{x}_1$ and $\boldsymbol{x}_2$ should be combined to form two dimensional space that represents the relation between $\boldsymbol{x}_1$ and $\boldsymbol{x}_2$.

In Fig. 1.3, we had to omit the forth scholar, fish, in order to represent the geometrical relationship between price ($\boldsymbol{x}_1$) and weight ($\boldsymbol{x}_2$). Nevertheless, we can represent the relation between price and weight in the plane using $\boldsymbol{u}^{\pm}$. From Eq. (2.47), we can get

$$\left(\lambda^{+}\boldsymbol{u}^{+}, \lambda^{-}\boldsymbol{u}^{-}\right) = X_1 \left(\boldsymbol{v}^{+}, \boldsymbol{v}^{-}\right) . \tag{2.48}$$

Then

$$X_1 = \left(\lambda^{+}\boldsymbol{u}^{+}, \lambda^{-}\boldsymbol{u}^{-}\right)\left(\boldsymbol{v}^{+}, \boldsymbol{v}^{-}\right)^T = \left(v_1^{+}\lambda^{+}\boldsymbol{u}^{+} + v_1^{-}\lambda^{-}\boldsymbol{u}^{-}, v_2^{+}\lambda^{+}\boldsymbol{u}^{+} + v_2^{-}\lambda^{-}\boldsymbol{u}^{-}\right) \tag{2.49}$$

since $\left(\boldsymbol{v}^{+}, \boldsymbol{v}^{-}\right)$ is orthogonal matrix. Thus in the plane spanned by $\lambda^{\pm}\boldsymbol{u}^{\pm}$, price ($\boldsymbol{x}_1$) and weight ($\boldsymbol{x}_2$) can be two points having coordinates $\left(v_1^{+}, v_1^{-}\right)$ and $\left(v_2^{+}, v_2^{-}\right)$, respectively.

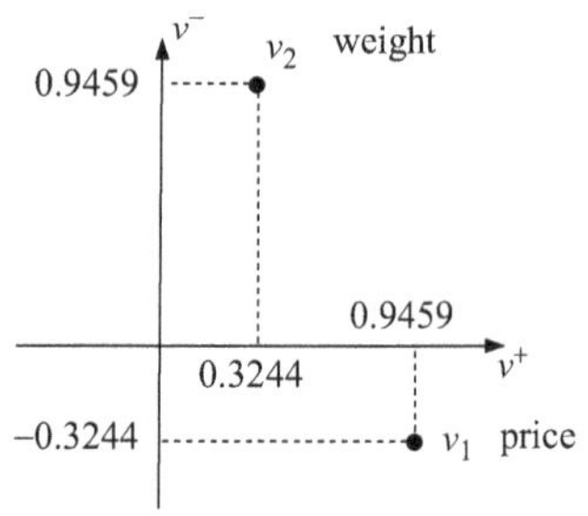

Fig. 2.1 A geometrical interpretation of price and weight originally shown in Fig. 1.3. $\boldsymbol{v}_1$: price, $\boldsymbol{v}_2$: weight

Figure 2.1 shows the geometrical interpretation of price and weight using the results given by SVD. In contrast to Fig. 1.3 where fish must be inevitably omitted, Fig. 2.1 does not omit anything but keep all information. Instead of that, it is difficult to interpret the meaning of axes, each of which simply represents foods: bread, beef, pork, in Fig. 1.3. Two axes in Fig. 2.1 represent a linear combination of foods represented as the four-dimensional vectors, $\lambda^+\boldsymbol{u}^+$ and $\lambda^-\boldsymbol{u}^-$, respectively, although we do not write down them here because they are at most confusing and are not helpful for our understanding at all.

Thus, it turns out that there is a trade-off; if we would like to keep interpretability of axes, we cannot represent the relation of features in the easily visible lower dimensional space. On the other hand, if we would like to have geometrical representation that can be easy to understand as shown in Fig. 2.1, we cannot avoid to lose the interpretability of axes. In some sense, the purpose of data science is to make balance between these two problems, i.e., interpretability of axes or that of relation of features. Most of the popular methods ever proposed are aiming to achieve this purpose. The fact that so many methods are proposed definitely suggests that there is still not a unique (the best) solution for this problem. The purpose of this book is also to add yet another solution to solve this problem effectively.

2.3 Principal Component Analysis

In the previous section, we demonstrated that SVD can give the plane that can represent the relation between two features, price and weight, in lower dimensional space, which is more easily interpreted than original four-dimensional space spanned by four foods: bread, pork, beef, and fish. It is also shown that SVD can be performed via diagonalization of matrix products, X^TX or XX^T. Apparently, although they seem to be nothing but mathematical or technical relationships, they actually do not. Diagonalization of these two matrix products is deeply related to principal component analysis (PCA) [2].

PCA is mathematically defined as the diagonalization of covariance matrix $S_{ii'} \in \mathbb{R}^{N\times N}$,

$$S_{ii'} = \left\langle \left(x_{ij'} - \left\langle x_{ij} \right\rangle_j \right) \cdot \left(x_{i'j'} - \left\langle x_{i'j} \right\rangle_j \right) \right\rangle_{j'} \tag{2.50}$$

$$= \left\langle x_{ij'} x_{i'j'} - \left\langle x_{ij} \right\rangle_j x_{i'j'} - x_{ij'} \left\langle x_{i'j} \right\rangle_j + \left\langle x_{ij} \right\rangle_j \left\langle x_{i'j} \right\rangle_j \right\rangle_{j'} \tag{2.51}$$

$$= \left\langle x_{ij'} x_{i'j'} \right\rangle_{j'} - \left\langle \left\langle x_{ij} \right\rangle_j x_{i'j'} \right\rangle_{j'} - \left\langle x_{ij'} \left\langle x_{i'j} \right\rangle_j \right\rangle_{j'} + \left\langle \left\langle x_{ij} \right\rangle_j \left\langle x_{i'j} \right\rangle_j \right\rangle_{j'} \tag{2.52}$$

$$= \left\langle x_{ij'} x_{i'j'} \right\rangle_{j'} - \left\langle x_{ij} \right\rangle_j \left\langle x_{i'j'} \right\rangle_{j'} - \left\langle x_{ij'} \right\rangle_{j'} \left\langle x_{i'j} \right\rangle_j + \left\langle x_{ij} \right\rangle_j \left\langle x_{i'j} \right\rangle_j \tag{2.53}$$

$$= \left\langle x_{ij'} x_{i'j'} \right\rangle_{j'} - \left\langle x_{ij} \right\rangle_j \left\langle x_{i'j'} \right\rangle_{j'}, \tag{2.54}$$

where

$$\left\langle x_{ij} x_{i'j} \right\rangle_j = \frac{1}{M} \sum_j x_{ij} x_{i'j} \tag{2.55}$$

$$\left\langle x_{ij} \right\rangle_j = \frac{1}{M} \sum_j x_{ij} \tag{2.56}$$

and $x_{ij} \in \mathbb{R}^{N \times M}$.

It is obvious that Eq. (2.50) is equivalent to XX^T if $\langle x_{ij} \rangle_j = 0$. Thus, PCA is equivalent to SVD in special cases.

Exercise

2.4 Apply PCA to Eq. (2.28).

2.4 Equivalence Between PCA and SVD

As can be seen in the previous section, the difference between SVD and PCA is simply if $\left\langle x_{ij} \right\rangle_j = 0$ or not. Nonetheless, it is not frequently discussed from the view point of data science how the difference affects the outcome. Suppose that S is the matrix whose component is $S_{ii'}$ given in Eq. (2.54). We also define vectors $\langle \boldsymbol{S}_i \rangle$,

$$\langle \boldsymbol{S}_i \rangle = \left(\left\langle x_{1j} \right\rangle_j, \left\langle x_{2j} \right\rangle_j, \ldots \left\langle x_{ij} \right\rangle_j, \ldots \left\langle x_{Nj} \right\rangle_j \right), \tag{2.57}$$

whose components are columnwise mean of X. Then using Eq. (2.10), S can be decomposed as

$$S = \frac{XX^T}{M} - \langle \boldsymbol{S}_i \rangle \times^0 \langle \boldsymbol{S}_i \rangle = \frac{1}{M} U \Sigma V^T \left(U \Sigma V^T \right)^T - \langle \boldsymbol{S}_i \rangle \times^0 \langle \boldsymbol{S}_i \rangle$$

$$= U \frac{\Sigma^2}{M} U^T - \langle \boldsymbol{S}_i \rangle \times^0 \langle \boldsymbol{S}_i \rangle. \tag{2.58}$$

On the other hand, applying PCA to S, we should get

$$SU' = U'\Lambda, \tag{2.59}$$

where $U' \in \mathbb{R}^{N\times N}$ is an orthogonal matrix and $\Lambda \in \mathbb{R}^{N\times N}$ is a diagonal matrix. Then

$$S = U'\Lambda U'^T. \tag{2.60}$$

Thus generally $U \neq U'$, and there are no ways to compute U' directly from U and $\langle \boldsymbol{S}_i \rangle$.

SVD can also be performed by the diagonalization of $X^T X$. Covariance matrix $S_{jj'} \in \mathbb{R}^{M\times M}$ is redefined as

$$S_{jj'} = \left\langle x_{ij} x_{ij'} \right\rangle_i - \left\langle x_{ij} \right\rangle_i \left\langle x_{ij'} \right\rangle_i . \tag{2.61}$$

Then using

$$\langle \boldsymbol{S}_j \rangle = \left(\langle x_{i1} \rangle_i , \langle x_{i2} \rangle_i , \ldots , \left\langle x_{ij} \right\rangle_i \ldots \langle x_{iM} \rangle_i \right), \tag{2.62}$$

we get

$$\begin{aligned} S = \frac{X^T X}{N} - \left\langle \boldsymbol{S}_j \right\rangle \times^0 \left\langle \boldsymbol{S}_j \right\rangle &= \frac{1}{N} \left(U \Sigma V^T \right)^T U \Sigma V^T - \left\langle \boldsymbol{S}_j \right\rangle \times^0 \left\langle \boldsymbol{S}_j \right\rangle \\ = V \frac{\Sigma^2}{N} V^T - \left\langle \boldsymbol{S}_j \right\rangle \times^0 \left\langle \boldsymbol{S}_j \right\rangle . \end{aligned} \tag{2.63}$$

Applying PCA to $S = X^T X \in \mathbb{R}^{M\times M}$, we get

$$SV' = V'\Lambda, \tag{2.64}$$

where $V' \in \mathbb{R}^{M\times M}$ is an orthogonal matrix and $\Lambda \in \mathbb{R}^{M\times M}$ is a diagonal matrix. Then

$$S = V'\Lambda V'^T. \tag{2.65}$$

Again, generally $V \neq V'$, and there are no ways to generate V' only from the information of V and $\langle \boldsymbol{S}_j \rangle$.

Thus, although diagonalization of $X^T X$ is equivalent to that of $X X^T$ in SVD, this does not stand for PCA because of columnwise or rowwise mean extraction. Once mean is extracted from a matrix X columnwisely, it is impossible to reproduce original matrix X or rowwisely mean extracted matrix. Since PCA is more frequently employed than SVD in data science, this inequality between PCA applied to $S_{ii'}$ and $S_{jj'}$ should be taken care of. From the data science point of views, if columnwise or rowwise mean extraction should be performed is not easy to decide in advance. It cannot be determined without the knowledge about the data set to which PCA is applied. This knowledge is often quoted as domain knowledge, which is often considered to be "untouched" by data scientists. Nonetheless, even when simple linear algebra like PCA is considered, domain knowledge cannot be avoided as shown in the above.

Exercise
2.5 Compare the solutions of Problems 2.3 and 2.4.

2.5 Geometrical Representation of PCA

In contrast to SVD, PCA is often discussed from the geometrical point of views. In this section, I would like to summarize some of geometrical interpretation of PCA, since it is also benefitable to interpret the geometrical representation of SVD.

2.5.1 *PCA Selects the Axis with the Maximal Variance*

Suppose that $U \in \mathbb{R}^{N \times N}$ is an orthogonal matrix. $X \in \mathbb{R}^{N \times M}$ is considered to be M N-dimensional vectors as in Eq. (2.5). Next, we apply columnwise mean extraction, i.e.,

$$\bar{X} = X - \underbrace{\left(\langle S_i \rangle^T, \ldots, \langle S_i \rangle^T\right)}_{M}, \tag{2.66}$$

where the second term of the right hand side is $N \times M$ matrix. Multiplying U to $\bar{X}$, we get a new matrix X' as

$$X' = U^T \bar{X}. \tag{2.67}$$

Thus

$$\frac{X' X'^T}{M} = \frac{1}{M} U^T \bar{X} (U^T \bar{X})^T = U^T \frac{\bar{X}\bar{X}^T}{M} U = U^T S U, \tag{2.68}$$

where $\frac{\bar{X}\bar{X}^T}{M} = S \in \mathbb{R}^{N \times N}$ is covariance matrix. If we can choose U such that $\frac{X'X'^T}{M}$ is diagonal, Λ, this is nothing but PCA, Eq. (2.59).

In this calculation, Eq. (2.67) can be considered to be coordinate transformation since

$$x'_{ij} = \sum_{i'} u_{ii'} \bar{x}_{i'j}. \tag{2.69}$$

What does the requirement that $X'X^T$ should be diagonal correspond to? As can be seen in the below, it is equivalent to the condition that x'_{ij} should have maximal variances, S_{ii}. Because of mean extraction defined in Eq. (2.66),

$$\left\langle \bar{x}_{ij} \right\rangle_j = 0. \tag{2.70}$$

Thus,

$$\left\langle x'_{ij} \right\rangle_j = 0 \tag{2.71}$$

as well. Then $S_{ii} = \left\langle x'_{ij} x'_{ij} \right\rangle_j$ and maximizing S_{ii} is equivalent to maximizing

$$\sum_j x'^2_{ij} - \lambda \left(\sum_{i'} u^2_{ii'} - 1 \right) = \sum_{i_1} \sum_{i_2} u_{ii_1} u_{ii_2} \left(\sum_j \bar{x}_{i_1 j} \bar{x}_{i_2 j} - \lambda \delta_{i_1 i_2} \right) + \lambda, \tag{2.72}$$

where Eq. (2.69) is substituted. The terms multiplied by λ are required such that

$$\boldsymbol{u}_i = (u_{i1}, u_{i2}, \ldots, u_{iN}) \tag{2.73}$$

is a unit vector; this requirement must be fulfilled in order that U is orthogonal. In order that, $u_{ii'}$ should satisfy

$$\frac{\partial}{\partial u_{ii_1}} \left\{ \sum_j x'^2_{ij} - \lambda \left(\sum_{i'} u^2_{ii'} - 1 \right) \right\} = \sum_{i_2} u_{ii_2} \left(\sum_j \bar{x}_{i_1 j} \bar{x}_{i_2 j} - \lambda \delta_{i_1 i_2} \right) = 0. \tag{2.74}$$

In order to have solutions other than the trivial solution, $\boldsymbol{u}_i = 0$, we need to solve the eigen value problem,

$$\sum_{i_2} u_{ii_2} \sum_j \bar{x}_{i_1 j} \bar{x}_{i_2 j} = \lambda u_{ii_1}. \tag{2.75}$$

Or equivalently,

$$NSU = U\Lambda, \tag{2.76}$$

which is nothing but PCA, Eq. (2.59), excluding a prefactor N in the left hand side. Thus applying PCA is nothing but generating the new feature x'_{ij} from $\bar{x}_{ij}$ so as to have maximum variance along the new axis.

Eigen value problem gives us more than one eigen value. The largest one corresponds to the maximal S_{ii}. We would like to discuss what the second largest eigen value corresponds to. In the subspace to the eigen vector $\boldsymbol{u}_i$ that corresponds to the largest eigen values, try to find direction $\boldsymbol{u}'_i$ along which the largest variance is given. This can be achieved by maximizing

$$\sum_j x_{ij}'^2 - \lambda \left(\sum_{ii'} u_{ii'}'^2 - 1 \right) - \alpha \left(\sum_{i'} u_{ii'} u'_{ii'} \right) \tag{2.77}$$

$$= \sum_{i_1} \sum_{i_2} u'_{ii_1} u'_{ii_2} \left(\sum_j \bar{x}_{i_1 j} \bar{x}_{i_2 j} - \lambda \delta_{i_1 i_2} \right) + \lambda - \alpha \left(\sum_{i'} u_{ii'} u'_{ii'} \right) \tag{2.78}$$

$$= \sum_{i_1} u'_{ii_1} \sum_{i_2} \left(u'_{ii_2} \sum_j \bar{x}_{i_1 j} \bar{x}_{i_2 j} - u'_{ii_2} \lambda \delta_{i_1 i_2} - \alpha u_{ii_2} \delta_{i_1 i_2} \right) + \lambda. \tag{2.79}$$

The last term in Eq. (2.78) is required such that $\boldsymbol{u}_i \perp \boldsymbol{u'}_i$.

Maximization is performed by

$$\frac{\partial}{\partial u'_{ii_1}} \left\{ \sum_{i_1} u'_{ii_1} \sum_{i_2} \left(u'_{ii_2} \sum_j \bar{x}_{i_1 j} \bar{x}_{i_2 j} - u'_{ii_2} \lambda \delta_{i_1 i_2} - \alpha u_{ii_2} \delta_{i_1 i_2} \right) + \lambda \right\} \tag{2.80}$$

$$= \sum_{i_2} \left(u'_{ii_2} \sum_j \bar{x}_{i_1 j} \bar{x}_{i_2 j} - u'_{ii_2} \lambda \delta_{i_1 i_2} - \alpha u_{ii_2} \delta_{i_1 i_2} \right) = 0 \tag{2.81}$$

$$\sum_{i_2} u'_{ii_2} \sum_j \bar{x}_{i_1 j} \bar{x}_{i_2 j} - \lambda u'_{ii_1} = \alpha u_{ii_1}. \tag{2.82}$$

Multiplying u_{ii_1} and taking summation of i_1, we get

$$\sum_{i_2} u'_{ii_2} \sum_{i_1} \sum_j \bar{x}_{i_1 j} \bar{x}_{i_2 j} u_{ii_1} - \lambda \sum_{i_1} u_{ii_1} u'_{ii_1} = \alpha \sum_{i_1} u_{ii_1}^2 = \alpha. \tag{2.83}$$

Because of Eq. (2.75), we get

$$\sum_{i_2} u'_{ii_2} \lambda u_{ii_2} - \lambda \sum_{i_1} u_{ii_1} u'_{ii_1} = \alpha = 0. \tag{2.84}$$

Thus Eq. (2.82) is now

$$\sum_{i_2} u'_{ii_2} \sum_j \bar{x}_{i_1 j} \bar{x}_{i_2 j} = \lambda u'_{ii_1}. \tag{2.85}$$

This is the same eigen value problem as PCA. Because

$$\frac{\partial}{\partial \alpha} \left\{ \sum_j x'^2_{ij} - \lambda \left(\sum_{ii'} u'^2_{ii'} - 1 \right) - \alpha \left(\sum_{i'} u_{ii'} u'_{ii'} \right) \right\} = \sum_{i'} u_{ii'} u'_{ii'} = 0 \tag{2.86}$$

$\boldsymbol{u'}_i \perp \boldsymbol{u}_i$. This is satisfied by restricting eigen vectors other than the first eigen vector. Since the second eigen vector has maximum eigen values among those other than the first eigen vector, the second eigen vector represents the direction that is both associated with the maximum variance and perpendicular to the first eigen vectors. As such, the nth eigen values, λ_n, are always equivalent to the maximal variance along the axis included in the subspace perpendicular to all eigen vectors $\boldsymbol{u}_i, i < n$.

Thus, if we employ the first n eigen vectors, $\boldsymbol{u}_i, i \leq n$, in order to represent samples or features, it is the geometrical representation to include maximal variance that can be expressed within n-dimensional space. In this sense, PCA can be considered to be a most effective (i.e., minimum loss of information) geometrical representation of given data set expressed as a matrix.

Exercise

2.6 Compute variances along the directions parallel to the eigen vectors given in Problems 2.4.

2.5.2 *PCA Selects the Axis with Minimum Residuals*

In the previous section, it was shown that PCA can give us the most effective geometrical representation within a given number of dimensions n. In this section, though it is equivalent, the geometrical representation given by PCA supports minimum residuals.

In order to compute residuals when Eq. (2.69) is employed, we need to find projection of $\bar{\boldsymbol{x}}_j = (\bar{x}_{1j}, \bar{x}_{2j}, \ldots, \bar{x}_{Nj})$ onto $\boldsymbol{u}_i$. The ith component of the projection is computed as

$$\sum_{i'} u_{ii'} \bar{x}_{i'j}. \tag{2.87}$$

Thus, the projection itself is defined as

$$\left(\boldsymbol{u}_i^T \cdot \bar{\boldsymbol{x}}_j \right) \boldsymbol{u}_i. \tag{2.88}$$

Then squared residual R^2 can be computed as

$$R^2 = \sum_j \left\{ \bar{\boldsymbol{x}}_j - \left(\boldsymbol{u}_i^T \cdot \bar{\boldsymbol{x}}_j \right) \boldsymbol{u}_i \right\}^2 \tag{2.89}$$

$$= \sum_j \left\{ \bar{\boldsymbol{x}}_j^T \bar{\boldsymbol{x}}_j - 2 \left(\boldsymbol{u}_i^T \cdot \bar{\boldsymbol{x}}_j \right)^2 + \left(\boldsymbol{u}_i^T \cdot \bar{\boldsymbol{x}}_j \right)^2 \boldsymbol{u}_i^T \cdot \boldsymbol{u}_i \right\} \tag{2.90}$$

$$= \sum_j \left\{ \bar{\boldsymbol{x}}_j^T \bar{\boldsymbol{x}}_j - \left(\boldsymbol{u}_i^T \cdot \bar{\boldsymbol{x}}_j \right)^2 \right\}. \tag{2.91}$$

Since $\bar{\boldsymbol{x}}_j^T \bar{\boldsymbol{x}}_j$ is constant, minimizing R^2 is equivalent to maximizing $\sum_j \left(\boldsymbol{u}_i^T \cdot \bar{\boldsymbol{x}}_j \right)^2$. Because of Eq. (2.69), $\sum_j \left(\boldsymbol{u}_i^T \cdot \bar{\boldsymbol{x}}_j \right)^2 = \sum_j \boldsymbol{x}'^T_j \boldsymbol{x}'_j = N S_{ii}$. Since PCA is proven to maximize S_{ii}, PCA is now proven to minimizing residuals, too. Thus, also in this sense, PCA can be considered to be a most effective (i.e., minimum loss of information) geometrical representation of given data set expressed as a matrix, too.

Exercise
2.7 Compute residuals around the directions of eigen vectors given in Problems 2.4.

2.5.3 Nonequivalence Between Two PCAs

In the previous two subsections, I have shown two equivalent geometrical interpretations of low dimensional representation given by the PCA, in the senses:

1. The geometrical space spanned by n principal components, $\boldsymbol{u}_i$, represents the those with the maximum variance.
2. The geometrical space spanned by n principal components, $\boldsymbol{u}_i$, represents the those with the minimum residuals.

On the other hand, in contrast to SVD, since PCA diagonalizes covariance matrix, applying PCA to X and X^T differs. This is because $S_{ii'}$ defined by Eq. (2.54) differs from $S_{jj'}$ defined by Eq. (2.61). Then the next question is how these two n-dimensional representations if $S_{ii'}$ or $S_{jj'}$ is employed differ with each other.

Generally speaking, it is completely unpredictable. It is very easy to add some matrix X^0 that satisfies

$$\langle x_{ij}^0 \rangle_j = 0 \tag{2.92}$$

$$\langle x_{ij}^0 \rangle_i \neq 0, \tag{2.93}$$

i.e., a matrix with zero columnwise mean and nonzero rowwise mean, to matrix X. This procedure does not affect $S_{ii'}$ at all, while it changes $S_{jj'}$. Thus they do toward

n-dimensional representation, too. Therefore, we cannot expect any equivalence between two PCAs diagonalizing $S_{ii'}$ or $S_{jj'}$. This often matters in data sciences. In contrast to the physical or social sciences where the target of study is clear, in data science, even what should be targeted is decided in the data-driven way. In Fig. 2.1, the relation between weight and price can be viewed only after applying SVD. It is impossible to decide how we apply PCA to data set in advance.

2.6 PCA as a Clustering Method

Usually, PCA is considered to be a kind of embedding method that represents the relationship among objects as geometrical fashion as demonstrated in the previous sections. Nonetheless, PCA can also be considered as a sort of clustering analysis that represents the relationship between objects by grouping [1]. Although there are many methods that cluster data points, clustering method whose equivalence with PCA is proven is K-means. K-means is one of the so-called centroid methods that define multiple centroids to be used as centers of generated clusters. K-means requires to find centroids, $\boldsymbol{m}_k \in \mathbb{R}^N, k = 1, \ldots, K$ when matrix $X \in \mathbb{R}^{N \times M}$ is considered to be a set of M N-dimensional vectors, $\boldsymbol{x}_j \in \mathbb{R}^N, j = 1, \ldots, M$, that minimizes

$$J_K = \sum_{k=1}^{K} \sum_{j \in C_k} \left(\boldsymbol{x}_j - \boldsymbol{m}_k\right)^2, \tag{2.94}$$

where

$$\boldsymbol{m}_k = \frac{1}{n_k} \sum_{j \in C_k} \boldsymbol{x}_j \tag{2.95}$$

with n_k being the number of $j \in C_k$. Equation (2.94) represents squared summation of deviations between centroids and $\boldsymbol{x}_j$ within each cluster C_k. Here each j is supposed to belong to C_k whose centroid $\boldsymbol{m}_k$ is the nearest to $\boldsymbol{x}_j$. Thus the task is to identify a set of $(\boldsymbol{m}_k, C_k), k = 1, \ldots, K$.

Suppose we define centroid subspace as that spanned by K centroids. Then the projection to centroids can be defined as:

Definition 2.1 The projection of any vector $\boldsymbol{x}$ to centroid subspace is

$$S_b \boldsymbol{x} = \sum_{k=1}^{K} n_k (\boldsymbol{m}_k^T \cdot \boldsymbol{x}) \boldsymbol{m}_k, \tag{2.96}$$

where

$$S_b = \sum_{k=1}^{K} n_k \boldsymbol{m}_k \times^0 \boldsymbol{m}_k \tag{2.97}$$

is the between center scattered matrix with n_k being the number of $j \in C_k$. The centroid subspace is generally considered to be the subspace in which K clusters are visibly well separated. Thus, obtaining centroid subspace is essential to see how K clusters are separated with each other.

In order to demonstrate that the projection onto the centroid subspace exhibits the clustered structure, we applied it to artificial data set. This data set consists of a matrix $X \in \mathbb{R}^{10\times 30}$. All components x_{ij} obey normal distribution, $\mathcal{N}(\mu, \sigma)$, with the mean of μ and the standard deviation of σ; $\sigma = 1$, while mean, μ, varies as follows:

$$\mu = \begin{cases} \sqrt{2}, & 1 \le j \le 10, \quad 1 \le i \le 5 \\ -1, & 11 \le j \le 20, \\ -1, & 21 \le j \le 30, \quad 1 \le i \le 5 \\ 1, & 21 \le j \le 30, \ 6 \le i \le 10 \\ 0, & \text{otherwise.} \end{cases} \tag{2.98}$$

This says that js are divided into three clusters as $C_1 = \{1 \le j \le 10\}$, $C_2 = \{11 \le j \le 20\}$, and $C_3 = \{21 \le j \le 30\}$ (see also Fig. 2.2). Although no $\boldsymbol{x}_i$ represents clear separation between three clusters, C_1, C_2 and C_3 (see Fig. 2.3), it is rather obvious that $S_b \boldsymbol{x}_i$, $1 \le i \le 10$ shown in Fig. 2.4 exhibits the more pronounced cluster structure than $\boldsymbol{x}_i$ (see also Appendix).

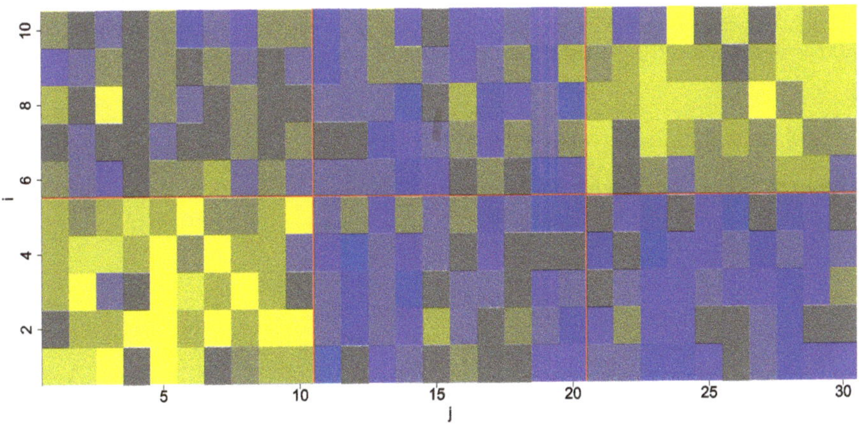

Fig. 2.2 Visualization of $x_{ij} \sim \mathcal{N}(\mu, 1)$, where μ is given by Eq. (2.98). Vertical red lines represent boundary between clusters, C_1, C_2 and C_3. The horizontal red line indicates the boundary between $1 \le i \le 5$ and $6 \le i \le 10$. Yellow (Blue) corresponds to larger (smaller) values

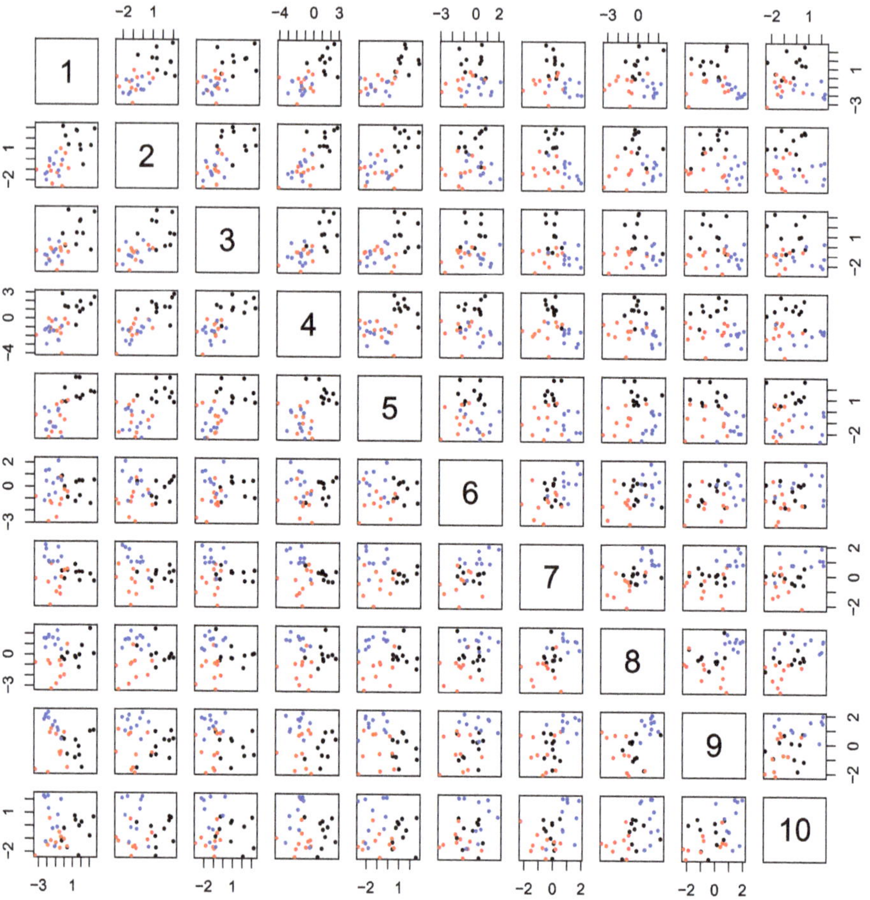

Fig. 2.3 Pairwise scatter plot of $\boldsymbol{x}_i \sim \mathcal{N}(\mu, 1), 1 \leq i \leq 10$ where μ is defined in Eq. (2.98). js that belong to clusters C_1, C_2, and C_3 are represented in black, red, and blue

Now we would like to relate PCA to K-means.

Theorem 2.1 *Cluster centroid subspace is spanned by the first $K-1$ principal directions, i.e.,*

$$S_b = \sum_{k=1}^{K-1} \lambda_k \boldsymbol{u}_k \times^0 \boldsymbol{u}_k, \tag{2.99}$$

where $\boldsymbol{u}_k \in \mathbb{R}^N$ is the kth principal component (PC) given by PCA.

Proof See Appendix.

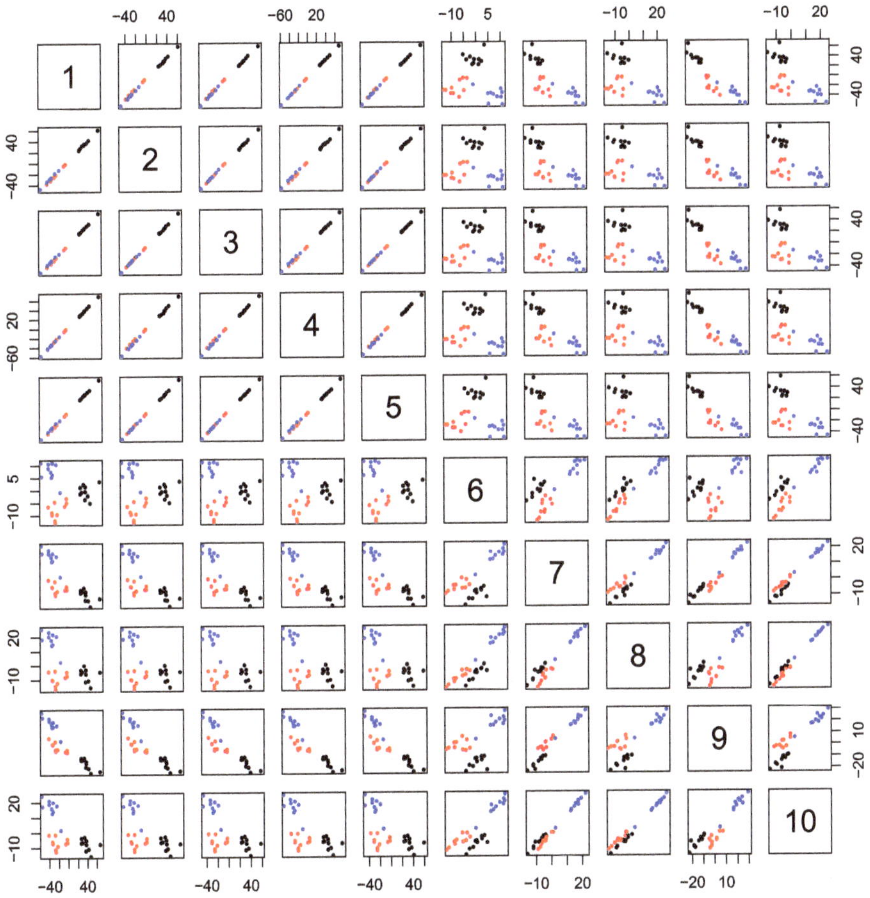

Fig. 2.4 Pairwise scatter plot of $S_b \boldsymbol{x}_i$, $1 \le i \le 10$ using S_b defined in Eq. (2.97). js that belong to clusters C_1, C_2, and C_3 are represented in black, red, and blue

In order to show equivalence of S_b defined in Eq. (2.99) presented by the Theorem 2.1 and that defined in Eq. (2.97), we have shown the pairwise scatter plot of $S_b \cdot \boldsymbol{x}_i$ using S_b computed by Eq. (2.99) in Fig. 2.5. It is also obvious that scatter plots in Fig. 2.5 are coincident with the three clusters. In order to further emphasize the equivalence between Eqs. (2.97) and (2.99), we have shown the scatter plot between $N^2 = 100$ elements of S_bs defined by Eqs. (2.97) and (2.99), respectively. The lack of complete coincidence is because proof of Theorem 2.1 requires complete clustering, while it can never be fulfilled in the real data set.

Anyway, it is obvious that PCA can be used for cluster realization when there are more or less clear clusters. In the general data science course, it is usually taught that embedding methods including PCA can visualize something different from those by clustering method. However, as we could see here, it is not very true since PCA can also visualize clustering if there are clusters, by projecting data onto the space.

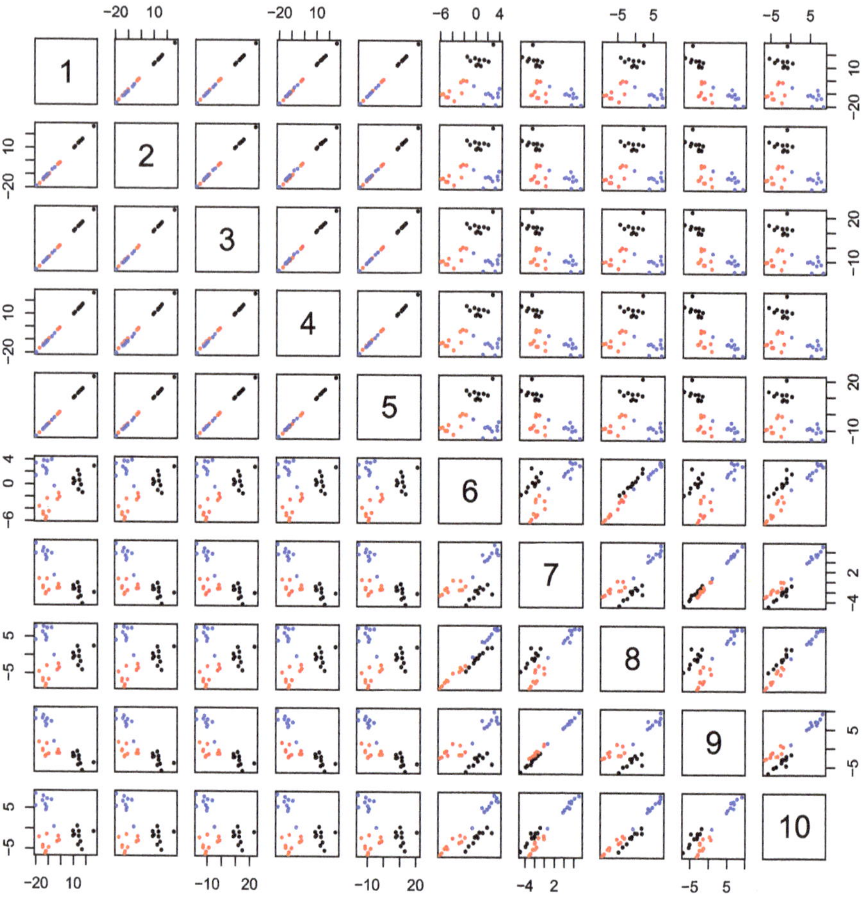

Fig. 2.5 Pairwise scatter plot of $S_b\boldsymbol{x}_i$, $1 \le i \le 10$ using S_b defined in Eq. (2.99). js that belong to clusters C_1, C_2, and C_3 are represented in black, red, and blue

Exercise

2.8 Apply PCA to

$$
X = \begin{pmatrix}
1 & 1 & 1 & 0 & 0 & 0 & 0 & 0 & 0 \\
1 & 1 & 1 & 0 & 0 & 0 & 0 & 0 & 0 \\
1 & 1 & 1 & 0 & 0 & 0 & 0 & 0 & 0 \\
0 & 0 & 0 & 1 & 1 & 1 & 0 & 0 & 0 \\
0 & 0 & 0 & 1 & 1 & 1 & 0 & 0 & 0 \\
0 & 0 & 0 & 1 & 1 & 1 & 0 & 0 & 0 \\
0 & 0 & 0 & 0 & 0 & 0 & 1 & 1 & 1 \\
0 & 0 & 0 & 0 & 0 & 0 & 1 & 1 & 1 \\
0 & 0 & 0 & 0 & 0 & 0 & 1 & 1 & 1
\end{pmatrix} \tag{2.100}
$$

and see if three clusters can be seen. Use any kinds of script language, e.g., R or python, to do this if necessary.

Appendix

Proof of Theorem 2.1

If we define vectors

$$\boldsymbol{h}_k = (0, \ldots, 0, \overbrace{1, \ldots, 1}^{n_k}, 0, \ldots, 0)^T / n_k^{1/2}, \tag{2.101}$$

which represent the members that belong to kth cluster, $\boldsymbol{m}_k$ can be rewritten as

$$\boldsymbol{m}_k = \frac{1}{n_k} \sum_{j \in C_k} \boldsymbol{y}_j = \frac{1}{\sqrt{n_k}} \sum_j \boldsymbol{h}_k(j) \boldsymbol{y}_j = \frac{1}{\sqrt{n_k}} Y \boldsymbol{h}_k \tag{2.102}$$

with defining

$$\boldsymbol{y}_j = \boldsymbol{x}_j - \langle \boldsymbol{x}_j \rangle_j \tag{2.103}$$

and

$$Y = \left(\boldsymbol{y}_1, \ldots, \boldsymbol{y}_M \right). \tag{2.104}$$

In the above, we redefined $\boldsymbol{m}_k$ using $\boldsymbol{y}_j$ instead of $\boldsymbol{x}_j$ in order to relate K-means to PCA more easily. Then S_b can also be rewritten as

$$S_b = \sum_{k=1}^{K} Y \boldsymbol{h}_k \times^0 Y \boldsymbol{h}_k = Y \left(\sum_{k=1}^{K} \boldsymbol{h}_k \times^0 \boldsymbol{h}_k \right) Y^T, \tag{2.105}$$

and J_k, Eq. (2.94), can be rewritten with Eq. (2.95) as

$$J_k = \sum_{k=1}^{K} \sum_{j \in C_k} \left(\boldsymbol{x}_j - \frac{1}{n_k} \sum_{j' \in C_k} \boldsymbol{x}_{j'} \right)^2 \tag{2.106}$$

$$= \sum_{k=1}^{K} \sum_{j \in C_k} \left(\boldsymbol{x}_j^2 - \frac{2}{n_k} \sum_{j' \in C_k} \boldsymbol{x}_j \boldsymbol{x}_{j'} + \frac{1}{n_k^2} \sum_{j', j'' \in C_k} \boldsymbol{x}_{j'} \boldsymbol{x}_{j''} \right) \tag{2.107}$$

$$
= \sum_{k=1}^{K} \sum_{j \in C_k} \boldsymbol{x}_j^2 - \frac{2}{n_k} \sum_{k=1}^{K} \sum_{j,j' \in C_k} \boldsymbol{x}_j \boldsymbol{x}_{j'}
$$

$$
+ \sum_{k=1}^{K} \left(\frac{\sum_{j \in C_k} 1}{n_k^2} \right) \sum_{j',j'' \in C_k} \boldsymbol{x}_{j'} \boldsymbol{x}_{j''} \tag{2.108}
$$

$$
= \sum_{j} \boldsymbol{x}_j^2 - \frac{1}{n_k} \sum_{k=1}^{K} \sum_{j,j' \in C_k} \boldsymbol{x}_j \boldsymbol{x}_{j'}. \tag{2.109}
$$

Using

$$
X = \left(\boldsymbol{x}_1, \ldots, \boldsymbol{x}_j, \ldots, \boldsymbol{x}_M \right) \tag{2.110}
$$

$$
H_K = \left(\boldsymbol{h}_1, \ldots, \boldsymbol{h}_K \right), \tag{2.111}
$$

J_k can be represented as

$$
J_k = \mathrm{Tr}\left(X^T X \right) - \mathrm{Tr}\left(H_K^T X^T X H_K \right). \tag{2.112}
$$

Since H_k that minimizes J_k is not altered even if X is replaced with Y as

$$
J_k = \mathrm{Tr}\left(Y^T Y \right) - \mathrm{Tr}\left(H_K^T Y^T Y H_K \right), \tag{2.113}
$$

H_k that minimizes J_k maximizes $\mathrm{Tr}\,(S_b)$, which is also represented as

$$
\mathrm{Tr}\,(S_b) = \mathrm{Tr}\left(H_K^T Y^T Y H_K \right). \tag{2.114}
$$

There is a theorem:

Theorem 2.2 *Let A be a symmetric matrix with eigenvalues $\lambda_1 \geq \cdots \geq \lambda_n$ and corresponding eigenvectors $(\boldsymbol{v}_1, \ldots, \boldsymbol{v}_n)$. The maximization of $Tr(QAQ)$ subject to constraints $Q^T Q = I_K$ has the solution $Q = (\boldsymbol{v}_1, \ldots, \boldsymbol{v}_K)R$, where R is an arbitrary $K \times K$ orthogonal matrix. And max $Tr(QAQ) = \lambda_1 + \cdots + \lambda_K$.*

Thus, max S_b can be given as

$$
\max_{H_K} \mathrm{Tr}\,(S_b) = \mathrm{Tr}\left(V^T Y^T Y V \right) = \mathrm{Tr}\left(\sum_{k=1}^{K-1} \lambda_k \boldsymbol{u}_k \times^0 \boldsymbol{u}_k \right) \tag{2.115}
$$

since $Y\boldsymbol{v}_k = \lambda_k^{1/2} \boldsymbol{u}_k$, where $V = (\boldsymbol{v}_1, \ldots, \boldsymbol{v}_K)$. This completes proof.

Although Ding and He [1] insist that it proves Eq. (2.99), it does not, because it does not guarantee that there is an R such that $RV = H_K$. Although all components

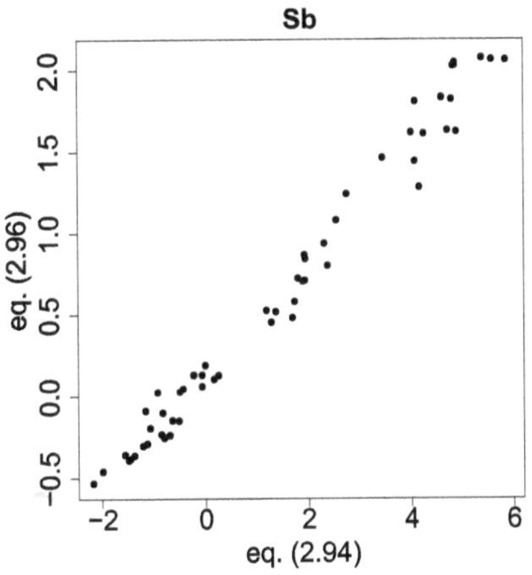

Fig. 2.6 Pairwise scatter plot of $N^2 = 100$ elements of S_bs defined in Eqs. (2.97) and (2.99), respectively

of H_K must be 0 or 1 in order that H_K represents clusters given by K-means, Theorem 2.2 does not have such a restriction that Q must be represented as $H_K R^T$. Actually, as can be seen in Fig. 2.6, S_b given by K-means, Eq. (2.97), does not completely match with S_b given by PCA, Eq. (2.99), but deviates from S_b given by PCA. Thus S_b given by PCA should be considered as not an alternative derivation, but at most a good approximation of S_b given by K-means.

References

1. Ding, C., He, X.: K-means clustering via principal component analysis. In: Proceedings of the Twenty-first International Conference on Machine Learning, ICML '04, pp. 29–. ACM, New York (2004). https://doi.org/10.1145/1015330.1015408. http://doi.acm.org/10.1145/1015330.1015408
2. Jolliffe, I.T.: Principal Component Analysis. Springer, Berlin (2002). https://doi.org/10.1007/b98835

Chapter 3
Tensor Decomposition

I painted her as an unapproachable enigma and never even tried to see her for who she was.
Ichigo, Darling in the FranXX, Season 1, Episode 16

3.1 Three Principal Realizations of TD

As has been mentioned in the previous sections, among huge realizations of TD [2], we discuss three most popular ones: canonical polyadic (CP) decomposition, Tucker decomposition and tensor train decomposition.[1] These three decomposition of tensor $\mathcal{X} \in \mathbb{R}^{N \times M \times K}$ whose element is x_{ijk} are expressed as CP decomposition

$$\mathcal{X} = \sum_{\ell=1}^{L} \lambda_\ell \boldsymbol{u}_\ell^{(i)} \times^0 \boldsymbol{u}_\ell^{(j)} \times^0 \boldsymbol{u}_\ell^{(k)} \tag{3.1}$$

and L is positive integer, λ_ℓ is weight, $\boldsymbol{u}_\ell^{(i)} \in \mathbb{R}^N$, $\boldsymbol{u}_\ell^{(j)} \in \mathbb{R}^M$, and $\boldsymbol{u}_\ell^{(k)} \in \mathbb{R}^K$.

Using Tucker decomposition,

$$\mathcal{X} = G \times_{\ell_1} U^{(i)} \times_{\ell_2} U^{(j)} \times_{\ell_3} U^{(k)}, \tag{3.2}$$

where $U_{\ell_1}^{(i)} \in \mathbb{R}^{N \times N}$,$U_{\ell_2}^{(j)} \in \mathbb{R}^{M \times M}$,$U_{\ell_3}^{(k)} \in \mathbb{R}^{K \times K}$ are singular value vectors, and $G \in \mathbb{R}^{N \times M \times K}$ is a core tensor. The components of $U^{(i)}$, $U^{(j)}$,$U^{(k)}$, and G are denoted as $u_{\ell_1 i}^{(i)}$, $u_{\ell_2 j}^{(j)}$, $u_{\ell_3 k}^{(k)}$, and $G(\ell_1, \ell_2, \ell_3)$

[1] Although the detailed algorithms of individual TDs will be presented in the later sections, readers might feel that they would like to try them in advance with reading prior sections that demonstrate examples. In that case, see Appendix A where I list some of implementations on various platformes.

Y-h. Taguchi, *Unsupervised Feature Extraction Applied to Bioinformatics*, Unsupervised and Semi-Supervised Learning,
https://doi.org/10.1007/978-3-031-60982-4_3

Using tensor train decomposition,

$$\mathcal{X} = G^{(i)} \times_{\ell_1} G^{(j)} \times_{\ell_2} G^{(k)}, \tag{3.3}$$

where $G^{(i)} \in \mathbb{R}^{N \times R_1}$, $G^{(j)} \in \mathbb{R}^{M \times R_1 \times R_2}$, and $G^{(k)} \in \mathbb{R}^{K \times R_2}$ with R_1 and R_2 being positive integer. $G^{(i)}$s' components, $G^{(j)}$s' components, $G^{(k)}$s' components are denoted as $G^{(i)}(i, \ell_1)$, $G^{(j)}(j, \ell_1, \ell_2)$, and $G^{(k)}(k, \ell_2)$. Although we employed three mode tensor, x_{ijk}, in the above, the extension to the higher mode should be straightforward.

All of these are in some sense the extension of SVD. In SVD, matrix $X \in \mathbb{R}^{N \times M}$ is represented as

$$X = \sum_{\ell=1}^{L} \lambda_\ell \boldsymbol{u}_\ell \times^0 \boldsymbol{v}_\ell, \tag{3.4}$$

where $L = \min(N, M)$ and $\boldsymbol{u}_\ell \in \mathbb{R}^N$, $\boldsymbol{v}_\ell \in \mathbb{R}^M$. It is obvious that CP decomposition is straight extension of SVD toward higher mode tensors. One problem of CP decomposition is that there are no ways to determine L in Eq. (3.1) a priori.

Tucker decomposition Eq. (3.2) also can be considered to be the extension of SVD to higher dimension, since Eq. (3.2) can also be represented as

$$\mathcal{X} = \sum_{\ell_1} \sum_{\ell_3} \sum_{\ell_2} G(\ell_1, \ell_2, \ell_3) \boldsymbol{u}_{\ell_1}^{(i)} \times^0 \boldsymbol{u}_{\ell_2}^{(j)} \times^0 \boldsymbol{u}_{\ell_3}^{(k)}, \tag{3.5}$$

where

$$\boldsymbol{u}_\ell^{(i)} = (u_{\ell 1}, \ldots, u_{\ell i}, \ldots, u_{\ell N}) \tag{3.6}$$

$$\boldsymbol{u}_\ell^{(j)} = (u_{\ell 1}, \ldots, u_{\ell j}, \ldots, u_{\ell M}) \tag{3.7}$$

$$\boldsymbol{u}_\ell^{(k)} = (u_{\ell 1}, \ldots, u_{\ell k}, \ldots, u_{\ell K}) \tag{3.8}$$

Only difference from CP decomposition is that individual vectors appear more than once in the right hand side.

Tensor train decomposition can be interpreted as an extension of SVD because Eq. (3.3) can be rewritten as

$$\mathcal{X} = \sum_{\ell_1=1}^{R_1} \sum_{\ell_2=1}^{R_2} \boldsymbol{G}_{\ell_1}^{(i)} \times^0 \boldsymbol{G}_{\ell_1,\ell_2}^{(j)} \times^0 \boldsymbol{G}_{\ell_2}^{(k)}, \tag{3.9}$$

where $\boldsymbol{G}_{\ell_1}^{(i)} \in \mathbb{R}^N$, $\boldsymbol{G}_{\ell_1 \ell_2}^{(j)} \in \mathbb{R}^M$, and $\boldsymbol{G}_{\ell_2}^{(k)} \in \mathbb{R}^K$ are

$$\boldsymbol{G}_{\ell_1}^{(i)} = \left(G^{(i)}(1, \ell_1), \ldots, G^{(i)}(N, \ell_1) \right) \tag{3.10}$$

$$G^{(j)}_{\ell_1\ell_2} = \left(G^{(j)}(1, \ell_1, \ell_2), \ldots, G^{(j)}(M, \ell_1, \ell_2)\right) \tag{3.11}$$

$$G^{(k)}_{\ell_2} = \left(G^{(k)}(1, \ell_2), \ldots, G^{(k)}(K, \ell_2)\right) \tag{3.12}$$

Thus all of three tensor decomposition listed in the above can be considered to be linear combinations of vector product, $\boldsymbol{a} \times^0 \boldsymbol{b} \times^0 \boldsymbol{c}$, although the number of terms differs; L for CP decomposition, $N \times M \times K$ for Tucker decomposition and $R_1 \times R_2$ for tensor train decomposition. These three TDs have their own pros and cons. CP decomposition has an advantage of interpretability; individual vectors in the right hand of Eq. (3.1) appear only once, thus it is easy to understand what each term means. A disadvantage of CP decomposition is that obtaining CP decomposition is non-deterministic polynomial time (NP) hard. Thus, no one knows how long it takes until convergence. Tucker decomposition does not have this disadvantage; it is expected to converge within polynomial time. Disadvantages of Tucker decomposition are two fold. The first disadvantage is that it is hard to interpret; because individual vectors in the right hand side of Eq. (3.2) appear multiple times, it is unclear what each vector represents. The second disadvantage is non-uniqueness. In actuality, we may use any orthogonal matrix, $R \in \mathbb{R}^{N \times N}$ whose components are denoted as $R_{\ell\ell_1}$, that satisfies $R^T R = I$ with the components R^T being denoted as $R_{\ell'_1\ell}$, where I is a unit matrix whose components are $\delta_{\ell'_1\ell_1}$, Eq. (3.2) is rewritten as, with denoting the components of G as $G(\ell'_1, \ell_2, \ell_3)$,

$$\mathcal{X} = G \times_{\ell'_1} I \times_{\ell_1} U^{(i)} \times_{\ell_2} U^{(j)} \times_{\ell_3} U^{(k)} \tag{3.13}$$

$$= G \times_{\ell'_1} \left(R^T \times_{\ell} R\right) \times_{\ell_1} U^{(i)} \times_{\ell_2} U^{(j)} \times_{\ell_3} U^{(k)} \tag{3.14}$$

$$= \left\{G \times_{\ell'_1} R^T\right\} \times_{\ell} \left\{R \times_{\ell_1} U^{(i)}\right\} \times_{\ell_2} U^{(j)} \times_{\ell_3} U^{(k)}. \tag{3.15}$$

If we define

$$G'(\ell, \ell_2, \ell_3) = \sum_{\ell'_1} G(\ell'_1, \ell_2, \ell_3) R_{\ell'_1\ell} \tag{3.16}$$

$$u'^{(i)}_{\ell i} = \sum_{\ell_1} R_{\ell\ell_1} u^{(i)}_{\ell_1 i} \tag{3.17}$$

Eq. (3.2) can be expressed as

$$\mathcal{X} = G' \times_{\ell} U'^{(i)} \times_{\ell_2} U^{(j)} \times_{\ell_3} U^{(k)} \tag{3.18}$$

which is nothing but an alternative representation of Tucker decomposition. It is also obvious that there are infinitely many solutions of Tucker decomposition since we can employ any orthogonal matrix R to derive alternative representations of Tucker decomposition.

Similarly, tensor train decomposition, Eq. (3.3), does not have uniqueness, either. Using R, Eq. (3.3) can be rewritten as, with denoting the components of $G^{(i)}$ as $G^{(i)}(i, \ell'_1)$,

$$\mathcal{X} = G^{(i)} \times_{\ell'_1} I \times_{\ell_1} G^{(j)} \times_{\ell_2} G^{(k)} \tag{3.19}$$

$$= G^{(i)} \times_{\ell'_1} \left(R^T \times_{\ell} R\right) \times_{\ell_1} G^{(j)} \times_{\ell_2} G^{(k)} \tag{3.20}$$

$$= \left\{G^{(i)} \times_{\ell'_1} R^T\right\} \times_{\ell} \left\{R \times_{\ell_1} G^{(j)}\right\} \times_{\ell_2} G^{(k)} \tag{3.21}$$

If we define

$$G'^{(i)}(i, \ell) = \sum_{\ell'_1} G^{(i)}(i, \ell'_1) R_{\ell'_1 \ell} \tag{3.22}$$

$$G'^{(j)}(j, \ell, \ell_2) = \sum_{\ell_1} R_{\ell \ell_1} G^{(j)}(j, \ell_1, \ell_2) \tag{3.23}$$

Eq. (3.3) can be rewritten

$$\mathcal{X} = G'^{(i)} \times_{\ell} G'^{(j)} \times_{\ell_2} G^{(k)} \tag{3.24}$$

which is nothing but an alternative representation of tensor train decomposition. It is also obvious that there are infinitely many solutions of tensor train decomposition since we can employ any orthogonal matrix R to derive alternative representations of tensor train decomposition.

The advantage of tensor train decomposition is the small number of parameters. For CP decomposition, Eq. (3.1), the number of parameters must be decided is $(N + M + K + 1)L$ and for Tucker decomposition, Eq. (3.2), the number of parameters that must be determined is $NMK + N^2 + M^2 + M^2$. On the other hand, in tensor train decomposition, Eq. (3.3), the number of parameters that must be decided is as many as $NR_1 + MR_1R_2 + KR_2$. In other words, the number of parameters that must be determined in tensor train decomposition is much smaller than the number of parameters that must be decided for CP and Tucker decomposition. This means, if we need to obtain the tensor decomposition of higher order modes, computational time and memory required is logarithmically small. This does not mean unfortunately that tensor train decomposition is always superior to CP decomposition and Tucker decomposition. There is no free lunch. In contrast to CP decomposition and Tucker decomposition, the order of suffix must be fixed in tensor train decomposition prior to executing tensor decomposition. In Eq. (3.3), the order of suffix in the left hand side is $i \to j \to k$ and is not commutable. This restriction of the suffix order does not exist in either CP decomposition or Tucker decomposition. This restriction might prevent tensor train decomposition from getting optimal solutions that can be obtained by CP decomposition or Tucker decomposition. At the moment, there are no guide lines on how to order suffix in order to get optimal solutions in tensor train decomposition; if the parameter space searched is narrow, the opportunity to get optimal solution is limited, too.

Exercise

3.1 Get CP decomposition, Tucker decomposition and tensor train decomposition of 3-mode tensor, $x_{ijk} = 1 \in \mathbb{R}^{3\times3\times3}$, though it might be trivial.

3.2 Performance of TDs as Tools Reducing the Degrees of Freedoms

In contrast to the MF that are associated with geometrical representations, TD generally lacks the interpretations based upon geometrical representation. Thus, it is important how TD can help us to interpret complex data set from the data science point of views. As an intuitive example that demonstrates the usefulness of TD as data mining tools, we consider the following simple case

$$x_{ijk} = i + j + k. \tag{3.25}$$

In principle, we do not need any complicated procedures like TD to understand this simple three mode tensor. Because we know what the tensor, Eq. (3.25), represents, it is also easy for us to understand how TD works when it is applied to this simple tensor. For the simplicity, I use only the case $x_{ijk} \in \mathbb{R}^{3\times4\times5}$. However, the essential result obtained by this assumption will be kept for larger tensors, too.

3.2.1 *Tucker Decomposition*

We start this analysis with applying Tucker decomposition, Eq. (3.2), to the tensor shown in Eq. (3.25). HOSVD algorithm (detailed explanation will be given later) is employed to obtain Tucker decomposition. Excluding those having essentially zero values with considering numerical accuracy, $G(\ell_1, \ell_2, \ell_3)$s in Eq. (3.2) are in Table 3.1. Thus, although the total number of G is $3 \times 4 \times 5 = 60$, as little as eight Gs have non zero values. Therefore, singular value vectors that can contribute to the decomposition Eq. (3.2) are limited to $1 \le \ell_1, \ell_2, \ell_3 \le 2$. The number of them is only six. Because $\boldsymbol{u}^{(i)}_{\ell_1}, \boldsymbol{u}^{(j)}_{\ell_2}$, and $\boldsymbol{u}^{(k)}_{\ell_3}$ have three, four and five components, these six vectors have in total $3 \times 2 + 4 \times 2 + 5 \times 2 = 24$ components. As a result, the total number of real numbers composed of Tucker decomposition, Eq. (3.2), applied to the tensor Eq. (3.25) is $8 + 24 = 32$. This number, 32, is about half of the number of elements of original tensor, $3 \times 4 \times 5 = 60$. This means, TD is effective to reduce the degrees of freedom in tensor, although it is not necessary because Eq. (3.25) is easy to understand without any kind of data reduction.

Table 3.1 Core tensors having nonzero values when Tucker decomposition Eq. (3.2) is applied to the tensor Eq. (3.25)

$G(\ell_1, \ell_2, \ell_3)$				
	$\ell_1 = 1$		$\ell_1 = 2$	
	$\ell_2 = 1$	$\ell_2 = 2$	$\ell_2 = 1$	$\ell_2 = 2$
$\ell_3 = 1$	-60.04	5.06×10^{-3}	-8.57×10^{-3}	-1.13
$\ell_3 = 2$	6.32×10^{-3}	1.57	-0.88	-0.32

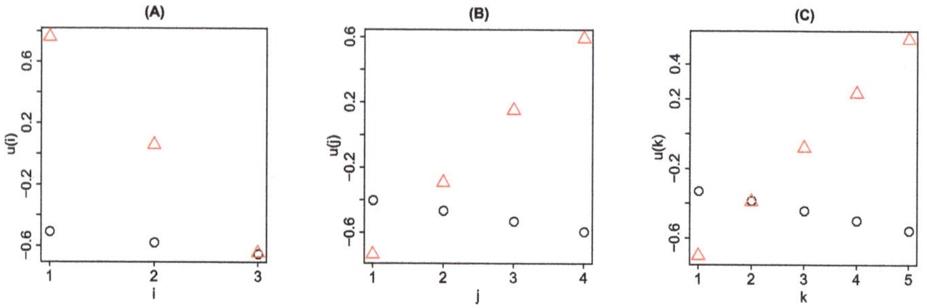

Fig. 3.1 Singular value vectors computed by applying Tucker decomposition, Eq. (3.2), to the tensor Eq. (3.25). **(a)** $\boldsymbol{u}^{(i)}_{\ell_1}$ **(b)** $\boldsymbol{u}^{(j)}_{\ell_2}$ **(c)** $\boldsymbol{u}^{(k)}_{\ell_3}$. $\circ : \ell_1, \ell_2, \ell_3 = 1$, $\triangle : \ell_1, \ell_2, \ell_3 = 2$

It is also important to see how the tensor Eq. (3.25) is decomposed by Tucker decomposition (Fig. 3.1). Firstly, all of these vectors represent monotonic dependence upon i, j or k. This suggests that TD can capture fundamental dependence of x_{ijk} in Eq. (3.25) upon i, j or k, since Eq. (3.25) shows the monotonic dependence upon i, j or k as well.

In addition to this, TD can also be used as an approximation to the tensor. As can be seen in Table 3.1, $G(1, 1, 1)$ has the maximum absolute values among eight G with nonzero values. Moreover, considering that Gs play a role of weight factors in Eq. (3.2), $G(1, 1, 1)$ have most of contributions since $\frac{G(1,1,1)^2}{\sum_{\ell_1,\ell_2,\ell_3} G(\ell_1,\ell_2,\ell_3)^2} = 0.998$. In actual, the scatter plot between x_{ijk} and the right hand side of Eq. (3.2) with only considering $\ell_1 = \ell_2 = \ell_3 = 1$ shows almost complete reproduction (Fig. 3.2).

In conclusion, Tucker decomposition Eq. (3.2) has ability to reduce the degrees of freedoms (about half of them) with keeping essential dependence upon i, j, k (monotonic dependence).

Exercise

3.2 Draw something that corresponds to Fig. 3.2 with employing more terms than $\ell_1 = \ell_2 = \ell_3 = 1$.

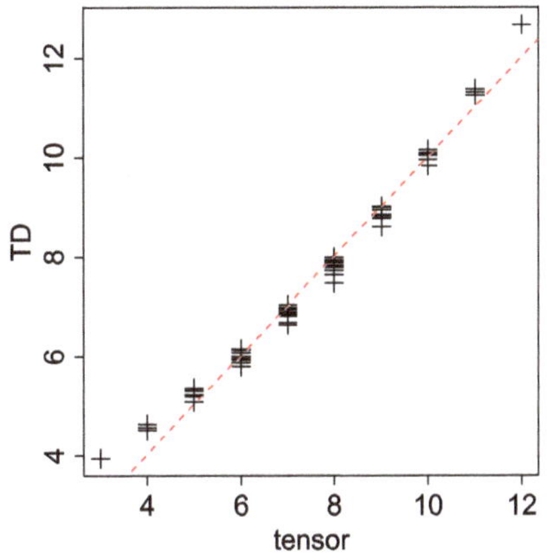

Fig. 3.2 Comparison between x_{ijk} in Eq. (3.25) and the recomputation from Tucker decomposition Eq. (3.2) with considering only $\ell_1 = \ell_2 = \ell_3 = 1$. Red broken line represents diagonal line (i.e., complete agreement)

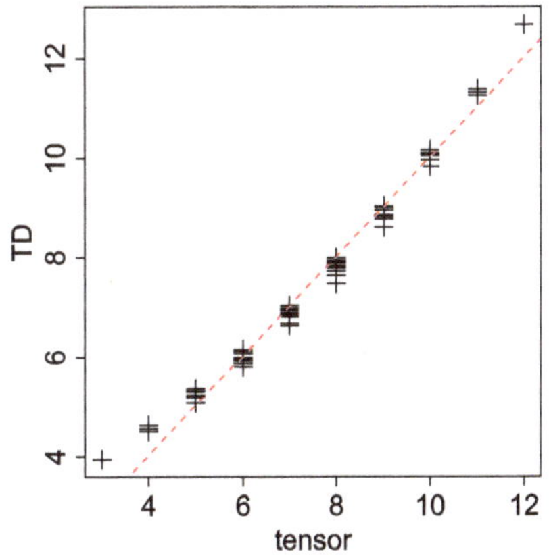

Fig. 3.3 Comparison between x_{ijk} in Eq. (3.25) and the recomputation from CP decomposition Eq. (3.1) with $L = 1$. Red broken line represents diagonal line (i.e., complete agreement)

3.2.2 CP Decomposition

Next, we consider CP decomposition, Eq. (3.1). It is usual that CP decomposition, Eq. (3.1), is more interpretable than Tucker decomposition, Eq. (3.2). This is because CP decomposition is a simple linear combination of tensor product of individual vectors while individual vectors are repeatedly used in Tucker decomposition, Eq. (3.2). Thus, apparently CP decomposition has more ability to relate vectors one by one; it is expected to make interpretation easier than Tucker decomposition.

Since we know that x_{ijk} in Eq. (3.25) can be well approximated by the single term in the right hand side of Eq. (3.2), we try to check if CP decomposition, Eq. (3.1), can represent x_{ijk} in Eq. (3.25) with $L = 1$. Figure 3.3 show the comparison between x_{ijk} in Eqs. (3.25) and (3.1) with $L = 1$ when CP decomposition is applied to x_{ijk} in Eq. (3.25). Figures 3.2 and 3.3 look identical; these two are really identical within numerical accuracy. Thus, CP decomposition can approximate x_{ijk} in Eq. (3.25) as well as Tucker decomposition did.

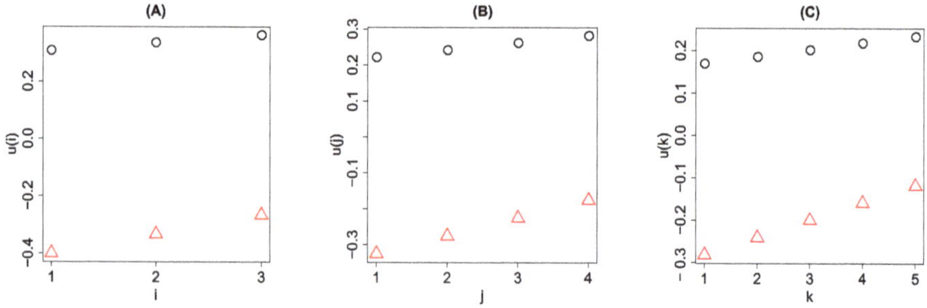

Fig. 3.4 Singular value vectors computed by applying CP decomposition Eq. (3.1) to the tensor Eq. (3.25) with $L = 2$. **(a)** $\boldsymbol{u}_\ell^{(i)}$ **(b)** $\boldsymbol{u}_\ell^{(j)}$ **(c)** $\boldsymbol{u}_\ell^{(k)}$. $\circ : \ell = 1$, $\triangle : \ell = 2$

Table 3.2 λ_ℓs with $L = 1$ and $L = 2$ when CP decomposition Eq. (3.1) is applied to the tensor Eq. (3.25)

λ_ℓ	$L = 1$	$L = 2$
$\ell = 1$	450.6	533.7
$\ell = 2$	–	83.7

In order to estimate the degrees of freedom that CP decomposition can represents x_{ijk} in Eq. (3.25) not approximately but completely, we try to find minimum L that can perform complete CP decomposition. Then we found that $L = 4$ is minimum. Thus, total number of real numbers required is $(3 + 4 + 5) \times 4 = 48$. Because this number is larger than 34 which is the number of real values to obtain Tucker decomposition that can perform complete decomposition, CP decomposition has less ability to reduce the degrees of freedom than Tucker decomposition.

The difference between Tucker decomposition and CP decomposition takes place when considering the second term. One might expect that Eq. (3.1) with $L = 2$ might be identical to summation of two termes composed of singular value vectors shown in Fig. 3.1. Figure 3.4 shows $\boldsymbol{u}_\ell^{(i)}$, $\boldsymbol{u}_\ell^{(j)}$, and $\boldsymbol{u}_\ell^{(k)}$ for $\ell = 1, 2$. In contrast to the expectation, Fig. 3.4 does not look like Fig. 3.1. In contrast to Fig. 3.1 where singular value vectors associated with distinct ℓ_1, ℓ_2, ℓ_3 values look different, those with distinct ℓ look similar excluding parallel vertical displacements in Fig. 3.4.

Table 3.2 shows the λ_ℓs with $L = 1$ and $L = 2$. The absolute ratio, $\left|\frac{\lambda_2}{\lambda_1}\right|$, of weights between the first term, λ_1, and the second term, λ_2, when $L = 2$ is comparatively larger than that between terms with the first and the second largest absolute values in Table 3.1, $\left|\frac{G(1,2,2)}{G(1,1,1)}\right|$. It is coincident with the fact that singular value vectors with $\ell = 1, 2$ in CP decomposition does not look distinct, because similar singular value vectors unlikely have very distinct weights. On the other hand, this suggests that CP decomposition fails to compute the additional small correction with keeping the contribution from the main term as large as Tucker decomposition did.

Although actual numerical algorithms to execute various TDs are not yet explained (see later part of this chapter), CP decomposition is not guaranteed to converge to the unique solution (see the following sections). Thus, in contrast to the apparent interpretability of CP decomposition, because TD itself is not unique but depends upon the initial values for the iterative computation, CP decomposition cannot be considered to have superior interpretability to Tucker decomposition.

Since the Tucker decomposition is easier to compute and has more converging algorithm, I prefer Tucker to CP in the approximations shown in the following application examples mentioned in this book in spite of the apparent interpretability of CP decomposition.

Exercise

3.3 Draw something that corresponds to Fig. 3.3 with employing more terms than $L = 1$.

3.2.3 Tensor Train Decomposition

Finally, I apply tensor train decomposition, Eq. (3.3), to x_{ijk} in Eq. (3.25). The result is

$$G^{(i)}_{\ell_1,i} = (i, 1) \tag{3.26}$$

$$G^{(j)}_{\ell_1,\ell_2,j} = \begin{pmatrix} 1 & 0 \\ j & 1 \end{pmatrix} \tag{3.27}$$

$$G^{(k)}_{\ell_2,k} = (1, k) \tag{3.28}$$

because

$$G^{(i)}_{\ell_1,i} \times_{\ell_1} G^{(j)}_{\ell_1,\ell_2,j} \times_{\ell_2} G^{(k)}_{\ell_2,k} = (i, 1) \begin{pmatrix} 1 & 0 \\ j & 1 \end{pmatrix} \begin{pmatrix} 1 \\ k \end{pmatrix} = (i + j, 1) \begin{pmatrix} 1 \\ k \end{pmatrix} = i + j + k. \tag{3.29}$$

The number of $G^{(i)}$ is three, that of $G^{(j)}$ is four and that of $G^{(k)}$ is five, thus the total number of real numbers that compose tensor train decomposition is $2 \times 3 + 4 \times 4 + 2 \times 5 = 32$. Since this number is smaller than 34 and 48, which are the minimum degrees of freedom to execute complete decomposition when Tucker and CP decomposition are applied to x_{ijk} in Eq. (3.25), respectively, tensor train decomposition has superior ability to reduce the degrees of freedom. In this example, the amount of superiority might look small, but if we consider tensors with the higher dimensions or modes, this difference matters.

On the other hand, tensor train decomposition has some disadvantages. The first disadvantage is that Eq. (3.3) is not invariant when the order of i, j, k are exchanged. It is obvious that i, j, k must be exchanged when the order of i, j, k in Eqs. (3.26)–

(3.28) are exchanged. In actual, the ability of reducing the number of freedoms itself is also altered. If the order of i, j, k is modified as j, i, k such that the number of matrices used is minimized, the total number of real numbers required decreases from 32 to $2 \times 4 + 4 \times 3 + 2 \times 5 = 30$. This might be problematic for the application of data science that requires interpretation of the obtained singular value vectors. If the order of i, j, k matters, we have to decide this order in advance, or select the best order after investigating the results. This is really problematic because the number of possibility on how to order i, j, k grows exponentially if we have to consider tensors with more number of modes. Selecting one of them might not be easy.

The second disadvantage is that tensor train decomposition does not have weight, by which we can know the primary terms in decomposition as in the cases of CP decomposition and Tucker decomposition. In the case of tensor train decomposition, we have no ways to know which combination among Eq. (3.3) is dominant. For the application of TD toward real data sets, it is not an ignorable point. Thus, in the application that will be discussed in the later parts of this text, I do not employ tensor train decomposition, either, as CP decomposition is not employed.

Exercise
3.4 Draw something that corresponds to Figs. 3.2 or 3.3 for tensor train decomposition.

3.2.4 TDs Are Not Always Interpretable

When applying TDs to x_{ijk} in Eq. (3.25), no matter how many degrees of freedom are required, three TDs, CP decomposition, Tucker decomposition, and tensor train decomposition can acquire essential feature of the tensor, i.e., monotonic dependence upon i, j, k. Although readers might trust the usefulness of these TDs as the tool for the application in data science, the situation is actually not so straightforward. Instead of the x_{ijk} in Eq. (3.25) we consider the tensor

$$x_{ijk} = \left(i - \frac{N+1}{2}\right) + \left(j - \frac{M+1}{2}\right) + \left(k - \frac{K+1}{2}\right) \tag{3.30}$$

such that average over either i, j, or k is equal to zero. Although this may not seem to dramatically change the results of TD, it actually does. Table 3.3 shows the list of Gs with nonzero values when Tucker decomposition is applied to x_{ijk} defined in Eq. (3.30). Compared with Table 3.1, although the number of Gs with nonzero values is eight which is the same as that in Table 3.1, individual absolute values of Gs are larger excluding $G(1, 1, 1)$. This suggests that $G(1, 1, 1)$ cannot acquire most of contributions in contrast to Table 3.3 but other Gs have substantial contributions. Figure 3.5 shows the singular value vectors, which are very different from those in Fig. 3.1 that represent monotonic dependence upon i, j, k. Singular value vectors in Fig. 3.5 have lost monotonic dependence upon

Table 3.3 Core tensors having nonzero values when Tucker decomposition, Eq. (3.2), is applied to the tensor Eq. (3.30)

$G(\ell_1, \ell_2, \ell_3)$				
	$\ell_1 = 1$		$\ell_1 = 2$	
	$\ell_2 = 1$	$\ell_2 = 2$	$\ell_2 = 1$	$\ell_2 = 2$
$\ell_3 = 1$	15.67	1.70	−1.71	5.69
$\ell_3 = 2$	1.86	−3.69	3.39	−3.04

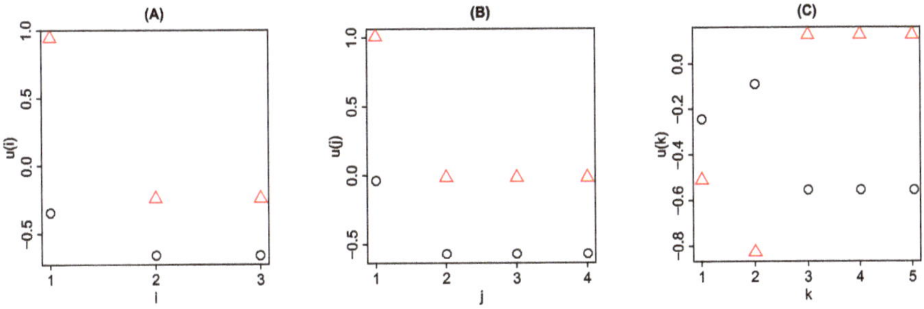

Fig. 3.5 Singular value vectors computed by applying Tucker decomposition Eq. (3.2) to the tensor Eq. (3.30). (**a**) $\boldsymbol{u}_{\ell_1}^{(i)}$ (**b**) $\boldsymbol{u}_{\ell_2}^{(j)}$ (**c**) $\boldsymbol{u}_{\ell_3}^{(k)}$. ○ : $\ell_1, \ell_2, \ell_3 = 1$, △:$\ell_1, \ell_2, \ell_3 = 2$

i, j, k in spite of that x_{ijk} itself in Eq. (3.30) still keeps monotonic dependence upon i, j, k as in Eq. (3.25). This drastic change is caused because x_{ijk}s in Eq. (3.30) take both negative and positive values while those in Eq. (3.25) take positive values only. Because the product between two negative values results in positive values, expressing the distinct signs of x_{ijk} with the products of vectors is not straightforward. Thus, singular value vectors inevitably lost the simple monotonic dependence upon i, j, k.

Thus, from the point of data science, tensors whose elements are both negatively and positively signed are not easy to be dealt with TDs. For the cases of matrix factorization, extraction of means affected the outcomes in unpredictable ways (see Sect. 2.5.3). Similarly, the outcomes of TDs are affected by whether means are extracted or not, because of the effect discussed in the above. How to extract means is also a key on the application of TD to real data sets, although this point is rarely emphasized.

Exercise

3.5 Draw something that corresponds to Fig. 3.2 for Tucker decomposition applied to Eq. (3.30).

3.3 Various Algorithms to Compute TDs

In this section, I introduce various algorithm to derive various TDs.

3.3.1 CP Decomposition

Firstly, I introduce how to compute CP decomposition, Eq. (3.1). Before introducing algorithm, I would like to mention about non-uniqueness of approximation of tensor by CP decomposition as demonstrated when CP decomposition, Eq. (3.1), with $L = 2$ is applied to x_{ijk} Eq. (3.25) in the previous section. In §3.3 of Kolda and Bader [2], there is an example of non-uniqueness when a specific three mode tensor is decomposed by CP decomposition with $L = 2$. The tensor $X \in \mathbb{R}^{N \times M \times K}$ has the form of

$$X = \boldsymbol{a}_1 \times^0 \boldsymbol{b}_1 \times^0 \boldsymbol{c}_2 + \boldsymbol{a}_1 \times^0 \boldsymbol{b}_2 \times^0 \boldsymbol{c}_1 + \boldsymbol{a}_2 \times^0 \boldsymbol{b}_1 \times^0 \boldsymbol{c}_1 \tag{3.31}$$

where $A = (\boldsymbol{a}_1, \boldsymbol{a}_2) \in \mathbb{R}^{N \times 2}$, $B = (\boldsymbol{b}_1, \boldsymbol{b}_2) \in \mathbb{R}^{M \times 2}$, and $C = (\boldsymbol{c}_1, \boldsymbol{c}_2) \in \mathbb{R}^{K \times 2}$. Then consider the specific form of CP decomposition with $L = 2$ as

$$Y = \alpha \left(\boldsymbol{a}_1 + \frac{\boldsymbol{a}_2}{\alpha} \right) \times^0 \left(\boldsymbol{b}_1 + \frac{\boldsymbol{b}_2}{\alpha} \right) \times^0 \left(\boldsymbol{c}_1 + \frac{\boldsymbol{c}_2}{\alpha} \right) - \alpha \boldsymbol{a}_1 \times^0 \boldsymbol{b}_1 \times^0 \boldsymbol{c}_1. \tag{3.32}$$

Then

$$\begin{aligned} ||X - Y|| = \frac{1}{\alpha} \Bigg|\Bigg| & \boldsymbol{a}_2 \times^0 \boldsymbol{b}_2 \times^0 \boldsymbol{c}_1 + \boldsymbol{a}_2 \times^0 \boldsymbol{b}_1 \times^0 \boldsymbol{c}_2 + \boldsymbol{a}_1 \times^0 \boldsymbol{b}_2 \times^0 \boldsymbol{c}_2 \\ & - \frac{1}{\alpha} \boldsymbol{a}_2 \times^0 \boldsymbol{b}_2 \times^0 \boldsymbol{c}_2 \Bigg|\Bigg| \end{aligned} \tag{3.33}$$

can be made arbitrarily small. Thus, CP decomposition with $L = 2$ for Eq. (3.31) can never be unique. The reason why this can happen is because two oppositely signed arbitrarily large terms can results in small value with canceling with each other. In this sense, no matter what algorithm is employed for CP decomposition, there is not unique approximation using CP decomposition.

Consequently, the algorithm of CP decomposition is inevitably empirical and does not guarantee neither uniqueness nor convergence. Here I introduce a specific algorithm that employs alternating least square (ALS). ALS is a general algorithm that minimizes multi arguments functions by alternating one argument with fixing other arguments. Suppose the case that minimization of the function $f(x, y, z)$ is difficult while $f(x, y_0, z_0)$ with fixing y_0 and z_0 is easy (this also stands for y and z). Then ALS algorithm repeatedly minimizes $f(x, y_0, z_0)$, $f(x_0, y, z_0)$ and $f(x_0, y_0, z)$ in turn until convergence. For example, let us consider the minimization of $f(x, y, z) = x^2 + y^2 + z^2$ with starting $x = y = z = 1$. Applying ALS to this problem is as follows. At first, try to minimize $f(x, 1, 1) = x^2 + 2$. It is obvious that $x = 0$ minimizes $x^2 + 2$. Then, x is decided to be 0. Then, we try to minimize $f(0, y, 1) = y^2 + 1$. It is again obvious $y = 0$ does. Then, y is decided to be 0. Finally, we try to minimize $f(0, 0, z) = z^2$. We get $z = 0$. The minimum value $f(0, 0, 0) = 0$ can be obtained by ALS algorithm.

In order to apply ALS to obtain CP decomposition, Eq. (3.1), we need some mathematics [2]. At first, Eq. (3.1) needs to be rewritten in the unfolded matrix form, $X^{i\times(jk)}$, of tensor $\mathcal{X} \in \mathbb{R}^{N\times M\times K}$

$$X^{i\times(jk)} = \hat{U}^{(i)} \times_{\ell} \left(U^{(j)} \times^{\ell} U^{(k)}\right)^{\ell\times(jk)} \tag{3.34}$$

with introducing matrices $\hat{U}^{(i)} = \left(\lambda_1 \boldsymbol{u}_1^{(i)}, \lambda_2 \boldsymbol{u}_2^{(i)}, \ldots, \lambda_L \boldsymbol{u}_L^{(i)}\right) \in \mathbb{R}^{N\times L}$, $U^{(j)} = \left(\boldsymbol{u}_1^{(j)}, \boldsymbol{u}_2^{(j)}, \ldots, \boldsymbol{u}_L^{(j)}\right) \in \mathbb{R}^{M\times L}$ and $U^{(k)} = \left(\boldsymbol{u}_1^{(k)}, \boldsymbol{u}_2^{(k)}, \ldots, \boldsymbol{u}_L^{(k)}\right) \in \mathbb{R}^{K\times L}$ and $\left(U^{(j)} \times^{\ell} U^{(k)}\right)^{\ell\times(jk)} \in \mathbb{R}^{L\times MK}$ is an unfolding of the tensor, $U^{(j)} \times^{\ell} U^{(k)} \in \mathbb{R}^{L\times M\times K}$. Then, we try to find $\hat{U}^{(i)}$ with fixing $U^{(j)}$ and $U^{(k)}$ such that

$$\min_{\hat{U}^{(i)}} \left\| X^{i\times(jk)} - \hat{U}^{(i)} \times_{\ell} \left(U^{(j)} \times^{\ell} U^{(k)}\right)^{\ell\times(jk)} \right\|_F , \tag{3.35}$$

where $||\cdots||_F$ is the Frobenius norm which is defined as the root of the squared summation of matrix elements. This is the same as liner regression problem with having NL elements of $\hat{U}^{(i)}$ as variables.

The solution of Eq. (3.35) can be obtained to compute Moore-Penrose pseudoinverse as

$$\hat{U}^{(i)} = X^{i\times(jk)} \times_{jk} \left[\left(U^{(j)} \times^{\ell} U^{(k)}\right)^{\ell\times(jk)}\right]^{\dagger} , \tag{3.36}$$

where $A^{\dagger}$ is the Moore-Penrose pseudoinverse of a matrix A. Moore-Penrose pseudoinverse is known to give the solution of $A\boldsymbol{x} = \boldsymbol{b}$ as the form $\boldsymbol{x} = A^{\dagger}\boldsymbol{b}$ including the cases that A is not a square matrix. Computing $A^{\dagger}$ from A is implemented in various application softwares, thus it is not discussed in details here.[2] After getting $\hat{U}^{(i)}$ with Eq. (3.36), we normalize the columns of $\hat{U}^{(i)}$ to get $U^{(i)}$. Then, $U^{(i)}$ is replaced with either $U^{(j)}$ or $U^{(k)}$ which can be obtained by repeating the above procedure until the convergence.

In order to see how ALS works for CP decomposition, we apply this algorithm to the simplest case. X is supposed to be a matrix instead of tensor as

$$X = \begin{pmatrix} 1 & 2 & 3 \\ 4 & 5 & 6 \\ 7 & 8 & 9 \end{pmatrix} . \tag{3.37}$$

[2] See Appendix for more details about Moore-Penrose pseudoinverse. Alternatively, one can simply execute linear regression analysis inear regression analysis, Eq. (3.35).

In CP decomposition with $L = 1$, X is supposed to be decomposed as

$$X = \boldsymbol{a} \times^0 \boldsymbol{b}, \tag{3.38}$$

where $\boldsymbol{a}, \boldsymbol{b} \in \mathbb{R}^3$. Although it is nothing but SVD, since we simply would like to demonstrate the usefulness of CP decomposition, it does not matter. Then we get

$$a_1 b_1 = 1 \tag{3.39}$$

$$a_1 b_2 = 2 \tag{3.40}$$

$$a_1 b_3 = 3 \tag{3.41}$$

$$a_2 b_1 = 4 \tag{3.42}$$

$$a_2 b_2 = 5 \tag{3.43}$$

$$a_2 b_3 = 6 \tag{3.44}$$

$$a_3 b_1 = 7 \tag{3.45}$$

$$a_3 b_2 = 8 \tag{3.46}$$

$$a_3 b_3 = 9 \tag{3.47}$$

In order to perform ALS, we need to express $\boldsymbol{a}$ by $\boldsymbol{b}$ and $\boldsymbol{b}$ by $\boldsymbol{a}$. This can be done by performing Eq. (3.39) + Eq. (3.40) + Eq. (3.41), Eq. (3.42)+Eq. (3.43)+Eq. (3.44), Eq. (3.45)+Eq. (3.46)+Eq. (3.47), Eq. (3.39)+Eq. (3.42)+Eq. (3.45), Eq. (3.40)+Eq. (3.43)+Eq. (3.46), and Eq. (3.41)+Eq. (3.44)+Eq. (3.47). This results in

$$\boldsymbol{a} = \frac{1}{\sum_i b_i} \begin{pmatrix} 6 \\ 15 \\ 24 \end{pmatrix} \tag{3.48}$$

$$\boldsymbol{b} = \frac{1}{\sum_i a_i} \begin{pmatrix} 12 \\ 15 \\ 18 \end{pmatrix} \tag{3.49}$$

ALS can be performed, by computing $\boldsymbol{a}$ by Eq. (3.48) then $\boldsymbol{b}$ by Eq. (3.49) and repeat them iteratively.

Starting from $\boldsymbol{b} = \begin{pmatrix} 1 \\ 1 \\ 1 \end{pmatrix}$, after one iteration, we get

$$\boldsymbol{a} = \begin{pmatrix} 2 \\ 5 \\ 8 \end{pmatrix} \tag{3.50}$$

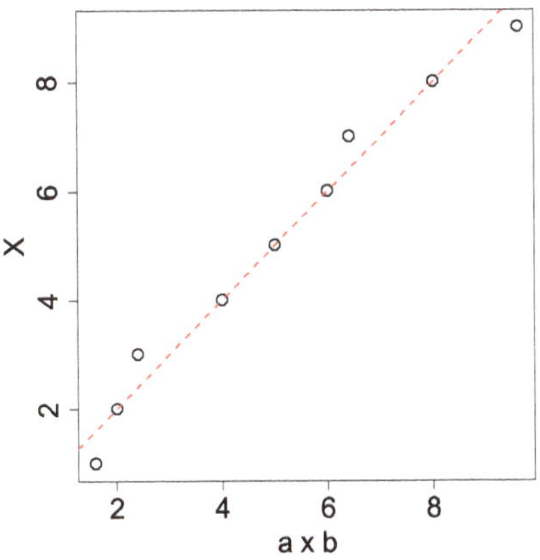

Fig. 3.6 Scatter plot of X, Eq. (3.37), and the approximation by CP decomposition, $\boldsymbol{a} \times^0 \boldsymbol{b}$ where $\boldsymbol{a}$ and $\boldsymbol{b}$ are given as Eqs. (3.50) and (3.51). Red broken lines indicates complete match

$$\boldsymbol{b} = \begin{pmatrix} 0.8 \\ 1.0 \\ 1.2 \end{pmatrix} \tag{3.51}$$

This satisfies Eqs. (3.48) and (3.49). Thus, they are converged solutions.

Next we would like to see how good it is. Figure 3.6 shows the comparison between X, Eq. (3.37), and $\boldsymbol{a} \times^0 \boldsymbol{b}$. It is obvious that they are highly coincident. Thus, CP decomposition implemented using ALS works well.

Here readers should notice that we need initial values of $U^{(i)}$, $U^{(j)}$, and $U^{(k)}$ in CP decomposition implemented using ALS (In the above example, we needed to initialize $\boldsymbol{b}$). Since uniqueness of approximate solution by CP decomposition is not guaranteed as demonstrated in Eq. (3.33), CP decomposition cannot give unique approximation but generally gives various approximation dependent upon initial values. From this point of view, employing CP decomposition for data science is not recommended because data science requires interpretation of obtained decomposition. If the results of CP decomposition has initial value dependence, it is not easy to interpret the outcome uniquely.

In order to extend the above calculation to tensors, $\mathcal{X} \in \mathbb{R}^{N_1 \times N_1 \times \cdots \times N_m}$ with arbitrary number of modes m, Eq. (3.1) is generalized as

$$\mathcal{X} = \sum_{\ell=1}^{L} \lambda_\ell \boldsymbol{u}_\ell^{(i_1)} \times^0 \boldsymbol{u}_\ell^{(i_2)} \times^0 \cdots \times^0 \boldsymbol{u}_\ell^{(i_m)}. \tag{3.52}$$

Figure 3.7 shows the generalized algorithm of CP decomposition aiming tensors with arbitrary number of modes m, which is the straight extension of ALS based CP decomposition algorithm described for the three mode tensor in the above.

Procedure CP decomposition
  Initialize $U^{(i_\alpha)}, \alpha \in [1,m]$
  repeat
    do $\alpha \in [1,m]$
      $\hat{U}^{(i_\alpha)} \longleftarrow X^{i_\alpha \times (i_1 i_2 \ldots i_{\alpha-1} i_{\alpha+1} \ldots i_m)} \times_{(i_1 i_2 \ldots i_{\alpha-1} i_{\alpha+1} \ldots i_m)}$
      $\left[\left(U^{(i_1)} \times^\ell U^{(i_2)} \times^\ell \cdots \times^\ell U^{(i_{\alpha-1})} \times^\ell U^{(i_{\alpha+1})} \times^\ell \cdots \times^\ell U^{(i_m)} \right)^{\ell \times (i_1 i_2 \cdots i_{\alpha-1} i_{\alpha+1} \cdots i_m)} \right]^\dagger$
      normalize columns of $\hat{U}^{(i_\alpha)}$ (storing norms as λ)
    end do
  until fit ceases to improve or maximum iterations exhausted
  return $\lambda, U^{(i_\alpha)}, \alpha \in [1,m]$
end procedure

Fig. 3.7 Algorithm of CP decomposition for tensors with arbitrary number of modes m

Exercise
3.6 Apply CP decomposition implemented using ALS to the tensor $X \in \mathbb{R}^{2\times 2\times 2}$

$$X_{ij1} = \begin{pmatrix} 1 & 2 \\ 3 & 4 \end{pmatrix} \tag{3.53}$$

$$X_{ij2} = \begin{pmatrix} 5 & 6 \\ 7 & 8 \end{pmatrix} \tag{3.54}$$

3.3.2 *Tucker Decomposition*

Tucker decomposition, Eq. (3.2), is not as popular as CP decomposition that has apparent ease to apply to data set. As discussed in the previous section, this apparent ease is not always true. Since I found that Tucker decomposition has numerous advantages in spite of its unpopularity and I can almost always make use of it in the applications described in the later part of this text book, I would like to discuss about it in more details in this section.

There are two popular implementations of Tucker decomposition, ALS based one and SVD based one. Since Tucker decomposition does not have uniqueness at all as discussed in the above, these two distinct implementations generally give distinct outcomes. The first one that makes use of ALS is named higher orthogonal iteration of tensors (HOOI). HOOI, as its name says, computes TD iteratively with orthogonalizing column vectors, because Tucker decomposition requires the orthogonal matrices as outcomes, although CP decomposition does not always require orthogonality between obtained singular value vectors. Using $U^{(i)} \in \mathbb{R}^{L_1 \times N}$, $U^{(j)} \in \mathbb{R}^{L_2 \times M}$, and $U^{(k)} \in \mathbb{R}^{L_3 \times K}$ defined in the previous subsection, Eq. (3.2) can be rewritten as

$$\mathcal{X} = G \times_{\ell_1} U^{(i)} \times_{\ell_2} U^{(j)} \times_{\ell_3} U^{(k)}. \tag{3.55}$$

In order to perform ALS, we need to express $U^{(i)}$ with $U^{(j)}$ and $U^{(k)}$. Since $U^{(i)}$, $U^{(j)}$, and $U^{(k)}$ are orthogonal matrices, it can be easily done as follows. First, we need to define a tensor $\mathcal{Y} \in \mathbb{R}^{N \times \ell_2 \times \ell_3}$

$$\mathcal{Y} = \mathcal{X} \times_j U^{(j)} \times_k U^{(k)}. \tag{3.56}$$

Since $U^{(j)}$ and $U^{(k)}$ are orthogonal matrices, $U_{(j)} \times_j U_{(j)} = I$ and $U_{(k)} \times_k U_{(k)} = I$. Then we get

$$\mathcal{Y} = G \times_{\ell_1} U^{(i)}. \tag{3.57}$$

Applying SVD to unfolded matrix $Y^{i \times (\ell_2 \ell 3)}$, we get

$$Y^{i \times (\ell_2 \ell_3)} = G^{\ell_1 \times (\ell_2 \ell_3)} \times_{\ell_1} U^{(i)}. \tag{3.58}$$

Thus Eqs. (3.56)–(3.58) give the procedure to compute $U^{(i)}$ from $U^{(k)}$ and $U^{(j)}$. Based upon ALS, we can repeated compute either of $U^{(i)}$, $U^{(j)}$ and $U^{(k)}$ from the other two of them until these are converged. After the convergence, we can compute G as

$$G = \mathcal{X} \times_i U^{(i)} \times_j U^{(j)} \times_k U^{(k)} \tag{3.59}$$

because $U^{(i)}$, $U^{(j)}$ and $U^{(k)}$ are orthogonal matrices.

One might notice that HOOI also needs the initialization of $U^{(i)}$, $U^{(j)}$, and $U^{(k)}$. In contrast to CP decomposition that has no ways to perform initialization uniquely, Tucker decomposition can have unique way to decide the initialization. It is called as higher order singular value decomposition (HOSVD). In order to perform HOSVD, we apply SVD to unfolded matrix $X^{i \times (jk)}$ in order to obtain $U^{(i)}$, because we get $U^{(i)}$ through getting the tensor $\mathcal{Y} \in \mathbb{R}^{L_1 \times M \times K}$ and its unfolded matrix $Y^{i \times (jk)}$ as

$$X^{i \times (jk)} = Y^{\ell_1 \times (jk)} \times_{\ell_1} U^{(i)} \tag{3.60}$$

$$\mathcal{Y} = G \times_{\ell_2} U^{(j)} \times_{\ell_3} U^{(k)} \tag{3.61}$$

Similarly, $U^{(j)}$ and $U^{(k)}$ can be obtained with applying SVD to unfolded matrices $X^{j \times (ik)}$ and $X^{k \times (ij)}$, respectively. Finally, using obtained $U^{(i)}$, $U^{(j)}$ and $U^{(k)}$, we can compute G as

$$G = \mathcal{X} \times_i U^{(i)} \times_j U^{(j)} \times_k U^{(k)}. \tag{3.62}$$

In order to extend the above computations to tensors with arbitrary modes m, Eq. (3.2) is extended as

$$\mathcal{X} = G \times_{\ell_1} U^{(i_1)} \times_{\ell_2} U^{(i_2)} \times_{\ell_3} \cdots \times_{\ell_m} U^{(i_m)}, \tag{3.63}$$

Procedure HOSVD
do $i_\alpha, \alpha \in [1, m]$
compute $U^{(i_\alpha)}$ with applying SVD to $X^{i_\alpha \times (i_1 i_2 \cdots i_{\alpha-1} i_{\alpha+1} \cdots i_m)}$ as
$X^{i_\alpha \times (i_1 i_2 \cdots i_{\alpha-1} i_{\alpha+1} \cdots i_m)} = Y^{\ell_\alpha \times (i_1 i_2 \cdots i_{\alpha-1} i_{\alpha+1} \cdots i_m)} \times_{\ell_\alpha} U^{(i_\alpha)}$
end do
$G = \mathcal{X} \times_{i_1} U^{(i_1)} \times_{i_2} U^{(i_2)} \times_{i_3} \cdots \times_{i_m} U^{(i_m)}$
return $G, U^{(i_\alpha)}, \alpha \in [1, m]$
end procedure

Fig. 3.8 Algorithm of HOSVD for tensors with arbitrary number of modes m

Procedure HOOI
Initialize $U^{(i_\alpha)}, \alpha \in [1, m]$ with HOSVD
repeat
do $i_\alpha \in [1, m]$
compute $U^{(i_\alpha)}$ with applying SVD to $Y^{i_\alpha \times (\ell_{i_1} \ell_{i_2} \cdots \ell_{\alpha-1} \ell_{\alpha+1} \cdots \ell_m)}$ as
$Y^{i_\alpha \times (\ell_1 \ell_2 \cdots \ell_{\alpha-1} \ell_{\alpha+1} \cdots \ell_m)} = G^{\ell_\alpha \times (\ell_1 \ell_2 \cdots \ell_{\alpha-1} \ell_{\alpha+1} \cdots \ell_m)} \times_{\ell_\alpha} U^{(i_\alpha)}$
with $\mathcal{Y} = \mathcal{X} \times_{i_1} U^{(i_1)} \times_{i_2} U^{(i_2)} \times_{i_3} \cdots \times_{i_{\alpha-1}} U^{(i_{\alpha-1})} \times_{i_{\alpha+1}} U^{(i_{\alpha+1})} \times_{i_{\alpha+2}} \cdots \times_{i_m} U^{(i_m)}$
end do
until fit ceases to improve or maximum iterations exhausted
$G = \mathcal{X} \times_{i_1} U^{(i_1)} \times_{i_2} \cdots \times_{i_m} U^{(i_m)}$
return $G, U^{(i_\alpha)}, \alpha \in [1, m]$
end procedure

Fig. 3.9 Algorithm of HOOI for tensors with arbitrary number of modes. "…" means the operation over modes excluding the selected ith mode for do loop

where $\mathcal{X}, G \in \mathbb{R}^{N_1 \times N_2 \times \cdots \times N_m}$ and $U^{(i_\alpha)} \in \mathbb{R}^{N_\alpha \times N_\alpha}, 1 \le \alpha \le m$.

Figure 3.8 shows the HOSVD algorithm for tensors with general number of modes and Fig. 3.9 shows the HOOI algorithm for tensors with general number of modes starting from initialization by HOSVD. In these definitions, we can get two algorithm to obtain Tucker decomposition, Eq. (3.2), for tensors with general number of modes. They are also free from arbitrary initialization in contrast to CP decomposition, because HOSVD does not need initialization while HOOI can be initialized uniquely with HOSVD.

I would like to mention some additional comments for these two algorithm. In Figs. 3.8 and 3.9, we do not specify the dimensions of $U^{(i)}, U^{(j)}, \ldots$. If we employ full rank, i.e., $U^{(i)} \in \mathbb{R}^{N \times N}$, $U^{(j)} \in \mathbb{R}^{M \times M}$, $U^{(k)} \in \mathbb{R}^{K \times K}$ …, HOSVD and HOOI do not differ from each other, since initialization using HOSVD gives complete solution thus it is no ways for HOOI to optimize. If we assign smaller dimensions to $U^{(i)}, U^{(j)}, \ldots$, there are possibilities that HOOI can optimize the results by HOSVD. If HOOI differs from HOSVD is completely data dependent. For the tensor Eq. (3.25), $U^{(i)}, U^{(j)}, \ldots$ whose ranks are much smaller than full rank can give complete solution. Thus, we cannot say that assignment of smaller dimensions to $U^{(i)}, U^{(j)}, \ldots$ always results in more optimal results by HOOI that by HOSVD.

One should also notice that HOSVD has superiority to CP decomposition (Fig. 3.7) and HOOI (Fig. 3.9) because arbitrary $U^{(i)}$ can be computed independent of others. Although anyway we cannot avoid computing other singular matrices, $U^{(j)}, U^{(k)}, \ldots$ because we cannot get G without computing all $U^{(i)}, U^{(j)}, U^{(k)}, \ldots$, it is a great advantage of HOSVD when considering applications.

Exercise

3.7 Apply HOSVD to the tensor $X \in \mathbb{R}^{2\times2\times2}$

$$X_{ij1} = \begin{pmatrix} 1 & 2 \\ 3 & 4 \end{pmatrix} \tag{3.64}$$

$$X_{ij2} = \begin{pmatrix} 5 & 6 \\ 7 & 8 \end{pmatrix} \tag{3.65}$$

3.3.3 Tensor Train Decomposition

After recognizing how to compute Tucker decomposition, it is relatively easy to understand how to compute tensor train decomposition [3] as well. Essentially, it is iterative SVD applied to unfolded matrices of a tensor. In order to show the algorithm that computes tensor train decomposition for the tensor with arbitrary number of modes m, tensor train decomposition, Eq. (3.3), is generalized as

$$\mathcal{X} = G^{(i_1)} \times_{\ell_1} G^{(i_2)} \times_{\ell_2} \cdots \times_{\ell_{\alpha-1}} G^{(i_\alpha)} \times_{\ell_\alpha} \cdots \times_{\ell_{m-1}} G^{(i_m)}, \tag{3.66}$$

where $G^{(i_1)} \in \mathbb{R}^{N_1\times R_1}$, $G^{(i_2)} \in \mathbb{R}^{N_2\times R_1\times R_2}$, $\ldots$, $G^{(i_\alpha)} \in \mathbb{R}^{N_\alpha\times R_{\alpha-1}\times R_\alpha}$, $\ldots$, $G^{(i_m)} \in \mathbb{R}^{N_m\times R_{m-1}}$. The components of $G^{(i_1)}$, $G^{(i_\alpha)}$, and $G^{(i_m)}$ are denoted as $G^{(i_1)}(i_1, \ell_1)$, $G^{(i_\alpha)}(i_\alpha R_{\alpha-1}, R_\alpha)$, and $G^{(i_m)}(i_m, R_{m-1})$.

Figure 3.10 shows the tensor train decomposition algorithm applied to tensor with arbitrary number of modes m. In order to perform the algorithm shown in Fig. 3.10, we need to know $R_\alpha, \alpha \in [1, m-1]$ in advance ($R_0 = 1$). It is known that [3]

```
Procedure Tensor train decomposition
    C ← X
    do i_α, α ∈ [1, m − 1]
        compute G^(i_α) ∈ R^{N_α×R_{α−1}×R_α}
            with applying SVD to unfolded matrix C^{(i_α ℓ_{α−1})×(i_{α+1}···i_m)} as
        C^{(i_α ℓ_{α−1})×(i_{α+1}···i_m)} = [G^(i_α)]^{(i_α ℓ_{α−1})×ℓ_α} ×_{ℓ_α} Y^{ℓ_α×(i_{α+1}···i_m)}
            with Y ∈ R^{R_α×N_{α+1}×···×N_m}
        C ← Y
    end do
    G^(i_m) ← C
    return Gs
end procedure
```

Fig. 3.10 Algorithm of tensor train decomposition for tensors with arbitrary number of modes m

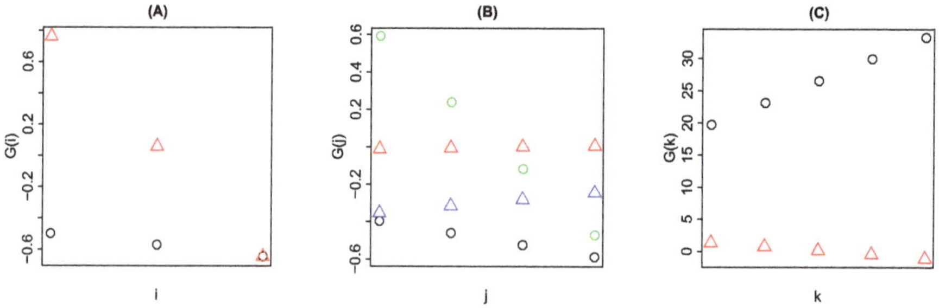

Fig. 3.11 Singular value vectors computed by applying tensor train decomposition (Fig. 3.10) to the tensor Eq. (3.25). (**a**) $G^{(i)}(i, \ell_1)$, $\circ : \ell_1 = 1$, $\triangle$:$\ell_1 = 2$ (**b**) $G^{(j)}(j, \ell_1, \ell_2)$, $\circ : (\ell_1, \ell_2) = (1, 1)$, $\triangle$:$(\ell_1, \ell_2) = (2, 1)$, $\circ$: $(\ell_1, \ell_2) = (1, 2)$, $\triangle$:$(\ell_1, \ell_2) = (2, 2)$ (**c**) $G^{(k)}(k, \ell_2)$, $\circ : \ell_2 = 1$, $\triangle$:$\ell_2 = 2$

$$R_\alpha = \text{rank}\left[X^{(i_1 i_2 \cdots i_\alpha)\times(i_{\alpha+1}\cdots i_m)}\right]. \tag{3.67}$$

In order to see how well the algorithm shown in Fig. 3.10 works, it is applied to the tensor Eq. (3.25) with $(R_0, R_1, R_2, R_3) = (1, 2, 2, 1)$ (Fig. 3.11). Here R_1 and R_2 are estimated by Eq. (3.67) with applying SVD to unfolded matrices, $X^{i\times(jk)}$ and $X^{(ij)\times k}$, respectively. If Fig. 3.11 is compared with Eqs. (3.26)–(3.28), it is a little bit more complicated. However, if it is inserted to Eq. (3.3), it turns out that Eq. (3.25) is reproduced completely. Thus, as far as the reduction of degrees of freedom is considered, algorithm shown in Fig. 3.10 has the solution Fig. 3.11 with the same performance as Eqs. (3.26)–(3.28).

Using tensor train decomposition to obtain approximation is easy. Simply employing R_α which is smaller than Eq. (3.67) such that SVD is truncated up to the first R_α components. Of course, this truncated tensor train decomposition is not guaranteed to be an optimal solution with fixed $R_\alpha, \alpha \in [1, m]$. If the truncated tensor train decomposition does not work well, further optimization might be required. In this case, it is rather straight forward to apply ALS to tensor train decomposition computed by the algorithm shown in Fig. 3.10. In order that, we need to introduce frame matrix $G^{(\neq i_\alpha)} \in \mathbb{R}^{(N_1\cdots N_{\alpha-1}N_{\alpha+1}\cdots N_m)\times(R_{\alpha-1}R_\alpha)}$ as

$$G^{(\neq i_\alpha)} = \left[G^{(i_1)} \times_{\ell_1} \cdots \times_{\ell_{\alpha-2}} G^{(i_{\alpha-1})} \times^{\ell_{\alpha-1}\ell_\alpha} G^{(i_{\alpha+1})} \cdots \right.$$
$$\left. \times_{\ell_{m-1}} G^{(i_m)}\right]^{(i_1 i_2\cdots i_{\alpha-1} i_{\alpha+1}\cdots i_m)\times(\ell_{\alpha-1}\ell_\alpha)}. \tag{3.68}$$

Then ALS that optimizes $G^{(i_\alpha)}$ with fixing $G^{(i_{\alpha'})}$, $\alpha' \neq \alpha$ can be done as

$$G^{(i_\alpha)} = \mathcal{X} \times_{i_1 i_2\cdots i_{\alpha-1} i_{\alpha+1}\cdots i_m} [G^{(\neq i_\alpha)}]^\dagger. \tag{3.69}$$

Procedure Tensor train decomposition with ALS
initialize $G^{(i_\alpha)}, \alpha \in [1, m]$ with the algorithm shown in Fig. 3.10
repeat
do $i_\alpha \in [1, m]$
compute $G^{(i_\alpha)}$ with eq. (3.69)
end do
until fit ceases to improve or maximum iterations exhausted
return $G^{(i_\alpha)}, \alpha \in [1, m]$
end procedure

Fig. 3.12 Algorithm of tensor train decomposition with ALS for tensors with arbitrary number of modes m

Figure 3.12 shows the algorithm of tensor train decomposition with ALS. Unfortunately, the algorithm shown in Fig. 3.12 still does not guarantee globally optimal solution.

Exercise
3.8 Apply tensor train decomposition to the tensor $X \in \mathbb{R}^{2\times2\times2}$

$$X_{ij1} = \begin{pmatrix} 3 & 4 \\ 4 & 5 \end{pmatrix} \tag{3.70}$$

$$X_{ij2} = \begin{pmatrix} 4 & 5 \\ 5 & 6 \end{pmatrix} \tag{3.71}$$

3.4 Interpretation Using TD

In order to demonstrate how effective use of TD is to interpret the date set, we apply TDs to dada set shown in Table 1.6. We have already applied SVD to data set with $k = 1$ in Table 1.6 in order to visualize the relation between price and weight (Fig. 2.1). Here we show the integrated analysis of $k = 1$ and $k = 2$ with TD. Because ks represent two shops, the integrated analysis by TD should represents how similar two shops are as well as how much they differ from each other. Figure 3.13 shows the results of SVD applied to data sets shown in Table 1.6 ($k = 2$). The difference between Figs. 2.1 and 3.13 represents the distinction between two shops ($k = 1$ and $k = 2$). v^+ axis represents the same contribution to price and weight while v^- axis represents the opposite contribution to them. $k = 1$ (shop 1, Fig. 2.1) has more distinct contribution between v^+ and v^- while that of $k = 2$ (shop 2, Fig. 3.13) is less. This represents the primary difference between $k = 1$ and $k = 2$ (two shops).

Next we apply Tucker decomposition (HOSVD algorithm, Fig. 3.8) to the data set shown in Table 1.6 that is formatted as $\mathcal{X} \in \mathbb{R}^{4\times2\times2}$ where is stand for goods, js stand for price ($j = 1$) and weight ($j = 2$), and ks stand for shops. Figure 3.14

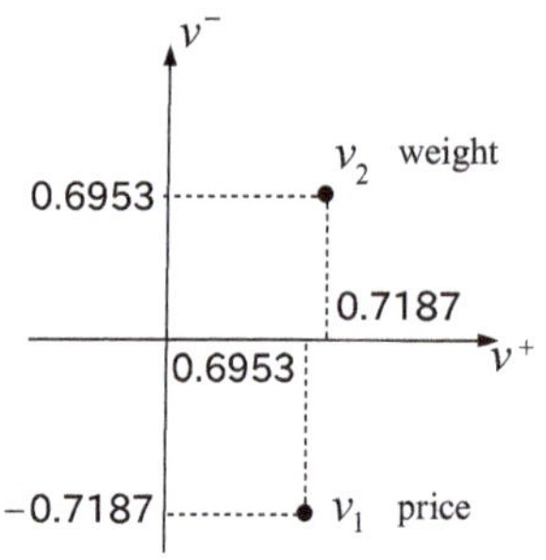

Fig. 3.13 A geometrical interpretation of price and weight originally shown in Table 1.6, $k = 2$. $\boldsymbol{v}_1$: price, $\boldsymbol{v}_2$: weight

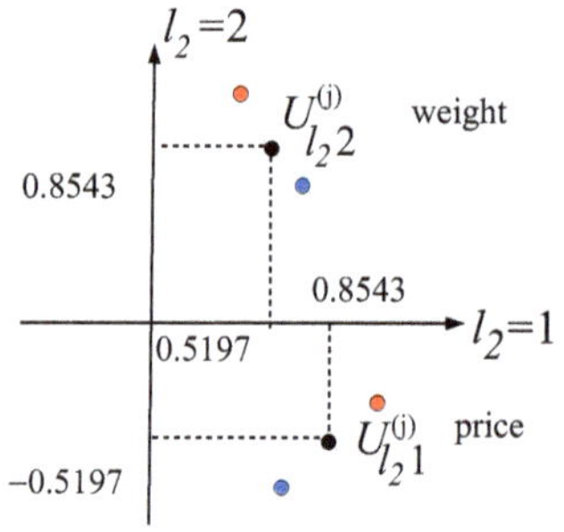

Fig. 3.14 A geometrical interpretation of price and weight originally shown in Table 1.6 with applying HOSVD (Fig. 3.8). $U^{(j)}_{\ell_2 1}$: price, $U^{(j)}_{\ell_2 2}$: weight. Red and blue dots correspond to the location of price and weight in Figs. 2.1 and 3.13, respectively

Table 3.4 $G(\ell_1, \ell_2, \ell_3)$s computed by HOSVD (Fig. 3.8 applied to data set shown in Table 1.6. ℓ_1: foods (Bread, Beef, Pork, and Fish), ℓ_2: properties (price and weight), ℓ_3:two shops

	$\ell_2 = 1$		$\ell_2 = 2$	
ℓ_3	1	2	1	2
ℓ_1				
1	1964	−283	46	−208
2	−25	−275	1316	427
3	18	13	−42	141
4	17	126	28	−5

shows that scatter plot of $U^{(j)}_{\ell_2 j}$; $U^{(j)}_{1j}$ and $U^{(j)}_{2j}$ correspond to v^+ and v^- in Figs. 2.1 and 3.13, respectively. It is rather obvious that Fig. 3.14 represents, in some sense, "in between" feature of Figs. 2.1 and 3.13, because $U^{(j)}_{11}$ is larger than v^+ for price in Fig. 3.13 and smaller that in Fig. 2.1, $U^{(j)}_{21}$ is larger than v^- for price in Fig. 3.13 and smaller than that in Fig. 2.1, $U^{(j)}_{12}$ is smaller than v^+ for price in Fig. 3.13 and larger than that in Fig. 2.1, and $U^{(j)}_{21}$ is larger than v^- for price in Fig. 3.13 and smaller than that in Fig. 2.1. Thus, integrated analysis using TD of two shops' data is seemingly successful.

The next question is how these factors, i.e., simultaneous and opposite effects between price and weight, affect data set. It can be also understood by investigating Gs that represents the interaction between distinct three features, i.e., foods (Bread, Beef, Pork, and Fish), properties (price and weight), and two shops. Table 3.4 shows Gs. Gs associated with larger absolute valued correspond to the combination of foods, properties and shops singular vectors that contribute more to the right hand

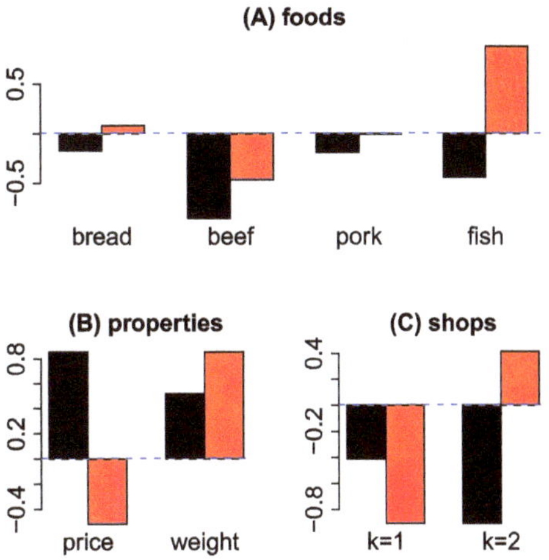

Fig. 3.15 $U^{(i)}$, $U^{(j)}$ and $U^{(k)}$ when HOSVD (Fig. 3.8) is applied to data set originally shown in Table 1.6. (**a**) $U^{(i)}$, foods, (**b**) $U^{(j)}$, properties, (**c**) $U^{(k)}$, shops. Black: $\ell_1 = \ell_2 = \ell_3 = 1$, red: $\ell_1 = \ell_2 = \ell_3 = 2$

side of Eq. (3.55). It is obvious that two combinations, $(\ell_1\ell_2, \ell_3) = (1, 1, 1)$ and $(2, 2, 1)$, outperform other combinations. First, we investigate what $U^{(k)}$ represents, because $U^{(k)}_{2k}$ does not seem to be important at all. As can be seen in Fig. 3.15c, $\ell_3 = 2$ represents the distinction between two shops because $U^{(k)}_{21}$ and $U^{(k)}_{22}$ are oppositely signed. Thus, the smaller absolute values of Gs associated with $\ell_3 = 2$ suggests the unimportance of the difference between two shops. In Fig. 3.14, we demonstrated that the difference between two shops is comparatively smaller than mean between two shops with comparing the results obtained by SVD and those by HOSVD. Nevertheless, even without comparison between SVD and HOSVD, by investigating the results by SVD and HOSVD independently, we can easily recognize the unimportance of the difference between two shops as shown here.

Next, we try to understand what the combinations $(\ell_1, \ell_2) = (1, 1)$ and $(2, 2)$ mean. $\ell_2 = 1$ and $\ell_2 = 2$ (Fig. 3.15b) correspond to the coincidence and distinction between price and weight as shown in Fig. 3.14. Thus, $\ell_1 = 1$ and $\ell_2 = 1$ (Fig. 3.15a) show that as well. $U^{(i)}_{1i}$ shows the coincidence between four foods. On the other hand, $U^{(i)}_{2i}$ shows the distinct signs between four foods, especially opposite signs between fish and beef. From these analysis, we can understand that applying TD to the data set enables us to understand many characteristic features hidden in data set. This ability of TD will be further demonstrated in the application of TDs to more extensive data set in the following sections.

One might wonder how other TDs work as well. In order to see if CP decomposition works as well, we apply CP decomposition to data set shown in Table 1.6. The reason why we use $L = 2$ because we know that there are only two important combinations of singular value vectors when HOSVD is applied to the same data set in the above. Figure 3.16 shows $\boldsymbol{u}^{(i)}_\ell, \boldsymbol{u}^{(j)}_\ell, \boldsymbol{u}^{(k)}_\ell$. It is rather obvious that they are coincident with Fig. 3.15 excluding some reversed signs that are not critical. This might suggests that CP decomposition works as well, but we need to remind that we assumed $L = 2$ based upon the results by HOSVD. In order

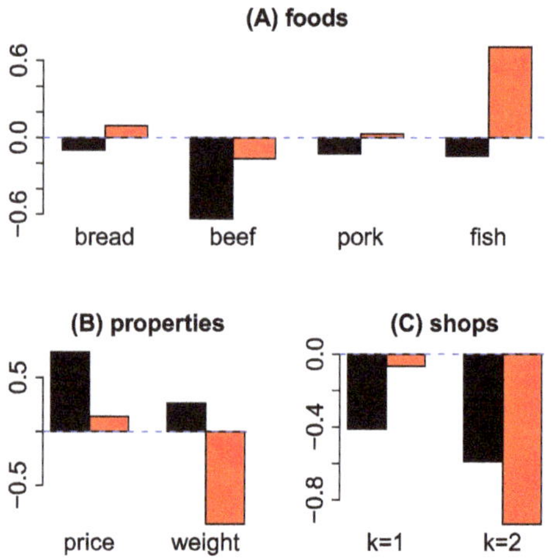

Fig. 3.16 $U^{(i)}$, $U^{(j)}$ and $U^{(k)}$ when CP decomposition with $L = 2$ is applied to data set originally shown in Table 1.6. (**a**) $\boldsymbol{u}^{(i)}{}_{\ell}$, foods, (**b**) $\boldsymbol{u}_{\ell}^{(j)}$, properties, (**c**) $\boldsymbol{u}_{\ell}^{(k)}$, shops. Black: $\ell = 1$, red: $\ell = 2$

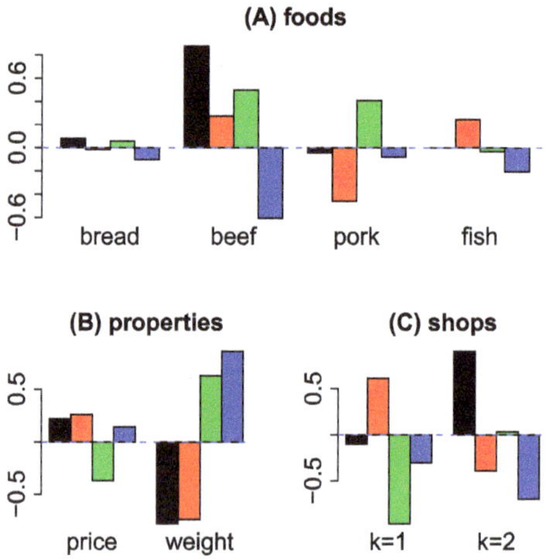

Fig. 3.17 $U^{(i)}$, $U^{(j)}$ and $U^{(k)}$ when CP decomposition with $L = 4$ is applied to data set originally shown in Table 1.6. (**a**) $\boldsymbol{u}^{(i)}{}_{\ell}$, foods, (**b**) $\boldsymbol{u}_{\ell}^{(j)}$, properties, (**c**) $\boldsymbol{u}_{\ell}^{(k)}$, shops. Black: $\ell = 1$, red: $\ell = 2$, green: $\ell = 3$, blue: $\ell = 4$

to see if we can identify that $L = 2$ is enough without the support of HOSVD, we apply CP decomposition with $L = 4$ to the same data set (Fig. 3.17). The result is rather disappointing. Not only it is not easily understood, but also there are no ways to identify which ℓs are important. Since λ_ℓs are 6052 ($\ell = 1$), 4109 ($\ell = 2$), 3810 ($\ell = 3$) and 9771.689 ($\ell = 4$), there are not outstandingly important ones in contrast to Gs in Table 3.4 where only $(\ell_1, \ell_2, \ell_3) = (1, 1, 1)$ and $(2, 2, 1)$ have outstandingly large contributions. Thus, CP decomposition has less ability to identify fewer number of important singular value vectors.

Finally, we apply tensor train decomposition, Fig. 3.10, to the same data set with $R_1 = R_2 = 2$ that enables us to retain two supposedly important two combinations. In this set up, although js (properties, i.e., price and weight) must be associated with $R_1 \times R_2 = 4$ singular value vectors, there are no ways to restrict number of singular value vectors attributed to j to two in the tensor train frame work. Figure

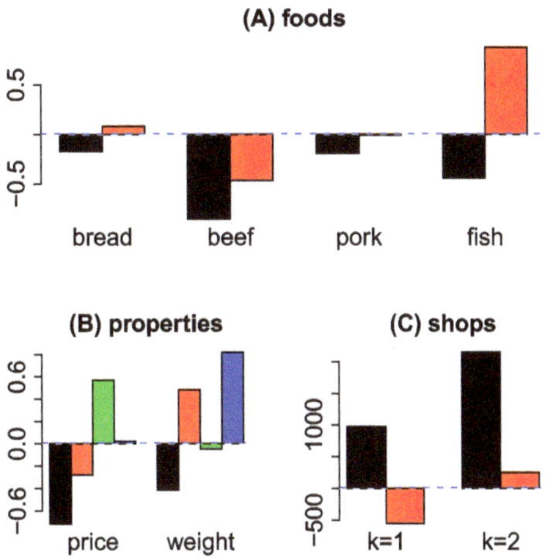

Fig. 3.18 $G^{(i)}(i, \ell_1)$, $G^{(j)}(j, \ell_1, \ell_2)$ and $G^{(k)}(k, \ell_2)$ when tensor train decomposition with $R_1 = R_2 = 2$ is applied to data set originally shown in Table 1.6. (**a**) $G^{(i)}(i, \ell_1)$, foods, (**b**) $G^{(j)}(j, \ell_1, \ell_2)$, properties, (**c**) $G^{(k)}(k, \ell_2)$, shops. Black: (**a**) $\ell_1 = 1$, (**b**) $(\ell_1, \ell_2) = (1, 1)$, (**c**) $\ell_2 = 1$, red: (**a**) $\ell_1 = 2$, (**b**) $(\ell_1, \ell_2) = (1, 2)$, (**c**) $\ell_2 = 1$, green: (**b**) $(\ell_1, \ell_2) = (2, 1)$, blue: (**b**)$(\ell_1, \ell_2) = (2, 2)$

3.18 shows the results of tensor train decomposition. Figure 3.18 is also coincident with Fig. 3.15 where HOSVD is employed if excluding $G^{(j)}(j, \ell_1, \ell_2), (\ell_1, \ell_2) = (2, 1), (2, 2)$. Nevertheless, we cannot exclude these two without the knowledge from HOSVD, because there are no weight factors like λ_ℓs for CP decomposition and Gs for HOSVD that can be used for the selection of important terms in TDs. In this sense, tensor train decomposition is also inferer to HOSVD because it cannot select primarily important two combinations.

3.5 Summary

In this section, I have introduced three popular TD methods, CP decomposition, Tucker decomposition, and tensor train decomposition. All three TDs have their own advantaged and disadvantages.

3.5.1 CP Decomposition

3.5.1.1 Advantages

The advantages of CP decomposition are as follows.

- Easy interpretability. CP decomposition can result in one to one correspondence between singular value vectors. Thus, the interpretation is easier than other two methods.
- The number of terms in right hand side of decomposition can be decided freely without any restriction.

- Because of freely decidable number of decomposition terms, truncation is uniquely decided.
- It has weights, λ_ℓ, that can evaluate importance of each term.

3.5.1.2 Disadvantages

The disadvantages of CP decomposition are as follows.

- With using known algorithms, it is not guaranteed to converge to global optimum.
- In some worst cases, there are not global optimum in the sense complete solution (i.e., no residuals) can achieve the limit when the absolute values of each terms go to infinity.
- It needs to have initial values to start and it reaches the local minimums dependent upon initial values.
- No known algorithm to converge within polynomial times.

3.5.2 Tucker Decomposition

3.5.2.1 Advantages

The advantages of Tucker decomposition are as follows.

- There are algorithm that can converge in polynomial times (e.g., HOSVD), although convergence to the global minimum is not guaranteed.
- It has weight, G, that can evaluate importance of each term.
- ALS can be used to optimize the solution obtained by the method with the guarantee of convergence within polynomial time.
- Although it is limited to the product of truncated rank of each mode, i.e., $\prod_{\alpha=1}^m R_\alpha$ where R_α is the truncated rank of αth mode, truncation decomposition straight forward.
- We do not need to assign initial values to perform ALS since initial values can be computed by HOSVD which requires only polynomial time.

3.5.2.2 Disadvantages

The disadvantages of Tucker decomposition are as follows.

- Since all possible combination of singular value vectors are present, selection of important terms based upon weight G is inevitably subjective.
- In the full rank TD, i.e., $R_\alpha = N_\alpha$ where N_α is number of variables in αth mode, the number of degrees of freedom increase to $\prod_{\alpha=1}^m N_\alpha + \sum_{\alpha=1}^m N_\alpha^2$ from that of original tensor, $\prod_{\alpha=1}^m N_\alpha$.

- It does not have unique solutions, because applying unitary transformation does not alter the amount of residues.

3.5.3 Tensor Train Decomposition

3.5.3.1 Advantages

The advantages of tensor train decomposition are as follows.

- It has superior ability to reduce degrees of freedoms to other two TDs. In CP decomposition, degrees of freedoms is as many as $L \sum_{\alpha=1}^{m} N_\alpha$. That in Tucker decomposition is as many as $\prod_{\alpha=1}^{m} N_\alpha + \sum_{\alpha=1}^{m} N_\alpha^2$. On the other hand, that of tensor train decomposition is as many as $N_1 R_1 + \sum_{\alpha=1}^{m} R_{\alpha-1} R_\alpha N_\alpha + N_m R_{m-1}$. Thus, degrees of freedom increases as only proportional to logarithmic order of terms in decomposition.
- It has algorithm that converges in polynomial time.
- ALS can be applied to optimize the obtained solution.
- We do not need to assign initial values to perform ALS since initial values can be computed by algorithm which requires only polynomial time.

3.5.3.2 Disadvantages

- It does not have unique solutions, because applying unitary transformation does not alter the amount of residues.
- It does not have weight that evaluates importance of each term.

3.5.4 Superiority of Tucker Decomposition

Considering the advantages and disadvantages of three TDs, we decided to employ Tucker decomposition implemented by HOSVD to be applied to real problems in the following sections. It is primarily because it has weight to select relevant terms. Tensor train decomposition does not have weight, thus it is not suitable to employ the application that needs the interpretation of the outcome of TDs. In other application, e.g., image analysis, because it is not require do interpret TDs themselves, tensor train decomposition does not have to be excluded. Nevertheless, in this monograph, the application to the biological problem is the main topics. In the application to biological problems, interpretability is important. Tensor train decomposition that lacks the weight to evaluate each term is not suitable.

On the other hand, CP decomposition apparently has more interpretability than Tucker decomposition because it provides one to one correspondence between singular value vectors. The apparent superior interpretability of this method is not fully trustable because of heavy initial value dependence. It is not also ideal one because increasing L often results in distinct results obtained by smaller L. In this case, it is unsure how large L should be.

Because of the above reasons, HOSVD is considered to be the best method that can be applied to tensors when interpretability is important. In addition to this, because HOSVD is natural extension of SVD to higher mode tensors, we can discuss the application of SVD (or PCA) and that of HOSVD in the integrated manner.

Appendix

Moore-Penrose Pseudoinverse

Moore-Penrose pseudoinverse [1], which is denoted as $A^\dagger$, of matrix A satisfies the following conditions.

- $AA^\dagger A = A$
- $A^\dagger AA^\dagger = A^\dagger$
- $(A^\dagger A)^T = A^\dagger A$
- $(AA^\dagger)^T = AA^\dagger$

Suppose we need to find $\boldsymbol{x} \in \mathbb{R}^M$ that satisfies

$$A\boldsymbol{x} = \boldsymbol{b} \tag{3.72}$$

where $A \in \mathbb{R}^{N\times M}$ and $\boldsymbol{b} \in \mathbb{R}^N$. It is known that there is a unique solution only when $N = M$.

Moore-Penrose pseudoinverse can *solve* Eq. (3.72) because

$$\boldsymbol{x} = A^\dagger \boldsymbol{b} \tag{3.73}$$

gives

- the unique solution of Eq. (3.72) when $N = M$.
- the $\boldsymbol{x}$ that satisfies Eq. (3.72) with minimum $|\boldsymbol{x}|$ when $N < M$ (i.e., when no unique solutions are available).
- the $\boldsymbol{x}$ with minimum $|A\boldsymbol{x} - \boldsymbol{b}|$ when $N > M$ (equivalent to so called the linear regression analysis).

When $N < M$, there are infinitely large number of solutions that satisfy Eq. (3.72). Moore-Penrose pseudoinverse allows us to select one of them, which has minimum $|\boldsymbol{x}|$. On the other hand, when $N > M$, there are not always solutions that satisfy Eq. (3.72). Moore-Penrose pseudoinverse allows us to select the solution having the minimum $|A\boldsymbol{x} - \boldsymbol{b}|$, i.e., the smallest residuals. Thus, by computing

Moore-Penrose pseudoinverse, we can always compute $\boldsymbol{x}$ that satisfies Eq. (3.72) as much as possible in some sense.

How to compute $A^\dagger$ is as follows. Apply SVD to A as

$$A = U\Sigma V^T \tag{3.74}$$

$U \in \mathbb{R}^{N\times M}$, $\Sigma, V \in \mathbb{R}^{M\times M}$ for $N > M$ and $U, \Sigma \in \mathbb{R}^{N\times N}$, $V \in \mathbb{R}^{M\times N}$ for $N < M$. When U or V is not a square matrix, $U^TU = V^TV = I$, but $UU^T \neq I$ and $VV^T \neq I$. When U and V are square matrices, $U^TU = UU^T = V^TV = VV^T = I$.

Then $A^\dagger$ can be defined as

$$A^\dagger = V\Sigma^{-1}U^T \tag{3.75}$$

It is not difficult to show that $A^\dagger = V\Sigma^{-1}U^T$ satisfies the required conditions because

$$AA^\dagger = \left(U\Sigma V^T\right)\left(V\Sigma^{-1}U^T\right) = U\Sigma\Sigma^{-1}U^T = UIU^T = \begin{cases} UU^T, N > M \\ I, N \leq M \end{cases} \tag{3.76}$$

and

$$A^\dagger A = \left(V\Sigma^{-1}U^T\right)\left(U\Sigma V^T\right) = V\Sigma^{-1}\Sigma V^T = VIV^T = \begin{cases} I, N \geq M \\ VV^T, N < M \end{cases} \tag{3.77}$$

where $V^TV = I$ for $N > M$ and $U^TU = I$ for $N < M$ are used.

Then when $N > M$,

$$AA^\dagger A = A\left(A^\dagger A\right) = AI = A, \tag{3.78}$$

$$A^\dagger AA^\dagger = \left(A^\dagger A\right)A^\dagger = IA^\dagger = A^\dagger \tag{3.79}$$

$$\left(AA^\dagger\right)^T = \left(UU^T\right)^T = \left(U^T\right)^T U^T = UU^T = AA^\dagger \tag{3.80}$$

$$\left(A^\dagger A\right)^T = I^T = I = A^\dagger A \tag{3.81}$$

On the other hand, when $N < M$,

$$AA^\dagger A = \left(AA^\dagger\right)A = IA = A, \tag{3.82}$$

$$A^\dagger AA^\dagger = A^\dagger\left(AA^\dagger\right) = A^\dagger I = A^\dagger \tag{3.83}$$

$$\left(AA^\dagger\right)^T = I^T = I = AA^\dagger \tag{3.84}$$

$$\left(A^{\dagger}A\right)^{T} = \left(VV^{T}\right)^{T} = \left(V^{T}\right)^{T} V^{T} = VV^{T} = A^{\dagger}A \tag{3.85}$$

When $N = M$, these are obvious because $AA^{\dagger} = A^{\dagger}A = I$.

The reason why we can treat Eq. (3.72) using Moore-Penrose pseudoinverse as mentioned in the above is as follows. Define

$$\boldsymbol{x}_0 = A^{\dagger}\boldsymbol{b} + \left(I - A^{\dagger}A\right)\boldsymbol{w} \tag{3.86}$$

with arbitrary vector $\boldsymbol{w}$. Then because

$$A\boldsymbol{x}_0 = AA^{\dagger}\boldsymbol{b} + \left(A - AA^{\dagger}A\right)\boldsymbol{w} = AA^{\dagger}\boldsymbol{b} \tag{3.87}$$

when $AA^{\dagger} = I$, i.e., $N \leq M$, $A\boldsymbol{x}_0 = \boldsymbol{b}$, $\boldsymbol{x}_0$ is a solution of Eq. (3.72). This corresponds to the cases where there are no unique solutions because the number of variables, M, is larger than the number of equations, N. $\boldsymbol{x}_0$ can be a unique solution only when $A^{\dagger}A = I$ as well, i.e., $N = M$ because of Eq. (3.86). This corresponds to the cases where there is a unique solution because the number of variables, M, is equal to the number of equations, N.

Here one should notice that $A^{\dagger}\boldsymbol{b} \perp \left(I - A^{\dagger}A\right)\boldsymbol{w}$ because

$$\begin{aligned}\left(I - A^{\dagger}A\right)\boldsymbol{w} \cdot A^{\dagger}\boldsymbol{b} &= \left(\left(I - A^{\dagger}A\right)\boldsymbol{w}\right)^{T} A^{\dagger}\boldsymbol{b} = \boldsymbol{w}^{T}\left(I - A^{\dagger}A\right)^{T} A^{\dagger}\boldsymbol{b} \\ &= \boldsymbol{w}^{T}\left(I - A^{\dagger}A\right)A^{\dagger}\boldsymbol{b} = \boldsymbol{w}^{T}(A^{\dagger} - A^{\dagger}AA^{\dagger})\boldsymbol{b} \\ &= \boldsymbol{w}^{T}0\boldsymbol{b} = 0.\end{aligned} \tag{3.88}$$

Thus from Eq. (3.86)

$$|\boldsymbol{x}_0|^2 = \left|A^{\dagger}\boldsymbol{b}\right|^2 + \left|\left(I - A^{\dagger}A\right)\boldsymbol{w}\right|^2 \tag{3.89}$$

This means, $|\boldsymbol{x}_0| > \left|A^{\dagger}\boldsymbol{b}\right|$. Therefore, $A^{\dagger}\boldsymbol{b}$ is the solution that satisfies Eq. (3.72) and has the smallest $|x_0|$ (In other words, the solution with the $L2$ regulation term).

When $AA^{\dagger} \neq I$, i.e., $N > M$, there are no solutions. This corresponds to the cases where there are no solutions because the number of variables, M, is smaller than the number of equations, N. In this case, $\boldsymbol{x} = A^{\dagger}\boldsymbol{b}$ is known to be optimal (i.e., the solution with minimum $|A\boldsymbol{x} - \boldsymbol{b}|$). In order to prove this, first we need to compute $A^{T}(AA^{\dagger}\boldsymbol{b} - \boldsymbol{b})$ as

$$\begin{aligned}A^{T}\left(AA^{\dagger}\boldsymbol{b} - \boldsymbol{b}\right) &= A^{T}\left(\left(AA^{\dagger}\right)^{T}\boldsymbol{b} - \boldsymbol{b}\right) = \left(\left(AA^{\dagger}A\right)^{T} - A^{T}\right)\boldsymbol{b} \\ &= \left(AA^{\dagger}A - A\right)^{T}\boldsymbol{b} = 0\boldsymbol{b} = 0.\end{aligned} \tag{3.90}$$

With taking transposition of the above, we can also get

$$\left(AA^{\dagger}\boldsymbol{b}-\boldsymbol{b}\right)^{T}A=0. \tag{3.91}$$

Using these, we can show

$$|A\boldsymbol{x}-\boldsymbol{b}|^{2}=\left|\left(A\boldsymbol{x}-AA^{\dagger}\boldsymbol{b}\right)+\left(AA^{\dagger}\boldsymbol{b}-\boldsymbol{b}\right)\right|^{2} \tag{3.92}$$

$$=\left|A\boldsymbol{x}-AA^{\dagger}\boldsymbol{b}\right|^{2}+\left(A\boldsymbol{x}-AA^{\dagger}\boldsymbol{b}\right)^{T}\left(AA^{\dagger}\boldsymbol{b}-\boldsymbol{b}\right)$$
$$+\left(AA^{\dagger}\boldsymbol{b}-\boldsymbol{b}\right)^{T}\left(A\boldsymbol{x}-AA^{\dagger}\boldsymbol{b}\right)+\left|AA^{\dagger}\boldsymbol{b}-\boldsymbol{b}\right|^{2} \tag{3.93}$$

$$=\left|A\boldsymbol{x}-AA^{\dagger}\boldsymbol{b}\right|^{2}+\left(\boldsymbol{x}-A^{\dagger}\boldsymbol{b}\right)^{T}A^{T}\left(AA^{\dagger}\boldsymbol{b}-\boldsymbol{b}\right)$$
$$+\left(AA^{\dagger}\boldsymbol{b}-\boldsymbol{b}\right)^{T}A\left(\boldsymbol{x}-A^{\dagger}\boldsymbol{b}\right)+\left|AA^{\dagger}\boldsymbol{b}-\boldsymbol{b}\right|^{2} \tag{3.94}$$

$$=\left|A\boldsymbol{x}-AA^{\dagger}\boldsymbol{b}\right|^{2}+\left(\boldsymbol{x}-A^{\dagger}\boldsymbol{b}\right)^{T}0$$
$$+0\left(\boldsymbol{x}-A^{\dagger}\boldsymbol{b}\right)+\left|AA^{\dagger}\boldsymbol{b}-\boldsymbol{b}\right|^{2} \tag{3.95}$$

$$=\left|A\boldsymbol{x}-AA^{\dagger}\boldsymbol{b}\right|^{2}+\left|AA^{\dagger}\boldsymbol{b}-\boldsymbol{b}\right|^{2} \tag{3.96}$$

$$\geq\left|AA^{\dagger}\boldsymbol{b}-\boldsymbol{b}\right|^{2}. \tag{3.97}$$

This means that $\boldsymbol{x}=A^{\dagger}\boldsymbol{b}$ is an optimal solution of Eq. (3.72).

References

1. Barata, J.C.A., Hussein, M.S.: The Moore–Penrose pseudoinverse: A tutorial review of the theory. Braz. J. Phys. **42**(1), 146–165 (2012). https://doi.org/10.1007/s13538-011-0052-z
2. Kolda, T., Bader, B.: Tensor decompositions and applications. SIAM Rev. **51**(3), 455–500 (2009). https://doi.org/10.1137/07070111X
3. Oseledets, I.: Tensor-train decomposition. SIAM J. Sci. Comput. **33**(5), 2295–2317 (2011). https://doi.org/10.1137/090752286

Part II
Feature Extractions

Feature extraction is a generation of new feature in the data-driven way. In this part, two methods, PCA and TD, are extensively considered. Although both are supposed to be fully linear methods, because they decompose variables into products of new variables, it can include nonlinear transformation partly. In addition to this, both have ability to reduce degrees of freedom. They are discussed from the data science point of views, with the applications to the data sets, for the usage in the later chapters.

Chapter 4
PCA-Based Unsupervised FE

There is no sound that I do not need.
Rio Kazumiya, Sound of the Sky, Season 1, Episode 3

4.1 Introduction: Feature Extraction vs Feature Selection

In this chapter, I mainly discuss about the situation where feature extraction or feature selection is inevitable. When or under what kind of conditions, do we need either or both of two? Here are some examples of such situations.

- **Case 1**: The number of features attributed to individual samples is larger than the number of samples.
- **Case 2**: Features attributed to individual samples are not independent of one another.
- **Case 3**: Some of features attributed to samples are not related to some properties that we would like to relate features to.

Although these above three cases are not comprehensive, they are good examples by which we can discuss the reason why we need feature extraction and/or feature selection. An example of case 1 is liner equations that can be represented as $A\boldsymbol{x} = \boldsymbol{b}$ where $A \in \mathbb{R}^{N \times M}, \boldsymbol{x} \in \mathbb{R}^{M}, \boldsymbol{b} \in \mathbb{R}^{N}$ and $\boldsymbol{x}$ represents variables, A represents coefficients, and $\boldsymbol{b}$ represents constants. When $N < M$, not only there are no unique solutions, but also there are always solutions, even when A and $\boldsymbol{b}$ are purely random numbers. The fact that there are no unique solutions prevents us from interpreting outcome, because there can be multiple distinct unique solutions. The fact that there are always solutions means that there might be meaningless solutions. In this case, we need feature extraction and/or feature selection such that we can have limited number of features that is smaller the number of samples. An example of case 2 is multicollinearity. In this case, although apparently, $A\boldsymbol{x} = \boldsymbol{b}$ is uniquely solvable, it is actually not because coefficient matrix A is not regular (in other words, row vectors are not independent of one another). In this case, we need to apply feature extraction or feature selection in order to obtain reduced number of features that

Y-h. Taguchi, *Unsupervised Feature Extraction Applied to Bioinformatics*, Unsupervised and Semi-Supervised Learning,
https://doi.org/10.1007/978-3-031-60982-4_4

enables us to get unique solutions. An example of case 3 is that some elements of A are zero. Especially, if A includes column vectors totally filled with zero, variables that correspond to these columns are not related to $\boldsymbol{b}$ at all. When A is given, we can simply discard these variables. Nevertheless, when A is required to be inferred from $\boldsymbol{x}$ and $\boldsymbol{b}$ (e.g., linear regression analysis), it is impossible to exclude there variables in advance. This might result in the incorrect estimation of A. In this case, we need feature selection that enables us to exclude variables not related to $\boldsymbol{b}$ in advance.

From these examples, we can know that the needs of feature selection and feature extraction is very ubiquitous. So, the next question is which is better strategy to address these problems. Unfortunately, the answer is highly context dependent and cannot be decided based upon mathematical considerations. For example, let us consider image analysis, e.g., face recognition. In this case, it is rather obvious that not all pixels of digital images but only a limited number of them is useful for the purpose. If small number of features generated from large number of pixels work well, there are no needs to go further. On the other hands, if the problem is the inference about bankruptcy, in other words, the prediction of who will bankrupt. In this case, even if a newly generated feature composed of numerous personal information, e.g., income, age, education history, address, and so on, works pretty well, it might not be a final goal. This is because collecting these information might cost or is impossible at all. If another feature composed of more limited number of features works, even if the performance is a little bit less, another one might be employed because of easiness to use. Thus, it is inevitable to specify situation that we want to discuss.

As for the targeted field, I would like to say that the targeted field is bioinformatics as the title of this book says. In bioinformatics analysis, it is very usual that feature selection is more favorable than feature extraction because of the following reasons. In bioinformatic analysis (or in biology although it means the same), measuring individual features often costs. Thus, measuring less number of features can reduce the cost spent to individual observations. This results in the increased number of observations that often leads to better outcome. Even when the measuring individual features does not cost, e.g., in the case of high throughput measurements, feature selection is often better than feature extraction, because each feature has its own meaning. For example, if features are genes, the selected limited number of genes are more intepretable than features generated by the combination of large number of genes. Thus, in the following I assume the situation where feature selection is more favorable than feature extraction even if not explicitly denoted.

4.2 Various Feature Selection Procedures

Although there are various ways to classify numerous number of previously proposed feature selection procedures, I would like to employ the one shown in Table 4.1. Feature selection strategies can be classified into two groups in two ways. One way is supervised ones vs unsupervised ones. Not to mention, supervised ones

Table 4.1 Classification of feature selections

	One by one	Collective
Supervised	Statistical tests[a]	Random forest, LASSO
Unsupervised	Highly variable genes, bimodal genes	PCA-based unsupervised FE

[a] *t*-test, limma, SAM

are definitely more popular than unsupervised ones. This is because the purpose of feature selection is usually purpose oriented. For example, if the study aims to investigate diseases, it is natural to consider genes expressed differently between patients and healthy controls. If the study aims to predict who will bankrupt, it is reasonable to consider features related to something financial. On the other hand, unsupervised feature selection might sound self-discrepancy, because it is unlikely possible to select features without any clear purposes. In spite of that, unsupervised feature selection is still possible. For example, it is natural to select features with maximum variance, because large variance might reflect the ability of the feature that represents diverse categories hidden in the considered sample. Thus, although it is less popular, unsupervised feature selection is still possible. Another way to classify feature selection strategies is one by one vs collective. The former means that feature selection is performed without the consideration of interaction between features. For example, when conventional statistical tests are applied to a feature of samples composed of two categories, the P-value that rejects the null hypothesis that a feature of members of two samples obeys the same distribution is computed. Then, if P value is small enough, say less than 0.01, the feature is identified as distinct between two categories. This means that each P-value attributed to each feature is not affected by other features at all. On the other hand, the latter considers the interaction between features. For example, when dummy variables are attributed to each of two categories, we can make linear regression using arbitrary number of features to predict dummy variables. In this case, the interaction between features included into regression equation is considered. Then, features used to construct regression equation with good performance are selected.

In order to demonstrate how differently feature selections that belong to four categories listed in Table 4.1 work, I prepare two synthetic data sets. Both are matrices $x_{ij} \in \mathbb{R}^{N \times M}$ where i and j correspond to features' index and samples' index, respectively. In both data sets, the only first $N_1 (< N)$ features, $x_{ij}, i \leq N_1$, are distinct between two classes where $j \leq \frac{M}{2}$ and $j > \frac{M}{2}$ belong to the first and second class, respectively. x_{ij} is also drawn from Gaussian or mixed Gaussian distribution where $\mathcal{N}(\mu, \sigma)$ represents Gaussian distribution that has mean of μ and standard deviation σ, respectively.

- Data set 1:

$$x_{ij} \sim \begin{cases} \mathcal{N}(0, \sigma) & j \leq \frac{M}{2},\ i \leq N_1 \\ \mathcal{N}(\mu_0, \sigma) & j > \frac{M}{2},\ i \leq N_1 \\ \frac{1}{2}\mathcal{N}(0, \sigma) + \frac{1}{2}\mathcal{N}(\mu_0, \sigma) & i > N_1. \end{cases} \tag{4.1}$$

- Data set 2:

$$x_{ij} \sim \begin{cases} \mathcal{N}(0,\sigma) & j \le \frac{M}{2},\ i \le N_1 \\ \mathcal{N}(\mu_0,\sigma) & j > \frac{M}{2},\ i \le N_1 \\ \mathcal{N}(\mu_1,\sigma) & \qquad i > N_1. \end{cases} \tag{4.2}$$

Thus, the only difference between two synthetic data sets is if the $N - N_1$ features (i.e., $i > N_1$) not distinct between two classes are drawn from bimodal [Eq. (4.1)] or unimodal [Eq. (4.2)] distributions. Specifically, $N = 100$, $M = 20$, $\mu_0 = 4$, $\mu_1 = \frac{\mu_0}{2} = 2$, $N_1 = 10$ and $\sigma = 1$ in the following. Performance are averaged over one hundred independent trials. The number of features distinct between two categories, N_1, is assumed to be known in advance. μ_1 is selected such that the sample mean of ith feature, $\langle x_{ij} \rangle_j$ defined by Eq. (2.56), does not differ between two models.

The statistical tests used belong to either of four categories. t-test is employed as a representative of one by one, supervised feature selection. P values computed by t-test are attributed to individual features. Top N_1 features with smaller P values are selected. As a representative of collective supervised feature selection, linear regression is employed. The dummy variable $y_j \in [0,1]^M$ is given such that $y_j = 0, j \le \frac{M}{2}$ and $y_j = 1, j > \frac{M}{2}$. Then using regression coefficient vector, $a_i \in \mathbb{R}^N$, $X\boldsymbol{a} = \boldsymbol{y}$ is assumed. $\boldsymbol{a}$ is computed with $\boldsymbol{a} = X^\dagger \boldsymbol{y}$ using Moore-Penrose pseudoinverse, $X^\dagger$, because there are no unique solutions due to $N > M$. Top N_1 features with larger absolute a_i is selected. As for representatives of one by one, unsupervised feature selections, two methods are employed. One is highly variable features. Sample variance of each feature,

$$\frac{1}{M}\left(x_{ij} - \frac{1}{M}\sum_{j=1}^{M} x_{ij}\right)^2, \tag{4.3}$$

is computed and top $N_1 = 10$ features associated with larger variance are selected. Another is unimodal test. Unimodal test computes P-values that reject the null hypothesis that x_{ij}s with fixed i are drawn from unimodal distribution; Hartigan's Dip Test, which rejects the null hypothesis that the distribution is unimodal [1] is used for this purpose. Then top $N_1 = 10$ features associated with smaller P-values are selected. Finally, as a representative of collective unsupervised feature selections, we employ PCA. PCA is applied to x_{ij} such that kth PC score vectors, $\boldsymbol{u}_k \in \mathbb{R}^N$, are attributed to features. In other words, $\boldsymbol{u}_k$ is computed as the eigen vectors of $S_{ii'}$, Eq. (2.50), $S_{ii'}\boldsymbol{u}_k = \lambda_k \boldsymbol{u}_k$ where λ_k is eigen value. Then, top $N_1 = 10$ features associated with the larger absolute first PC score, $|u_{1i}|$, are selected (The reason why this procedure works as feature selection will be discussed later).

Table 4.2 shows the number of features that are distinct between two classes and are also selected by individual methods. When tests are applied to data sets 1 and 2, two supervised methods samely achieved well although the collective method achieved a little bit worse than one by one method. The performance achieved by

Table 4.2 Performance of statistical tests applied to two synthetic data set 1 defined by Eq. (4.1) and data set 2 defined by Eq. (4.2). Numbers represent mean number of features selected by each method, among N_1 features distinct between two classes, $i < N_1(= 10)$. Shuffled means that class labels are shuffled

	Supervised		Unsupervised		
	One by one	Collective	One by one		Collective
Data set	*t*-test	Linear regression	Variance	Unimodal test	PCA
1	10.00	9.88	1.20	1.68	8.75
2	10.00	9.79	9.99	5.68	10.00
1 (shuffled)	1.03	0.08	1.34	1.66	8.78
2 (shuffled)	0.94	0.89	10.00	5.76	10.00

unsupervised method is quite distinct between two data sets. Two unsupervised one by one methods fail when data set 1 is considered while they performed better for data set 2. This is reasonable because all N features obey the identical distribution if class labels are not considered. Thus, unsupervised methods have no ways to distinguish features with and without distinction between two classes. In this sense, it is remarkable that PCA, an unsupervised and collective method, can perform similarly well for both data sets 1 and 2.

One might wonder why unsupervised method must be considered, because supervised methods perform better. This impression changes once the class labels are shuffled. It is reasonable that no supervised methods work well. On the other hands, it is also reasonable that the performance by unsupervised method do not change because of class label shuffling. This suggests that unsupervised feature selections are better choices when class labels are not available or not trustable.

Unsupervised collective feature selection, PCA, is successful for data set 1, for which other unsupervised methods fail, and shuffled data set, for which supervised collective methods fail. It is important why it can happen. In order to see this, we investigate the first PC loading vectors, $\boldsymbol{v}_1 \in \mathbb{R}^M$, which is defined as $\boldsymbol{v}_1 = \frac{1}{\lambda_1} X^T \boldsymbol{u}_1$ (see Eq. (2.21)). Figure 4.1 shows the first PC loading vectors. For all cases, u_{ij}s with $j \leq \frac{M}{2}$ take positive values while u_{ij}s with $j > \frac{M}{2}$ take negative value. Since $\boldsymbol{u}_1 = \lambda_1 X \boldsymbol{v}_1$, u_{1i} reflects the difference between two classes. Thus, selecting is associated with absolutely larger u_{1i} can identify correctly features associated with distinction between two classes for all four cases. This is the reason why PCA can always perform well.

4.3 PCA Applied to More Complicated Patterns

In the previous section, feature selection with two classes was discussed. Nevertheless, it is the simplest case. There are many more complicated feature selections. One direction is to have more classes than two. Another direction is to have more than one classifications simultaneously. Here, let us discuss both together,

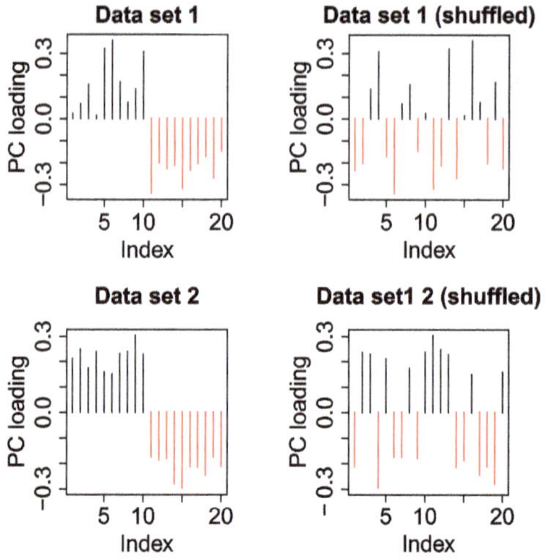

Fig. 4.1 The first PC loading vectors, $\boldsymbol{v}_1 \in \mathbb{R}^M$, for data set 1, shuffled data set 1, data set 2 and shuffled data set 2. Black and red bars correspond to classes 1 and 2, respectively

i.e., feature extraction under the conditions having more than one classification with more than two classes. In order to demonstrate feature selections under this condition, we extend data set 2, Eq. (4.2), as follows.

Data set 3

$$x_{ij} \sim \begin{cases} \mathcal{N}(0,\sigma) & j \le \frac{M}{2}, \quad i \le N_1 \\ \mathcal{N}(\mu_0,\sigma) & j > \frac{M}{2}, \quad i \le N_1 \\ \mathcal{N}(0,\sigma) & j \le \frac{M}{4}, \quad N_1 < i \le N_1+N_2 \\ \mathcal{N}(\mu_1,\sigma) & \frac{M}{4} < j \le \frac{M}{2}, \quad N_1 < i \le N_1+N_2 \\ \mathcal{N}(2\mu_1,\sigma) & \frac{M}{2} < j \le \frac{3M}{4}, \quad N_1 < i \le N_1+N_2 \\ \mathcal{N}(3\mu_1,\sigma) & j > \frac{3M}{4}, \quad N_1 < i \le N_1+N_2 \\ \mathcal{N}(\mu_2,\sigma) & i > N_1+N_2. \end{cases} \tag{4.4}$$

Features $i \le N_1$ are composed of two classes, those $N_1 < i \le N_1 + N_2$ are composed of four classes, and those $i > N_1 + N_2$ are composed of no classes. Thus the feature selection aims to identify which features are composed of how many classes.

Now the problem is more difficult. For example, simply trying to identify which features are composed of two classes does not help us to distinguish between features composed of two classes and those composed of four classes, because four classes can be also considered to be two classes if each two of four classes are considered as one class. Thus in order to perform feature selections under such a complicated condition, we usually need more detailed information about class labeling.

It is not very easy to adapt to this situation. Suppose we have already known 20 samples are classified into the four classes as

$$(A, A, A, A, A, B, B, B, B, B, C, C, C, C, C, D, D, D, D, D) \tag{4.5}$$

or into the two classes as

$$(E, E, E, E, E, E, E, E, E, E, F, F, F, F, F, F, F, F, F, F). \tag{4.6}$$

Even if this is the case, identification of features with four classes is not straightforward. Simple linear regression analysis is not applicable, because we know only that four classes differ from one another. In order to perform linear regression analysis, we need to assign numbers to each of four classes. If we do not know practical relationship between four classes, there are no ways to assign numbers to four classes. Pairwise comparison between four classes might be possible, but might not work well, because we need to integrate pairwise comparisons in order to rank features. Suppose we try to all possible six pairwise comparisons in Eq. (4.5), as

$$(A, B), (A, C), (A, D), (B, C), (B, D), (C, D). \tag{4.7}$$

If we consider this is occasionally applied to Eq. (4.6), they correspond to comparisons of

$$(E, E), (E, F), (E, F), (E, F), (E, F), (F, F). \tag{4.8}$$

Thus, in contrast to the expectation, four out of six comparisons will report that they differ. Thus, if difference between two classes, E and F, is greater than that between pairs in four classes, A, B, C, and D, integration of six pairwise comparison might report that Eq. (4.8) more fits to four classes than Eq. (4.7). In the following, we consider in which occasion integration of six pairwise comparisons occasionally report that Eq. (4.8) is more likely to be four classes than Eq. (4.7). For the simplicity, we assume that all pairwise comparisons (E,F) in Eq. (4.8) are higher ranked than all pairwise comparisons in Eq. (4.7). The requirement that difference between two classes among four classes should be smaller than that among two classes is not unrealistic. It is very usual that values of features have both upper and lower boundary. In this case, the distinction between two classes when samples are classified into two classes is that between the upper and the lower halves. On the other hand, the distinction between two classes when samples are classified into four classes is that between any pairs of four quantiles. If region is divided into two, the distinction is larger than that when region is divided into four. In this case, the following happens (Table 4.3). Four pairwise comparisons (E,F) in Eq. (4.8) is always higher ranked than corresponding four pairwise comparisons, (A,C), (A,D), (B,C) and (B,D) in Eq. (4.7). On the other hand, two pairwise comparisons (E,E) and (F,F) in Eq. (4.8) are always lower ranked than corresponding two pairwise comparisons, (A,B) and (C,D). There are N_1 features composed of two classes and N_2 features composed of four classes. Thus mean rank of pairs (A,B) and (C,D) are $\frac{N_2}{2}$ because N_2 features composed of four classes are ranked higher

Table 4.3 Mean (expected) rank, mean lowest rank and mean top ranks of pairwise comparisons. Integrated rank is summation of ranks of six pairwise comparisons

							Integrated rank
Pairs in Eq. (4.7)	(A,B)	(A,C)	(A,D)	(B,C)	(B,D)	(C,D)	$3N_2 + 4N_1$
mean rank	$\frac{N_2}{2}$	$N_1 + \frac{N_2}{2}$	$N_1 + \frac{N_2}{2}$	$N_1 + \frac{N_2}{2}$	$N_1 + \frac{N_2}{2}$	$\frac{N_2}{2}$	
Pairs in Eq. (4.8)	(E,E)	(E,F)	(E,F)	(E,F)	(E,F)	(F,F)	$N + 2N_1 + N_2$
mean rank	$\frac{N+N_2}{2}$	$\frac{N_1}{2}$	$\frac{N_1}{2}$	$\frac{N_1}{2}$	$\frac{N_1}{2}$	$\frac{N+N_2}{2}$	
Pairs in Eq. (4.7) mean lowest rank	(A,B) N_2	(A,C) $N_1 + N_2$	(A,D) $N_1 + N_2$	(B,C) $N_1 + N_2$	(B,D) $N_1 + N_2$	(C,D) N_2	$4N_1 + 6N_2$
Pairs in Eq. (4.8)	(E,E)	(E,F)	(E,F)	(E,F)	(E,F)	(F,F)	$N - N_1 + N_2 + 4$
mean top rank	$\frac{N_2+N-N_1}{2}$	1	1	1	1	$\frac{N_2+N-N_1}{2}$	

than other features. Mean rank of (A,C), (A,D), (B,C) and (B,D) are $\frac{N_2}{2} + N_1$ because N_1 features composed of two classes are always ranked higher than N_2 features composed of four classes. Mean rank of four pairs (E,F) in Eq. (4.8) is $\frac{N_1}{2}$ because N_1 features composed of two classes are higher ranked than other features. Mean rank of two pairs (E,E) and (F,F) are $\frac{N+N_2}{2}$ because N_2 features composed of four classes are higher ranked than others. Next, integrated rank is computed as the summation over six pairwise comparisons. Then, integrated rank of features composed of four classes is

$$2 \times \frac{N_2}{2} + 4 \times \left(N_1 + \frac{N_2}{2}\right) = 4N_1 + 3N_2 \tag{4.9}$$

and integrated rank of features composed of two classes is

$$2 \times \frac{N + N_2}{2} + 4 \times \frac{N_1}{2} = N + 2N_1 + N_2. \tag{4.10}$$

In order that N_2 features composed of four classes are higher ranked than N_1 features composed of two classes based upon integrated rank in average, Eq. (4.9) < Eq. (4.10). Thus

$$\text{Eq. (4.10)} - \text{Eq. (4.9)} > 0 \tag{4.11}$$

$$N + 2N_1 + N_2 - (4N_1 + 3N_2) > 0 \tag{4.12}$$

$$N - 2N_1 - 2N_2 > 0 \tag{4.13}$$

$$N > 2(N_1 + N_2) \tag{4.14}$$

is required. Otherwise, integrated rank based upon six pairwise comparisons, Eq. (4.7), cannot select N_2 features composed of four classes more likely than N_1 features composed of two classes. This means that total number of features distinct between any pairs of classes must not exceed the half of total number of features. This requirement is unlikely fulfilled always.

Equation (4.14) that cannot always be expected to be satisfied is only for average. Even if Eq. (4.14) stands, at most only half of selected features is correctly composed of four classes. If we require that there should not be any false positives, requirement can become more strict (Table 4.3). In order that, we have to require that top-ranked features among those composed of two classes must be always ranked lower than the lowest ranked features among those composed of four classes. The rank of bottom ranked feature among those composed of four classes by the two pairwise comparison (A,B) and (C,D) in Eq. (4.7) is N_2 because there are N_2 features that are composed of four classes and are ranked higher than other features. The rank of feature ranked as bottom by the four pairwise comparisons (A,C), (A,D), (B,C) and (B,D) in Eq. (4.7) among those composed of four classes is $N_1 + N_2$ because N_1 features that are composed of two classes and are ranked higher than

N_2 features composed of four classes. On the other hand, features ranked as top by two pairwise comparisons (E,E) and (F,F) in Eq. (4.8) among those composed of two classes are ranked uniformly between N_2 and $N - N_1$. This is because N_2 features composed of four classes are higher ranked than N_1 features composed of two classes and there are N_1 features ranked lower than top-ranked features among those composed of two classes. Thus, mean top-ranked features among those composed of two classes by two pairwise comparisons (E,E) and (F,F) in Eq. (4.8) is $\frac{N-N_1+N_2}{2}$. The rank of feature ranked as top by four pairwise comparisons (E,F) in Eq. (4.8) among those composed of two classes is 1, because N_2 features composed of two classes are higher ranked than other features. Thus integrated bottom rank among N_2 features composed of four classes is

$$2 \times N_2 + 4 \times (N_1 + N_2) = 4N_1 + 6N_2 \tag{4.15}$$

while integrated top rank among N_1 features composed of two classes is

$$2 \times \left(\frac{N - N_1 + N_2}{2} \right) + 4 = N - N_1 + N_2 + 4. \tag{4.16}$$

In order that there are no false positives, i.e., N_2 features composed of four classes is always ranked higher than N_1 features composed of two classes, Eq. (4.16) > Eq. (4.15),

$$\text{Eq. (4.16)} - \text{Eq. (4.15)} > 0 \tag{4.17}$$

$$N - N_1 + N_2 + 4 - (4N_1 + 6N_2) > 0 \tag{4.18}$$

$$N - 5N_1 - 5N_2 + 4 > 0 \tag{4.19}$$

$$N + 4 > 5(N_1 + N_2). \tag{4.20}$$

This means that the number of features composed of two classes and that of four classes must be less than ten percentages of N if $N_1 = N_2$. This is a less likely fulfilled requirement than Eq. (4.14). Thus integration of six pairwise comparisons unlikely correctly identifies N_2 features composed of four classes when features composed of two classes coexist with them.

Because pairwise comparisons are not expected to work well to identify features composed of multiple classes when more than two kinds of multiple classes coexist, e.g., Eq. (4.4), usually any other alternative strategies are recommended to employ; ones of such alternative strategies are categorical regressions. In categorical regression, class labels are converted to dummy variables, δ_{kj} that takes 1 when jth sample belongs to kth class otherwise 0. Then, categorical regression analysis of x_{ij} is

$$x_{ij} = a_i + \sum_k b_{ik} \delta_{kj}, \tag{4.21}$$

Table 4.4 δ_{kj} in categorical regression, Eq. (4.21), assuming either four classes, Eq. (4.5), and two classes, Eq. (4.6), respectively

	4 classes				2 classes	
k	$1 \le j \le 5$,	$6 \le j \le 10$,	$11 \le j \le 15$,	$16 \le j \le 20$,	$1 \le j \le 10$,	$11 \le j \le 20$
1	1	0	0	0	1	0
2	0	1	0	0	0	1
3	0	0	1	0		
4	0	0	0	1		

where a_i and b_{ik} are the regression coefficients specific to ith feature. Pairwise comparisons that assume four classes could not distinguish features composed of four classes from those composed of two classes well. This problem does not exist in categorical regression analysis anymore. Suppose the simplest cases correspond to two classes, Eq. (4.6), and four classes, Eq. (4.5), as

$$(1, 1, 1, 1, 1, 1, 1, 1, 1, 1, 2, 2, 2, 2, 2, 2, 2, 2, 2, 2) \tag{4.22}$$

and

$$(1, 1, 1, 1, 1, 2, 2, 2, 2, 2, 3, 3, 3, 3, 3, 4, 4, 4, 4, 4) \tag{4.23}$$

respectively. It is obvious that there are no residual errors when Eq. (4.21) assuming four classes (Table 4.4) is applied to Eq. (4.22) if $a_i = \frac{3}{2}$, $b_{i1} = b_{i2} = -\frac{1}{2}$, $b_{i3} = b_{i4} = \frac{1}{2}$. Because there are no residual errors when Eq. (4.21) assuming four classes (Table 4.4) is applied to Eq. (4.23) as well if $a_i = \frac{5}{2}$, $b_{i1} = -\frac{3}{2}$, $b_{i2} = -\frac{1}{2}$, $b_{i3} = \frac{1}{2}$, and $b_{i4} = \frac{3}{2}$, this cannot discriminate four classes from two classes. Nevertheless, Eq. (4.21) assuming two classes (Table 4.4) can discriminate two classes from four classes. If $a_1 = \frac{3}{2}$, $b_{i1} = -\frac{1}{2}$ and $b_{i2} = \frac{1}{2}$, there are no residual errors for Eq. (4.22). On the other hand, there are no solutions with no residual errors when Eq. (4.21) assuming two classes (Table 4.4) is applied to Eq. (4.23). Thus, integration of categorical regression analyses assuming four classes and two classes can identify features composed of two classes and those composed of four classes successfully.

In order to see if categorical regression analysis, Eq. (4.21), can identify features composed of two classes and those composed of four classes simultaneously, we apply categorical regression, Eq. (4.21), to data set 3, Eq. (4.4), as follows. First we apply categorical regression, Eq. (4.21), assuming four classes to data set 3. Because categorical regression assuming four classes are simultaneously coincident with features composed of four classes and those composed of two classes, we select top ranked N_2 features, which is the total number of features that are composed of either two or four classes, i.e., $i \le N_1 + N_2$. Then, we apply categorical regression assuming two classes to data set 3. Because categorical regression assuming two classes are coincident with only features composed of two classes,

Table 4.5 Performance of statistical tests applied to synthetic data sets 3 defined by Eq. (4.4). Numbers represent mean number of features distinct between two calsses, $i < N_1 (= 10)$, and four classes, $N_1 < i \le N_1 + N_2$, among N_1 features selected by each method, respectively

Categorical regression		PCA-based unsupervised FE	
Two classes	Four classes	Two classes	Four classes
10.00	10.00	9.97	9.97

we select top ranked N_1 features, which is the total number of features that are composed of two classes, i.e., $i \le N_1$. Features selected by categorical regression assuming two classes are considered as features composed of two classes. On the other hand, features selected by categorical regression assuming four classes but not selected by categorical regression assuming two classes are considered as features composed of four classes. Table 4.5 shows the performance of this integrated categorical regression assuming two classes and four classes when $N = 100, M = 20, \mu_0 = 8, \mu_1 = \mu_2 = \frac{\mu_0}{2} = 2, N_1 = 10, N_2 = 10$ and $\sigma = 1$ in data set 3, Eq. (4.4). Performance is averaged over one hundred independent trials. Categorical regression can identify features composed of two classes and four classes completely.

In order to see if PCA-based unsupervised FE is applicable, it is applied to the same data set, too. In this case, we selected top 10 features and the second top 10 features (i.e., ranked between 11th and 20th). Since we do not know which one correspond to two classes or four classes, after investigating coincidence, we assign top 10 to four classes and the second top 10 to two classes. PCA-based unsupervised FE is also successful (Table 4.5). Only disadvantages of PCA-based unsupervised FE is that it cannot find the correspondence between selected sets of features and the number of classes in advance.

In order to see this, we can observe the first PC loading vector, $\boldsymbol{v}_1$ (Fig. 4.2). It is obvious the first PC loading vector is coincident with four classes. This is the reason why the top ranked 10 features are coincident with, not two classes, but four classes. Although we do not repeat the application to shuffled data, it is obvious that categorical regression does not work toward shuffled data because feature selection is performed with class labeling. PCA-based unsupervised FE is not affected by shuffling, because PC score vectors, $\boldsymbol{u}_k$s, which is used for feature selection, are not affected by the order of samples, thus are not affected by the class labeling as well. Thus, in this complicated situation, i.e., coexistence of features composed of two classes and four classes, PCA-based unsupervised FE is the most favorable method.

4.4 Identification of Non-Sinusoidal Periodicity By PCA-Based Unsupervised FE

Identification of periodicity, no matter whether it is spatial or temporal, has ever been central issue of data science. In order to identify periodicity, sinusoidal

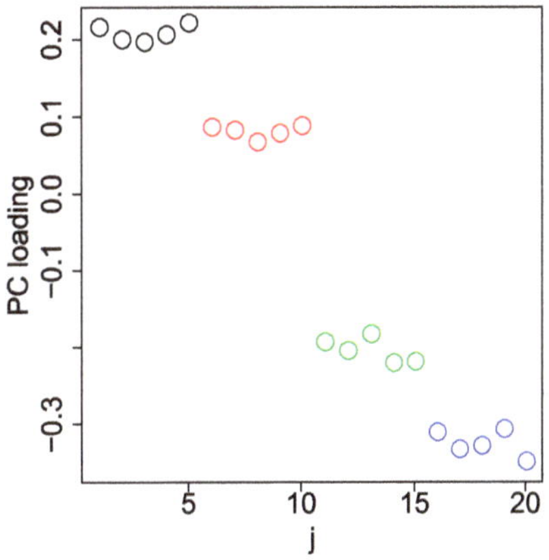

Fig. 4.2 The first PC loading vectors, $\boldsymbol{v}_1 \in \mathbb{R}^M$, for data set 3

regression is often used. Sinusoidal regression is defined as

$$x_{ij} = a_i + b_i \sin\left(\frac{2\pi}{T} j\right) + c_i \cos\left(\frac{2\pi}{T} j\right), \tag{4.24}$$

where a_i, b_i, c_i are regression coefficients specific to ith feature and T is period. In the following, for the simplicity, $T \in \mathbb{N}$. There are multiple practical problems on regression analysis. At first, we need to know period T in advance in order to apply regression analysis to data set. Of course, it is possible to estimate T from the data set with considering T to be a fitting parameter as well. Nevertheless, there are no known algorithm to find best T values, because any minimization algorithm applied to residues might fall in local minimum that differ from true T. Second, and more critical problem is that not all periodicity is sinusoidal. Only requirement of x_{ij} to be periodic with the period T is

$$x_{ij} = x_{ij+T} \tag{4.25}$$

which does not restrict functional forms to be sinusoidal at all.

In order to see how well sinusoidal regression, Eq. (4.24), can work, we apply it to the data set 4 with period of T

Data set 4

$$x_{ij} = \begin{cases} f_{(i+j) \bmod T} + a\varepsilon_{ij} & i \le N_1 \\ a\varepsilon_{ij} & i > N_1 \end{cases} \tag{4.26}$$

where $f_j \in \mathbb{R}^T$ and $\varepsilon_{ij} \in \mathbb{R}^{N \times M}$ are drawn from normal distribution $\mathcal{N}(0, \sigma)$, mod is modulo operation and $0 < a < 1$ is the coefficient that represents signal noise ration. Because of the term $(i + j) \bmod T$, $\left\{x_{ij} \middle| 1 \le j \le M\right\}$s have distinct phases from one another. Performance is averaged over 100 independent trials. Table 4.6 shows the performance when $N = 100, M = 50, T = 10, a = 0.1, \sigma = 1, N_1 =$

Table 4.6 Performance of statistical tests applied to synthetic data sets 4 defined by Eq. (4.26). Numbers represent mean number of features with period T, $i \leq N_1(=10)$ among N_1 features selected by each method, respectively

Sinusoidal regression	PCA-based unsupervised FE
5.72	10

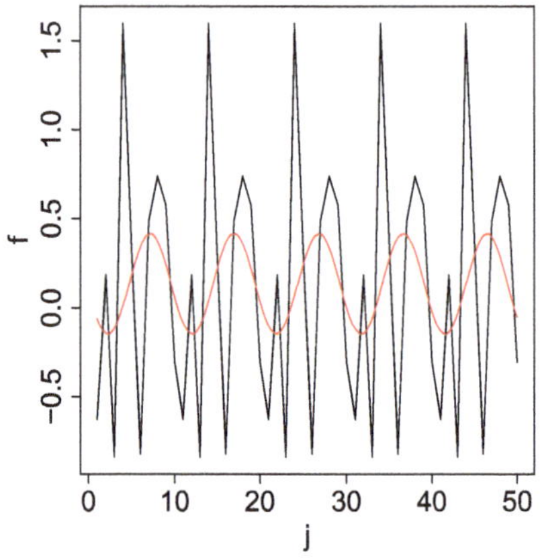

Fig. 4.3 Typical $f_{j \bmod T} \in \mathbb{R}^M (M = 50, T = 10)$ in Eq. (4.26) (black) and its sinusoidal regression, Eq. (4.24) (red)

10. It is as small as 5.72 which is hardly said to be a good performance. This low performance is because of f_j's non-sinusoidal functional form (Fig. 4.3).

Next, we apply PCA-based unsupervised FE to data set 4, Eq. (4.26), as in Sect. 4.2 excluding one point; Instead of ranking features based on the absolute value of the first PC score, $|u_{1i}|$, features are ranked based upon squared sum of the first and second PC scores $u_{1i}^2 + u_{2i}^2$. Table 4.6 shows the performance which is as large as 10, i.e., no errors.

The reason why we need to employ, not only the first PC score, u_{1i}, but also the second PC score, u_{2i}, can be seen in Fig. 4.4. As can be seen in Fig. 4.4(A), the first and second PC loading represent periodic function of period $T(=10)$. And the first 10 pairs of the first and the second PC scores, $u_{ki}, i \leq N_1(=10), k \leq 2$, form circular trajectory in the plain spanned by the first and the second PC (Fig. 4.4b). This is because of the term $(i + j) \bmod T$ in Eq. (4.26), that generates phase shift between features $x_{ij}, i \leq N_1(=10)$. In some cases, the corresponding PC loading, v_{1j} and v_{2j}, represent not the period T, but the period $\frac{T}{2}$ or $\frac{T}{3}$. Nevertheless, in data set 4, Eq. (4.26), only features $i \leq N_1(=10)$ can be coincident with higher modes, $\frac{T}{2}$ or $\frac{T}{3}$. Thus, these cases also can identify periodic features $i \leq N_1(=10)$ correctly.

Although in the above explanation, we use circular trajectory shown in Fig. 4.4b to reasons why we need to employ the first two PC scores for feature selection. In the practical application, the order of analysis can be reversed. First, we might observe the pairwise scatter plots of PC scores to identify which pairs of features have periodicity because periodic features should draw circular trajectory. Next, we can see individual PC loading as in Fig. 4.4a in order to see period T. This

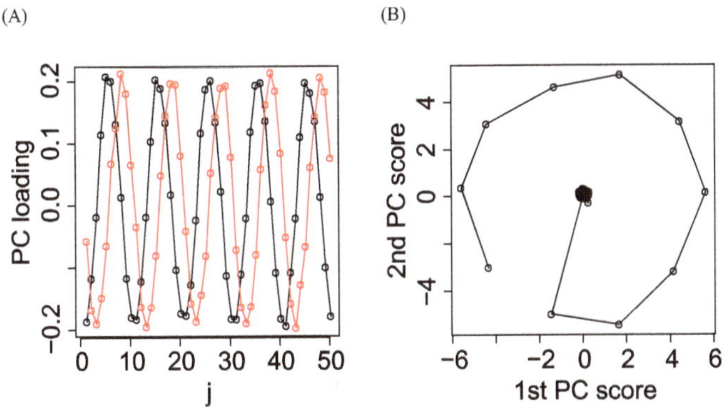

Fig. 4.4 (**a**) A typical first PC loading (black), v_{1j}, and second PC loading (red), v_{2j}. (**b**) Scatter plot of typical first PC score, u_{1i}, and the second PC score, u_{2i}, that correspond to PC loading shown in (**a**)

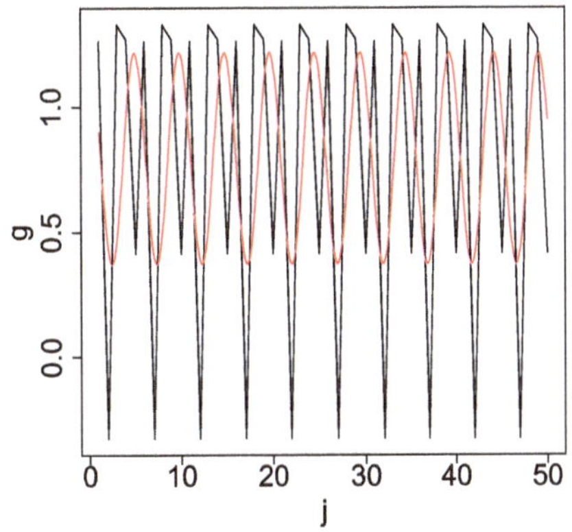

Fig. 4.5 Typical $g_{j \bmod T'} \in \mathbb{R}^M (T' = 5)$ in Eq. (4.27) (black) and its sinusoidal regression, Eq. (4.24) with $T = T' = 5$ (red)

is possible because it is unsupervised method that assumes no specific periodic functional forms in advance. In this sense, PCA-based unsupervised FE is superior to the sinusoidal regression to select periodic features.

In order to see if PCA-based unsupervised FE can recognize periodicity under the more complicated situation, I modified data set 4, Eq. (4.26), such that cycles with two period, T and T', coexist, i. e.

Data set 5

$$x_{ij} = \begin{cases} f_{(i+j) \bmod T} + a\varepsilon_{ij} & i \le N_1 \\ g_{(i+j) \bmod T'} + a\varepsilon_{ij} & N_1 < i \le N_2 \\ a\varepsilon_{ij} & i > N_2 \end{cases} \tag{4.27}$$

where $g_j \in \mathbb{R}^{T'}$ is drawn from normal distribution $\mathcal{N}(0, \sigma)$. Figure 4.5 shows the typical g that is far from sinusoidal profile ($T' = 5, N_2 = 20$, other parameters are the same as those in Eq. (4.26)). Figure 4.6 shows the typical first to forth PC

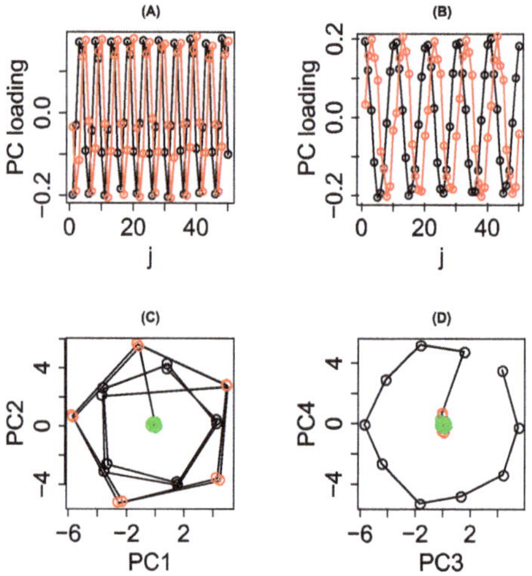

Fig. 4.6 (A) Typical the first PC loading (black), v_{1j}, and the second PC loading (red), v_{2j}. (B) Typical the third PC loading (black), v_{3j}, and the fourth PC loading (red), v_{4j}. (C) Scatter plot of typical the first PC score, u_{1i}, and the second PC score, u_{2i}, that correspond to PC loading shown in (A). (D) Scatter plot of typical the third PC score, u_{3i}, and the fourth PC score, u_{4i}, that correspond to PC loading shown in (B). Black open circles: $j \leq N_1(=10)$, red open circles: $N_1 < j \leq N_2(=20)$, green open circles: $N_2 < j$

scores, $\boldsymbol{u}_k$, $1 \leq k \leq 4$, and PC lading, $\boldsymbol{v}_k$, $1 \leq k \leq 4$. It is obvious that Fig. 4.6a and c corresponds to period $T' = 5$ and Fig. 4.6b and d corresponds to period $T = 10$, respectively. Thus, PCA-based unsupervised FE basically has ability to identify features with two distinct periods even when they coexist. The problem is that the first four PCs do not always correspond to two periods, $T' = 5$ and $T = 10$, but other four PCs, e.g., the second, third, seventh and eighth PCs, correspond to these two period, in contrast to data set 4, Eq. (4.26), where the first two PC loading always correspond to period $T = 10$. Thus, in order to make use of PCA to identify features with two distinct periods, we need to identify which PC loading correspond to two periods, $T = 10$ and $T' = 5$, respectively by applying sinusoidal regression, Eq. (4.24) with $T = 10$ and $T = T' = 5$. Thus, detailed procedure is as follows.

1. Apply PCA to data set 5, x_{ij} (Eq. (4.27)).
2. Apply sinusoidal regression, Eq. (4.24), with $T = T' = 5$ to PC loading, $\boldsymbol{v}_k$ and select top two, k_1 and k_2.
3. Apply sinusoidal regression, Eq. (4.24), with $T = 10$ to PC loading, $\boldsymbol{v}_k$ and select top two, k'_1 and k'_2.
4. Select top ranked $N_2(=20)$ features using squared sum of two v_{ki}s, $v^2_{k'_1 i} + v^2_{k'_2 i}$, selected in step 3 (This is because PC score, $\boldsymbol{u}_k$ with period $T' = 5$, identify features with periods $T' = 5$ and $T = 10$ as can be seen in Fig. 4.6c).

Table 4.7 Performance of statistical tests applied to synthetic data sets 5 defined by Eq. (4.27). Numbers represent mean number of features with period $T = 10$, $i \leq N_1(= 10)$ among N_1 features selected by each method and that of features with period $T = T' = 5$, $N_1 < i \leq N_2(= 20)$ among $N_2 - N_1(= 10)$ features selected by each method, respectively

Sinusoidal regression		PCA-based unsupervised FE	
$T = 10$	$T = T' = 5$	$T = 10$	$T = T' = 5$
6.32	6.75	9.73	9.99

5. Select top ranked $N_1(= 10)$ features using squared sum of two v_{ki}s, $v_{k_1 i}^2 + v_{k_2 i}^2$, selected in step 2 (This is because PC score, $\boldsymbol{u}_k$ with period $T = 10$, identify only features with periods $T = 10$ as can be seen in Fig. 4.6d).
6. Identify features selected in step 5 as those with period $T = 10$.
7. Identify features selected in step 4 but not in step 5 as those with period $T = 5$.

Performance is averaged over 100 independent trials (Table 4.7). PCA-based unsupervised FE obviously can identify features with two distinct periods almost completely.

In order to see if sinusoidal regressions, Eq. (4.24) with $T = 10$ and $T = T' = 5$, can perform as well as PCA-based unsupervised FE, we applied sinusoidal regression to data set 5, Eq. (4.27), too. Top $10(= N_1 = N_2 - N_1)$ features were selected with $T = 10$ and $T = T' = 5$, respectively (Table 4.7). Sinusoidal regression is clearly inferior to PCA-based unsupervised FE, possibly because of non-sinusoidal nature of f_j (Fig. 4.3) and g_j (Fig. 4.5) in Eq. (4.27).

4.5 Null Hypothesis

In the above examples, the number of features considered, e.g., those composed of multiple classes or those with specific period, are known in advance. Nevertheless, in the real application, it is unrealistic to assume that the number of features that should be selected is known in advance. In this case, usually P-values are attributed to individual features. These P-values represent the possibility that observation can happen accidentally under the null hypothesis that represents something opposite to the nature that selected features should obey.

For example, when we search features composed of two classes, the P-values represent the possibility that absolute difference of means between two classes can become accidentally larger than observed values when all observations are drawn from the same distribution (e.g., normal distribution with the same mean and standard deviation). If P-values are small enough, we can consider these features to be those composed of two classes, because the observed difference can unlikely appear if there are no classes.

There are some issues in this strategy. The first one is how we can select the null hypothesis. P-values are obviously dependent upon the selection of null hypothesis. Thus, it is important to select "correct" null hypothesis to address proper P-values to features. Unfortunately, there is no known established strategy to select the correct null hypothesis. Null hypothesis, which should be rejected, cannot be observable. Even if majority of features does not always follow null hypothesis, it might simply mean that most of features are associated with properties searched. Therefore, only requirement is to present clearly null hypothesis together with the P-values attributed to features.

Another issue is how small P-values should be. Generally, P-values are considered to be false ratio. In other words, if we select n features associated with P-values smaller than p, there can be at most np features selected wrongly in spite of that they obey the null hypothesis. Thus, ideally, p should be as small as $\frac{1}{n}$ such that there are no false positives. Nonetheless, it is often unrealistic to require $p < \frac{1}{n}$ especially when n is large and data is noisy. Therefore, practically, p is set to be 0.01 or 0.05, because it is enough if the 99 % or 95 % of selected feature are correct, for the usual purpose.

The third, and the most critical issue is the problem of multiple comparisons. When there are N features to which P-values are attributed, P-values can be accidentally as small as $\frac{1}{N}$. When N is large, e.g., $N \sim 10^4$, it causes a problem. Even if some features have P-values as small as 10^{-4}, we cannot reject null hypothesis. Thus, we cannot select these features as those associated with properties searched, e.g., composed of two classes. In spite of that, it is often unrealistic to require that P-values should be as small as 10^{-4}. Although there are many ways to address this difficulty, we employ Benjamini Hochberg (BH) criterion, because it is known to work practically well, in the applications described in the following chapters.

The basic idea of BH criterion is very simple. If the features obeys null hypothesis completely, e.g., apparently two classes features are drawn from the same distribution, e.g., normal distribution, the distribution of P-values should be uniform distribution $\in [0, 1]$, because this is the definition of probability. Thus, if we order P-values in ascending order, the ith largest P-value should be as large as $\frac{i}{N}$. In other words, if the ith largest P-value is smaller than $\frac{i}{N}$, it unlikely occurs under the null hypothesis.

Considering these discussions, BH criterion is as follows.

1. Order P-values attributed to ith feature, P_i, in ascending order.
2. Find the smallest i_0 such that $P_{i_0} > \frac{i_0}{N} p$ where p is threshold P-values.
3. Select features, $i \leq i_0$, such that their attributed P-values are practically supposed to be less than p.

Throughout the remaining part of this book, we employ this criterion to adjust P-values with considering multiple comparisons as many as the number of features, N.

4.6 Feature Selection with Considering P-Values

In order to perform feature selection with considering P-values, we select null hypothesis for the distribution of PC score, u_{ki}, as normal distribution. In order to assign P-values to features, we employ χ^2 distribution as

$$P_i = P_{\chi^2}\left[> \sum_k \left(\frac{u_{ki}}{\sigma_k} \right)^2 \right], \quad (4.28)$$

where $P_{\chi^2}[> x]$ is the cumulative probability that the argument is larger than x. The summation is taken over PCs selected for identification of ith feature that fulfills desired condition. The degrees of freedom of χ^2 distribution is equal to the number of PCs included in the summation. σ_k is the standard deviation of u_{ki}. Then features associated with adjusted P-values less than 0.01 are selected.

Other methods compared with PCA-based unsupervised FE in the previous section can also attribute P-values to individual features. Using these P-values, features associated with adjusted P-values less than 0.01 can be selected. This enables us to compare performance between the various methods.

At first, we perform analysis shown in Table 4.2 with replacing identification of features based upon top ranked $N_1(= 10)$ features with that based upon features associated with adjusted P-values less than 0.01. Unfortunately, not all tests shown in Table 4.2 can derive P-values. Evaluation based upon variance has no ways to attribute to P-values, because no null hypothesis can exist. Regression analysis cannot either, because complete fitting is always possible because the number of features, N, is larger than number of samples, M. Thus, only remaining three, t-test unimodal test, and PCA-based unsupervised FE can be employed. We do not apply shuffling in this case, because the effect of shuffling was presented in Table 4.2.

Evaluations based upon adjusted P-values do not always give us N_1 features selected. Thus, instead of presenting the number of correctly selected features as in Table 4.2, we need to present confusion matrix, which is demonstrated in Table 4.8. Suppose that there are two classes, positive set and negative set (in the case of feature selection, positive corresponds to features with considered properties, e.g., those composed of two classes, and negative corresponds to features without considered properties, e.g., those without any classes). The number of positives predicted as positive is true positive (TP). The number of positives predicted as not positive is

Table 4.8 Confusion matrix

	Real	
prediction	positive	negative
positive	TP	FP
negative	FN	TN

TP true positive, *FP* false positive, *FN* false negative, *TN* true negative

Table 4.9 Confusion matrices when statistical tests are applied to synthetic data sets 1 defined by Eq. (4.1) and features associated with adjusted P-values less than 0.01 are selected. $N_1 = 10$

	t-test		unimodal test		PCA	
	$i \le N_1$	$N_1 < i$	$i \le N_1$	$N_1 < i$	$i \le N_1$	$N_1 < i$
Data set 1: $N = 100, M = 20$						
selected	10.00	0.10	0.03	0.08	0.00	0.00
not selected	0.00	89.90	9.97	89.92	10.00	90.00
Data set 1: $N = 100, M = 10$						
selected	5.96	0.18	0.01	0.06	0.00	0.00
not selected	4.04	89.82	9.99	89.94	10.00	90.00
Data set 1: $N = 1000, M = 20$						
selected	9.98	0.2	0.0	0.2	0.00	0.00
not selected	0.02	989.8	10	989.8	10.00	990.00
Data set 1: $N = 1000, M = 10$						
selected	1.16	0.2	0.0	0.04	0.00	0.00
not selected	8.84	989.8	10	989.96	10.00	990.00

Table 4.10 Confusion matrices when statistical tests are applied to synthetic data sets 2 defined by Eq. (4.2) and features associated with adjusted P-values less than 0.01 are selected. $N_1 = 10$

	t-test		unimodal test		PCA	
	$i \le N_1$	$N_1 < i$	$i \le N_1$	$N_1 < i$	$i \le N_1$	$N_1 < i$
Data set 2: $N = 100, M = 20$						
selected	10.00	0.07	0.0	0.07	0.00	0.01
not selected	0.00	89.93	10.00	89.93	10.00	89.99
Data set 2: $N = 100, M = 10$						
selected	6.08	0.06	0.00	0.00	0.00	0.00
not selected	3.92	89.94	10.00	90.00	10.00	90.00
Data set 2: $N = 1000, M = 20$						
selected	9.98	0.1	0.00	0.00	9.97	0.07
not selected	0.02	989.9	10.00	990.0	0.03	989.03
Data set 2: $N = 1000, M = 10$						
selected	1.09	0.01	0.0	0.04	9.4	0.00
not selected	8.91	989.99	10	989.96	0.6	990.0

false negative (FN). The number of negatives predicted as positive is false positive (FP). The number of negatives predicted as not positive is true negative (TN). If FN=FP=0, it is complete prediction.

Confusion matrices when three statistical tests are applied to data set 1, Eq. (4.1), and data set 2, Eq. (4.2), are shown in Tables 4.9 and 4.10, respectively. The performance is averaged over 100 independent trials. t-test performs almost equally between data sets 1 and 2, although the performance decreases as M decreases or N increase. PCA-based unsupervised FE totally fails for data set 1, while it is successful for larger N in data set 2. Unimodel test has never been successful. One remarkable point is that PCA-based unsupervised FE can outperform t-test when $N = 1000$ and $M = 10$. This suggests that PCA-based unsupervised FE might be the best when $N \gg M$; the situation $N \gg M$ is very usual in the bioinformatics. This is the basic motivation that this text book is written.

In spite of that PCA-based unsupervised FE is an unsupervised method that does not fully make use of available information while t-test is a supervised method that fully makes use of available information, the reason why PCA-based unsupervised FE can outperform t-test when $N \gg M$ is as follows. In t-test, P-values increase as M decrease (i.e., less significant). On the other hand, the correction of P-values considering multiple comparisons is enhanced as N increase. Thus, adjusted P-values become larger (less significant) as N increases. This means, if $N \gg M$, t-test hardly computes small enough P-values. On the other hand, in PCA-based unsupervised FE where P-values are computed by u_{1i} which is less affected by varying M, P-values are less dependent on M. In Table 4.10, TPs computed by PCA-based unsupervised FE do not change much between $M = 10$ and $M = 20$ when $N = 1000$. In addition to this, in this set up, N_1 that represents the number of positives remains unchanged while N increase. This means, the number of negatives increases. Generally, negatives are associated with smaller absolute values of u_{1i} because u_{1i} is associated with v_{1j} that represents distinction between two classes (Fig. 4.1). P-values are computed based upon normalized u_{1i}, Eq. (4.28), thus absolute values u_{1i} attributed to positives become relatively larger as the number of negatives increase. This process has tendency that increasing the number of negatives reduces P-values attributed to positives (i.e., more significant). Because of that, in Table 4.10, PCA-based unsupervised FE is successful only when $N = 1000$.

This is the reason why PCA-based unsupervised FE is employed for the feature selection in bioinformatics where $N \gg M$ is quite usual. P-values computed by PCA-based unsupervised FE is less affected by M that is typically small in bioinformatics while P-values decrease for larger N that is typically very large in bioinformatics. Thus, PCA-based unsupervised FE is very fitted to the problems in bioinformatics.

One might be interested in what will happen if selection based upon adjusted P-values is applied to other examples discussed in the above. The answer is that it is dependent upon various parameters. In the examples analyzed in this section, PCA-based unsupervised FE can outperform t-test only when $N = 1000$ and $M = 10$. Thus, if it works well when it is applied to real data set is also dependent upon the properties of data sets. The general tendency that PCA-based unsupervised FE works well only when $N \gg M$ is universal independent of the data sets considered. Thus, the discussion about in which situation PCA-based unsupervised FE that selects features based upon adjusted P-values works well is postponed to the later chapters where PCA-based unsupervised FE is applied to real data sets. The readers can see many examples where PCA-based unsupervised FE works well or not in these later chapters.

4.7 Stability

Weaker sensitivity of PCA-based unsupervised FE on the number of samples, M, naturally results in the stability of feature selection. The stability of feature selection is defined as the robustness of feature selection when samples changes. Suppose that samples are drawn from some distributions. If selected features vary every time samples are drawn from distribution, it is problematic in biology where individual features, e.g., genes, have meanings.

In PCA-based unsupervised FE, P-values are less dependent upon the number of samples. In other words, every time we select half of samples among the available samples, P-values attributed to individual features do not change. If P-values attributed to individual features do not change, the selected features do not change, either. This is definitely equivalent to the stability. In the applications of PCA-based unsupervised FE to real data sets described in the following chapters, readers will see many examples that PCA-based unsupervised FE outperforms other methods from the point of stability. This is yet another reason why PCA-based unsupervised FE is a recommended method to be used in bioinformatics.

4.8 Summary

In this chapter, I proposed to make use of PCA as a tool of feature selection. PCA-based unsupervised FE can identify features composed of multiple classes better than conventional supervised methods, e.g., *t*-test and categorical regression. When it is applied to identification of non-sinusoidal periodic features, PCA-based unsupervised FE can outperform another conventional method, sinusoidal regression. With attributing P-values to features under the null hypothesis that PC scores obey χ^2 distribution, PCA-based unsupervised FE correctly identifies features composed of two classes only when $N \gg M$, i.e., the number of features is much larger than the number of samples.

Reference

1. Hartigan, J.A., Hartigan, P.M.: The dip test of unimodality. Ann. Stat. **13**(1), 70–84 (1985). https://doi.org/10.1214/aos/1176346577

Chapter 5
TD-Based Unsupervised FE

Although our world might have no reason to exist, it sounds fantastic, because we can make the reason for ourselves.
Filicia Heideman, Sound of the Sky, Season 1, Spisode 7

5.1 TD as a Feature Selection Tool

In this chapter, I would like to make use of TD as a feature selection tool. Suppose that $x_{ijk} \in \mathbb{R}^{N \times M \times K}$ represents the value of the ith feature of the samples having jth and kth properties as

Data set 6:

$$x_{ijk} \sim \begin{cases} \mathcal{N}(\mu, \sigma), & i \leq N_1,\ j \leq \frac{M}{2},\ k \leq \frac{K}{2} \\ \mathcal{N}(0, \sigma), & \text{otherwise} \end{cases} \tag{5.1}$$

In this example, j and k are supposed to be classified into two classes, $j \leq \frac{M}{2}, K \leq \frac{M}{2}$ and $j > \frac{M}{2}$ or $j > \frac{K}{2}$ for $i \leq N_1$. Then, x_{ijk} is drawn from normal distribution, $\mathcal{N}(\mu, \sigma)$, with positive mean, $\mu > 0$, only when $j \leq \frac{M}{2}, k \leq \frac{K}{2}$ otherwise $\mu = 0$. The purpose of feature selection is to find N_1 features associated with two classes shown in Eq. (5.1).

Tucker decomposition, Eq. (3.2), with HOSVD algorithm, Fig. 3.8, is applied to data set 6, Eq. (5.1), with $N = 1000$, $M = K = 6$, $N_1 = 10$, $\mu = 2$, $\sigma = 1$, as

$$x_{ijk} = \sum_{\ell_1=1}^{N} \sum_{\ell_2=1}^{M} \sum_{\ell_3=1}^{K} G(\ell_1, \ell_2, \ell_3) u_{\ell_1 i}^{(i)} u_{\ell_2 j}^{(j)} u_{\ell_3 k}^{(k)}, \tag{5.2}$$

where $\boldsymbol{u}_{\ell_1}^{(i)} \in \mathbb{R}^N$, $\boldsymbol{v}_{\ell_2}^{(i)} \in \mathbb{R}^M$, $\boldsymbol{u}_{\ell_3}^{(k)} \in \mathbb{R}^K$, $G(\ell_1, \ell_2, \ell_3) \in \mathbb{R}^{N \times M \times K}$. Figure 5.1a and b show a typical realization of $\boldsymbol{u}_1^{(j)}$ and $\boldsymbol{u}_1^{(k)}$, respectively. It is obvious that these two correctly reflect the distinction between $j > \frac{M}{2}, k > \frac{K}{2}$ and $j \leq \frac{M}{2}, k \leq \frac{K}{2}$.

Y-h. Taguchi, *Unsupervised Feature Extraction Applied to Bioinformatics*, Unsupervised and Semi-Supervised Learning,
https://doi.org/10.1007/978-3-031-60982-4_5

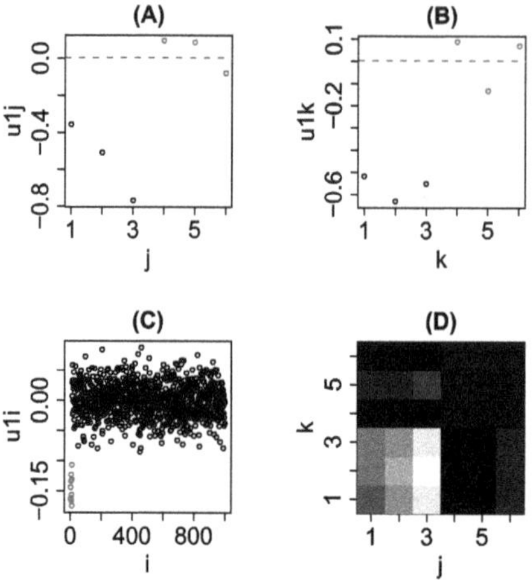

Fig. 5.1 A typical realization of $u_1^{(i)}, u_1^{(j)}, u_1^{(k)}$ when Tucker decomposition, Eq. (3.2), with HOSVD algorithm, Fig. (3.8) is applied to data set 6, Eq. (5.1) with $N = 1000, M = K = 6, N_1 = 10, \mu = 2, \sigma = 1$. **(a)** $u_1^{(j)}$, **(b)** $u_1^{(k)}$, black and red circles correspond to $j \leq \frac{M}{2}, k \leq \frac{K}{2}$ and $j > \frac{M}{2}, k > \frac{K}{2}$, respectively. Red broken lines show base line. **(c)** $u_1^{(i)}$. Red open circle correspond to $i \leq N_1$, i.e., features associated with j, k dependence. **(d)** $u_1^{(j)} \times^0 u_1^{(k)}$. Brighter squares have larger values

Table 5.1 $G(\ell_1, 1, 1)$s that correspond to Fig. 5.1

ℓ_1	1	4	2	6
$G(\ell_1, 1, 1)$	−35.484412	2.137686	1.748955	−1.705922

Next, we would like to identify which $u_{\ell_1}^{(i)}$ can be used for feature selection. In contrast to PCA-based unsupervised FE, it is not clear which $u_{\ell_1}^{(i)}$ should be used, because there is no one to one correspondence among $u_{\ell_1}^{(i)}, u_{\ell_2}^{(j)}, u_{\ell_3}^{(k)}$; instead of that, their relationship is represented through the core tensor, G.

In order to see this relationship, we order $G(\ell_1, 1, 1)$ with descending order of absolute values; Table 5.1 shows the core tensors, $G(\ell_1, 1, 1)$, sorted in this order. Table 5.1 suggests that $u_1^{(i)}$ is most likely associated with $u_1^{(j)}$ and $u_1^{(k)}$, because $G(1, 1, 1)$ has the largest absolute value among $G(\ell_1, 1, 1)$. Actually, $u_1^{(i)}$ shown in Fig. 5.1c obviously has larger absolute values for $i \leq N_1$ than others. Thus, the strategy proposed here, i.e., first find singular value vectors attributed to samples and associated with desired class dependence, then identify singular value vectors, attributed to features, that share G having larger absolute values with them, can identify features with not known in advance j, k dependence in fully unsupervised manner. The reason why it works so well is obvious. If we see $u_{\ell_2}^{(j)} \times^0 u_{\ell_3}^{(k)}$ that is shown in Fig. 5.1d, it is fully associated with the j, k dependence defined in Eq. (5.1) that means only $j, k < \frac{M}{2}$ are drawn from normal distribution with positive mean while others are drawn from those with zero mean.

Next issue might be if TD-based unsupervised FE can outperform conventional methods. As a representative of conventional methods, we employ again categorical regression analysis, Eq. (4.21), that is modified to be adapted to co-existence of two kinds of classes,

$$x_{ijk} = a_i + \sum_{s=1}^{2} b_{is}\delta_{sj} + \sum_{s=1}^{2} c_{is}\delta_{sk}, \tag{5.3}$$

where a_i, b_{is}, c_{is} are regression coefficients. δ_{sj} and δ_{sk} are the function that takes 1 only when sample j or k belong to the sth class otherwise 0.

In order to perform feature selection, P-values need to be addressed to features. For categorical regression analysis, P-values computed by categorical regression analysis is used as it is. For TD-based unsupervised FE,

$$P_i = P_{\chi^2}\left[> \left(\frac{u_{1i}^{(i)}}{\sigma_1}\right)^2 \right] \tag{5.4}$$

is used to attribute$P$-values to features where σ_1 is the standard deviation of $u_{1i}^{(i)}$. Both P-values, i.e., computed with TD-based unsupervised FE and categorical regression analysis, are corrected by BH criterion and features associated with adjusted P-values less than 0.01 are selected. Table 5.2 shows the performances achieved by TD-based unsupervised FE and categorical regression, Eq. (5.3). Performance is averaged over 100 independent examples. In contrast to TD-based unsupervised FE that can identify more than 60% of features associated with searched j, k dependence, categorical regression, Eq. (5.3), could identify almost no features. The cause of this drastic low performance is obvious. Eq. (5.3) assumes four classes, because j and k are composed of two classes, respectively. Thus, two classes times two classes are equal to four classes. Nevertheless, Eq. (5.1) obviously admit two classes, i.e., $j \le \frac{M}{2}, k \le \frac{K}{2}$ versus others. This not proper assumption in the model (categorical regression analysis) results in poor performance. In actual, if we employ categorical regression as

$$x_{ijk} = a_i + \sum_{s=1}^{2} b_{is}\delta_{sjk}, \tag{5.5}$$

where δ_{sjk} is a function that takes 1 only when

$s = 1$: $\quad j \le \frac{M}{2}$ and $k \le \frac{K}{2}$
$s = 2$: $\quad j > \frac{M}{2}$ or $k > \frac{K}{2}$

otherwise 0 and a_i, b_{sjk} are regression coefficients, categorical regression can outperform TD-based unsupervised FE as expected (Table 5.2). The only problem is that it is usually impossible to assume two classes in spite of that there are four

Table 5.2 Confusion matrices when statistical tests are applied to synthetic data sets 6 and 7 defined by Eq. (5.1) and features associated with adjusted P-vales less than 0.01 are selected. $N_1 = 10$. "categorical test(two classes)" corresponds to Eq. (5.3), "categorical test(four classes)" corresponds to Eq. (5.5) and "categorical test(nine classes)" corresponds to Eq. (5.7)

Data set 6	TD-based unsupervised FE		Categorical test(four classes)		Categorical test(two classes)	
	$i \le N_1$	$N_1 < i$	$i \le N_1$	$N_1 < i$	$i \le N_1$	$N_1 < i$
selected	6.34	0.00	0.63	0.00	7.35	0.00
not selected	3.66	990	9.37	990	2.65	990
Data set 7	TD-based unsupervised FE		Categorical test(nine classes)		Categorical test(two classes)	
	$i \le N_1$	$N_1 < i$	$i \le N_1$	$N_1 < i$	$i \le N_1$	$N_1 < i$
selected	8.73	0.00	4.58	0.00	10.0	0.00
not selected	1.27	990	5.42	990	0.00	990

classes based upon the apparent category. In this case, unsupervised method can outperform supervised method.

In order to confirm these tendencies, we prepare additional synthetic data.

Data set 7:

$$x_{ijk} \sim \begin{cases} \mathcal{N}(\mu, \sigma), & i \leq N_1, \frac{M}{3} < j \leq \frac{2M}{3}, \frac{K}{3} < k \leq \frac{2K}{3} \\ \mathcal{N}(0, \sigma), & \text{otherwise} \end{cases}. \tag{5.6}$$

Equation (5.3) is modified as

$$x_{ijk} = a_i + \sum_{s=1}^{3} b_{is}\delta_{sj} + \sum_{s=1}^{3} c_{is}\delta_{sk} \tag{5.7}$$

with three classes, $1 \leq j \leq \frac{M}{3}$ or $1 \leq k \leq \frac{K}{3}$ for $s = 1$, $\frac{M}{3} < j \leq \frac{2M}{3}$ or $\frac{K}{3} < k \leq \frac{2K}{3}$ for $s = 2$, and $\frac{2M}{3} < j \leq M$ or $\frac{2K}{3} < k \leq K$ for $s = 3$. On the other hand, Eq. (5.5) remains unchanged although δ_{sjk} takes 1 only when

$s = 1$: $\frac{M}{3} < j \leq \frac{2M}{3}$ and $\frac{K}{3} < k \leq \frac{2K}{3}$
$s = 2$: $j \leq \frac{M}{3}$ or $j > \frac{2M}{3}$ or $k \leq \frac{K}{3}$ or $k > \frac{2K}{3}$

otherwise 0. $M = K = 12$ and other parameters remain unchanged. As expected (Table 5.2), the performances of categorical regressions applied to set 7 are improved from those applied to data set 6, because the number of samples, MK, increases while the number of features, N, unchanges. In spite of these improved performances of categorical regression analyses, TD-based unsupervised FE still outperforms three classes × three classes = nine classes categorical regression analysis, Eq. (5.7) (see Table 5.2). Thus, as far as apparent categories that do not correctly reflect true category are considered, TD-based unsupervised FE can outperform supervised method. It is very usual in genomic data analysis that it is unclear if apparent categories are coincident with true, but unknown, classes. This is possibly the reason why TD-based unsupervised FE often outperforms supervised methods in the applications to bioinformatics that will be introduced in the later part of this book.

It should be also emphasized that TD-based unsupervised FE can outperform supervised methods only when $N \gg MK$, i.e., the number of features is much larger than the number of samples. Although we do not demonstrate this using more synthetic data sets, one should remember this point when one would like to employ TD-based unsupervised FE.

5.2 Comparisons with Other TDs

Here I employed only Tucker decomposition, Eq. (3.2), with HOSVD algorithm, Fig. 3.8, for feature selection. Since I have already argued the superiority of

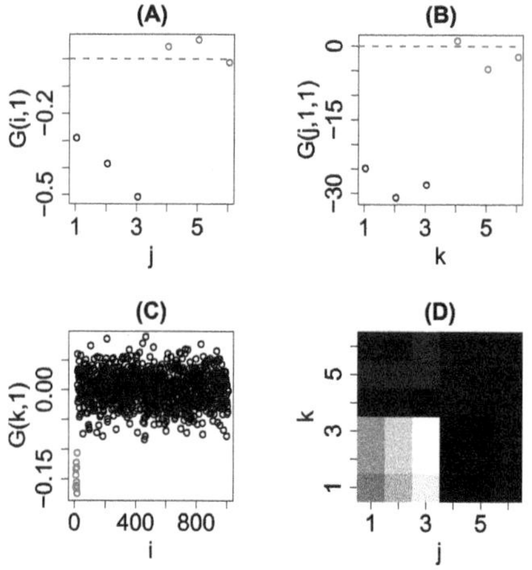

Fig. 5.2 $G^{(j)}(j, 1, 1), G^{(k)}(k, 1), G^{(i)}(i, 1)$ when tensor train decomposition, Eq. (3.3), with $R_1 = R_2 = M = K = 6$ is applied to data set 6, Eq. (5.1) whose results obtained by Tucker decomposition is shown in Fig. 5.1. (**a**) $G^{(j)}(j, 1, 1)$, (**b**) $G^{(k)}(k, 1)$, black and red circles correspond to $j \leq \frac{M}{2}, k \leq \frac{K}{2}$ and $j > \frac{M}{2}, k > \frac{K}{2}$, respectively. Red broken lines show base line. (**c**) $G^{(i)}(i, 1)$. Red open circle correspond to $i \leq N_1$, i.e., features associated with j, k dependence. (**d**) $G^{(j)}(j, 1, 1) \cdot G^{(k)}(k, 1)$. Brighter squares have larger values

Tucker decomposition toward other two TDs, CP decomposition and tensor train decomposition, it might not be necessary to demonstrate superiority of Tucker decomposition to other two TDs. Nevertheless, it is not meaningless to see what we can get when the other two TDs are applied to data set 6.

First, tensor train decomposition, Eq. (3.3), with $R_1 = R_2 = M = K = 6$ is applied to data set 6, whose results obtained by Tucker decomposition is shown in Figs. 5.1 and 5.2. Figure 5.2 looks very similar to Fig. 5.1. In spite of that, tensor train decomposition is still inferior to Tucker decomposition. First of all, we have no idea how we should choose R_is that decide the rank of tensor train decomposition. In the present case, we can try to find R_is that result in the same result as that in Fig. 5.1. If not, we can have no ways to decide R_is. Second, we do not know how to relate $G^{(j)}(j, 1, 1)$, $G^{(k)}(k, 1)$, and $G^{(i)}(i, 1)$ with one another, because there are no core tensor that plays the role to connect singular vectors in Tucker decomposition (Table 5.1) where we know what I should search. If not as in the present case, i.e., tensor train decomposition, we have no ideas which core tensors given by tensor tran decomposition are selected for the feature selection.

Next, we apply CP decomposition, Eq. (3.1), with $L = 1$ to data set 6, whose results obtained by Tucker decomposition is shown in Fig. 5.1. Figure 5.3 represents the two independent results staring from different initial values (One should remember that CP decomposition need to be given by initial values from which computation starts). At first, they clearly differ from each other. Second, the second realizations, (B), (D), and (F), do not correspond to the distinction between two classes and fail to identify features with not known in advance j, k dependence,

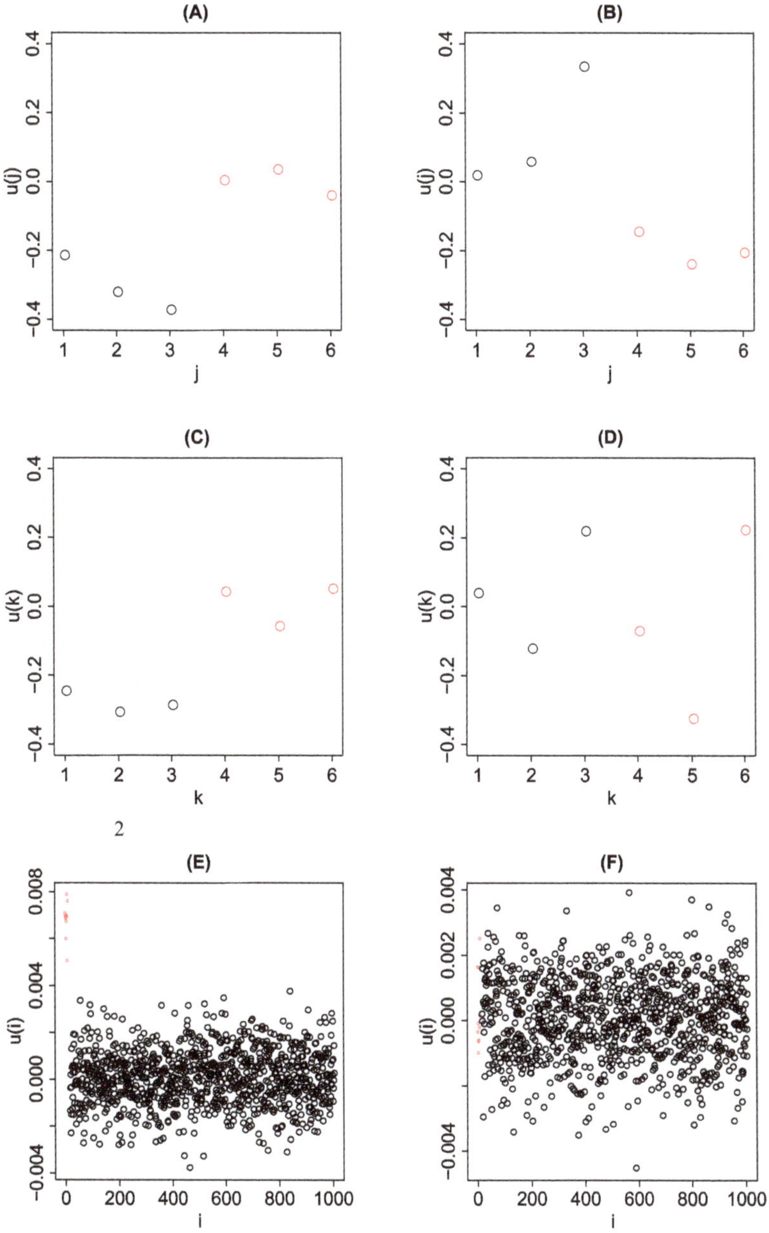

Fig. 5.3 Two typical convergent realizations starting from different initial values of CP decomposition, Eq. (3.1), with $L = 1$ applied to data set 6, Eq. (5.1), whose results obtained by Tucker decomposition is shown in Fig. 5.1. (**a**) and (**b**) $\boldsymbol{u}_1^{(j)}$, black and red circles correspond to $j \leq \frac{M}{2}$ and $j > \frac{M}{2}$, respectively. (**c**) and (**d**) $\boldsymbol{u}_1^{(k)}$, black and red circles correspond to $k \leq \frac{K}{2}$ and $k > \frac{K}{2}$, respectively. (**e**) and (**f**) $\boldsymbol{u}_1^{(i)}$. Red open circle correspond to $i \leq N_1$, i.e., features associated with j, k dependence

$i \leq N_1$. Thus, CP decomposition is inferior to Tucker decomposition because of initial condition dependence as discussed earlier.

These comparisons suggest that Tucker decomposition is superior to tensor train decomposition and CP decomposition as a tool of feature selection.

5.3 Generation of a Tensor from Matrices

In the previous section, we showed that TD-based unsupervised FE can outperform conventional supervised feature selection, categorical regression analysis, when the number of features is much larger than the number of samples and true classification is a complex function of apparent labeling. Although TD-based unsupervised FE is shown to be effective, it is unfortunately not so frequent that there are data sets formatted as tensor, because getting tensor requires more observation than matrices. In order to get $N \times M$ matrix that represents M samples with N features, required number of observations is as many as the number of samples, i.e., M. On the other hand, in order to get $N \times M \times K$ tensors that correspond to N features observed under the combination of M times and K times measurements, the required number of observation is as many as $K \times M$. If we need to have tensors with more modes, the number of observation will increase, too. Thus, even if TD-based unsupervised FE is an effective method, we usually cannot have data set formatted as tensors, to which TD-based unsupervised FE is applicable.

In order to have more opportunities to which we can apply TD-based unsupervised FE, we can propose to generate tensors from matrices [1], which are obtained more easily than tensors. Suppose that we have two matrices, $x_{ij} \in \mathbb{R}^{N \times M}$ and $x_{ik} \in \mathbb{R}^{N \times K}$, which represents i features under the jth experimental conditions and the kth experimental conditions, respectively. A typical observation is that N health conditions, blood pressure, body mass, body temperature, hight, weight, etc are observed M individuals in Japan and K individuals in USA. Then we can get tensor $x_{ijk} \in \mathbb{R}^{N \times M \times K}$ by simply multiply x_{ij} and x_{ik},

$$x_{ijk} = x_{ij} x_{ik} \tag{5.8}$$

TD can be applied to x_{ijk} as usual. It does not have to be restricted to the product of two matrices. We can generate $m + 1$ mode tensor by multiplying m matrices, $x_{ij_1}, x_{ij_2}, \cdots, x_{ij_m}$ as

$$x_{ij_1 j_2 \cdots j_m} = \prod_{s=1}^{m} x_{ij_s}. \tag{5.9}$$

On the other hand, we can consider the alternative cases where not features but samples are common between two matrices. Suppose that for K individuals two distinct N and M observations are performed and are recorded as matrices form,

$x_{ik} \in \mathbb{R}^{N \times K}$ and $x_{jk} \in \mathbb{R}^{M \times K}$. A typical example is that there are N goods in kth shop and x_{ik} represents a price of ith good in kth shop. On the other hand, x_{jk} represents the number of customers at jth time point at kth shop. We can generate tensor $x_{ijk} \in \mathbb{R}^{N \times M \times K}$ as

$$x_{ijk} = x_{ik} x_{jk}. \tag{5.10}$$

Again we can employ more matrices as

$$x_{i_1 i_2 \cdots i_m j} = \prod_{s=1}^{m} x_{i_s j}. \tag{5.11}$$

From the mathematical point of views, although there are no needs to distinguish between equations Eqs. (5.11) and (5.9), they should be considered separately from the data science point of views. Then hereafter we denote Eq. (5.11), i.e., the cases sharing samples, as case I while Eq. (5.9), i.e., the cases sharing features, as case II, respectively.

5.4 Reduction of Number of Dimensions of Tensors

It is possible to produce tensors from matrices. However, it increases number of features. When two matrices, $x_{ij} \in \mathbb{R}^{N \times M}$ and $x_{ik} \in \mathbb{R}^{N \times K}$ are multiplied in order to generate a tensor $x_{ijk} \in \mathbb{R}^{N \times M \times K}$ (case II), the number of features increases from $N \times (M + K)$ to $N \times M \times K$. Thus, we need some way to reduce the number of dimensions of generated tensors. Here we propose taking summation of shared features, i.e.,

$$\tilde{x}_{i_1 i_2 \cdots i_m} = \sum_j x_{i_1 i_2 \cdots i_m j} \tag{5.12}$$

$$\tilde{x}_{j_1 j_2 \cdots j_m} = \sum_i x_{i j_1 j_2 \cdots j_m}. \tag{5.13}$$

Then the number of dimensions increases from $N \times (M + K)$ not to $N \times M \times K$ but to $M \times K$ for case II while from $(N + M) \times K$ not to $N \times M \times K$ but to $N \times M$ for case I.

One might wonder how we can compute singular value matrices that correspond to indices of which are taken summation when TD is applied to $\tilde{x}_{i_1 i_2 \cdots i_m}$ or $\tilde{x}_{j_1 j_2 \cdots j_m}$. These missing singular value matrices are recovered by the following computations,

$$\boldsymbol{u}_\ell^{(i;j_s)} = X^{(ij_s)} \times_{j_s} \boldsymbol{u}_\ell^{(j_s)} \tag{5.14}$$

$$\boldsymbol{u}_\ell^{(j;i_s)} = X^{(ji_s)} \times_{i_s} \boldsymbol{u}_\ell^{(i_s)}, \tag{5.15}$$

where $X^{(ij_s)} \in \mathbb{R}^{N \times M_s}$ and $X^{(ji_s)} \in \mathbb{R}^{M \times N_s}$, respectively. Thus, we have m singular value matrices that correspond to i_s or j_s, instead of one singular value matrix. This might look problematic. Nevertheless, practically, if m singular value matrices obtained are mutually highly correlated, it is not practically problematic. Thus, case to case, we might employ this approximate strategy. In order to distinguish these tensors from the previous one, we call those generated after the partial summation of index, Eqs. (5.12) and (5.13) as type II while those without partial summation, Eqs. (5.9) and (5.11), as type I. Table 5.3 summarizes the distinction between cases and types.

5.5 Identification of Correlated Features Using Type I Tensor

The purpose of introduction of tensors summarized in Table 5.3 is simply because we would like to make use of TD-based unsupervised FE when no tensors are available. Nevertheless, we can make use of tensors listed in Table 5.3 for the additional alternative purpose as bi-product: identification of mutually correlated features. Suppose we have two sets of observations to K samples formatted as matrices, $x_{ik} \in \mathbb{R}^{N \times K}$ and $x_{jk} \in \mathbb{R}^{M \times K}$. The question is to search pairs of features between two sets.

The standard strategy is to compute pairwise correlation between x_{ik} and x_{jk},

$$r_{ij} = \frac{\frac{1}{K}\sum_k \left(x_{ik} - \frac{1}{K}\sum_{k'} x_{ik'}\right)\left(x_{jk} - \frac{1}{K}\sum_{k'} x_{jk'}\right)}{\sqrt{\frac{1}{K}\sum_k \left(x_{ik} - \frac{1}{K}\sum_{k'} x_{ik'}\right)^2 \frac{1}{K}\sum_k \left(x_{jk} - \frac{1}{K}\sum_{k'} x_{jk'}\right)^2}} \tag{5.16}$$

and to identify pairs of i and j associated with significant correlation. In the following, we will show some synthetic data set where pairwise computation of correlation does not work well while TD applied to a tensor generated from the product of two matrices, $x_{ijk} = x_{ik}x_{jk}$, can identify correlated pairs successfully.

In order for this purpose, we prepare data set 8 as follows.

Data set 8:

Table 5.3 Distinction between cases and types

	Type I		Type II	
Case I	$x_{i_1 i_2 \cdots i_m j} = \prod_{s=1}^{m} x_{i_s j}$	Eq. (5.11)	$\tilde{x}_{i_1 i_2 \cdots i_m} = \sum_j x_{i_1 i_2 \cdots i_m j}$	Eq. (5.12)
Case II	$x_{i j_1 j_2 \cdots j_m} = \prod_{s=1}^{m} x_{i j_s}$	Eq. (5.9)	$\tilde{x}_{j_1 j_2 \cdots j_m} = \sum_i x_{i j_1 j_2 \cdots j_m}$	Eq. (5.13)

Table 5.4 Confusion matrices when statistical tests are applied to synthetic data sets 8 defined by Eqs. (5.17) and (5.18) and features associated with adjusted P-vales less than 0.05 are selected for pairwise correlation and 0.1 for TD-based unsupervised FE

Data set 8	Pairwise correlation		TD-based unsupervised FE			
	$i \le N_1$ & $j \le M_1$	otherwise	$i \le N_1$	$N_1 < i$	$j \le M_1$	$M_1 < j$
selected	15.47	38.49	6.20	0.00	6.14	0.00
not selected	84.53	9861.51	3.80	90.00	3.86	90.00

$$x_{ik} \sim \begin{cases} k + \mathcal{N}(\mu, \sigma) & i \le N_1 \\ \mathcal{N}(\mu, \sigma) & \text{otherwise} \end{cases} \tag{5.17}$$

$$x_{jk} \sim \begin{cases} k + \mathcal{N}(\mu, \sigma) & j \le M_1 \\ \mathcal{N}(\mu, \sigma) & \text{otherwise} \end{cases} \tag{5.18}$$

This means, only features $i \le N_1$ and $j \le M_1$ shares the k dependence while no other pairs are correlated. In this set up, the number of positive (correlated) pairs is $N_1 \times M_1$ among total number of pairs, $N \times M$.

In order to see if pairwise correlation analysis can identify correlated pairs, we compute Pearson's correlation coefficients between all $N \times M$ pairs, x_{ik} and x_{jk}. Then computed correlation coefficient, r_{ij}, is converted to t_{ij} as

$$t_{ij} = \frac{r_{ij}(K-2)}{\sqrt{1-r^2}} \tag{5.19}$$

that is known to obey t distribution with the degrees of freedom of $K - 2$. Then P-values are computed using t distribution and are attributed to all of $N \times M$ pairs. These P-values are corrected by BH criterion and pairs associated with adjusted P-values less than 0.05 is considered to be correlated. Table 5.4 shows the confusion matrix averaged over 100 independent trials when $N = M = 100, N_1 = M_1 = 10, K = 6, \mu = \sigma = 1$. In this set up, the number of positive pairs is $N_1 \times M_1 = 100$. It is obvious that there are more false positives (38.49) than true positives (15.47). Thus, it unlikely works well. Next, we apply TD based unsupervised FE to data set 8 with generating case I type I tensor (Table 5.4) as Eq. (5.10). We apply HOSVD algorithm, Fig. 3.8, to data set 8. Figure 5.4a and b show typical $\boldsymbol{u}_1^{(i)}$ and $\boldsymbol{u}_1^{(j)}$ obtained when HOSVD is applied to data set 8, respectively. These two have obviously larger absolute values for $i \le N_1$ and $j \le M_1$ than $i > N_1$ and $j > M_1$, respectively. This suggests that $\boldsymbol{u}_1^{(i)}$ and $\boldsymbol{u}_1^{(j)}$ can successfully identify features with correlations ($i \le N_1$ or $j \le M_1$) from those without correlations ($i > N_1$ or $j > M_1$). How it comes to be possible can be understood by observing $\boldsymbol{u}_1^{(k)}$ (Fig. 5.5). $\boldsymbol{u}_1^{(k)}$ clearly reflects the dependence upon k shown in Eqs. (5.17) and (5.18). Since $G(1, 1, 1)$ is the largest among $G(\ell_1, \ell_2, 1)$, $\boldsymbol{u}_1^{(i)}$ and $\boldsymbol{u}_1^{(j)}$ naturally assign larger

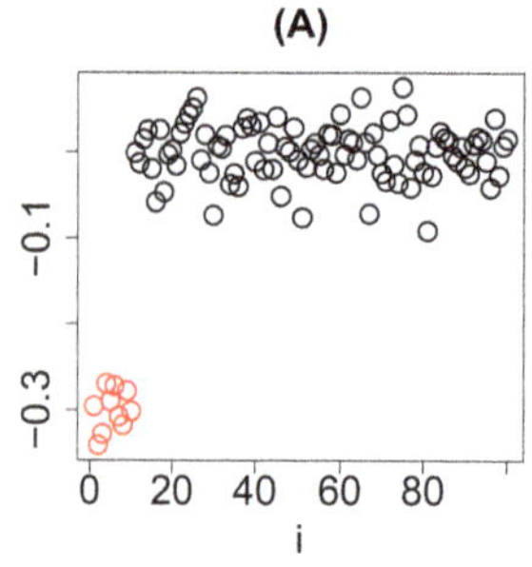

Fig. 5.4 A typical realization of $\boldsymbol{u}_1^{(i)}$ and $\boldsymbol{u}_1^{(j)}$ when Tucker decomposition, Eq. (3.2), with HOSVD algorithm, Fig. 3.8 is applied to data set 8, Eqs. (5.17) and (5.18) with $N = M = 100, N_1 = M_1 = 10, K = 6, \mu = \sigma = 1$. (**a**) $\boldsymbol{u}_1^{(i)}$, red and black open circles correspond to $i \leq N_1$ and $i > N_1$, respectively. (**b**) $\boldsymbol{u}_1^{(j)}$, red and black open circles correspond to $j \leq M_1$ and $j > M_1$, respectively

Fig. 5.5 $\boldsymbol{u}_1^{(k)}$ that corresponds to $\boldsymbol{u}_1^{(i)}$ and $\boldsymbol{u}_1^{(j)}$ shown in Fig. 5.4

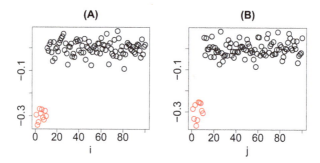

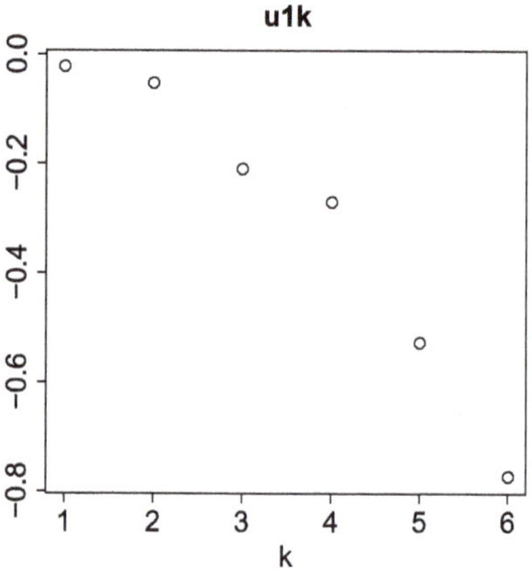

absolute values to $u_{1i}^{(i)}$ and $u_{1j}^{(j)}$ that shares embedded k dependence, i.e., $i \leq N_1$ or $j \leq M_1$.

In order to see if $u_{1i}^{(i)}$ and $u_{1j}^{(j)}$ are useful for the feature selection, P-values are attributed to i as Eq. (5.4) and j as

$$P_j = P_{\chi^2}\left[> \left(\frac{u_{1j}^{(j)}}{\sigma_1'} \right)^2 \right], \tag{5.20}$$

where σ_1' is the standard deviation of $u_{1j}^{(j)}$. Then is and js associated with adjusted P-value less than 0.1 are selected (Performances are averaged over 100 independent trials). Table 5.4 shows the corresponding confusion matrices. Although the performance cannot be said vary good, it is remarkable that there are no FP which are as many as 38.49 in pairwise correlation analysis (Table 5.4). TD-based unsupervised FE also has more TPs than correlation analysis; 6.20 or 6.14 TPs among 10 positives versus 15.47 TP among 100 positives.

Only from this specific example, we cannot conclude that TD-based unsupervised FE can always outperform the conventional methods. Nevertheless, in the application to the real data set that will be shown in the later, we will see that TD based unsupervised FE can achieve better performances than conventional supervised methods.

5.6 Identification of Correlated Features Using Type II Tensor

In the previous section, we can see that TD-based unsupervised FE can correctly recognize the features with mutual correlation that cannot be recognized by conventional pairwise correlation analysis. In this section, we would like to see if type II tensor, Eq. (5.12), can samely identify features with mutual correlations using the same data set 8, Eqs. (5.17) and (5.18). In the present specific case, type II tensor can be defined as

$$\tilde{x}_{ij} = \sum_{k=1}^{K} x_{ijk} = \sum_{k=1}^{K} x_{ik} x_{jk}. \tag{5.21}$$

TD, or essentially it is SVD because HOSVD is equivalent to SVD when it is applied to matrix, is applied to $\tilde{x}_{ij}$. Figure 5.6 shows the comparison of $\boldsymbol{u}_1^{(i)}$ and $\boldsymbol{u}_1^{(j)}$ between type I and type II tensors. Although slight deviation can be observed, they are coincident enough to recognize features with mutual correlations, i.e., $i \leq N_1$ and $j \leq M_1$, respectively. Thus as long as considering feature selection, replacing type I tensor with type II tensor does not cause any problems.

Then we need to see if two vectors,

$$\boldsymbol{u}_1^{(k;i)} = X^{(ik)} \times_i \boldsymbol{u}_1^{(i)} \tag{5.22}$$

$$\boldsymbol{u}_1^{(k;j)} = X^{(jk)} \times_j \boldsymbol{u}_1^{(j)} \tag{5.23}$$

are coincident with each other and reflects k dependence when $\boldsymbol{u}_1^{(i)}$ and $\boldsymbol{u}_1^{(j)}$ are computed from type II tensor (matrix), Eq. (5.21). Figure 5.7 shows $\boldsymbol{u}_1^{(k:i)}$ and $\boldsymbol{u}_1^{(k:j)}$. They are not only coincident with each other, but also reflecting k dependence in Eqs. (5.17) and (5.18), respectively. Thus, replacing type I tensor with type II, at least in the present case, does not likely cause any problems.

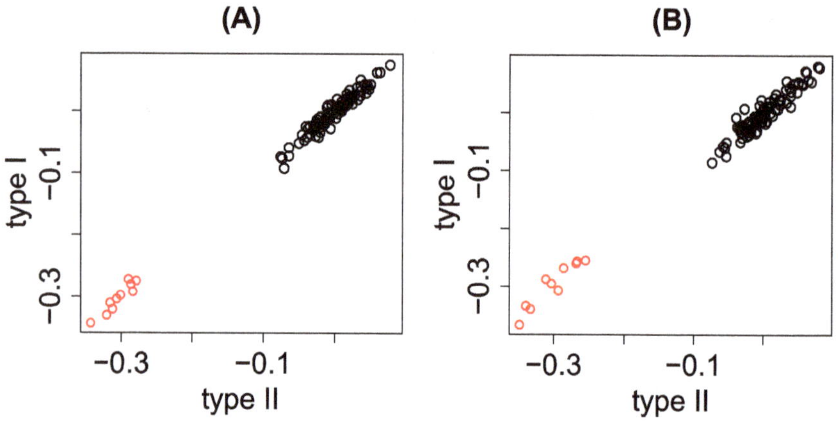

Fig. 5.6 Comparison between $\boldsymbol{u}_1^{(i)}$ and $\boldsymbol{u}_1^{(j)}$ in Fig. 5.4 and those when SVD is applied to type II tensor (matrix), $\tilde{x}_{ij}$, defined in Eq. (5.21). (**a**) $\boldsymbol{u}_1^{(i)}$, red and black open circles correspond to $i \leq N_1$ and $i > N_1$, respectively. (**b**) $\boldsymbol{u}_1^{(j)}$, red and black open circles correspond to $j \leq M_1$ and $j > M_1$, respectively

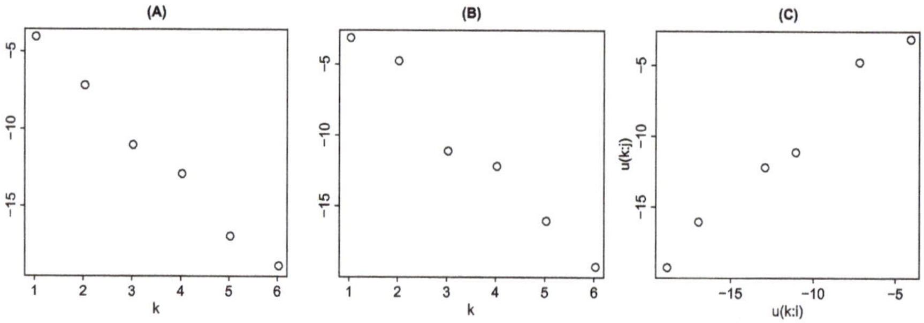

Fig. 5.7 Comparison between $\boldsymbol{u}_1^{(k:i)}$ and $\boldsymbol{u}_1^{(k:j)}$ computed by Eqs. (5.22) and (5.23), respectively. (**a**) $\boldsymbol{u}_1^{(k:i)}$ (**b**) $\boldsymbol{u}_1^{(k:j)}$, (**c**) scatter plot of (**a**) and (**b**)

5.7 Feature Selection with Integrating Multiple Profiles

5.7.1 Samples Sharing Cases

We can introduce yet another way [4] to integrate multiple profiles other than those shown in Table 5.3. Suppose that we have K profiles $x_{i_k jk} \in \mathbb{R}^{N_k \times M \times K}$ which corresponds to Case I in Table 5.3, i.e., K profiles share samples associated with the feature i_k whose number is N_k. In order to integrate them, we generate matrices by multiplying $x_{i_k jk}$ with itself as

$$x_{jj'k} = \sum_{i_k=1}^{N_k} x_{i_k jk} x_{i_k j'k}, \tag{5.24}$$

where $x_{jj'k} \in \mathbb{R}^{M \times M \times K}$. Then HOSVD is applied to $x_{jj'k}$ to get

$$x_{jj'k} = \sum_{\ell_1=1}^{M} \sum_{\ell_2=1}^{M} \sum_{\ell_3=1}^{K} G(\ell_1 \ell_2 \ell_3) u_{\ell_1 j}^{(j)} u_{\ell_2 j'}^{(j')} u_{\ell_3 k}^{(k)} \tag{5.25}$$

where $\boldsymbol{u}_{\ell_1}^{(j)}, \boldsymbol{u}_{\ell_2}^{(j')} \in \mathbb{R}^M$, $\boldsymbol{u}_{\ell_3}^{(k)} \in \mathbb{R}^K$, $G(\ell_1 \ell_2 \ell_3) \in \mathbb{R}^{M \times M \times K}$. Although singular value matrices attributed to i_ks are not obtained due to taking summation of i_k in Eq. (5.24), they can be recovered as in Eq. (5.14),

$$\boldsymbol{u}_{\ell_1}^{(i_k;j)} = X^{(i_k j)} \times_j \boldsymbol{u}_{\ell_1}^{(j)} \tag{5.26}$$

where $\boldsymbol{u}_{\ell_1}^{(i_k;j)} \in \mathbb{R}^{N_k}$ and $X^{(i_k j)} \in \mathbb{R}^{N_k \times M}$ whose components are $x_{i_k jk}$. Once the $\boldsymbol{u}_{\ell_1}^{(i_k;j)}$s are recovered and the $\boldsymbol{u}_{\ell_1}^{(j)}$ of interest is decided, the corresponding $u_{\ell_1 i_k}^{(i_k;j)}$ is replaced by $u_{\ell i}^{(i)}$ in Eq. (5.4) and P_{i_k} is attributed to the i_kth feature. The remaining process is the same as the preceding sections. Since the size of the tensor to which TD is applied is as small as $M \times M \times K$, this implementation can drastically reduce the memory size to compute TD.

In order to see how well it works, we applied the above procedure to a synthetic data set.

Data set 9:

$$x_{i_k jk} = \begin{cases} \mathcal{N}(\mu, \sigma), & i \leq N_1^k, j \leq \frac{M}{2} \\ \mathcal{N}(0, \sigma), & \text{otherwise} \end{cases} \tag{5.27}$$

When we applied the proposed procedure for $K = 3, N_1 = 10^4, N_2 = 2 \times 10^4, N_3 = 3 \times 10^4, N_1^k = 10^2, \mu = 3, \sigma = 1, M = 10$ with one hundred ensembles (Table 5.5) where $\ell_1 = 1$ is always used to select features since $\ell_1 = 1$ is associated with distinction between two classes (Fig. 5.8), features associated with distinction between two classes are almost always selected whereas there are

Table 5.5 Confusion matrices when two statistical tests, TD-based unsupervised FE and t-test, are applied to the synthetic data set 9 defined by Eq. (5.27) and features associated with adjusted P-values less than 0.01 are selected

		TD-based unsupervised FE		t-test	
Data set 9	k	$i \leq N_1^k$	$i > N_1^k$	$i \leq N_1^k$	$i > N_1^k$
Selected	1	97.08	0.03	0.22	0.00
Not selected		2.92	9899.97	99.78	9900.00
Selected	2	98.53	0.17	0.08	0.00
Not selected		1.47	19899.83	99.92	19900.00
Selected	3	98.67	0.30	0.07	0.02
Not selected		1.33	29899.70	99.93	29899.98

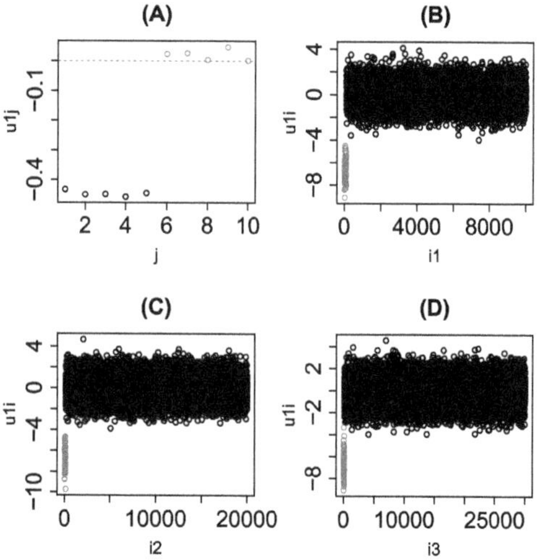

Fig. 5.8 A typical realization of $\boldsymbol{u}_1^{(j)}$ and $\boldsymbol{u}_1^{(i_k;j)}$, $k = 1, 2, 3$ for Eqs. (5.25) and (5.26) when data set 9 with $K = 3, N_1 = 10^4, N_2 = 2 \times 10^4, N_3 = 3 \times 10^4, N_1^k = 10^2, \mu = 3, \sigma = 1, M = 10$ is used. (**a**) $\boldsymbol{u}_1^{(j)}$, black and red circles correspond to $j \leq \frac{M}{2}$ and $j > \frac{M}{2}$, respectively. The red broken line shows the baseline. (**b**) $\boldsymbol{u}_1^{(i_1;j)}$, (**c**) $\boldsymbol{u}_1^{(i_2;j)}$, and (**d**) $\boldsymbol{u}_1^{(i_3;j)}$. The red open circles correspond to $i \leq N_1$, that is, the features associated with the dependence on j

essentially no false positives. On the other hand, if we apply the t-test to the same data set, essentially no true positives are identified (Table 5.5). Although we cannot insist that our proposed strategy can always outperform the t-test, at least there are some *large p small n* situations where our proposed strategy, TD-based unsupervised FE, can outperform t-test. Thus, our proposed strategy is worthwhile trying.

5.7.2 Features Sharing Cases

We can also employ a similar strategy when features are shared among multiple profiles. Suppose that we have K profiles $x_{ij_kk} \in \mathbb{R}^{N \times M_k \times K}$ which corresponds to Case II in Table 5.3, i.e., K profiles share features associated with samples j_k whose number is M_k. In order to integrate them, we generate matrices by multiplying x_{ij_kk} with itself as

$$x_{ii'k} = \sum_{j_k=1}^{M_k} x_{ij_kk} x_{i'j_kk}, \tag{5.28}$$

where $x_{ii'k} \in \mathbb{R}^{N \times N \times K}$. Nevertheless, it is not as effective as in the case where features are shared among multiple profiles, since it requires as many as N^2 computer memories, which can be huge because N is a large number. HOSVD is applied to $x_{ii'k}$ to get

$$x_{ii'k} = \sum_{\ell_1=1}^{N} \sum_{\ell_2=1}^{N} \sum_{\ell_3=1}^{K} G(\ell_1 \ell_2 \ell_3) u_{\ell_1 i}^{(i)} u_{\ell_2 i'}^{(i')} u_{\ell_3 k}^{(k)}, \tag{5.29}$$

where $\boldsymbol{u}_{\ell_1}^{(i)}, \boldsymbol{u}_{\ell_2}^{(i')} \in \mathbb{R}^N$, $\boldsymbol{u}_{\ell_3}^{(k)} \in \mathbb{R}^K$, $G(\ell_1\ell_2\ell_3) \in \mathbb{R}^{N \times N \times K}$. In order to evaluate which ℓ_1 is associated with classification for individual profiles, we compute singular value vectors attributed to samples for the kth profiles as in the Eq. (5.15),

$$\boldsymbol{u}_{\ell_1}^{(j_k;i)} = X^{(ij_k)} \times_i \boldsymbol{u}_{\ell_1}^{(i)}, \tag{5.30}$$

where $\boldsymbol{u}_{\ell_1}^{(j_k;i)} \in \mathbb{R}^{M_k}$ and $X^{(ij_k)} \in \mathbb{R}^{N \times M_k}$ whose components are x_{ij_kk}. Once $\boldsymbol{u}_{\ell_1}^{(j_k;i)}$s are recovered and $u_{\ell_1 j_k}^{(j_k;i)}$ of interest is decided, the corresponding $\boldsymbol{u}_{\ell_1}^{(i)}$ is substituted to $u_{\ell i}^{(i)}$ in Eq. (5.4) and P_i is attributed to the ith feature. The remaining process is the same as the preceding sections.

In order to see how well it works, we applied the above procedure to a synthetic data set.

Data set 10:

$$x_{ij_kk} = \begin{cases} \mathcal{N}(\mu, \sigma), & N_1^k \le i \le N_2^k, j_k \le \frac{M_k}{2} \\ \mathcal{N}(0, \sigma), & \text{otherwise.} \end{cases} \tag{5.31}$$

When we applied the proposed procedure for $K = 3, N = 10^3, M_1 = 8, N_1^1 = 1, N_2^1 = 10, M_2 = 10, N_1^2 = 11, N_2^2 = 20, M_3 = 12, N_1^3 = 21, N_2^3 = 30, \mu = 3, \sigma = 1$ with one hundred ensembles (Table 5.6) where $\ell_1 = 1$ for $k = 3$, $\ell_1 = 2$ for $k = 2$ and $\ell_1 = 3$ for $k = 1$ are used to select features since the selected i_ks using $\boldsymbol{u}_{\ell_1}^{(i_k;j)}$ computed with Eq. (5.30) are well coincident with $N_1^k \le i \le N_2^k$ (Fig. 5.9). Although their performances are inferior to those TD-based unsupervised FE achieved in Table 5.5, if they are compared with the performances achieved by t-test (Table 5.6), the performances achieved by TD-based unsupervised FE are still much better than t-test. Although we cannot insist that our proposed strategy can always outperform t-test, at least there are some *large p small n* situations where our proposed strategy can outperform t-test. Thus, our proposed strategy is worthwhile trying.

Table 5.6 Confusion matrices when two statistical tests, TD-based unsupervised FE and t-test, are applied to the synthetic data set 10 defined by Eq. (5.31) and features associated with adjusted P-values less than 0.01 are selected

		TD-based unsupervised FE		t-test	
Data set 10	k	$N_1^k \le i \le N_2^k$	otherwise	$N_1^k \le i \le N_2^k$	otherwise
Selected	1	5.82	1.98	0.03	0.00
Not selected		4.18	988.02	9.97	990.00
Selected	2	5.49	3.33	0.16	0.01
Not selected		4.51	988.67	9.84	989.99
Selected	3	7.87	1.85	0.07	0.04
Not selected		2.13	988.15	9.08	989.96

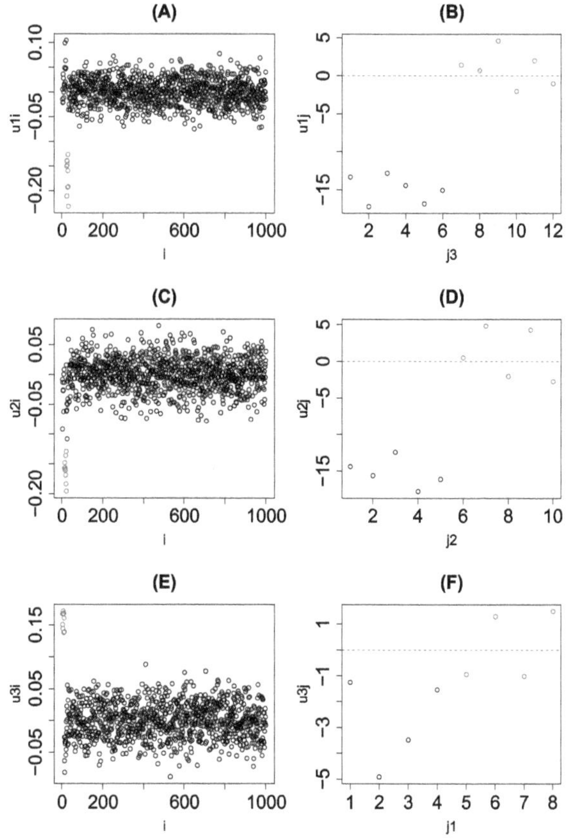

Fig. 5.9 A typical realization of $u^{(j)}_{\ell_1 j}$ and $\boldsymbol{u}^{(i_k;j)}_{\ell_1}$, $k = 1, 2, 3$ for Eqs. (5.29) and (5.30) when data set 10 with $K = 3, N = 10^3, M_1 = 8, N^1_1 = 1, N^1_2 = 10, M_2 = 10, N^2_1 = 11, N^2_2 = 20, M_3 = 12, N^3_1 = 21, N^3_2 = 30, \mu = 3, \sigma = 1$ is used. (**a**) $u^{(i)}_{1i}$ and (**b**) $\boldsymbol{u}^{(j_3;i)}_1$ (for $k = 3$), (**c**) $u^{(i)}_{2i}$ and (**d**) $\boldsymbol{u}^{(j_2;i)}_2$ (for $k = 2$), (**e**) $u^{(i)}_{3i}$ and (**f**) $\boldsymbol{u}^{(j_1;i)}_3$ (for $k = 1$). Red open circles in (**a**), (**c**), and (**e**) correspond to $N^k_1 \leq i \leq N^k_2$. Black and red circles in (**b**), (**d**), and (**f**) correspond to $j_k \leq \frac{M_k}{2}$ and $j_k > \frac{M_k}{2}$, respectively. Red broken line shows baseline

We should also note that $\ell_1 = 1, 2, 3$ corresponds to $k = 3, 2, 1$ (Fig. 5.9). In contrast to data set 9, features in data set 10 are distinct from each other and features with distinction between two classes, $N^k_1 \leq i \leq N^k_2$, differ between ks. Thus, ℓ_1 that corresponds to k varies dependent upon k.

Since it is a very memory-consuming strategy, although it has not yet been applied to real bioinformatics problem, the strategy itself might have some possibility to be applied to real data set.

5.8 Feature Selection with Integrating Multiple Profiles Using Projection

Although we have introduced the alternative strategy in the previous section to integrate multiple profiles other than those in Table 5.3, the above strategy has some limitation; it requires huge memory when features are shared between profiles. In this section, we introduce yet another strategy to integrate multiple profiles. In this

Table 5.7 Confusion matrices when statistical tests are applied to synthetic data sets 9 defined by Eq. (5.27) and features associated with adjusted P -values less than 0.01 and 0.05 are selected

		TD-based unsupervised FE			
		$P = 0.01$		$P = 0.05$	
Data set 9	k	$i \le N_1^k$	$i > N_1^k$	$i \le N_1^k$	$i > N_1^k$
Selected	1	11.92	0.02	19.23	0.27
Not selected		88.08	9899.98	80.77	9899.73
Selected	2	12.73	0.10	18.99	0.60
Not selected		87.27	19899.90	81.01	19899.40
Selected	3	11.63	0.09	17.32	0.56
Not selected		88.37	29899.91	82.68	29899.44

newly proposed strategy, individual profiles are embedded into low dimensional space prior to integration. In contrast to the impression, this strategy seems to work pretty well.

5.8.1 Samples Sharing Cases

As in the section Sect. 5.7.1, we first consider sample sharing cases. The profiles are again $x_{i_k jk} \in \mathbb{R}^{N_k \times M \times K}$. Before integrating profiles, SVD is applied to individual profiles as

$$x_{i_k jk} = \sum_{\ell=1}^{L} \lambda_\ell u_{\ell i_k}^k v_{\ell j}^k, \tag{5.32}$$

where $\boldsymbol{u}_\ell \in \mathbb{R}^N$, $\boldsymbol{v}_\ell \in \mathbb{R}^M$. Up to L singular value vectors retrieved from K profiles are formatted as a tensor.

$$x_{\ell jk} = v_{\ell j}^k = \sum_{\ell_1=1}^{L} \sum_{\ell_2=1}^{M} \sum_{\ell_3=1}^{K} G(\ell_1 \ell_2 \ell_3) u_{\ell_1 \ell}^{(\ell)} u_{\ell_2 j}^{(j)} u_{\ell_3 k}^{(k)}, \tag{5.33}$$

where $\boldsymbol{u}_{\ell_1}^{(\ell)} \in \mathbb{R}^L$, $\boldsymbol{u}_{\ell_2}^{(j)} \in \mathbb{R}^M$, and $\boldsymbol{u}_{\ell_3}^{(k)} \in \mathbb{R}^K$. After identifying $\boldsymbol{u}_{\ell_2}^{(j)}$ of interest, singular value vectors attributed to features are computed as in the Eq. (5.14),

$$\boldsymbol{u}_{\ell_2}^{(i_k;j)} = X^{(i_k j)} \times_j \boldsymbol{u}_{\ell_2}^{(j)}, \tag{5.34}$$

where $\boldsymbol{u}_{\ell_2}^{(i_k;j)} \in \mathbb{R}^{N_k}$ and $X^{(i_k j)} \in \mathbb{R}^{N_k \times M}$ whose components are $x_{i_k jk}$. The $u_{\ell_2 i_k}^{(i_k;j)}$ is substituted for $u_{\ell i}^{(i)}$ in Eq. (5.4) and P_{i_k} is attributed to the i_kth feature. The remaining process is the same as the preceding sections. Since the size of tensor to which TD is applied is as small as $L \times M \times K$, this implementation can drastically reduce the memory size to compute TD.

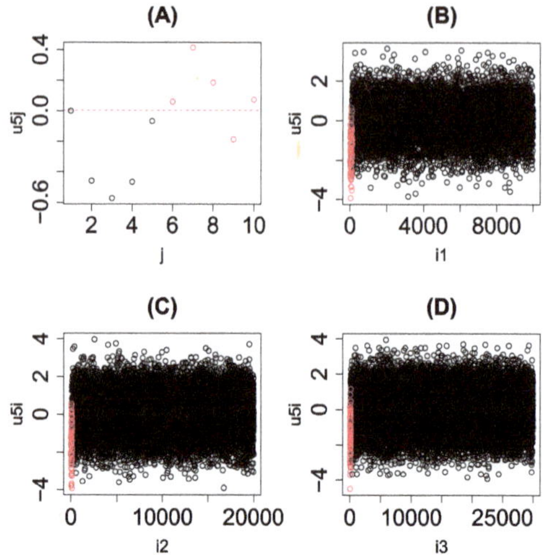

Fig. 5.10 A typical realization of $u_{5j}^{(j)}$ and $\boldsymbol{u}_{\ell_2}^{(i_k;j)}, k = 1, 2, 3$ for Eqs. (5.33) and (5.34) when data set 9 with $K = 3, N_1 = 10^4, N_2 = 2 \times 10^4, N_3 = 3 \times 10^4, N_1^k = 10^2, \mu = 3, \sigma = 1, M = 10, L = 10$ is used. (**a**) $u_{5j}^{(j)}$, black and red circles correspond to $j \leq \frac{M}{2}$ and $j > \frac{M}{2}$, respectively. Red broken line shows baseline. (**b**) $\boldsymbol{u}_5^{(i_1;j)}$, (**c**) $\boldsymbol{u}_5^{(i_2;j)}$, (**d**) $\boldsymbol{u}_5^{(i_3;j)}$. Red open circles correspond to $i \leq N_1$, i.e., features associated with j dependence. In this specific realization, $\ell_2 = 5$

In order to evaluate the performance of this strategy, we applied the proposed strategy to data set 9 with the same set up as in Table 5.5 with $L = 10$ (Table 5.7). Although the performances substantially decreased from those TD-based unsupervised FE achieved in Table 5.5 even if we increase the threshold P-values to 0.05, they are still much better than those t-test achieved in Table 5.5. Since we do not know which $\ell_2 (\leq L = 10)$ is the most coincident with the distinction between the two classes (Fig. 5.10), we selected ℓ_2 associated with the smallest P-values when categorical regression was applied to each of one hundred ensembles. In addition to this, the present strategy has an advantage that the strategy proposed in Sect. 5.7.1 lacks; since the required computational memory is as small as $L \times M \times K$, when M is large, the present strategy can reduce the required computational memory than the strategy proposed in Sect. 5.7.1 that requires the computational memory as large as $M \times M \times K$. Although we assume *large p small n* problems, i.e., $N \gg M$, if N is very large, M still can be very large (i.e., $M \gg 1$). In actual, the present strategy was applied to single cell multiomics profiles (see Sect. 7.12.4) whose number of samples is huge since it is equal to the number of single cells [3].

5.8.2 Features Sharing Cases

In contrast to the samples sharing case discussed in the previous subsection that has little advantages compared with those in Sect. 5.7.1, features sharing case discussed in this subsection has more advantages than those in Sect. 5.7.2; it can drastically reduce the required computational memories. As in Sect. 5.7.2. we assume that we have K profiles $x_{ij_kk} \in \mathbb{R}^{N \times M_k \times K}$. As in the previous subsection, SVD is applied to x_{ij_kk} as

$$x_{ij_kk} = \sum_{\ell=1}^{L} \lambda_\ell u^k_{\ell i} v^k_{\ell j_k}. \tag{5.35}$$

Then $u^k_{\ell i}$s are bundled to have a tensor

$$x_{i\ell k} = u^k_{\ell i} = \sum_{\ell_1=1}^{N} \sum_{\ell_2=1}^{L} \sum_{\ell_3=1}^{K} G(\ell_1 \ell_2 \ell_3) u^{(i)}_{\ell_1 i} u^{(\ell)}_{\ell_2 \ell} u^{(k)}_{\ell_3 k} \tag{5.36}$$

where $\boldsymbol{u}^{(i)}_{\ell_1} \in \mathbb{R}^N$, $u^{(\ell)}_{\ell_2 \ell} \in \mathbb{R}^L$, $\boldsymbol{u}^{(k)}_{\ell_3} \in \mathbb{R}^K$, $G(\ell_1 \ell_2 \ell_3) \in \mathbb{R}^{N \times L \times K}$. In order to evaluate which ℓ_1 is associated with classification for individual profiles, we compute singular value vectors attributed to samples for the kth profiles as in Eq. (5.30). Once $\boldsymbol{u}^{(j_k;i)}_{\ell_1}$ is recovered as in the Eq. (5.15),

$$\boldsymbol{u}^{(j_k;i)}_{\ell_1} = X^{(ij_k)} \times_i \boldsymbol{u}^{(i)}_{\ell_1}. \tag{5.37}$$

where $\boldsymbol{u}^{(j_k;i)}_{\ell_1} \in \mathbb{R}^{M_k}$ and $X^{(ij_k)} \in \mathbb{R}^{N \times M_k}$ whose component are x_{ij_kk} and $u^{(j_k;i)}_{\ell_1 j_k}$ of interest is decided, the corresponding $\boldsymbol{u}^{(i)}_{\ell_1}$ is substituted to $u^{(i)}_{\ell i}$ in Eq. (5.4) and P_i is attributed to ith feature. The remaining process is the same as the above.

In order to evaluate the performance of this strategy, we applied the proposed strategy to the data set 10 with the same set up as in Table 5.6 with $L = 100$ (Table 5.8). Although the performances substantially decreased from those TD-based unsupervised FE achieved in Table 5.6 even if we increase the threshold P-values to 0.05, they are still much better than those t-test achieved in Table 5.6. Since we do not know which $\ell_1 (\leq L = 100)$ is the most coincident with the distinction two classes (Fig. 5.11), we selected ℓ_1 associated with the smallest P-values when categorical regression was applied to each of one hundred ensembles. The most remarkable thing in this strategy is the reduction of computational memory. In Sect. 5.7.2, it requires the computational memory as large as $N \times N \times K$. On the other, in the present strategy, the computational memory required is as small as $N \times L \times K$, which is much smaller than $N \times N \times K$. Thus, even if the performance decreases from Sect. 5.7.2, the strategy proposed in this section is worth while trying.

Table 5.8 Confusion matrices when statistical tests are applied to synthetic data sets 10 defined by Eq. (5.31) and features associated with adjusted P-values less than 0.01 and 0.05 are selected

		TD-based unsupervised FE			
		$P = 0.01$		$P = 0.05$	
Data set 10	k	$N_1^k \le i \le N_2^k$	otherwise	$N_1^k \le i \le N_2^k$	otherwise
Selected	1	6.07	0.33	6.44	0.37
Not selected		3.93	989.67	3.56	989.63
Selected	2	2.19	0.33	2.27	0.74
Not selected		7.81	989.67	7.73	989.26
Selected	3	1.02	0.00	1.03	0.00
Not selected		9.89	990.00	8.97	990.00

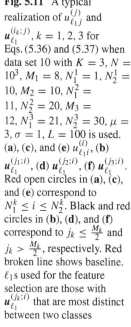

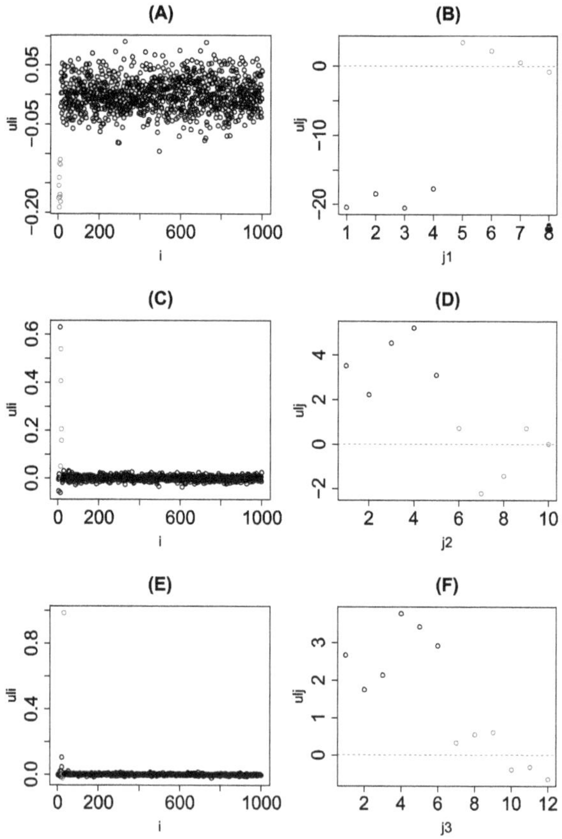

Fig. 5.11 A typical realization of $u_{\ell_1 j}^{(j)}$ and $\boldsymbol{u}_{\ell_1}^{(i_k;j)}$, $k = 1, 2, 3$ for Eqs. (5.36) and (5.37) when data set 10 with $K = 3$, $N = 10^3$, $M_1 = 8$, $N_1^1 = 1$, $N_2^1 = 10$, $M_2 = 10$, $N_1^2 = 11$, $N_2^2 = 20$, $M_3 = 12$, $N_1^3 = 21$, $N_2^3 = 30$, $\mu = 3$, $\sigma = 1$, $L = 100$ is used. (**a**), (**c**), and (**e**) $u_{\ell_1 i}^{(i)}$, (**b**) $\boldsymbol{u}_{\ell_1}^{(j_1;i)}$, (**d**) $\boldsymbol{u}_{\ell_1}^{(j_2;i)}$, (**f**) $\boldsymbol{u}_{\ell_1}^{(j_3;i)}$. Red open circles in (**a**), (**c**), and (**e**) correspond to $N_1^k \le i \le N_2^k$. Black and red circles in (**b**), (**d**), and (**f**) correspond to $j_k \le \frac{M_k}{2}$ and $j_k > \frac{M_k}{2}$, respectively. Red broken line shows baseline. ℓ_1s used for the feature selection are those with $\boldsymbol{u}_{\ell_1}^{(j_k;i)}$ that are most distinct between two classes

5.9 Kernel Tensor Decomposition

Although we have developed many ways to apply TD to various data analyses, all of analyses were restricted to liner analysis. Although basically we are not interested in nonlinear analyses since nonlinear analyses prevent us from directly interpreting the results, we introduced some methods to extend TD to nonlinear region; kernel tensor decomposition (KTD), to demonstrate that TD-based unsupervised FE can be expected to the non-linear analyses.

Now we can introduce the concept of KTD [2]. Starting from the standard representation of Tucker decomposition,

$$x_{ijk} = \sum_{\ell_1=1}^{N} \sum_{\ell_2=1}^{M} \sum_{\ell_3=1}^{K} G(\ell_1 \ell_2 \ell_3) u_{\ell_1 i}^{(i)} u_{\ell_2 j}^{(j)} u_{\ell_3 k}^{(k)} \tag{5.38}$$

we can "kernelize" it as

$$\begin{aligned} x_{jkj'k'} &= \sum_{i=1}^{N} x_{ijk} x_{ij'k'} \\ &= \sum_{\ell_1=1}^{N} \sum_{\ell_2=1}^{M} \sum_{\ell_3=1}^{K} \sum_{\ell_1'=1}^{N} \sum_{\ell_2'=1}^{M} \sum_{\ell_3'=1}^{K} G(\ell_1 \ell_2 \ell_3) G(\ell_1' \ell_2' \ell_3') \\ &\quad \times \left(\sum_{i=1}^{N} u_{\ell_1 i}^{(i)} u_{\ell_1' i}^{(i)} \right) u_{\ell_2 j}^{(j)} u_{\ell_3 k}^{(k)} u_{\ell_2' j'}^{(j')} u_{\ell_3' k'}^{(k')}. \end{aligned} \tag{5.39}$$

Since

$$\sum_{i=1}^{N} u_{\ell_1 i}^{(i)} u_{\ell_1' i}^{(i)} = \delta_{\ell_1 \ell_1'} \tag{5.40}$$

we can get

$$\sum_{i=1}^{N} x_{ijk} x_{ij'k'} = \sum_{\ell_1=1}^{N} \sum_{\ell_2=1}^{M} \sum_{\ell_3=1}^{K} \sum_{\ell_2'=1}^{M} \sum_{\ell_3'=1}^{K} G(\ell_1 \ell_2 \ell_3) G(\ell_1 \ell_2' \ell_3') u_{\ell_2 j}^{(j)} u_{\ell_3 k}^{(k)} u_{\ell_2' j'}^{(j')} u_{\ell_3' k'}^{(k')}. \tag{5.41}$$

If we can introduce

$$G(\ell_2 \ell_3 \ell_2' \ell_3') = \sum_{\ell_1=1}^{N} G(\ell_1 \ell_2 \ell_3) G(\ell_1 \ell_2' \ell_3') \tag{5.42}$$

we can get

$$x_{jkj'k'} = \sum_{\ell_2=1}^{M} \sum_{\ell_3=1}^{K} \sum_{\ell_2'=1}^{M} \sum_{\ell_3'=1}^{K} G(\ell_2 \ell_3 \ell_2' \ell_3') u_{\ell_2 j}^{(j)} u_{\ell_3 k}^{(k)} u_{\ell_2' j'}^{(j')} u_{\ell_3' k'}^{(k')}. \tag{5.43}$$

One can notice that the above is nothing but TD of linear kernel, $x_{jkj'k'}$. Thus we can successfully reformat TD as TD of kernel. This allows to apply TD to nonlinear kernels, too, as

$$K(x_{ijk}, x_{ij'k'}) = \exp\left\{-\frac{\sum_{i=1}^{N}\left(x_{ijk} - x_{ij'k'}\right)^2}{\sigma^2}\right\}$$
$$= \sum_{\ell_2=1}^{M}\sum_{\ell_3=1}^{K}\sum_{\ell_2'=1}^{M}\sum_{\ell_3'=1}^{K} G(\ell_2\ell_3\ell_2'\ell_3') u_{\ell_2 j}^{(j)} u_{\ell_3 k}^{(k)} u_{\ell_2' j'}^{(j')} u_{\ell_3' k'}^{(k')} \quad (5.44)$$

which is known as radial basis function (RBF) kernel or

$$K(x_{ijk}, x_{ij'k'}) = \left(\sum_{i=1}^{N} x_{ijk}x_{ij'k'} + 1\right)^d$$
$$= \sum_{\ell_2=1}^{M}\sum_{\ell_3=1}^{K}\sum_{\ell_2'=1}^{M}\sum_{\ell_3'=1}^{K} G(\ell_2\ell_3\ell_2'\ell_3') u_{\ell_2 j}^{(j)} u_{\ell_3 k}^{(k)} u_{\ell_2' j'}^{(j')} u_{\ell_3' k'}^{(k')}, \quad (5.45)$$

which is known as polynomial kernel.

To see if nonlinear KTD can capture nonlinear structure, we applied KTD with RBF kernel (Eq. 5.44) and polynomial kernel (Eq. 5.45) to so called "Swiss roll" structure (Fig. 5.12a and b) as follows. "Swiss roll' is formatted as a tensor, $x_{ijk} \in \mathbb{R}^{3\times M\times K}$ as

$$p_j = 2\sqrt{\frac{j}{M}} \quad (5.46)$$

$$x_{1jk} = p_j \cos(2\pi p_j) \quad (5.47)$$

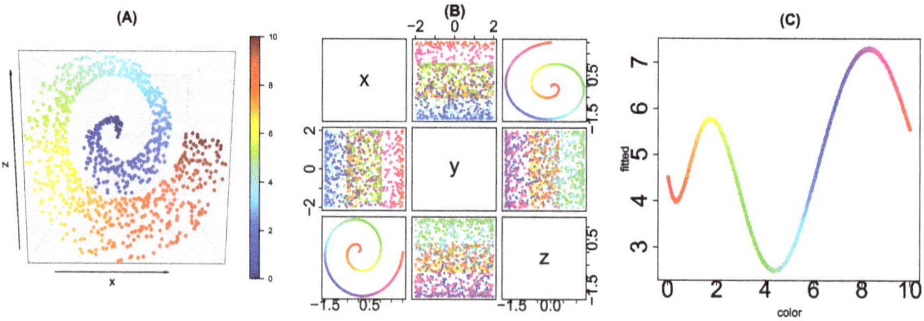

Fig. 5.12 "Swiss roll" structure. Color graduation shows positions along the surface, j. (**a**) Perspective view, (**b**) projections, and (**c**) results of linear regression. Vertical and horizontal axes are fitted values and distance along the surface, j, respectively. Correlation coefficient is 0.277

$$x_{2jk} = \epsilon \tag{5.48}$$

$$x_{3jk} = p_j \sin(2\pi p_j), \tag{5.49}$$

where $M = 10^3$, $K = 10$ and ϵ is the uniform random real number in [-2,2]. This means that x_{ijk} is a bundle of K "Swiss roll"s associated with distinct random numbers. To evaluate how much it can represent the distance along surface, j, in the linear dimensions, we perform the linear regression

$$j = \sum_{k=1}^{K} \sum_{i=1}^{3} a_i x_{ijk} + b. \tag{5.50}$$

The Pearson correlation coefficient between distance along the surface, j, and fitted value is as small as 0.277 (Fig. 5.12c). Thus, as expected, although there is a significant correlation between the distance along the surface, j, and the coordinates, the correlation is poor.

Now we apply KTD with RBF kernel (Eq. (5.44), $\sigma^2 = 10^3$) and that with polynomial kernel (Eq. (5.45), $d = 2$) to x_{ijk} (Figs. 5.13 and 5.14). Since the results of the linear regression using the first three singular value vectors,

$$j = \sum_{\ell_2=1}^{3} a_{\ell_2} u_{\ell_2 j}^{(j)} + b, \tag{5.51}$$

are excellent (Figs. 5.13c and 5.14c) and the Pearson's correlation coefficients between the distance along surface, j, and fitted values are as large as 0.896 and 0.786, respectively, KTDs with RBF kernel and polynomial kernel successfully unfold nonlinear structure in "Swiss roll" onto the linear coordinates.

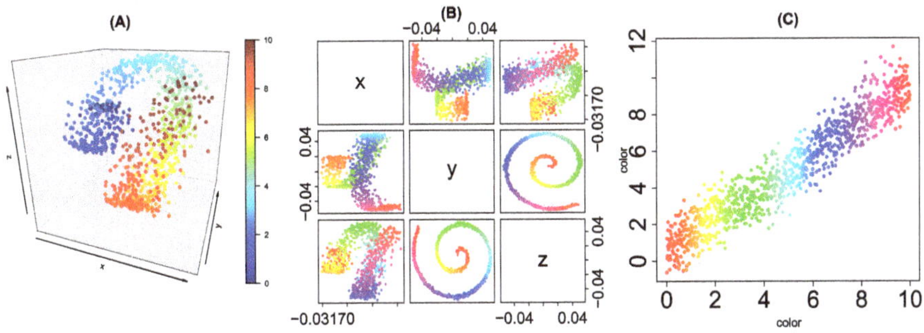

Fig. 5.13 KTD with RBF kernel applied to "Swiss roll" structure. Color graduation shows positions along the surface, j. (**a**) Perspective view, (**b**) projections, and (**c**) the results of linear regression. Vertical and horizontal axes are fitted values and distance along the surface, j, respectively. The correlation coefficient is 0.896

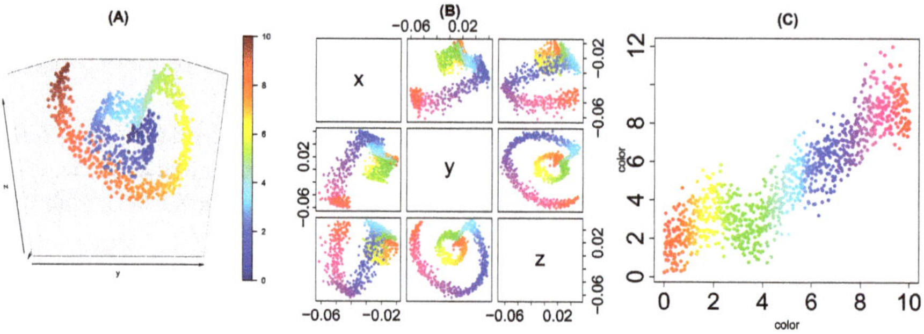

Fig. 5.14 KTD with polynomial kernel to "Swiss roll" structure. Color graduation shows positions along the surface, j. (**a**) Perspective view, (**b**) projections, and (**c**) the results of linear regression. Vertical and horizontal axes are fitted values and distance along the surface, j, respectively. Correlation coefficient is 0.786

Although we successfully implemented a nonlinear kernel into TD, there is an issue; feature index, i, was missing during the process to have kernel, $K(x_{ijk}, x_{ij'k'})$. For linear kernel, although missing singular value vectors attributed to feature, $u^{(i)}_{\ell_1\ell_2 i}$, can be retrieved as

$$u^{(i)}_{\ell_1\ell_2 i} = \sum_{j=1}^{M}\sum_{k=1}^{K} x_{ijk} u^{(j)}_{\ell_2 j} u^{(k)}_{\ell_3 k} \tag{5.52}$$

and feature (i) selection can be performed as usual, it is not the case for nonlinear kernels.

We prepare $x_{ijk} \in \mathbb{R}^{N \times M \times 3}$ as

$$p_j = 2\sqrt{\frac{j}{M}} \tag{5.53}$$

$$x_{ij1} = p_j \cos(2\pi p_j) \tag{5.54}$$

$$x_{ij2} = \epsilon \tag{5.55}$$

$$x_{ij3} = p_j \sin(2\pi p_j) \tag{5.56}$$

for $i \le N_1 \le N$ (associated with "Swiss roll' structure) and x_{ijk}s are uniform random numbers in $[-1, 1]$ for $i > N_1$ (not associated with "Swiss roll" structure). The task is to identify $i \le N_1$ associated with "Swiss roll" structure among N features. Then, we shall do as follows. After excluding one specific i, we perform KTD and evaluate the decrease of coincidence between $u^{(j)}_{\ell_2 j}$ and/or $u^{(k)}_{\ell_3 k}$ with class labels. As a result we can select top features ranked based upon largeness of decrease as a criterion. Although it is very time consuming process since we have to repeat whole computation every time we exclude one i, it works.

Suppose that we compute the correlation coefficient between js and the fitted values with Eq. (5.51) using $u^{(j)}_{\ell_2 j}$ given by Eq. (5.44) or (5.45) where $N_1 = 10$, $N = 1000$, and $M = 100$ excluding one i one by one. Then if we selected top N_1 is whose removal reduces absolute values of correlation coefficients more, the selected i are always identical to those $i \leq N_1$ (the results are not shown here, since it is useless if the correct is are always selected). Thus, we can make use of this strategy, i.e., selecting is whose removal decreases the coincidence with "Swiss roll" structure, to identify is associated with x_{ijk} having coincidence with distance along surface, j, i.e., those associated with "Swiss roll" structure.

5.10 Summary

In this chapter, we proposed feature section using TD, named TD-based unsupervised FE. TD-based unsupervised FE can outperform conventional supervised method when the number of samples is much less than the number of features and true classification is a complex function of apparent labeling. We also further extended the concept of tensor such that we can make use of TD-based unsupervised FE even when only matrices are given. As a bi-product, we come to be able to select features with mutual correlations even when conventional pairwise correlation analysis fails. Furthermore, we have proposed alternative ways to integrate multiple profiles when samples or features are shared among multiple profiles. We also extended our methods to kernel methods to deal with nonlinear structures. Nothing shown in this chapter are proven, but are only demonstrated by synthetic data set. Nonetheless, we will see that TD-based unsupervised FE can work very well when it is applied to real examples, i.e., the applications toward bioinformatics in the later part of this book.

References

1. Taguchi, Y.H.: Tensor decomposition-based unsupervised feature extraction applied to matrix products for multi-view data processing. PLoS One **12**(8), e0183933 (2017). https://doi.org/10.1371/journal.pone.0183933
2. Taguchi, Y.H., Turki, T.: Mathematical formulation and application of kernel tensor decomposition based unsupervised feature extraction. Knowl.-Based Syst. **217**, 106834 (2021). https://doi.org/10.1016/j.knosys.2021.106834. https://www.sciencedirect.com/science/article/pii/S0950705121000976
3. Taguchi, Y.h., Turki, T.: Tensor-decomposition-based unsupervised feature extraction in single-cell multiomics data analysis. Genes **12**(9), 1442 (2021). https://doi.org/10.3390/genes12091442. https://www.mdpi.com/2073-4425/12/9/1442
4. Taguchi, Y.h., Turki, T.: Novel feature selection method via kernel tensor decomposition for improved multi-omics data analysis. BMC Med. Genet. **15**(1), 37 (2022). https://doi.org/10.1186/s12920-022-01181-4

Part III
Applications to Bioinformatics

In this part, I explain the main purpose of this book, i.e., applications of PCA- or TD-based unsupervised FE to bioinformatics. Although I know that readers might not be very much interested in biology, because data analyzed are real numbers formatted as matrices and tensors, methods themselves should be able to be applied to other data sets than biology. Starting from the introduction of fundamental knowledge required in order to understand examples presented in this part, various applications of PCA- and TD-based unsupervised FE will be discussed.

Chapter 6
Applications of PCA-Based Unsupervised FE to Bioinformatics

I do not need any other reason to be born than being your friend.
Rikka Takarada, SSSS.GRIDMAN, Season 1, Episode 11

6.1 Introduction

PCA is an old technology. It has been invented more than a 100 years ago. Thus many might think that it is nothing but text book level matter (in other words, nothing new to be worthwhile being investigated might not exist anymore). Especially, because modernized many nonlinear methods have been proposed, people think that no linear methods can have any superiority toward these advanced technologies. In spite of such a general belief, PCA still might be an effective method. In this chapter, I propose to make use of PCA for unsupervised feature selection. The word "unsupervised feature selection" might sound like discrepancy. If so, please look at the following. I am sure that you can understand what I mean by this word and might start to think that PCA is still an effective method and is worthwhile investigating further.

6.2 Some Introduction to Genomic Science

Although it is unrealistic to fully explain fundamental genome biology required to understand the contents in this chapter, I will try to outline very basic points here to help readers to understand. Readers who would like to understand the contents more deeply should consult with fully mentioned text books, e.g., Gnome 4th Edition [7].

Y-h. Taguchi, *Unsupervised Feature Extraction Applied to Bioinformatics*, Unsupervised and Semi-Supervised Learning,
https://doi.org/10.1007/978-3-031-60982-4_6

6.2.1 Central Dogma

Although it has become a little bit old fashioned, the so-called central dogma still remains a gold standard in genome science. Central dogma says that any functional proteins are translated from mRNA (translation) which is transcribed from DNA (transcription), not vice versa. Thus, fundamentally, study of DNA is an essential part of genome science. DNA is an abbreviation of deoxyribonucleic acid, which is a long chain of sequence composed of only four kinds of molecules: cytosine, guanine, adenine, and thymine, which are often abbreviated as using the first letters of words: C, G, A, and T. DNA is located in nuclei of cells, which functions as information transporter between generations as well as acts as information resources from which cellar activities are decided. The way by which DNA stores information is digital and is in some sense similar to digital information in modernized computer or information science; it is fitted to be investigated with information science. This sympathy between DNA and information science is a key factor from which bioinformatics was born.

DNA takes the form of double helix, which is generated by the pairing two complementary DNA sequences. Among four molecules that form DNA, T, C, G, and A, A binds to T and G binds to C. Thus two DNA sequences that form double helix store the same information, like positive films and negative films in photography.

The objects fitted to be studied by information science in genome biology are not only DNA but also mRNA and protein as well. mRNA is an abbreviation of messenger RNA, and RNA is an abbreviation of ribonucleic acid. RNA is a partial copy of lengthy DNA sequence whose total length is as many as 3 billion (3×10^9) base pairs (in the case of human being). Only difference between DNA and RNA other than length is that uracil (U) is used instead of T in RNA; thus, four molecules that form RNA are A, U, G, and C. Protein is a long polymer composed of 20 amino acids, each of which is coded using triplet of nucleic acids in RNA and DNA sequence; these triplets are called as codon. Because the number of total codons is as many as $4^3 = 64$, while the total number of amino acids used for generating proteins is as small as 20, multiple codons code the same amino acid. Some of codons are also used as terminator that marks the point where transcription ends.

Proteins translated from mRNA form complex structure to function as blocks of organisms, enzyme that accelerates the chemical reactions and vehicle that transport something within body. The three dimensional (tertiary) structures are believed to be dependent upon only amino acid sequence, although the strict relationship between amino acid sequence and protein tertiary structure has not yet been known.

6.2.2 Regulation of Transcription

The mostly focused process in this and the next chapter is transcription [55]. There are multiple reasons why transcription is mostly focused:

- Because of high throughput technologies including microarray and high throughput sequencing (HTS), RNA and DNA have become the easiest parts to be measured.
- In contrast to the proteins whose tertiary structure is heavily related to the functions, the functions of RNA and DNA are primarily related to sequence only. Thus, we do not need to perform downstream analysis about the structures about RNA and DNA so much.
- Noncoding RNAs which do not have any proteins translated from them have many functions to regulate transcription itself.

Thus, DNA sequence, RNA sequence, and the amount of transcript (i.e., RNA) are mainly measured and studied extensively. Thus, the application of PCA-based unsupervised FE is much easier to be applied to DNA and RNA sequence as well as the mount of transcripts.

6.2.3 The Technologies to Measure the Amount of Transcript

Measuring the amount of RNA is essentially the count of the number of RNAs transcribed from DNA. There are two major methods of high throughput measurements of the amount of RNA [54]: one is microarray and another is high throughput sequencing. Microarray is the technology that prepares numerous probes that specifically bind to individual RNAs; the amount of RNA that binds to probe is measured by photo emission from the fluorescent molecules that decorate RNA. HTS employs more direct strategy to measure RNA. HTS tries to count the number of RNAs by sequencing each RNA. One point that must be taken into account is that HTS can measure only fragments of individual RNAs, not a full length one. Thus, after counting the number of fragments, each fragment must be annotated in the reference of external resources (e.g., as a part of known gene). Although there are numerous ways to sequence RNA, we are not willing to explain the details of sequencing technology. Only essential point to be explained to understand the contents in the book is that the output from microarray is real number, while the output from HTS is integer. In addition to this, although output from HTS is guaranteed to be proportional to the amount of transcript, that from microarray is not. Output of microarray is often translated to logarithmic values because logarithmic values more likely obey Gaussian distribution, which is believed to be a natural outcome.

6.2.4 Various Factors That Regulate the Amount of Transcript

Although it is not fully understood yet, various factors regulate transcription [54, 55]. In this subsection, I try to explain how some factors control gene expression via regulation of transcription. Most important known factors that regulate transcription are undoubtedly transcription factors (TFs). TFs are usually composed of proteins that bind to DNA region known as prompter. TFs primarily control the transcription initiation. One possible problem from the point of data analysis is that TFs are proteins. As mentioned in the above, the most easy factor to be measured is not protein but RNA. Because of that, TFs are not frequently to be targets of investigations in this and the next chapter. The second major factor that regulates transcription is methylation. If promoter region is methylated, TF is forbidden to bind to the region. As a result, the transcription of RNAs whose promoter is methylated is disturbed. In addition to this, DNA methylation is heritated during DNA replication that takes place in cell duplication. Thus DNA methylation can affect transcription for longer period than other factors. The additional factor that can regulate transcription in the posttranscription process is microRNA (miRNA). miRNAs are RNAs not translated to proteins but have functions; their functions are to destroy RNAs before translation takes place. Each miRNA can identify each target mRNA by the complementary binding to 8 bp length seed region within 3' untranslated region (UTR) of mRNA (the term 3' is used to identify which edge of mRNA is targeted). UTR is region of mRNA that is not translated to protein. Although there are more factors that can affect transcription, DNA methylation and miRNA expression are mostly analyzed factors in this and the next chapters.

6.2.5 Other Factors to be Considered

Some other factors that can affect transcription will be discussed time to time. Single-nucleotide polymorphism (SNP) is the replacement of single nucleotide in DNA. Although there are many reasons that cause SNP, it is primarily caused by miss-duplication of DNA during cell division. Until now, although SNP is primarily considered to alter amino acid sequence of protein, it can also affect transcription. For example, SNP in promoter region or 3' UTR region can affect the binding between TF and DNA or that between miRNA and mRNA, thus affecting the transcription. Although it is not extensively investigated, SNP and transcription are mutually interacted.

Histone modification is another factor that regulates transcription. Histone is protein around which DNA winds. This process is necessary in order that long DNA chain does not get tangled up. DNA that tightly winds around histone cannot be transcribed because no TFs can bind to it. Because histone modification that means that small molecules bind to histone tail can affect how tightly DNA can wind round histone, histone modification is additional factor that can affect transcription.

Finally, although it is not so frequently analyzed, proteome and metabolome can be treated. Proteome is a set of proteins translated. Their amount can be causes or outcome of transcription. Metabolome is a set of compounds generated as a consequence of chemical reaction to which some proteins take place as enzyme. Thus, metabolome is, indirectly, affected by transcription.

Integrated analysis of these factors is often annotated as multiomics data analysis, because it integrates genome, transcriptome, proteome, metabolome, i.e., several "-ome" data sets.

6.3 Biomarker Identification

Although PCA-based unsupervised FE is applied to various bioinformatics topics, we would like to start from identification of biomarker; biomarker is a kind of disease marker that can tell you about your healthy status. When you can take health check, various factors are measured from blood and urine. You will be warned if some of the measured components have nonstandard values. Identification of biomarkers is very important to facilitate diagnosis, as medical knowledge is not required. In this sense, most readers are not considered to be medical professionals, so identifying biomarkers is likely to be the most understandable to readers. Thus, beginning to explain the identification of biomarkers that require less medical knowledge, readers who are not medical experts may not be as stressful.

6.3.1 Biomarker Identification Using Circulating miRNA

Circulating miRNA means miRNAs that circulate in the body. For example, blood miRNA is a typical circulating miRNA. The reason why biomarkers are searched within circulating miRNA is as follows. At first, obtaining circulating miRNA is less painful than getting tissue miRNAs that are expected to be more likely directly related to diseases than blood miRNA. In order to obtain tissue miRNA, one needs surgery or needle biopsy that inevitably injures patients' body and results in some pain. In order to get blood, there need to be also needle but with less pain. Thus, if we can find useful disease biomarker in the blood, it is very convenient. On the other hand, identification of disease biomarker using circulating miRNA is more challenging than that using tissue miRNA, because circulating miRNA reflects whole body state that is not always related to specific disease. Thus, in data science point of views, identification of disease biomarker using circulating miRNA might be challenging and interesting.

6.3.1.1 Biomarker Identification Using Serum miRNA

As the first example, we consider serum miRNAs [58]. Serum is the liquid component of blood that does not include either blood vessels or clotting factors. Serum is also supposed to contain all proteins (other than those contributing to blood clotting), electrolytes, antibodies, antigens, hormones, and any exogenous substances. Thus, it is suitable to search biomarker in it.

Table 6.1 List of serum miRNAs samples

k	Group	Number of samples (M_k)
0	Controls	70
1	Lung cancer	32
2	Prostate cancer	23
3	Melanoma	35
4	Wilms tumors	5
5	Ovarian cancer	15
6	Gastric cancers	13
7	Pancreatic ductal adenocarcinoma	45
8	Other pancr. tumors and diseases	48
9	Pancreatitis	38
10	Chronic Obstructive Pulmonary Disease	24
11	Periodontitis	18
12	Sarcoidosis	45
13	Acute myocardial infarction	20
14	Multiple sclerosis	23

Here we make use of serum miRNA data set that is publically available [24]. It includes serum miRNAs measurements for 14 diseases and healthy controls (Table 6.1). Although it does not always include enough number of samples in individual diseases in the recent standards (because it was 6 years ago when I performed this study), because I believe that it is a good intuitive example from which the explanation starts, I introduce the application of PCA-based unsupervised FE to them. The expression of miRNAs was measured by microarray technology. It includes only 863 human miRNAs because it is an old study. Nowadays, more number of human miRNAs are identified. For example, the most recent miRBase [25] (http://www.mirbase.org/index.shtml, version 22), which is a primary miRNA database that is periodically updated, includes as many as 1917 pre-miRNAs, each of which usually includes two complementary miRNAs (thus, the most updated version includes c.a. 4000 miRNAs). The full data set of used gene expression profiles is available from Gene Expression Omnibus (GEO) [44] with GEO ID, GSE31568.

Although I am not interested in describing how to retrieve gene expression from GEO, I briefly introduce about it. GEO includes multiple format of gene expression: typically processed (normalized) one and raw one. The former is the one after the correction assuming some hypothesis, e.g., background correction. As mentioned in the above, microarray measures gene expression with light emission. Thus, measurement of gene expression by microarray is quite indirect. In order to compensate the errors and biases introduced by this technology, gene expression is often corrected based upon some assumption. Nevertheless, as can be seen in the below, PCA-based unsupervised FE can make use of raw (unprocessed) data quite successfully (throughout the application of PCA- and TD-based unsupervised FE, it will not be very usual to use processed data).

Then, also in this case, I downloaded raw data set, GSE31568_raw.txt.gz.[1] This file includes all miRNA expressions listed in Table 6.1 as one file. The first row excluding header line that includes GEO ID of individual samples annotates distinction between controls and diseases. Then we generated gene expression profiles as a form of matrix, $x_{ij}^{(k)} \in \mathbb{R}^{863 \times (M_0+M_k)}$, that represents ith miRNA expression of jth sample where $M_0(=70)$ and M_k, $1 \leq k \leq 14$ are the number of controls and the kth disease samples, respectively.

In order to apply PCA-based unsupervised FE to $x_{ij}^{(k)}$, PCA is applied to it such that PC scores, $\boldsymbol{u}_\ell^{(k)} \in \mathbb{R}^{863}$, are attributed to miRNAs and PC loading, $\boldsymbol{v}_\ell^{(k)} \in \mathbb{R}^{M_0+M_k}$, are attributed to samples. Unfortunately, we cannot identify PC loadings associated with distinction between healthy controls and patients. Then, empirically, we employ the following strategy to select miRNAs used for biomarkers. At first, compute length, r_i, of PC score as

$$r_i = \sqrt{\sum_{\ell=1}^{2} u_{\ell i}^2}. \tag{6.1}$$

Then top-ranked 10 miRNAs with larger r_i are selected. Table 6.2 shows the list of miRNAs selected for each of 14 pairs composed of controls and patients of one of 14 diseases. Interestingly, miRNAs selected in each of 14 pairs are heavily overlapped. In spite of that 140 miRNAs are selected in total, there are only twelve miRNAs. Nine out of twelve miRNAs are selected in all of fourteen pairs of control and patients.

Although it is interesting that selected miRNAs are highly overlapped between fourteen diseases, because no PC loading exhibits distinction between controls and diseases, it might not be related to biology at all. In order to see this, we try to make use of these miRNAs selected in order to discriminate between controls and diseases. If they can, they are considered to be disease biomarkers.

In order that, we employ the following strategy. Instead of full-size miRNA expression profile matrix, $x_{ij}^{(k)} \in \mathbb{R}^{863 \times (M_0+M_k)}$, we prepare reduced miRNA expression profile, $x_{ij}^{(k)\prime} \in \mathbb{R}^{10 \times (M_0+M_k)}$, that includes selected 10 miRNAs only. Then PCA is applied to $x_{ij}^{(k)\prime}$ again in order to get PC loading, $\boldsymbol{v}_\ell^{(k)\prime} \in \mathbb{R}^{M_0+M_k}$, attributed to samples. Then the first L PC loading, $\boldsymbol{v}_\ell^{(k)\prime}, \ell \leq L$, are used to linear discriminant analysis (LDA) in order to discriminate between controls and diseases.

Here LDA is a classical method to discriminate between multiple classes using linear algebra. In order to perform LDA, we need to compute several variables. Then new variable $y_j \in \mathbb{R}^{M_0+M_k}$ is defined as

[1] ftp://ftp.ncbi.nlm.nih.gov/geo/series/GSE31nnn/GSE31568/suppl/GSE31568_raw.txt.gz.

Table 6.2 Twelve miRNAs selected by applying PCA-based unsupervised FE to each of pairs of controls and the kth disease in Table 6.1. o: selected, ×: not selected. Suffix suggested that coincident with most recent miRBase v. 22

k (diseases)	1	2	3	4	5	6	7	8	9	10	11	12	13	14	Suffix
miR-425	o	o	o	o	o	o	o	o	o	o	o	o	o	o	5p
miR-191	o	o	o	o	o	o	o	o	o	o	o	o	o	×	5p
miR-185	o	o	o	o	o	o	o	o	o	o	o	o	o	o	5p
miR-140-3p	o	o	o	o	o	o	o	o	o	o	o	o	o	o	3p
miR-15b	o	o	o	o	o	o	o	o	o	o	o	o	o	o	5p
miR-16	o	o	o	o	o	o	o	o	o	o	o	o	o	o	5p
miR-320a	o	o	o	o	o	o	o	o	o	o	o	o	o	o	
miR-486-5p	o	o	o	o	o	o	o	o	o	o	o	o	o	o	5p
miR-92a	o	o	o	o	o	o	o	o	o	o	o	o	o	o	3p
miR-19b	o	×	o	×	×	×	o	×	×	×	×	×	×	o	3p
miR-106b	×	o	×	o	o	o	×	o	o	o	o	×	o	o	5p
miR-30d	×	×	×	×	×	×	×	×	×	×	×	o	×	×	5p

$$y_j = \boldsymbol{a} \times_\ell \left(\boldsymbol{v}_j^{(k)'} - \left\langle \boldsymbol{v}_j^{(k)'} \right\rangle_j \right) = \boldsymbol{a} \times_\ell \delta \boldsymbol{v}_j^{(k)'} \tag{6.2}$$

with deciding $\boldsymbol{a} \in \mathbb{R}^L$ such that y_j discriminates controls and diseases where

$$\boldsymbol{v}_j^{(k)'} = \left(v^{(k)'}{}_{1j}, v^{(k)'}{}_{2j}, \cdots, v^{(k)'}{}_{Lj} \right). \tag{6.3}$$

In order to decide $\boldsymbol{a}$, we need to compute in-class centroid as

$$\langle y_j \rangle_j^{(k)} = \begin{cases} \frac{1}{M_k} \sum_{j=1}^{M_k} y_{M_0+j} & k \neq 0 \\ \frac{1}{M_0} \sum_{j=1}^{M_0} y_j & k = 0 \end{cases}, \tag{6.4}$$

which is also written as

$$\langle y_j \rangle_j^{(k)} = \begin{cases} \boldsymbol{a} \times_\ell \left\langle \delta \boldsymbol{v}_j^{(k)'} \right\rangle_j^{(k)} & k \neq 0 \\ \boldsymbol{a} \times_\ell \left\langle \delta \boldsymbol{v}_j^{(0)'} \right\rangle_j^{(0)} & k = 0 \end{cases}. \tag{6.5}$$

The task is maximizing the difference between in-class centroid

$$\begin{aligned} \Delta^{(k)} &= \left(\langle y_j \rangle_j^{(k)} - \langle y_j \rangle_j^{(0)} \right)^2 & (6.6) \\ &= \boldsymbol{a} \times_\ell \left[\left\{ \left\langle \delta \boldsymbol{v}_j^{(k)'} \right\rangle_j^{(k)} - \left\langle \delta \boldsymbol{v}_j^{(0)'} \right\rangle_j^{(0)} \right\} \times^0 \left\{ \left\langle \delta \boldsymbol{v}_j^{(k)'} \right\rangle_j^{(k)} - \left\langle \delta \boldsymbol{v}_j^{(0)'} \right\rangle_j^{(0)} \right\} \right] \times_{\ell'} \boldsymbol{a} & (6.7) \\ &\equiv \boldsymbol{a} \times_\ell \Sigma_B \times_{\ell'} \boldsymbol{a} & (6.8) \end{aligned}$$

relative to summation of in-class variances

$$\Delta^{(0,k)} = \left\langle \left(y_j - \langle y_{j'} \rangle_{j'}^{(k)} \right)^2 \right\rangle_j^{(k)} + \left\langle \left(y_j - \langle y_{j'} \rangle_{j'}^{(0)} \right)^2 \right\rangle_j^{(0)} \tag{6.9}$$

$$= \boldsymbol{a} \times_\ell \left[\left\{ \delta \boldsymbol{v}_j^{(k)\prime} - \left\langle \delta \boldsymbol{v}_j^{(k)\prime} \right\rangle_j^{(k)} \right\} \times^0 \left\{ \delta \boldsymbol{v}_j^{(k)\prime} - \left\langle \delta \boldsymbol{v}_j^{(k)\prime} \right\rangle_j^{(k)} \right\} \right] \times_{\ell'} \boldsymbol{a} \tag{6.10}$$

$$+ \boldsymbol{a} \times_\ell \left[\left\{ \delta \boldsymbol{v}_j^{(0)\prime} - \left\langle \delta \boldsymbol{v}_j^{(0)\prime} \right\rangle_j^{(0)} \right\} \times^0 \left\{ \delta \boldsymbol{v}_j^{(0)\prime} - \left\langle \delta \boldsymbol{v}_j^{(0)\prime} \right\rangle_j^{(0)} \right\} \right] \times_{\ell'} \boldsymbol{a} \tag{6.11}$$

$$\equiv \boldsymbol{a} \times_\ell \Sigma_W \times_{\ell'} \boldsymbol{a}. \tag{6.12}$$

It is known that this task can be performed by maximizing

$$L(\boldsymbol{a}, \lambda) = \Delta^{(k)} - \lambda \left(\Delta^{(0,k)} - 1 \right) \tag{6.13}$$

with respect to $\boldsymbol{a}$ and λ, which is also known as method of Lagrange multipliers. It is equivalent to maximize $\Delta^{(k)}$ with keeping $\Delta^{(0,k)} = 1$. In order to find $\boldsymbol{a}$ that maximizes $L(\boldsymbol{a}, \lambda)$, we require that derivatives of $L(\boldsymbol{a}, \lambda)$ with respect to $\boldsymbol{a}$ must be zero.

$$\frac{\partial L(\boldsymbol{a}, \lambda)}{\partial \boldsymbol{a}} = 2 \left(\Sigma_B \times_{\ell'} \boldsymbol{a} - \lambda \Sigma_W \times_{\ell'} \boldsymbol{a} \right) = 0. \tag{6.14}$$

This is equivalent to eigen value problem

$$\Sigma_W^{-1} \Sigma_B \boldsymbol{a} = \lambda \boldsymbol{a}, \tag{6.15}$$

and $\boldsymbol{a}$ can be obtained as the first eigen vector. LDA can be performed to find y_0 such that the distinction between two sets, $y_j < y_0$ and $y_j > y_0$, is maximally coincident with distinction between controls and diseases.

Table 6.3 shows the performance measured using Leave One Out Cross Validation (LOOCV). In LOOCV, one of $M_0 + M_k$ samples is removed from computing LDA, and y_j for removed one is computed by Eq. (6.4) using obtained $\boldsymbol{a}$. Then, it is discriminated if $y_j > y_0$ or $y_j < y_0$. This procedure is repeated for all $M_0 + M_k$ samples.

The reason why LOOCV is required is as follows. If LDA is performed considering all samples, it cannot be said to be true prediction since we know the classification of the sample whose classification is tried to predict. In real situation, biomarker must predict sample classification without knowing the answer. Thus LDA should be performed excluding a sample of which classification is tried to be predicted.

In Table 6.3, there are many performance measures. Here I briefly explain them based upon confusion matrix (Table 4.8) as

Table 6.3 Various performance achieved by LDA (with LOOCV) using PC loading computed by 10 miRNAs selected by PCA-based unsupervised FE

k	Group	L	Accuracy	Specificity	Sensitivity	Precision
1	Lung cancer	5	0.784	0.800	0.750	0.632
2	Prostate cancer	5	0.806	0.800	0.826	0.576
3	Melanoma	10	0.867	0.857	0.886	0.756
4	Wilms tumors	7	0.867	0.886	0.600	0.273
5	Ovarian cancer	6	0.800	0.786	0.867	0.464
6	Gastric cancers	9	0.806	0.800	0.826	0.576
7	Pancreatic ductal adenocarcinoma	2	0.765	0.743	0.800	0.667
8	Other pancr. tumors and diseases	7	0.814	0.771	0.875	0.724
9	Pancreatitis	8	0.933	0.786	0.921	0.700
10	Chronic obstructive pulmonary disease	2	0.713	0.671	0.833	0.465
11	Periodontitis	10	0.807	0.814	0.778	0.519
12	Sarcoidosis	10	0.835	0.800	0.889	0.741
13	Acute myocardial infarction	7	0.789	0.900	0.757	0.964
14	Multiple sclerosis	10	0.892	0.871	0.957	0.710

$$\text{Accuracy} = \frac{TP + TN}{TP + TN + FP + FN}, \tag{6.16}$$

$$\text{Specificity} = \frac{TN}{TN + FP}, \tag{6.17}$$

$$\text{Sencitivity} = \frac{TP}{TP + FN}, \tag{6.18}$$

$$\text{Precision} = \frac{TP}{TP + FP}. \tag{6.19}$$

Accuracy measures the ratio of the number of correctly predicted samples to the number of total samples. Specificity measures the ratio of the number of correctly predicted control samples to the number of control samples. Sensitivity measures the ratio of the number of correctly predicted disease samples to the number of disease samples. Precision measures the ratio of the number of correctly predicted disease samples to the number of samples predicted as diseases.

Generally, the performance in Table 6.3 is quite well if we consider the performance is achieved in the fully unsupervised manner. This suggests that PCA-based unsupervised FE can be useful even when it is applied to the real applications.

One might wonder if these performances must be evaluated based upon the comparisons with other methods. Before starting to discuss this point, I would like to point out one important point specific to this application. In this application, selecting miRNAs as small as possible is important because measuring more miRNAs costs more in practical applications. In addition to this, the selected miRNAs should not be dependent upon sets of samples considered. If the best selected miRNAs vary from samples to samples, it becomes useless. From this point of view, PCA-based unsupervised FE is superior to other supervised methods. As mentioned in the above, we can select miRNAs using all samples even including samples whose classification is tried to predict because we did not use classification of samples at all. In the real application, we can do as follows. Suppose we have both samples with known classification and those without known classification. Then apply PCA to all samples including both. Select top 10 miRNAs using distance computed the first and the second PC scores shown in Eq. (6.1). PC loading is recomputed using selected 10 miRNAs. These all processes can be done without knowing sample classifications at all.

Actually, Keller et al. [24] failed to select a set of fixed 10 miRNAs because they needed to exclude samples to be predicted. This results in a distinct set of miRNAs selected dependent upon the samples excluded. Thus, before comparing performance by other methods with those by PCA-based unsupervised FE, primarily we have to know if other methods can select stable feature selection without strong sample dependency.

In order to evaluate stability, we define the stability test as follows:

1. Select randomly 90 % of the samples within control M_0 samples and disease M_k samples, respectively.
2. Apply PCA to $x_{ij} \in \mathbb{R}^{N \times 0.9(M_0+M_k)}$.
3. Select top 10 miRNAs using r_i defined in Eq. (6.1).
4. Repeat the process over independent 100 trials and count miRNAs selected in all 100 trials.

When applying this stability test to PCA-based unsupervised FE, although there are 140 miRNAs selected for pairwise 14 discrimination where 10 miRNAs are selected, 129 miRNAs are always selected when the above stability test is applied to PCA-based unsupervised FE. Thus, feature selection with PCA-based unsupervised FE is quite stable.

Next we apply stability test to *t*-test which Keller et al. [24] employed. In this case, in step 3, P values computed by *t*-test are used for selecting miRNAs instead of r_i. Then only 40 miRNAs among 140 selected miRNAs are always selected. This means that *t*-test is quite inferior to PCA-based unsupervised FE from the point of stability.

In order to confirm the superiority of PCA-based unsupervised FE toward other supervised methods from the point of stability, we also apply stability test to significance analysis of microarrays (SAM) [65], gene selection based on a mixture of marginal distributions (gsMMD) [35], and ensemble recursive feature elimination (RFE) [1]. The performance of these advanced supervised methods from the point of stability is quite disappointing. Among 140 miRNAs selected, only 30, 5, 1, 1, 0 miRNAs are always selected when stability test is applied to SAM, up- and downregulation by gsMMD, RFE, ensemble RFE. It is quite obvious that more advanced methods proposed to achieve better discrimination are inferior to PCA-based unsupervised FE from the point of stability.

We cannot emphasize too much the importance of stability of feature selection here, although it is generally overlooked. As one can see in the below, PCA-based unsupervised FE is always outstanding over the conventional supervised methods from this point of view, stability.

In order to clarify if this superiority of PCA-based unsupervised FE is because of its unsupervised nature, we try here additional unsupervised method, unsupervised feature filtering (UFF) [67]. UFF is SVD based unsupervised method. Because SVD is in some sense equivalent to PCA as mentioned in the earlier part of this book, UFF has similar theoretical base to that of PCA-based unsupervised FE. UFF makes use of entropy computed by SVD. Entropy H is defined as

$$\rho_i = \frac{\lambda_i^2}{\sum_{i=1}^N \lambda_i^2} \tag{6.20}$$

$$H = -\frac{1}{\log N} \sum_{i=1}^N \rho_i \log \rho_i, \tag{6.21}$$

where N is the number of features (in this example, the number of miRNAs). λ_i is singular value when SVD is applied to x_{ij}. H represents how complicated structure x_{ij} has. When λ_i is constant, i.e., there are no structures at all, $\rho_i = \frac{1}{N}$ and $H = 1$. On the other hand, when $\lambda_i \neq 0$ only for one specific i, because $\rho_i = 1$ and $\rho_{i'} = 0, i' \neq i$, $H = 0$. In UFF we compute H without ith feature as

$$H_i = -\frac{1}{\log N} \sum_{i' \neq i} \rho_{i'} \log \rho_{i'}. \tag{6.22}$$

Then we have selected top 10 miRNAs having larger $\Delta H_i = H - H_i$. Interestingly, stability test applied to UFF results in 111 always selected miRNAs among 140 miRNAs. This number is comparative with 129 miRNAs achieved by PCA-based unsupervised FE. Thus, the reason why PCA-based unsupervised FE outperformed other conventional supervised methods from the point of stability is likely because of unsupervised nature.

Finally, we discuss the difference between two unsupervised methods: PCA-based unsupervised FE and UFF. In UFF, SVD must be repeated as many as times equal to the number of features, which can often become 10^4 in the case that mRNAs are considered as a feature. On the other hand, PCA-based unsupervised FE requires PCA only once. Thus computationally, UFF is far more challenging than PCA-based unsupervised FE. Thus as far as these two methods achieve competitively, there are no needs to employ UFF than PCA-based unsupervised FE.

Although the readers might be primarily interested in statistical methods themselves, not in biology, I briefly explain how we can evaluate the outcome also using domain knowledge, i.e., the knowledge is outside the statistical analysis and only in the biology. It is also important that the outcome driven from statistical methods is coincident with domain knowledge from the biological point of views because the coincidence with domain knowledge supports the reliability of statistical methods employed.

Although it is not mathematical, the so-called literature search is a powerful method. Simply searching database for the coincidence, one can easily get evidences that support outcome. Among the diseases listed in Table 6.1 there are multiple cancers included (lung cancer, prostate cancer, melanoma, Wilms tumor, ovarian cancer, gastric cancer, pancreatic ductal adenocarcinoma). Thus, it is not a bad idea to seek database with the words, e.g., "cancer" and the name of specific miRNAs. The most useful database for this purpose is pubmed[2] that corrects titles and abstract of the papers published in major biological journals. With the search of "cancer" and "miR-425," the readers can easily find that the miR-425 is known to be oncogenic, i.e., expressive in cancer. Thus, inclusion of miR-425 into one of biomarkers for diseases including many cancers is reasonable.

On the other hand, it is not so straightforward. Since we have employed PCA and LDA that are linear methods in order to select and construct biomarker, we can easily evaluate if the expression of miR-425 contributes positively or negatively to identify disease samples besides normal control.

From Eq. (2.21),

$$\boldsymbol{v}_\ell^{(k)'} = \frac{1}{\lambda'_\ell} X^{(k)'^T} \boldsymbol{u}_\ell^{(k)'}, \tag{6.23}$$

where λ'_ℓ is the singular value which is obtained by square root of the ℓth eigen value computed by PCA. $X^{(k)'}$ is the matrix whose component is $x_{ij}^{(k)'} \in \mathbb{R}^{10\times(M_0+M_k)}$, and $\boldsymbol{u}_\ell^{(k)'} \in \mathbb{R}^{10}$ is PC score computed by applying PCA to X'.

$$v_{\ell j}^{(k)'} = \frac{1}{\lambda'_\ell} \boldsymbol{x}_j^{(k)'} \times_i \boldsymbol{u}_\ell^{(k)'}, \tag{6.24}$$

where $\boldsymbol{x}_j^{(k)'} = \left(x_{1j}^{(k)'}, \cdots, x_{10j}^{(k)'}\right)$.

$$y_j = \boldsymbol{a} \times_\ell \left\{ \left(\boldsymbol{x}_j^{(k)'} - \left\langle \boldsymbol{x}_j^{(k)'} \right\rangle_j \right) \times_i U^{(k)'} \right\}, \tag{6.25}$$

where

$$U^{(k)'} = \left(\frac{\boldsymbol{u}_1^{(k)'}}{\lambda'_1}, \cdots, \frac{\boldsymbol{u}_L^{(k)'}}{\lambda'_L} \right) \in \mathbb{R}^{10\times L}. \tag{6.26}$$

Then we can compute the contribution from ith miRNA to y_j as

[2] https://www.ncbi.nlm.nih.gov/pubmed/.

$$y_{ij} = \left(\boldsymbol{a} \times_\ell \boldsymbol{u}_i^{(k)\prime}\right) \cdot \left(x_{ij}^{(k)\prime} - \left\langle x_{ij}^{(k)\prime} \right\rangle_j\right), \tag{6.27}$$

where

$$\boldsymbol{u}_i^{(k)\prime} = \left(\frac{u_{1i}^{(k)\prime}}{\lambda_1'}, \cdots, \frac{u_{Li}^{(k)\prime}}{\lambda_L'}\right) \in \mathbb{R}^L. \tag{6.28}$$

When $y_j > y_0$ corresponds to disease, i with $\boldsymbol{a} \times_\ell \boldsymbol{u}_i^{(k)\prime} > 0$ is considered to contribute to disease positively because upregulation of ith miRNA in jth sample enhances the tendency that the jth sample is identified as disease sample.

Figure 6.1 shows $\boldsymbol{a} \times_\ell \boldsymbol{u}_i^{(k)\prime}$ whose signs are assigned such that upregulation of ith miRNA in jth sample enhances the tendency that the jth sample is identified as disease sample. In contrast to expectation, miR-425 identified as oncogenic has mainly negative values. Thus, although inclusion of miR-425 as disease biomarker is reasonable, the effect is opposite to the expectation. This suggests that consulting to domain knowledge is very useful to validate the outcome from statistical analysis. This observation can be a start point why serum miRNA can have opposite sign to that in tissues which really contributes to diseases. Although we are not willing to discuss this point comprehensively, after further literature search, miR-486 is tumor suppressor, and miR-92a and miR-106b are oncogenic.

I would like to emphasize that the present strategy that relates $\boldsymbol{a} \times_\ell \boldsymbol{u}_i^{(k)\prime}$ to the outcome is helpful to interpret the functions of individual features. Thus, it should be employed in any other data science research.

Another strategy that validates outcome obtained statistically is more biology oriented in the sense that it makes use of biological knowledge fully. As mentioned in the above, miRNAs have their own targets whose numbers range from tens to hundreds. Because targeted mRNAs have their own functions, miRNAs can be validated along the biological concepts if their target mRNAs functions are evaluated. It is called enrichment analysis. Suppose that in total there are N genes among which N_1 genes' mRNAs are targeted by a specific miRNA. On the other hand, suppose that there is a set of N_2 genes that share some specific function. In this situation, suppose that there are $N_{12}(\leq N_1, N_2)$ genes that are not only targeted by the considered miRNAs but also have the specific function. Then Fisher's exact test checks if the number of overlaps is more than that of those by accident or not. In order to perform Fisher's exact test, we need to make the table that represents this situation. Assuming that whether an mRNA is targeted by the miRNA or not is not related to whether the mRNA has the function or not, Fisher [14] found that the situation shown in Table 6.4 occurs with the probability

$$P(N_{12}) = \frac{(N - N_1)!N_1!(N - N_2)!N_2!}{(N - N_1 - N_2 + N_{12})!(N_2 - N_{12})!(N_1 - N_{12})!N_{12}!}. \tag{6.29}$$

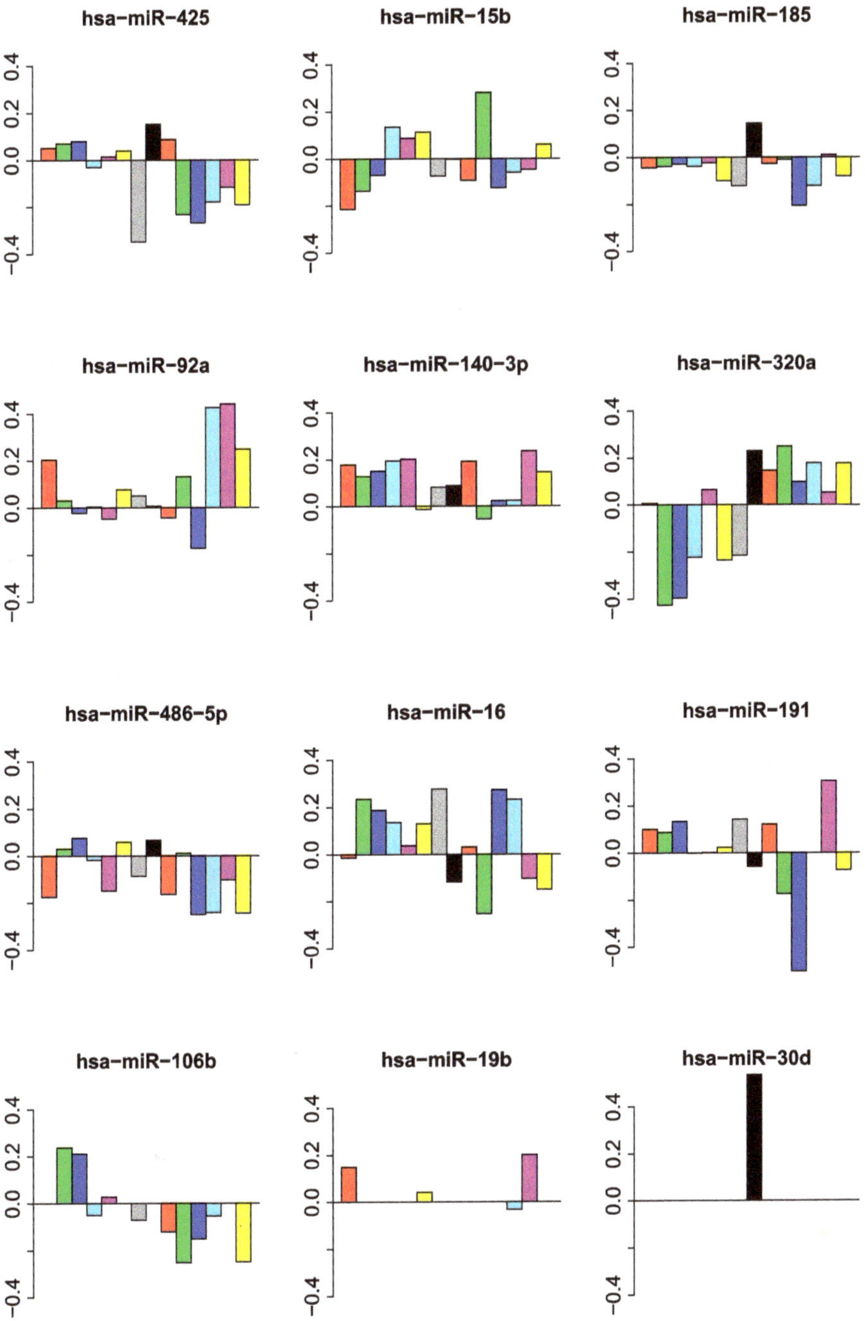

Fig. 6.1 The contribution of the ith miRNA in jth sample toward each disease, $\boldsymbol{a} \times_\ell \boldsymbol{u}_i^{(k)\prime}$, whose sign is assigned such that upregulation contributes to identification that jth sample is identified as disease. From left to right, diseases are lung cancer (red), other pancreatic tumors and diseases (green), pancreatitis (blue), ovarian cancer (cyan), COPD (pink), ductal pancreatic cancer (yellow), gastric cancer (gray), sarcoidosis (black), prostate cancer (red), acute myocardial infarction (green), periodontitis (blue), multiple sclerosis (cyan), melanoma (pink), and Wilm's tumor (yellow)

Table 6.4 Various performance achieved by LDA (with LOOCV) using PC loading computed by 10 miRNAs selected by PCA-based unsupervised FE

k	mRNAs without a function	mRNAs with a function	Total
mRNAs not targeted by an miRNA	$N - N_1 - N_2 + N_{12}$	$N_2 - N_{12}$	$N - N_1$
mRNAs targeted by an miRNA	$N_1 - N_{12}$	N_{12}	N_1
Total	$N - N_2$	N_2	N

P values can be computed by summing up probability Eq. (6.29) for N_{12} larger than real observation. Fisher's exact test can evaluate the accidental probability that the miRNA's target mRNAs are associated with the specific function. If P values corrected with multiple comparison criterion (e.g., BH criterion) are small enough, we can insist that the specific miRNA is likely related to this function. This can be done by uploading a set of miRNAs to DIANA-miRPath [68], a web server that automatically evaluates these probabilities. In order to upload each miRNA listed in Table 6.1, one needs to add suffix such that it is adapted to the most recent miRBase and "hsa" to specify species. For example, instead of uploading "miR-425," one must upload the name of "miR-425-5p." "miR-320a" must be uploaded as without suffix, "miR-320a." The option that specifies miRNA target gene database is "Tarbase."

Table 6.5 is the result when considering Kyoto Encyclopedia of Genes and Genomes (KEGG) [23] pathway that evaluates genes based upon metabolic paths that describe chemical reactions mediated via proteins coded by genes. Among 68 pathways associated with adjusted P values less than 0.05, as many as 17 cancer related pathways are included. This also supports the reliability of selected 12 miRNAs by PCA-based unsupervised PCA.

6.3.2 *Circulating miRNAs as Universal Disease Biomarker*

In this section, occasionally, miRNAs are used for biomarker that identifies if multiple diseases are highly overlapped. It is the next question if it is occasional or not. In order to see this, we need to see if miRNAs selected in this section can diagnose other diseases. For this purpose, we have collected miRNA expression of other diseases from various studies [59]. We have collected seven blood miRNA expressions from the GEO: Alzheimer's disease (AD) (GSE46579) [26], carcinoma (GSE37472) [30], coronary artery disease (CAD) (GSE49823), nasopharyngeal carcinoma (NPC) (GSE43329), HCC (GSE50013) [41], breast cancer (BC) (GSE41922) [8], and acute myeloid leukemia (AML) (GSE49665) [38] (Table 6.6). This is really heterogeneous data set. Not only collected diseases but also resources as well as methods include multiple ones. Thus it is suitable to check if 12 miRNAs can work as robust universal disease biomarker.

Table 6.5 KEGG pathway enrichment analysis by DIANA-miRPath for 12 miRNAs listed in Table 6.1. Twelve cancer related pathways among total 61 pathways detected are listed. The number of genes is that of genes included in the union set of genes targeted by at least one of 12 miRNAs. The number of miRNAs is that of miRNAs whose target genes are included in the pathway

Rank	KEGG pathway	Adjusted *P* value	The number of genes	The number of miRNAs
1.	Proteoglycans in cancer (hsa05205)	2.96×10^{-14}	120	12
12.	Prostate cancer (hsa05215)	1.35×10^{-6}	60	12
14.	Glioma (hsa05214)	9.41×10^{-6}	41	12
15.	Chronic myeloid leukemia (hsa05220)	1.12×10^{-5}	48	12
16.	Renal cell carcinoma (hsa05211)	1.36×10^{-5}	45	12
22.	Pathways in cancer (hsa05200)	4.49×10^{-5}	201	12
25.	Colorectal cancer (hsa05210)	1.34×10^{-4}	39	12
26.	Small cell lung cancer (hsa05222)	1.78×10^{-4}	54	12
31.	Pancreatic cancer (hsa05212)	3.59×10^{-4}	43	12
34.	Non-small cell lung cancer (hsa05223)	4.24×10^{-4}	35	12
36.	Central carbon metabolism in cancer (hsa05230)	4.83×10^{-3}	39	12
37.	Endometrial cancer (hsa05213)	5.62×10^{-3}	31	12
40.	Melanoma (hsa05218)	8.04×10^{-3}	39	12
43.	Transcriptional misregulation in cancer (hsa05202)	9.42×10^{-3}	90	12
44.	Bladder cancer (hsa05219)	9.77×10^{-3}	25	12
60.	Acute myeloid leukemia (hsa05221)	2.67×10^{-2}	33	11
61.	Thyroid cancer (hsa05216)	3.00×10^{-2}	17	12

The procedure is almost similar. Excluding identification of 10 miRNAs using PCA-based unsupervised FE, 12 miRNAs are considered to be chosen in common for seven diseases. Then, PC loading $\boldsymbol{v}_\ell^{(k)\prime}$ for kth disease is computed with applying PCA to $x_{ij}{}' \in \mathbb{R}^{12\times(M_0+M_k)}$. Then optimal top L $\boldsymbol{v}_\ell^{(k)\prime}$ is used for discriminating patients from controls with LDA. Performance is evaluated by LOOCV. Table 6.7 shows the performance toward seven diseases. The disease-wise mean performance (accuracy = 0.791, sensitivity = 0.785, specificity = 0.800) is almost the same as (even a little bit better than) that in the previous study (accuracy = 0.784, sensitivity = 0.750, specificity = 0.800) [58]. This suggests that identification of 12 miRNAs universally for 12 diseases is not accidental, but they are truly useful for identification of a wide range of diseases from healthy people. Thus, I named them as universal disease biomarker (UDB). The possibility of UDB is not frequently recognized. Nevertheless, because blood miRNAs can reflect whole body status, they can be UDB that can diagnose multiple diseases. Actually, we very recently [60] identified 107 blood miRNAs that can successfully discriminate familial amyotrophic lateral sclerosis, sporadic amyotrophic lateral sclerosis, healthy controls, and gene mutation holders. Among twelve miRNAs identified here, as many as nine miRNAs (miR-30d, miR-19b, miR-106b, miR-425, miR-185, miR-191, miR-92a, miR-16, and miR-140-3p) are included in the

Table 6.6 List of blood miRNA expression profiles used in validation for 12 miRNAs in Table 6.1 as a universal disease biomarker

Diseases	Alzheimer	Carcinoma	CAD	NPC
GEO ID	GSE46579	GSE37472	GSE49823	GSE43329
Number of miRNAs	502	565	746	886
Total samples	70	56	26	50
Disease samples	48	30	13	31
Healthy control samples	22	26	13	19
Methodology	HTS	qPCR	RT-PCR	Microarray
Source	Whole blood	Peripheral serum	Plasma sample	Plasma sample
Diseases	HCC	BC	AML	
GEO ID	GSE50013	GSE41922	GSE49665	
Number of miRNAs	231	274	128	
Total samples	40	54	65	
Disease samples	20	32	52	
Healthy control samples	20	22	13	
Methodology	RT-PCR	RT-PCR	Microarray	
Source	Plasma sample	Preoperative serum	Peripheral blood	

Table 6.7 Performance of PCA-based LDA using 12 miRNAs in Table 6.1 toward seven diseases with LOOCV. The "Mean of previous study" corresponds to the mean over the performance in Table 6.3. L: optimal number of PC loading used for LDA

Diseases	Accuracy	Sensitivity	Specificity	L
AD	0.829	0.833	0.818	8
Carcinoma	0.768	0.730	0.800	11
CAD	0.846	0.846	0.846	3
NPC	0.740	0.806	0.632	12
HCC	0.700	0.700	0.700	9
BC	0.870	0.813	0.955	3
AML	0.784	0.769	0.846	8
Mean	0.791	0.785	0.800	–
Mean of previous study [58]	0.784	0.750	0.800	–

107 miRNAs. Nine out of twelve might not look like large enough, because 107 miRNAs are selected from as many as 3391 miRNAs [60], and the fact that selected 107 miRNAs have nine overlaps with twelve miRNAs is highly significant ($P = 4.5 \times 10^{-4}$ by Fisher's exact test). Identification of UDB using circulating miRNAs should be searched more extensively and seriously.

6.3.3 Biomarker Identification Using Exosomal miRNAs

In the previous subsubsection, we have shown that serum biomarker can discriminate various diseases from normal controls. In this section, we would like to demonstrate that blood miRNA can even work as disease progression biomarker. miRNAs considered are those in exosome.

Exosome is a small vehicle composed of lipid bilayer membrane. It is released from cells and includes various compounds originated from cells inside. As a result, exosome is a good target by which we can know the state inside cells. The functions of exosome are not yet fully understood. Although some reports say that exosome is used to transfer some compounds from cells to cells, what the purpose is specifically is not yet understood.

Recently, exosome is considered to be a candidate as biomarker, because it can carry something out of cell insides. It is also reported that some cancers make use of functions of exosome. These suggest that compounds in exosome can reflect the change inside cells coincident with disease initiation and progression. Exosome also includes miRNAs originated from cells and is circulating in the blood. Thus, exosomal miRNAs are good targets from which disease biomarkers are generated.

The targeted diseases are liver diseases, which are classified as hepatitis. Hepatitis is a kind of chronic inflammation disease caused by various causes. Hepatitis itself is not lethal, but it becomes cirrhosis of the liver earlier or later, and finally results in deadly liver cancer. Thus, treatment of hepatitis before cirrhosis of the liver is critically important. Thus inference of hepatitis progression is very important. On the other hand, because therapy changes dependent upon disease causes, it is also very important to diagnose disease cause. Therefore, the aim of constructing biomarker is not only discrimination between healthy controls and hepatitis, but also constructing biomarker that can discriminate hepatitis having different causes and progression stages. Thus, the more advanced biomaker than that we have identified in the previous subsubsection is required.

The data set is downloaded from GEO with GEO ID GSE33857 [33] (Table 6.8). This data set includes three hepatitis: one caused by hepatitis B virus (HBV) infection, one caused by hepatitis C virus (HCV) infection, and nonalcoholic steatohepatitis (NASH). The microarray used for these measurements includes 887 miRNAs. Because each miRNA is measured by multiple probes, the number of probes included is 14192. Because feature selection below is performed not in

Table 6.8 List of exosome miRNA expression profiles used in this chapter. CHB: chronic hepatitis B, CHC: chronic hepatitis C, NASH:Nonalcoholic steatohepatitis

CHC	CHB	NASH	Normal
Primary sets			
64	4	12	24
Validation sets			
31	12	8	–

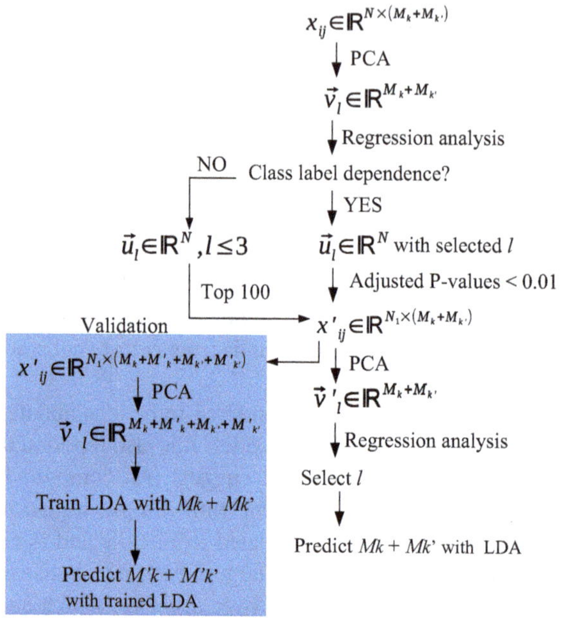

Fig. 6.2 Flowchart of discrimination between diseases using exosomal miRNAs. The performance obtained with following this flowchart is in Table 6.10

individual miRNAs base but in probe base, it is obvious that the number of features (14192) is much larger than the number of samples (104 for primary set).

Figure 6.2 illustrates the analysis flow of discrimination between CHB, CHC, NASH, and healthy controls (unshaded part) for the primary set:

1. Expression profiles, $x_{ij} \in \mathbb{R}^{N \times (M_k + M_{k'})}$, of ith miRNA and jth samples with kth and k'th disease or normal sample with the number of samples M_k and $M_{k'}$, respectively.
2. Apply PCA to x_{ij} such that PC loading, $\boldsymbol{v}_\ell \in \mathbb{R}^{M_k + M_{k'}}$, is attributed to samples.
3. Apply categorical regression analysis

$$v_{\ell j} = a_\ell + \sum_{s \in \{k,k'\}} b_{\ell s} \delta_{sj} \tag{6.30}$$

   to $\boldsymbol{v}_\ell$ and attribute P values to ℓs. P values are adjusted by BH criterion and ℓs associated with adjusted P values less than 0.05 are selected (the cases without ℓs that pass this filtering will be discussed later).
4. Attribute P value to ith miRNA as

$$P_i = P_{\chi^2}\left[> \sum_\ell \left(\frac{u_{\ell i}}{\sigma_\ell} \right)^2 \right], \tag{6.31}$$

   where the summation is taken over ℓs associated with adjusted P values less than 0.05 and σ_ℓ is the standard deviation of $u_{\ell i}$.

Table 6.9 Frequencies of PC loading selected via Eqs. (6.30) and (6.32) when one of the samples is sequentially excluded among 100 trials. ℓs not shown are not selected. PC loading shown in bold is selected and used for selection of miRNA probes

CHC vs. Normal						
ℓ	**1**	**2**	**3**	27		
Eq. (6.30)	88	88	88	1		
Eq. (6.32)	88	50	88	–		
CHB vs. CHC						
ℓ	**2**	**3**				
Eq. (6.30)	20	68				
ℓ	**1**	3	4	**5**	6	12
Eq. (6.32)	68	1	1	67	3	1
NASH vs. CHC						
ℓ	**1**	**2**	**3**			
Eq. (6.30)	12	76	76			
ℓ	**1**	**3**	**4**			
Eq. (6.32)	76	76	76			

5. ith miRNAs associated with adjusted P values less than 0.01 are selected. With using only selected N_1 miRNAs, expression profile, $x_{ij}' \in \mathbb{R}^{N_1 \times (M_k + M_{k'})}$, is composed.
6. PCA is again applied to x_{ij}' and $\boldsymbol{v}_\ell' \in \mathbb{R}^{(M_k + M_{k'})}$ are obtained. Categorical regression is applied to $\boldsymbol{v}_\ell'$ as

$$v_{\ell j}' = a_\ell' + \sum_{s \in \{k,k'\}} b_{\ell s}' \delta_{sj} \tag{6.32}$$

and ℓs with adjusted P values less than 0.05 are used for LDA.
7. kth and k'th diseases or normal samples are discriminated with LDA using ℓs selected. LOOCV is employed for the cross validation.

There is one problem in the above process. When features are selected, it is forbidden to use the information of samples to be discriminated. Nevertheless, in Eqs. (6.30) and (6.32), all of the class labeling are used.

In order to see if exclusion of one sample to be discriminated (because we employ LOOCV) can alter the selected miRNAs, we perform the following. When PC loading is selected using Eqs. (6.30) and (6.32), one of the $M_k + M_{k'}$ samples is removed and P values are recomputed and adjusted to select PC loading. Table 6.9 is the result for 100 trials. As can be seen, only three pairs, CHC vs. normal, CHB vs. CHC, and NASH vs. CHC, out of six possible pairs among four classes, CHB, CHC, NASH, and normal controls, have nonzero selected PC loading in Eq. (6.30) (please note if no PC loading is selected in Eq. (6.30), the process is terminated and Eq. (6.32) is not performed). As expected, PC loading is selected with high stability for these three pairs. Thus, we decide to select PC loading shown in bold in Table 6.9 for these three pairs. Table 6.10 shows the confusion matrices of the discrimination between CHC, CHB, NASH, and normal controls. For these three pairs, CHC vs. normal, CHB vs. CHC, and NASH vs. CHC, performance is relatively well.

Table 6.10 Confusion matrix of discrimination between CHB, CHC, NASH, and normal controls. Columns: True, Rows: Prediction. P values are computed by Fisher's exact test

	Primary set[a]								
	Predict	CHC	Normal	Predict	CHC	CHB	Predict	CHC	NASH
	CHC	64	4	CHC	62	1	CHC	63	2
	Normal	0	20	CHB	2	3	NASH	1	10
P values		3.46×10^{-16}			7.80×10^{-4}			7.40×10^{-10}	
Odds ratio		∞			71.4			234.2	
	Primary set[b]								
	Predict	CHB	NASH	Predict	CHB	Normal	Predict	NASH	Normal
	CHB	2	3	CHB	2	9	NASH	9	7
	NASH	2	9	Normal	2	15	Normal	2	17
P values		5.55×10^{-1}			1.00×10^{0}			8.79×10^{-3}	
Odds ratio		2.78			1.63			10.1	
	Validation set								
	Predict	CHB	NASH	Predict	CHC	CHB	Predict	CHC	NASH
	CHB	18	6	CHC	74	3	CHC	73	8
	NASH	4	12	CHB	21	17	NASH	22	12
P values		3.21×10^{-3}			1.90×10^{-7}			2.24×10^{-3}	
Odds ratio		8.42			19.3			4.89	

[a] N_1 probes are selected based upon significance.
[b] Top $N_1(= 100)$ probes with smaller P values are selected without considering significance.

For other three pairs without CHC, i.e., CHB versus NASH, CHB versus normal, NASH versus normal, because there are no PC loading associated with adjusted P values less than 0.05 (the "NO" branch to the question "Class label dependence?" in Fig. 6.2), we cannot select miRNAs using Eq. (6.31). Instead, we attribute P values to miRNAs with Eq. (6.31) using $1 \leq \ell \leq 3$. Then top 100 miRNAs with smaller P values are selected even if they are not significantly small. PC loading, $\boldsymbol{v}_\ell{}'$, with applying PCA to $x_{ij}{}' \in \mathbb{R}^{10 \times (M_k + M_{k'})}$. Using $\boldsymbol{v}_\ell{}'$, $1 \leq \ell \leq 3$, LDA is performed. The results for these three pairs are also shown in Table 6.10. Performance for CHB is not good.

In Table 6.11, we list miRNAs selected. As in the case of serum miRNAs, they are highly overlapped. Thus, as miRNAs in serum, exosomal miRNAs have potential to be UDB, too.

Finally, we try to validate the suitability of selected miRNAs in Table 6.11 using validation set in Table 6.8. The procedure is as follows (shaded part in Fig. 6.2):

1. We construct expression profiles of selected N_1 probes listed in Table 6.11, $x_{ij}{}' \in \mathbb{R}^{N_1 \times (M_k + M_k' + M_{k'} + M_{k'}')}$, where M_k' and $M_{k'}'$ are the number of samples in validation set (Table 6.8) of kth and k'th disease or control samples.
2. PC loading $\boldsymbol{v}_\ell{}'$ is computed with applying PCA to $x_{ij}{}'$.

Table 6.11 List of miRNAs included in N_1 probes selected in order to compute $\boldsymbol{v}_\ell{}'$ (Fig. 6.2) with applying PCA to $x_{ij}{}' \in \mathbb{R}^{N_1 \times (M_k + M_{k'})}$

miRNAs	CHC vs. normal	CHB vs. CHC	NASH vs. CHC	CHB vs. NASH	CHB vs. normal	NASH vs. normal
N_1	176	140	170	100	100	100
miR-638	○	○	○	○	○	○
miR-320c	○	○	○	○	○	○
miR-486-5p	○	○	○	○	○	○
miR-451	○	○	○	○	○	○
miR-1974_v14.0	○	○	○	○	○	○
miR-1246	○	○	○	○	○	○
miR-720	○	○	○	○	○	
miR-762	○	○	○		○	○
miR-630	○	○	○	○		○
miR-92a	○	○	○		○	
miR-1275	○	○	○			
miR-1225-5p	○		○			
miR-1207-5p	○		○			
miR-1202	○					
miR-22		○		○		
miR-532-3p		○				
miR-1202			○		○	
miR-122		○				
miR-1306		○				
miR-34b		○				
miR-16		○				
miR-1					○	
miR-1271					○	

3. $\boldsymbol{a}$ in Eq. (6.2) is computed using only $\boldsymbol{v}_\ell{}'$ (ℓs used are the same as listed in Table 6.9 for Eq. (6.32)) of $M_k + M_{k'}$ samples in primary set. In other words, $\Delta^{(k)}$ and $\Delta^{(0,k)}$ are computed using only $M_k + M_{k'}$ samples in primary set.
4. Using obtained $\boldsymbol{a}$, y_j for $M_k' + M_{k'}'$ samples in validation set is computed using Eq. (6.2).
5. $M_k' + M_{k'}'$ samples in validation set are discriminated between k and k' using obtained y_j.

The exclusion of $M_k' + M_{k'}'$ samples in validation set for computing $\boldsymbol{a}$ is required in the step 3 because we should not use any information about to which category samples in validation set belong; this information is not available in the real situation. On the other hand, their expression profiles themselves are allowed to be used for computing $\boldsymbol{v}_\ell{}'$ because we have miRNA expression of validation set in advance even if we do not know about labeling.

Table 6.10 also shows the results for these validation sets. The performance is pretty good. Interestingly, CHB samples that cannot be well discriminated in primary samples are well discriminated in the validation sample, in spite of that it is usually expected that performance in validation set decreases than training set. This suggests that probe selection by PCA-based supervised FE can work pretty well even if the number of samples available is small as demonstrated in synthetic data set in the previous chapter. In conclusion, exosomal miRNA has ability to diagnose not only disease but also cause of diseases, because it can discriminate among CHB, CHC, and NASH, which are hepatitis caused by different causes.

Next, we would like to see if exosomal miRNA can diagnose hepatitis progression. There are two features that describe hepatitis progression, inflammation and fibrosis. Because hepatitis is chronic inflammation, there might be no need to explain why inflammation describes hepatitis progression. On the other hand, fibrosis is not so direct measure. As mentioned in the above, hepatitis develops to cirrhosis. Cirrhosis is fibrosis of liver. Thus, it is reasonable to consider fibrosis progression to be disease progression measure of hepatitis. Both inflammation and fibrosis are diagnosed for some CHC samples using integer grade.

Table 6.12 shows the frequency of inflammation and fibrosis grade levels diagnosed. In order to infer these levels using exosomal miRNAs, we do as follows:

1. PCA is applied to $x_{ij} \in \mathbb{R}^{N \times M_k}$ where M_k is the total number of CHC samples with inflammation or fibrosis diagnose (Table 6.12). Unfortunately, we cannot identify any PC loading, $\boldsymbol{v}_\ell \in \mathbb{R}^{M_k}$, that are significantly associated with inflammation or fibrosis levels.
2. After attributing P_is to miRNAs using Eq. (6.31) with $\ell \leq 3$, top 100 probes with smaller P_is are selected.
3. PCA is applied to miRNA expression profile including only selected 100 probes, $x_{ij}{}' \in \mathbb{R}^{100 \times M_k}$.
4. The obtained $\boldsymbol{v}_\ell{}', \ell \leq 2$ as well as patient ages are used for LDA. In this case, ages must be considered together with miRNA expression in order to get significant results.

Table 6.13 shows the confusion matrices between predicted and true inflammation and fibrosis. Although the number of used miRNAs increases to twice, it is still as many as $\sim$10, which is less than 5% of the total number of miRNAs in the array, 887. In order to see if they are significant, we compute correlation coefficient between true and predicted inflammation and fibrosis levels. Although the correlation is not very high, they are associated with significant P values ($P = 0.02$, see page 113 for how to compute P values attributed to correlation

Table 6.12 Number of samples having specific inflammation and fibrosis levels in CHC samples

Inflammation	1	2	3	Fibrosis	0	1	2	3
Number	36	18	9	Number	3	33	16	12

Table 6.13 Confusion matrices between true (columns) and predicted (rows) inflammation and fibrosis levels. Below the table, correlation coefficients with associated P values as well as 21 miRNAs to which top 100 probes are attributed are listed

True								
Inflammation				Fibrosis				
Prediction	1	2	3	Prediction	0	1	2	3
1	28	10	4	0	1	2	0	0
2	5	6	2	1	2	23	9	6
3	3	2	3	2	0	3	5	2
				3	0	5	2	4
Corr. = 0.29, P=0.02				Corr. = 0.30, P=0.02				

miR-1225-5p miR-1275 miR-638 miR-320c miR-197_v14.0 miR-194* miR-630 miR-720 miR-300 miR-1179 miR-373 miR-1181 miR-1246 miR-320d miR-532-3p miR-518d-3p miR-34b miR-664 miR-668 miR-147 miR-664*

coefficients). Thus, we can expect that exosomal miRNAs can diagnose hepatitis progression together with patients' ages.

Multi-class LDA Here we need to explain how LDA for two classes can be extended to multi-classes because discrimination of inflammation or fibrosis levels require multi-classes discrimination. At first, $\triangle^{(k)}$ in Eq. (6.6) and $\triangle^{(0,k)}$ in Eq. (6.9) are replaced with

$$\triangle^{\text{inter}} = \sum_{k \neq k'} \left(\langle y_j \rangle_j^{(k)} - \langle y_j \rangle_j^{(k')} \right)^2 \tag{6.33}$$

and

$$\triangle^{\text{intra}} = \sum_k \left\langle \left(y_j - \langle y_{j'} \rangle_{j'}^{(k)} \right)^2 \right\rangle_j^{(k)}, \tag{6.34}$$

respectively. Then we get Eq. (6.15) with modified Σ_W and Σ_B. In contrast to the two classes discrimination, we need the first S eigen vectors, $\boldsymbol{a}_p, p \leq S$, for S classes discrimination. Then y_j^ps are attributed to jth sample with substituting $\boldsymbol{a}_p$ to $\boldsymbol{a}$ in Eq. (6.2). $\boldsymbol{y}_j$ is defined as

$$\boldsymbol{y}_j = \left(y_j^1, y_j^2, \cdots, y_j^S \right), \tag{6.35}$$

and kth class centroid vector $\langle \boldsymbol{y}_j \rangle_j^{(k)}$ is computed as in Eq. (6.4). The distance between jth sample and kth centroid

$$d_{kj} = \left| \boldsymbol{y}_j - \langle \boldsymbol{y}_j \rangle_j^{(k)} \right| \tag{6.36}$$

is computed. Finally, jth sample is classified into the kth class having the smallest d_{jk} among S classes. □

In conclusion, we have found that exosomal miRNAs cannot only discriminate between healthy controls and hepatitis patients, but also diagnose hepatitis progression. The advantages of the usage of PCA-based unsupervised FE is that we can reduce the number of probes down to $\sim 10^2$ from 14192. As for the number of miRNAs, it is ~ 10 among 887 total miRNAs. Considering that it is the result of unsupervised methods, it is remarkable and demonstrates usefulness of PCA-based unsupervised FE in the real application.

6.4 Integrated Analysis of mRNA and miRNA Expression

In the previous section, circulating miRNA can be an effective biomarker. As mentioned in the above, miRNAs can affect the biological processes through targeting mRNAs. Thus, considering miRNA and miRNA expression might be more effective to understand biological systems.

6.4.1 *Understanding Soldier's Heart from the mRNA and miRNA*

Soldier's heart means that veterans often have heart problems without any physiological abnormalities. Thus, it is believed to be post-traumatic stress disorder (PTSD) driven disorder. PTSD is a mental disorder caused by life-threading stresses, e.g., experiences in battlefields or encounters to disaster. Even after the stresses passed out, human beings sometimes have mental problems, not to be fully relaxed. PTSD not only affects mental sides, but also affects physical sides. In this sense, it is critically important to know how life-threading stress causes gene expression anomaly in heart. In this subsection, we would like to fulfill this requirement by analyzing miRNA and mRNA expression profiles in stressed mice hearts [56].

Table 6.14 lists 48 samples for which mRNA and miRNA expression profiles are measured. These are downloaded from GEO ID GSE52875; the file GSE52875_RAW.tar including individual 48 raw data files is downloaded. Individual 48 files whose file names start from "GSM" are loaded into R via read.csv func-

Table 6.14 Number of samples in miRNA and mRNA expression profiles of stressed mice hearts

Rest period	1 day				1 day	10 days	
Stress expose period	1 day	2 days	3 days	10 days	5 days		42 days
Control/Stressed	0/4	4/4	0/4	4/4	4/4	4/4	4/4

tion, and "gProcessedSignal" columns in each file are collected as one dataframe. All probes having ControlType=0 are excluded for the further analyses.

Here we would like to emphasize the difficulty of feature selection in this case. In the case of discrimination between diseases and healthy control, miRNAs should be expressed distinctly between two classes. Even diagnosing disease progression, the direction among multiple classes is clear; inflammation and fibrosis should increase as disease progressed. In contrast to these cases where how expression is expected to differ is more or less clear, how miRNA expression differs between 12 classes (5 controls and 7 stressed samples) is unclear. Of course, although miRNA expression should differ between corresponding controls and stressed samples, pairwise comparisons between five pairs of stressed samples and controls might not be a good idea because individual comparison might identify nonoverlapping sets of mRNAs and miRNAs. Because the aim of this chapter is to identify disease causing mRNAs and miRNAs, identification of sets of nonoverlapping mRNAs and miRNAs is not desirable. Thus, how we can identify mRNAs and miRNAs that contribute to diseases is not an easy problem.

Here, we apply PCA-based unsupervised FE to mRNA expression and miRNA expression separately. We denote mRNA and miRNA expression as $x_{ij}^{\text{mRNA}} \in \mathbb{R}^{59305 \times 48}$ and $x_{kj}^{\text{miRNA}} \in \mathbb{R}^{2640 \times 48}$, respectively. Although the total number of mRNAs measured is as many as 37890, since some mRNAs are measured by multiple probes, the total number of probes, 59305, is much larger than the total number of mRNAs. Although there are only 660 miRNAs measured, because they are measured by as many as four probes, the total number of probes is 2640. Although x_{ij}^{mRNA} is standardized, i.e., $\sum_i x_{ij}^{\text{mRNA}} = 0, \sum_i \left(x_{ij}^{\text{mRNA}}\right)^2 = 59305$, x_{kj}^{miRNA} are not. PCA is applied to x_{ij}^{mRNA} and x_{kj}^{miRNA} such that PC loading $\boldsymbol{v}_\ell^{\text{mRNA}} \in \mathbb{R}^{48}$ and $\boldsymbol{v}_\ell^{\text{miRNA}} \in \mathbb{R}^{48}$ are attributed to 48 samples.

In order to see which PC loading is associated with distinction among 12 classes, $1 \leq s \leq 12$, we apply categorical regression analysis

$$v_{\ell j}^{\text{mRNA}} = a_\ell^{\text{mRNA}} + \sum_{s=1}^{12} b_{\ell s}^{\text{mRNA}} \delta_{sj} \tag{6.37}$$

$$v_{\ell j}^{\text{miRNA}} = a_\ell^{\text{miRNA}} + \sum_{s=1}^{12} b_{\ell s}^{\text{miRNA}} \delta_{sj}, \tag{6.38}$$

where $a_\ell^{\text{mRNA}}, b_{\ell s}^{\text{mRNA}}, b_{\ell s}^{\text{miRNA}}, b_\ell^{\text{miRNA}}$ are the regression coefficients. P values are attributed to PC loading and corrected by BH criterion. As a result, PC loading with $\ell = 1, 2, 4, 10$ for mRNA and $\ell = 1, 2$ for miRNA has adjusted P values less than 0.05. Figures 6.3 and 6.4 show the boxplots of PC loading, $\boldsymbol{v}_\ell^{\text{mRNA}}$, $\ell = 1, 2, 4, 10$, and those of PC loading, $\boldsymbol{v}_\ell^{\text{miRNA}}$, $\ell = 1, 2$, respectively. It is obvious that PCA can successfully identify PC loading associated with dependence upon 12 classes for mRNA and miRNA. It might be difficult to identify such complicated dependence

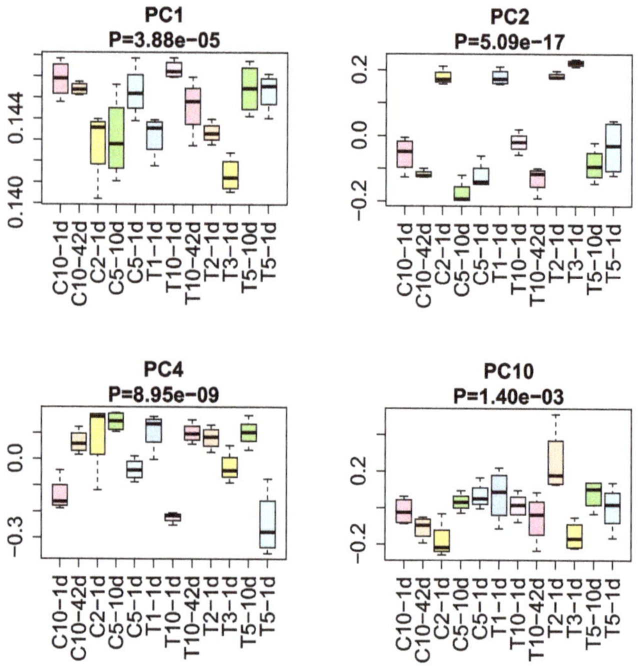

Fig. 6.3 The first, second, fourth, and tenth PC loading, $\boldsymbol{v}_\ell^{\text{mRNA}}$, $\ell = 1, 2, 4, 10$, to which adjusted P values less than 0.05 are attributed. P values above each plot are adjusted ones. Labels of classes: C:control, T:stressed, the numbers adjusted to T or C: period of stress, XXd: days of rest. See Table 6.14, too. Coloring is just for visibility and does not correspond to experimental conditions

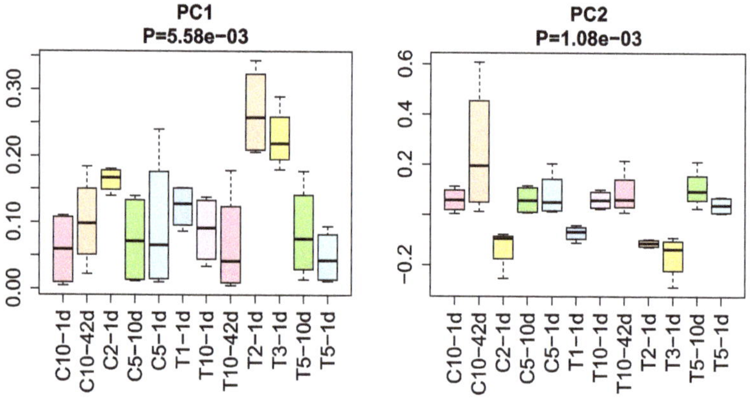

Fig. 6.4 The first and second PC loading, $\boldsymbol{v}_\ell^{\text{miRNA}}$, $\ell = 1, 2$, to which adjusted P values less than 0.05 are attributed. P values above each plot are adjusted ones. Labels of classes: C:control, T:stressed, the numbers adjusted to T or C: period of stress, XXd: days of rest. See Table 6.14, too. Coloring is just for visibility and does not correspond to experimental conditions

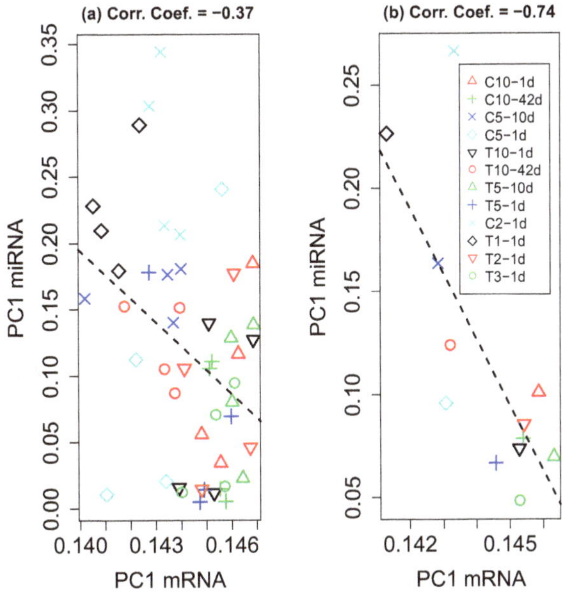

Fig. 6.5 Scatter plot between PC loading, (a) v_{1j}^{mRNA} and v_{1j}^{miRNA} (b) $\left\langle v_{1j}^{\mathrm{mRNA}} \right\rangle_j^{(s)}$ and $\left\langle v_{1j}^{\mathrm{miRNA}} \right\rangle_j^{(s)}$. Correlation coefficients are (a) -0.37, $P = 9.54 \times 10^{-3}$, (b) -0.74, $P = 6.28 \times 10^{-3}$

with conventional supervised methods, because we need to specify the dependence upon class labeling in advance, for supervised methods.

Although PCA can identify PC loading that is coincident with twelve classes, the problem is if these coincidences are biological reasonable or not. In order to discuss this point, we need domain knowledge. As mentioned in the above, primary function of miRNA is to destroy mRNA. Thus, miRNA expression should be negatively correlated to mRNA expression. Hence, PC loading that corresponds to these two should as well. Fig. 6.5a shows the scatter plot of the first PC loading between miRNA and mRNA attributed to 48 samples. The correlation coefficients are -0.37 with associated P value of 9.54×10^{-3}. Thus, as expected, v_{1j}^{mRNA} and v_{1j}^{miRNA} is significantly correlated. In order to further confirm the biological reliability, summation over four replicates is taken. Then the correlation between

$$\left\langle v_{1j}^{\mathrm{mRNA}} \right\rangle_j^{(s)} = \frac{1}{4} \sum_{j \in s} v_{1j}^{\mathrm{mRNA}} \tag{6.39}$$

and

$$\left\langle v_{1j}^{\mathrm{miRNA}} \right\rangle_j^{(s)} = \frac{1}{4} \sum_{j \in s} v_{1j}^{\mathrm{miRNA}} \tag{6.40}$$

is computed (Fig. 6.5b). If the negative correlation between miRNA and mRNA is biologically reliable, taking summation over four replicates within each class, k, should enhance the negative correlation. As expected, the correlation coefficient

Table 6.15 Correlation coefficients between $v^{\text{mRNA}}_{\ell j}$, $\ell = 1, 2, 4, 10$ and $v^{\text{miRNA}}_{\ell j}$, $\ell = 1, 2$, and those between $\left\langle v^{\text{mRNA}}_{\ell j} \right\rangle^{(s)}_j$, $\ell = 1, 2, 4, 10$ and $\left\langle v^{\text{miRNA}}_{\ell j} \right\rangle^{(s)}_j$, $\ell = 1, 2$

$v^{\text{mRNA}}_{\ell j}$	$v^{\text{miRNA}}_{\ell j}$		$\left\langle v^{\text{mRNA}}_{\ell j} \right\rangle^{(s)}_j$	$\left\langle v^{\text{miRNA}}_{\ell j} \right\rangle^{(s)}_j$	
ℓ	1	2	ℓ	1	2
1	−0.37	0.65	1	−0.73	0.78
2	0.64	−0.62	2	0.80	−0.83
4	0.12	−0.21	4	0.30	−0.26
10	0.13	0.07	10	0.15	0.09

decreases (absolute value increases) from −0.37 to −0.71, while associated P value decreases ($P = 6.28 \times 10^{-3}$); significance increases. Because the number of observations decreases from 48 to 12, it is reasonable even if P value associated with correlation coefficient increases (becomes less significant). In spite of that, P value actually decreases; this suggests that negative correlation between miRNA and mRNA is likely biologically reliable. Table 6.15 shows the correlation coefficients between other PC loading. Other than pairs including v^{mRNA}_{10j}, all pairs of PC loading between miRNA and mRNA have at least one pair associated with negative correlation. This suggests that PCA has ability to identify expected negative correlations between miRNA and mRNA, in spite of that no requirements for negative correlations are assumed during the selection of PC loading. This suggests that PCA has ability to identify biologically reasonable PC loading in an unsupervised manner.

Next, in order to identify mRNAs and miRNAs that contributed to PTSD-mediated heart disease we show scatter plots of $\boldsymbol{u}^{\text{mRNA}}_{\ell}$ and $\boldsymbol{u}^{\text{miRNA}}_{\ell}$ (Fig. 6.6), because the first two PC loading has stronger mutual correlations between miRNA and mRNA (Table 6.15) that are coincident the miRNA function that destroys mRNA. It is obvious that the first PC scores have more contributions. Thus, I decided to select miRNA and mRNA using the first PC scores. Nevertheless, the second PC loading are positively correlated with the first ones (Table 6.15), and mRNAs and miRNAs having larger contribution to the second one should be excluded. The problem is that miRNAs and mRNAs having how large contribution to the second PC score should be excluded. In order to estimate this, we select top 100 mRNAs and miRNAs simultaneously having the first PC score, u^{mRNA}_{1j} or u^{miRNA}_{1j}, whose absolute values are larger (i.e., highly ranked), and the second PC score, u^{mRNA}_{2j} or u^{miRNA}_{2j}, whose absolute values are less than threshold value D. Figure 6.7 shows how D affects the selection of top 100 mRNAs and miRNAs. We can see that too small D heavily affects the selection, while large enough D affects less. Then we decide to select $D = 20$ for mRNAs and $D = 5000$ for miRNAs, respectively. As a result, 27 miRNAs (mmu-miR-451, -22, -133b, -709, -126-3p, -30c, -29a, -143, -24, -23b, -133a, -378, -30b, -29b, -125b-5p, -675-5p, -16, -26a, -30e, -1983, -691, -23a, -690, -207, and -6691, and mmu-let-7b and -7g) and 59 mRNAs (Table 6.16) are associated with at least one of the selected probes.

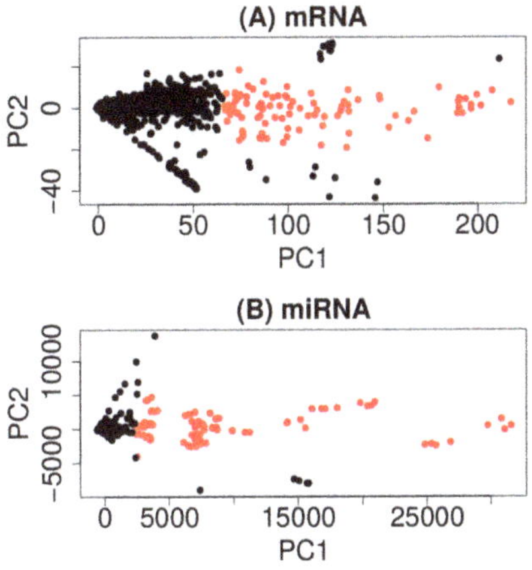

Fig. 6.6 Scatter plots of PC scores attributed to (**a**) mRNA, $\boldsymbol{u}_\ell^{\text{mRNA}}, \ell = 1, 2$, and (**b**) miRNA, $\boldsymbol{u}_\ell^{\text{miRNA}}, \ell = 1, 2$. Red dots are selected for further analysis

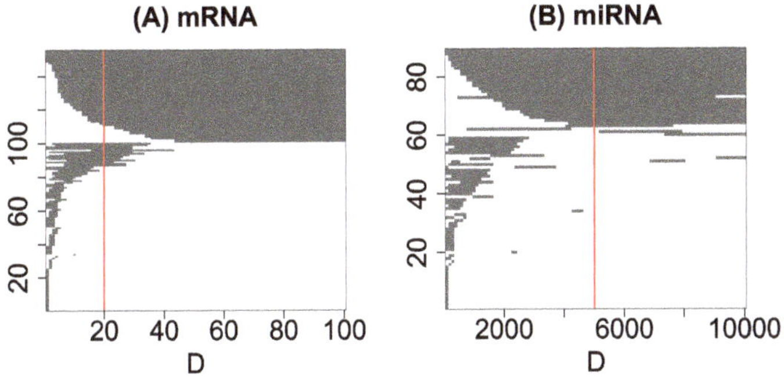

Fig. 6.7 Dependence of selected (**a**) mRNAs and (**b**) miRNAs upon D, which is a threshold value to exclude mRNAs and miRNAs having too large contribution to the second PC score. Horizontal axis: D, Vertical axis: arbitrary gene ID. Gray: selected, white: not selected. Vertical red lines indicate employed "D"

We use seed matching to identify miRNA target genes, with the so-called 7mer-m8 [48] detecting exact matches to positions 2–8 of mature miRNAs (seed + position 8). Among the 59 mRNAs, 24 are targeted by at least one of the 27 selected miRNAs. In addition, 47 pairs of miRNAs and miRNA target genes are identified. In total, there are 45/47 negative correlation coefficients between miRNAs and miRNA target genes. We also examine significance of correlation coefficients, with 26/47 pairs (more than half) associated with significant correlations (significance is judged if P values adjusted by BH criterion are less than 0.05 or not. Two positive correlations are judged insignificant because only negative correlation is biologically meaningful; see page 113 for how to compute P values attributed to correlation coefficients) and confirm negative correlation between miRNAs and miRNA target genes.

Next, we try to see if mRNAs and miRNAs selected are distinctly expressed between stressed and control samples. Because only five experimental conditions

Table 6.16 Selected 59 mRNAs

Refseq	Gene symbol	Refseq	Gene symbol
NM_010859	Myl3	NM_007450	SLC25A4
NM_001083955	Hba-a2	NM_001164248	Tpm1
NM_001164171	Myh6	NM_010861	Myl2
NM_009943	Cox6a2	NM_008218	Hba-a1
NM_013593	Mb	NM_001177307	Aldoa
NM_011619	Tnnt2	NM_144886	Exosc2
NM_008084	Gapdh	NM_009944	Cox7a1
NM_175329	Chchd10	NM_001033435	Milr1
NM_010174	Fabp3	NM_001161419	Atp5g1
NM_016774	Atp5b	NM_008617	Mdh2
NM_024166	Chchd2	NM_011540	Tcap
NM_007747	Cox5a	NM_024223	CRIP2
NM_194341	SYNRG	NM_175015	Atp5g3
NM_008220	Hbb-bt	NM_009941	Cox4i1
NM_009429	TPT1	NM_008653	Mybpc3
NM_027519	medag	NM_027862	Atp5h
NM_009964	Cryab	NM_009406	Tnni3
NM_001100116	1700047I17Rik2	NM_026701	Pbld1
NM_023312	Ndufa13	NM_026614	Tnni3
NM_025352	Uqcrq	NM_198415	Ckmt2
NM_170759	Zfp628	NM_029816	2610028H24Rik
NM_019883	UBA52	NM_025641	Uqcrh
NM_007505	Atp5a1	NM_177369	Myh8
NM_010888	Ndufs6	NM_007751	Cox8b
NM_010239	Fth1	NM_010212	Fhl2
NM_173011	Idh2	NM_007475	Rplp0
NM_023374	Sdhb	NM_053071	Cox6c
NM_025983	Atp5e	NM_080633	Aco2
NM_018858	Pebp1	NM_031165	Hspa8
NM_020582	Atp5j2		

have both stressed and control samples (Table 6.14), we consider only these five conditions. For mRNA, logarithmic ratios between stressed and control samples

$$\log\left(\frac{x_{ij_q}^{\mathrm{mRNA}}}{x_{ij'_q}^{\mathrm{mRNA}}}\right) \tag{6.41}$$

for ith mRNA in selected 59 mRNAs are computed. Here $(j_q, j'_q), q \leq 5$ are the five pairs of stressed and control samples. Then, we apply t-test to a set of 59 ×

Table 6.17 P values computed by t-tests applied to logarithmic ration, Eq. (6.41)

Rest period	2 days	5 days		10 days	
Stress expose period	1 day	1 day	10 days	1 day	42 days
PCA-based unsupervised FE					
Control < Stress	6.79×10^{-14}	7.41×10^{-8}	0.77	1.0	0.03
Control > Stress	1.0	1.0	0.22	2.35×10^{-8}	0.97
Categorical regression					
Control < Stress	6.28×10^{-9}	5.71×10^{-23}	0.80	1.0	7.37×10^{-6}
Control > Stress	1.0	1.0	0.20	2.01×10^{-4}	1.00
BAHSIC					
Control < Stress	1.00	1.00	1.00	1.0	1.05×10^{-3}
Control > Stress	6.97×10^{-4}	7.98×10^{-10}	5.45×10^{-15}	5.43×10^{-4}	1.00

$5 = 295$ computed logarithmic ratios to see if their mean value is significantly positive or negative. Table 6.17 shows the P values. In not all but some conditions, between control and stressed samples, mRNA expressions are expressed differently for selected 59 mRNAs. Because PCA-based unsupervised FE does not require the distinct expression between control and stressed samples, these suggest that PCA-based unsupervised FE can identify genes expressed distinctly between control and stressed samples in unsupervised manner.

Unfortunately, logarithmic ratio, Eq. (6.41), does not work well for miRNA expression profile. Thus, I propose alternative. First, we apply t-test to kth miRNA in one of five experimental conditions between four control samples and four stressed samples and compute P values, P_k, that reject the null hypothesis that means of four replicates are equal between control samples and stressed samples toward alternative hypothesis that means of four replicates in control samples are less than means of four replicates in stressed samples. Then, a set of logarithmic P values, $\log P_k$, are compared between selected 27 miRNAs and other miRNAs by to test if mean $\log P_k$ is distinct between selected 27 miRNAs and others (Table 6.18). This test can address significant P values to all five experimental conditions for which both stressed and control samples are available.

PCA-based unsupervised FE successfully identified mRNAs and miRNAs, which are negatively and mutually correlated and distinctly expressed between some pairs of control and stressed samples. Nevertheless, if other methods can perform same, the usefulness of PCA-based unsupervised FE is limited. Thus, it might be important to compare the performance with other popular or conventional methods.

The first conventional method tried is categorical regression analysis.

$$x_{ij}^{\text{mRNA}} = a_i + \sum_{s=1}^{12} b_{is}\delta_{sj} \tag{6.42}$$

and

Table 6.18 P values computed by t-tests between 27 selected miRNAs and other miRNAs applied to logarithmic P value, $\log P_k$, where P_k is also computed by t-test that rejects null hypothesis that means of four replicates are equal between control and stressed samples toward the alternative hypothesis that mean of four replicates in control samples is less than that of stressed samples. If mean $\log P_k$ in selected 27 miRNAs is less than that in other miRNAs, the amount by which mean miRNA expressions of control samples are less than stressed samples in 27 miRNAs is greater than that in other miRNAs and vice versa

Rest period	2 days	5 days		10 days	
Stress expose period	1 day	1 day	10 days	1 day	42 days
PCA-based unsupervised FE					
Control > Stress	1.98×10^{-5}	1.0	1.00	1.0	1.0
Control < Stress	1.0	3.15×10^{-13}	4.39×10^{-4}	1.02×10^{-3}	5.67×10^{-4}
Categorical regression					
Control > Stress	0.98	0.07	0.49	2.21×10^{-3}	0.4
Control < Stress	0.02	0.93	0.51	0.99	0.6
BAHSIC					
Control > Stress	2.93×10^{-3}	0.98	0.49	0.91	0.96
Control < Stress	1.00	0.02	0.51	0.09	0.04

$$x_{kj}^{\text{miRNA}} = a_k + \sum_{s=1}^{12} b_{ks}\delta_{sj}, \tag{6.43}$$

where summation is taken over twelve classes, $1 \leq s \leq 12$. Top-ranked 100 probes with smaller P values computed by categorical regression are selected. 23 mRNAs and 73 miRNAs are associated with top-ranked 100 probes, respectively. Although there are 181 pairs of miRNAs and miRNA target genes identified by seed match, only 37 pairs are associated with significant negative correlations. 37/181 is much less than that for PCA-based unsupervised FE, 45/47. Table 6.17 shows the results for t-test applied to a logarithmic ratio of 23 mRNAs selected by categorical regression analysis. The performance is, at most, comparative with PCA-based unsupervised FE in Table 6.17. Nonetheless, performance for identification of miRNA expressed distinctly between control and stressed samples is obviously less significant than that of PCA-based unsupervised FE (Table 6.18). Thus, in average, ability of categorical regression analysis to identify negatively correlated pairs of miRNAs and miRNAs that are expressed distinctly between control and stressed samples is less than that of PCA-based unsupervised FE.

In addition to the comparison with categorical regression analysis, I try another more advanced FE, backward elimination using Hilbert–Schmidt norm of the cross-covariance operator (BAHSIC) [45]. HSIC is the evaluation of coincidence between features and class labeling based upon inner product. Inner product of feature vectors between jth and j'th samples is defined as

$$\boldsymbol{x}_j \times_i \boldsymbol{x}_{j'}, \tag{6.44}$$

where

$$\boldsymbol{x}_j = \left(x_{1j}, x_{2j}, \cdots, x_{ij}, \cdots, x_{Nj}\right) \tag{6.45}$$

and

$$\boldsymbol{x}_{j'} = \left(x_{1j'}, x_{2j'}, \cdots, x_{ij}, \cdots, x_{Nj'}\right). \tag{6.46}$$

Similarly, inner product of class labeling vectors,

$$\boldsymbol{\delta}_j = \left(\delta_{1j}, \delta_{2j}, \cdots, \delta_{sj}, \cdots, \delta_{Sj}\right) \tag{6.47}$$

and

$$\boldsymbol{\delta}_{j'} = \left(\delta_{1j'}, \delta_{2j'}, \cdots, \delta_{sj'}, \cdots, \delta_{Sj'}\right), \tag{6.48}$$

can be defined as

$$\boldsymbol{\delta}_j \times_s \boldsymbol{\delta}_{j'}. \tag{6.49}$$

Coincidence between $\boldsymbol{x}_j \times_i \boldsymbol{x}_{j'}$ and $\boldsymbol{\delta}_j \times_s \boldsymbol{\delta}_{j'}$ means that larger (smaller) $\boldsymbol{x}_j \times_i \boldsymbol{x}_{j'}$ should be associated with larger (smaller) $\boldsymbol{\delta}_j \times_s \boldsymbol{\delta}_{j'}$. HSIC can qualitatively evaluate this coincidence as

$$\begin{aligned} \| C_{sj} \|_{HS}^2 &= \left\langle \boldsymbol{x}_j \times_i \boldsymbol{x}_{j'} \cdot \boldsymbol{\delta}_j \times_s \boldsymbol{\delta}_{j'} \right\rangle_{jj'} \\ &+ \left\langle \boldsymbol{x}_j \times_i \boldsymbol{x}_{j'} \right\rangle_{jj'} \left\langle \boldsymbol{\delta}_j \times_s \boldsymbol{\delta}_{j'} \right\rangle_{jj'} - 2\left\langle \left\langle \boldsymbol{x}_j \times_i \boldsymbol{x}_{j''} \right\rangle_{j''} \left\langle \boldsymbol{\delta}_{j'} \times_s \boldsymbol{\delta}_{j''} \right\rangle_{j''} \right\rangle_{jj'}, \end{aligned} \tag{6.50}$$

where the last term is added such that $\| C_{sj} \|_{HS}^2 = 0$ when features and class labeling are totally independent. $\langle \cdot \rangle_{jj'}$ is the average over j and j'. BAHSIC makes use of HSIC for FE. One of the N features is excluded from the computation when HSIC is computed. Then, a feature associated with the least decreased HSIC is removed. Then the process is repeated until the desired number of features remains. In order to accelerate the process, not one but more features (e.g., top 10 %) are eliminated before HSIC is recomputed for further feature elimination. BAHSIC is applied to miRNA and miRNA expression by eliminating top 10 % until 100 features remain.

The 100 probes selected by BAHSIC are associated with 37 mRNAs and 47 miRNAs, respectively. Although there are 169 pairs of miRNAs and miRNA target genes identified by seed match, only 73 pairs are associated with significant negative correlations. Although 73/169 is better than 37/181 by categorical regression, it is much less than that for PCA-based unsupervised FE, 45/47. Table 6.17 shows the results for t-test applied to logarithmic ratio of 37 mRNAs selected by BAHSIC. The performance is better than PCA-based unsupervised FE in Table 6.17. Nonetheless, performance for identification of miRNA expressed distinctly between control and stressed samples is obviously less significant than that of PCA-based unsupervised FE (Table 6.18). Thus, in average, ability of BAHSIC to identify negatively correlated pairs of miRNAs and miRNAs that are expressed distinctly between control and stressed samples is less than that of PCA-based unsupervised FE.

The advantage of PCA-based unsupervised FE toward categorical regression and BAHSIC is in some sense obvious. Categorical regression and BAHSIC can identify mRNAs and miRNAs with significant category dependence. Thus, negative correlation between miRNAs and mRNAs cannot be guaranteed. On the other hand, PCA-based unsupervised FE can provide PC loading by which we can see if negative correlation is persisted. Thus, although PCA-based unsupervised FE itself is unsupervised method, because there are opportunities that we can screen mRNAs and miRNAs with considering additional information (in this case, negative correlation), PCA-based unsupervised FE is more manageable method than other two methods.

Next, we compare the stability of FE among these three methods (Table 6.19). Because all 12 classes (Table 6.14) are composed of four replicates, stability test is performed by eliminating one of the four replicates randomly for all 12 classes. Then, stability test is applied to three FEs, and outcomes are summed up over 100 independent trials. For PCA-based unsupervised FE, 78 probes associated with mRNAs and 27 probes associated miRNAs are always selected, respectively. There are only ten probes associated with mRNAs not always selected. In addition to this, no other probes associated with miRNAs are selected. This means, independent of the ensemble, 27 miRNAs are always selected. In contrast to the performance achieved by PCA-based unsupervised FE, for categorical regression analysis, 24 probes associated mRNAs and eight probes associated miRNAs are always selected, respectively. Nevertheless, there are as many as 122 probes associated with mRNAs and selected by only once. 33 and 29 probes associated with miRNAs were selected only once and twice, respectively. Thus, it is obvious that stability of categorical regression as feature selection tool is much less than PCA-based unsupervised FE. For BAHSIC, no probes for mRNAs and 63 probes for miRNAs are always selected, respectively. There are 31 probes associated with miRNAs not selected always. Thus, it is again obvious that stability of BAHSIC as feature selection tool is much less than PCA-based unsupervised FE. As a result, from the point of stability, PCA-based unsupervised FE outperforms categorical regression analysis and BAHSIC.

Finally, we evaluate selected genes biologically. Because readers might not be so interested in biological background, I present here only one evaluation. As has been done in biological validation of circulating miRNAs biomarker, enrichment analysis is easy way to evaluate obtained mRNAs. In contrast to the evaluation of miRNAs, we have a list of genes. Thus we can upload genes to the Database for Annotation, Visualization and Integrated Discovery (DAVID) [21] that evaluates sets of genes by enrichment analysis. We upload 24, 21, and 37 mRNAs that are selected by PCA-based unsupervied FE, categorical regression analysis, and BAHSIC and are also simultaneously targeted by 27, 73, and 47 miRNAs selected by these individual three methods (that is, the most confident set of mRNAs based upon the integrated analysis of mRNA and miRNA expression by these three methods). Table 6.20

Table 6.19 The frequency of probes associated with mRNAs and miRNAs, selected by either PCA-based unsupervised FE, categorical regression, or BAHSIC, among 100 independent trials. Bold numbers are the number of probes always selected (i.e., 100 times)

Stability analysis of PCA-based unsupervised FE											
mRNA											
Frequency	1	8	32	36	41	43	73	89	98	99	100
Number of probes	1	1	1	1	1	1	1	1	1	1	**78**
miRNA											
Frequency	100										
Number of probes	**27**										

Stability analysis of categorical regression-based FE																				
mRNA																				
Frequency	1	2	3	4	5	6	7	8	9	10	11	12	13	14	15	16	17	18	19	20
Number of probes	122	61	25	17	15	18	8	9	10	6	6	13	7	6	7	4	4	2	5	1
Frequency	21	22	23	24	25	26	27	28	29	30	31	32	33	34	36	37	39	40	42	43
Number of probes	7	1	6	5	2	4	6	3	5	5	2	2	3	3	5	3	4	2	4	1
Frequency	44	45	46	47	48	49	50	51	55	57	58	60	62	63	65	67	68	71	72	74
Number of probes	1	1	4	2	1	3	2	2	1	1	1	1	1	1	1	1	1	1	1	1
Frequency	75	77	78	79	82	83	84	87	88	90	94	95	96	97	99	100				
Number of probes	1	1	1	3	1	4	1	1	2	1	2	1	1	2	3	**24**				
miRNA																				
Frequency	1	2	3	4	5	6	7	8	9	10	11	12	13	14	15	16	17	18	19	20
Number of probes	33	29	15	17	3	10	5	8	9	7	2	1	2	4	1	5	2	2	6	1
Frequency	21	22	23	24	25	26	27	28	29	30	31	32	33	35	36	37	38	39	40	41
Number of probes	2	4	1	5	3	2	4	2	2	4	3	1	1	3	1	1	1	4	3	5
Frequency	42	43	44	46	47	48	49	50	52	53	55	56	57	59	60	61	66	67	68	70
Number of probes	2	1	2	2	3	1	2	2	2	1	2	3	2	1	4	1	2	2	2	1
Frequency	71	74	76	77	78	80	81	82	83	84	86	87	88	89	91	93	95	100		
Number of probes	1	2	1	1	1	1	1	1	2	2	1	5	1	2	3	2	2	**8**		

Stability analysis of BAHSIC																
mRNA																
Frequency	1	2	3	4	5	6	7	8	9	10	11	12	13	14		
Number of probes	2133	886	474	280	197	136	84	55	41	14	7	6	4	1		
miRNA																
Frequency	1	3	4	20	21	25	32	40	41	43	45	60	61	62	69	70
Number of probes	2	1	1	1	3	1	1	1	1	1	1	1	1	1	1	1
Frequency	73	74	75	77	79	82	86	87	90	93	95	96	97	98	99	100
Number of probes	3	2	1	3	6	1	1	1	1	1	1	1	3	2	3	**68**

Table 6.20 KEGG pathway enriched by 24 mRNAs targeted by 27 miRNAs, identified by DAVID

KEGG pathway	Number of genes	%	*P* values	Adjusted *P* values
Cardiac muscle contraction	7	30.4	2.30×10^{-9}	3.20×10^{-8}
Parkinson's disease	7	30.4	5.80×10^{-8}	4.10×10^{-7}
Oxidative phosphorylation	6	26.1	2.30×10^{-6}	1.10×10^{-5}
Alzheimer's disease	6	26.1	1.20×10^{-5}	4.20×10^{-5}
Huntington's disease	6	26.1	1.20×10^{-5}	3.50×10^{-5}

The number of genes is genes included in pathway, and % is the ratio genes included in pathway among 24 mRNAs. *P* values and those adjusted by BH criterion were provided by DAVID.

shows the results for mRNAs selected by PCA-based unsupervised FE. These are KEGG pathways related heart disease and neurodegenerative diseases. It is quite reasonable because PTSD-mediated heart disease should be associated with both heart and brain problems. On the other hand, no KEGG pathway enrichment is obtained by uploading mRNAs selected by categorical regression or BAHSIC. These results suggest that PCA-based unsupervised FE can outperform categorical regression and BAHSIC also from biological point of views.

6.4.2 *Identifications of Interactions Between miRNAs and mRNAs in Multiple Cancers*

In the previous subsection, we decided to select 100 probes in advance. As a result, this decision works pretty well. On the other hand, it is also possible to decide the number of features selected in fully data-driven way. In actual, when genes (or miRNAs) are selected based upon expression profile, it is very usual to select genes based upon if these are differentially expressed genes (DEGs) or not. Although there are no definition about what DEGs are, two criteria are often employed:

Statistical significance Several statistical tests are applied to check if genes are differently expressed between two classes.

Fold change (FC) The ratio of the amount of expressions between two classes.

Because most of statistical tests are scale invariant, i.e., even if amount expression is globally doubled, significance does not change. This scale invariance is unlikely true because gene expressed twice more should have more important functions. In order to compensate this difficulty, FC is employed. Generally speaking, DEGs associated with more FC and more significance are better to be selected. On the other hand, the employment of two independent criteria can cause uncertainty. There can be several choices on how to balance two criteria.

This problem is critical for the selection of pairs of miRNAs and mRNAs. When identifying mRNAs and miRNAs pairs, we require:

Table 6.21 A part of significant DEG identification for mRNAs and miRNAs. Preceding studies

Cancer	Significance criteria		References
	miRNA	mRNA	
HCC	FDR ≤ 0.01; $\log_2$ FC ≥ 1		[13]
NSCLC	FDR < 0.1 by SAM		[29]
ESCC	From preceding studies	FC > 1.5	[69]
	FDR < 0.05	FC > 3; FDR < 0.001	[71]
	FDR < 0.05		[31]
PC	None		[72]
CRC	FDR < 0.05		[15]
CC	FC > 1.2; FDR< 0.1		[27]
BC	miRtest [4]	No description	[6]
PDA	FDR* < 0.05; $\mid \log \text{FC} \mid > 1$		[28]

HCC: Hepatocellular carcinoma, NSCLC: non-small cell lung cancer, ESCC: esophageal squamous cell carcinomas, PC: prostate cancer, CRC: Colorectal cancer, CC: Colon cancer, BC: breast cancer, PDA: Pancreatic Ductal Adenocarcinoma, *: Bonferroni's correction-adjusted *P* value.

- miRNAs should be DEG between control and treated samples.
- mRNAs should be DEG between control and treated samples.
- miRNAs and mRNAs should be mutually negatively correlated.

Because these three requirements are independent, finding miRNAs and mRNAs pairs fulfilling these three conditions is not an easy task. Simply applying these three criteria parallelly to mRNAs and miRNAs might result in no intersections between those satisfying each of three conditions. Especially, significant negative correlations are hard to achieve because of too many pairs. Typically, the number of miRNAs is 10^3, while the number of mRNAs is 10^4. Thus, the number of pairs is as many as 10^7. This means that, if multiple comparison correction is considered, P values must be smaller than $10^{-2} \times 10^{-7} = 10^{-9}$, which is unlikely satisfied especially when a large enough number of samples are not available. Nevertheless, if the number of candidate mRNAs and miRNAs is reduced in advance by identifying DEGs for miRNAs and mRNAs, required P values associated with correlation between miRNAs and mRNAs can be much larger. For example, we can reduce the number of miRNAs and mRNAs down to 10^2 and 10^3, respectively, and required minimum P values can increase up to $10^{-2} \times 10^{-5} = 10^{-7}$. Then the combination of P values and FC for DEGs identification is often optimized without any proper reasons such that a desired number of negatively correlated miRNAs and mRNAs pairs can be identified. Table 6.21 is a partial list of identification criteria of DEGs for mRNAs and miRNAs. It is obvious that there is no de facto standard.

In this subsection, I would like to show [50] that employment of PCA-based unsupervised FE enables us to identify mutually negatively correlated pairs of miRNAs and mRNAs that are expressive differently between controls and treated samples with the unified criterion for multiple cancers, in contrast to the various criteria dependent upon cancers as shown in Table 6.21. Figure 6.8 illustrates how

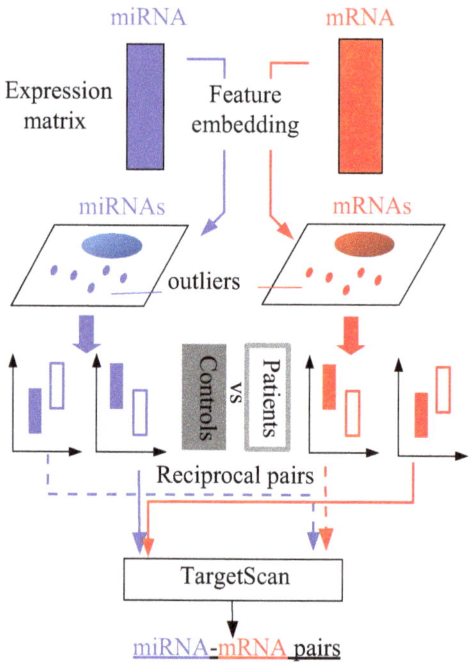

Fig. 6.8 Schematic figure that illustrates how to identify miRNAs–mRNAs pairs from expression profile using PCA-based unsupervised FE. PCA is applied to miRNAs and mRNAs profiles such that PC scores, $\boldsymbol{u}_\ell^{\mathrm{miRNA}}$ and $\boldsymbol{u}_\ell^{\mathrm{mRNA}}$, are attributed to miRNAs and mRNAs, respectively. P values are computed using χ^2 distribution as in Eqs. (6.52) and (6.51). mRNAs and miRNAs associated with adjusted P values less than 0.01 are selected. Then, mRNAs and miRNAs expressed distinctly between normal tissues and cancers are selected as outliers. miRNA–mRNA pairs are identified by TargetScan among these mRNA and miRNAs with reciprocal relationship (i.e., mRNAs upregulated and miRNAs downregulated in cancer or mRNAs downregulated and miRNAs upregulated in cancer)

to identify miRNA–mRNA pairs from expression profiles. Table 6.22 summarizes the results identified by PCA-based unsupervised FE.

The more detailed procedure is as follows. Suppose we have mRNA expression profile, $\boldsymbol{x}^{\mathrm{mRNA}} \in \mathbb{R}^{N_1 \times M_1}$, and miRNA expression profile, $\boldsymbol{x}^{\mathrm{miRNA}} \in \mathbb{R}^{N_2 \times M_2}$. Here, N_1 and N_2 are the number of mRNAs (probes) and miRNAs (probes), respectively. M_1 and M_2 are the number of samples of mRNAs profiles and miRNAs profiles, respectively. PCA is applied to mRNAs and miRNAs profiles, and PC loading, $\boldsymbol{v}_\ell^{\mathrm{mRNA}} \in \mathbb{R}^{N_1}$ and $\boldsymbol{v}_\ell^{\mathrm{miRNA}} \in \mathbb{R}^{N_2}$, are obtained. Then $\boldsymbol{v}_\ell^{\mathrm{mRNA}} \in \mathbb{R}^{N_1}$ and $\boldsymbol{v}_\ell^{\mathrm{miRNA}} \in \mathbb{R}^{N_2}$ associated with significant distinction between normal tissues and cancers (P values computed by t-test must be less than 0.05) are selected as shown in "ℓ(P value)s used for FE" of Table 6.22. After identifying PC loading used for FE, P values are attributed to ith mRNA and miRNAs as

Table 6.22 Summary of the investigated mRNA and miRNA expressions. Probes identified and not identified by PCA-based unsupervised FE are denoted as selected and nonselected, respectively. ℓs are PC scores used for computation of P values, Eqs. (6.51) and (6.52). P values associated with ℓs are computed by applying t-test to PC loading to test if it is distinct between normal tissues and cancers

		Number of samples		Number of probes		
Cancers	GEO ID	Tumors	Controls	Selected	Non-selected	$\ell(P$ value)s used for FE
HCC						
mRNA	GSE45114	24	25	269	22963	2 (7.5×10^{-9}), 3 (7.2×10^{-5}), 4 (2.1×10^{-6})
miRNA	GSE36915	68	21	58	1087	1 (9.6×10^{-8}), 3 (3.2×10^{-16}), 4 (8.5×10^{-10})
NSCLC						
mRNA	GSE18842	46	45	1098	53504	1 (5.5×10^{-10}), 2 (3.9×10^{-30}), 3 (1.04×10^{-2})
miRNA	GSE15008	187	174	268	3428	1 (8.0×10^{-5}), 2 (2.4×10^{-10}), 3 (1.3×10^{-2}), 4 (1.4×10^{-20}), 5 (4.6×10^{-30}), 6 (3.0×10^{-2})
ESCC						
mRNA	GSE38129	30	30	189	22088	3 (2.1×10^{-18})
miRNA	GSE13937	76	76	37	1217	2 (2.8×10^{-5}), 3 (3.9×10^{-2}), 4 (7.8×10^{-3}), 5 (2.0×10^{-4}), 6 (3.7×10^{-6}), 7 (4.2×10^{-2})
Prostate cancer						
mRNA	GSE21032	150	29	399	43020	3 (5.4×10^{-15})
miRNA	GSE64318	27	27	23	700	1 (2.0×10^{-2}), 2 (9.3×10^{-3}), 4 (1.4×10^{-3})
Colon/colorectal cancer						
mRNA	GSE41258	186	54	309	21974	1 (6.2×10^{-4}), 2 (2.1×10^{-2}), 3 (3.7×10^{-2}), 4 (5.1×10^{-23}), 5 (2.1×10^{-2})
miRNA	GSE48267	30	30	12	839	5 (2.2×10^{-15})
Breast cancer						
mRNA	GSE29174	110	11	980	33600	2 (3.3×10^{-20}), 3 (8.0×10^{-21}), 4 (1.1×10^{-6}), 5 (2.5×10^{-2})
miRNA	GSE28884	173	16	18	2258	1 (4.9×10^{-10}), 2 (4.0×10^{-11})

$$P_i^{\text{mRNA}} = P_{\chi^2}\left[> \sum_{\ell} \left(\frac{u_{\ell i}^{\text{mRNA}}}{\sigma_{\ell}^{\text{mRNA}}} \right)^2 \right] \tag{6.51}$$

and

$$P_i^{\text{miRNA}} = P_{\chi^2}\left[> \sum_{\ell} \left(\frac{u_{\ell i}^{\text{miRNA}}}{\sigma_{\ell}^{\text{miRNA}}} \right)^2 \right] \tag{6.52}$$

where summation is taken over ℓs selected (Table 6.22). miRNAs and mRNAs associated with BH criterion adjusted P values less than 0.01 are selected.

In order to identify reliable mRNAs and miRNAs pairs among selected mRNAs and miRNAs, the following procedures are further performed. In order to fulfill the requirement of reciprocal relationship between miRNAs and mRNAs expression, mRNAs and miRNAs up/downregulated in cancers compared with normal tissue are identified. This has been done by applying t-test. Obtained P values are adjusted by BH criterion, and mRNAs and miRNAs associated with adjusted P values less than 0.05 are selected. Reciprocal pairs of miRNAs and miRNAs, i.e., upregulated miRNAs and downregulated mRNAs or upregulated mRNAs and downregulated miRNAs, are compared with pairs included in TargetScan [2] that stores a list of miRNA target mRNAs. In order to do this, Predicted_Targets_Info file that is supposed to include all of human conserved targets is obtained, and all pairs included in this file remain as final candidate miRNAs and mRNAs. Table 6.23 summarizes the biological validation of identified pairs. Most of the pairs in seven cancers other than PC are composed of miRNAs and mRNAs that are previously reported to be related to cancers to which miRNAs and mRNAs are identified. We also check if pairs are in starbase [70] that stores the information of miRNA–mRNA pair. Generally, half of pairs are included in starbase. This suggests that PCA-based unsupervised FE can identify a limited number of mRNAs and miRNAs between which biologically reliable reciprocal pairs can be identified. In this regard, PCA-based unsupervised FE is more effective than the standard strategy that requires combinatorial usage of statistical test and FC. In addition to this, we can employ unified criterion that adjusted P values must be less than 0.01. To my knowledge, no other methods can perform identification of a reliable number of miRNAs and mRNAs by the unified criterion valuable for as many as six cancers, i.e., six cancers listed in Table 6.23 other than PC.

6.5 Integrated Analysis of Methylation and Gene Expression

As can be seen in the previous section, integrated analysis of mRNAs and miRNAs can give us the more reliable identification of mRNAs than selecting mRNAs based upon only the criterion of DEG. Thus, it is better to consider something other than

Table 6.23 Summary of the biological validation of identified pairs. "Pairs with previous studies" suggest that mRNA and miRNA are reported to be related to cancers to which these pairs are identified. "Number of pairs in starbase" suggests the number of pairs included in any cancers in starbase. More detailed information is available in Tables S1 to S18 [50]

Cancer	HCC	NSCLC	ESCC	PC	CCC/CC	BC
Number of pairs	21	311	4	32	8	37
Pairs with previous studies	19	270	4	19	7	32
Number of pairs in starbase	9	144	2	12	3	17

Table 6.24 Samples used in this chapter. pre: before metastasis, post: after metastasis. Numbers are the number of biological replicates. mRNAs expression profiles and promoter methylation profiles are obtained via GEO ID: GSE52143 and GSE52144, respectively

Cell lines	A549		HTB56		
Metastasis	pre	post	pre	post	Total
mRNA	3	3	3	3	12
Methylation	2	2	2	2	8

miRNA expression together with mRNA expression. One possible candidate is DNA methylation which is known to suppress mRNA expression.

6.5.1 *Aberrant Promoter Methylation and Expression Associated with Metastasis*

Metastasis is a developed stage of cancer. After metastasis takes place, cancer cell starts to leave from original location where cancer initiates, to migrate to all over the body and to grow there. Thus, once metastasis starts, therapy of cancer becomes drastically difficult. Therefore, suppressing progression to metastasis is critically important in cancer therapy. In this regard, we try to identify critical genes for cancer progression to metastasis based upon the integrated analysis of mRNA expression and promoter methylation [66]. Table 6.24 shows the number of samples used in this chapter (Files, GSE52143_series_matrix.txt.gz and GSE52144_series_matrix.txt.gz in series matrix session in GEO are used for mRNA expression profiles and promoter methylation profiles, respectively). There are two cell lines for which pre-/post-metastasis samples available. Thus, there are four classes. Two and three biological replicates are for methylation and mRNA, respectively. Before starting analysis, I would like to emphasize the difficulty of the analysis. First of all, there are only three and two biological replicates for mRNA expression profiles and promoter methylation profiles, respectively, while there are as many as 33297 probes and 27578 probes for mRNA expression profiles and methylation profiles, respectively. This means that identification of DEG and differentially methylated site (DMS) is not easy.

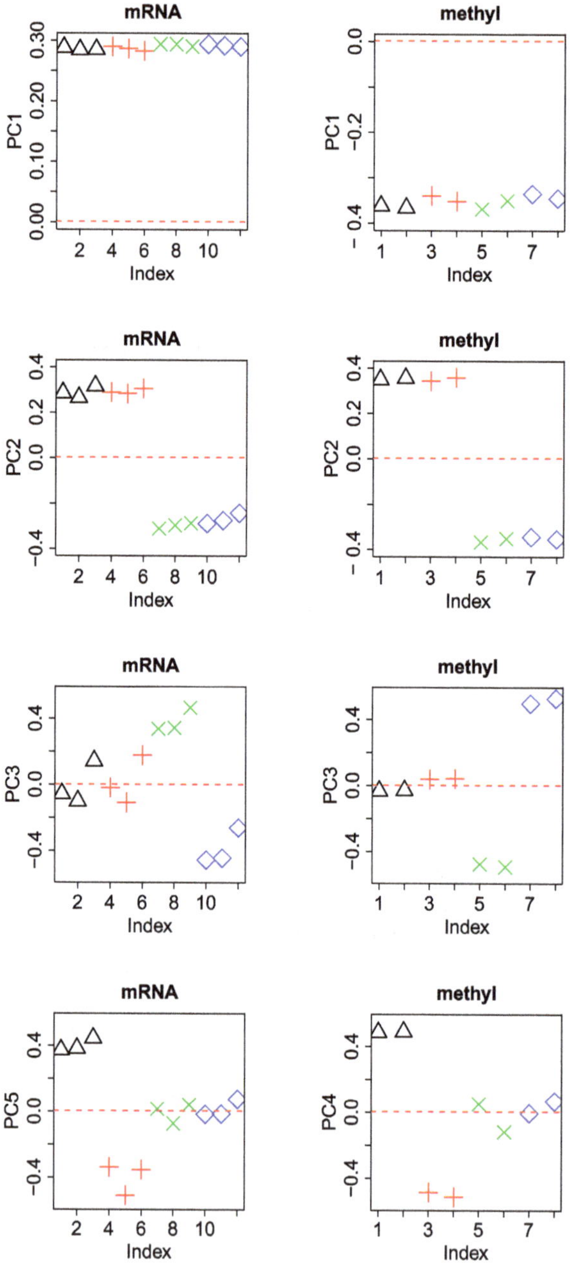

Fig. 6.9 PC loading obtained by applying PCA to mRNA expression and promoter methylation. △: A549 cell line pre-metastasis, +: A549 cell line post-metastasis, ×: HTB56 cell line pre-metastasis, ◇: HTB56 cell line post-metastasis. Left column: the first, second, third, and fifth PCs for mRNA. Right column: the first, second, third, and fourth PCs for promoter methylation

We apply PCA to mRNA expression profile, $x_{ij}^{\text{mRNA}} \in \mathbb{R}^{33297 \times 12}$, and promoter methylation profiles, $x_{kj}^{\text{methyl}} \in \mathbb{R}^{27578 \times 8}$. Then we get PC loading attributed to mRNA and miRNA samples, $\boldsymbol{v}_{\ell}^{\text{mRNA}}$ and $\boldsymbol{v}_{\ell}^{\text{methyl}}$, respectively. Figure 6.9 shows the obtained PC

Table 6.25 The number of probes selected using P values computed by Eqs. (6.53) and (6.54). ℓ: PCs used for FE

	mRNA			methylation		
ℓ	2	3	5	2	3	4
Number of selected probes	422	261	248	512	369	270

loading. The first PC loading does not exhibit any sample dependence. Thus, it is not useful to identify DEG and DMS. The second PC loading exhibits some sample dependence. Nevertheless, it is coincident with the distinction between cell lines. The third PC loading exhibits the distinction between pre- and post-metastasis as expected, but only for HTB56 cell lines. The fifth PC loading for mRNA and the fourth PC loading for methylation exhibit the distinction between pre- and post-metastasis as expected, but again only for A549 cell lines.

We attribute P values to probe using corresponding PC scores using χ^2 distribution as

$$P_i^\ell = P_{\chi^2}\left[> \left(\frac{u_{\ell i}^{\text{mRNA}}}{\sigma_\ell^{\text{mRNA}}}\right)^2\right], \ell = 2, 3, 5 \tag{6.53}$$

and

$$P_k^\ell = P_{\chi^2}\left[> \left(\frac{u_{\ell k}^{\text{miRNA}}}{\sigma_\ell^{\text{miRNA}}}\right)^2\right], \ell = 2, 3, 4. \tag{6.54}$$

P values are adjusted by BH criterion. Then probes associated with adjusted P values less than 0.01 are selected. Table 6.25 lists the number of probes selected. At least, for all cases, PCA-based unsupervised FE can identify probes with significant P values.

Because we are aiming to perform integrated analysis of mRNA expression and promoter methylation, it is important to see if genes selected for mRNA expression and promoter methylation are significantly overlapped. In order to do this, mRNA probes and methylation probes are converted into list of genes to which probes are attributed (the correspondence between probes and genes is available as files GPL6244-24073.txt and GPL8490-65.txt in GEO). Table 6.26 shows the confusion matrix and the results of Fisher's exact test. For all three cases, PCA-based unsupervised FE identifies mRNAs and promoter methylation sites with significant overlaps. On the other hand, the overlap of genes associated with the progression from pre- to post-metastasis is much less than those associated with distinction between two cell lines. PCA-based unsupervised FE is powerful enough method to detect this slight overlap.

Figure 6.10 shows the scatter plot of PC loading shown in Fig. 6.9, which is averaged over within each of four classes. PC loading other than the first PC loading are mutually correlated between mRNA and methylation. This is coincident with

Table 6.26 Confusion matrices and associated with P values and odds ratio

Distinction between cell lines					
	Methylation ($\ell = 2$)	Not selected	Selected	P value	Odds ratio
mRNA	Not selected	13065	340	1.39×10^{-24}	7.12
($\ell = 2$)	selected	286	53		
Distinction between pre- and post-metastasis for HTB56 cell line					
	Methylation ($\ell = 3$)	Not selected	Selected	P value	Odds ratio
mRNA	Not selected	13252	313	0.04	2.24
($\ell = 3$)	selected	170	9		
Distinction between pre- and post-metastasis for A549 cell line					
	Methylation ($\ell = 4$)	Not selected	Selected	P value	Odds ratio
mRNA	Not selected	13402	232	0.01	3.33
($\ell = 5$)	Selected	104	6		

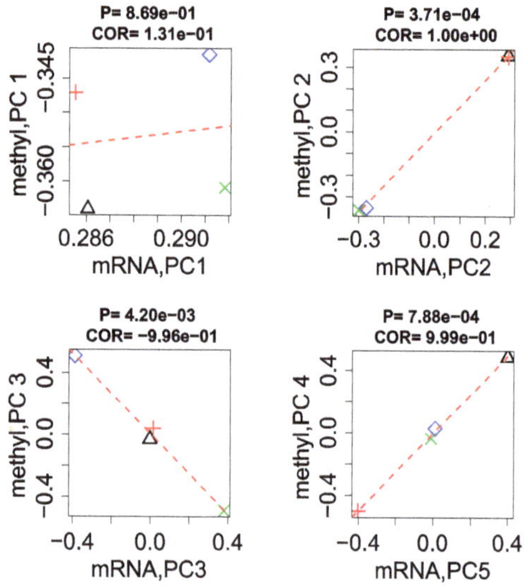

Fig. 6.10 Scatter plot of PC loading in Fig. 6.9 averaged within four classes. Broken straight lines are liner regression. △: A549 cell line pre-metastasis, +: A549 cell line post-metastasis, ×: HTB56 cell line pre-metastasis, ◇: HTB56 cell line post-metastasis

Table 6.26 where significant overlap of selected genes between mRNA expression and promoter methylation is detected.

In order to confirm the superiority of PCA-based unsupervised FE toward conventional supervised methods, we apply t-test as follows:

1. Six samples in A549 cell lines vs. six samples in HTB56 cell lines for mRNA expression
2. Four samples in A549 cell lines vs. four samples in HTB56 cell lines for promoter methylation
3. Three samples in pre-metastasis vs. three samples in post-metastasis for A549 mRNA expression
4. Two samples in pre-metastasis vs. two samples in post-metastasis for A549 promoter methylation

5. Three samples in pre-metastasis vs. three samples in post-metastasis for HTB56 mRNA expression
6. Two samples in pre-metastasis vs. two samples in post-metastasis for HTB56 promoter methylation

Among the above comparisons, (3), (4), and (6) have no probes associated with adjusted P values less than 0.01. Comparison 5. has only five probes associated with adjusted P values less than 0.01. Thus, t-test is useless for the identification of mRNAs and promoter methylation distinct between pre- and post-metastasis. Comparisons 1. and 2. have 7074 and 2186 probes associated with adjusted P values less than 0.01. Nevertheless, there are not significant overlaps between genes to which these probes are attributed ($P = 0.57$ and odds ratio is as small as 0.97). In addition to this, we also apply categorical regression analysis assuming four classes to mRNA expression and promoter methylation independently. It identifies 7501 and 6573 probes, respectively. Nevertheless, odds ratio of overlap detection between genes associated with identified probes is 0.67, which is even smaller than the expectation, 1.0, for random selections. Thus, PCA-based unsupervised FE can outperform conventional supervised methods.

Finally, we need to validate identified genes biologically as usual. We upload 15 genes shown in Table 6.27 to the molecular signatures database (MSigDB) [47] with specifying "C2: curated gene sets," which is composed of "CGP: chemical

Table 6.27 List of nine genes chosen for the distinction between metastasis of HTB56 cell lines in common between mRNA expression and promoter methylation, and six genes chosen for the distinction between metastasis of A549 cell lines in common between mRNA expression and promoter methylation

Refseq ID	Gene symbol	Gene Name
Distinction between pre- and post-metastasis for HTB56 cell line		
NM_153608	ZNF114	Zinc finger protein 114
NM_152457	ZNF597	Zinc finger protein 597
NM_152753	scube3	Signal peptide, CUB domain, and EGF like domain containing 3
NM_000793	DIO2	Deiodinase, Iodothyronine type II
NM_002145	HOXB2	Homeobox B2
NM_032040	CCDC8	Coiled-coil domain containing 8
NM_004613	TGM2	Transglutaminase 2
NM_001275	CHGA	Chromogranin A)
NM_006762	LAPTM5	Lysosomal protein transmembrane 5
Distinction between pre- and post-metastasis for A549 cell line		
NM_000201	ICAM1	Intercellular adhesion molecule 1
NM_005562	LAMC2	Laminin subunit gamma 2
NM_002996	CX3CL1	C-X3-C motif chemokine ligand 1
NM_020182	PMEPA1	Prostate transmembrane protein, androgen induced 1
NM_004633	IL1R2	Interleukin 1 receptor type 2
NM_022164	TINAGL1	Tubulointerstitial nephritis antigen like 1

and genetic perturbations," "CP: Canonical pathways," "CP:BIOCARTA: BioCarta gene sets, " " CP:KEGG: KEGG gene sets," and "CP:REACTOME: Reactome gene sets." Table 6.28 show the results. In total, as many as 46 gene sets are significantly overlapped with uploaded 15 genes (false discovery rate (FDR) q values are less than 0.05). Twenty seven out of 46 gene sets are directly related to tumors and cancers (asterisked). The fact that more than half of identified gene sets are cancer and tumor related demonstrates the ability of PCA-based unsupervised FE that can select biologically reliable gene sets. In addition to this, it is rare that as small as 15 genes have such a huge number of enriched terms, because P values computed by enrichment analysis have tendency to increase, i.e., to become less significant, as the number of genes is smaller. This suggests that PCA-based unsupervised FE has ability to identify a small number of critical genes also from biological point of views.

6.5.2 *Epigenetic Therapy Target Identification Based upon Gene Expression and Methylation Profile*

As mentioned in the previous subsection, cancer therapy is always difficult. In order to challenge this difficult task, other than usual therapies, epigenetic therapy recently collects many researchers' interest, because epigenetic is expected to affect cancer initiation and progression [40]. Thus, conversely, modifying epigenetic profile might contribute to the cancer therapy [3]. One possible difficulty of epigenetic therapy is identification of target genes. In contrast to small molecule drug that has target proteins to which small molecule binds, epigenetic therapy generally targets the alteration of epigenetic profiles, e.g., promoter methylation and histone modification. Thus, it is unclear which genes are targeted by individual epigentic therapy. Because PCA-based unsupervised FE has ability to identify DEGs associated with methylation alteration, PCA-based unsupervised FE is expected to be fitted to detect genes targeted by epigenetic therapy.

In this data set, we analyze mRNA expression and methylation profiles before and after reprogramming, which means to add pluripotency to differentiated cells, of various cancer cell lines. Here, pluripotency is the ability of cells that can differentiate into any kind of cells. The reason why we analyze gene expression profiles of reprogrammed cells is because methylation profiles altered during reprogramming and associated with altered mRNA expression are potential targets of epigenetic therapy. The data sets we analyze [57] are taken from GEO with ID GSE35913. They consist of eight cell lines, H1 (ES cell), H358 and H460 (NSCLC), IMR90 (Human Caucasian fetal lung fibroblast), iPCH358, iPCH460, iPSIMR90 (reprogrammed cell lines), and piPCH358 (redifferentiated iPCH358) with three biological replicates. In total, there were three replicates $\times$ 8 cell lines $\times$ 2 properties (gene expression and promoter methylation) = 48 samples. It is a typical multi-class

Table 6.28 Enrichment analysis by MSigDB. #1 : # Genes in Gene Set (K) #2: # Genes in Overlap (k), P: p value, Q: FDR q value. Cancer or tumor related terms are asterisked. "XE-Y" means $X \times 10^{-Y}$

Gene set name	#1	Description	#2	k/K	P	Q
BOYAULT_LIVER_CANCER _SUBCLASS_G5_DN	27	* Downregulated genes in hepatocellular carcinoma (HCC) subclass G5, defined by unsupervised clustering.	3	0.1111	2.97E-08	1.42E-04
KHETCHOUMIAN_TRIM24 _TARGETS_UP	47	* Retinoic acid-responsive genes upregulated in hepatocellular carcinoma (HCC) samples of TRIM24 knockout mice.	3	0.0638	1.64E-07	3.92E-04
ONDER_CDH1_TARGETS _2_DN	464	* Genes downregulated in HMLE cells (immortalized nontransformed mammary epithelium) after E-cadhedrin (CDH1) knockdown by RNAi.	4	0.0086	3.20E-06	2.78E-03
RASHI_RESPONSE _TO_IONIZING_RADIATION_2	127	Cluster 2: late ATM dependent genes induced by ionizing radiation treatment.	3	0.0236	3.35E-06	2.78E-03
KRISHNAN_FURIN _TARGETS_UP	12	Genes upregulated in naive T lymphocytes lacking FURIN : Cre-Lox knockout of FURIN in CD4+ cells.	2	0.1667	3.43E-06	2.78E-03
KIM_RESPONSE_TO _TSA_AND_DECITABINE_UP	129	* Genes upregulated in glioma cell lines treated with both decitabine and TSA .	3	0.0233	3.51E-06	2.78E-03
DARWICHE_SQUAMOUS _CELL_CARCINOMA_UP	146	* Genes upregulated in squamous cell carcinoma (SCC) compared to normal skin.	3	0.0205	5.09E-06	3.09E-03
DARWICHE_PAPILLOMA _RISK_HIGH_UP	147	* Genes upregulated during skin tumor progression from normal skin to high-risk papilloma.	3	0.0204	5.19E-06	3.09E-03

(continued)

Table 6.28 (continued)

Gene set name	#1	Description	#2	k/K	P	Q
PETROVA_ENDOTHELIUM _LYMPHATIC_VS_BLOOD_DN	162	Genes downregulated in BEC (blood endothelial cells) compared to LEC (lymphatic endothelial cells)	3	0.0185	6.95E-06	3.68E-03
SMID_BREAST_CANCER _BASAL_UP	648	* Genes upregulated in basal subtype of breast cancer samples.	4	0.0062	1.19E-05	5.69E-03
AFFAR_YY1_TARGETS _UP	214	Genes upregulated in MEF cells (embryonic fibroblast) expressing 25% of YY1.	3	0.014	1.60E-05	6.52E-03
BOQUEST_STEM _CELL_DN	216	Genes downregulated in freshly isolated CD31- (stromal stem cells from adipose tissue) versus the CD31+ (non-stem) counterparts.	3	0.0139	1.64E-05	6.52E-03
WONG_ADULT_TISSUE _STEM_MODULE	721	The “adult tissue stem” module: genes coordinately upregulated in a compendium of adult tissue stem cells.	4	0.0055	1.82E-05	6.65E-03
SENESE_HDAC1_AND _HDAC2_TARGETS_UP	238	* Genes upregulated in U2OS cells (osteosarcoma) upon knockdown of both HDAC1 and HDAC2 by RNAi.	3	0.0126	2.19E-05	7.17E-03
HOLLERN_SOLID_NODULAR _BREAST_TUMOR_DN	30	* Genes that have low expression in mammary tumors of solid nodular histology.	2	0.0667	2.26E-05	7.17E-03
KOINUMA_TARGETS_OF _SMAD2_OR_SMAD3	824	Genes with promoters occupied by SMAD2 or SMAD3 in HaCaT cells (keratinocyte) according to a ChIP-chip analysis.	4	0.0049	3.06E-05	8.58E-03
SHIN_B_CELL _LYMPHOMA_CLUSTER_8	36	* Cluster 8 of genes distinguishing among different B lymphocyte neoplasms.	2	0.0556	3.27E-05	8.64E-03
HELLER_SILENCED_BY _METHYLATION_UP	282	* Genes upregulated in at least one of three multiple myeloma (MM) cell lines treated with the DNA hypomethylating agent decitabine (5-aza-2’-deoxycytidine).	3	0.0106	3.64E-05	9.12E-03

PHONG_TNF_RESPONSE _NOT_VIA_P38	337	* Genes whose expression changes in Calu-6 cells (lung cancer) by TNF were not affected by p38 inhibitor LY479754.	3	0.0089	6.17E-05	1.44E-02
VILIMAS_NOTCH1 _TARGETS_UP	52	Genes upregulated in bone marrow progenitors by constitutively active NOTCH1.	2	0.0385	6.86E-05	1.44E-02
SCHUETZ_BREAST_CANCER _DUCTAL_INVASIVE_UP	351	* Genes upregulated in invasive ductal carcinoma (IDC) relative to ductal carcinoma in situ (DCIS, noninvasive).	3	0.0085	6.97E-05	1.44E-02
KRIEG_HYPOXIA _VIA_KDM3A	53	* Genes dependent on KDM3A for hypoxic induction in RCC4 cells (renal carcinoma) expressing VHL.	2	0.0377	7.13E-05	1.44E-02
NABA_MATRISOME	1028	Ensemble of genes encoding extracellular matrix and extracellular matrix-associated proteins	4	0.0039	7.25E-05	1.44E-02
PID_TXA2PATHWAY	57	Thromboxane A2 receptor signaling	2	0.0351	8.25E-05	1.57E-02
PHONG_TNF _TARGETS_UP	63	* Genes upregulated in Calu-6 cells (lung cancer) at 1 h time point after TNF treatment.	2	0.0317	1.01E-04	1.85E-02
PID_INTEGRIN1 _PATHWAY	66	Beta1 integrin cell surface interactions	2	0.0303	1.11E-04	1.93E-02
SATO_SILENCED_BY _METHYLATION_IN _PANCREATIC_CANCER_1	419	* Genes upregulated in the pancreatic cancer cell lines (AsPC1, Hs766T, MiaPaCa2, Panc1) but not in the non-neoplastic cells (HPDE) by decitabine (5-aza-2'-deoxycytidine).	3	0.0072	1.18E-04	1.93E-02
TURASHVILI_BREAST _NORMAL_DUCTAL_VS _LOBULAR_UP	68	Genes upregulated in normal ductal and normal lobular breast cells.	2	0.0294	1.18E-04	1.93E-02
DELYS_THYROID _CANCER_UP	443	* Genes upregulated in papillary thyroid carcinoma (PTC) compared to normal tissue.	3	0.0068	1.39E-04	2.20E-02

(continued)

Table 6.28 (continued)

Gene Set Name	#1	Description	#2	k/K	P	Q
RADMACHER_AML _PROGNOSIS	78	* The "Bullinger validation signature" used to validate prediction of prognostic outcome of acute myeloid leukemia (AML) patients with a normal karyotype.	2	0.0256	1.55E-04	2.38E-02
HELLEBREKERS_SILENCED _DURING_TUMOR_ANGIOGENESIS	80	* Genes downregulated in tumor-conditioned vs. quiescent endothelial cells and upregulated upon treatment with decitabine and TSA.	2	0.025	1.63E-04	2.38E-02
LIEN_BREAST_CARCINOMA _METAPLASTIC_VS_DUCTAL_UP	83	* Genes upregulated between two breast carcinoma subtypes: metaplastic (MCB) and ductal (DCB).	2	0.0241	1.75E-04	2.38E-02
SANA_TNF _SIGNALING_UP	83	Genes upregulated in five primary endothelial cell types (lung, aortic, iliac, dermal, and colon) by TNF.	2	0.0241	1.75E-04	2.38E-02
KIM_GLIS2 _TARGETS_UP	84	Partial list of genes upregulated in the kidney of GLIS2 knockout mice compared to the wild type.	2	0.0238	1.80E-04	2.38E-02
RASHI_RESPONSE_TO _IONIZING_RADIATION_6	84	Cluster 6: late responding genes activated in ATM deficient but not in the wild type tissues.	2	0.0238	1.80E-04	2.38E-02
KANG_IMMORTALIZED _BY_TERT_UP	89	Upregulated genes in the signature of adipose stromal cells (ADSC) immortalized by forced expression of telomerase (TERT).	2	0.0225	2.02E-04	2.60E-02
ENK_UV_RESPONSE _KERATINOCYTE_UP	530	Genes upregulated in NHEK cells (normal epidermal keratinocytes) after UVB irradiation.	3	0.0057	2.35E-04	2.94E-02
BASSO_CD40 _SIGNALING_UP	101	* Gene upregulated by CD40 signaling in Ramos cells (EBV negative Burkitt lymphoma).	2	0.0198	2.60E-04	3.17E-02
SMID_BREAST _CANCER_LUMINAL_B_DN	564	* Genes downregulated in the luminal B subtype of breast cancer.	3	0.0053	2.82E-04	3.36E-02
GHANDHI_DIRECT _IRRADIATION_UP	110	Genes significantly (FDR $< 10\%$) upregulated in IMR-90 cells (fibroblast) in response to direct irradiation.	2	0.0182	3.08E-04	3.58E-02

WANG_ESOPHAGUS _CANCER_VS_NORMAL_UP	121	* Upregulated genes specific to esophageal adenocarcinoma (EAC) relative to normal tissue.	2	0.0165	3.72E-04	4.22E-02
BUYTAERT_PHOTODYNAMIC _THERAPY_STRESS_DN	637	* Genes downregulated in T24 (bladder cancer) cells in response to the photodynamic therapy (PDT) stress.	3	0.0047	4.03E-04	4.46E-02
ZHONG_SECRETOME_OF _LUNG_CANCER_AND_FIBROBLAST	132	* Proteins secreted in coculture of LKR-13 tumor cells (non-small cell lung cancer, NSCLC) and MLg stroma cells (fibroblasts).	2	0.0152	4.43E-04	4.79E-02
PANGAS_TUMOR _SUPPRESSION_BY_SMAD1 _AND_SMAD5_UP	134	* Genes upregulated in ovarian tumors from mouse models for the BMP SMAD signaling (gonad specific double knockout of SMAD1 and SMAD5).	2	0.0149	4.56E-04	4.83E-02
PROVENZANI _METASTASIS_DN	136	* Genes downregulated in polysomal and total RNA samples from SW480 cells (primary colorectal carcinoma, CRC) compared to the SW620 cells (lymph node metastasis from the same individual).	2	0.0147	4.70E-04	4.86E-02

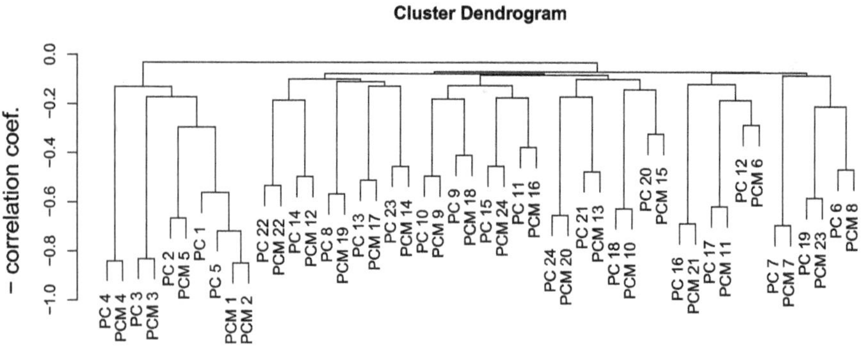

Fig. 6.11 Hierarchical clustering (UPGMA) applied to set of 24 $\boldsymbol{v}_\ell^{\text{mRNA}}$ (labeled as PCℓ) and 24 $\boldsymbol{v}_\ell^{\text{methyl}}$ (labeled as PCMℓ) with using negative signed Pearson's correlation coefficients as distance. $\boldsymbol{v}_\ell^{\text{mRNA}}$ and $\boldsymbol{v}_\ell^{\text{methyl}}$ for $\ell = 3$ and 4, i.e., four edges on the left end, are paired with high correlations

data set because there are no clear one to one correspondence. Then we apply PCA-based unsupervised FE as well as categorical regression analysis to this data set.

First, PCA is applied to mRNA expression, $x_{ij}^{\text{mRNA}} \in \mathbb{R}^{47321\times 24}$, and methylation profiles, $x_{kj}^{\text{methyl}} \in \mathbb{R}^{25728\times 24}$, and compute PC loading, $\boldsymbol{v}_\ell^{\text{mRNA}} \in \mathbb{R}^{24}$ and $\boldsymbol{v}_\ell^{\text{methyl}} \in \mathbb{R}^{24}$, which are attributed to samples. In order to perform integrated analysis of mRNA expression and methylation profile, we need to know pairs of PC loading of mRNA expression and promoter methylation associated with reciprocal relationship. For this purpose, we apply unweighted pair group method using arithmetic average (UPGMA) to set of 24 $\boldsymbol{v}_\ell^{\text{mRNA}}$ and 24 $\boldsymbol{v}_\ell^{\text{methyl}}$ with using negative signed Pearson's correlation coefficients as distance. Figure 6.11 shows the result of UPGMA. It is obvious that $\boldsymbol{v}_3^{\text{mRNA}}$ and $\boldsymbol{v}_3^{\text{methyl}}$, and $\boldsymbol{v}_4^{\text{mRNA}}$ and $\boldsymbol{v}_4^{\text{methyl}}$, are paired with high correlations (correlation coefficient $\sim$ 0.9). Figure 6.12 shows the selected PC loading. It is obvious that they have dependence upon eight classes. Especially, it is remarkable that four classes that represent reprogrammed cells ("iPCH358," "iPCH460," "iPSIMR90," and "piPCH358") have almost same values. Thus, mRNAs and methylation associated with these PC loading likely exhibit the distinction between pre- and post-reprogramming.

The algorithm of UPGMA is as follows. Suppose there are N features to be clustered and pairwise distances $d_{ii'} \in \mathbb{R}^{N\times N}$ between ith and i'th features are available:

1. Find a pair i and i' with minimum distance $d_{ii'}$.
2. Merge i and i' into a newly generated *pseudo* feature i''.
3. Compute pairwise distance between i'' and $i_0 \neq i, i'$s as

$$d_{i''i_0} = \frac{d_{ii_0} + d_{i'i_0}}{2}. \tag{6.55}$$

4. If there are more than one feature, go back to step 1.

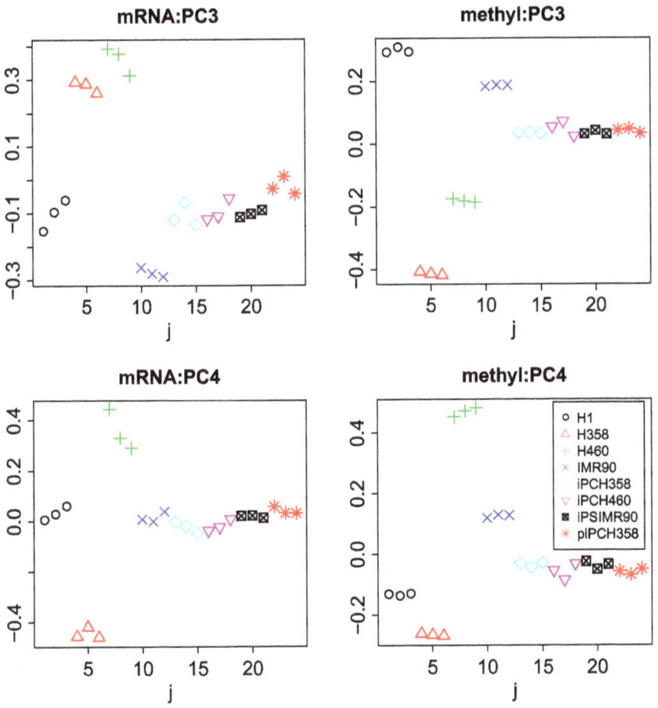

Fig. 6.12 PC loading, $\boldsymbol{v}_\ell^{\mathrm{mRNA}}$ and $\boldsymbol{v}_\ell^{\mathrm{methyl}}$ for $\ell = 3, 4$

Table 6.29 The number of probes selected using P values computed by Eqs. (6.53) and (6.54)

	mRNA		Methylation	
ℓ	3	4	3	4
Number of selected probes	283	310	200	199

In order to confirm the stability of the selection of pairs $\ell = 3, 4$, we systematically remove one of the 24 samples and apply UPGMA to 23 samples. Although $\ell = 3, 4$ are not always clustered together, four PC loading that are most similar to PC loading $\ell = 3$ or 4 when all 24 samples are used are always clustered together. Thus, the selection of $\ell = 3, 4$ as features clustered together is robust.

Then we attribute P values to probes using corresponding PC scores using χ^2 distribution as Eqs. (6.53) and (6.54) where ℓs listed beside these equations are replaced with $\ell = 3, 4$. P values are adjusted by BH criterion. Then probes associated with adjusted P values less than 0.05 are selected. Table 6.29 lists the number of probes selected. At least, for all cases, PCA-based unsupervised FE can identify probes with significant P values.

Because we are aiming to perform integrated analysis of mRNA expression and promoter methylation, it is important to see if genes selected for mRNA expression and promoter methylation are significantly overlapped. In order to do this, mRNA probes and methylation probes are converted into the list of genes to which probes

Table 6.30 Confusion matrices and associated with P values and odds ratio

	Methylation	Not selected	Selected	P value	Odds ratio
PCA-based unsupervised FE: $\ell = 3$					
mRNA	Not selected	13118	191	0.04	2.25
	Selected	274	9		
PCA-based unsupervised FE: $\ell = 4$					
mRNA	Not selected	13092	190	0.05	2.06
	Selected	301	9		
Categorical regression					
mRNA	Not selected	1180	6294	3.79×10^{-3}	0.87
	Selected	1080	5038		

are attributed (the correspondence between probes and genes is available as files GPL8490-65.txt in GEO). Table 6.30 shows the confusion matrix and the results of Fisher's exact test. For all three cases, PCA-based unsupervised FE identifies mRNAs and promoter methylation sites with significant overlaps. We also apply categorical regressions and found that 11332 and 5038 probes are associated with adjusted P values less than 0.05 for mRNA expression and promoter methylation. On the other hand, the overlap of genes associated with categorical regression analysis has odds ratio less than 1.0 (0.87), which suggests that overlaps are less than random selection (the small P value assigned means that overlap is significantly *less* than that expected when the selection is random). Thus, categorical regression analysis fails to identify genes associated with aberrant mRNA expression and promoter methylation simultaneously. PCA-based unsupervised FE is a powerful enough method to detect this slight overlap.

Finally, we need to validate identified genes biologically. We upload 18 genes shown in Table 6.31 to MSigDB [47] with specifying "C2: curated gene sets." In total, as many as 85 gene sets are significantly overlapped with uploaded 18 genes (false discovery rate (FDR) q values are less than 0.05). It takes three tables to display 46 gene sets (Table 6.28 show the results). If we list all 85 gene sets here, it will take more than six tables which is simply annoying. Thus, we are not willing to list all of them here. Forty five out of 86 gene sets are listed because they are directly related to tumors and cancers. "C2: curated gene sets" is composed of "CGP: chemical and genetic perturbations," "CP: Canonical pathways," "CP:BIOCARTA: BioCarta gene sets," "CP:KEGG: KEGG gene sets," and "CP:REACTOME: Reactome gene sets." The fact that more than half of identified gene sets are cancer and tumor related demonstrates the ability of PCA-based unsupervised FE that can select biologically reliable gene sets. In addition to this, P values computed by enrichment analysis generally have tendency to increase as the number of genes is smaller. Thus, it is rare that as small as 18 genes have such a huge number of enriched gene sets. This suggests that PCA-based unsupervised FE has ability to identify a small number of critical genes also from a biological point of views.

Table 6.31 List of nine genes chosen by PCA-based unsupervised FE with $\ell = 3, 4$ in common between mRNA expression and promoter methylation (Table 6.30)

Refseq ID	Gene symbol	Gene name
$\ell = 3$		
NM_213606	SLC16A12	Solute carrier family 16 member 12
NM_004321	KIF1A	Kinesin family member 1A
NM_015881	DKK3	Dickkopf WNT signaling pathway inhibitor 3
NM_014220	TM4SF1	Transmembrane 4 L six family member 1
NM_003012	SFRP1	Secreted frizzled related protein 1
NM_019102	HOXA5	Homeobox A5
NM_001458	FLNC	Filamin C
NM_201525	ADGRG1	Adhesion G-protein-coupled receptor G1
NM_001992	F2R	Coagulation factor II thrombin receptor
$\ell = 4$		
NM_000393	COL5A2	Collagen type V alpha 2 chain
NM_002727	SRGN	Serglycin
NM_005558	LAD1	Ladinin 1
NM_012307	EPB41L3	Erythrocyte membrane protein band 4.1 like 3
NM_005562	LAMC2	Laminin subunit gamma 2
NM_000993	RPL31	Ribosomal protein L31
NM_201525	ADGRG1	Adhesion G-protein-coupled receptor G1
NM_004360	CDH1	Cadherin 1
NM_002354	EPCAM	Epithelial cell adhesion molecule

Before closing this subsection, we would like to add a few biological supportive evidences that 18 genes in Table 6.31 likely include genes targeted by epigenetic therapy more directly.

The first evidence is the comparison with cell lines resistant to epigenetic therapy [32]. Histone deacetylase (HDAC) inhibitor is one of promising epigenetic therapy. Histone acetylation is generally believed to accelerate gene transcription. Thus, deacetylation is supposed to deactivate genes. In this regard, HDAC inhibitor suppresses the deactivation of genes by histone deacetylase. Miyanaga et al. [32] compared various cell lines to determine whether they were resistant to HDAC inhibitors. We investigated SFRP1 expression, which is in Table 6.31, between HDAC inhibitor-resistant cell lines and nonresistant cell lines for adenocarcinoma and squamous cell carcinoma and found different levels of SFRP1 expression (Table 6.32). SFRP1 expression is likely targeted by HDAC inhibitor because its expression decreases in cells resistant to HDAC inhibitor that should reactivate target genes. On the other hand, DKK3 which is also in Table 6.30 is not consistently affected by HDAC inhibitor. Thus, SFRP1 is more likely to be a HDAC inhibitor target in cancer therapy than DKK3 although both are in selected 18 genes (Table 6.31).

The second evidence is the alteration of histone acetylation by HDAC inhibitor shown in Table 6.32; HDAC inhibitor reduces the histone acetylation of SFRP1 [61]

Table 6.32 Gene expression difference between nonresistant and resistant cell lines to HDAC inhibitor as well as alteration of histone acetylation treated by HDAC inhibitor

Gene expression				
Adenocarcinoma				
		P value	Nonresistant cell lines	Resistant cell lines
SFRP1		4.64×10^{-4}	611.06	>92.60
DKK3		6.73×10^{-2}	263.27	> 30.59
Squamous cell carcinoma				
SFRP1		7.42×10^{-3}	304.53	>49.53
DKK		4.61×10^{-1}	261.38	< 506.25
Histone modification (H3K9K14ac)				
SFRP1	Cell line	*P* value	0 h	2 h
	(A549)	2.90×10^{-2}	−1.29	< −0.52
	(H1299)	4.06×10^{-2}	−2.51	<−1.85
	(CL1-1)	8.71×10^{-1}	−1.38	<−1.34
DKK3	(A549)	6.19×10^{-1}	−1.17	< −1.01
	(H1299)	1.98×10^{-3}	−1.70	<−0.48
	(CL1-1)	1.48×10^{-1}	−0.59	>−1.13
SALL4	(A549)	1.71×10^{-3}	−2.44	< −1.05
	(H1299)	5.23×10^{-1}	−2.62	>−2.86
	(CL1-1)	1.03×10^{-4}	0.997	> −0.59

for A549 and H1299 cell lines that are generated from non-small cell lung cancer (NSCLC), from which H358 and H640 whose gene expression and methylation level are analyzed in this chapter are generated. DKK3 and SALL4 are less consistently affected by HDAC inhibitor than SFRP1 for these two cell lines. On the other hand, when HDAC inhibitor is used for CL1-1 cell lines that are generated from cervix are not consistent at all for SFRP1, SALL4, and DKK3. Thus, SFRP1 is most likely a target of epigenetic therapy toward NSCLC. In conclusion, PCA-based unsupervised FE is an effective method to integrate methylation profile and mRNA expression as in the integrated analysis of mRNA and miRNA expression.

6.5.3 Identification of Genes Mediating Transgenerational Epigenetics Based upon Integrated Analysis of mRNA Expression and Promoter Methylation

Transgenerational epigenetics (TGE) [63] is one of recently established but important topics on evolution. Because of central dogma, it is generally believed that only heritable information is that stored in DNA sequence. On the other hand, there might be some other ways that transfer information intergenetically. Epigenetics that means alteration of genome without changing DNA sequence might transfer

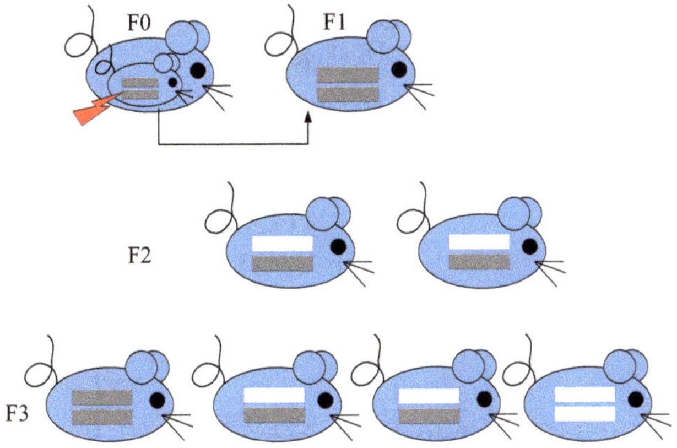

Fig. 6.13 F1 generation is exposed to endocrine disruptor during F0 generation is pregnant (red thunder mark). In F1 generation, both chromosome is directly affected by endocrine disruptor (gray rectangular). In F2 generation, all chromosomes is a pair of chromosome affected directly by endocrine disruptor (gray rectangular) and that not (white rectangular). In F3 generation, one fourth individuals (right end one) have no chromosomes affected directly by endocrine disruptor (white rectangular)

Table 6.33 List of files used in this chapter

GSE43559 (gene expression)		GSE59510 (promoter methylation)	
GEO ID	Description	GEO ID	Description
GSM1065332	PGC E13 F3-Control biological rep1	GSM1438556	E16-Vip2/Cip2
GSM1065333	PGC E13 F3-Control biological rep2	GSM1438557	E13-Vip2/Cip1
GSM1065334	PGC E13 F3-Vinclozolin biological rep1	GSM1438558	E13-Vip1/Cip1
GSM1065335	PGC E13 F3-Vinclozolin biological rep2	GSM1438559	E16-Vip1/Cip1
GSM1065336	PGC E16 F3-Control biological rep1	GSM1438560	E16-Vip2/Cip1
GSM1065337	PGC E16 F3-Control biological rep2	GSM1438561	E13-Vip2/Cip2
GSM1065338	PGC E16 F3-Vinclozolin biological rep1		
GSM1065339	PGC E16 F3-Vinclozolin biological rep2		

information intergenetically. If so, it can be an alternative important factor that can contribute to evolution. In spite of that, TGE is not completely understood.

One possible way to study TGE is to study the effect of endocrine disruptors toward embryo. The reason is as follows. First of all, expose of embryo to endocrine disruptors is known to cause some disease. Thus, at least, we can expect to detect the effect of it no matter what it is. Secondary, by preparing the clone animals, we can guarantee that expose to endocrine disruptors does not alter DNA sequence. Third, by considering F3 generation, we can expect that DNA is not directly exposed to endocrine disruptor (Fig. 6.13). Skinner et al [43] performed this kind of experiments. We apply PCA-based unsupervised FE to their data set [49]. Data set analyzed is downloaded from GEO with GEO ID GSE59511. Table 6.33 shows the list of files used in this chapter. E13 and E16 correspond to 13 days and 16 days after the fertilization, respectively. Eight mRNA expression profiles, $x_{ij}^{\text{mRNA}} \in \mathbb{R}^{27342\times 8}$, are further converted into $\tilde{x}_{ij}^{\text{mRNA}} \in \mathbb{R}^{27342\times 8}$ as

$$\tilde{\boldsymbol{x}}_i^{\text{mRNA}} = \begin{pmatrix} \tilde{x}_{i1}^{\text{mRNA}} \\ \tilde{x}_{i2}^{\text{mRNA}} \\ \tilde{x}_{i3}^{\text{mRNA}} \\ \tilde{x}_{i4}^{\text{mRNA}} \\ \tilde{x}_{i5}^{\text{mRNA}} \\ \tilde{x}_{i6}^{\text{mRNA}} \\ \tilde{x}_{i7}^{\text{mRNA}} \\ \tilde{x}_{i8}^{\text{mRNA}} \end{pmatrix} = \begin{pmatrix} \frac{\text{E13 control rep 1}}{\text{E13 treated rep 1}} \\ \frac{\text{E13 control rep 2}}{\text{E13 treated rep 2}} \\ \frac{\text{E13 control rep 2}}{\text{E13 treated rep 1}} \\ \frac{\text{E13 control rep 1}}{\text{E13 treated rep 2}} \\ \frac{\text{E16 control rep 1}}{\text{E16 treated rep 1}} \\ \frac{\text{E16 control rep 2}}{\text{E16 treated rep 2}} \\ \frac{\text{E16 control rep 2}}{\text{E16 treated rep 1}} \\ \frac{\text{E16 control rep 1}}{\text{E16 treated rep 2}} \end{pmatrix} \tag{6.56}$$

because we cannot get PC loading distinct between E13 and E16 otherwise and promoter methylation is provided as ratio by the original studies' researchers. Methylation profiles, $x_{kj}^{\text{methyl}} \in \mathbb{R}^{14162\times 6}$, are used as it is, because it is provided as ratio between control and treated samples.

PCA is applied to $\tilde{x}_{ij}^{\text{mRNA}}$ and x_{kj}^{methyl} and PC loading $\boldsymbol{v}_\ell^{\text{mRNA}} \in \mathbb{R}^8$ and $\boldsymbol{v}_\ell^{\text{methyl}} \in \mathbb{R}^6$ attributed to samples are obtained. After investigation of obtained PC loading, we find that $\boldsymbol{v}_2^{\text{mRNA}}$ and $\boldsymbol{v}_1^{\text{methyl}}$ have distinction between E13 and E16 (Fig. 6.14).

P values are attributed to probes using χ^2 distribution as

$$P_i = P_{\chi^2}\left[> \left(\frac{u_{2i}^{\text{mRNA}}}{\sigma_2^{\text{mRNA}}} \right)^2 \right], \tag{6.57}$$

and

$$P_k = P_{\chi^2}\left[> \left(\frac{u_{1k}^{\text{methyl}}}{\sigma_1^{\text{methyl}}} \right)^2 \right]. \tag{6.58}$$

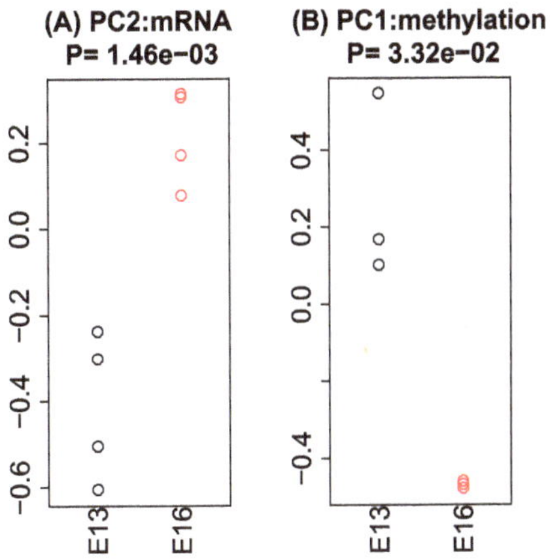

Fig. 6.14 (**a**) v_2^{mRNA} (**b**)v_1^{methyl}. P values are computed by t-test. Black open circle: E13, red open circle: E16

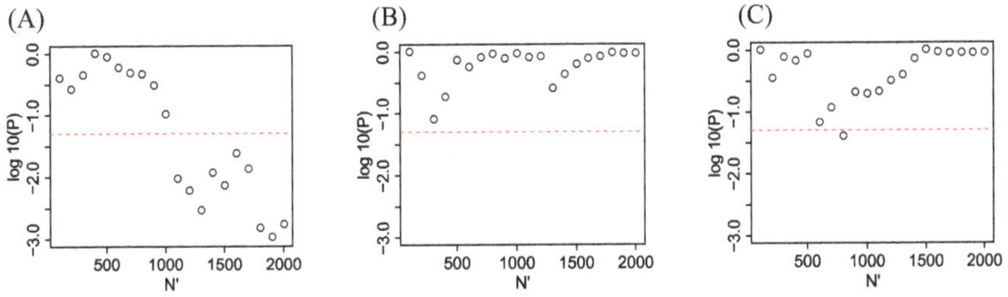

Fig. 6.15 P values computed by Fisher's exact that evaluates overlap between top N' selected genes of mRNA and methylation by (**a**) PCA-based unsupervised FE (**b**) t-test (**c**) limma. Horizontal broken line indicates $P = 0.05$

Unfortunately, no probes for methylation are associated with adjusted P values less than 0.05. Thus, we give up evaluating probes with P values. Instead, we decide to evaluate probes using overlap between mRNA and methylation. That is, we select top N' probes with smaller P values computed by Eqs. (6.57) and (6.58) for mRNA and methylation, respectively. Then compute P values with applying Fisher's exact test to evaluate overlaps of genes to which top N' probes are attributed for mRNA and methylation. Annotation of mRNA probes is available in the file GPL6247-249.txt. Annotation of methylation probes is in methylation profile files themselves.

In order to evaluate the number of genes chosen in common between mRNA and methylation, we also compute P values attributed to probes by two additional methods: t-test and Linear Models for Microarray Data (limma) [37]. Although limma is a simple linear regression analysis using logarithmic values, it is known to work pretty well for DEG identification. Figure 6.15 shows the dependence of P values computed by Fisher's exact test upon N' that is the number of top-ranked genes with smaller P values computed by PCA-based unsupervised FE, t-test, and limma. In contrast to PCA-based unsupervised FE which has P values less than 0.05 for $N' > 1000$, t-test does not fulfill this criterion $N' \leq 2000$, while more advanced limma can have only one N' associated with $P < 0.05$. Thus, from the point of integrated analysis of mRNA and methylation, PCA-based unsupervised FE outperforms other two conventional or popular methods for DEG identification.

Although PCA-based unsupervised FE successfully integrates mRNA expression and methylation profiles, if genes chosen in common between mRNA expression and methylation are not biologically valid, it is useless. In order to evaluate genes chosen in common between mRNA expression and methylation profiles, we upload 63 genes (Table 6.34) selected by PCA-based unsupervised FE when $N' = 1100$, which is minimum N' associated with $P < 0.05$, to DAVID. Table 6.35 lists gene ontology (GO) terms identified by DAVID. Here, GO terms [62] are composed of a human curated list of genes supposed to have biological concepts about biological process (BP), cellular components (CC), and molecular function (MF). Most enriched GO terms are related to olfactory receptor activity, which is known to be related to TGE [42]. Thus, not only PCA-based unsupervised FE can identify common genes between mRNA and methylation, but also identified GO terms are reasonable. PCA-based unsupervised FE is successful from the biological point of views, too.

6.6 Time Development Analysis

Analysis of temporal data set is another important topic of not only data science but also bioinformatics. For example, periodic motion often plays a critical role in biology. Typical examples where periodic motions play critical roles include heartbeats, circadian rhythm, and cell division cycle. For all of them, keeping the stability of periodicity is critically important. Thus, identification of genes that can contribute to periodic motion is also critical. Another example is development or disease progression. It is also important which genes drive these processes. From the view point of feature selection, the task is similar to those mentioned in the previous sections: to identify genes having time dependence. The only difference is that there are no clear definition of what the time dependence. In some sense, it is very closed to clustering. If we can find a set of genes that share similar time dependence, it might be the evidence that these are critical time dependence. The definition of periodicity is also unclear. Only definition of periodicity is that some function of time t, $f(t)$, should satisfy the condition that $f(t+T) = f(t)$ for all t in order to be a periodic function of period T. Nevertheless, because the time points measured are limited, it is usual that there are no pairs of points between whose time interval is exactly T. In this case, sinusoidal regression is often employed, in spite of that it is not guaranteed to capture all kinds of periodic motions because not all periodic functions are sinusoidal.

In the following subsections, we will demonstrate how effective is to employ PCA-based unsupervised FE in order to identify genes with time dependence. As mentioned in the above, it is quite difficult to assume the time dependent functional form to identify time dependent genes in advance. Because of this difficulty,

Table 6.34 63 mRNAs selected by PCA-based unsupervised FE

Refseq mRNA	Gene symbol	Description
NM_021866	CCR2	C-C motif chemokine receptor 2
NM_001000650	Olr624	Olfactory receptor 624
NM_001109617	PRAMEF27	PRAME family member 27
NM_017061	LOX	Lysyl oxidase
NM_012523	CD53	Cd53 molecule
NM_001033998	ITGAL	Integrin subunit alpha L
NM_022866	SLC13A3	Solute carrier family 13 member 3
NM_001109383	ANGPTL1	Angiopoietin-like 1
NM_001109118	ELOVL2	ELOVL fatty acid elongase 2
NM_001111269	LOC689064	Beta-globin
NM_001000551	Olr218	Olfactory receptor 218
NM_001107660	CAR1	Carbonic anhydrase I
NM_023968	npy2r	Neuropeptide Y receptor Y2
NM_053994	PDHA2	Pyruvate dehydrogenase (lipoamide) alpha 2
NM_001111321	Vom2r80	Vomeronasal 2 receptor, 80
NM_020104	MYL1	Myosin, light chain 1
NM_001000646	Olr635	Olfactory receptor 635
NM_001001071	Olr862	Olfactory receptor 862
NM_001000648	Olr633	Olfactory receptor 633
NM_001109218	RGD1565355	Similar to fatty acid translocase/CD36
NM_001000600	Olr796	Olfactory receptor 796
NM_001013952	LOC300308	Similar to hypothetical protein 4930509O22
NM_013025	CCL3	C-C motif chemokine ligand 3
NM_001000566	Olr542	Olfactory receptor 542
NM_022218	CMKLR1	Chemerin chemokine-like receptor 1
NM_013158	DBH	Dopamine beta-hydroxylase
NM_001109374	LRRTM1	Leucine rich repeat transmembrane neuronal 1
NM_021853	kcnt1	Potassium sodium-activated channel subfamily T member 1
NM_175586	TAAR7B	Trace amine-associated receptor 7b
NM_001008946	Vom1r29	Vomeronasal 1 receptor 29
NM_001047891	RGD1310507	Similar to RIKEN cDNA 1300017J02
NM_001008947	Vom1r34	Vomeronasal 1 receptor 34
NM_020071	FGB	Fibrinogen beta chain
NM_001080938	Tas2r124	Taste receptor, type 2, member 124
NM_012909	AQP2	Aquaporin 2
NM_030856	LRRN3	Leucine rich repeat neuronal 3
NM_001099492	Vom2r19	Vomeronasal 2 receptor, 19
NM_013149	AHR	Aryl hydrocarbon receptor
NM_001011892	SERPINF2	Serpin family F member 2
NM_001012224	NFE2	Nuclear factor, erythroid 2
NM_001013177	Sult1c2a	Sulfotransferase family, cytosolic, 1C, member 2a
NM_053843	FCGR2A	Fc fragment of IgG, low-affinity IIa, receptor

(continued)

Table 6.34 (continued)

Refseq mRNA	Gene symbol	Description
NM_001106056	TRIM52	Tripartite motif-containing 52
NM_001000523	Olr1381	Olfactory receptor 1381
NM_001007729	PF4	Platelet factor 4
NM_001000080	Olr1583	Olfactory receptor 1583
NM_001107036	MPO	Myeloperoxidase
NM_022696	HAND2	Heart and neural crest derivatives expressed 2
NM_001001053	Olr545	Olfactory receptor 545
NM_001024805	Hbe2	Hemoglobin, epsilon 2
NM_001000384	Olr408	Olfactory receptor 408
NM_001001362	Olr1059	Olfactory receptor 1059
NM_138537	LOC171573	Spleen protein 1 precursor
NM_001000896	Olr1726	Olfactory receptor 1726
NM_134326	ALB	Albumin
NM_001001017	Olr1143	Olfactory receptor 1143
NM_017105	BMP3	Bone morphogenetic protein 3
NM_012893	ACTG2	Actin, gamma 2, smooth muscle, enteric
NM_001000619	Olr727	Olfactory receptor 727
NM_001012112	ANKRD9	Ankyrin repeat domain 9
NM_001001114	Olr1701	Olfactory receptor 1701
NM_001108651	HEBP1	Heme binding protein 1
NM_001014222	Dmrtc1c1	DMRT-like family C1c1

unsupervised nature of PCA-based unsupervised FE works quite well. Let us start to try to identify genes that can drive cell division cycle.

6.6.1 Identification of Cell Division Cycle Genes

Cell division cycle is a primary process of living organisms. There are not living organisms that do not perform cell division, because only through cell division, living organism can develop or make the descendants. Thus, maintenance of stability of cell division cycle is quite critical. In this sense, identification of cell division cycle genes is very important for understanding living materials. Fortunately, studying cell division cycle is not difficult, because cell division can be observed even using unicellular organisms, which can often be cultured in Petri dish. Although cell structure of unicellular organisms like bacteria generally differs from that of multicellular organisms like human beings, fortunately there are some unicellular organisms that share the cell structure with multicellular organisms, e.g., yeast. Because of this ease of experiment, there is a long history of study of cell division cycles.

Table 6.35 GO terms detected by DAVID. %: the ratio of genes annotated, Benjamini: *P* values corrected by BH criterion

Category	Term	Count	%	*P* value	Benjamini
GOTERM_BP_DIRECT	GO:0007186 G-protein-coupled receptor signaling pathway	26	41.3	8.52×10^{-10}	3.57×10^{-7}
GOTERM_BP_DIRECT	GO:0050911 detection of chemical stimulus involved in sensory perception of smell	16	25.4	3.00×10^{-5}	6.26×10^{-3}
GOTERM_CC_DIRECT	GO:0016021 integral component of membrane	33	52.4	2.35×10^{-4}	2.07×10^{-2}
GOTERM_CC_DIRECT	GO:0072562 blood microparticle	5	7.94	5.80×10^{-4}	2.55×10^{-2}
GOTERM_MF_DIRECT	GO:0004984 olfactory receptor activity	16	25.4	5.83×10^{-5}	6.91×10^{-3}
GOTERM_MF_DIRECT	GO:0004930 G-protein-coupled receptor activity	17	27.0	7.26×10^{-5}	4.31×10^{-3}

In this subsection, we try to reanalyze gene expression profiles of yeast during mitotic cell division cycle (i.e., normal cell division cycle that is not related to reproductive processes) [51]. One possible obstacle of this experiment is synchronization. In natural state, individual yeasts perform cell division cycle with randomized phase. In other words, cell division always takes place in some individual yeast cells. Under such a condition, measuring gene expression might not exhibit any periodicity at all. In order to avoid such a situation, all yeast cells must be synchronized before experiments start. And almost only one possible way to perform synchronization is arresting cell cycle [22]. There are multiple ways to arrest cell cycle, e.g., the usage of mutant or cutting off the food supply. Cell cycle arresting has one problem; after cell cycle releasing arresting, cell cycles start to be desynchronized; living organisms have no benefits for cell cycle synchronization, which gradually vanishes and returns to randomized phase. If desynchronization is rapid, there are no ways to observe gene expression of cell division cycle for longer period. This results in again typical large p small n problem, i.e., small number of time points (often less than 100) versus a huge number of genes (a few thousands).

The first data set we analyze is yeast metabolic cycle [64]. Gene expression profile, $x_{ij} \in \mathbb{R}^{9335 \times 36}$, composed of 36 times points and 9335 genes. Thirty six time points are supposed to be composed of three cycles based upon external observations. Thus, it corresponds to the observation over three cycles. It can be downloaded as a file GSE3431_series_matrix.txt from GEO ID GSE3431. PCA is applied to x_{ij} and PC loading $\boldsymbol{v}_\ell \in \mathbb{R}^{36}$ is attributed to time points. Figure 6.16

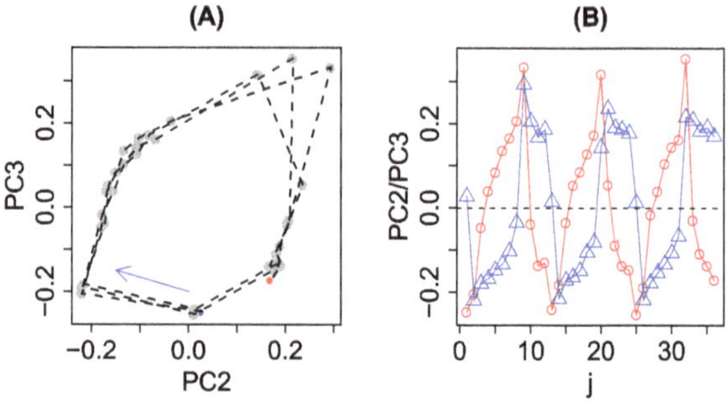

Fig. 6.16 (**a**) The scatter plot of $\boldsymbol{v}_\ell, \ell = 2, 3$. Blue filled circle, $j = 1$, red filled circle: $j = 36$, gray filled circle: $1 < j < 36$. Horizontal axis: $\ell = 2$ and vertical axis: $\ell = 3$. (**b**) Time dependence of PC loading $\boldsymbol{v}_\ell, \ell = 2, 3$. Blue: $\ell = 2$, red: $\ell = 3$

shows the $\boldsymbol{v}_\ell, \ell = 2, 3$. As expected, they are coincident with three periodic cycles. Although we never use the assumption that they are periodic motions, PCA correctly identifies periodic motions. In addition to this, time dependence of periodic motion is far from sinusoidal motion that is usually assumed. Furthermore, the functional shapes of $\boldsymbol{v}_2$ and $\boldsymbol{v}_3$ differ from each other so much. Although individual genes are expected to have time dependence of functional form of a liner combination of $\boldsymbol{v}_2$ and $\boldsymbol{v}_3$, these apparently differ from each other because of complete different functional forms of $\boldsymbol{v}_2$ and $\boldsymbol{v}_3$. Thus it is completely different from sinusoidal function for which each gene shares same functional form excluding the time shift. This suggests the limitation of employment of sinusoidal function to recognize periodic nature within gene expression. No fitting of specific functional forms to gene expression cannot identify periodic genes correctly.

In order to identify cell cycle regulated genes, we attribute P values to ith genes using χ^2 distribution as

$$P_i = P_{\chi^2}\left[> \sum_{\ell=2}^{3} \left(\frac{u_{\ell i}}{\sigma_\ell}\right)^2\right], \tag{6.59}$$

P_is are adjusted by BH criterion. Then we found that 298 probes are associated with adjusted P values less than 0.01 (Fig. 6.17).

It is not easy to evaluate selected genes without biological consideration because the functional form of PC loading, $\boldsymbol{v}_\ell, \ell = 2, 3$, is not a simple mathematical function. One possible evaluation is linear regression analysis that tests if selected genes are periodic or not. Then we perform two linear regression analyses,

$$x_{ij} = a_i + \sum_{\ell=2}^{3} b_{i\ell} v_{\ell j} \tag{6.60}$$

and

$$x_{ij} = a'_i + b'_{i1} \sin\left(\frac{2\pi j}{12}\right) + b'_{i2} \cos\left(\frac{2\pi j}{12}\right) \tag{6.61}$$

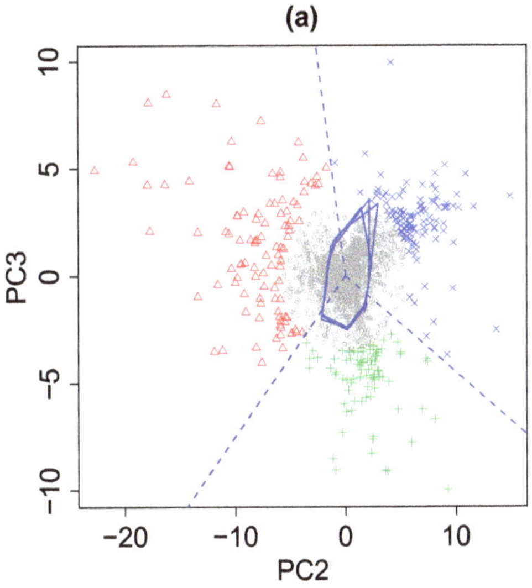

Fig. 6.17 Scatter plot of PC scores, $\boldsymbol{u}_\ell, \ell = 2, 3$. Points colored other than gray are selected 298 probes. Points displayed with red, green, and blue marks are three clusters identified by K-means assuming three clusters to $\boldsymbol{u}_\ell, \ell = 2, 3$

to selected 298 probes where $a_i, b_{i\ell}, a'_i, b'_{i1}, b'_{i2}$ are regression coefficients. Attributed P values are corrected by BH criterion, and the largest (i.e., the least significant) adjusted P values are 0.004 and 0.01, respectively. Thus, PCA-based unsupervised FE can correctly identify genes having period 12 in fully unsupervised manner.

One might wonder how we identify $\ell = 2, 3$ for periodic function without assuming periodicity. We plot $\boldsymbol{v}_\ell$ and $\boldsymbol{v}_{\ell'}$ and identify if they form limit cycle. In order to evaluate the amount that each trajectory is limit cycle, we compute winding number, which counts how many times each orbit moves round the origin anticlockwise direction (Fig. 6.18). It is obvious that $\boldsymbol{v}_\ell, \ell = 2, 3$ exhibit most clear limit cycle. Because identification of limit cycle does not require the knowledge about the periodicity in advance, we can identify $\boldsymbol{v}_\ell, \ell = 2, 3$ in fully unsupervised manner. One possible bi-product of winding number analysis is identification of periodic motion other than period of 12. $\boldsymbol{v}_\ell, \ell = 2, 4$ exhibit period doubling (eight letter shaped). Sinusoidal regression that assumes specific period cannot identify these motions. It is another advantage of using unsupervised methods.

It is also possible to select genes based upon linear regression analysis. We compare the performance of linear regression analysis with that by PCA-based unsupervised FE, by applying Eqs. (6.60) and (6.61) to not selected 298 but all gene expression profiles. P values are attributed to all genes with linear regression analysis, and obtained P values are adjusted by BH criterion. Then we select gene associated adjusted P values less than 0.01. This results in as many as 5598 genes by Eq. (6.60) and 4676 genes by Eq. (6.61), respectively, both of which are more than half of the total number of genes, 9335. Because it is too many, it is better to be screened based upon additional criterion. Nevertheless, no suitable criteria as FC when two classes are clearly defined are known for the detection of periodic motion.

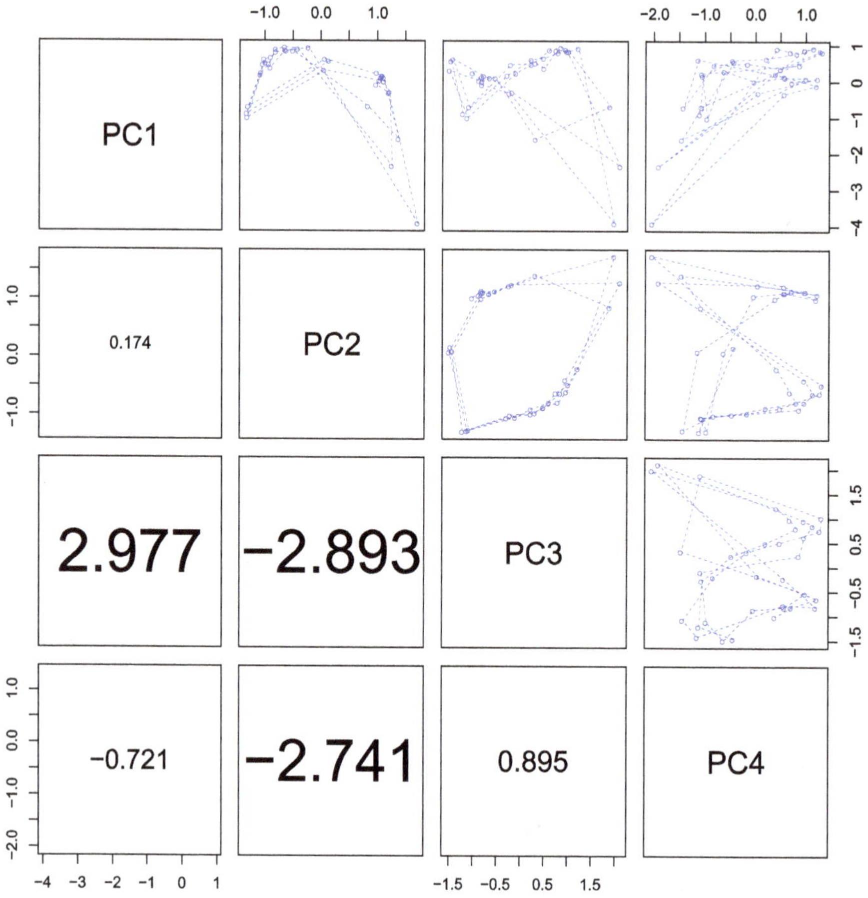

Fig. 6.18 Upper triangle: Scatter plot of PC loading, $\boldsymbol{v}_\ell$, $1 \leq \ell \leq 4$. Lower triangle: winding number (anticlockwise direction is positive)

Thus, reduction of the number of genes is not straightforward. Because PCA-based unsupervised FE can identify as small as 298 probes only with P values, it is more convenient than gene selection based upon liner regression analysis.

Finally, we would like to evaluate genes associated with the selected 298 probes biologically. Tu et al. [64] identified cell cycle regulated genes using linear regression analysis, Eq. (6.61), which they call sinusoidal regression. Then, classify them into three classes based upon the similarity of functional forms, with visual inspection of time course expression. They also recognized that these three classes are associated with specific biological function. In order to see if these three classes are reproduced in the present results, we apply K-means to selected 298 probes with $\boldsymbol{u}_\ell, \ell = 2, 3$. Three colored clusters in Fig. 6.17 correspond to three clusters obtained by K-means. Genes associated with probes in three clusters are separately uploaded to DAVID. Table 6.36 lists the significant KEGG pathway

Table 6.36 GO CC terms detected for genes in Fig. 6.17 by DAVID. %: the ratio of genes annotated, Benjamini: P values corrected by BH criterion

Category	Term	Count	%	P value	Benjamini
Red triangles in Fig. 6.17					
GOTERM_CC_DIRECT	GO:0005739 mitochondrion	36	42.4	4.11×10^{-6}	2.40×10^{-4}
GOTERM_CC_DIRECT	GO:0005886 plasma membrane	21	24.7	8.24×10^{-6}	2.43×10^{-4}
GOTERM_CC_DIRECT	GO:0005618 cell wall	8	9.4	4.27×10^{-5}	8.40×10^{-4}
GOTERM_CC_DIRECT	GO:0005777 peroxisome	8	9.4	9.57×10^{-5}	1.40×10^{-3}
GOTERM_CC_DIRECT	GO:0005576 extracellular region	9	10.6	1.46×10^{-4}	1.72×10^{-3}
GOTERM_CC_DIRECT	GO:0005741 mitochondrial outer membrane	8	9.4	3.82×10^{-4}	3.75×10^{-3}
GOTERM_CC_DIRECT	GO:0071944 cell periphery	13	15.3	3.96×10^{-4}	$3.33. \times 10^{-3}$
GOTERM_CC_DIRECT	GO:0005782 peroxisomal matrix	4	4.7	7.07×10^{-4}	5.02×10^{-3}
GOTERM_CC_DIRECT	GO:0009277 fungal-type cell wall	8	9.4	1.07×10^{-3}	7.03×10^{-3}
Green crosses in Fig. 6.17					
GOTERM_CC_DIRECT	GO:0005739 mitochondrion	48	71.6	1.27×10^{-19}	8.64×10^{-18}
GOTERM_CC_DIRECT	GO:0005758 mitochondrial intermembrane space	17	25.4	3.13×10^{-16}	1.13×10^{-14}
GOTERM_CC_DIRECT	GO:0005743 mitochondrial inner membrane	20	29.9	3.08×10^{-12}	6.99×10^{-11}
GOTERM_CC_DIRECT	GO:0070469 respiratory chain	6	9.0	5.71×10^{-7}	9.70×10^{-6}
GOTERM_CC_DIRECT	GO:0005759 mitochondrial matrix	11	16.4	1.12×10^{-6}	1.53×10^{-5}
GOTERM_CC_DIRECT	GO:0005750 mitochondrial respiratory chain complex III	5	7.5	4.29×10^{-6}	4.86×10^{-5}
GOTERM_CC_DIRECT	GO:0005749 mitochondrial respiratory chain complex II, succinate dehydrogenase complex (ubiquinone)	4	6.0	7.01×10^{-5}	6.81×10^{-4}
GOTERM_CC_DIRECT	GO:0000788 nuclear nucleosome	4	6.0	1.04×10^{-4}	8.86×10^{-4}
GOTERM_CC_DIRECT	GO:0042645 mitochondrial nucleoid	5	7.5	1.24×10^{-4}	9.36×10^{-4}
GOTERM_CC_DIRECT	GO:0005618 cell wall	6	9.0	8.74×10^{-4}	5.92×10^{-3}
GOTERM_CC_DIRECT	GO:0045261 proton-transporting ATP synthase complex, catalytic core F(1)	3	4.5	1.19×10^{-3}	7.31×10^{-3}

(continued)

Table 6.36 (continued)

Category	Term	Count	%	*P* value	Benjamini
GOTERM_CC_DIRECT	GO:0005576 extracellular region	7	10.4	1.28×10^{-3}	7.24×10^{-3}
GOTERM_CC_DIRECT	GO:0031298 replication fork protection complex	4	6.0	3.15×10^{-3}	1.64×10^{-3}
GOTERM_CC_DIRECT	GO:0000786 nucleosome	3	4.48	4.14×10^{-3}	2.00×10^{-2}
Blue crosses in Fig. 6.17					
GOTERM_CC_DIRECT	GO:0030529 intracellular ribonucleoprotein complex	54	65.1	6.94×10^{-49}	3.89×10^{-47}
GOTERM_CC_DIRECT	GO:0005840 ribosome	52	62.7	2.29×10^{-46}	6.42×10^{-45}
GOTERM_CC_DIRECT	GO:0022625 cytosolic large ribosomal subunit	31	37.3	1.27×10^{-32}	2.36×10^{-31}
GOTERM_CC_DIRECT	GO:0005622 intracellular	37	44.6	6.58×10^{-32}	9.21×10^{-31}
GOTERM_CC_DIRECT	GO:0022627 cytosolic small ribosomal subunit	21	25.3	1.22×10^{-21}	1.37×10^{-20}
GOTERM_CC_DIRECT	GO:0005737 cytoplasm	65	78.3	1.73×10^{-12}	1.62×10^{-11}
GOTERM_CC_DIRECT	GO:0015935 small ribosomal subunit	9	10.8	4.05×10^{-11}	3.24×10^{-10}
GOTERM_CC_DIRECT	GO:0030687 preribosome, large subunit precursor	10	12.1	4.07×10^{-6}	2.85×10^{-5}
GOTERM_CC_DIRECT	GO:0030686 90S preribosome	9	10.8	1.33×10^{-5}	8.29×10^{-5}
GOTERM_CC_DIRECT	GO:0031429 box H/ACA snoRNP complex	3	3.6	1.88×10^{-3}	1.05×10^{-3}

enrichment. Green and blue crosses in Fig. 6.17 correspond to two classes to which mitochondrial and ribosomal GO cellular component (CC) terms are enriched, respectively. Red triangles in Fig. 6.17 are not very clear, but it is enriched by the GO CC term of cell walls, which is deeply related to cell division. Thus, it is coincident. Therefore, K-means clustering applied to 298 probes identified by PCA-based unsupervised FE reproduces the biological clusters of genes reported by Tu et al. [64].

Another example of yeast cell division cycle to which PCA-based unsupervised FE is applied is data set stored in Cyclebase [39]. Cyclebase collected gene expression profiles of four species, one plant (*Arabidopsis thaliana*), two yeasts (*Saccharomyces cerevisiae*, *Schizosaccharomyces pombe*), and human (*Homo sapiens*). For the yeast that we analyze in the present chapter, *S. cerevisiae*, there are eight time course data sets available, one of which is excluded because there are only prescreened genes included. Profiles are available as a file budding_experiments.tsv, which is downloadable from Cyclebase. Because PCA-based unsupervised FE tries to identify genes as outliers, we need all genes before screening; otherwise we cannot identify outliers because of a lack of non-outlier genes. As a result, we apply PCA-based unsupervised FE to remaining seven gene expression profiles (Figs. 6.19 and 6.20).

In contrast to Fig. 6.18, not all trajectories exhibit clear limit cycles. In that case, we select pairs of PC loading associated with largest absolute winding numbers for gene selection (the pairs of PCs selected are shown in captions in Figs. 6.19 and 6.20). Then, using selected pairs of PC loading, (ℓ, ℓ'), P values are attributed to ith gene as

$$P_i = P_{\chi^2}\left[> \sum_{\ell_1=\ell,\ell'} \left(\frac{u_{\ell_1 i}}{\sigma_{\ell_1}}\right)^2\right]. \tag{6.62}$$

P_i is adjusted by BH criterion. Genes associated with P values less than 0.05 are selected (Fig. 6.21). For each of seven expression profiles, PCA-based unsupervised FE identifies more than 100 genes are selected. In order to evaluate selected genes, we see how much they are overlapped, because the selected genes should be largely overlapped between seven sets of genes if they are biologically valid. As in Fig. 6.21, 37 genes are chosen in common at least six among seven experiments. If considering that the total number of genes is at least several thousands (dependent on microarrays used in individual experiments), this coincident is too strong to occur accidentally. Thus, as far as coincidence, PCA-based unsupervised FE is successful.

In order to see if other supervised methods can similarly achieve sufficient coincidence between seven experiments, we apply sinusoidal regression to gene expression profile, x_{ij}, as

$$x_{ij} = a_i + b_{i1} \sin\left(\frac{2\pi j}{T}\right) + b_{i2} \cos\left(\frac{2\pi j}{T}\right). \tag{6.63}$$

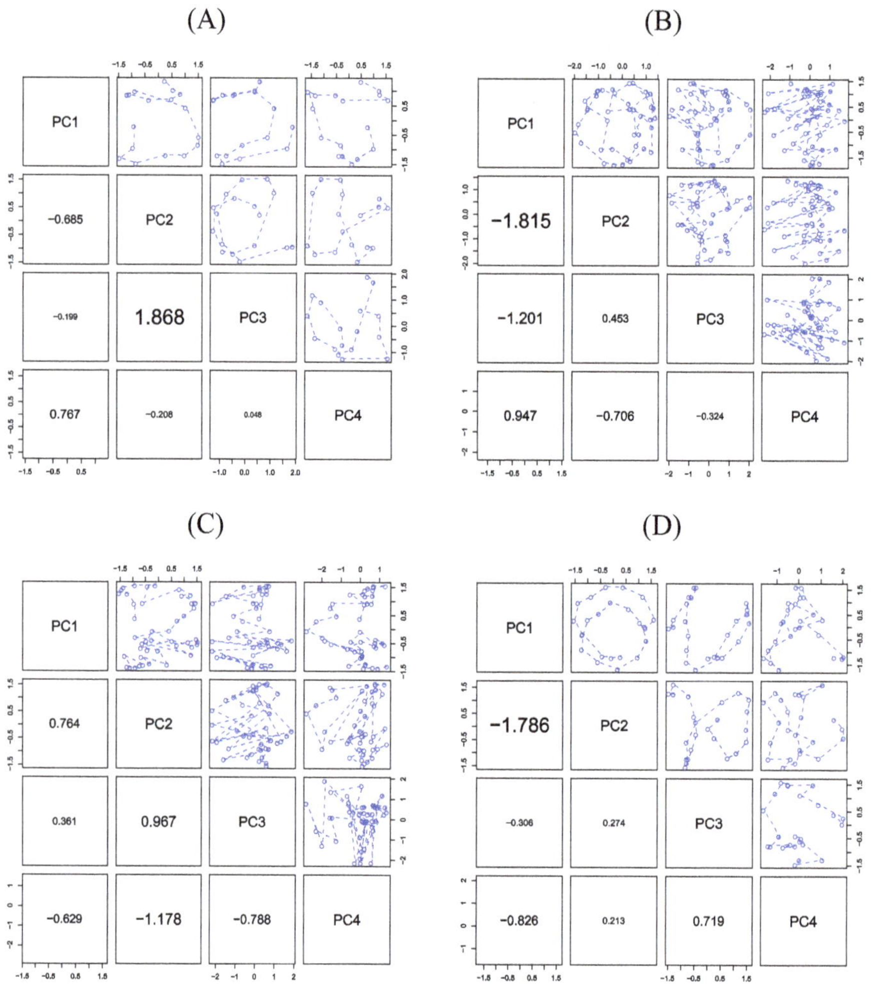

Fig. 6.19 Scatter plot of PC loading, $\boldsymbol{v}_\ell$, $1 \leq \ell \leq 4$ for Cyclebase. (ℓ, ℓ') denotes PC used for gene selection. (**a**) Cho et al. [11], $(\ell, \ell') = (2, 3)$. (**b**) Granovskaia et al. [17], G1 phase arrest by α-factor,$(\ell, \ell') = (1, 2)$. (**c**) Granovskaia et al. [17], G1 phase arrest by temperature-sensitive cdc28-13 mutant cells, $(\ell, \ell') = (2, 4)$. (**d**) Pramila et al. [34], α-Factor synchronization, $(\ell, \ell') = (1, 2)$. Other notations are the same as Fig. 6.18

One problem is the decision of period T. For five out of seven experiments, because the periods are denoted in Table 6.1 [16], we employ these values. For Fig. 6.19b and c, because no information is available, we decide T with visual inspection of $\boldsymbol{v}_\ell$ and $\boldsymbol{v}_{\ell'}$. P values obtained by sinusoidal regression are adjusted by BH criterion. Then genes associated with adjusted P values less than 0.05 are selected (Table 6.37). Although sinusoidal regression can also identify a large enough number of cell cycle regulated genes, the number of selected genes varies

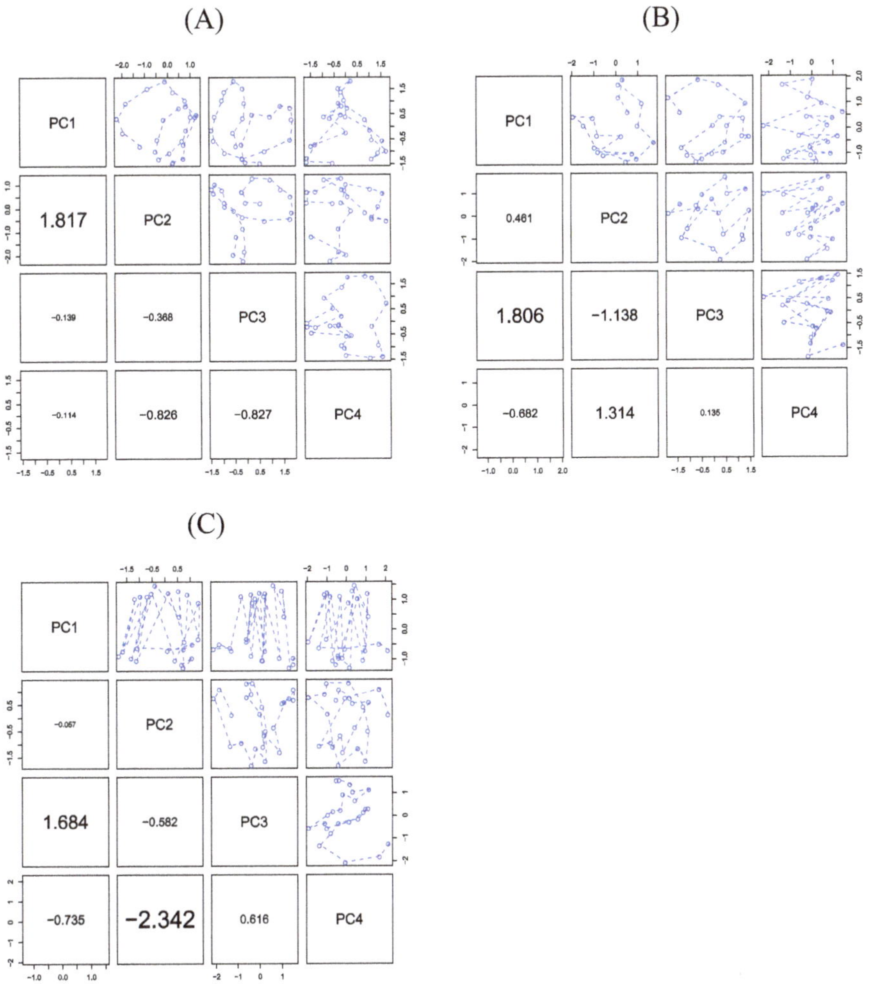

Fig. 6.20 (**a**) Pramila et al. [34], α-Factor synchronization: 38 data, $(\ell, \ell') = (1, 2)$. (**b**) Spellman et al. [46], α factor arrest, $(\ell, \ell') = (1, 3)$. (**c**) Spellman et al. [46], arrest of a cdc15 temperature-sensitive mutant. $(\ell, \ell') = (2, 4)$. Other notations are the same as Fig. 6.19

from experiments to experiments. In order to evaluate the amount of coincidence among seven experiments, we select top 150 genes with smaller P values. Then, as small as 12 genes are selected in at least six among seven experiments. Thus, PCA-based unsupervised FE can identify more common genes among seven experiments.

Finally, we evaluate 37 genes (Table 6.38) selected by PCA-based unsupervised FE biologically. Because DAVID does not identify any significant terms when 37 genes selected by PCA-based unsupervised FE are uploaded, we instead employ YeastMine [5] (Table 6.39), which includes more carefully curated biological terms specifically for yeast. For comparisons, 36 top-ranked genes in Cyclebase

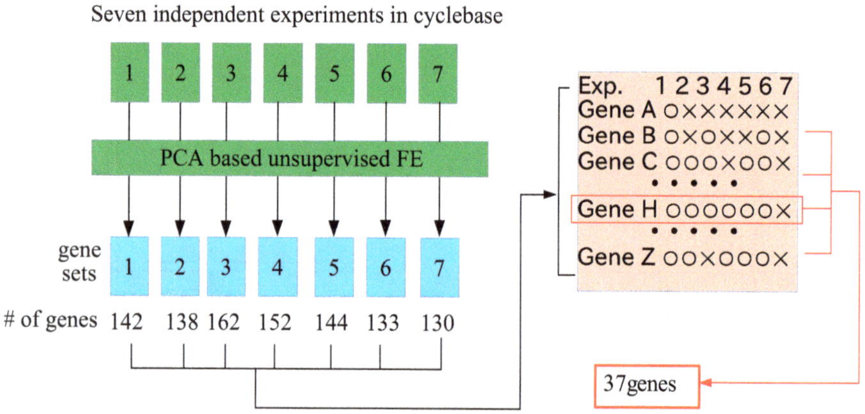

Fig. 6.21 The results of PCA-based unsupervised FE applied to seven yeast cell division cycle gene expressions in Cyclebase. Experiments one to seven correspond to Figs. 6.19a–d and 6.20e–g, in this order

Table 6.37 The number of genes selected using P values computed by Eq. (6.63)

Experiments	1	2	3	4	5	6	7
Number of genes measured	6214	6378	6378	6102	6145	6075	5673
Number of time points	17	41	44	24	24	18	24
Period T	8	14	20	12	12	9	10
Number of selected genes	364	3624	1164	1790	1951	298	354

Table 6.38 Genes selected by PCA-based unsupervised FE, cyclebase, and sinusoidal regression. Genes in bold are 15 genes chosen in common over three methods

Methods	Genes
PCA-based unsupervised FE	AIM34, ALK1, **AXL2**, CDC5, CLB1, CLB2, CLB6, CLN1, **CLN2**, **CSI2**, EGT2, **GAS3**, GIN4, **HHO1**, HHT1, HOF1, HST3, **HTA1**, **HTA2**, **KCC4**, MCD1, MMR1, **MNN1**, MSH2, **MSH6**, **PDS1**, PHO3, **POL30**, **PRY2**, RAD27, **RFA1**, RNR1, SFG1, SRC1, SWI5, **TOS4**, YOX1
Cyclebase	AIM34, **AXL2**, BUD3, **CLN2**, **CSI2**, **GAS3**, **HHO1**, HHT1, HMLALPHA2, **HTA1**, **HTA2**, HTB1, HTB2, **KCC4**, MATALPHA2, MCD1, **MNN1**, MRC1, **MSH6**, NRM1, **PDS1**, **POL30**, **PRY2**, **RFA1**, SRC1, SVS1, **TOS4**, TOS6, YRF1-1, YRF1-2, YRF1-3, YRF1-4, YRF1-5, YRF1-6, YRF1-7, YRF1-8
Sinusoidal regression	ALK1, ASF1, **AXL2**, BUD3, CDC21, CDC9, **CLN2**, **CSI2**, DSN1, ERP3, **GAS3**, HHF1, HHF2, **HHO1**, HHT2, **HTA1**, **HTA2**, HTB1, HTB2, **KCC4**, MCM5, **MNN1**, MSA1, MSH2, **MSH6**, NRM1, NUF2, **PDS1**, **POL30**, **PRY2**, RAD27, **RFA1**, RFA2, RSR1, SGO1, SML1, SPC98, TOF2, **TOS4**, WTM2

(Tables 6.38 and 6.40) and 40 genes selected by sinusoidal regression, Eq. (6.63), in at least five among seven experiments (Tables 6.38, 6.39, 6.40, and 6.41) are also uploaded.

Table 6.39 Top five GO BP term/publication enrichments reported by YeastMine in 37 genes identified by PCA-based unsupervised FE. #: number of genes associated with GO BP terms or mentioned in the publications. PMID: PubMed ID

PCA-based unsupervised FE			
GO BP term		p value	#
Cell cycle	[GO:0007049]	5.32×10^{-10}	24
Cell cycle process	[GO:0022402]	3.08×10^{-8}	21
Mitotic cell cycle	[GO:0000278]	4.45×10^{-8}	17
Mitotic cell cycle process	[GO:1903047]	2.23×10^{-7}	16
Cell division	[GO:0051301]	1.02×10^{-6}	15
Publication	PMID	p Value	#
Clustering time-varying gene expression profiles using scale-space signals	[16452778]	9.74×10^{-24}	20
Serial regulation of transcriptional regulators in the yeast cell cycle	[11572776]	6.14×10^{-17}	16
Identification of a core set of signature cell cycle genes whose relative order of time to peak expression is conserved across species	[22135306]	6.34×10^{-12}	10
Identification of sparsely distributed clusters of cis-regulatory elements in sets of co-expressed genes	[15155858]	3.71×10^{-10}	9
Computational reconstruction of transcriptional regulatory modules of the yeast cell cycle	[17010188]	4.17×10^{-10}	12

Venn diagram of three gene sets is in Fig. 6.22. At most, one third of genes are chosen in common. Thus, these three gene sets are quite distinct. It is also obvious that 37 genes selected by PCA-based unsupervised FE are most significantly enriched by the cell cycle related genes. Thus, from the biological point of views, PCA-based unsupervised FE outperforms cyclebase as well as conventional sinusoidal regression.

In conclusion, although applications are limited to yeast cell division cycle, PCA-based unsupervised FE obviously has the superior ability to identify periodic genes in fully unsupervised manner.

6.6.2 *Identification of Disease Driving Genes*

As can be seen in the section that describes biomarker identification (Sect. 6.3), disease alters gene expression. Gene expression is also associated with disease

Table 6.40 Top five GO BP term/publication enrichments reported by YeastMine in 36 top-ranked genes by Cyclebase. #: number of genes associated with GO BP terms or mentioned in the publications. PMID: PubMed ID

Cyclebase			
GO BP term		p value	#
Chromosome organization	[GO:0051276]	1.13×10^{-8}	20
Telomere maintenance via recombination	[GO:0000722]	3.34×10^{-8}	8
DNA metabolic process	[GO:0006259]	3.50×10^{-8}	19
Telomere maintenance	[GO:0000723]	2.07×10^{-6}	9
Anatomical structure homeostasis	[GO:0060249]	2.07×10^{-6}	9
Publication	PMID	p Value	#
Genome-wide array-CGH analysis reveals YRF1 gene copy number variation that modulates genetic stability in distillery yeasts	[26384347]	9.74×10^{-24}	20
Transcriptional effects of the potent enediyne anticancer agent Calicheamicin gamma(I)(1)	[11880039]	1.11×10^{-11}	7
Linking DNA replication checkpoint to MBF cell cycle transcription reveals a distinct class of G1/S genes	[22333912]	2.32×10^{-11}	11
Mcm1p-induced DNA bending regulates the formation of ternary transcription factor complexes	[12509445]	2.35×10^{-11}	8
A genetic screen for yeast genes induced by sustained osmotic stress	[12868060]	1.82×10^{-10}	7

progression (Sect. 6.3.3). In this sense, it is not surprising even if we can identify a set of genes that describes disease progression well. In this subsection, we try to identify genes that discriminate time developments between dengue fever (DF) and dengue hemorrhagic fever (DHF) [18]. DF is usually nonlethal mosquito-borne virus disease. Nevertheless, it rarely develops to lethal DHF. Because DHF usually develops only after remission of DF, it is supposed that if DHF develops from DF is dependent upon the patients' status. In spite of that, it is unclear what kind of difference decides if DHF develops after patients start to recover from DF. In this subsection, we try to apply PCA-based unsupervised FE to patients' blood gene expression profiles in order to identify which genes are related to DHF developments.

In this subsection, we apply PCA-based unsupervised FE to five DF patients' blood gene expression profiles in order to identify genes that make DHF develop. Five data sets analyzed are shown in Table 6.42 (data sets 1 to 5) [52]. These five data sets are quite distinct. In data set 1, which is composed of four classes,

Table 6.41 Top five GO BP term/publication enrichments reported by YeastMine in 40 genes identified by sinusoidal regression, Eq. (6.63). #: number of genes associated with GO BP terms or mentioned in the publications. PMID: PubMed ID

Sinusoidal regression			
GO BP Term		p Value	#
Cell cycle	[GO:0007049]	4.45×10^{-10}	26
Cell cycle process	[GO:0051276]	3.27×10^{-8}	22
Chromosome organization	[GO:0006259]	5.07×10^{-8}	21
Cellular response to DNA damage stimulus	[GO:0006974]	5.85×10^{-8}	17
Chromatin assembly or disassembly	[GO:0006333]	7.32×10^{-7}	9
Publication	PMID	p value	#
Clustering time-varying gene expression profiles using scale-space signals.	[16452778]	8.69×10^{-23}	20
Histone h3 exerts a key function in mitotic checkpoint control.	[19917722]	3.69×10^{-15}	9
Regulation of cell cycle-dependent gene expression in yeast.	[2201678]	5.846×10^{-15}	11
Molecular biology. Nucleosomes help guide yeast gene activity.	[15961637]	6.85×10^{-14}	8
Brownian dynamics simulation of directional sliding of histone octamers caused by DNA bending.	[16802969]	6.85×10^{-14}	8

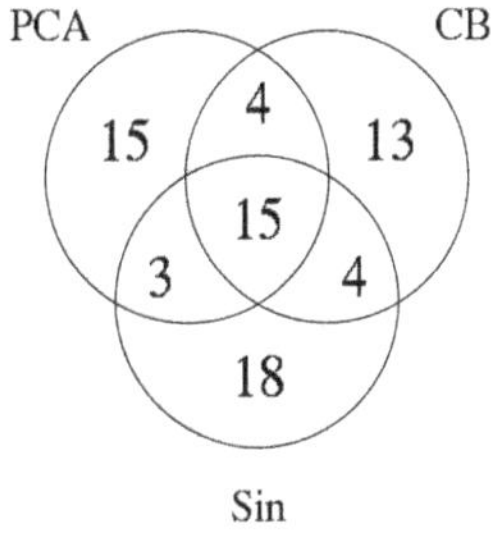

Fig. 6.22 Venn diagram of genes in Table 6.38. PCA: PCA-based unsupervised FE, CB: Cyclebase, Sin: sinusoidal fitting

there are healthy controls (HC), acute patients (AC), DF and DHF patients. On the other hand, in data set 2, which is also composed of four classes, three are Acute and Convalescent patients, both of which are composed of DF and DHF patients. In the following, we would like to demonstrate that starting the analysis of these quite distinct two data sets, genes chosen in common between two data sets can describe distinct time development between DF and DHF. In order to show this, we apply PCA-based unsupervised FE to data sets 1 and 2 and identify genes for both

Table 6.42 List of samples included in data sets 1, 2, 3, 4, and 5. DSS: Dengue Shock Syndrome. GSE51808: RMA normalization was performed using Expression Console software. GSE13052: Intensity was acquired using Beadstudio software Intensity was background normalized (subtract the background value). GSE25001: Data was normalized by Beadstudio software. GSE43777: RMA normalization was performed using Expression Console software. For more details see papers that reported these data sets

Data set 1 (GSE51808)	Affymetrix HT HG-U133+ PM Array Plate			
	Healthy controls (HC)	Acute patients (AC)	DF	DHF
	9	19	18	10
Data set 2 (GSE13052)	Sentrix HumanRef-8 Expression BeadChip			
	Acute	Convalescent		
uncomplicated (DF)	10	5		
DSS* (DHF)	9	6		
Data set 3 (GSE25001)	Illumina humanRef-8 v2.0 expression beadchip			
	Acute	0-1	Disease (Fever)	Follow-up
DF	56	32	31	16
DHF	24	12	20	18

	Affymetrix Human Genome U133 Plus 2.0 Array							
Data set 4 (GSE43777-GPL570)	G0	G1	G2	G3	G4	G5	G6	G7
DF	0	2	5	8	9	5	11	12
DHF	0	0	3	8	10	5	11	12
Data set 5 (GSE43777-GPL201)	Affymetrix Human HG-Focus Target Array							
DF	2	5	21	18	22	22	24	45
DHF	0	0	0	1	3	1	1	3

data sets. Then, using genes chosen in common, we try to see how selected genes describe distinct time developments between DF and DHF, using data sets 3 to 5.

At first, we apply PCA to data sets 1, $x_{ij} \in \mathbb{R}^{54715\times 56}$, and 2, $x_{ij} \in \mathbb{R}^{23454\times 30}$, after standardization, $\sum_i x_{ij} = 0$ and $\sum_i x_{ij}^2 = N$, where N is the number of probes, 54715 and 23454, respectively. Selection of PCs used for gene selection is not straightforward, because no single PC can discriminate four classes in Table 6.42. Upper triangles of Fig. 6.23 show the PC loading, $\boldsymbol{v}_\ell$, $1 \leq \ell \leq 4$ for data sets 1 and 2. With visual inspection, we decide to employ $\ell = 2, 3$ for both data sets because this combination is most coincident with clear clusters coincident with class labels. On the other hand, it is obvious that there are no PC loading that discriminate between DHF and DF. Two clusters are coincident with only the distinction between sample with and without symptom. In order to support this decision quantitatively, we perform LDA, Eq. (6.2), assuming two classes, and compute accuracy, Eq. (6.16) (lower triangles of Fig. 6.23). It is obvious that $\ell = 2, 3$ achieves the highest accuracy.

Then we attribute P values to probes assuming χ^2 distribution as

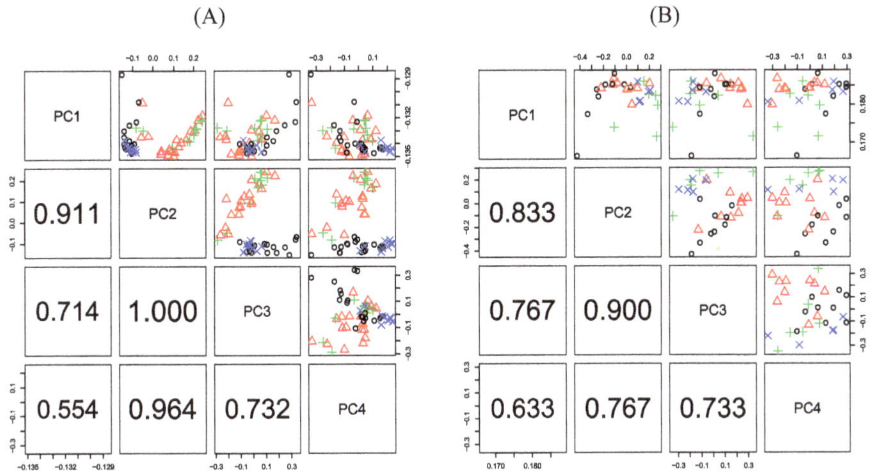

Fig. 6.23 Upper triangle: Scatter plot of PC loading, $\boldsymbol{v}_\ell, 1 \leq \ell \leq 4$, for (**a**) data set 1, ○: Convalescent Patient, △: Dengue Fever, +: Dengue Hemorrhagic, ×: Healthy control, (**b**) Data set 2, ○: acute DSS, △: acute uncomplicated, +: convalescent DSS, ×: convalescent uncomplicated. Lower triangle: accuracy of two classes (with and without symptom) discrimination. (**a**) ○+× vs. △++, (**b**) ○+△ vs. ++×

$$P_i = P_{\chi^2}\left[> \sum_{\ell=2}^{3} \left(\frac{u_{\ell i}}{\sigma_\ell}\right)^2 \right]. \tag{6.64}$$

P values are adjusted by BH criterion, and probes associated with adjusted P values less than 0.01 are selected. As a result, 879 and 275 probes are selected for data sets 1 and 2, respectively (Fig. 6.24). Considering the fact that the total number of probes is 10^4, while the number of selected $\sim 10^2$, the regions where selected probes (red dots) distribute are very huge and selected probes are really outliers. The number of genes included in common in both sets of selected genes as many as 46 (Table 6.43). In order to check if as many as 46 chosen in common genes can occur accidentally, we apply Fisher's exact test (Table 6.44). It is obvious that 46 genes are too large to occur accidentally.

In order to see if other conventional supervised feature selection methods can work same, we test three methods, limma, categorical regression analysis and significance analysis of microarrays (SAM) [65], which is t-test modified so as to be fitted to microarray analysis. These three tests are performed under the two assumptions of either four classes or two classes. Two classes are assumed to be those with and without symptom, as shown in the caption of Fig. 6.23. Probes with adjusted P values less than 0.01 are selected. Table 6.45 shows the results. Because the numbers of probes identified in data set 1 are too large, no methods are useful to select a small enough number of genes chosen in common between two data sets. Although this definitely suggests the superiority toward these three conventional methods, we can make use of them with taking into account the results of PCA-

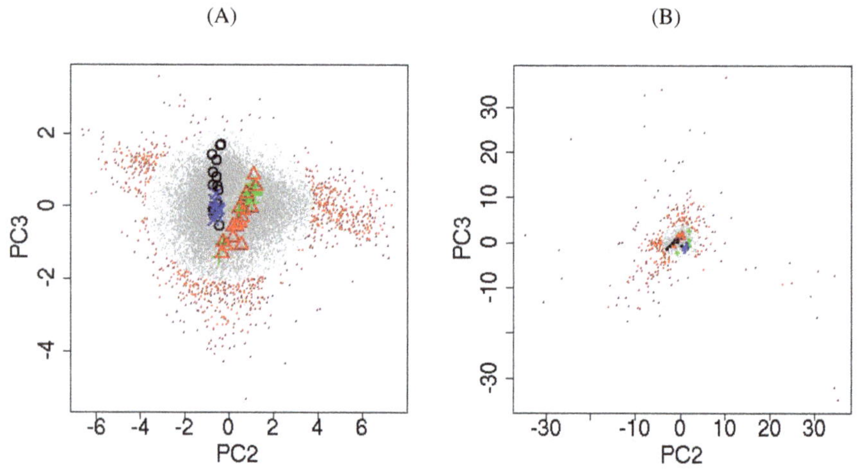

Fig. 6.24 PC score, $u_\ell, \ell = 2, 3$, for (**a**) data set 1 and (**b**) data set 2. Gray dots: not selected probes, red dots: selected dots. Colored marks are PC loading, as shown in Fig. 6.23

Table 6.43 Forty six and forty one genes common between 879 and 275 genes selected by PCA-based unsupervised FE and categorical regression. Bold genes are common

46 genes associated with probes chosen in common by PCA-based unsupervised FE
FBXO7 MX1 LY6E **IFI27** TNFSF10 OAS1 CDC20 GYPC **PI3** FCGR3A HBA1 HBA2 HBG1 HBG2 IFI44L IFIT3 CCR1 FPR1 STAT2 ISG15 OASL CD38 TNFRSF17 CXCR1 ZBP1 HBB IFI35 MKRN1 APOBEC3A ALAS2 IL1RN RSAD2 ASCC2 IFIT2 **ADIPOR1** SLC25A37 OAS3 SDF2L1 TMEM140 FKBP11 HERC5 ITM2C TXNDC5 STRADB **SLC25A39** EPSTI1
41 genes associated with probes chosen in common by categorical regression
FBXO7 PSMB2 LGALS1 NMT1 TMX2 LRRC41 IDH3A BAG1 **IFI27** UBE2S ATOX1 **PI3** BAK1 MRPL28 CHAF1B HAGH PSMD11 XPNPEP1 TSPAN5 GART RTN1 YARS SLC43A3 **ADIPOR1** DCXR MRPS18A SIL1 DPP3 GPN2 TESC KCTD14 GMPPB CAMK1D TACO1 OSBP2 STRADB **SLC25A39** EHD4 TRIM69 HAVCR2 SESN3

Table 6.44 Confusion matrices and associated P values and odds ratio

	Data set 1	Not selected	Selected	P value	Odds ratio
PCA-based unsupervised FE					
Data set 2	Not selected	13574	186	2.17×10^{-22}	7.51
	Selected	447	46		
Categorical regression (2 classes)					
Data set 2	Not selected	13680	185	5.73×10^{-16}	5.50
	Selected	551	41		

based unsupervised FE. PCA-based unsupervised FE has already shown that 879 and 275 probes are large enough to have a reasonable number of common genes between two data sets. Thus, we select top-ranked 879 and 275 probes based upon P values computed by one of three methods in order to compare the performance

Table 6.45 Number of genes identified by sam, limma, and categorical regression. Two classes mean "DHF+DF" vs. "CP+HC" for data set 1 (GSE51808) and "Acute" vs. "Convalescent" for data set 2 (GSE13052). *:all probes

	Sam		Limma		Categorical regression	
Data set	Two classes	Four classes	Two classes	Four classes	Two classes	Four classes
1	17680	16647	54715*	13506	15447	13941
2	2427	865	21795	20629	679	581

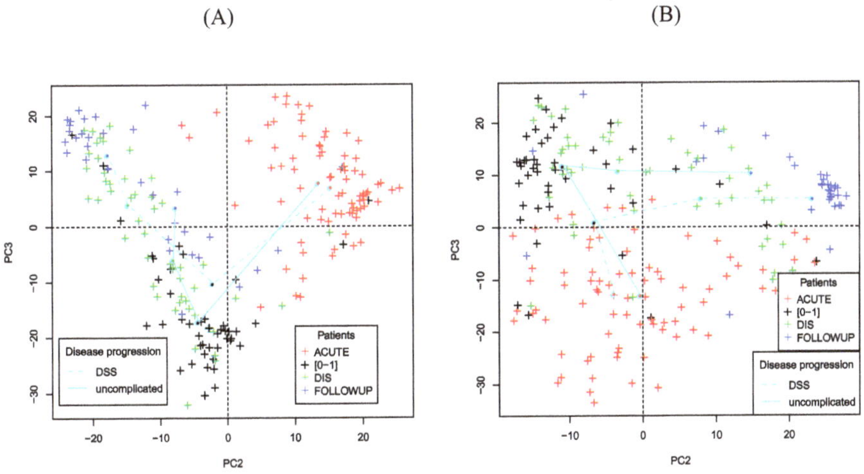

Fig. 6.25 Scatter plot of PC loading $\boldsymbol{v}_\ell$, $\ell = 2, 3$, obtained by applying PCA to data set 3 using only genes selected by PCA-based unsupervised FE or categorical regression to data sets 1 and 2. (**a**) PCA-based unsupervised FE. (**b**) Categorical regression

with PCA-based unsupervised FE. SAM attributed $P = 0$ to more than 879 genes in data set 1. Thus we cannot select top-ranked 870 genes in data set 1 by SAM. Although limma allowed us to identify specified top-ranked genes, the number of common probes is a few. Thus, neither SAM not limma can compete with PCA-based unsupervised FE. Categorical regression analysis identifies 32 and 41 probes in common between 879 and 275 probes for two classes and four classes cases, respectively. As many as 41 genes (Table 6.43) chosen in common is highly significant (Table 6.44). Thus, categorical regression is comparable with PCA-based unsupervised FE.

These two sets of genes selected by PCA-based unsupervised FE and by categorical regression analysis are quite distinct. There are only five genes chosen in common between two sets. In order to see which gene set is better, we need to evaluate them biologically. For this purpose, we employ data set 3 (Table 6.42). We apply PCA to either 46 genes selected by PCA-based unsupervised FE or 41 genes selected by categorical regression in data set 3. If selected genes are reasonable, samples with distinct labels should be separately located in the plane spanned by PC loading. Figure 6.25 shows that scatter plot of PC loading obtained by applying PCA

Table 6.46 *P* values computed by *t*-test applied to the distinction of the second and the third PC scores between "DSS" and "uncomplicated" patients in Fig. 6.25

ACUTE	[0-1]	DIS	FOLLOWUP	
PCA-based unsupervised FE				
PC2	2.14×10^{-1}	5.62×10^{-1}	7.87×10^{-3}	4.15×10^{-3}
PC3	7.23×10^{-1}	1.07×10^{-1}	6.41×10^{-3}	9.73×10^{-3}
Categorical regression				
PC2	1.24×10^{-1}	2.78×10^{-1}	8.84×10^{-4}	4.00×10^{-2}
PC3	$9, 16 \times 10^{-1}$	1.26×10^{-2}	5.48×10^{-2}	6.49×10^{-2}

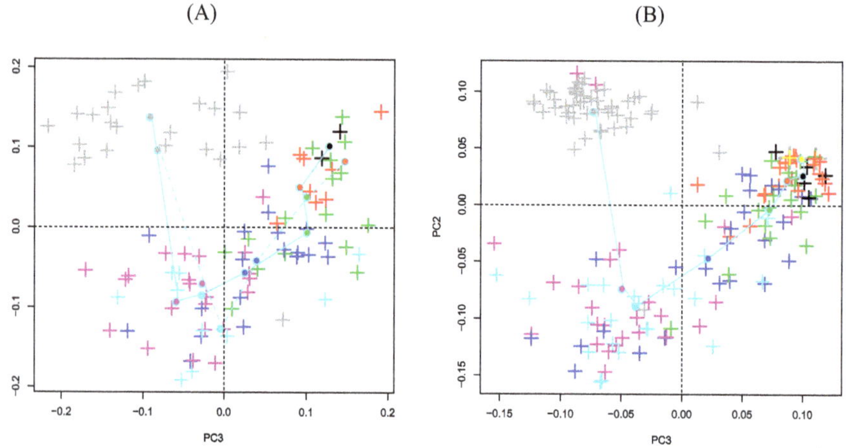

Fig. 6.26 Scatter plot of PC loading, $\boldsymbol{v}_\ell$, $\ell = 2, 3$, obtained by applying PCA to data sets 4 (**a**) and 5 (**b**), using only 46 genes (Table 6.43) selected by PCA-based unsupervised FE applied to data sets 1 and 2. The correspondence between the colored crosses (+) and disease progression are black (stage G1), red (stage G2), green (stage G3), blue (stage G4), cyan (stage G5), magenta (stage G6), and gray (stage G7). Cyan solid and broken lines correspond to DF and DHF, respectively

to either 46 genes selected by PCA-based unsupervised FE or 41 genes selected by categorical regression in data set 3. There are four disease stages in data set 3: acute, [0-1] (0 or 1 days after the symptom), DIS (disease), and FOLLOWUP (after remission). For both cases, disease progression can be seen in this order. It is also interesting, in later stage, DHF and DF are distinct to some extent. In order to further validate two gene sets, we apply *t*-test to see how distinct DHF and DF are quantitatively. Table 6.46 shows the result. DHF and DF are more distinct when genes selected by PCA-based unsupervised FE are used. Thus, PCA-based unsupervised FE selected more reasonable genes than categorical regression.

Basically, we believe that the above performance is good enough to demonstrate the superiority of PCA-based unsupervised FE over the conventional supervised methods. Nevertheless, we would like to emphasize the robustness of the selected 46 genes by applying PCA to additional data set, data sets 4 and 5 (Table 6.42). Figure 6.26 shows the scatter plots of PC loading, $\boldsymbol{v}_\ell$, $\ell = 2, 3$, obtained by applying

PCA to data sets 4 and 5 with only 46 genes (Table 6.43) selected with PCA-based unsupervised FE applied to data sets 1 and 2. The V letter shapes seen in Fig. 6.25 are conserved for two data sets, too, although the distinction between DHF and DF is weaker. This is possibly because the number of samples is smaller (12 samples in G7 stage of data set 4, while 16 or 18 samples in follow-up stage of data set 3). Nevertheless, in G7 stage, the second PC loading, v_{2j}, is still significantly distinct (P=0.05 and 0.04 with a t-test and Wilcoxon signed-rank sum test, respectively). Because 46 genes selected for data set 1 and data set 2 are valid to describe disease progression in three independent data sets, data sets 3 to 5, 46 genes in Table 6.43 should be key genes that describe the distinction between DF and DHF.

In conclusion, PCA-based unsupervised FE has superior ability to identify a limited number of genes that can describe DHF progression.

6.7 Gene Selection for Single-Cell RNA-seq

Single-cell RNA high throughput sequencing (scRNA-seq) is a newly developed technology. scRNA-seq can measure RNA expression in single-cell base [20]. In contrast to the usual HTS that can measure gene expression profile only in tissue base, scRNA-seq can measure gene expression profile within individual cells.

From the data science point of views, scRNA-seq is distinct from conventional tissue-based gene expression measurements in the following two points:

Larger number of samples In contrast to the conventional tissue-based measurements, because the number of samples is as many as the number of cells, it can be as large as 10^3.

Missing labeling Because geometrical information of individual cell within tissue is missing during the process of library preparation, samples (cells) are not basically labeled.

Because of these two primary differences, applying PCA-based unsupervised FE to scRNA-seq is challenging. As emphasized many times, PCA-based unsupervised FE is invented to be applied to large p small n problem. It is interesting to see if PCA-based unsupervised FE is useful even if the number of samples increases up to 10^3. On the other hand, missing labeling is advantageous to PCA-based unsupervised FE, because it is designed to be fitted to unlabeled samples. Because of these pros and cons, it is very unclear if PCA-based unsupervised FE is applicable to scRNA-seq.

In this section, we apply PCA-based unsupervised FE to a scRNA-seq data set [53]. scRNA-seq data is downloaded from GEO with GEO ID GSE76381. It includes human embryo ventral midbrain cells between 6 and 11 weeks of gestation, mouse ventral midbrain cells at six developmental stages between E11.5 and E18.5, Th+ neurons at P19-P27, and fluorescence activated cell sorting (FACS)-sorted putative dopaminergic neurons at P28-P56 from Slc6a3-Cre/tdTomato mice. That is, it includes data set of brain development of human and mouse. The purpose of the analysis is to understand what is common between human and mouse brain development, based upon gene expression analysis. PCA-based unsupervised FE is applied to these data sets separately. Here E denotes the number of days after fertilization, while P denotes postnatal days.

Usually, the first step is to identify which PC loading exhibits the desired class dependence. Nevertheless, for scRNA-seq data, because no labeling is available for samples, it is impossible to find which PC loading reflect something to be searched. Then, we decide to select the first L PC scores, $\boldsymbol{u}_\ell$, $\ell \leq L$, and attribute P values to gene i assuming χ^2 distribution as

$$P_i = P_{\chi^2}\left[> \sum_{\ell=1}^{L} \left(\frac{u_{\ell i}}{\sigma_\ell}\right)^2 \right]. \tag{6.65}$$

P values are corrected by BH criterion, and genes associated with adjusted P values less than 0.01 are selected. The problem is how to decide L. In this case, we select the smallest L such that selected genes are as many as a few hundreds. Then, we find that 116 and 118 genes are selected if $L = 2$ and $L = 3$ for human and mouse, respectively.

The first evaluation of these two sets of selected genes is the amount of overlap between them. Among 116 selected human genes and 118 mouse genes, as many as 53 genes are chosen in common (Table 6.47). Unfortunately, in contrast to the microarray, there are no definite number of "total genes" for scRNA-seq. Thus, tentatively, we assume that there are 20,000 genes for mouse and human. Table 6.48 shows the confusion matrix and the result of Fisher's exact test. In any case, overlap is highly significant.

In order to compare the performance with other methods, first we consider highly variable genes [10]. The procedure of how to perform highly variable genes is as follows. First we perform regression analysis

$$\log_{10}\left(\frac{\sqrt{\langle x_{ij}^2\rangle_j - \langle x_{ij}\rangle_j^2}}{\langle x_{ij}\rangle_j}\right) = \frac{1}{2}\log_{10}\left(\frac{\beta}{\langle x_{ij}\rangle_j} + \alpha\right) + \epsilon_i, \tag{6.66}$$

where α and β are regression coefficients and ϵ_i is residual. P values are attributed to genes, i, assuming χ^2 distribution as

$$P_i = P_{\chi^2}\left[> \left(\frac{\epsilon_i}{\sigma'}\right)^2 \right]. \tag{6.67}$$

P values are corrected by BH criterion. Genes associated with adjusted P values less than 0.01 are selected. We identify 168 human genes and 171 mouse genes between which 44 genes are chosen in common (Table 6.47). Although it is a little bit less significant than PCA-based unsupervised FE, it is still highly significant (Table 6.48).

Table 6.47 Fifty three genes common between 116 human and 118 mice genes selected by PCA-based unsupervised FE

53 genes associated with probes chosen in common by PCA-based unsupervised FE
ACTB ACTG1 ATP5E CALM2 COX7C EEF1A1 FAU FTH1 H2AFZ H3F3B HMGB1 HMGB2 HMGN2 HSP90AA1 HSP90AB1 MALAT1 MAP1B MARCKS MEG3 NGFRAP1 PABPC1 PPIA PTMA RPL18A RPL23 RPL24 RPL32 RPL35 RPL38 RPL39 RPL4 RPLP1 RPLP2 RPS12 RPS13 RPS16 RPS20 RPS24 RPS25 RPS28 RPS29 RPS3 RPS5 RPS6 RPSA SPARC STMN1 STMN2 SUMO2 TMSB10 TMSB4X TUBA1A TUBA1B
44 genes associated with probes chosen in common by highly variable genes
ALDH1A1 APLN CARTPT CAV1 CCL2 CCL3 CCL4 CD93 CLDN5 COL4A1 COL4A2 CRABP1 CSF1R CX3CR1 DBH FLT1 FN1 HPGDS ICAM2 IGFBP3 IGFBP7 IL1B ITM2A KDR NEFM NPY NTS P2RY12 PF4 PLEK RGS5 SLC2A1 SLC38A5 SLC6A2 SLC6A4 SLC7A1 SLC7A5 SNCG SPARCL1 SPP1 SRGN SST TPH2 VWF
21 genes associated with probes chosen in common by bimodal genes
AP2B1 AP2M1 ASH1L EIF4B FXR1 G3BP1 HNRNPH2 IK ILF3 MIDN NMT1 OCIAD1 PNN PPIG PRPF6 PSMD11 RPS15 SETD5 SRP72 TAX1BP1 WAC
76 genes associated with probes chosen in common by dpFeature
ACTG1 ALDH1A1 ANK3 ARL6IP1 ATP1A2 ATP5E B2M BASP1 BGN CALD1 CALM2 CCL3 CCL4 CCNB1 CDK1 CELF4 CENPF CKS1B CKS2 CLDN5 COL4A1 COL4A2 COX7A2 COX7C CRMP1 CST3 CYR61 DCX DPYSL2 DPYSL3 DNRB EEF1A1 ELAVL2 ELAVL4 ESAM ETS1 FABP5 FABP7 FAU FGFBP3 FLT1 FN1 FOS FSTL1 GAP43 GNB2L1 GNG11 GPM6A GPM6B GRIA2 GSTP1 H2AFZ H3F3B HES1 HMGB1 HMGB2 HMGN2 HN1 HSP90AA1 IGFBP7 INA ITM2A KCNQ1OT1 KIF5C KPNA2 LGALS1 MALAT1 MAP1B MAP2 MEG3 MIAT MLLT11 MYL12A MYL6 NCAM1 NDUFA4

Next, we compare PCA-based unsupervised FE with bimodal genes [12]. Bimodal genes are selected based upon the P values computed by Hartigan's Dip Test, which rejects the null hypothesis that the distribution is unimodal [19]. The concept behind bimodal genes is that if gene expresssion of a gene is unimodal, it is unlikely that the expression is coincident with the distinction between two classes. We attribute P values to genes using this test. P values are adjusted by BH criterion. Then genes associated with adjusted P values less than 0.01 are selected. As a result, 11344 human and 10849 mouse genes are selected. Thus, it is obvious that bimodel genes are too many. In order to see the coincident between the two gene sets, we select top-ranked 200 human and mouse genes based upon P value computed by Hartigan's Dip Test. This results in as small as 21 genes chosen in common (Table 6.47). Thus, as far as consistency between human and mouse, bimodel genes are inferior to either PCA-based unsupervised FE or highly variable genes (Table 6.48).

Finally, we compare PCA-based unsupervised FE with dpFeature [36], which was proposed very recently as an advanced tool to select genes in scRNA-seq. It selects 13775 human and 13362 mouse genes. Thus it cannot select a reasonable number of genes. In order to verify consistency between human and mouse, we selected top-ranked 200 genes and compare them. Then there are 76 common genes (Table 6.47). The significance is comparable with PCA-based unsupervised FE (Table 6.48).

Table 6.48 Confusion matrices and associated P values and odds ratio

PCA-based unsupervised FE					
	Mouse	Not selected	Selected	P value	Odds ratio
Human	Not selected	19819	63	2.21×10^{-91}	255.00
	Selected	65	53		
Highly variable genes					
	Data set 1	Not selected	Selected	P value	Odds ratio
Data set 2	Not selected	19705	124	7.13×10^{-54}	54.97
	Selected	127	44		
Bimodal genes					
	Data set 1	Not selected	Selected	P value	Odds ratio
Data set 2	Not selected	19621	179	1.00×10^{-15}	12.85
	Selected	179	21		
dpFeature					
	Data set 1	Not selected	Selected	P value	Odds ratio
Data set 2	Not selected	19676	124	1.03×10^{-105}	96.98
	Selected	124	76		

Although biological validations of selected genes using enrichment analysis are available elsewhere [53], I am not willing to discuss about it in detail here, because coincidence between mouse and human can be a biological evaluation to some extent; the results by enrichment analysis also support the superiority of PCA-based unsupervised FE to other three methods. In addition to this, Chen et al. [9] evaluated biologically multiple gene selection methods applicable to scRNA-seq using enrichment analysis; they concluded that PCA-based unsupervised FE is at least competitively good with other compared methods.

In conclusion, PCA-based unsupervised FE is at least comparable with other popular or conventional methods.

6.8 Summary

In this chapter, I demonstrated how we can make use of PCA-based unsupervised FE in the application to bioinformatics, especially, in the field of feature selection. In all application examples, PCA is applied such that PC loading, $\boldsymbol{v}_\ell$, is attributed to samples, while PC score, $\boldsymbol{u}_\ell$, is attributed to features (genes, miRNAs). The next step is to investigate PC loading, $\boldsymbol{v}_\ell$, in order to identify ℓs used for computing P values. This step is the most difficult. The simplest case is to apply liner regression analysis to PC loading and to identify which PC loading is coincident with class labeling. Nevertheless, it is not always possible. When class labeling is missing or no PC loading is coincident with class labeling, simply $\ell \leq L$ is employed (in this case, there are no definite ways to decide L). In some case, we need to find a set of ℓs that are coincident with class labeling. In this case, we need to draw scatter plot of PC loading, $\boldsymbol{v}_\ell$. When we aim to perform integrated analysis, e.g., miRNAs and mRNAs or methylation and mRNAs, the coincidence of PC loading between these two must be investigated. When samples are shared between two, correlation coefficients and hierarchical clustering using correlation coefficients are useful. If samples are not shared, we need to investigate PC loading with more additional information, e.g., down/upregulated in treated samples toward control samples simultaneously between these two features. Thus, although PCA-based unsupervised FE is powerful method, in contrast to other machine learning techniques, we need more deep understanding of data to be analyzed. This can be pros or cons of PCA-based unsupervised FE.

References

1. Abeel, T., Helleputte, T., de Peer, Y.V., Dupont, P., Saeys, Y.: Robust biomarker identification for cancer diagnosis with ensemble feature selection methods. Bioinformatics **26**(3), 392–398 (2009). https://doi.org/10.1093/bioinformatics/btp630
2. Agarwal, V., Bell, G.W., Nam, J.W., Bartel, D.P.: Predicting effective microRNA target sites in mammalian mRNAs. eLife **4**, e05005 (2015). https://doi.org/10.7554/elife.05005
3. Ahuja, N., Sharma, A.R., Baylin, S.B.: Epigenetic therapeutics: a new weapon in the war against cancer. Annu. Rev. Med. **67**(1), 73–89 (2016). https://doi.org/10.1146/annurev-med-111314-035900
4. Artmann, S., Jung, K., Bleckmann, A., Beissbarth, T.: Detection of simultaneous group effects in microRNA expression and related target gene sets. PLoS One **7**(6), e38365 (2012)
5. Balakrishnan, R., Park, J., Karra, K., Hitz, B.C., Binkley, G., Hong, E.L., Sullivan, J., Micklem, G., Michael Cherry, J.: Yeastmine—an integrated data warehouse for saccharomyces cerevisiae data as a multipurpose tool-kit. Database **2012**, bar062 (2012). http://dx.doi.org/10.1093/database/bar062
6. Bleckmann, A., Leha, A., Artmann, S., Menck, K., Salinas-Riester, G., Binder, C., Pukrop, T., Beissbarth, T., Klemm, F.: Integrated miRNA and mRNA profiling of tumor-educated macrophages identifies prognostic subgroups in estrogen receptor-positive breast cancer. Mol. Oncol. **9**(1), 155–166 (2015)
7. Brown, T.A.: Genomes, vol. 4, 4 edn. Garland Science, New York (2017). https://www.crcpress.com/Genomes-4/Brown/p/book/9780815345084
8. Chan, M., Liaw, C.S., Ji, S.M., Tan, H.H., Wong, C.Y., Thike, A.A., Tan, P.H., Ho, G.H., Lee, A.S.G.: Identification of circulating MicroRNA signatures for breast cancer detection. Clin. Cancer Res. **19**(16), 4477–4487 (2013). https://doi.org/10.1158/1078-0432.ccr-12-3401
9. Chen, B., Lau, K.S., Herring, C.A.: pyNVR: investigating factors affecting feature selection from scRNA-seq data for lineage reconstruction. Bioinformatics (2018). https://dx.doi.org/10.1093/bioinformatics/bty950
10. Chen, H.I.H., Jin, Y., Huang, Y., Chen, Y.: Detection of high variability in gene expression from single-cell rna-seq profiling. BMC Genomics **17**(7), 508 (2016). https://doi.org/10.1186/s12864-016-2897-6
11. Cho, R.J., Campbell, M.J., Winzeler, E.A., Steinmetz, L., Conway, A., Wodicka, L., Wolfsberg, T.G., Gabrielian, A.E., Landsman, D., Lockhart, D.J., Davis, R.W.: A genome-wide transcriptional analysis of the mitotic cell cycle. Mol. Cell **2**(1), 65–73 (1998). https://doi.org/10.1016/s1097-2765(00)80114-8
12. DeTomaso, D., Yosef, N.: Fastproject: a tool for low-dimensional analysis of single-cell rna-seq data. BMC Bioinf. **17**(1), 315 (2016). https://doi.org/10.1186/s12859-016-1176-5

13. Ding, M., Li, J., Yu, Y., Liu, H., Yan, Z., Wang, J., Qian, Q.: Integrated analysis of miRNA, gene, and pathway regulatory networks in hepatic cancer stem cells. J. Transl. Med. **13**, 259 (2015)
14. Fisher, R.A.: On the interpretation of χ^2 from contingency tables, and the calculation of p. J. R. Stat. Soc. **85**(1), 87 (1922). https://doi.org/10.2307/2340521
15. Fu, J., Tang, W., Du, P., Wang, G., Chen, W., Li, J., Zhu, Y., Gao, J., Cui, L.: Identifying microRNA-mRNA regulatory network in colorectal cancer by a combination of expression profile and bioinformatics analysis. BMC Syst. Biol. **6**, 68 (2012)
16. Gauthier, N.P., Larsen, M.E., Wernersson, R., de Lichtenberg, U., Jensen, L.J., Brunak, S., Jensen, T.S.: Cyclebase.org—a comprehensive multi-organism online database of cell-cycle experiments. Nucleic Acids Res. **36**(suppl_1), D854–D859 (2008). http://dx.doi.org/10.1093/nar/gkm729
17. Granovskaia, M.V., Jensen, L.J., Ritchie, M.E., Toedling, J., Ning, Y., Bork, P., Huber, W., Steinmetz, L.M.: High-resolution transcription atlas of the mitotic cell cycle in budding yeast. Genome Biol. **11**(3), R24 (2010). https://doi.org/10.1186/gb-2010-11-3-r24
18. Gubler, D.J.: Dengue and dengue hemorrhagic fever. Clin. Microbiol. Rev. **11**(3), 480–496 (1998). https://doi.org/10.1128/cmr.11.3.480
19. Hartigan, J.A., Hartigan, P.M.: The dip test of unimodality. Ann. Stat. **13**(1), 70–84 (1985). https://doi.org/10.1214/aos/1176346577
20. Hwang, B., Lee, J.H., Bang, D.: Single-cell RNA sequencing technologies and bioinformatics pipelines. Exp. Mol. Med. **50**(8), 1–14 (2018). https://doi.org/10.1038/s12276-018-0071-8
21. Jiao, X., Sherman, B.T., Huang, D.W., Stephens, R., Baseler, M.W., Lane, H.C., Lempicki, R.A.: DAVID-WS: a stateful web service to facilitate gene/protein list analysis. Bioinformatics **28**(13), 1805–1806 (2012). https://doi.org/10.1093/bioinformatics/bts251
22. Juanes, M.A.: Methods of Synchronization of Yeast Cells for the Analysis of Cell Cycle Progression, pp. 19–34. Springer, New York (2017). https://doi.org/10.1007/978-1-4939-6502-1_2
23. Kanehisa, M., Furumichi, M., Tanabe, M., Sato, Y., Morishima, K.: KEGG: new perspectives on genomes, pathways, diseases and drugs. Nucleic Acids Res. **45**(D1), D353–D361 (2016). https://doi.org/10.1093/nar/gkw1092
24. Keller, A., Leidinger, P., Bauer, A., ElSharawy, A., Haas, J., Backes, C., Wendschlag, A., Giese, N., Tjaden, C., Ott, K., Werner, J., Hackert, T., Ruprecht, K., Huwer, H., Huebers, J., Jacobs, G., Rosenstiel, P., Dommisch, H., Schaefer, A., Müller-Quernheim, J., Wullich, B., Keck, B., Graf, N., Reichrath, J., Vogel, B., Nebel, A., Jager, S.U., Staehler, P., Amarantos, I., Boisguerin, V., Staehler, C., Beier, M., Scheffler, M., Büchler, M.W., Wischhusen, J., Haeusler, S.F.M., Dietl, J., Hofmann, S., Lenhof, H.P., Schreiber, S., Katus, H.A., Rottbauer, W., Meder, B., Hoheisel, J.D., Franke, A., Meese, E.: Toward the blood-borne miRNome of human diseases. Nat. Methods **8**(10), 841–843 (2011). https://doi.org/10.1038/nmeth.1682
25. Kozomara, A., Griffiths-Jones, S.: miRBase: annotating high confidence microRNAs using deep sequencing data. Nucleic Acids Res. **42**(D1), D68–D73 (2014). http://dx.doi.org/10.1093/nar/gkt1181
26. Leidinger, P., Backes, C., Deutscher, S., Schmitt, K., Mueller, S.C., Frese, K., Haas, J., Ruprecht, K., Paul, F., Stähler, C., Lang, C.J., Meder, B., Bartfai, T., Meese, E., Keller, A.: A blood based 12-miRNA signature of alzheimer disease patients. Genome Biol. **14**(7), R78 (2013). https://doi.org/10.1186/gb-2013-14-7-r78
27. Li, X., Gill, R., Cooper, N.G., Yoo, J.K., Datta, S.: Modeling microRNA-mRNA interactions using PLS regression in human colon cancer. BMC Med. Genet. **4**, 44 (2011)
28. Liu, P.F., Jiang, W.H., Han, Y.T., He, L.F., Zhang, H.L., Ren, H.: Integrated microRNA-mRNA analysis of pancreatic ductal adenocarcinoma. Genet. Mol. Res. **14**(3), 10288–10297 (2015)
29. Ma, L., Huang, Y., Zhu, W., Zhou, S., Zhou, J., Zeng, F., Liu, X., Zhang, Y., Yu, J.: An integrated analysis of miRNA and mRNA expressions in non-small cell lung cancers. PLoS One **6**(10), e26502 (2011)
30. MacLellan, S.A., Lawson, J., Baik, J., Guillaud, M., Poh, C.F.Y., Garnis, C.: Differential expression of miRNAs in the serum of patients with high-risk oral lesions. Cancer Med. **1**(2), 268–274 (2012). https://doi.org/10.1002/cam4.17

31. Meng, X.R., Lu, P., Mei, J.Z., Liu, G.J., Fan, Q.X.: Expression analysis of miRNA and target mRNAs in esophageal cancer. Braz. J. Med. Biol. Res. **47**(9), 811–817 (2014)
32. Miyanaga, A., Gemma, A., Noro, R., Kataoka, K., Matsuda, K., Nara, M., Okano, T., Seike, M., Yoshimura, A., Kawakami, A., Uesaka, H., Nakae, H., Kudoh, S.: Antitumor activity of histone deacetylase inhibitors in non-small cell lung cancer cells: development of a molecular predictive model. Mol. Cancer Ther. **7**(7), 1923–1930 (2008). DOI 10.1158/1535-7163.MCT-07-2140. http://mct.aacrjournals.org/content/7/7/1923
33. Murakami, Y., Toyoda, H., Tanahashi, T., Tanaka, J., Kumada, T., Yoshioka, Y., Kosaka, N., Ochiya, T., Taguchi, Y.H.: Comprehensive miRNA expression analysis in peripheral blood can diagnose liver disease. PLoS One **7**(10), e48366 (2012). https://doi.org/10.1371/journal.pone.0048366
34. Pramila, T., Wu, W., Miles, S., Noble, W.S., Breeden, L.L.: The forkhead transcription factor hcm1 regulates chromosome segregation genes and fills the s-phase gap in the transcriptional circuitry of the cell cycle. Genes Dev. **20**(16), 2266–2278 (2006). DOI 10.1101/gad.1450606. http://genesdev.cshlp.org/content/20/16/2266.abstract
35. Qiu, W., He, W., Wang, X., Lazarus, R.: A marginal mixture model for selecting differentially expressed genes across two types of tissue samples. Int. J. Biostat. **4**(1) (2008). https://doi.org/10.2202/1557-4679.1093
36. Qiu, X., Mao, Q., Tang, Y., Wang, L., Chawla, R., Pliner, H.A., Trapnell, C.: Reversed graph embedding resolves complex single-cell trajectories. Nat. Methods **14**(10), 979–982 (2017). https://doi.org/10.1038/nmeth.4402
37. Ritchie, M.E., Phipson, B., Wu, D., Hu, Y., Law, C.W., Shi, W., Smyth, G.K.: limma powers differential expression analyses for rna-sequencing and microarray studies. Nucleic Acids Res. **43**(7), e47 (2015). http://dx.doi.org/10.1093/nar/gkv007
38. Rommer, A., Steinleitner, K., Hackl, H., Schneckenleithner, C., Engelmann, M., Scheideler, M., Vlatkovic, I., Kralovics, R., Cerny-Reiterer, S., Valent, P., Sill, H., Wieser, R.: Overexpression of primary microRNA 221/222 in acute myeloid leukemia. BMC Cancer **13**(1), 1–12 (2013). https://doi.org/10.1186/1471-2407-13-364
39. Santos, A., Wernersson, R., Jensen, L.J.: Cyclebase 3.0: a multi-organism database on cell-cycle regulation and phenotypes. Nucleic Acids Res. **43**(D1), D1140–D1144 (2015). http://dx.doi.org/10.1093/nar/gku1092
40. Sharma, S., Kelly, T.K., Jones, P.A.: Epigenetics in cancer. Carcinogenesis **31**(1), 27–36 (2009). https://doi.org/10.1093/carcin/bgp220
41. Shen, J., Wang, A., Wang, Q., Gurvich, I., Siegel, A.B., Remotti, H., Santella, R.M.: Exploration of genome-wide circulating MicroRNA in hepatocellular carcinoma: MiR-483-5p as a potential biomarker. Cancer Epidemiol. Biomarkers Prev. **22**(12), 2364–2373 (2013). https://doi.org/10.1158/1055-9965.epi-13-0237
42. Skinner, M.K.: Environmental stress and epigenetic transgenerational inheritance. BMC Med. **12**(1), 1–5 (2014). https://doi.org/10.1186/s12916-014-0153-y
43. Skinner, M.K., Haque, C.G.B.M., Nilsson, E., Bhandari, R., McCarrey, J.R.: Environmentally induced transgenerational epigenetic reprogramming of primordial germ cells and the subsequent germ line. PLoS One **8**(7), 1–15 (2013). https://doi.org/10.1371/journal.pone.0066318
44. Soboleva, A., Yefanov, A., Evangelista, C., Robertson, C.L., Lee, H., Kim, I.F., Phillippy, K.H., Marshall, K.A., Tomashevsky, M., Holko, M., Serova, N., Zhang, N., Sherman, P.M., Ledoux, P., Davis, S., Wilhite, S.E., Barrett, T.: NCBI GEO: archive for functional genomics data sets–update. Nucleic Acids Res. **41**(D1), D991–D995 (2012). https://dx.doi.org/10.1093/nar/gks1193
45. Song, L., Smola, A., Gretton, A., Bedo, J., Borgwardt, K.: Feature selection via dependence maximization. J. Mach. Learn. Res. **13**(May), 1393–1434 (2012)
46. Spellman, P., Sherlock, G., Zhang, M., Iyer, V., Anders, K., Eisen, M., Brown, P., Botstein, D., Futcher, B.: Comprehensive identification of cell cycle-regulated genes of the yeast saccharomyces cerevisiae by microarray hybridization. Mol. Biol. Cell **9**(12), 3273–3297 (1998)

47. Subramanian, A., Tamayo, P., Mootha, V.K., Mukherjee, S., Ebert, B.L., Gillette, M.A., Paulovich, A., Pomeroy, S.L., Golub, T.R., Lander, E.S., Mesirov, J.P.: Gene set enrichment analysis: A knowledge-based approach for interpreting genome-wide expression profiles. Proc. Natl. Acad. Sci. **102**(43), 15545–15550 (2005). https://doi.org/10.1073/pnas.0506580102
48. Taguchi, Y.H.: Inference of target gene regulation by miRNA via mirage server. In: Wan, J. (ed.) Introduction to Genetics: DNA Methylation, Histone Modification and Gene Regulation, chap. 9, pp. 175–200. iConcept Press, New York (2013)
49. Taguchi, Y.H.: Identification of aberrant gene expression associated with aberrant promoter methylation in primordial germ cells between e13 and e16 rat f3 generation vinclozolin lineage. BMC Bioinf. **16**(18), S16 (2015). https://doi.org/10.1186/1471-2105-16-S18-S16
50. Taguchi, Y.H.: Identification of more feasible MicroRNA–mRNA interactions within multiple cancers using principal component analysis based unsupervised feature extraction. Int. J. Mol. Sci. **17**(5), 696 (2016). https://doi.org/10.3390/ijms17050696
51. Taguchi, Y.H.: Principal component analysis based unsupervised feature extraction applied to budding yeast temporally periodic gene expression. BioData Mining **9**(1), 22 (2016). https://doi.org/10.1186/s13040-016-0101-9
52. Taguchi, Y.H.: Principal components analysis based unsupervised feature extraction applied to gene expression analysis of blood from dengue haemorrhagic fever patients. Sci. Rep. **7**(1) (2017). https://doi.org/10.1038/srep44016
53. Taguchi, Y.H.: Principal component analysis-based unsupervised feature extraction applied to single-cell gene expression analysis. In: Intelligent Computing Theories and Application, pp. 816–826. Springer International Publishing, Berlin (2018). https://doi.org/10.1007/978-3-319-95933-7_90
54. Taguchi, Y.H.: Comparative transcriptomics analysis. In: Ranganathan, S., Gribskov, M., Nakai, K., Schönbach, C. (eds.) Encyclopedia of Bioinformatics and Computational Biology, pp. 814–818. Academic Press, Oxford (2019). https://doi.org/10.1016/B978-0-12-809633-8.20163-5. http://www.sciencedirect.com/science/article/pii/B9780128096338201635
55. Taguchi, Y.H.: Regulation of gene expression. In: Ranganathan, S., Gribskov, M., Nakai, K., Schönbach, C. (eds.) Encyclopedia of Bioinformatics and Computational Biology, pp. 806–813. Academic Press, Oxford (2019). https://doi.org/10.1016/B978-0-12-809633-8.20667-5. http://www.sciencedirect.com/science/article/pii/B9780128096338206675
56. Taguchi, Y.H., Iwadate, M., Umeyama, H.: Principal component analysis-based unsupervised feature extraction applied to in silico drug discovery for posttraumatic stress disorder-mediated heart disease. BMC Bioinf. **16**(1), 139 (2015). https://doi.org/10.1186/s12859-015-0574-4
57. Taguchi, Y.H., Iwadate, M., Umeyama, H.: SFRP1 is a possible candidate for epigenetic therapy in non-small cell lung cancer. BMC Med. Genet. **9**(1), 28 (2016). https://doi.org/10.1186/s12920-016-0196-3
58. Taguchi, Y.H., Murakami, Y.: Principal component analysis based feature extraction approach to identify circulating microRNA biomarkers. PLoS One **8**(6), e66714 (2013). https://doi.org/10.1371/journal.pone.0066714
59. Taguchi, Y.H., Murakami, Y.: Universal disease biomarker: can a fixed set of blood microRNAs diagnose multiple diseases? BMC. Res. Notes **7**(1), 581 (2014). https://doi.org/10.1186/1756-0500-7-581
60. Taguchi, Y.H., Wang, H.: Exploring microrna biomarker for amyotrophic lateral sclerosis. Int. J. Mol. Sci. **19**(5), 1318 (2018). DOI 10.3390/ijms19051318. http://www.mdpi.com/1422-0067/19/5/1318
61. Tang, Y.A., Wen, W.L., Chang, J.W., Wei, T.T., Tan, Y.H.C., Salunke, S., Chen, C.T., Chen, C.S., Wang, Y.C.: A novel histone deacetylase inhibitor exhibits antitumor activity via apoptosis induction, f-actin disruption and gene acetylation in lung cancer. PLoS One **5**(9), e12417 (2010). https://doi.org/10.1371/journal.pone.0012417
62. The Gene Ontology Consortium: The Gene Ontology Resource: 20 years and still GOing strong. Nucleic Acids Res. **47**(D1), D330–D338 (2018). https://dx.doi.org/10.1093/nar/gky1055

63. Tollefsbol, T. (ed.): Transgenerational Epigenetics. Elsevier, Amsterdam (2014). https://doi.org/10.1016/c2012-0-02853-0
64. Tu, B.P.: Logic of the yeast metabolic cycle: temporal compartmentalization of cellular processes. Science **310**(5751), 1152–1158 (2005). https://doi.org/10.1126/science.1120499
65. Tusher, V.G., Tibshirani, R., Chu, G.: Significance analysis of microarrays applied to the ionizing radiation response. Proc. Natl. Acad. Sci. **98**(9), 5116–5121 (2001). https://doi.org/10.1073/pnas.091062498
66. Umeyama, H., Iwadate, M., Taguchi, Y.H.: TINAGL1 and B3GALNT1 are potential therapy target genes to suppress metastasis in non-small cell lung cancer. BMC Genomics **15**(9), S2 (2014). https://doi.org/10.1186/1471-2164-15-S9-S2
67. Varshavsky, R., Gottlieb, A., Horn, D., Linial, M.: Unsupervised feature selection under perturbations: meeting the challenges of biological data. Bioinformatics **23**(24), 3343–3349 (2007). http://dx.doi.org/10.1093/bioinformatics/btm528
68. Vlachos, I.S., Zagganas, K., Paraskevopoulou, M.D., Georgakilas, G., Karagkouni, D., Vergoulis, T., Dalamagas, T., Hatzigeorgiou, A.G.: DIANA-miRPath v3.0: deciphering microRNA function with experimental support. Nucleic Acids Res. **43**(W1), W460–W466 (2015). https://doi.org/10.1093/nar/gkv403
69. Wu, B., Li, C., Zhang, P., Yao, Q., Wu, J., Han, J., Liao, L., Xu, Y., Lin, R., Xiao, D., Xu, L., Li, E., Li, X.: Dissection of miRNA-miRNA interaction in esophageal squamous cell carcinoma. PLoS One **8**(9), e73191 (2013)
70. Yan, X., Chen, X., Liang, H., Deng, T., Chen, W., Zhang, S., Liu, M., Gao, X., Liu, Y., Zhao, C., Wang, X., Wang, N., Li, J., Liu, R., Zen, K., Zhang, C.Y., Liu, B., Ba, Y.: miR-143 and miR-145 synergistically regulate ERBB3 to suppress cell proliferation and invasion in breast cancer. Mol. Cancer **13**(1), 220 (2014). https://doi.org/10.1186/1476-4598-13-220
71. Yang, Y., Li, D., Yang, Y., Jiang, G.: An integrated analysis of the effects of microRNA and mRNA on esophageal squamous cell carcinoma. Mol. Med. Rep. **12**(1), 945–952 (2015)
72. Zhang, W., Edwards, A., Fan, W., Flemington, E.K., Zhang, K.: miRNA-mRNA correlation-network modules in human prostate cancer and the differences between primary and metastatic tumor subtypes. PLoS One **7**(6), e40130 (2012)

Chapter 7
Application of TD-Based Unsupervised FE to Bioinformatics

May my wish never come true.
Rikka Takarada, SSSS.GRIDMAN, Season 1, Episode 12

7.1 Introduction

Because of continuous price reduction of multiomics data measurements, including gene expression, promoter methylation, single-nucleotide polymorphism (SNP), histone modification, and miRNA expression, more number of experimental conditions need to be considered. For example, if gene expression is measured for various tissues of patients, gene expression has to be formatted, not in matrix, but in tensor, as patients vs. tissue vs. genes. In this case, TD rather than PCA is a suitable technology to apply. On the other hand, in the previous chapter, we aimed various integrated analyses, e.g., miRNA and mRNA expression, mRNA expression and methylation, and mRNA expression of two species. If genes or features are shared in the integrated analysis, generation of case I or II tensor and application of TD to it is a suitable treatment. In the following, we introduce some applications of TD-based unsupervised FE to either of the cases.

7.2 PTSD-Mediated Heart Diseases

The first example to be processed as a tensor form is post-traumatic stress disorder (PTSD)-mediated heart diseases. Although this disease has already been analyzed in the previous chapter (Sect. 6.4.1), the data set analyzed there includes only one tissue, the heart. Nonetheless, if one would like to understand how PTSD mediates heart disease, we need to know the gene expression of both heart and brain. Fortunately, there is such a kind of data set. In this section, I would like to demonstrate the usefulness of TD-based unsupervised FE applied to gene expression

Y-h. Taguchi, *Unsupervised Feature Extraction Applied to Bioinformatics*,
Unsupervised and Semi-Supervised Learning,
https://doi.org/10.1007/978-3-031-60982-4_7

Table 7.1 Samples used in this study. Numbers before/after comma are control/treated samples. h: hours, w: weeks

Stress, days	5		10			5		10	
rest period	24 h	1.5 w	24 h	6 w		24 h	1.5 w	24 h	6 w
AY	3,2	5,4	3,4	3,4	HC	3,5	4,5	5,4	4,5
MPFC	4,5	5,5	3,4	4,4	SE	3,2	2,3	3,3	3,3
ST	5,5	5,5	5,4	4,4	VS	5,5	5,5	3,4	5,4
Blood	5,5	5,5	4,5	4,5	Heart	5,5	4,5	5,5	5,5
Hemibrain	5,5	4,5	5,5	5,5	Spleen	5,5	5,5	5,4	5,5

AY amygdala, *HC* hippocampus, *MPFC* medial prefrontal cortex, *SE* septal nucleus, *ST* striatum, *VS* ventral striatum

of multiple tissues aiming to understand PTSD-mediated heart disease based upon the recent publication [94].

The data set analyzed is composed of the following samples (Table 7.1): It includes ten tissues under eight experimental conditions. This data set is formatted as a five-mode tensor, $x_{ij_1j_2j_3j_4} \in \mathbb{R}^{43,699\times 2\times 10\times 2\times 3}$, of the ith probe, subjected to j_1th treatment ($j_1 = 1$: control, $j_1 = 2$: treated [stress-exposed] samples), in the j_2th tissue [$j_2 = 1$: amygdala (AY), $j_2 = 2$: hippocampus (HC), $j_2 = 3$: medial prefrontal cortex (MPFC), $j_2 = 4$: septal nucleus (SE), $j_2 = 5$: striatum (ST), $j_2 = 6$: ventral striatum (VS), $j_2 = 7$: blood, $j_2 = 8$: heart, $j_2 = 9$: hemibrain, $j_2 = 10$: spleen], with the j_3th stress duration ($j_3 = 1$: 10 days, $j_3 = 2$: 5 days) and j_4th rest period after application of stress ($j_4 = 1$: 1.5 weeks, $j_4 = 2$: 24 h, $j_4 = 3$: 6 weeks). Zero values are assigned to missing observations (e.g., measurements at 6 weeks after a 5-day period of stress are not available).

The higher-order singular value decomposition (HOSVD) algorithm (Fig. 3.8) is applied to $x_{ij_1j_2j_3j_4}$ as

$$x_{ij_1j_2j_3j_4} = \sum_{\ell_5=1}^{43,699}\sum_{\ell_1=1}^{2}\sum_{\ell_2=1}^{10}\sum_{\ell_3=1}^{2}\sum_{\ell_4=1}^{3} G(\ell_1,\ell_2,\ell_3,\ell_4,\ell_5) u_{\ell_1 j_1}^{(j_1)} u_{\ell_2 j_2}^{(j_2)} u_{\ell_3 j_3}^{(j_3)} u_{\ell_4 j_4}^{(j_4)} u_{\ell_5 i}^{(i)} \tag{7.1}$$

where $u_{\ell_5 i}^{(i)} \in \mathbb{R}^{43,699\times 43,699}$, $u_{\ell_1 j_1}^{(j_1)} \in \mathbb{R}^{2\times 2}$, $u_{\ell_2 j_2}^{(j_2)} \in \mathbb{R}^{10\times 10}$, $u_{\ell_3 j_3}^{(j_3)} \in \mathbb{R}^{2\times 2}$, and $u_{\ell_4 j_4}^{(j_4)} \in \mathbb{R}^{3\times 3}$ are singular value vectors and $G(\ell_1,\ell_2,\ell_3,\ell_4,\ell_5) \in \mathbb{R}^{43,699\times 2\times 10\times 2\times 3}$ is the core tensor.

We need to specify which singular value vectors attributed to genes, $\boldsymbol{u}_{\ell_1}^{(i)}$, is used for gene selection. For this purpose, we investigate other singular value vectors, $\boldsymbol{u}_{\ell_k}^{(j_k)}$, $1 \le k \le 4$. One of the important points is tissue specificity. What I would like to find is a set of genes expressive in common between heart and brain. Because $1 \le j \le 6$ and $j = 9$ correspond to brain and $j = 8$ corresponds to heart, we need to find $\boldsymbol{u}_{\ell_2}^{(j_2)}$ expressive in common $j = 1, 2, \ldots, 6, 8, 9$. Figure 7.1 shows the singular

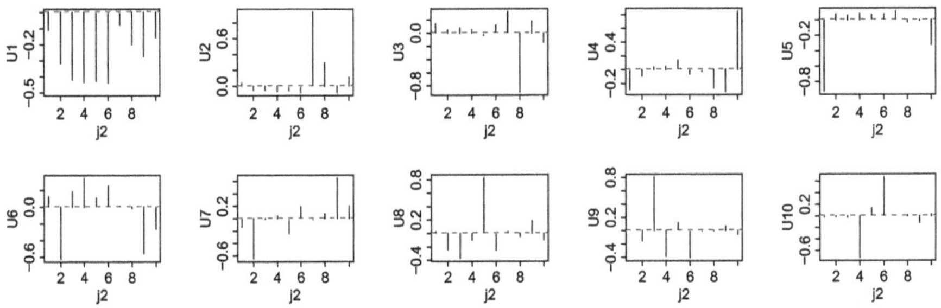

Fig. 7.1 Singular value vectors, $\boldsymbol{u}_{\ell_2}^{(j_2)}$, $1 \leq \ell_2 \leq 10$. Red horizontal broken lines show baseline

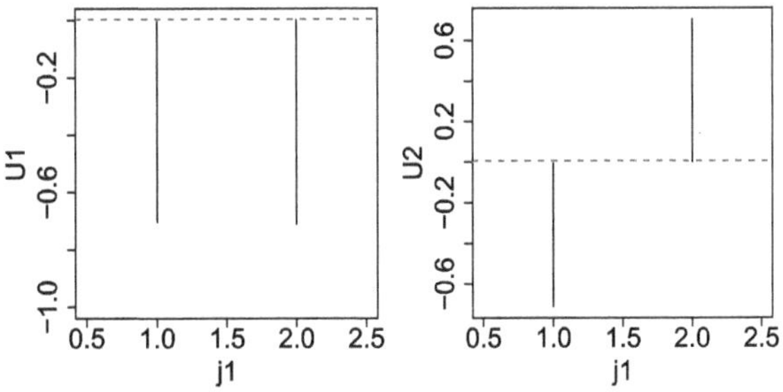

Fig. 7.2 Singular value vectors, $\boldsymbol{u}_{\ell_1}^{(j_1)}$, $\ell_1 = 1, 2$. Red horizontal broken lines show baseline

value vectors, $\boldsymbol{u}_{\ell_2}^{(j_2)}$, $1 \leq \ell_2 \leq 10$. Although no $\boldsymbol{u}_{\ell_2}^{(j_2)}$ fully satisfies this requirement, $\boldsymbol{u}_{4}^{(j_2)}$ relatively fulfills this requirement. $\boldsymbol{u}_{4}^{(j_2)}$ are negatively signed in common for $j = 1, 2, 8, 9$ that correspond to AY, HC, heart, and hemibrain. Especially, because AY and HC are very important in PTSD [67], it is promising that we can get singular value vector expressive in common AY, HC, and heart.

The next important requirement is that control and stressed samples should be oppositely expressive. This means $u_{\ell_1 1}^{(j_1)} = -u_{\ell_1 2}^{(j_1)}$. This requirement is easy to fulfill because $u_{\ell_1 1}^{(j_1)} = -u_{\ell_1 2}^{(j_1)}$ or $u_{\ell_1 1}^{(j_1)} = u_{\ell_1 2}^{(j_1)}$ must be satisfied when there are only two classes and the mean is zero. Figure 7.2 shows the singular value vectors, $\boldsymbol{u}_{\ell_1}^{(j_1)}$, $\ell_1 = 1, 2$. As expected, $\ell_1 = 2$ corresponds to the reversed sign between control and stressed samples.

Because there are no known predefined desirable properties for experimental conditions, i.e., stress and rest period, we should find $G(2, 4, \ell_3, \ell_4, \ell_5)$ with the larger absolute values. Table 7.2 shows the top-ranked G with larger absolute values. Then we can find that $\ell_5 = 1, 4, 11$ are associated with $G(2, 4, \ell_3, \ell_4, \ell_5)$ with the larger absolute values. Thus, we decided to attribute P values using $\ell_5 = 1, 4, 11$ by assuming χ^2 distribution as

$$P_i = P_{\chi^2}\left[> \sum_{\ell_5=1,4,11} \left(\frac{u_{\ell_5 i}}{\sigma_{\ell_5}} \right)^2 \right]. \tag{7.2}$$

Table 7.2 Top-ranked $G(\ell_1 = 2, \ell_2 = 4, \ell_3, \ell_4, \ell_5)$ with greater absolute values

ℓ_3	ℓ_4	ℓ_5	$G(2, 4, \ell_3, \ell_4, \ell_5)$
1	1	11	−35.0
1	1	1	−30.8
2	2	1	−30.3
2	3	4	−30.0
2	3	1	28.7
2	2	4	28.5

Table 7.3 Thirteen combinations of tissues and experimental conditions where the selected 801 probes are differentially expressed between stress-exposed and control samples. MPFC: medial prefrontal cortex. ○: associated with P values that are computed by t-test, adjusted by BH criterion and less than 0.01

Stress duration	10 days		5 days	
rest period	24 h	6 weeks	24 h	1.5 weeks
AY		○		○
HC		○	○	○
MPFC		○		
Heart	○			○
Hemibrain			○	○
Spleen		○	○	○

P values are corrected by BH criterion, and 801 probes associated with adjusted P values less than 0.01 are selected.

The first validation of selected 801 probes is to see if these are expressed distinctly between control and stressed samples, selectively on only the heart and brain. In order to confirm this, we apply t-test to the selected 801 probes between control and stressed samples for all combinations of tissues, rest and stressed periods. P values are corrected by BH criterion and conditions associated with adjusted P values less than 0.01 are considered to be expressed distinctly and significantly between control and stressed samples. Table 7.3 shows the results. The selected 801 genes are expressed distinctly between control and stressed samples, selectively in heart, HC, and AY (It is also in spleen, because it is oppositely expressed toward heart, HC, and AY as shown in Fig. 7.1).

Here we would like to emphasize the difficulty of gene selection in this data set. As mentioned above, what we aim for is quite abstract, i.e., "genes expressive in common between brain and heart as well as distinctly between control and stressed samples." As a result, we realize that common expression between AY, HC, and heart is possible (with the investigation of $\boldsymbol{u}_4^{(j_2)}$ in Fig. 7.1). Generally, it is impossible to know this combination in advance. When no clear purpose is given in advance, supervised methods cannot perform well while unsupervised methods can.

In order to see how well other conventional supervised methods perform, we test three methods, SAM, limma, and categorical regression analysis. The first example to be compared with TD-based unsupervised FE is categorical regression analysis. For the data set shown in Table 7.1, the only possible way to apply categorical

Table 7.4 Results of gene selection based on categorical regression. P values are adjusted by BH criterion

Adjusted P values	$P > 0.01$	$P < 0.01$	$P > 0.05$	$P < 0.05$	$P > 0.1$	$P < 0.1$
Number of probes	2222	41,157	1986	41,713	1839	41,860

Table 7.5 Results by SAM. p0 is the ratio of the null hypothesis, FDR corresponds to the adjusted P values. Called is the number of genes that break the null hypothesis. Expected number of false positives is False × FDR × p0

	Delta	p0	False	Called	FDR
1	0.1	0.011	38,538.08	43,379	0.0094
2	11.4	0.011	0.02	5424	3.9e–08
3	22.7	0.011	0	323	0
4	34.0	0.011	0	40	0
5	45.2	0.011	0	7	0
6	56.5	0.011	0	4	0
7	67.8	0.011	0	2	0
8	79.1	0.011	0	1	0
9	90.3	0.011	0	1	0
10	101.6	0.011	0	1	0

regression is to treat it as 80 classes (ten tissues vs. four experimental conditions vs. control and stressed samples). Although it is better to consider the pair of control and stressed samples, it is impossible. Typically, although a ratio might be taken, because it is not paired samples, i.e., there is no one-to-one correspondence, and we cannot take a ratio. Table 7.4 shows the result of categorical regression analysis. Because of treatment as 80 classes, genes associated with any kind of distinction are detected (i.e., associated with significantly small adjusted P values). As a result, almost all genes are judged as distinct between some combinations. It is obvious that this result is not desirable for our purpose, "genes expressive in common between brain and heart distinctly between control and stressed samples," at all, because of lack of specificity. To screen these genes, we need some additional criterion that TD-based unsupervised FE does not require. Thus, TD-based unsupervised FE is more fitted to the present purpose than categorical regression.

Next, we apply SAM with assuming 80 classes to the data set shown in Table 7.1. Table 7.5 shows the result of SAM. $p0$, which represents the contribution of null hypothesis that no distinction among 80 classes, is 1%. This means that almost all genes are distinctly expressive in either of the combinations. Although FDR corresponds to the adjusted P values, it is clear that all genes are associated with FDR less than 0.01. Although this conclusion itself is coincident with that of categorical regression, in this sense, SAM is not useful to select "genes expressive in common between brain and heart distinctly between control and stressed samples," either.

Finally, we apply limma to the data set shown in Table 7.1. Fortunately, limma enables us to select genes that are distinct between any pairs of controls and samples.

Table 7.6 Results of gene selection based on limma. P values are adjusted by limma itself

	Case A : not considering differential expression					
Adjusted P values	$P > 0.01$	$P < 0.01$	$P > 0.05$	$P < 0.05$	$P > 0.1$	$P < 0.1$
Number of probes	0	43,379	0	43,379	0	43,379
Adjusted P values	Case B: considering differential expression					
	$P > 0.01$	$P < 0.01$	$P > 0.05$	$P < 0.05$	$P > 0.1$	$P < 0.1$
Number of probes	25,992	17,387	17,745	25,634	13,542	29,837

Table 7.7 KEGG pathway enrichment by the 457 genes identified by TD-based unsupervised FE. Adjusted P values are by BH criterion

Category	Term	Gene count	%	P Value	Adjusted P value
KEGG_PATHWAY	Ribosome	57	12.8	8.4×10^{-58}	1.0×10^{-55}
KEGG_PATHWAY	Parkinson's disease	48	10.8	3.6×10^{-33}	2.2×10^{-31}
KEGG_PATHWAY	Oxidative phosphorylation	47	10.5	1.7×10^{-32}	6.9×10^{-31}
KEGG_PATHWAY	Alzheimer's disease	50	11.2	2.5×10^{-28}	7.5×10^{-27}
KEGG_PATHWAY	Huntington's disease	48	10.8	3.6×10^{-26}	8.6×10^{-25}
KEGG_PATHWAY	Cardiac muscle contraction	30	6.7	2.4×10^{-21}	4.8×10^{-20}
KEGG_PATHWAY	Glycolysis/gluconeogenesis	10	2.2	1.5×10^{-3}	2.6×10^{-2}

Thus, we apply limma in two ways. One assumes 80 classes (case A in Table 7.6) and another assumes 40 classes (case B in Table 7.6) composed of 40 (ten tissues vs. four experimental conditions) pairwise combinations between control an stress samples. Possibly because of its advanced feature, limma successfully denies the detection of genes expressive distinct among any pairs of 80 classes (case A). Nevertheless, limma still detects too many positives in 40 pairwise comparisons (case B). As expected, because of the lack of well-defined screening criterion, three supervised methods are useless to find "genes expressive in common between brain and heart as well as distinctly between control and stressed samples." In conclusion, none of the three conventional supervised methods are as useful as TD-based unsupervised FE for the present purpose.

Although TD-based unsupervised FE successfully identifies genes expressive distinct between control and stressed samples in tissue-specific manner (Table 7.3), if it is biologically useless, it cannot be considered to be successful. In order to evaluate selected probes biologically, we try to identify protein-coding genes associated with these 801 probes. Then, we find 457 genes (because of lack of space, we cannot list all of 457 genes, which are available as Additional file 5 [94], if the readers are particularly interested in them). We upload 457 genes to DAVID. The result is quite promising. Table 7.7 shows the enriched KEGG pathway associated with adjusted P values less than 0.05. They include four neurodegenerative diseases as well as one cardiac problem. Thus, they are quite suitable to be candidate genes that cause PTSD-mediated heart diseases as those in Table 6.20 where PTSD-mediated heart disease is investigated by PCA-based unsupervised FE.

7.3 Drug Discovery from Gene Expression

Drug discovery is a time-consuming and expensive process. It starts from preparing as many small molecules as possible. Then, it is tried to find one effective way to target diseases by exhaustive search. The number of initially prepared molecules can be 10^4; testing these many number of compounds is expensive and time-consuming. If we can reduce the number of initial candidate small molecules to one tenth, it benefits so much to reduce time and cost required.

In this sense, so-called in silico drug discovery develops with much expectation to fulfill this requirement. in silico drug discovery aims to identify candidate small molecules without *wet* experiments. By making full use of recently developed computational power, including CPU with high-speed computing, storage that can store massive information, as well as recently developed machine learning technique, in silico drug discovery enables us to prepare a set of more promising candidate small molecules as drugs.

Traditionally, there are two main streams of in silico drug discovery. One is ligand-based drug design [1] (LBDD) and another is structure-based drug design [5] (SBDD). LBDD aims to identify new candidate drug compounds based on their similarity with known drugs. LBDD has huge varieties dependent upon how similarity is defined. The advantage of LBDD is that it has more trust, i.e., larger probability of finding true drug compounds, and requires smaller computational resources than SBDD. The disadvantage of LBDD is that it requires the information of known drugs and fails to find new drug candidates that lack similarity with known drugs. On the contrary, SBDD has the advantage that it can predict new candidate drugs without the information of known drugs. The disadvantage of SBDD is that it requires massive computation, because it must execute docking simulation between drug candidate compounds and target proteins. Another disadvantage of SBDD is that it needs a protein tertiary structure to which individual candidate drug compounds must bind. Experimental measurements of protein tertiary structure itself are difficult tasks. Although it has become much easier because of the invention of cryo-electron microscopy [41] than before, it still needs to pay a huge amount of money and long time. When there are no protein tertiary structures available, the protein tertiary structure itself must be computationally predicted [28]. The prediction inevitably has inaccuracy that affects the prediction of the binding affinity of small molecules.

In order to compensate for these disadvantages of LBDD and SBDD, the third option is recently proposed: drug design from gene expression [25]. Posttreatment gene expression can be used to screen candidate compounds for their ability to induce the target phenotype. This approach is very useful once posttreatment gene expression is available. In this section, we try to make use of TD-based unsupervised FE to predict new drug target by analyzing posttreatment gene expression [97].

Posttreatment gene expression is obtained from LINCS [90]. L1000 is highly reproducible, comparable to RNA sequencing, and suitable for computational inference of the expression levels of 81% of non-measured transcripts. Gene expression

Table 7.8 The number of the inferred compounds and inferred genes associated with significant dose-dependent activity. The target genes predicted by means of the comparison with the data showing upregulation of the expression of individual genes (predicted targets) are also shown

Tumor Cell lines	BT20	HS578T	MCF10A	MCF7	MDAMB231	SKBR3
Tumor	Breast					
Inferred genes	41	57	42	55	41	46
Inferred compounds	4	3	2	6	5	6
All compounds	110	106	106	108	108	106
Predicted targets	418	576	476	480	560	423
Tumor Cell lines	A549	HCC515	HA1E	HEPG2	HT29	PC3
Tumor	Lung	Lung	Kidney	Liver	Colon	Prostate
Inferred genes	45	46	48	54	50	63
Inferred compounds	8	5	7	2	2	9
All compounds	265	270	262	269	270	270
Predicted targets	428	352	423	396	358	439
Tumor Cell lines	A375					
Tumor	Melanoma					
Inferred genes	43					
Inferred compounds	6					
All compounds	269					
Predicted targets	421					

The full list of inferred genes and predicted targets is available in Additional file 7 [97]. Inferred compounds are presented in Table 7.9. 'All compounds' rows represent the total number of compounds used for the treatment of each cell line

profile is available in GEO with GEO ID GSE70138. Table 7.8 summarizes the gene expression profiles. They include 13 cell lines to which 100–300 compounds (denoted as "all compounds") are treated. One problem of this data set is that it includes only 978 genes' expression profiles, because it is measured by Luminex scanners. Gene expression profiles in individual cell lines are formatted as tensor, $x_{ijk} \in \mathbb{R}^{978\times 6\times K}$; i denotes gene (probe), j denotes dose density of drug compound, and k stands for individual compounds among K total number of compounds that correspond to "all compounds" in Table 7.8. The HOSVD algorithm (Fig. 3.8) is applied as

$$x_{ijk} = \sum_{\ell_1=1}^{978}\sum_{\ell_2=1}^{6}\sum_{\ell_3=1}^{K} G(\ell_1,\ell_2,\ell_3) u_{\ell_1 i}^{(i)} u_{\ell_2 i}^{(j)} u_{\ell_3 k}^{(k)} \tag{7.3}$$

where $\boldsymbol{u}_{\ell_1}^{(i)} \in \mathbb{R}^{978}$, $\boldsymbol{u}_{\ell_2}^{(j)} \in \mathbb{R}^{6}$, $\boldsymbol{u}_{\ell_3}^{(k)} \in \mathbb{R}^{K}$, are the singular value vectors, and $G(\ell_1,\ell_2,\ell_3) \in \mathbb{R}^{978\times 6\times K}$ is the core tensor.

The first step is to identify genes whose expression is altered by drug treatment. Thus, we try to identify which $\boldsymbol{u}^{(j)}$ has monotonic dependence upon dose density. Figure 7.3 shows $\boldsymbol{u}_{\ell_2}^{(j)}$, $1 \le \ell_2 \le 3$ for 13 cell lines listed in Table 7.8. It is obvious

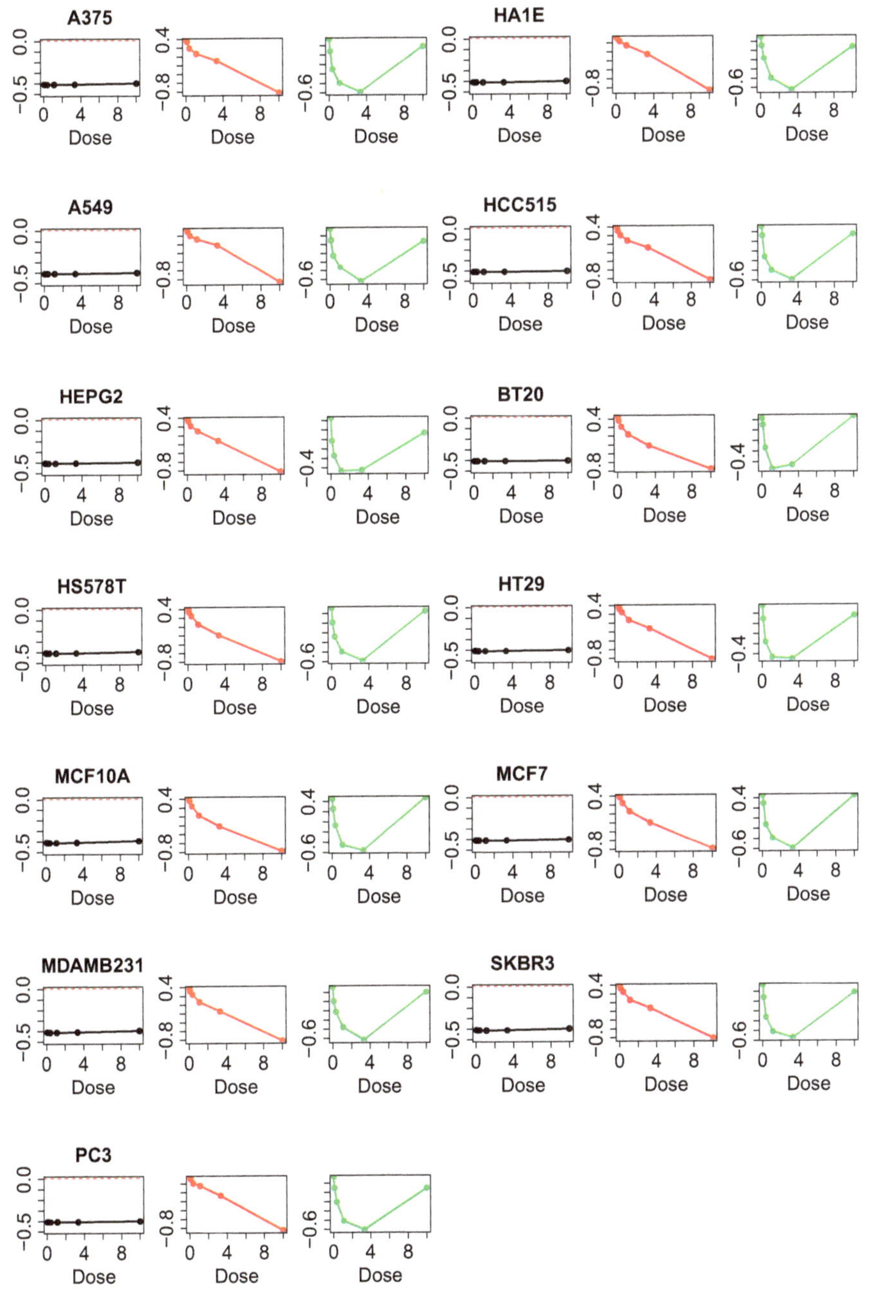

Fig. 7.3 Singular value vectors, $\boldsymbol{u}_{\ell_2}^{(j)}, 1 \leq \ell_2 \leq 3$. Red horizontal broken lines show baseline. Black: $\ell_2 = 1$, red: $\ell_2 = 2$, green: $\ell_2 = 3$

that $\boldsymbol{u}_2^{(j)}$ shows almost linear dependence upon dose independent of cell lines. The next task is to identify $G(\ell_1, 2, \ell_3)$ with larger absolute values in order to decide which $\boldsymbol{u}_{\ell_1}^{(i)}$ and $\boldsymbol{u}_{\ell_3}^{(k)}$ are used for selecting the combinations of genes and compounds that commit linear dose dependence. Because

$$G(\ell_1 \leq 6, \ell_2 \leq 6, \ell_3 \leq 6) = \frac{\sum_{\ell_1 \leq 6, \ell_2 \leq 6, \ell_3 \leq 6} G(\ell_1, \ell_2, \ell_3)^2}{\sum_{\ell_1, \ell_2, \ell_3} G(\ell_1, \ell_2, \ell_3)^2} \tag{7.4}$$

exceeds 0.95 for almost all cell lines, it is decided to employ $(\ell_1 \leq 6, \ell_2 = 2, \ell_3 \leq 6)$ components for FE. Nonetheless, in the case of PC3 cells, $(\ell_1 \leq 8, \ell_2 = 2, \ell_3 \leq 8)$, as an exception, are used for FE because the eighth component is found to have non-negligible contributions in this cell line.

To identify the genes and compounds associated with a significant dose-dependent activity, it is assumed that $u_{\ell_1 \leq 6, i}$ and $u_{\ell_3 \leq 6, k}$ follow independent normal distributions and P values are attributed to the ith gene and the kth compounds using a χ^2 distribution,

$$P_i = P_{\chi^2}\left[> \sum_{\ell_1 \leq 6} \left(\frac{u_{\ell_1 i}^{(i)}}{\sigma_{\ell_1}} \right)^2 \right] \tag{7.5}$$

and

$$P_k = P_{\chi^2}\left[> \sum_{\ell_3 \leq 6} \left(\frac{u_{\ell_3 k}^{(k)}}{\sigma_{\ell_3}} \right)^2 \right] \tag{7.6}$$

where σ_{ℓ_1} and σ_{ℓ_3} are standard deviations of $u_{\ell_1 i}^{(i)}$ and $u_{\ell_3 k}^{(k)}$, respectively. For PC3 cells, $\ell_1 \leq 8$ and $\ell_3 \leq 8$ are used in the above equations. $P_{\chi^2}[> x]$ is the cumulative probability that the argument is greater than x assuming a χ^2 distribution with eight degrees of freedom for PC3 cell lines and with six degrees of freedom for other cell lines. P_i and P_k are adjusted by means of the BH criterion, and compounds and genes associated with the adjusted P value lower than 0.01 are selected as those associated with a significant dose-dependent cellular response. The number of selected genes and compounds are listed as "inferred genes" and "inferred compounds" in Table 7.8, respectively. The above process is illustrated in Fig. 7.4.

The next task is to identify proteins to which selected compounds bind. "Inferred genes" in Table 7.8 do not correspond to the proteins to which selected compounds bind, because they are the genes whose mRNA expression is altered because of drug treatment. Usually, mRNA expression of proteins to which selected compounds bind is not altered because of drug treatment. Thus, we need to infer the protein targeted by drug treatment. Thus, we need additional external information that lists the genes whose mRNA expression is altered because of a gene perturbation. Then

Fig. 7.4 Starting from gene expression profile formatted as tensor, x_{ijk}, singular value vectors, $\boldsymbol{u}_{\ell_1}^{(i)}, \boldsymbol{u}_{\ell_2}^{(j)}$, and $\boldsymbol{u}_{\ell_3}^{(k)}$, are obtained. After identifying $\ell_2 = 2$ as associated with linear dose dependence (see Fig. 7.3), $\ell_1 \leq 6$ and $\ell_3 \leq 6$ are decided to be used for FE because of larger contribution defined in Eq. (7.4). Genes i and compounds k are selected using $\boldsymbol{u}_{\ell_1}^{(i)}, \ell_1 \leq 6, \boldsymbol{u}_{\ell_3}^{(k)}, \ell_3 \leq 6$

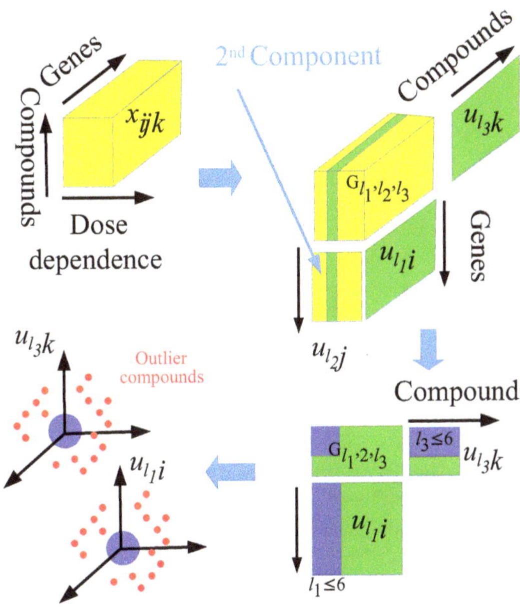

if "inferred genes" matched with genes whose mRNA expression is altered because of the gene perturbation, we infer the perturbed gene as target protein (Fig. 7.5). There can be multiple resources from which we can retrieve the list of genes whose mRNA expression is altered because of single-gene perturbation. Here we employ Enrichr [49] that collects multiple data resources in order to perform various enrichment analyses. After uploading "inferred genes" to Enrichr, we list genes associated with adjusted P values less than 0.01 in the category of "Single-gene Perturbations from GEO up." Their number corresponds to the number of "predicted targets" in Table 7.8. This strategy is especially efficient for LINCS data set that includes only the expression of 978 genes. Employing the strategy in Fig. 7.5, we can identify target proteins not included in these 978 genes.

Next we would like to evaluate if our prediction is correct, i.e., if "inferred compounds" bind to "predicted targets." In principle, it is impossible to check the accuracy of our prediction without experiments. Thus, instead of executing experiments, we compare our prediction with known list of target proteins of drug compounds. For this purpose, we employ two information resources, drug2gene.com [81] and DSigDB [124]. Table 7.9 shows the results of Fisher's exact test that evaluates overlaps between "predicted targets" and known target proteins of "inferred compounds." If P value computed by Fisher's exact test is less than 0.05, it is significant (no correction considering multiple comparisons). It is obvious that in most of the cases, our prediction significantly overlaps with known target proteins of drug compounds. Thus, TD-based unsupervised FE can be used for in silico drug discovery from gene expression.

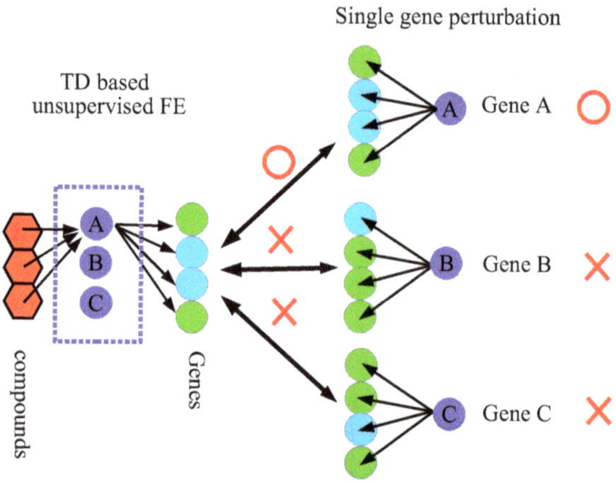

Fig. 7.5 After the drug (red hexagon) treatments, we can detect mRNAs with altered expression (filled cyan circle) along with those without altered expression (filled green circle). We have no information about proteins (circled A, B and C). List of genes with altered expression can be compared with genes with altered expression when genes A, B or C is perturbed. Then, we can identify compounds might bind to protein A, because the list of genes whose mRNA expression is altered are common

It is also interesting that "inferred compounds" are largely overlapped among cell lines. Because two to nine compounds are identified in each of the 13 cell lines, the total number of identified compounds can be several tens. Nevertheless, the number of compounds listed in Table 7.9 is as small as 19. In some sense, it might be a piece of evidence that our strategy is correct. It is reasonable that anticancer drugs are effective for multiple cancers. Thus, large overlap of "inferred compounds" between distinct cell lines makes sense. On the other hand, analyses based upon distinct gene expression profiles unlikely result in largely overlapped results without any biological reasons. Possibly, the results shown in Table 7.9 are trustable.

Although we employed single-gene perturbation to infer target proteins from the list of genes with altered expression caused by drug treatment, any other database that can describe gene interaction should be usable. As an alternative, we try "PPI hub proteins" in Enrichr instead of "single-gene perturbations from GEO up." The primary difference between "PPI hub proteins" and "single-gene perturbations from GEO up" is the number of genes included. "PPI hub proteins" includes only a few hundred genes, while "single-gene perturbations from GEO up" includes a few thousand genes. This suggests that the results using "PPI hub proteins" might be less significant. Table 7.10 lists the results of Fisher's exact test of the comparison between predicted targets based upon "PPI hub proteins" and drug2gene.com database. In contrast to the expectation, all cases have significant overlap with drug2gene.com. This supports our expectation that any kind of gene–gene interaction is usable together with TD-based unsupervised FE for in sicilo drug discovery from gene expression.

Table 7.9 Compound–gene interactions presented in Table 7.8 that significantly overlap with interactions described in two data sets. For each compound in the table, the upper row the drug2gene.com data set is used for comparisons [81], the lower row the DSigDB data set is used for comparisons [124]. Columns represent cell lines used in the analysis: (1) BT20, (2) HS578T, (3) MCF10A, (4) MCF7, (5) MDAMB231, (6) SKBR3, (7) A549, (8) HCC515, (9) HA1E,(10) HEPG2, (11) HT29, (12) PC3, (13) A375

Compounds	(1)	(2)	(3)	(4)	(5)	(6)	(7)	(8)	(9)	(10)	(11)	(12)	(13)
Dabrafenib													○
													○
Dinaciclib							○	○	○	○	○	○	○
							○	○	○	○	○	○	○
CGP-60474			○	○	○	○	○		○			○	○
			×	×	×	×	×		×			×	○
LDN-193189	○				○								○
	○				○								○
OTSSP167							—	—		—		—	—
							○	○		○		○	○
WZ-3105		—		—	—		—	—	—			—	—
		○		○	○		○	○	○			○	○
AT-7519				○		○		○	○			○	
				○		○		○	○			○	
BMS-387032				○		○	○		○				
				○		○	○		○				
JNK-9L									○				
									○				
Alvocidib	○	○	○	○	○	○			○				
	—	—	—	—	—	—			—				
GSK-2126458							—					—	
							—					—	
NVP-BEZ235							○					○	
							×					×	
Torin-2							×					×	
							○					○	
NVP-BGT226					—			—			—	—	
					—			—			—	—	
QL-XII-47	—												
	—												
Celastrol	○												
	—												
A443654		○		○									
		○		○									
NVP-AUY922					×	○							
					—	—							
Radicicol						○							
						—							

○ a significant overlap between the data sets ($P < 0.05$), × no significant overlap between the data sets, — no data, *Blank* no significant dose–response relation is identified. The confusion matrix and a full list of genes chosen in common are available in Additional file 3 [97]

Table 7.10 A significant overlap demonstrated between compound–target interactions presented in Table 7.8 and drug2gene.com. In this case, the "PPI Hub Proteins" category in Enrichr is used. Labels (1)–(13) represent the same cell lines as described in Table 7.9

Compounds	(1)	(2)	(3)	(4)	(5)	(6)	(7)	(8)	(9)	(10)	(11)	(12)	(13)
										○	○	○	
CGP-60474			○	○	○	○	○		○			○	○
LDN-193189					○								
AT-7519				○		○		○	○			○	
BMS-387032				○		○	○		○				
Alvocidib		○	○	○	○	○			○				
NVP-BEZ235												○	
Celastrol	○												
A443654		○		○									
NVP-AUY922					○	○							
Radicicol						○							

The full list of confusion matrices and genes chosen in common is available in Additional file 3 [97]

It might be useful to demonstrate how more direct and simple approach fails. One possible simpler alternative is to apply linear regression

$$x_{ijk} = a_{ik} + b_{ik} D_j \tag{7.7}$$

where D_j is the jth dose density and a_{ik} and b_{ik} are regression coefficients. Then simply select i and k associated with more significant P values as in the case of TD-based unsupervised FE. In order to show that it cannot give us the reasonable set of is and ks, we apply Eq. (7.7) to A375 cell lines ((13) in Tables 7.8, 7.9 and 7.10) as an example. After correcting P values that Eq. (7.7) gives by BH criterion, we find that all compounds have adjusted P values less than 0.01 with at least one of genes while all genes have adjusted P values less than 0.01 with at least one of the compounds. Thus, by simply requesting "adjusted P values less than 0.01" as in the case of TD-based unsupervised FE, we cannot screen either genes or compounds. We can still try to select "top-ranked" genes or compounds. In order to show that this cannot work well either, we apply two distinct criteria to select "top-ranked" compounds as

- Select top ten–ranked compounds having a larger number of genes associated with adjusted P values less than 0.01.
- Suppose P_{ik} is P value that Eq. (7.7) gives. Select top ten–ranked compounds having smaller $\sum_i \log P_{ik}$.

These two criteria rank compounds with more significant correlation with genes through dose density in some sense. The result is a bit disappointing (Table 7.11). Only three of the top ten compounds are chosen in common. This suggests that it is not easy to select compounds in a robust way simply based upon P values that

Table 7.11 Compounds selected by P values that Eq. (7.7) gives, for A375 cell line ((13) in Tables 7.8, 7.9 and 7.10). Bold ones are chosen in common

Compounds selected
Criterion 1
Chelerythrine chloride, TGX-221, lapatinib, AS-601245, PIK-93, **canertinib**, LDN-193189, MK-2206, PF-04217903, **DCC-2036**
Criterion 2
ALW-II-49-7, AZ20, BI-2536, **canertinib**, celastrol, **chelerythrine chloride**, CHIR-99021, **DCC-2036**, dovitinib, GSK-1904529A

Eq. (7.7) gives. Thus, TD-based unsupervised FE is a much better strategy without any additional criterion than adjusted P values than selection based upon P values that Eq. (7.7) gives.

Before closing this section, I would like to mention briefly why the results of TD-based unsupervised FE differ from that based upon linear regression, Eq. (7.7), so much in spite that both TD-based unsupervised FE and linear regression try to find the combinations of genes and compounds associated with dose dependence. As can be seen in Fig. 7.3, $\boldsymbol{u}_2^{(j)}$ used for FE is not simple linear function of dose density. In spite of that, the dependence of $\boldsymbol{u}_2^{(j)}$ upon dose density is quite universal, in other words, independent of cell lines. TD is the only method that can successfully identify this universal (independent of cell lines) functional form. There are no other ways to find it in advance. This cannot be achieved by any other supervised method, because any supervised methods cannot avoid assuming something contradictory to this universal functional form. Because of this superiority, TD-based unsupervised FE can achieve good performance shown in Tables 7.9 and 7.10.

7.4 Universality of miRNA Transfection

miRNA transfection is a popular method that finds miRNA target genes experimentally. Nevertheless, some doubt arises if transfected miRNA can work similarly to endogenous miRNAs [40], because it causes various unexpected effects that cannot be seen by the upregulation of endogenous miRNAs. Because the aim of miRNA transfection experiments is to find miRNA target genes, only genes downregulated by the transfection are searched. Nevertheless, it is quite usual to find that many mRNAs are upregulated because of transfection. These upregulated mRNAs are usually ignored, because it is not interpretable from the knowledge about conventional miRNA functions. On the other hand, Jin et al. [40] argued that miRNA transfection can cause nonspecific changes in gene expression. To my knowledge, there are no studies that try to identify these nonspecific effects from a more positive point of view.

In this section, using TD-based unsupervised FE, we aim to study how universal these nonspecific gene expression alterations by miRNA transfections are. Thus, we collect multiple studies where multiple miRNA transfection experiments are performed. In individual studies, genes whose expression are altered in common over multiple miRNA transfection experiments are tried to be identified. Then it is checked if genes identified in individual studies are common over multiple studies. If so, sequence-nonspecific off-target regulation of mRNA does really exist and might also play some critical roles in biology.

The identification of genes altered in common by sequence-nonspecific off-target regulation caused by miRNA transfection can be performed by TD-based unsupervised FE as follows [96]: In usual application of TD-based unsupervised FE, singular value vectors associated with desired sample dependence, e.g., distinction between patients and healthy controls, are searched to identify genes associated with such a dependence. On the contrary, in the present application, we aim to seek singular value vectors "not" associated with the distinction between transfected miRNAs, because lack of transfected miRNA dependence might be the evidence that gene expression alteration caused by miRNA expression toward these genes are because of sequence-nonspecific off-target regulation, no matter what the biological reasons that cause it are. Table 7.12 lists 11 studies including the gene expression profiles collected for the analysis in this study. It is obvious that they are quite diverse. Not only used cell lines but also transfected miRNAs differ from experiment to experiment. Both KO (knockout) and OE (overexpression) are considered. Thus, if there are genes chosen in common among these 11 studies, it is quite likely caused by sequence-nonspecific off-target regulation.

Because of their diversity, not only TD-based unsupervised FE but also PCA-based unsupervised FE is used. If the number of samples used for individual transfection in individual experiments does not match with one another, multiple experiments in which distinct miRNAs are transfected are hardly formulated in tensor forms. In these cases, PCA-based unsupervised FE is employed instead. In the following, individual data sets and how to format them in either matrix or tensor are discussed in a more detail in appendix.

Table 7.13 shows the results. In spite of the heterogeneous data sets analyzed, they are highly consistent with one another. Thus, there might be some universal mechanisms that cause sequence-nonspecific off-target regulation.

From the data science point of view, it is important to see if other methods can derive the set of genes associated with the same amount of consistency among 11 studies listed in Table 6.12. For comparison, we select t-test. What we aim is essentially to find genes expressed distinctly between control and transfected samples. This kind of two-class comparisons can be done by t-test too. In order to see if t-test is inferior to TD- and PCA-based unsupervised FE, t-test is applied to 11 studies. In this analysis, samples in individual studies are divided into two classes: samples to which no miRNAs (or mock miRNA) were transfected and samples to which miRNAs were transfected. Two-sided t-test is applied to 11 individual studies. Then, obtained P values are adjusted by BH criterion. Then, probes associated with adjusted P values less than 0.01 are selected (Table 7.14). The result is a little disappointing. For 5 out of 11 studies, t-test cannot identify any differently expressed genes. On the other hand, the numbers of selected genes vary from 35

Table 7.12 Eleven studies were conducted for this analysis. More detailed information on how to process individual experiments in these 11 studies is available in appendix. Methods: PCA- or TD-based unsupervised FE is used

Exp	GEO ID	Cell lines (cancer)	miRNA	Misc	Methods
1	GSE26996	BT549 (breast cancer)	miR-200a/b/c		PCA
2	GSE27431	HEY (ovarian cancer)	miR-7/128	mas5	PCA
3	GSE27431	HEY (ovarian cancer)	miR-7/128	Plier	PCA
4	GSE8501	Hela (cervical cancer)	miR-7/9/122a/128a/132/133a /142/148b/181a		TD
5	GSE41539	CD1 mice	cel-miR-67, hsa-miR-590-3p, hsa-miR-199a-3p		PCA
6	GSE93290	multiple	miR-10a-5p, 150-3p/5p, 148a-3p/5p, 499a-5p, 455-3p		TD
7	GSE66498	multiple	miR-205/29a/144-3p/5p, 210,23b,221/222/223		TD
8	GSE17759	EOC 13.31 microglia cells	miR-146a/b	(KO/OE)	TD
9	GSE37729	HeLa	miR-107/181b	(KO/OE)	TD
10	GSE37729	HEK-293	miR-107/181b	(KO/OE)	TD
11	GSE37729	SH-SY5Y	181b	(KO/OE)	TD

to 11,060, which is in contrast to the range of the number of genes selected by PCA- or TD-based unsupervised FE $\sim 10^2$ (Table 7.13). These numbers are unlikely biologically trustable. This possibly shows the failure of the methodology.

In order to further demonstrate the inferiority of *t*-test to TD- or PCA-based unsupervised FE, we try to reproduce the results of PCA- or TD-based unsupervised FE in Table 7.13. Since the number of genes selected by *t*-test are often 0 (Table 7.14), the same number of top-ranked genes with smaller *P* values as those in PCA- or TD-based unsupervised FE are selected in individual experiments based on *P* values computed by *t*-test even though *P* values are not significant. It is obvious that the selected genes by *t*-test are less coincident with each other than the selected genes by PCA- or TD-based unsupervised FE (Table 7.13) because odds ratios are smaller and *P* values are larger. Thus, also from the point of coincidence between 11 studies, *t*-test is inferior to TD- or PCA-based unsupervised FE.

Although PCA- or TD-based unsupervised FE successfully identify sets of genes highly coincident between heterogeneous 11 studies, if they are not biologically reasonable, they are useless. In order to see the biological values of selected genes, we here show one evaluation, although many evaluations were performed in my published paper [96] (I am not willing to show all of them here, because it might be simply boring).

Table 7.13 Fisher's exact tests for coincidence among 11 miRNA transfection studies for PCA-or TD-based unsupervised FE and *t*-test. Upper triangle: *P* value; lower triangle: odds ratio; #: number of genes selected in individual studies. "Xe-Y" means that "X $\times 10^{-Y}$"

PCA- or TD-based unsupervised FE												
Exp.		1	2	3	4	5	6	7	8	9	10	11
	#	232	711	747	441	123	292	246	873	113	104	120
1	232		4.14e–19	6.59e–22	3.96e–41	4.12e–71	9.41e–70	2.90e–60	1.34e–17	1.15e–27	6.84e–26	2.66e–07
2	711	7.68		0.00	1.89e–18	4.93e–27	5.59e–20	2.69e–32	4.62e–13	9.23e–16	8.66e–12	1.37e–03
3	747	8.30	345.52		3.63e–20	7.96e–21	5.70e–12	1.82e–27	9.52e–12	1.18e–14	1.01e–12	3.90e–06
4	441	18.23	5.19	5.34		6.14e–41	1.01e–34	1.44e–69	4.61e–11	2.16e–30	4.09e–28	1.35e–10
5	123	53.86	9.04	7.27	17.48		2.9e–179	1.27e–63	6.24e–15	3.16e–25	2.37e–17	4.69e–09
6	292	61.50	8.15	5.52	17.71	204.39		3.53e–53	2.57e–15	6.65e–22	1.65e–12	5.60e–05
7	246	20.27	5.35	4.67	12.39	20.11	22.03		6.91e–42	1.77e–36	4.50e–31	2.78e–14
8	873	18.61	7.22	6.51	8.29	15.61	18.53	20.73		1.81e–07	1.37e–06	2.76e–02
9	113	39.34	9.87	8.77	25.98	32.44	34.90	21.94	16.02		3.7e–125	9.27e–18
10	104	40.29	8.22	8.27	26.64	23.34	20.86	21.56	15.18	517.87		6.82e–16
11	120	10.15	3.19	4.43	9.19	11.55	8.11	8.28	4.92	19.57	18.70	
t-test												
Exp.		1	2	3	4	5	6	7	8	9	10	11
	#	232	711	747	441	123	292	246	873	113	104	120
1	232		4.96e–04	8.49e–01	2.59e–01	6.35e–01	1.00e+00	5.40e–01	1.00e+00	4.08e–01	6.45e–01	6.68e–01
2	711	2.56		6.40e–69	1.38e–02	1.25e–01	1.55e–01	9.36e–03	1.00e+00	1.00e+00	3.76e–01	1.00e+00
3	747	0.80	10.49		8.65e–01	5.28e–01	3.76e–01	2.47e–01	7.79e–01	7.75e–01	5.30e–01	1.00e+00
4	441	1.55	1.90	0.89		6.58e–01	1.00e+00	4.31e–01	1.26e–01	2.71e–01	2.56e–01	1.00e+00
5	123	0.00	0.00	0.36	1.39		1.13e–22	1.00e+00	3.86e–01	1.00e+00	1.00e+00	1.00e+00
6	292	0.77	1.83	0.32	0.72	27.05		3.71e–01	1.00e+00	1.00e+00	1.00e+00	1.00e+00
7	246	1.16	0.48	0.71	1.22	0.67	0.31		4.47e–01	1.83e–01	7.60e–02	2.04e–01
8	873	0.64	1.00	1.17	2.15	2.09	0.00	0.46		1.59e–01	4.54e–01	1.27e–03
9	113	0.00	0.81	0.60	0.00	0.00	0.00	0.25	2.91		1.18e–03	4.07e–01
10	104	0.00	0.32	0.35	1.75	0.00	0.00	0.00	1.68	5.56		6.37e–01
11	120	1.31	0.78	0.88	0.97	0.00	0.00	1.69	6.87	0.00	0.00	

Table 7.14 The number of genes selected by t-test. Two numbers beside colon are the number of control and transfected samples, respectively

Studies	1	2	3	4	5	6	7	8	9	10	11
Samples	6:6	3:4	6:4	18:18	2:2	16:16	19:19	18:18	6:12	6:12	4:4
Selected genes	11,060	0	0	0	0	35	280	55	5949	5730	0

Table 7.15 is the result for KEGG pathway enrichment by uploading selected genes to Enrichr. It is obvious that not only there are many significant enrichments but also they are highly coincident between 11 studies. Thus, the coincidence of selected genes between 11 studies shown in Table 7.13 is also biologically reasonable. In this sense, PCA- or TD-based unsupervised FE can identify biologically meaningful genes chosen in common between heterogeneous studies including various miRNAs transfected to various cell lines. Universal nature detected has seemingly biological importance too.

7.5 One-Class Differential Expression Analysis for Multiomics Data Set

In general, there are two kinds of biological experiments, in vivo and in vitro. In vivo means real biological experiments using living organisms, e.g., animals and plants. Nevertheless, in vivo cannot be said as very economical, because it wastes the whole body even when we are interested in a specific tissue. For example, even if you are interested in liver disease, in vivo experiments require to cultivate a whole body. You may wonder if only liver can be separately cultivated, it would be more effective. in vivo experiments recently have a tendency to be avoided from the ethical point of view, too, because they kill numerous animals. in vitro experiments can fulfill these requirements more or less. in vitro makes use of cell lines, which are immortalized cells that are often made out of cancer cells. Once a cell line is established, you can do any kind of experiments in vitro using cell lines. Because cell lines can be cultivated even in a dish, it is definitely cost-effective and does not kill any animal.

One possible problem of in vitro is the lack of control samples. It is known that cell lines differ from the tissue cells from which cell lines are established. Thus, usually, cell lines are compared between not treated and treated ones. Characterizing immortalized cell lines themselves is not an easy task.

In this section, we propose a method that can characterize cancer cell lines from gene expression without comparing with something [92]. In this criterion, genes that are expressed in common over multiple cancer subtypes are searched and are considered to be characteristic gene expressions of cancer cell lines. In this regard, TD-based unsupervised FE used to identify expressed genes in common over multiple miRNAs transfection studies in the previous section is employed again.

Table 7.15 In each of 11 studies, 20 top-ranked significant KEGG pathways whose associated genes significantly match some genes selected for each experiment are identified. Thus, the following KEGG pathways are most frequently ranked within the top 20. "Xe-Y" means that "X $\times 10^{-Y}$"

Exp.	#	(i)	(ii)	(iii)	(iv)	(v)	(vi)	(vii)	(viii)	(ix)	(x)
1	(232)	31/137	7/168	10/142	6/133	9/55	9/193			7/169	8/203
	[10]	3.69e–29	3.18e–02	1.66e–04	3.45e–02	1.02e–06	6.85e–03			3.18e–02	3.01e–02
2	(711)	36/137	18/168	14/142	12/133	13/55				16/169	18/203
	[12]	3.43e–19	1.48e–03	1.05e–02	3.20e–02	5.92e–06				8.12e–03	8.12e–03
3	(747)	23/137	15/168			14/55				18/169	19/203
	[15]	3.58e–07	1.94e–02			1.20e–06				2.02e–03	4.78e–03
4	(441)	50/137	15/168	19/142	18/133	6/55	19/193	7/78	12/151	9/169	
	[10]	2.92e–45	1.91e–04	3.97e–08	6.42e–08	2.49e–02	3.40e–06	2.74e–02	4.44e–03	1.29e–01	
5	(123)	9/137						8/78		6/169	8/203
	[23]	2.97e–06						6.08e–07		4.29e–03	3.03e–04
6	(292)	45/137	20/168	19/142	18/133	4/55	19/193	11/78	12/151		
	[14]	1.35e–46	3.32e–11	2.27e–11	4.00e–11	7.95e–02	2.24e–09	4.90e–07	4.87e–05		
7	(246)	40/137	9/168	10/142	9/133		11/193	4/78	7/151		6/203
	[6]	5.61e–42	6.60e–03	5.80e–04	1.32e–03		1.16e–03	2.57e–01	6.31e–02		4.52e–01
8	(873)	75/137	30/168	32/142	32/133		36/193	14/78	24/151	25/169	
	[24]	5.59e–63	2.09e–09	9.32e–13	1.89e–13		7.51e–12	1.39e–04	1.62e–06	3.11e–06	
9	(113)	18/137	11/168	12/142	10/133	6/55	12/193	4/78	11/151		
	[20]	8.24e–18	7.10e–08	1.66e–09	8.42e–08	8.85e–06	2.96e–08	6.64e–03	2.96e–08		
10	(104)	11/137	8/168	9/142	8/133	5/55	10/193		8/151		
	[20]	1.98e–08	6.68e–05	3.23e–06	1.71e–05	1.56e–04	3.23e–06		3.60e–05		
11	(120)	6/137		4/142		5/55					5/203
	[3]	9.04e–03		8.49e–02		2.98e–03					6.83e–02

(i) Ribosome:hsa03010, (ii) Alzheimer's disease:hsa05010, (iii) Parkinson's disease:hsa05012, (iv) Oxidative phosphorylation:hsa00190, (v) Pathogenic *Escherichia coli* infection:hsa05130, (vi) Huntington's disease:hsa05016, (vii) Cardiac muscle contraction:hsa04260, (viii) Nonalcoholic fatty liver disease (NAFLD):hsa04932, (ix) Protein processing in endoplasmic reticulum:hsa04141, and (x) Proteoglycans in cancer:hsa05205. (numbers):gene, [numbers]:KEGG pathways associated with adjusted P values less than 0.01. Upper rows in each exp: (the number of genes coinciding with the genes selected for each experiment)/(genes listed in Enrichr in each category). Lower rows in each exp: adjusted P values provided by Enrichr

In addition to this, TD-based unsupervised FE is used as a tool that integrates omics data. The data set used is downloaded from database of transcriptional start sites (DBTSS) [91], which is a database of transcriptional start sites (TSS), and includes RNA-seq, TSS-seq, and ChIP-seq (histone modification, H3K27ac). These are observed in 26 NSCLC subtype cell lines using HTS technology; DBTSS also stores various omics data set measured on various cell lines and living organisms.

Before starting the analysis, we briefly explain the difference among TSS-seq, RNA-seq, and ChIP-seq. As its name says, TSS-seq tries to sequence RNA transcribed from the region around TSS. Thus, TSS-seq basically counts how many times transcription starts. On the other hand, RNA-seq counts the fragments taken from any part of the whole RNA. In this sense, RNA-seq counts the total amount of RNA transcribed. Generally, TSS-seq and RNA-seq are positively correlated, although there are no known functional forms that relate between these two, because the function is affected by many factors, e.g., individual genes have various lengths, and some genes are long while others are short. If longer genes are transcribed, the ratio RNA-seq to TSS-seq becomes larger. In addition to this, individual genes have isoforms, each of which has a different length. This mechanism is called an alternative splicing. If more number of longer isoforms are transcribed from each gene, it also contributes to the increased RNA-seq/TSS-seq ratio. Although there are many detailed points that must be considered in order to relate RNA-seq to TSS-seq, there is one clear point; TSS-seq and RNA-seq should be positively correlated. Thus, seeking genes associated with both more TSS-seq counts and RNA-seq counts can reduce the possibility that genes are wrongly identified as being upregulated or downregulated, e.g., because of technical issues like miss amplification.

ChIP-seq is a different technology that detects to which part of DNA the protein binds. Although I do not explain the details of the relationship between DNA and proteins that bind to it, basically, DNA-binding protein can control the rate of transcription. ChIP-seq can study this relationship by considering DNA-binding protein. Histone modification is a more advanced feature. In order to suppress the self-entanglements of lengthy DNA, long DNA string is wrapped around a protein core called histone. Because tightly wrapped DNA is hardly transcribed, how tightly DNA is wrapped around histone can affect the amount of transcription drastically. On the other hand, affinity between histone and DNA can be affected by the chemical modification of histone. Among various histone modifications, acetylation of histone tail is supposed to enhance the transcription by reducing the affinity between DNA and histone. As a result, considering histone modification (H3K27ac) together with RNA-seq and TSS-seq can further reduce the possibility of wrongly identified up/downregulated genes. In the following, we try to seek genes simultaneously associated with the increased TSS-seq, RNA-seq, and ChIP-seq measured H3K27ac counts.

When formatting RNA-seq, TSS-seq, and ChIP-seq measurement data into tensor form, how we practically perform this is a problem. Fundamentally, although it is possible to perform it in a single nucleotide base, it results in too huge a tensor that requires too large a memory to manage. In this case, it is better to employ a coarse-graining approach that takes average over local chromosome regions. The problem is how long regions should be. If the length of the region is too large, each region includes more than one (protein-coding) gene. Then, increased or decreased counts within each region might reflect more than one gene. This will result in low interpretability. On the other hand, if the length of the region is too short, individual (protein-coding) genes are expressed over multiple regions. It again results in low interpretability. Thus, there should be a somewhat optimal length of region. In this section, I try 25,000 nucleotides as a length of region. Generally, the average length of protein is $\sim 10^2$. Because one amino acid is coded by three nucleotides (codon), the length of region that codes individual protein-coding genes should be at most $\sim 10^3$. The regions that code protein-coding genes are typically composed of both exon and intron, which correspond to translated and non-translated regions, respectively. Thus, the region of DNA that codes individual genes might be doubled. It is still expected not to exceed $\sim 10^3$ so much. Actually, some literature reported that the average length of DNA regions that code human protein-coding genes is still a little bit shorter than $\sim 10^4$ [38]. Nevertheless, if the region over which TSS-seq, RNA-seq, and ChIP-seq count data is averaged is as long as the expected length of the DNA region that codes individual protein-coding genes, boundaries between the averaging regions might frequently fall into the middle of the DNA region that codes individual protein-coding region. Thus, the length of region that averages counts data should be a few times longer than the expected length of the DNA region that codes individual protein-coding region. Based upon these considerations, a 25,000-nucleotide region over which TSS-seq, RNA-seq, and ChIP-seq counts are averaged is proposed.

In the data set having a type "human lung adenocarcinoma cell line 26 cell line" in in-house data category, RNA-seq, TSS-seq, and ChIP-seq data are used. Among ChIP-seq data, only the H3K27ac is used (H3K27ac means that K27 position of the third histone (H3) is acethlyated). Counts are averaged over chromosomal regions fragmented to regions of the length of 25,000 nucleotides. Tensors are generated for each chromosome separately. Then, tensor is the form of $x_{ijk} \in \mathbb{R}^{N \times 26 \times 3}$, where N is the total number of regions of the length of 25,000 nucleotides within each chromosome, j stands for 26 cell lines, and k stands for counts of TSS-seq, RNA-seq, and ChIP-seq. HOSVD algorithm, Fig. 3.8, is applied to x_{ijk} as

$$x_{ijk} = \sum_{\ell_1=1}^{N} \sum_{\ell_2=1}^{26} \sum_{\ell_3=1}^{3} G(\ell_1, \ell_2, \ell_3) u_{\ell_1 i}^{(i)} u_{\ell_2 j}^{(j)} u_{\ell_3 k}^{(k)} \tag{7.8}$$

where $u_{\ell_1 i}^{(i)} \in \mathbb{R}^{N \times N}$, $u_{\ell_2 j}^{(j)} \in \mathbb{R}^{26 \times 26}$, and $u_{\ell_3 k}^{(k)} \in \mathbb{R}^{3 \times 3}$ are singular value matrices and $G(\ell_1, \ell_2, \ell_3) \in \mathbb{R}^{N \times 26 \times 3}$ is a core tensor.

First, we need to find $\boldsymbol{u}_{\ell_2}^{(j)}$ that is independent of 26 cell lines and $\boldsymbol{u}_{\ell_3}^{(k)}$ that is independent of RNA-seq, TSS-seq, and ChIP-seq. Figure 7.6 shows $\boldsymbol{u}_1^{(j)}$. Excluding the X chromosome, it is highly independent of 26 cell lines. Then, we decide to employ $\ell_2 = 1$. Figure 7.7 shows $\boldsymbol{u}_1^{(k)}$. They are highly independent of TSS-seq, RNA-seq, and ChIP-seq. Then, we decide to employ $\ell_3 = 1$.

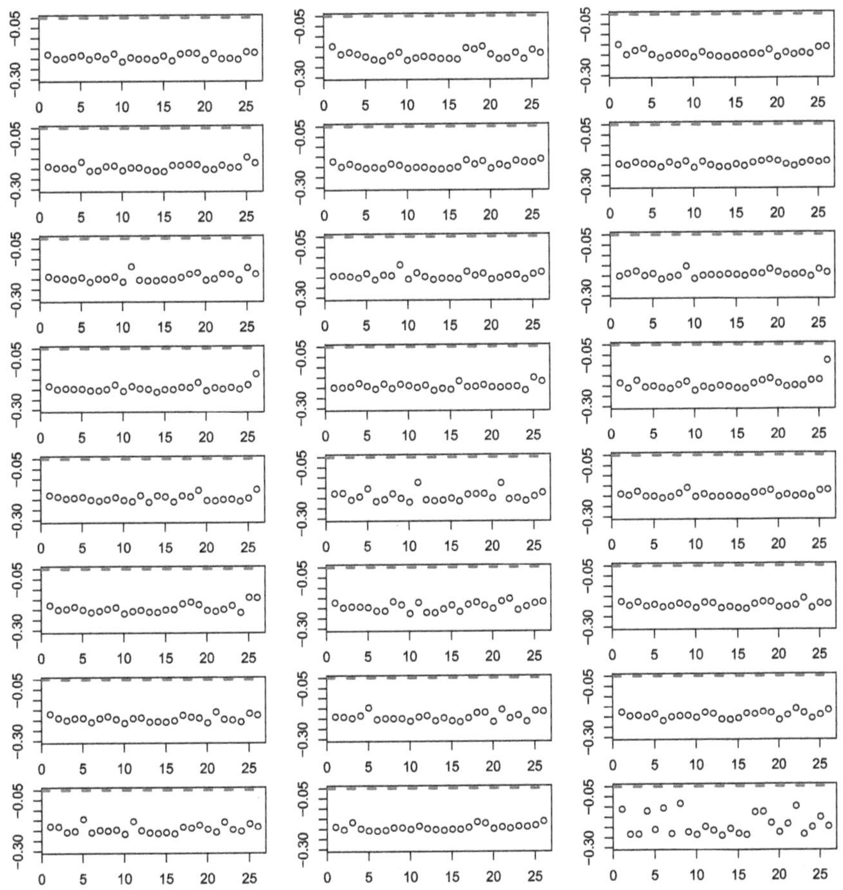

Fig. 7.6 $u_1^{(j)}$. The first row, from left to right, chromosomes 1, 2, and 3; the second row, from left to right, chromosomes 4, 5, 6, and so on; and the last row, from left to right, chromosomes 22, X, and Y. The red broken line is the baseline

Then, we try to find which $G(\ell_1, 1, 1)$ has the largest absolute value and find that $G(1, 1, 1)$ has always the largest absolute values independent of chromosome. Thus, $u_1^{(i)}$ is used to attributed P value to regions as

$$P_i = P_{\chi^2}\left[> \left(\frac{u_{1i}^{(i)}}{\sigma_1}\right)^2\right]. \tag{7.9}$$

P values are collected from 24 chromosomes and are corrected by BH criterion. Then, 826 regions associated with adjusted P values less than 0.01 are selected. 826 is very small compared with the total number of regions, because the total number of regions is about $3 \times 10^9 / 2.5 \times 10^4 \sim 10^5$ where 3×10^9 is the total length of

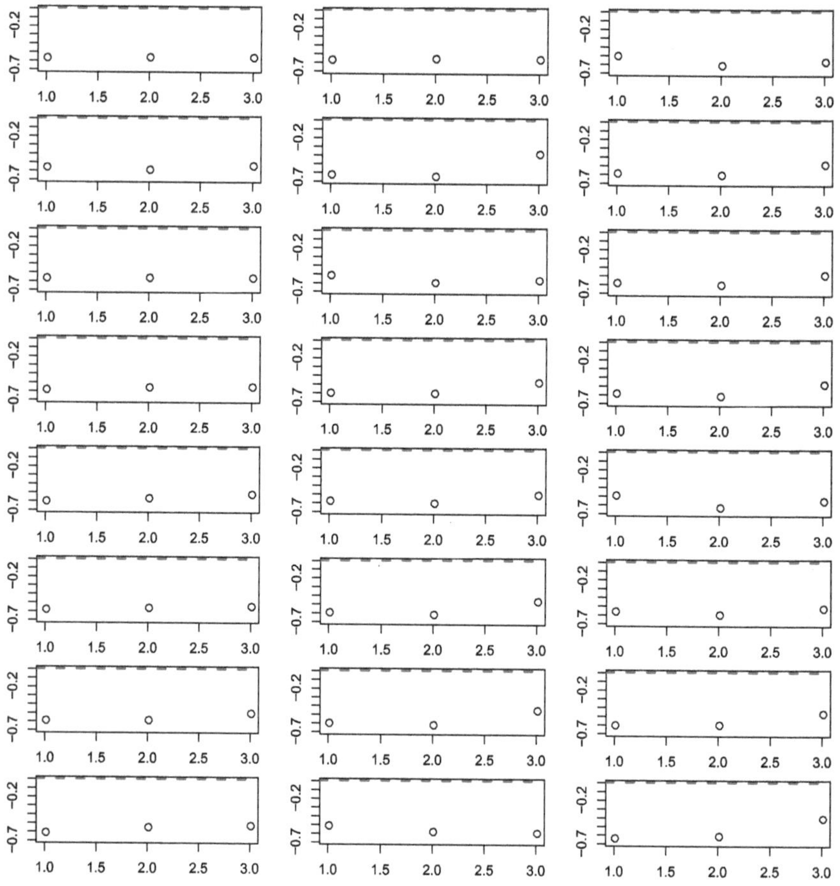

Fig. 7.7 $u_1^{(k)}$. The first row, from left to right, chromosomes 1, 2, and 3; the second row, from left to right, chromosomes 4, 5, 6, and so on; and the last row, from left to right, chromosomes 22, X, and Y. The red broken line is the baseline

human genome while 2.5×10^4 is the length of individual regions; 826 corresponds to as little as 0.8% of regions. This is reasonable because only a less percentage of genome codes protein-coding genes.

In order to validate these selected regions, we upload 1741 Entrez genes associated with these 826 regions to DAVID. Entrez gene is manually curated gene unique ID that is integer number [59]. Table 7.16 lists the KEGG pathway enrichment associated with adjusted P values less than 0.05. At a glance, they do not look like related to cancers. Nevertheless, some of them are cancer-related terms. For example, the relationship between "Antigen processing and presentation" and cancer is often discussed [8]. Parkinson's disease is often reported to be related to lung cancer [119]. Although we are not willing to discuss fully the relations between

Table 7.16 KEGG pathway enrichment by the 1741 Entrez genes identified by TD-based unsupervised FE. Adjusted P values are by BH criterion

Category	Term	Gene count	%	P value	Adjusted P value
KEGG_PATHWAY	Ribosome	73	4.2	9.8×10^{-38}	2.7×10^{-35}
KEGG_PATHWAY	Spliceosome	39	2.2	6.2×10^{-10}	8.4×10^{-8}
KEGG_PATHWAY	Protein processing in endoplasmic reticulum	41	2.4	8.0×10^{-8}	7.3×10^{-6}
KEGG_PATHWAY	Antigen processing and presentation	22	1.3	8.0×10^{-6}	5.5×10^{-4}
KEGG_PATHWAY	Pathogenic Escherichia coli infection	17	1.0	1.7×10^{-5}	9.2×10^{-4}
KEGG_PATHWAY	Parkinson's disease	30	1.7	9.6×10^{-5}	4.3×10^{-3}
KEGG_PATHWAY	Biosynthesis of antibiotics	39	2.2	1.6×10^{-4}	6.3×10^{-3}
KEGG_PATHWAY	Oxidative phosphorylation	26	1.5	1.0×10^{-3}	3.5×10^{-2}
KEGG_PATHWAY	Bacterial invasion of epithelial cells	18	1.0	1.2×10^{-3}	3.6×10^{-2}
KEGG_PATHWAY	Alzheimer's disease	30	1.7	1.7×10^{-3}	4.6×10^{-2}

the detected KEGG pathway enrichment and NSCLC, it is obvious that TD-based unsupervised FE can detect a set of genes including those related to NSCLC.

Although it is better to evaluate the performance of TD-based unsupervised FE based upon the comparison with other methods, it is not easy because there are no control samples to be compared. Thus, alternatively, we select genes based upon the ratio of standard deviation to average over 26 cell lines, because the smaller ratio of variance to mean might suggest smaller variability between 26 cell lines. For each of TSS-seq, RNA-seq, and ChIP-seq, we select the top 5 regions with a smaller ratio. Then regions chosen in common among TSS-seq, RNA-seq, and ChIP-seq are collected; we find that 2041 Entrez genes are included in these regions chosen in common. This number, 2041, is compared with 1741, which is the number of Entrez genes selected by TD-based unsupervised FE. Thus, uploading these to DAVID is a suitable test to see if TD-based unsupervised FE is superior to this alternative method. Then we find that only two KEGG pathways, "spliceosome" and "ubiquitin-mediated proteolysis," are associated with adjusted P values less than 0.05. This suggests that TD-based unsupervised FE can identify a far more biologically reasonable set of genes than this alternative approach.

7.6 General Examples of Case I and II Tensors

Before demonstrating individual cases using case I and case II tensors in detail, we demonstrate various cases briefly based upon the recent publication [93]. As shown in Table 5.3, matrices or low-mode tensor can be combined to generate (higher-mode) tensor. In this section, we demonstrate how the combinations shown in Table 5.3 works to select genes critical to the diseases or phenomena considered.

7.6.1 Integrated Analysis of mRNA and miRNA

Integrated analysis of mRNA and miRNA was also performed by PCA-based unsupervised FE (Sect. 6.3), which is once applied to mRNA and miRNA separately. Then, the obtained two sets of PC loading attributed to the sample were investigated to seek those sharing a common nature between two sets. After that, corresponding PC scores attributed to mRNA and miRNA were used for FE. On the contrary, in the application of TD-based unsupervised FE to the integrated analysis of mRNA and miRNA, mRNA and miRNA expression profiles are integrated in advance.

The analyzed data set is composed of mRNA and miRNA profiles which were measured for multi-class breast cancer samples including normal breast tissues [27]. mRNA and miRNA expression profiles of multiomics data are downloaded from GEO using GEO ID GSE28884. At first, GSE28884_RAW.tar is downloaded and expanded. For mRNA, 161 files whose names ended by the string "c.txt.gz" are used. Each file is loaded into R by read.csv command and the second column named "M" is employed as mRNA expression values. Probes not associated with Human Genome Organisation (HUGO) gene names are discarded, and 13,393 probes remain. One hundred sixty-one files whose names end by the string "geo.txt.gz" are used for miRNA expression profiles; mRNA expression profiles of the corresponding samples are also used. Each file is loaded into R by read.csv command, and the second column ("Count") is summed using the same third column ("Annotation") values. If the resulting total sum is less than 10, it is discarded and not used for further analysis.

Because the 161 samples are shared between miRNA and mRNA expression profiles, the multiomics data corresponds to case I data (Table 5.3). TD-based unsupervised FE is applied to the data set in order to identify disease-critical genes and latent relations between miRNA and mRNA, whose expression profiles are $x_{i_1 j}^{\text{mRNA}} \in \mathbb{R}^{13,393 \times 161}$ and $x_{i_2 j}^{\text{miRNA}} \in \mathbb{R}^{755 \times 161}$, respectively. They can be formatted as case I tensor as

$$x_{i_1 i_2 j} = x_{i_1 j}^{\text{mRNA}} x_{i_2 j}^{\text{miRNA}}. \tag{7.10}$$

HOSVD, Fig. 3.8, is applied to $x_{i_1 i_2 j}$ as

$$x_{i_1 i_2 j} = \sum_{\ell_1=1}^{13,393} \sum_{\ell_2=1}^{755} \sum_{\ell_3=1}^{161} G(\ell_1, \ell_2, \ell_3) u_{\ell_1 i_1}^{(i_1)} u_{\ell_2 i_2}^{(i_2)} u_{\ell_3 j}^{(j)} \tag{7.11}$$

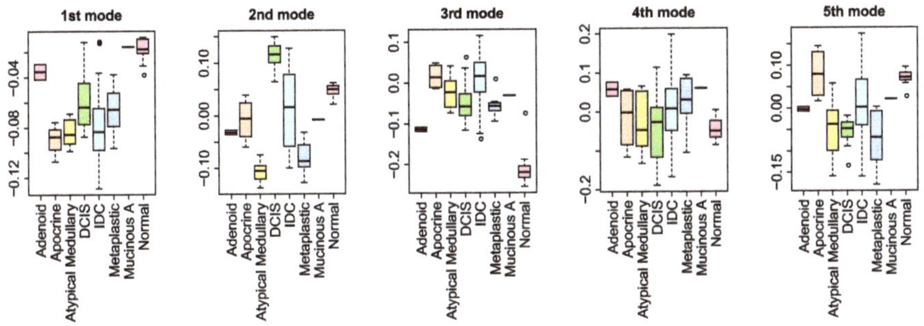

Fig. 7.8 Boxplot of $\boldsymbol{u}_{\ell_3}^{(j)}$, $1 \leq \ell_3 \leq 5$ when HOSVD is applied as Eq. (7.11). P values computed by categorical regression. First: 2.39×10^{-5}, second: 5.83×10^{-14}, third: 1.36×10^{-24}, fourth: 2.58×10^{-2}, fifth: 2.12×10^{-5}

Table 7.17 Top ten–ranked $G(\ell_1, \ell_2, 1 \leq \ell_3 \leq 5)$s with larger absolute values among $1 \leq \ell_1, \ell_2, \ell_3 \leq 10$ in Eq. (7.11)

ℓ_1	1	2	4	3	5
ℓ_2	1	1	1	1	1
ℓ_3	1	2	4	3	5
$G(\ell_1, \ell_2, \ell_3)$	1.67×10^5	-1.03×10^5	7.48×10^4	-6.64×10^4	6.23×10^4
ℓ_1	3	1	3	2	1
ℓ_2	2	2	1	2	2
ℓ_3	3	3	5	3	2
$G(\ell_1, \ell_2, \ell_3)$	3.00×10^4	-2.87×10^4	-2.33×10^4	-2.02×10^4	-1.48×10^4

where $u_{\ell_1 i_1}^{(i_1)} \in \mathbb{R}^{13,393 \times 13,393}$,$u_{\ell_2 i_2}^{(i_2)} \in \mathbb{R}^{755 \times 755}$, and $u_{\ell_3 j}^{(j)} \in \mathbb{R}^{161 \times 161}$ are singular value matrices and $G(\ell_1, \ell_2, \ell_3) \in \mathbb{R}^{13,393 \times 755 \times 161}$ is the core tensor.

First, we need to seek singular value vectors, $\boldsymbol{u}_{\ell_3}^{(j)} \in \mathbb{R}^{161}$, with significant cancer subtype dependence. Figure 7.8 shows boxplots of $\boldsymbol{u}_{\ell_3}^{(j)}$, $1 \leq \ell_3 \leq 5$; it is obvious that these singular value vectors have significant class (cancer subtypes) dependence. The next step is to find $G(\ell_1, \ell_2, 1 \leq \ell_3 \leq 5)$ with larger absolute values. Table 7.17 shows the top-ranked $G(\ell_1, \ell_2, 1 \leq \ell_3 \leq 5)$s; there are clearly only $1 \leq \ell_1 \leq 5$ and $1 \leq \ell_2 \leq 2$, respectively. Thus, P values are attributed to i_1 and i_2 using $u_{\ell_1 i_1}^{(i_1)}$, $1 \leq \ell_1 \leq 5$ and $u_{\ell_2 i_2}^{(i_2)}$, $1 \leq \ell_1 \leq 2$, respectively as

$$P_{i_1} = P_{\chi^2}\left[> \sum_{\ell_1=1}^{5} \left(\frac{u_{\ell_1 i_1}^{(i_1)}}{\sigma_{\ell_1}} \right)^2 \right], \tag{7.12}$$

$$P_{i_2} = P_{\chi^2}\left[> \sum_{\ell_2=1}^{2} \left(\frac{u_{\ell_2 i_2}^{(i_2)}}{\sigma_{\ell_2}} \right)^2 \right]. \tag{7.13}$$

Computed P values are adjusted by BH criterion; i_1s and i_2s associated with adjusted P values less than 0.01 are selected. Then, 426 mRNA probes and 7 miRNAs are selected, respectively.

In order to evaluate selected 426 mRNAs biologically, we upload these mRNAs to DAVID. Then we can find numerous enrichment. Tables 7.18 and 7.19 show the results of gene ontology (GO) term enrichment (adjusted P values less than 0.05). BP is related to biological features, CC is related to the location within the cell, and MF is a function of genes as molecules. Although we are not willing to summarize all of them, most of them are reasonably related to cancers, e.g., immune-related or cell surface enrichment. Thus, TD-based unsupervised FE is likely successful to identify cancer-related genes.

In order to demonstrate the superiority of type I tensor, we also employ type II tensor as

$$x_{i_1 i_2} = \sum_j x_{i_1 i_2 j}. \tag{7.14}$$

Applying SVD to $x_{i_1 i_2}$, we get singular value vectors $u^{(i_1)}_{\ell_1 i_1} \in \mathbb{R}^{13,393 \times 161}$ and $u^{(i_2)}_{\ell_2 i_2} \in \mathbb{R}^{755 \times 161}$. In order to select singular vector used for FE, we need to know the dependence upon classes (in this case, cancer subtype). Thus, we need singular value vectors attributed to samples. It is computed as Eqs (5.12) and (5.13),

$$u^{j;i_1}_{\ell_1 j} = \sum_{i_1=1}^{13,393} x_{i_1 j} u^{(i_1)}_{\ell_1 i_1} \tag{7.15}$$

$$u^{j;i_2}_{\ell_2 j} = \sum_{i_2=1}^{755} x_{i_2 j} u^{(i_2)}_{\ell_2 i_2} \tag{7.16}$$

Figure 7.9 shows a boxplot of $u^{j;i_1}_{\ell_1 j}$ and $u^{j;i_2}_{\ell_2 j}$ for $1 \leq \ell_3 \leq 5$. It is obvious that these singular value vectors have significant class (cancer subtypes) dependence.

Thus, P values are attributed to i_1 and i_2 using $u^{(i_1)}_{\ell_1 i_1}$ and $u^{(i_2)}_{\ell_2 i_2}$ for $1 \leq \ell_3 \leq 5$, respectively, as

$$P_{i_1} = P_{\chi^2}\left[> \sum_{\ell_1=1}^{5} \left(\frac{u^{(i_1)}_{\ell_1 i_1}}{\sigma_{\ell_1}} \right)^2 \right], \tag{7.17}$$

$$P_{i_2} = P_{\chi^2}\left[> \sum_{\ell_2=1}^{5} \left(\frac{u^{(i_2)}_{\ell_2 i_2}}{\sigma_{\ell_2}} \right)^2 \right]. \tag{7.18}$$

P values are adjusted by BH criterion. i_1 and i_2 associated with adjusted P values less than 0.01 are selected. Then, 374 mRNA probes and 21 miRNAs are selected.

Table 7.18 GO BP enrichment by the 426 emsembl genes identified by TD-based unsupervised FE. Adjusted *P* values are by BH criterion

Category	Term	Gene count	%	*P* value	Adjusted *P* value
GOTERM_BP_DIRECT	Immune response	36	11.4	2.7×10^{-14}	5.6×10^{-11}
GOTERM_BP_DIRECT	Signal transduction	57	18.1	5.1×10^{-12}	5.3×10^{-9}
GOTERM_BP_DIRECT	Type I interferon signaling pathway	10	3.2	1.8×10^{-6}	1.2×10^{-3}
GOTERM_BP_DIRECT	Collagen catabolic process	10	3.2	1.8×10^{-6}	1.2×10^{-3}
GOTERM_BP_DIRECT	Positive regulation of cell proliferation	25	7.9	3.2×10^{-6}	1.3×10^{-3}
GOTERM_BP_DIRECT	Cell–cell signaling	18	5.7	3.1×10^{-6}	1.6×10^{-3}
GOTERM_BP_DIRECT	Response to estradiol	11	3.5	4.8×10^{-6}	1.6×10^{-3}
GOTERM_BP_DIRECT	Defense response to virus	14	4.4	8.0×10^{-6}	2.3×10^{-3}
GOTERM_BP_DIRECT	B cell receptor signaling pathway	8	2.5	4.5×10^{-5}	1.1×10^{-2}
GOTERM_BP_DIRECT	Positive regulation of cAMP metabolic process	4	1.3	1.1×10^{-4}	2.4×10^{-2}
GOTERM_BP_DIRECT	Response to peptide hormone	7	2.2	1.2×10^{-4}	2.4×10^{-2}
GOTERM_BP_DIRECT	Negative regulation of apoptotic process	21	6.7	1.9×10^{-4}	2.6×10^{-2}
GOTERM_BP_DIRECT	Defense response	8	2.5	1.8×10^{-4}	2.6×10^{-2}
GOTERM_BP_DIRECT	T cell activation	7	2.2	1.7×10^{-4}	2.7×10^{-2}
GOTERM_BP_DIRECT	T cell differentiation	6	1.9	1.7×10^{-4}	2.8×10^{-2}
GOTERM_BP_DIRECT	Skeletal system development	11	3.5	1.7×10^{-4}	3.1×10^{-2}
GOTERM_BP_DIRECT	Chemokine-mediated signaling pathway	8	2.5	2.6×10^{-4}	3.3×10^{-2}
GOTERM_BP_DIRECT	Mast cell activation	4	1.3	2.9×10^{-4}	3.4×10^{-2}
GOTERM_BP_DIRECT	Adaptive immune response	11	3.5	3.1×10^{-4}	3.5×10^{-2}
GOTERM_BP_DIRECT	Cell surface receptor signaling pathway	15	4.8	4.0×10^{-4}	4.2×10^{-2}
GOTERM_BP_DIRECT	Cellular response to interferon-alpha	4	1.3	4.3×10^{-4}	4.3×10^{-2}
GOTERM_BP_DIRECT	Inflammatory response	18	5.7	4.5×10^{-4}	4.3×10^{-2}
GOTERM_BP_DIRECT	Apoptotic process	23	7.3	5.2×10^{-4}	4.5×10^{-2}
GOTERM_BP_DIRECT	Humoral immune response	7	2.2	5.0×10^{-4}	4.6×10^{-2}
GOTERM_BP_DIRECT	Positive regulation of neutrophil chemotaxis	5	1.6	5.5×10^{-4}	4.6×10^{-2}
GOTERM_BP_DIRECT	Collagen fibril organization	6	1.9	6.0×10^{-4}	4.8×10^{-2}
GOTERM_BP_DIRECT	Proteolysis	21	6.7	6.4×10^{-4}	4.9×10^{-2}

Table 7.19 GO CC and MF enrichment by the 426 emsembl genes identified by TD-based unsupervised FE. Adjusted P values are by BH criterion

Category	Term	Gene count	%	P value	Adjusted P value
GOTERM_CC_DIRECT	Extracellular space	84	26.7	1.60×10^{-26}	4.90×10^{-24}
GOTERM_CC_DIRECT	Extracellular region	82	26	3.10×10^{-20}	4.80×10^{-18}
GOTERM_CC_DIRECT	Extracellular exosome	97	30.8	9.00×10^{-13}	9.20×10^{-11}
GOTERM_CC_DIRECT	External side of plasma membrane	23	7.3	1.00×10^{-11}	7.70×10^{-10}
GOTERM_CC_DIRECT	Cell surface	23	7.3	1.20×10^{-4}	7.40×10^{-3}
GOTERM_CC_DIRECT	Extracellular matrix	15	4.8	4.80×10^{-4}	1.80×10^{-2}
GOTERM_CC_DIRECT	Multivesicular body	5	1.6	4.40×10^{-4}	1.90×10^{-2}
GOTERM_CC_DIRECT	Anchored component of membrane	9	2.9	6.20×10^{-4}	2.10×10^{-2}
GOTERM_CC_DIRECT	Cytosol	80	25.4	4.20×10^{-4}	2.10×10^{-2}
GOTERM_MF_DIRECT	Protein homodimerization activity	34	10.8	8.60×10^{-7}	4.90×10^{-4}
GOTERM_MF_DIRECT	RAGE receptor binding	5	1.6	2.90×10^{-5}	5.50×10^{-3}
GOTERM_MF_DIRECT	Chemokine activity	8	2.5	2.40×10^{-5}	6.70×10^{-3}
GOTERM_MF_DIRECT	CXCR3 chemokine receptor binding	4	1.3	5.40×10^{-5}	7.60×10^{-3}
GOTERM_MF_DIRECT	Receptor binding	18	5.7	2.00×10^{-4}	1.90×10^{-2}
GOTERM_MF_DIRECT	Serine-type endopeptidase activity	15	4.8	2.00×10^{-4}	2.20×10^{-2}
GOTERM_MF_DIRECT	Protein binding	187	59.4	2.90×10^{-4}	2.30×10^{-2}
GOTERM_MF_DIRECT	Identical protein binding	28	8.9	4.00×10^{-4}	2.80×10^{-2}

In order to validate selected 374 mRNAs, we upload these mRNAs to DAVID. Then we can find numerous enrichment. Table 7.20 shows the results of GO term enrichment (adjusted P values less than 0.05) as in Tables 7.18 and 7.19. Thus, although the number of enrichment decreases compared to the type I tensor, still there are many cancer-related GO terms. Thus, the type II tensor approach is still valid enough biologically.

Finally, in order to emphasize the superiority of TD-based unsupervised FE to conventional supervised methods, we apply categorical regression analysis to mRNAs expression,

$$x_{i_1 j} = a_{i_1} + \sum_s b_{i_1 s} \delta_{js} \tag{7.19}$$

where a_{i_1} and $b_{i_1 s}$ are regression coefficient. Based upon the results by categorical regression analysis, because too many 16,917 mRNAs probes are associated with

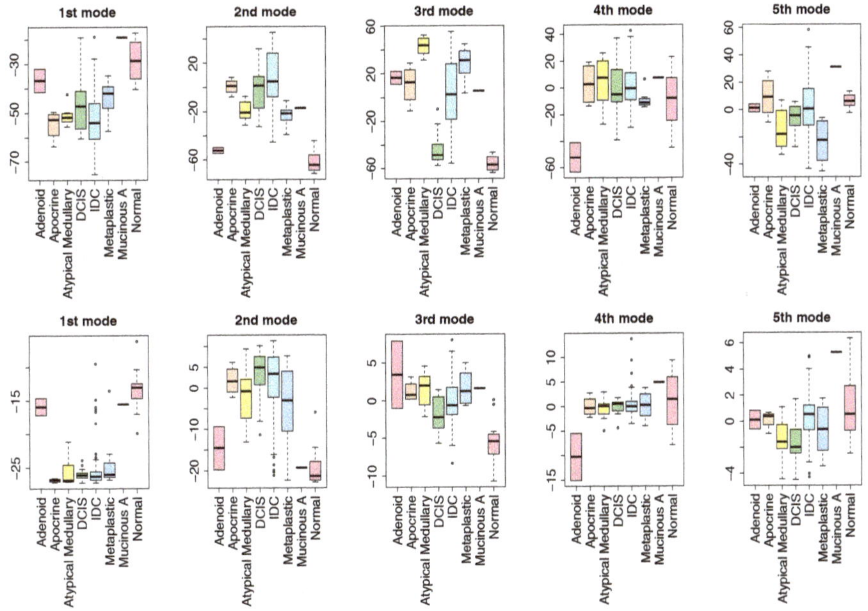

Fig. 7.9 Boxplot of $u^{j;i_1}_{\ell_1 j}$ (upper row) and $u^{j;i_2}_{\ell_2 j}$ (lower row) for $1 \le \ell_3 \le 5$ computed by Eqs. (7.15) and (7.16). P values computed by categorical regression. Upper, first: 4.07×10^{-11}, second: 4.36×10^{-22}, third: 2.03×10^{-23}, fourth: 4.14×10^{-4}, fifth: 1.57×10^{-4}. Lower, first: 3.36×10^{-27}, second: 3.91×10^{-13}, third: 7.39×10^{-9}, fourth: 9.32×10^{-5}, fifth: 2.82×10^{-5}

adjusted P values less than 0.01, we instead upload top-ranked 500 mRNAs with smaller P values to DAVID. As a result, only one GO CC enrichment, cytoplasm, associated with adjusted P values less than 0.05, 1.9×10^{-3}, is detected. Although more advanced methods than categorical regression analysis might achieve better performance, this drastic decrease of the number of detected GO terms enrichment demonstrates the superiority over the conventional supervised method. In this sense, TD-based unsupervised FE is outstanding, no matter whether type I or type II tensor is used.

7.6.2 Temporally Differentially Expressed Genes

Although type I and type II tensor approaches achieved good performance in integrated analysis of multi-class multiomics data set in the previous section, it is better if we can demonstrate yet another example in which TD-based unsupervised FE can achieve better performance. In this subsection, we try to identify genes temporally expressed distinctly between two classes.

Table 7.20 GO BP, CC, and MF enrichment by the 374 emsembl genes identified by TD-based unsupervised FE for type II tensor. Adjusted P values are by BH criterion

Category	Term	Gene count	%	P value	Adjusted P value
GOTERM_BP_DIRECT	Response to estradiol	15	5.1	2.50×10^{-10}	5.00×10^{-7}
GOTERM_BP_DIRECT	Collagen catabolic process	11	3.7	8.20×10^{-8}	5.50×10^{-5}
GOTERM_BP_DIRECT	Skeletal system development	15	5.1	5.60×10^{-8}	5.70×10^{-5}
GOTERM_BP_DIRECT	Positive regulation of cell proliferation	26	8.8	2.10×10^{-7}	1.10×10^{-4}
GOTERM_BP_DIRECT	Collagen fibril organization	8	2.7	2.90×10^{-6}	1.20×10^{-3}
GOTERM_BP_DIRECT	Extracellular matrix organization	15	5.1	4.40×10^{-6}	1.30×10^{-3}
GOTERM_BP_DIRECT	Extracellular matrix disassembly	10	3.4	4.10×10^{-6}	1.40×10^{-3}
GOTERM_BP_DIRECT	Ossification	10	3.4	6.20×10^{-6}	1.60×10^{-3}
GOTERM_BP_DIRECT	Signal transduction	40	13.5	1.50×10^{-5}	3.30×10^{-3}
GOTERM_BP_DIRECT	Cell–cell signaling	15	5.1	8.00×10^{-5}	1.50×10^{-2}
GOTERM_BP_DIRECT	Response to peptide hormone	7	2.4	7.60×10^{-5}	1.50×10^{-2}
GOTERM_BP_DIRECT	Regulation of branching involved in prostate gland morphogenesis	4	1.40	1.40×10^{-4}	2.20×10^{-2}
GOTERM_BP_DIRECT	Mammary gland alveolus development	5	1.7	1.40×10^{-4}	2.40×10^{-2}
GOTERM_BP_DIRECT	Cellular response to hypoxia	9	3.0	1.80×10^{-4}	2.50×10^{-2}
GOTERM_BP_DIRECT	Immune response	19	6.4	2.10×10^{-4}	2.80×10^{-2}
GOTERM_BP_DIRECT	Proteolysis	21	7.1	2.30×10^{-4}	2.80×10^{-2}
GOTERM_BP_DIRECT	Aging	11	3.7	4.00×10^{-4}	4.60×10^{-2}
GOTERM_CC_DIRECT	Extracellular space	89	30.1	6.10×10^{-33}	1.80×10^{-30}
GOTERM_CC_DIRECT	Extracellular region	80	27.0	2.60×10^{-21}	3.90×10^{-19}
GOTERM_CC_DIRECT	Extracellular matrix	27	9.1	8.60×10^{-13}	8.60×10^{-11}
GOTERM_CC_DIRECT	Extracellular exosome	91	30.7	1.90×10^{-12}	1.40×10^{-10}
GOTERM_CC_DIRECT	Proteinaceous extracellular matrix	21	7.1	7.00×10^{-9}	4.10×10^{-7}
GOTERM_CC_DIRECT	Cell surface	24	8.1	1.20×10^{-5}	6.20×10^{-4}
GOTERM_CC_DIRECT	Basement membrane	8	2.7	2.20×10^{-4}	9.20×10^{-3}
GOTERM_CC_DIRECT	Cytosol	75	25.3	4.20×10^{-4}	1.50×10^{-2}
GOTERM_MF_DIRECT	Growth factor activity	12	4.1	7.60×10^{-5}	6.70×10^{-3}
GOTERM_MF_DIRECT	Heparin binding	12	4.1	6.80×10^{-5}	7.20×10^{-3}
GOTERM_MF_DIRECT	Collagen binding	8	2.7	5.50×10^{-5}	7.20×10^{-3}
GOTERM_MF_DIRECT	Calcium ion binding	28	9.5	5.30×10^{-5}	9.30×10^{-3}
GOTERM_MF_DIRECT	Protein binding	178	60.1	3.90×10^{-5}	1.00×10^{-2}

(continued)

Table 7.20 (continued)

Category	Term	Gene count	%	P value	Adjusted P value
GOTERM_MF_DIRECT	RAGE receptor binding	5	1.7	2.10×10^{-5}	1.10×10^{-2}
GOTERM_MF_DIRECT	Protein homodimerization activity	26	8.8	4.40×10^{-4}	3.20×10^{-2}
GOTERM_MF_DIRECT	Identical protein binding	26	8.8	6.30×10^{-4}	3.20×10^{-2}
GOTERM_MF_DIRECT	Serine-type peptidase activity	7	2.4	5.70×10^{-4}	3.30×10^{-2}
GOTERM_MF_DIRECT	Insulin-like growth factor I binding	4	1.4	8.60×10^{-4}	3.40×10^{-2}
GOTERM_MF_DIRECT	Extracellular matrix structural constituent	7	2.4	8.00×10^{-4}	3.40×10^{-2}
GOTERM_MF_DIRECT	Metalloendopeptidase activity	9	3.0	5.50×10^{-4}	3.60×10^{-2}
GOTERM_MF_DIRECT	Fibronectin binding	5	1.7	8.00×10^{-4}	3.80×10^{-2}
GOTERM_MF_DIRECT	Serine-type endopeptidase activity	13	4.4	1.10×10^{-3}	3.90×10^{-2}
GOTERM_MF_DIRECT	Protein kinase binding	16	5.4	1.40×10^{-3}	4.90×10^{-2}

Table 7.21 List of samples in EGF treatment experiments

Time points (hours)	0	0.5	1	2	4	6	8	12	18	24	48
Control	3	1	1	1	1	1	0	0	0	2	3
EGF treated	0	2	1	1	1	1	1	1	1	3	3

The first data set analyzed is the comparison of NSCLC cell line H1975, with and without EGF treatment [4]. EGF is a gene supposed to accelerate cell growth and is known to be expressive frequently in cancers. Thus, EGF treatment is expected to activate cancer cell lines. The data set is composed of two mRNA expression profiles, $x_{ij_1}^{\text{control}} \in \mathbb{R}^{39,937\times 13}$ and $x_{ij_2}^{\text{EGF}} \in \mathbb{R}^{39,937\times 15}$, which are gene expressions of cell lines without and with EGF treatment, respectively. j_1 and j_2 represent time points (tps) after the treatment (Table 7.21). Because they share genes, $x_{ij_1}^{\text{control}}$ and $x_{ij_2}^{\text{EGF}}$ can be converted to case II type I tensor as

$$x_{ij_1j_2} = x_{ij_1}^{\text{control}} x_{ij_2}^{\text{EGF}}. \tag{7.20}$$

HOSVD, Fig. 3.8, is applied to $x_{ij_1j_2}$ as

$$x_{ij_1j_2} = \sum_{\ell_1=1}^{13} \sum_{\ell_2=1}^{15} \sum_{\ell_3=1}^{39,937} G(\ell_1, \ell_2, \ell_3) u_{\ell_1 j_1}^{(j_1)} u_{\ell_2 j_2}^{(j_2)} u_{\ell_3 i}^{(i)} \tag{7.21}$$

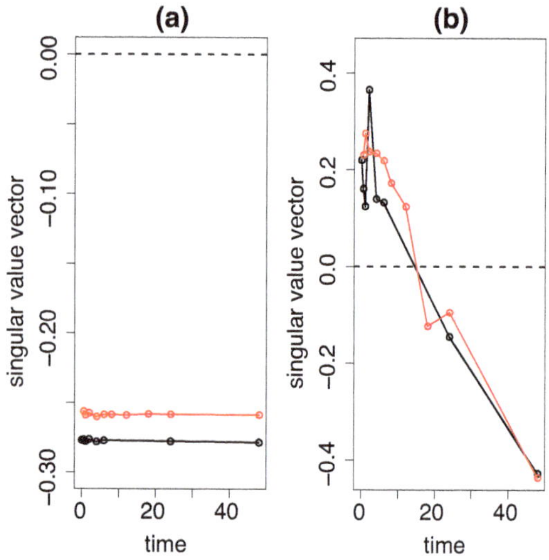

Fig. 7.10 Singular value vectors, Eq. (7.21). (**a**) $\boldsymbol{u}_1^{(j_1)}$ (black) and $\boldsymbol{u}_1^{(j_2)}$ (red). (**b**) $\boldsymbol{u}_2^{(j_1)}$ (black) and $\boldsymbol{u}_2^{(j_2)}$(red)

Table 7.22 Top ten–ranked $G(\ell_1, \ell_2, \ell_3)$s with larger absolute values in Eq. (7.21)

ℓ_1	1	2	1	3	1
ℓ_2	1	1	2	1	3
ℓ_3	1	2	2	3	4
$G(\ell_1, \ell_2, \ell_3)$	-4.03×10^4	-1.56×10^3	1.49×10^3	1.05×10^3	-5.79×10^2
ℓ_1	4	2	5	1	4
ℓ_2	1	1	1	4	1
ℓ_3	5	3	6	6	4
$G(\ell_1, \ell_2, \ell_3)$	4.24×10^2	4.16×10^2	3.25×10^2	3.19×10^2	-2.62×10^2

At first, we need to find singular value vectors $\boldsymbol{u}_{\ell_1}^{(j_1)} \in \mathbb{R}^{13}$ and $\boldsymbol{u}_{\ell_2}^{(j_2)} \in \mathbb{R}^{15}$ that exhibit distinct temporal expression between them. Figure 7.10 shows time development of $\boldsymbol{u}_{\ell_1}^{(j_1)}$ and $\boldsymbol{u}_{\ell_2}^{(j_2)}$ for $\ell_1 = \ell_2 = 1, 2$. Here, the components of singular value vectors sharing the time points are averaged within individual vectors, $u_{\ell_1 j_1}^{(j_1)}$. It is obvious that $\boldsymbol{u}_1^{(j_1)}$ and $\boldsymbol{u}_1^{(j_2)}$ do not exhibit any time dependence while $\boldsymbol{u}_2^{(j_1)}$ and $\boldsymbol{u}_2^{(j_2)}$ do. Thus, there is a possibility that genes associated with $\boldsymbol{u}_2^{(j_1)}$ and $\boldsymbol{u}_2^{(j_2)}$ also exhibit the temporal difference between control and EGF-treated cells.

In order to select genes associated with $\boldsymbol{u}_2^{(j_1)}$ and $\boldsymbol{u}_2^{(j_2)}$, we need to find $G(\ell_1, \ell_2, \ell_3)$, $\ell_1 = 2$ or $\ell_2 = 2$ having larger absolute values (Table 7.22); $G(2, 1, 2)$ and $G(1, 2, 2)$ have larger absolute values. Thus, we decide to use $\boldsymbol{u}_2^{(i)}$ for FE. P values are attributed to i as

$$P_i = P_{\chi^2}\left[> \left(\frac{u_{2i}^{(i)}}{\sigma_2}\right)^2 \right]. \tag{7.22}$$

P values are corrected by BH criterion and genes associated with adjusted P values less than 0.01 are selected. Then 552 mRNA probes are selected.

Next, we need to see if the selected 552 mRNA probes really exhibit the temporal difference between control and EGF-treated cells. For this purpose, we compute the correlation coefficient between

$$\left(x_{i1}^{\text{control}}, \ldots, x_{i13}^{\text{control}}, x_{i1}^{\text{EGF}}, \ldots, x_{i15}^{\text{EGF}}\right) \tag{7.23}$$

and

$$\left(u_{2,1}^{(j_1)}, \ldots, u_{2,13}^{(j_1)}, u_{2,1}^{(j_2)}, \ldots, u_{2,15}^{(j_2)}\right) \tag{7.24}$$

to see if 552 selected genes are coincident with $\boldsymbol{u}_2^{(j_1)}$ and $\boldsymbol{u}_2^{(j_2)}$. Figure 7.11a shows the histogram of correlation coefficients. Because there are two peaks at ± 1, it is obvious that the gene expressions of the selected 552 mRNA probes are highly coincident with $\boldsymbol{u}_2^{(j_1)}$ and $\boldsymbol{u}_2^{(j_2)}$.

Before comparing 552 genes directly between control and EGF-treated cells, we need shift and scale individual gene expression profiles such that they have the same baseline and amplitude. Thus, we apply the following linear regression:

$$u_{2j_1}^{(j_1)} = a_i x_{ij_1}^{\text{control}} + b_i \tag{7.25}$$

$$u_{2j_2}^{(j_2)} = a_i x_{ij_2}^{\text{EGF}} + b_i \tag{7.26}$$

where a_i and b_i are the regression coefficients. Because regression coefficients are shared between control and EGF-treated ones, this does not reduce the difference between these two. Then, we compare $a_i x_{ij_1}^{\text{control}} + b_i$ and $a_i x_{ij_2}^{\text{EGF}} + b_i$ of the selected 552 mRNA probes (Fig. 7.11b). Not all, but the comparisons of five out of seven time points excluding two time points, 4 and 24 h, after the EGF treatment are associated with P values less than 0.05. Thus, TD-based unsupervised FE has the ability to select genes associated with temporal distinction.

Next, we try to see if the type II tensor approach works as well. Because case II tensor shares the feature whose number is generally much larger than the number of samples, type II tensor where the shared dimension is summed up can result in a much smaller number of components. Type II tensor is defined as

$$x_{j_1 j_2} = \sum_{i=1}^{39,937} x_{ij_1 j_2}. \tag{7.27}$$

where $x_{ij_1 j_2}$ is defined in Eq. (7.20). The number of components in $x_{j_1 j_2} \in \mathbb{R}^{13 \times 15}$ is $13 \times 15 = 195$, which is as small as 1/39,937 of the number of components in $x_{ij_1 j_2} \in \mathbb{R}^{39,937 \times 13 \times 15}$. Thus, if the type II tensor approach works as well, it is very effective. SVD is applied to $x_{j_1 j_2}$ as

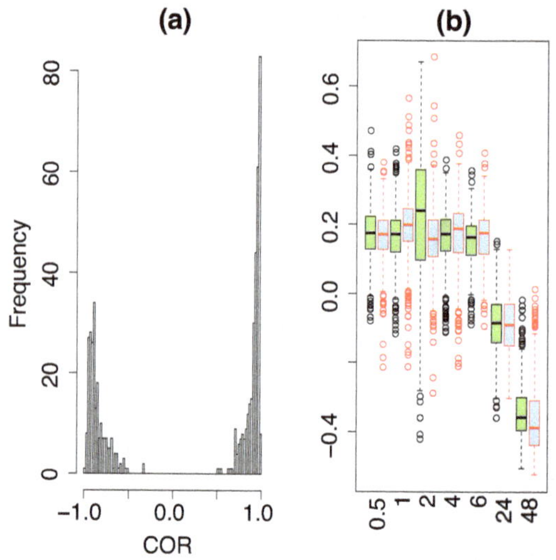

Fig. 7.11 (**a**) Histogram of correlation coefficients between Eqs. (7.23) and (7.24) for case II type I tensor, Eq. (7.20). (**b**) Boxplot of Eqs. (7.25) (black boxes filled with green) and (7.26) (red boxes filled with blue) for case II type I tensor, Eq. (7.20). P values computed by t-test: 0.5 h:2.83×10^{-2}, 1 h:6.81×10^{-8}, 2 h:5.63×10^{-12}, 4 h:3.5×10^{-1}, 6 h:4.83×10^{-2}, 24 h:5.0×10^{-1}, 48 h:1.70×10^{-6}

$$x_{j_1 j_2} = \sum_{\ell} \lambda_\ell u^{(j_1)}_{\ell j_1} u^{(j_2)}_{\ell j_2} \tag{7.28}$$

Figure 7.12 shows the $\boldsymbol{u}^{(j_1)}_\ell$ and $\boldsymbol{u}^{(j_2)}_\ell$ for $\ell = 1, 2$. Basically, it looks similar to Fig. 7.10. Thus, we decide to employ $\ell = 2$ for FE. Then, singular value vectors attributed to i can be computed as Eq. (5.14),

$$u^{i;j_1}_{\ell i} = \sum_{j_1=1}^{13} x^{\text{control}}_{i j_1} u^{(j_1)}_{\ell j_1} \tag{7.29}$$

$$u^{i;j_2}_{\ell i} = \sum_{j_2=1}^{15} x^{\text{EGF}}_{i j_2} u^{(j_2)}_{\ell j_2} \tag{7.30}$$

Thus, P values are also attributed to i in two ways as

$$P^{j_1}_i = P_{\chi^2}\left[> \left(\frac{u^{(i;j_1)}_{2i}}{\sigma_2} \right)^2 \right], \tag{7.31}$$

$$P^{j_2}_i = P_{\chi^2}\left[> \left(\frac{u^{(i;j_2)}_{2i}}{\sigma'_2} \right)^2 \right]. \tag{7.32}$$

P values are corrected by BH criterion. mRNA probes associated with adjusted P values less than 0.01 are selected. Then, 482 and 487 mRNA probes, between

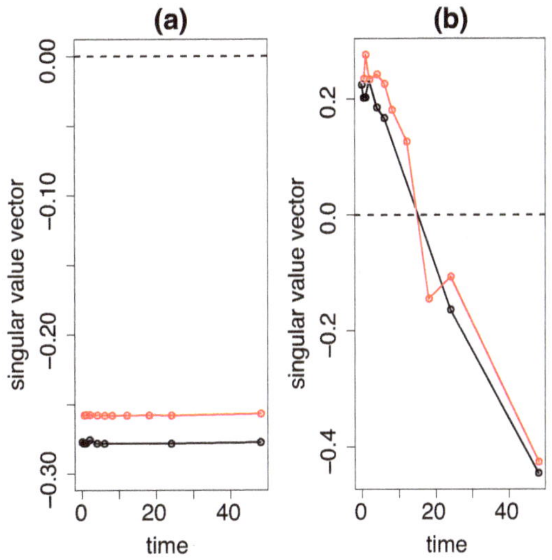

Fig. 7.12 Singular value vectors, Eq. (7.28). (**a**) $\boldsymbol{u}_1^{(j_1)}$ (black) and $\boldsymbol{u}_1^{(j_2)}$ (red). (**b**) $\boldsymbol{u}_2^{(j_1)}$ (black) and $\boldsymbol{u}_2^{(j_2)}$(red)

which 396 mRNA probes are chosen in common, are selected using $P_i^{j_1}$ and $P_i^{j_2}$, respectively. Thus, in some sense, the type II tensor approach can give the results coincident between two approximations of singular value vectors attributed to i using Eqs. (7.29) and (7.30).

Next, we need to see if the 396 mRNA probes chosen in common really exhibit temporal difference between control and EGF-treated cells as in the case of the type I tensor approach. The correlation coefficient between Eqs. (7.23) and (7.24) is computed again to see the coincidence between gene expression and singular value vectors (Fig. 7.13a). It is obvious that the peak at ± 1 is much steeper than that in Fig. 7.11a. This suggests that the type II tensor approach might be better than the type I tensor approach in spite of the smaller computational resources required.

In order to confirm the superiority of the type II tensor approach, we again apply linear regression Eqs. (7.25) and (7.26) replacing singular value vectors with those obtained by type II tensor (Fig. 7.13b). Because six among seven time points excluding 4 h after the EGF treatment are associated with P values less than 0.05, the type II tensor approach is superior to the type I tensor approach.

Finally, in order to validate 552 and 396 mRNA probes selected by type I and II tensor approches, respectively, we upload RefSeq mRNA IDs associated with these probes to DAVID. Table 7.23 lists the KEGG pathways identified by DAVID for type I and II tensor approaches. Although common five KEGG pathways are associated with adjusted P values less than 0.05, P values for the type II tensor approach are smaller than those for the type I tensor approach. Because P values are more likely smaller for more number of genes uploads, smaller P values attributed to KEGG pathways by the type II tensor approach where less number of genes are selected suggests the superiority of the type II tensor approach from the biological point of view.

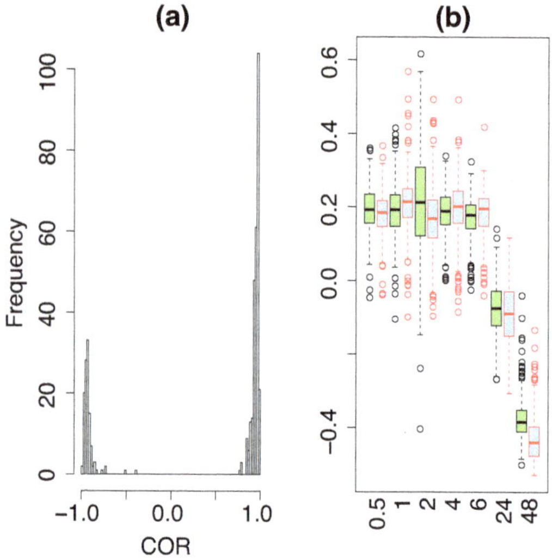

Fig. 7.13 (**a**) Histogram of correlation coefficients between Eqs. (7.23) and (7.24) for case II type II tensor, Eq. (7.27). (**b**) Boxplot of Eqs. (7.25) (black boxes filled with green) and (7.26) (red boxes filled with blue) for case II type II tensor, Eq. (7.27) P values computed by t-test: 0.5 h:1.68×10^{-2}, 1 h:2.56×10^{-5}, 2 h: 3.83×10^{-7}, 4 h:9.14×10^{-2}, 6 h:7.30×10^{-4}, 24 h:2.36×10^{-2}, 48 h:5.55×10^{-38}

Table 7.23 KEGG pathways identified by DAVID for genes associated with 552 (upper numbers) and 396 (lower numbers) miRNA probes selected using type I, Eq. (7.20), and II, Eq. (7.27), tensor approach. Adjusted P values are by BH criterion

Category	Term	Count	%	P value	Adjusted P value
KEGG_PATHWAY	Cell cycle	29 28	9.0 12.1	7.2×10^{-24} 3.7×10^{-29}	1.0×10^{-21} 3.2×10^{-27}
KEGG_PATHWAY	Oocyte meiosis	14 14	4.3 6.0	7.6×10^{-8} 1.4×10^{-10}	5.5×10^{-6} 5.8×10^{-9}
KEGG_PATHWAY	DNA replication	8 9	2.5 3.9	2.8×10^{-6} 3.2×10^{-9}	1.4×10^{-4} 9.3×10^{-8}
KEGG_PATHWAY	Progesterone-mediated oocyte maturation	8 9	2.5 3.9	9.2×10^{-4} 4.0×10^{-6}	3.3×10^{-2} 8.6×10^{-5}
KEGG_PATHWAY	p53 signaling pathway	7 6	2.2 2.6	1.2×10^{-3} 7.7×10^{-4}	3.5×10^{-2} 1.3×10^{-2}

Although the type II approach is better than the type I approach in this specific example, because it is highly dependent upon the data sets analyzed, it is difficult to know in advance which is better.

7.7 Gene Expression and Methylation in Social Insects

As the first example of the application of the case I tensor approach, we employ the multiomics analysis of social insects. Social insects, e.g., ants and bees, are known to have castes where distinct phenotypes appear in spite of shared genomes. Thus, it is interesting to know what drives differentiation between castes.

One possible scenario is the alteration of epigenome [115], because epigenome has a plasticity that can mediate differentiation between castes. The most typical caste is composed of a queen and workers. The former, queen, concentrates on reproduction while the latter, workers, serve to maintain colony. In spite of their strict difference in phenotype, they are often known to be relatives. Thus, they share the genome to some extent having distinct phenotype. This suggests that epigenome can play potential roles in the differentiation of caste.

In this section, we try to identify genes associated with differential expression and methylation between castes, especially queens and workers [95], because such genes are potential candidates that can mediate distinct phenotypes between castes. Thus, we employ TD-based unsupervised FE that can integrate multiomics data sets. The data set analyzed [72] is composed of two insect species, bee (*P. canadensis*) and ant (*Dinoponera quadriceps*). Table 7.24 shows the number of samples available from GEO with GEO ID GSE59525. As can be seen, it is a typical large p small n data set.

Because the amount of gene expression is measured by the unit of reads per kilobase of exon per million mapped reads (RPKM), it is used as it is. Because the gene expression profile of *P. canadensis* was $\log_2$-ratio converted, it is expanded to the original one as 2^x where x is gene expression. On the other hand, we would like to employ case II tensor format (Table 5.3) where genes are shared. Thus, we need to convert methylation profiles to be attributed to individual genes. Thus, assuming m_{s_1} and m_{s_2} are methylation and nonmethylation values respectively at locus s, the relative methylation within the ith gene can be defined as

$$\frac{\sum_{s \in i} m_{s_1}}{\sum_{s \in i} \left(m_{s_1} + m_{s_2}\right)} \tag{7.33}$$

where $\sum_{s \in i}$ is taken over s bases within DNA sequences corresponding to the ith gene body; the reason why methylation is not in the promoter region but in the gene body is summed up and is attributed to genes because gene body methylation is

Table 7.24 Number of samples in social insect study [72]

	Methylation			mRNA	
Caste	Control	Queen	Worker	Queen	Worker
Polymnia canadensis	1	3	3	4	6
Dinoponera quadriceps	1	3	3	7	6

believed to affect gene expression in insects [123]. Relative methylation profile is formatted as

$$x_{ik}^{\text{metyl, bee}} \in \mathbb{R}^{N \times 7}, \tag{7.34}$$

$$x_{ik}^{\text{metyl, ant}} \in \mathbb{R}^{N \times 7}, \tag{7.35}$$

where N is the number of genes. $k = 1$ corresponds to control samples. $2 \leq k \leq 4$ and $5 \leq k \leq 7$ correspond to queens and workers, respectively. On the other hand, mRNA expression is formatted as

$$x_{ij}^{\text{mRNA, bee}} \in \mathbb{R}^{N \times 10}, \tag{7.36}$$

$$x_{ij}^{\text{mRNA, ant}} \in \mathbb{R}^{N \times 13}. \tag{7.37}$$

where $1 \leq j \leq 4$ and $5 \leq j \leq 10$ for bee correspond to queens and workers, respectively, while $1 \leq j \leq 7$ and $8 \leq j \leq 13$ for ant correspond to queens and workers, respectively. Then case II tensor is generated as

$$x_{ijk}^{\text{bee}} = x_{ij}^{\text{mRNA, bee}} x_{ik}^{\text{metyl, bee}}, \tag{7.38}$$

$$x_{ijk}^{\text{ant}} = x_{ij}^{\text{mRNA, ant}} x_{ik}^{\text{metyl, ant}}, \tag{7.39}$$

where $x_{ijk}^{\text{bee}} \in \mathbb{R}^{N \times 10 \times 7}$ and $x_{ijk}^{\text{ant}} \in \mathbb{R}^{N \times 13 \times 7}$. HOSVD, Fig. 3.8, is applied to x_{ijk}^{bee} and x_{ijk}^{ant} as

$$x_{ijk}^{\text{bee}} = \sum_{\ell_1=1}^{N} \sum_{\ell_2=1}^{10} \sum_{\ell_3=1}^{7} G(\ell_1, \ell_2, \ell_3) u_{\ell_1 i}^{\text{bee}(i)} u_{\ell_2 j}^{\text{bee}(j)} u_{\ell_3 k}^{\text{bee}(k)} \tag{7.40}$$

$$x_{ijk}^{\text{ant}} = \sum_{\ell_1=1}^{N} \sum_{\ell_2=1}^{13} \sum_{\ell_3=1}^{7} G(\ell_1, \ell_2, \ell_3) u_{\ell_1 i}^{\text{ant}(i)} u_{\ell_2 j}^{\text{ant}(j)} u_{\ell_3 k}^{\text{ant}(k)} \tag{7.41}$$

where $u_{\ell_1 i}^{\text{bee}(i)} \in \mathbb{R}^{N \times N}$, $u_{\ell_2 j}^{\text{bee}(j)} \in \mathbb{R}^{10 \times 10}$, $u_{\ell_3 k}^{\text{bee}(k)} \in \mathbb{R}^{7 \times 7}$, $u_{\ell_1 i}^{\text{ant}(i)} \in \mathbb{R}^{N \times N}$, $u_{\ell_2 j}^{\text{ant}(j)} \in \mathbb{R}^{13 \times 13}$, and $u_{\ell_3 k}^{\text{ant}(k)} \in \mathbb{R}^{7 \times 7}$.

Next, as usual, we need to find which singular value vectors are coincident with the distinction between queens and workers. Figures 7.14a and b, 7.15a and b show singular value vectors associated with the highest distinction between queens and workers. Unfortunately, singular value vectors of methylation do not exhibit small enough P values to be significant. Nevertheless, because selected genes might exhibit significant distinct expression between queens and works, we continue the procedure. We seek $G(\ell_1, 1, 3)$ for *P. canadensis* and $G(\ell_1, 1, 5)$ for *D. quadriceps* with larger absolute values.

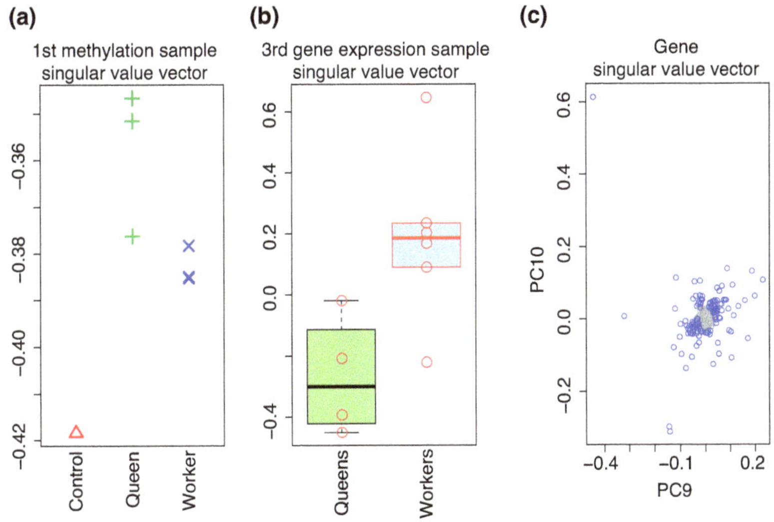

Fig. 7.14 Singular value vectors for *P. canadensis*. *P* values are computed by *t*-test between queens and workers. (**a**) $\boldsymbol{u}_1^{\text{bee}(k)}$, $P = 1.1 \times 10^{-1}$ (**b**) $\boldsymbol{u}_3^{\text{bee}(j)}$, $P = 1.65 \times 10^{-2}$ (**c**) $\boldsymbol{u}_{\ell_1}^{\text{bee}(i)}$, $\ell_1 = 9, 10$. The blue open circles are selected genes

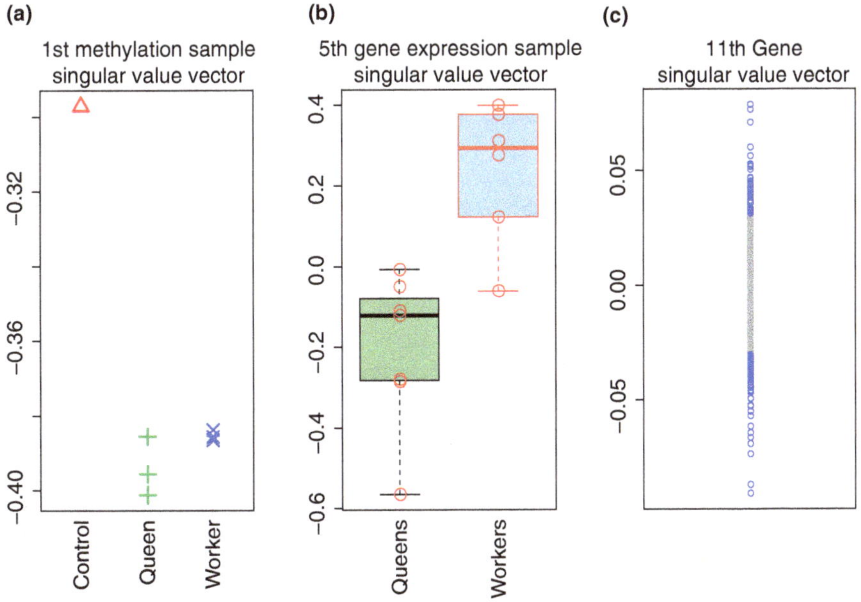

Fig. 7.15 Singular value vectors for *D. quadriceps*. *P* values are computed by *t*-test between queens and workers. (**a**) $\boldsymbol{u}_1^{\text{ant}(k)}$, $P = 1.9 \times 10^{-1}$ (**b**) $\boldsymbol{u}_5^{\text{ant}(j)}$, $P = 1.25 \times 10^{-3}$ (**c**) $\boldsymbol{u}_{11}^{\text{ant}(i)}$. The blue open circles are selected genes

Table 7.25 The top ten core tensors, G, with large absolute values

P. canadensis		*D. quadriceps*	
ℓ_1	$G(\ell_1, 1, 3)$	ℓ_1	$G(\ell_1, 1, 5)$
9	−79.8	11	−54.8
10	75.4	12	4.1
7	−61.4	25	3.4
11	38.4	2	−2.9
5	−23.4	23	2.8
4	−16.0	9	2.4
12	−11.9	20	−2.2
1	−5.4	8	2.2
13	5.4	10	−1.7
6	−4.5	22	−1.4

Table 7.25 lists the top-ranked Gs with larger absolute values. Then we decide that $\boldsymbol{u}_{\ell_1}^{\text{bee}(i)}, \ell_1 = 9, 10$ and $\boldsymbol{u}_{11}^{\text{ant}(i)}$ are used for FE (Figs. 7.14c and 7.15c). P values are attributed to the ith gene as

$$P_i^{\text{bee}} = P_{\chi^2}\left[> \sum_{\ell_1=9}^{10}\left(\frac{u_{\ell_1 i}^{\text{bee}(i)}}{\sigma_{\ell_1}}\right)^2\right], \tag{7.42}$$

and

$$P_i^{\text{ant}} = P_{\chi^2}\left[> \left(\frac{u_{11i}^{\text{ant}(i)}}{\sigma_{11}}\right)^2\right], \tag{7.43}$$

P values are adjusted by BH criterion. Genes associated with adjusted P values less than 0.01 are selected. As a result, 133 and 128 genes are selected for *P. canadensis* and *D. quadriceps*, respectively.

The point is selected genes are associated with distinct gene expression and methylation between queens and workers simultaneously. Then we apply three statistical tests to 133 genes and 128 genes between queens and workers (Table 7.26). Selected genes exhibit simultaneous distinct gene expression and methylation between queens and workers for *P. canadensis*, but not for *D. quadriceps*. Thus, selected genes can be potential factors that can mediate caste differentiation for *P. canadensis*, but not for *D. quadriceps*. Although we are not sure that the lack of detection for *D. quadriceps* is because of biological reason or failure of our methodology, at least, our purpose is achieved for *P. canadensis*. In order to clarify this point, we need to continue the research.

In order to see if conventional supervised methods can do this, we apply t-test to gene expression and promoter methylation to find genes that exhibit significant distinction between queens and workers. As a result, two genes for distinct gene expression between queens and workers for *D. quadriceps* are associated with

Table 7.26 Statistical tests of the differences (between queens and workers) in gene expression and methylation. The genes identified by TD-based unsupervised FE are analyzed by t (the t-test), Wilcox (the Wilcoxon rank sum test), and Kolmogorov–Sinai (KS) (the Kolmogorov–Sinai test), all two-sided

		t	Wilcox	KS
P. canadensis	Gene expression	1.71×10^{-3}	1.89×10^{-2}	0.08
	Methylation	1.74×10^{-4}	5.06×10^{-3}	1.02×10^{-3}
D. quadriceps	Gene expression	2.73×10^{-12}	9.05×10^{-12}	4.41×10^{-11}
	Methylation	0.3757	0.7163	0.4413

adjusted P vales less than 0.01. This poor performance is because of the small number of samples. Thus, TD-based unsupervised FE has the ability to find significant genes for large p small n problem, for which the conventional supervised method fails.

Before closing this section, we would like to validate selected genes from the biological point of view. Because these two insects are not included in popular enrichment servers, e.g., DAVID or Enrichr, we download a list of GO terms,[1] PCAN.v01.GO.tsv for *P. canadensis* and DQUA.v01.GO.tsv for *D. quadriceps*. Fisher's exact test is performed in order to evaluate enrichment and computed P values are corrected by BH criterion. GO terms associated with adjusted P values less than 0.05 are searched. There are three GO terms, Lipid transporter activity (GO:0005319), Lipid particle (GO:0005811), and Lipid transpor (GO:0006869) enriched in 133 genes selected for *P. canadensis*, while there are no GO terms enriched in 128 genes selected for *D. quadriceps*. This might be reasonable because 128 genes selected for *D. quadriceps* are not associated with distinct methylation between queens and workers (Table 7.26). Anyway, 133 genes selected for *P. canadensis*, which is simultaneously associated with distinct gene expression and methylation between queens and workers, are associated with a few GO term enrichment. Thus, at least for *P. canadensis*, TD-based unsupervised FE is useful also from the biological point of view.

7.8 Drug Discovery from Gene Expression: II

In Sect. 7.3, we have already shown that TD-based unsupervised FE successfully identifies compounds that affect gene expression in dose-dependent manner and their target proteins from only gene expression profiles in fully unsupervised manner. Nevertheless, it is strictly restricted to cancers because gene expression profiles are measured in cancer cell lines. Identifying drug compounds that are effective for other diseases requires additional gene expression profiles treated by compounds in specific diseases, e.g., model animals or cell lines originating from

[1] Paper Wasp and Denosaur Ant Project. Accessed 15 Jan. 2019. http://wasp.crg.eu/download.html.

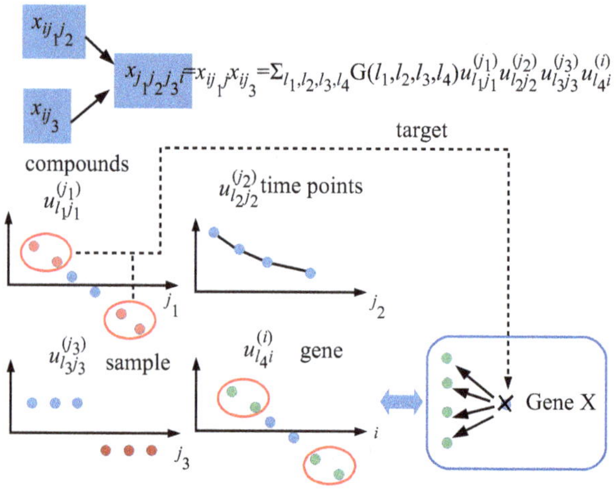

Fig. 7.16 Integrated analysis of gene expression profile of drug-treated animals, $x_{ij_1j_2}$ and human gene expression profiles of patients and healthy control, x_{ij_3}. i: genes, j_1: compounds, j_2: time point after the treatment, j_3: human samples

the disease. Thus, in the manner in Sect. 7.3, the effectiveness of methods is quite limited.

In this section, by using case II tensor where genes are shared between two matrices or tensors, we try to identify disease-effective drugs without measuring gene expression repeatedly for individual diseases. The study design is as follows (Fig. 7.16): $x_{ij_1j_2}$ is ith gene expression profiles of animals treated by j_1 compound at the time point j_2 after the treatment. x_{ij_3} is human gene expression profile of gene i at j_3th patients or healthy control. Case II tensor $x_{ij_1j_2j_3}$ is generated as

$$x_{ij_1j_2j_3} = x_{ij_1j_2}x_{ij_3} \tag{7.44}$$

HOSVD algorithm, Fig. 3.8, is applied to $x_{ij_1j_2j_3}$ as

$$x_{ij_1j_2j_3} = \sum_{\ell_1=1}^{N_1}\sum_{\ell_2=1}^{N_2}\sum_{\ell_3=1}^{N_3}\sum_{\ell_4=1}^{N_4} G(\ell_1,\ell_2,\ell_3,\ell_4)u_{\ell_1j_1}^{(j_1)}u_{\ell_2j_2}^{(j_2)}u_{\ell_3j_3}^{(j_3)}u_{\ell_4i}^{(i)} \tag{7.45}$$

Then, $\boldsymbol{u}_{\ell_2}^{(j_2)}$ that exhibits time dependence and $\boldsymbol{u}_{\ell_3}^{(j_3)}$ that exhibits distinction between healthy controls and patients are searched. After identifying ℓ_2 and ℓ_3, ℓ_1 and ℓ_4 associated with $G(\ell_1,\ell_2,\ell_3,\ell_4)$ with larger absolute values are selected. Once ℓ_1 and ℓ_4 are selected, P values are attributed to i and j_1 as

$$P_i = P_{\chi^2}\left[> \left(\frac{u_{\ell_4i}}{\sigma_{\ell_4}}\right)^2\right], \tag{7.46}$$

and

$$P_{j_1} = P_{\chi^2}\left[> \left(\frac{u_{\ell_1 j_1}}{\sigma_{\ell_1}}\right)^2 \right]. \tag{7.47}$$

P values are corrected by BH criterion and i and j_1 associated with adjusted P values less than 0.01 (filled pink circles and filled light green circles surrounded by pink oval in Fig. 7.16) are supposed to be selected. Target proteins are decided by the comparison with external databases (as shown in Fig. 7.5). This process results in the set of drug candidate compounds and candidate target proteins. Figure 7.17

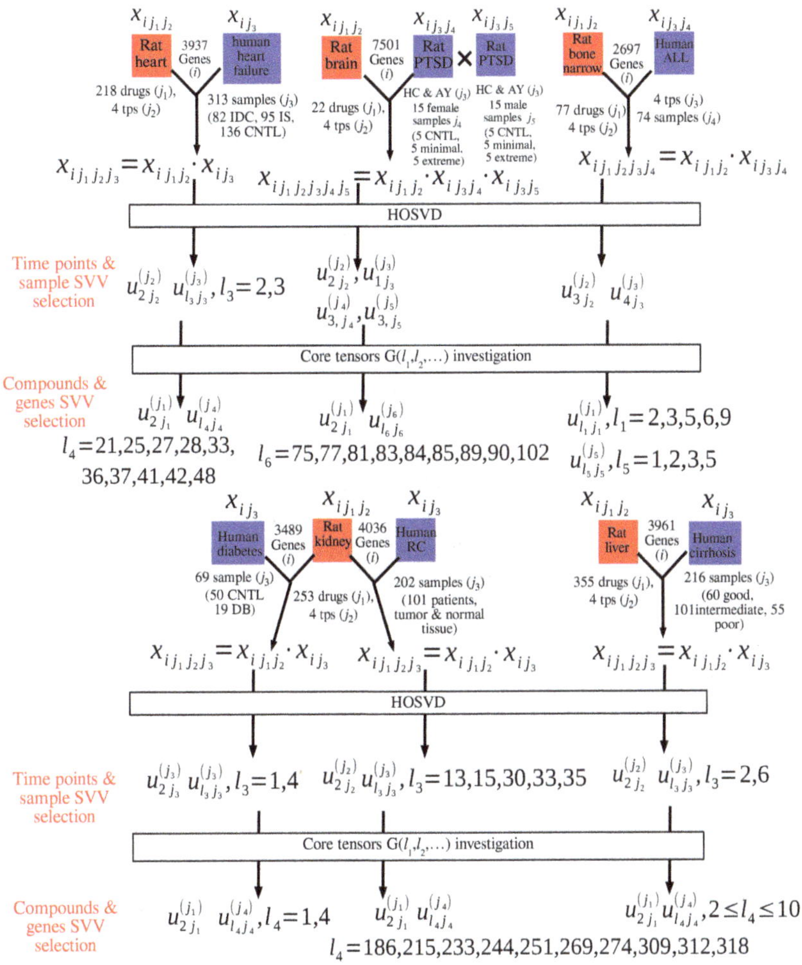

Fig. 7.17 Schematics that illustrate the procedure of TD-based unsupervised FE applied to various diseases and DrugMatrix data sets. SVV: singular value vector. Selected four time points (tps) are 1/4, 1, 3, and 5 days after treatment

and Table 7.27 summarize the process till the selection of singular value vectors are attributed to genes and compounds. Six diseases are analyzed, heart failure, PTSD, acute lymphoblastic leukemia (ALL), diabetes, renal carcinoma, and cirrhosis. In some cases, modes of case II tensors are more than four because human gene expression profiles are represented as tensors, not matrices.

Gene expression profiles of model animals are downloaded from DrugMatrix [68] where rats are treated as model animals and gene expression profiles of various tissues are extracted. Corresponding human or rat disease expression profiles are downloaded from GEO. For heart failure, human disease heart failure gene expression profiles and rat heart gene expression profiles treated by drugs are used. For PTSD, stressed mouse brain gene expression profiles and rat brain gene expression profiles treated by drugs are used. For ALL, drug-treated rat and ALL human patients' bone marrow gene expression profiles are used. For diabetes and renal carcinoma, drug-treated rat kidney gene expression profiles are used. Diabetes and renal carcinoma human patients' kidney gene expression profiles are used for diabetes and renal carcinoma, respectively. For cirrhosis, drug-treated rat liver gene expression profiles and cirrhosis human liver expression profiles are used. See appendix for more details.

After selecting genes and drugs, genes are uploaded to Enrichr for target protein identification. Genes enriched (adjusted P values less than 0.01) in "single-gene perturbation GEO up" and "single-gene perturbation GEO down" are selected as target proteins. This process is similar to that illustrated in Fig. 7.5. Table 7.28 summarizes the number of identified genes, compounds, and target proteins.

In order to validate the relationship between drugs and target proteins predicted, we compare them with DINIES [122] that stores known protein–drug interactions. We upload drugs one by one to DENIES with parameters "chemogenomic approach" and "with learning on all DBs" and can get the list of target proteins. They are merged into a list of proteins because individual proteins can be targeted by multiple drugs. The obtained set of target proteins is compared with predicted targets in Table 7.28. Here, the total proteins considered are limited to genes included in "Single_Gene_Perturbations_from_GEO_all_list" of Enrichr. Table 7.29 shows the results of evaluation by Fisher's exact test and χ^2 test. Ten out of twelve are evaluated as significant (P values less than 0.05) by either Fisher's exact test or χ^2 test. This suggests that TD-based unsupervised FE can be used for the prediction of target proteins and diseases of drugs only from gene expression profile, in a fully unsupervised manner in the sense that it does not require any pre-knowledge about disease–drug or protein–drug interactions.

Table 7.27 A summary of TDs and identification of various singular value vectors for the identification of candidate drugs and genes used to find genes encoding drug target proteins. In all cases, ℓ_1 stands for singular value vectors of compounds, whereas ℓ_k with the last (largest) k denotes gene singular value vectors. ℓ_2 stands for singular value vectors of time points in DrugMatrix data. The remaining singular value vectors correspond to sample singular value vectors dependent on the properties of gene expression profiles of diseases. See also Fig. 7.17 for the corresponding data

	Tensors				
Diseases	DrugMatrix	Disease	Generated	Core tensor	Singular value vectors
Heart	$x_{ij_1j_2}$	x_{ij_3}	$x_{ij_1j_2j_3}$	$G(\ell_1\ell_2\ell_3\ell_4)$	$u^{(j_k)}_{\ell_k,j_k}, k \le 3, u^{(i)}_{\ell_4,i} \in \mathbb{R}^{N_k \times N_k}$
failure	$\in \mathbb{R}^{N_4 \times N_1 \times N_2}$	$\in \mathbb{R}^{N_4 \times N_3}$	$\in \mathbb{R}^{N_4 \times N_1 \times N_2 \times N_3}$	$\in \mathbb{R}^{N_1 \times N_2 \times N_3 \times N_4}$	$(N_1, N_2, N_3, N_4) = (218, 4, 313, 3937)$
Selected				$\ell_1 = 2; \ell_2 = 2; \ell_3 = 2, 3; \ell_4 = 21, 25, 27, 28, 33, 36, 37, 41, 42, 48$	
PTSD	$x_{ij_1j_2}$	$x_{ij_3j_k}, k = 4, 5$	$x_{ij_1j_2j_3j_4j_5}$	$G(\ell_1\ell_2\ell_3\ell_4\ell_5\ell_6)$	$u^{(j_k)}_{\ell_k,j_k}, k \le 5, u^{(i)}_{\ell_6,i} \in \mathbb{R}^{N_k \times N_k}$
Rat	$\in \mathbb{R}^{N_6 \times N_1 \times N_2}$	$\in \mathbb{R}^{N_6 \times N_3 \times N_k}$	$\in \mathbb{R}^{N_6 \times N_1 \times N_2 \times N_3 \times N_4 \times N_5}$		$(N_1, N_2, N_3, N_4, N_5, N_6) = (22, 4, 2, 15, 15, 7501)$
Model				$\in \mathbb{R}^{N_1 \times N_2 \times N_3 \times N_4 \times N_5 \times N_6}$	
Selected				$\ell_1 = 2; \ell_2 = 2; \ell_3 = 1; \ell_4 = \ell_5 = 3; \ell_6 = 75, 77, 81, 83, 84, 85, 89, 90, 102$	
ALL	$x_{ij_1j_2}$	$x_{ij_3j_4}$	$x_{ij_1j_2j_3j_4}$	$G(\ell_1\ell_2\ell_3\ell_4\ell_5)$	$u^{(j_k)}_{\ell_k,j_k}, k \le 4, u^{(i)}_{\ell_5,i} \in \mathbb{R}^{N_k \times N_k}$
	$\in \mathbb{R}^{N_5 \times N_1 \times N_2}$	$\in \mathbb{R}^{N_5 \times N_3 \times N_4}$	$\in \mathbb{R}^{N_5 \times N_1 \times N_2 \times N_3 \times N_4}$		$(N_1, N_2, N_3, N_4, N_5) = (77, 4, 4, 74, 2597)$
Selected				$\in \mathbb{R}^{N_1 \times N_2 \times N_3 \times N_4 \times N_5}$ $\ell_1 = 2, 3, 5, 6, 9, 10; \ell_2 = 3; \ell_3 = 4; \ell_5 = 1, 2, 3, 5$	
Diabetes	$x_{ij_1j_2}$	x_{ij_3}	$x_{ij_1j_2j_3}$	$G(\ell_1\ell_2\ell_3\ell_4)$	$u^{(j_k)}_{\ell_k,j_k}, k \le 3, u^{(i)}_{\ell_4,i} \in \mathbb{R}^{N_k \times N_k}$
	$\in \mathbb{R}^{N_4 \times N_1 \times N_2}$	$\in \mathbb{R}^{N_4 \times N_3}$	$\in \mathbb{R}^{N_4 \times N_1 \times N_2 \times N_3}$	$\in \mathbb{R}^{N_1 \times N_2 \times N_3 \times N_4}$	$(N_1, N_2, N_3, N_4) = (253, 4, 69, 3489)$
Selected				$\ell_1 = 2; \ell_2 = 2; \ell_3 = 1, 4; \ell_4 = 1, 4$	
Renal	$x_{ij_1j_2}$	x_{ij_3}	$x_{ij_1j_2j_3}$	$G(\ell_1\ell_2\ell_3\ell_4)$	$u^{(j_k)}_{\ell_k,j_k}, k \le 3, u^{(i)}_{\ell_4,i} \in \mathbb{R}^{N_k \times N_k}$
Carcinoma	$\in \mathbb{R}^{N_4 \times N_1 \times N_2}$	$\in \mathbb{R}^{N_4 \times N_3}$	$\in \mathbb{R}^{N_4 \times N_1 \times N_2 \times N_3}$	$\in \mathbb{R}^{N_1 \times N_2 \times N_3 \times N_4}$	$(N_1, N_2, N_3, N_4) = (253, 4, 202, 4036)$
Selected			$\ell_1 = 2; \ell_2 = 2; \ell_3 = 13, 15, 30, 33, 35; \ell_4 = 186, 215, 233, 244, 251, 269, 274, 309, 312, 318$		
Cirrhosis	$x_{ij_1j_2}$	x_{ij_3}	$x_{ij_1j_2j_3}$	$G(\ell_1\ell_2\ell_3\ell_4)$	$u^{(j_k)}_{\ell_k,j_k}, k \le 3, u^{(i)}_{\ell_4,i} \in \mathbb{R}^{N_k \times N_k}$
	$\in \mathbb{R}^{N_4 \times N_1 \times N_2}$	$\in \mathbb{R}^{N_4 \times N_3}$	$\in \mathbb{R}^{N_4 \times N_1 \times N_2 \times N_3}$	$\in \mathbb{R}^{N_1 \times N_2 \times N_3 \times N_4}$	$(N_1, N_2, N_3, N_4) = (355, 4, 216, 3961)$
Selected				$\ell_1 = 2; \ell_2 = 2; \ell_3 = 2, 6; 2 \le \ell_4 \le 10$	

Table 7.28 The number of genes, drugs, and target proteins identified by TD-based unsupervised FE

Disease	Inferred genes	Inferred compounds	Predicted target	
			Up	Down
Heart failure	274	43	556	449
PTSD	374	6	578	548
ALL	24	2	91	57
Diabetes	65	14	186	140
Renal carcinoma	225	14	229	177
Cirrhosis	132	27	510	488

7.9 Integrated Analysis of miRNA Expression and Methylation

The unsupervised method is often useful when applied to something for which no pre-knowledge is available. For example, two kinds of omics data might be correlated with unknown reasons. To search for this kind of hidden (latent) relationship, the unsupervised method is critically useful. In this section, we propose the application of case I type II tensor to investigate the relationship between miRNA expression and methylation, between which no direct relationships are biologically expected.

Promoter methylation of genes targeted by miRNAs can of course affect the expression of these genes. Nevertheless, there seem to be no biological reasons that promoter methylation of genes targeted by miRNAs affects the expression of these miRNAs themselves and vice versa. Thus, if we can find any correlation between these two, it might be a starting point for finding a new biological point of view.

In this section, we make use of the cancer genome atlas (TCGA) data set [111]. The data set we analyze is composed of eight normal ovarian tissue samples and 569 tumor samples. Our data set includes expression data on 723 miRNAs as well as promoter methylation profiles of 24,906 genes. They are formatted as matrices

$$x_{ij}^{\text{methyl}} \in \mathbb{R}^{24,906\times 577} \tag{7.48}$$

$$x_{kj}^{\text{miRNA}} \in \mathbb{R}^{723\times 577} \tag{7.49}$$

They are converted to case I tensor because they share samples as

$$x_{ijk} = x_{kj}^{\text{miRNA}} x_{ij}^{\text{methyl}} \tag{7.50}$$

Usually, HOSVD, Fig. 3.8, is supposed to be applied to x_{ijk} as

Table 7.29 Fisher's exact test (P_F) and the uncorrected χ^2 test (P_{χ^2}) of known drug target proteins regarding the inference of the present study. Rows: known drug target proteins (DINIES). Columns: Inferred drug target proteins using "single-gene perturbations from GEO up" or "single-gene perturbations from GEO down." OR: odds ratio

		Single-gene perturbations from GEO up					Single-gene perturbations from GEO down				
		F	T	P_F	P_{χ^2}	RO	F	T	P_F	P_{χ^2}	RO
Heart	F	521	517	3.4×10^{-4}	3.9×10^{-4}	3.02	628	416	1.3×10^{-3}	7.3×10^{-4}	2.61
Failure	T	13	39				19	33			
PTSD	F	500	560	3.8×10^{-2}	3.1×10^{-2}	2.67	532	529	6.1×10^{-3}	4.5×10^{-3}	3.81
	T	6	18				5	19			
All	F	979	89	2.7×10^{-1}	3.0×10^{-1}	2.19	1009	57	1.0×10^{0}	–	–
	T	10	2				12	0			
Diabetes	F	889	177	1.2×10^{-2}	7.1×10^{-3}	3.00	936	130	3.6×10^{-4}	2.0×10^{-5}	5.13
	T	15	9				14	10			
Renal	F	847	219	2.0×10^{-2}	1.2×10^{-2}	2.75	895	169	4.3×10^{-2}	2.2×10^{-2}	2.64
Carcinoma	T	14	10				16	8			
Cirrhosis	F	572	490	1.1×10^{-2}	8.1×10^{-3}	2.91	595	467	1.6×10^{-3}	1.1×10^{-3}	3.81
	T	8	20				7	21			

$$x_{ijk} = \sum_{\ell_1=1}^{24,906} \sum_{\ell_2=1}^{577} \sum_{\ell_3=1}^{723} G(\ell_1, \ell_2, \ell_3) u_{\ell_1 i}^{(i)} u_{\ell_2 j}^{(j)} u_{\ell_3 k}^{(k)}. \tag{7.51}$$

Unfortunately, x_{ijk} is too huge to apply HOSVD directly. Thus, instead, we derive type II tensor as

$$x_{ik} = \sum_{j=1}^{577} x_{ijk}. \tag{7.52}$$

Now it is a matrix. Thus, we can apply PCA to it. Then we can have PC score $\boldsymbol{u}_\ell \in \mathbb{R}^{723}$ attributed to miRNA and PC loading $\boldsymbol{v}_\ell \in \mathbb{R}^{24,906}$ attributed to methylation. The singular value vectors attributed to sample j is computed in two ways as Eq. (5.15)

$$u_{\ell j}^{(j;k)} = \sum_k u_{\ell k} x_{kj}^{\text{miRNA}}, \tag{7.53}$$

$$u_{\ell j}^{(j;i)} = \sum_i v_{\ell i} x_{ij}^{\text{methyl}}. \tag{7.54}$$

The first things to check is if there are any ℓs such that $\boldsymbol{u}_\ell^{(j;k)} \in \mathbb{R}^{577}$ and $\boldsymbol{u}_\ell^{(j;i)} \in \mathbb{R}^{577}$ satisfy the following requirements simultaneously:

- $\boldsymbol{u}_\ell^{(j;i)}$ and $\boldsymbol{u}_\ell^{(j;k)}$ are significantly correlated.
- $\boldsymbol{u}_\ell^{(j;k)}$ is expressed distinctly between healthy controls ($j \leq 8$)and patients ($j > 8$).
- $\boldsymbol{u}_\ell^{(j;i)}$ is expressed distinctly between healthy controls ($j \leq 8$) and patients ($j > 8$).

In order to validate these requirements visually, we show a scatter plot for $1 \leq \ell \leq 9$ (Fig. 7.18). More or less all nine scatter plots look like satisfying the above requirements simultaneously. In order to select $\boldsymbol{u}_\ell$ and $\boldsymbol{v}_\ell$ used for miRNA and gene selection, respectively, we need to identify which ℓ satisfies the above requirements best. Thus, we propose several measures. First, we select miRNAs and genes. P values are attributed as

$$P_k = P_{\chi^2}\left[> \left(\frac{u_{\ell k}}{\sigma_\ell}\right)^2\right], \tag{7.55}$$

$$P_i = P_{\chi^2}\left[> \left(\frac{v_{\ell i}}{\sigma'_\ell}\right)^2\right]. \tag{7.56}$$

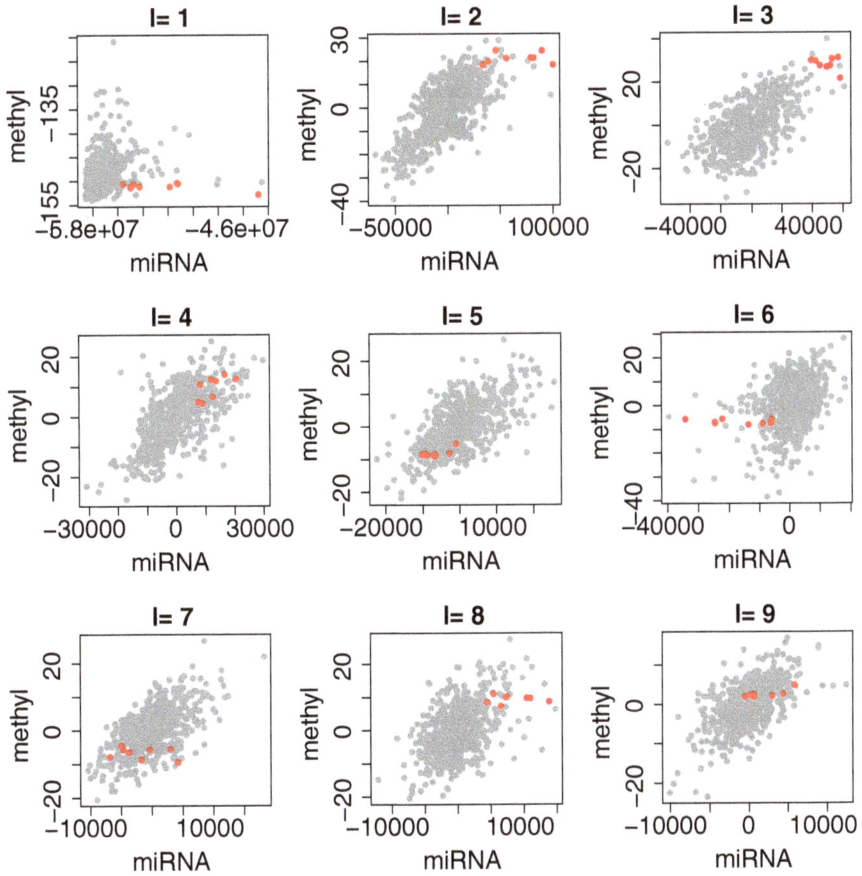

Fig. 7.18 Scatter plots of $\boldsymbol{u}_\ell^{(j;k)}$ (horizontal) and $\boldsymbol{u}_\ell^{(j;i)}$ (vertical) for $1 \le \ell \le 9$. Red-filled circle:eight normal controls ($j \le 8$), gray-filled circles:ovarian cancer patients ($j > 8$)

P values are adjusted by BH criterion, and i and k associated with adjusted P values less than 0.01 are selected. Then we require genes and miRNA selected similar to the above requirements as

- Selected genes and miRNAs are significantly correlated.
- Selected miRNAs are expressed distinctly between normal controls ($j \le 8$) and patients ($j > 8$).
- Selected genes are methylated distinctly between normal controls ($j \le 8$) and patients ($j > 8$).

Thus, we compute the following:

(a) Correlation coefficient between $\boldsymbol{u}_\ell^{(j;i)}$ and $\boldsymbol{u}_\ell^{(j;k)}$
(b) P value attributed to the above correlation coefficients

(c) P values computed by t-test that evaluates if $\boldsymbol{u}_\ell^{(j;k)}$ is distinct between normal control ($j \leq 8$) and patients ($j > 8$)
(d) P values computed by t-test that evaluates if $\boldsymbol{u}_\ell^{(j;i)}$ is distinct between normal control ($j \leq 8$) and patients ($j > 8$)
(e) Ratio of significantly correlated pairs of genes and miRNAs selected
(f) Ratio of miRNA associated with adjusted P values computed by t-test that evaluate if selected miRNAs are expressed distinctly between normal control ($j \leq 8$) and patients ($j > 8$)
(g) Ratio of genes associated with adjusted P values computed by t-test that evaluate if selected genes are methylated distinctly between normal control ($j \leq 8$) and patients ($j > 8$)
(h) The number of selected miRNAs
(i) The number of selected genes

Here, significant correlation is evaluated if associated BH criterion adjusted P values are less than 0.01 (See page 113 for how to compute P values attributed to correlation coefficients). Table 7.30 shows the result. $\ell = 3$ seems to be the best, because $\ell = 3$ is the best for the sixth and the seventh measures and the second best in the fifth measure; the fifth, sixth, and seventh measures are important because they are direct evaluations of selected genes and miRNAs. Because the number of selected genes and miRNAs do not vary dependent on ℓ so much, it is the best to select $\ell = 3$. Because more than 88% of genes and miRNAs and their pairs satisfy the desired requirements in the above (88% is the smallest ratio (percentage) among requirements from (e) to (g) in Table 7.30), TD-based unsupervised FE can be considered to have the ability to select miRNAs and genes satisfying desired requirements mentioned in the above.

In order to see if other supervised methods can identify set of genes and miRNAs satisfying desired requirements, i.e., selected genes are methylated distinctly between healthy control and patients, miRNAs selected are expressed distinctly between healthy controls and patients, selected genes and miRNAs are significantly correlated; we apply t-test to select genes methylated distinctly between healthy controls and patients and miRNA expressed distinctly between healthy controls and

Table 7.30 Measures that evaluate which ℓ satisfies the desired requirements best. The number in the first row corresponds to the alphabetical list in the main text

ℓ	(a)	(b)	(c)	(d)	(e)	(f)	(g)	(h)	(i)
1	0.187	6.35×10^{-6}	6.25×10^{-3}	4.42×10^{-7}	–	**1.000**	–	2	0
2	**0.718**	$\mathbf{1.95 \times 10^{-92}}$	1.28×10^{-4}	1.21×10^{-11}	**0.944**	0.571	0.834	7	241
3	0.628	1.49×10^{-64}	$\mathbf{3.06 \times 10^{-8}}$	5.55×10^{-10}	0.884	**1.000**	**0.905**	7	284
4	0.649	2.45×10^{-70}	6.15×10^{-5}	1.02×10^{-4}	0.539	0.714	0.597	7	273
6	0.348	6.76×10^{-18}	1.68×10^{-3}	$\mathbf{5.71 \times 10^{-17}}$	0.350	0.375	0.674	8	132
7	0.624	1.27×10^{-63}	2.00×10^{-1}	7.65×10^{-7}	0.365	0.400	0.758	5	293
8	0.500	8.60×10^{-38}	1.33×10^{-4}	5.89×10^{-13}	0.274	0.833	0.775	6	231
9	0.593	3.50×10^{-56}	6.44×10^{-2}	3.35×10^{-5}	0.182	0.667	0.681	3	251

patients. P values are attributed to miRNAs and genes and adjusted by BH criterion. Then, 214 miRNAs and 19,395 genes associated with adjusted P values less than 0.01 are selected. In order to see how much ratio of significantly correlated pairs among total $241 \times 19{,}395 = 4{,}829{,}355$ pairs is, we compute correlation coefficients between them and attribute P values to these pairs (See page 113 for how to compute P values attributed to correlation coefficients). P values are corrected by BH criterion, and 555,391 pairs are associated with adjusted P values less than 0.01. Because this is as small as 11.5% of 4,829,355 pairs, t-test is inferior to TD-based unsupervised FE to identify genes and miRNAs satisfying desired requirements.

This poor performance might be because of the too many genes and miRNAs selected. P values given by t-test have a strong tendency to reduce its value when many samples are available. In this example, because as many as 575 samples are available, even gene and miRNAs associated with small distinction are associated with small enough P values. In order to avoid this difficulty, we reduce the number of genes and miRNAs selected by t-test as many as those by TD-based unsupervised FE, by selecting top-ranked 7 miRNA and 284 methylation probes attributed to genes based upon P values computed by t-test. Then among $7 \times 284 = 1967$ pairs, as small as 50 pairs are associated with adjusted P values less than 0.01 attributed to correlation coefficient. Thus, only 2.5% of 1967 pairs are significantly correlated. Thus, the ratio decreases instead of increasing as opposed to the expectation.

It might be possible to select genes and miRNAs starting by identifying significantly correlated pairs before finding genes and miRNAs distinct between healthy control and patients. Then correlation coefficients are computed among all pairs of genes and miRNAs. P values are attributed to correlation coefficient (See page 113 for how to compute P values attributed to correlation coefficients) and are corrected by BH criterion. Then among $24{,}906 \times 723 = 18{,}007{,}038$ pairs, 1,197,772 pairs are associated with adjusted P values less than 0.01. Unfortunately, these pairs include all genes and miRNAs. Thus, starting from pairs significantly correlated is not an effective strategy. These poor performance achieved by t-test as well as correlation analysis demonstrate the difficulty of identifying gene and miRNAs satisfying the desired requirement, i.e., selected genes are methylated distinctly between healthy control and patients, miRNAs selected are expressed distinctly between healthy controls and patients, and selected genes and miRNAs are significantly correlated, which is easily achieved by TD-based unsupervised FE.

Before closing this section, genes and miRNA selected should be biologically evaluated too. First, 240 gene symbols associated with 284 probes are uploaded to DAVID (Table 7.31). At a glance, although it does not look deeply related to cancers, a detailed investigation can alter this impression. This data is about ovarian cancer. The most major subtype is surface epithelial-stromal tumor which is known to be associated with keratinization [65]. Thus, the detection of keratinization as the most enriched term is reasonable while the third enriched one is also related to keratinization. Because the fifth one, epidermis development, is the parent term of keratinization, it is also understandable.

Table 7.31 GO BP enrichment by the 274 gene symbols identified by TD-based unsupervised FE for ovarian cancer data from TCGA Adjusted *P* values are by BH criterion

Category	Term	Gene count	%	*P* value	Adjusted *P* value
GOTERM_BP_DIRECT	Keratinization	14	6.2	9.3E–15	1.1E–11
GOTERM_BP_DIRECT	Peptide cross-linking	14	6.2	1.7E–14	9.6E–12
GOTERM_BP_DIRECT	Keratinocyte differentiation	15	6.6	2.8E–13	1.1E–10
GOTERM_BP_DIRECT	Acute-phase response	7	3.1	6.4E–6	1.8E–3
GOTERM_BP_DIRECT	Epidermis development	9	4.0	8.0E–6	1.8E–3

Next, the selected seven miRNAs are uploaded to DIANA-mirpath for evaluation (Fig. 7.19). It is obvious that they are enriched with various cancers. Thus, the selected seven miRNAs are supposed to be related to cancers.

In conclusion, TD-based unsupervised FE successfully identifies reasonable genes and miRNAs also from the biological point of view.

7.10 Integrated Analysis of mRNA and miRNA II

In this subsection, we investigated yet another integration of mRNA and miRNA [69]. The purpose of this study is to integrate mRNA expression profiles and miRNA expression profiles that share samples based upon case I type II strategy (Table 5.3).

There are two data sets analyzed in this subsection.

- Data Set Sect. 7.10-1:This data set includes 324 samples composed of 253 kidney tumors and 71 normal kidney, respectively. 19,536 mRNA expression and 825 miRNA expression are measured. In the following subsection, we denote this data set as TCGA data set since we retrieved this data set from TCGA.
- Data Set Sect. 7.10-2: This data set includes 34 samples composed of 17 healthy controls and 17 patients (for the same type of kidney tumors), respectively. It measures 33,698 mRNA expression and 319 miRNA expression profiles. In the following subsection, we denote this data set as GEO data set since we retrieved this data set from GEO.

Roughly speaking, since the number of samples in the GEO data set is as small as one tenth of that of the TCGA data set, the GEO data set is regarded to be a kind of validation set. Both data sets are formatted as matrices as $x_{ij} \in \mathbb{R}^{N \times M}$ and $x_{kj} \in \mathbb{R}^{K \times M}$ that represent the expression of the ith mRNA or the kth miRNA of jth sample, respectively.

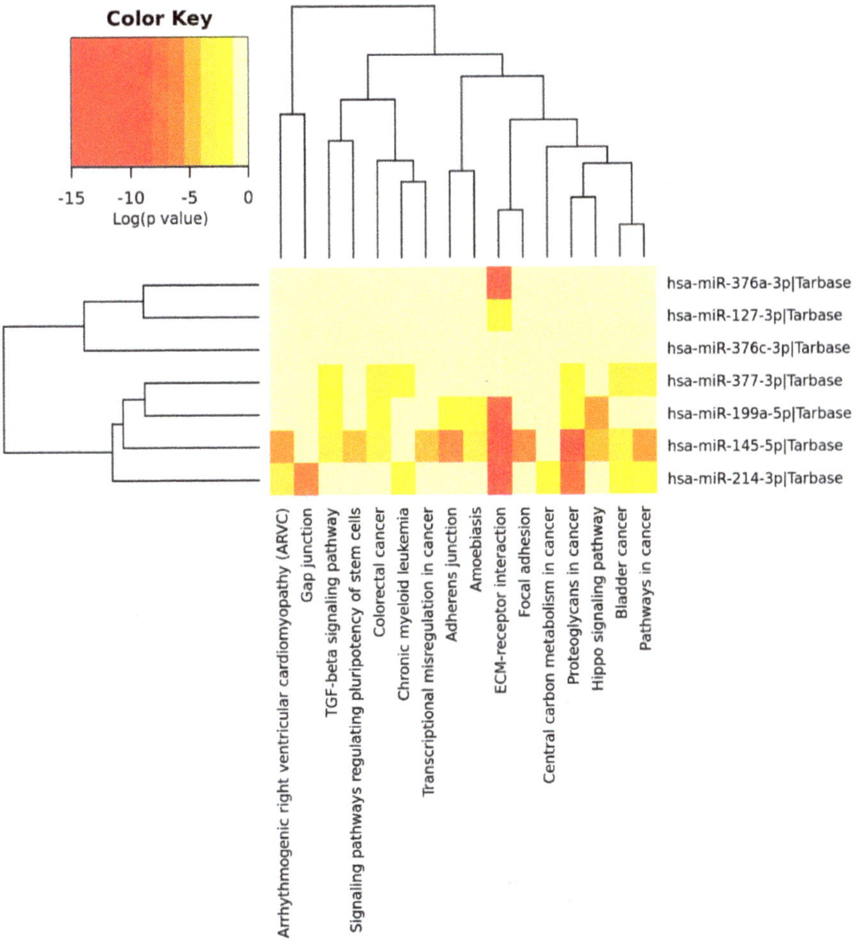

Fig. 7.19 Heatmap that summarizes the results of DIANA-mirpath for the selected seven miRNAs, specifying the "pathways union" option

Following the strategy of case I type II tensor, we generate

$$x_{ik} = \sum_{j=1}^{M} x_{ij} x_{kj} \in \mathbb{R}^{M \times K} \tag{7.57}$$

to which we apply SVD and get

$$x_{ik} = \sum_{\ell=1}^{L} \lambda_\ell u_{\ell i}^{(i)} u_{\ell k}^{(k)}. \tag{7.58}$$

Then we can recover singular value vectors attributed to samples as

$$v_{\ell j}^{(j;i)} = \sum_{i=1}^{N} x_{ij} u_{\ell i} \tag{7.59}$$

$$v_{\ell j}^{(j;k)} = \sum_{k=1}^{K} x_{kj} u_{\ell k}. \tag{7.60}$$

The first (mathematical) evaluation of the outcome is the coincidence between $v_{\ell j}^{(j;i)}$ and $v_{\ell j}^{(j;k)}$. Since $v_{\ell j}^{(j;i)}$ and $v_{\ell j}^{(j;k)}$ can be regarded to be sample dependence based upon mRNA expression and miRNA expression, respectively, if they are not coincident with each other, the analyses cannot be regarded as successful.

We have sought the pair of $v_{\ell j}^{(j;i)}$ and $v_{\ell j}^{(j;k)}$ that have the largest absolute Pearson correlation coefficient. For TCGA data set (Fig. 7.20), $\ell = 2$ has the largest absolute correlation coefficient (0.9049376, $P = 1.63 \times 10^{-121}$). For GEO data set (Fig. 7.20), although $\ell = 2$ has only the fifth largest absolute Pearson correlation coefficient (0.9306852, $P = 1.58 \times 10^{-15}$), since the value does not substantially smaller than the maximum value (0.9690192) and ℓs that have the first to the fourth largest absolute Pearson correlation coefficients are larger than 20 which unlikely have large enough contribution to x_{ik}, we also decided to employ $\ell = 2$ for GEO data set. It is obvious that $v_{2j}^{(j;i)}$ and $v_{2j}^{(j;k)}$ are highly correlated for TCGA and GEO data sets.

The second evaluation is if $v_{\ell j}^{(j;i)}$ and $v_{\ell j}^{(j;k)}$ are distinct between normal tissues and tumors. We applied t-test and found that they are highly distinct between normal

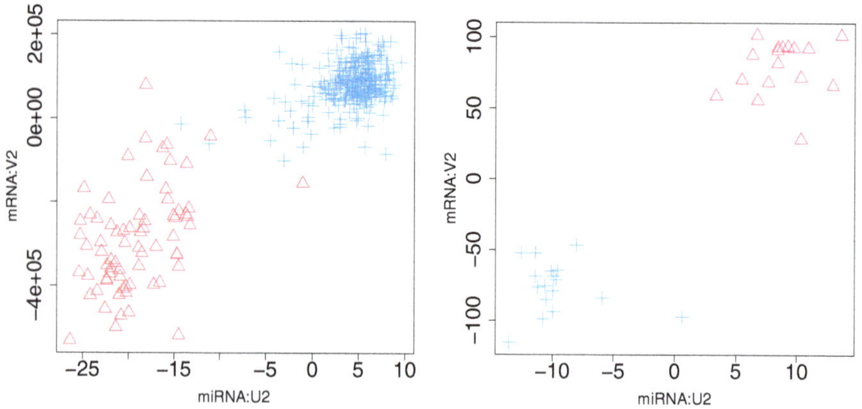

Fig. 7.20 Scatter plot of $v_{2j}^{(j;i)}$ (vertical axis) and $v_{2j}^{(j;k)}$ (horizontal axis). Red open triangles (blue crosses) correspond to normal (tumor) tissue. Pearson correlation coefficients are 0.9049376 ($P = 1.63 \times 10^{-121}$) and 0.9306852 ($P = 1.58 \times 10^{-15}$) for the left panel (TCGA data set) and the right panel (GEO data set), respectively

Table 7.32 P values obtained by applying t-test to $v_{\ell j}^{(j;i)}$ and $v_{\ell j}^{(j;k)}$, respectively, to detect distinction between tumors and healthy controls

	TCGA	GEO
mRNA	8.438412×10^{-39}	6.743853×10^{-22}
miRNA	3.625502×10^{-63}	2.544428×10^{-18}

Table 7.33 Confusion matrix between genes selected in TCGA and GEO data sets. P value computed by Fisher's exact test is 8.971×10^{-11}

		GEO	
		Not selected	Selected
TCGA	Not selected	17,209	160
	Selected	60	11

tissues and tumors (Table 7.32). It is again obvious that they are highly distinct between two classes (tumors and normal tissues).

Now we come to the stage of biological validation, although I am not willing to discuss outcomes biologically in an extensive manner since the majority of readers of this book are unlikely to be interested in biology. At first, as usual, we select genes using the following procedure. We attribute P values to i and k as

$$P_i = P_{\chi^2}\left[> \left(\frac{u_{2i}^{(i)}}{\sigma_2}\right)^2\right] \tag{7.61}$$

$$P_k = P_{\chi^2}\left[> \left(\frac{u_{2k}^{(k)}}{\sigma_2'}\right)^2\right]. \tag{7.62}$$

P_is and P_ks are corrected by BH criterion and mRNAs and miRNAs associated with adjusted P_i and P_k less than 0.01 are selected. Then 72 and 209 mRNAs are selected for TCGA and GEO data sets, respectively, and 11 and 3 miRNAs are selected for TCGA and GEO data sets, respectively. Next, we checked how many mRNAs and miRNAs are common between GEO and TCGA data sets. All three miRNAs selected in GEO data sets are in 11 miRNAs selected in the TCGA data set. On the other hand, among 72 mRNAs selected in TCGA data sets and 209 mRNAs selected in GEO data sets, there are 11 mRNAs that share gene symbols between TCGA and GEO data sets (Table 7.33). The reason why the total number of selected mRNAs and miRNAs changes is because we consider only mRNAs and miRNAs included in TCGA and GEO data sets to construct a confusion matrix. The P value determined using the Fisher's exact test was 8.97×10^{-11}, and the odds ratio was 19.7. Therefore, the coincidence of selected mRNAs and miRNAs by the proposed strategy between TCGA and GEO data sets is very well. Since these two data sets are independent, this overlap unlikely to occur accidentally.

Table 7.34 Top ten enriched terms in "Human Phenotype Ontology" category

Term	Overlap	*P* value	Adjusted *P* value
Polyuria (HP:0000103)	3/21	1.64×10^{-7}	7.92×10^{-6}
Abnormal urine output (HP:0012590)	3/25	2.83×10^{-7}	7.92×10^{-6}
Abnormality of renal excretion (HP:0011036)	3/29	4.49×10^{-7}	8.38×10^{-6}
Increased circulating renin level (HP:0000848)	2/8	7.69×10^{-6}	1.08×10^{-4}
Hypokalemic alkalosis (HP:0001949)	2/11	1.51×10^{-5}	1.40×10^{-4}
Tetany (HP:0001281)	2/11	1.51×10^{-5}	1.40×10^{-4}
Hypomagnesemia (HP:0002917)	2/12	1.81×10^{-5}	1.40×10^{-4}
Polydipsia (HP:0001959)	2/14	2.49×10^{-5}	1.40×10^{-4}
Abnormal drinking behavior (HP:0030082)	2/14	2.49×10^{-5}	1.40×10^{-4}
Abnormality of magnesium homeostasis (HP:0004921)	2/14	2.49×10^{-5}	1.40×10^{-4}

To confirm that this overlap is not accidental, we uploaded commonly selected 11 gene symbols (SLC12A3, NNMT, C7, SLC12A1, NDUFA4L2, ANGPTL4, UMOD, VEGFA, AIF1L, AQP2, and KNG1) to Enrihr [49] to evaluate their biological significance. Although we are not willing to discuss their enriched terms comprehensively, there are some remarkable points. For example, "Human Phenotype Ontology" category reports many reasonable enrichment (Table 7.34). The top three are kidney diseases. The fourth one reports the increased level of renin that is a protein generated in the kidney. The fifth one is hypokalemic alkalosis that can be caused also because of the treatment of diuretics. The seventh one, hypomagnesemia, and tenth one, abnormality of magnesium homeostasis, are similar to each other and are known to be related to kidney disease. The eighth one, polydipsia, and the ninth one, abnormal drinking behavior, mean too much drinking that is related to kidney because kidney controls blood water concentration. The common 11 genes are not likely to occur by chance, but rather because of some biological reasons.

In the following, we are trying to show that other methods cannot be superior to TD-based unsupervised FE when they are applied to the present data sets. At first, we consider PCA-based unsupervised FE that has worked very successfully toward various data sets in the previous chapter when PCA-based unsupervised FE is applied to profiles that share samples. Before applying PCA, the data sets were normalized as

$$\sum_{i=1}^{N} x_{ij} = \sum_{k=1}^{K} x_{kj} = 0 \tag{7.63}$$

$$\sum_{i=1}^{N} x_{ij}^2 = N \tag{7.64}$$

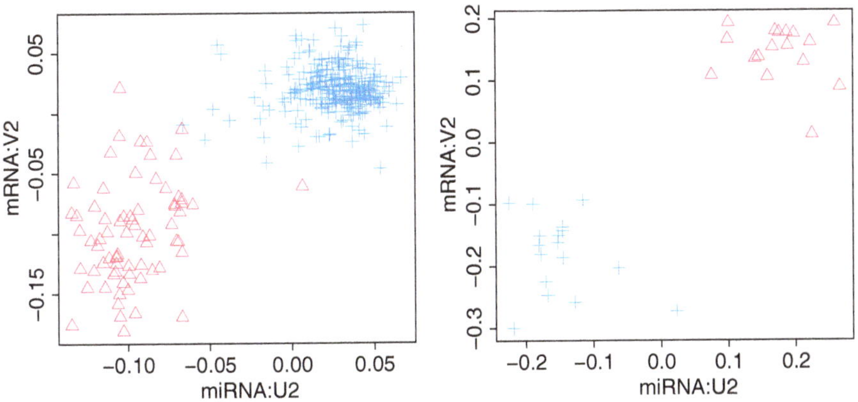

Fig. 7.21 Scatter plot of $\hat{v}_{2j}^{(j;i)}$ (vertical axis) and $\hat{v}_{2j}^{(j;k)}$ (horizontal axis). Red open triangles (blue crosses) correspond to normal (tumor) tissue. Pearson correlation coefficients are 0.8394552 ($P = 2.75 \times 10^{-87}$) and 0.8764739 ($P = 1.10 \times 10^{-11}$) for the left panel (TCGA data set) and the right panel (GEO data set), respectively (PCA-based unsupervised FE)

$$\sum_{k=1}^{K} x_{kj}^2 = K \tag{7.65}$$

Since PCA is equivalent to SVD when mean is extracted, instead of applying PCA, we apply SVD to x_{ij} and x_{kj} and get

$$x_{ij} = \sum_{\ell=1}^{L} \lambda_\ell \hat{u}_{\ell i}^{(i)} \hat{v}_{\ell j}^{(j;i)} \tag{7.66}$$

$$x_{kj} = \sum_{\ell=1}^{L} \lambda'_\ell \hat{u}_{\ell k}^{(k)} \hat{v}_{\ell j}^{(j;k)} \tag{7.67}$$

To distinguish those from singular value vectors obtained by TD above, we add circumflex (*hat*) on them. As in the case of TD, we first try to find which pairs of $\hat{v}_{\ell j}^{(j;i)}$ and $\hat{v}_{\ell j}^{(j;k)}$ have the largest absolute Pearson correlation coefficient. For TCGA data set and GEO data set (Fig. 7.21), $\ell = 2$ has the largest absolute correlation coefficient (0.8394552 for TCGA $P = 2.75 \times 10^{-87}$ and 0.8764739 $P = 1.10 \times 10^{-11}$). It is obvious that $\hat{v}_{2j}^{(j;i)}$ and $\hat{v}_{2j}^{(j;k)}$ are highly correlated for TCGA and GEO data sets, although correlation coefficients are a bit smaller than those obtained by TD.

The second evaluation is if $\hat{v}_{2j}^{(j;i)}$ and $\hat{v}_{2j}^{(j;k)}$ are distinct between normal tissues and tumors. We applied t-test and found that they are highly distinct between normal tissues and tumors (Table 7.35). It is again obvious that they are highly distinct between two classes (tumors and normal tissues).

Table 7.35 P values obtained by applying t-test to $\hat{v}_{2j}^{(j;i)}$ and $\hat{v}_{2j}^{(j;k)}$, respectively, to detect distinction between tumors and healthy controls (PCA-based unsupervised FE)

	TCGA	GEO
mRNA	2.397644×10^{-77}	2.040212×10^{-17}
miRNA	2.334593×10^{-36}	7.317031×10^{-17}

Table 7.36 Confusion matrix between genes selected in TCGA and GEO data sets. P value computed by Fisher's exact test is 3.84×10^{-6} (PCA-based unsupervised FE)

		GEO	
		Not selected	Selected
TCGA	Not selected	17,271	100
	Selected	63	6

Then is TD useless and is PCA enough? To see this point, we check the overlap of selected mRNAs and miRNAs between GEO and TCGA as has been done above. 69 and 106 mRNAs are selected for TCGA and GEO data sets, respectively, and ten and three miRNAs are selected for TCGA and GEO data sets, respectively. Next, we checked how many mRNAs and miRNAs are common between GEO and TCGA data sets. All three miRNAs selected in GEO data sets are in 11 miRNAs selected in the TCGA data set. On the other hand, among 70 mRNAs selected in the TCGA data set and 131 mRNAs selected in GEO data set, there are six mRNAs that share gene symbols between TCGA and GEO data sets (Table 7.36). The reason why the total number of selected mRNAs and miRNAs changes is we consider only mRNAs and miRNAs included in TCGA and GEO data sets to construct a confusion matrix. The P value determined using the Fisher's exact test was 3.842×10^{-6} and the odds ratio was 16.4. Thus, although overlap is still highly significant, since the total number of common genes is halved, it is inferior to TD (Table 7.33).

To confirm the inferiority of PCA toward TD, we uploaded six common genes (NNMT, SLC12A1, ANGPTL4, UMOD, AQP2, and KNG1) to Enrichr. Table 7.37 shows the enrichment of "Human Phenotype Ontology" category. If compared with Table 7.34, not only the significance decreases but also enriched terms are less related to kidney. Vomiting, constipation, and gout are unlikely directly related to kidney. Thus, PCA is inferior to TD from the biological point of view too.

Finally, we compared the performance of TD-based unsupervised FE with some state-of-the-art (SOTA) methods: t-test, SAM [113], and limma [79]. When we applied these SOTA methods to TCGA and GEO data sets and identified genes and miRNAs associated with adjusted P values less than 0.01, t-test identified 13,895 and 12,152 genes for TCGA and GEO, respectively. t-test also identified 339 and 74 miRNAs for TCGA and GEO, respectively. On the other hand, SAM identified 14,485 and 16,336 genes for TCGA and GEO, respectively, whereas SAM identified 441 and 108 miRNAs for TCGA and GEO, respectively. Finally, limma identified 18,225 and 28,524 genes for TCGA and GEO, respectively, and 663 and 319 miRNAs for TCGA and GEO, respectively. Thus, generally, these SOTA

Table 7.37 Top ten enriched terms in "Human Phenotype Ontology" category (PCA-based unsupervised FE)

Term	Overlap	*P* value	Adjusted *P* value
Polyuria (HP:0000103)	2/21	1.57×10^{-5}	5.16×10^{-4}
Abnormal urine output (HP:0012590)	2/25	2.24×10^{-5}	5.16×10^{-4}
Abnormality of renal excretion (HP:0011036)	2/29	3.03×10^{-5}	5.16×10^{-4}
Dehydration (HP:0001944)	2/56	1.15×10^{-4}	1.46×10^{-3}
Vomiting (HP:0002013)	2/108	4.27×10^{-4}	4.36×10^{-3}
Constipation (HP:0002019)	2/170	1.05×10^{-3}	8.95×10^{-3}
Hyperactive renin–angiotensin system (HP:0000841)	1/8	2.40×10^{-3}	1.11×10^{-2}
Increased circulating renin level (HP:0000848)	1/8	2.40×10^{-3}	1.11×10^{-2}
Gout (HP:0001997)	1/8	2.40×10^{-3}	1.11×10^{-2}
Metabolic alkalosis (HP:0200114)	1/10	3.00×10^{-3}	1.11×10^{-2}

Table 7.38 Confusion matrix between genes selected in TCGA and GEO data sets. (*t*-test and limma)

		GEO			
		t-test		limma	
		Not selected	Selected	Not selected	Selected
TCGA	Not selected	17,181	189	17,183	187
	Selected	70	0	70	0

methods failed to select limited number of genes and miRNAs although TD-based unsupervised FE could. Although we can conclude that SOTA methods are inferior to TD-based unsupervised FE only by this result, we further tried to confirm the superiority of TD-based unsupervised FE toward the SOTA methods by reducing the selected genes and miRNAs as many as TD-based unsupervised FE selected and counting the number of overlaps between TCGA and GEO. At first, we realized that it was not always possible to reduce the number of selection, since more number of genes (for SAM) and miRNAs (for SAM and limma) have $P = 0$ than those selected by TD-based unsupervised FE. Although we could count the overlap of genes and miRNAs selected by *t*-test between TCGA and GEO data sets, there are no genes selected between TCGA and GEO data sets (Table 7.38), and there is only one miRNA selected between TCGA and GEO data sets. On the other hand, these selected by limma have no commonly selected genes between TCGA and GEO data sets (Table 7.38). Thus, at least from the point of the ability that we can select common genes and/or miRNAs between two distinct data sets, SOTA methods are inferior to TD-based unsupervised FE.

In conclusion, not only TD-based unsupervised FE successfully identified genes and miRNAs associated with cancer, but it also outperformed SOTA methods.

7.11 Integrated Analysis of Multiple Profiles

7.11.1 The Effect of Vaccination by Integrating Multiple Profiles

In this section, we consider the effect of vaccination by integrating multiple profiles, one methylation profile, one mRNA expression profile, and two proteome profiles [109] that share the samples as the procedure described in Sect. 5.7.1, although we aim to integrate not matrices but tensors. Four profiles are formatted as tensors as

$$x_{i_k j_1 j_2} \in \mathbb{R}^{N_k \times 5 \times 15} \tag{7.68}$$

that represents the amount of i_kth feature at j_1th time points of j_2th individual of kth profiles ($1 \leq k \leq 4$). They are standardized as

$$\sum_{i_k=1}^{N_k} x_{i_k j_1 j_2} = 0 \tag{7.69}$$

$$\sum_{i_k=1}^{N_k} x_{i_k j_1 j_2}^2 = N_k. \tag{7.70}$$

To integrate four profiles represented as four tensors associated with distinct number of features, i_k, we get product sum of tensors as

$$x_{k j_1 j_2 j_1' j_2'} = \sum_{i_k=1}^{N} x_{i_k j_1 j_2 k} x_{i_k j_1' j_2' k} \in \mathbb{R}^{4 \times 5 \times 15 \times 5 \times 15}. \tag{7.71}$$

They are further normalized as

$$\sum_{j_1=1}^{5} \sum_{j_2=1}^{15} \sum_{j_1'=1}^{5} \sum_{j_2'=1}^{15} x_{k j_1 j_2 j_1' j_2'} = 5 \times 15 \times 5 \times 15. \tag{7.72}$$

HOSVD was applied to the above-normalized tensor and we get

$$x_{k j_1 j_2 j_1' j_2'} = \sum_{\ell_1=1}^{5} \sum_{\ell_2=1}^{15} \sum_{\ell_3=1}^{5} \sum_{\ell_4=1}^{15} \sum_{\ell_5=1}^{4} G(\ell_1 \ell_2 \ell_3 \ell_4 \ell_5) u_{\ell_1 j_1}^{(j_1)} u_{\ell_2 j_2}^{(j_2)} u_{\ell_3 j_1'}^{(j_1')} u_{\ell_4 j_2'}^{(j_2')} u_{\ell_5 k}^{(k)} \tag{7.73}$$

where $G \in \mathbb{R}^{5\times15\times5\times15\times4}$ is a core tensor, and $u^{(j_1)}_{\ell_1 j_1}, u^{(j_1')}_{\ell_3 j_1'} \in \mathbb{R}^{5\times5}$, $u^{(j_2)}_{\ell_2 j_2}, u^{(j_2')}_{\ell_4 j_2'} \in \mathbb{R}^{15\times15}$, and $u^{(k)}_{\ell_5 k} \in \mathbb{R}^{4\times4}$ are singular value matrices and orthogonal matrices.

Before selecting the features, we need to specify singular value vectors of interest, attributed to samples, j_2, time points, j_1, and k. The conditions of singular value vectors of interest are:

- $u^{(j_1)}_{\ell_1 j_1}$ and $u^{(j_1')}_{\ell_3 j_1'}$ should have j_1 dependence since features should change their values as time goes.
- $u^{(j_2)}_{\ell_2 j_2}$ and $u^{(j_2')}_{\ell_4 j_2'}$ are preferable to be constant, since we seek features whose values are independent of human subjects.
- $u^{(k)}_{\ell_5 k}$ is supposed to be constant since we would like to find features whose dependence upon j_1 and j_2 is independent of k.

We realized that $\ell_1 = 2$ and $\ell_2 = \ell_5 = 1$ satisfy the above requirements (Fig. 7.22). Using identified $u^{(j_1)}_{2j_1}$ and $u^{(j_2)}_{1j_2}$, we recover singular value vectors attributed to i_k as

$$u^{(i_k;j_1j_2)}_{21i_k} = \sum_{j_1=1}^{5}\sum_{j_2=1}^{15} x_{i_k j_1 j_2} u^{(j_1)}_{2j_1} u^{(j_2)}_{1j_2} \tag{7.74}$$

then P values are attributed to i_k as

$$P_{i_k} = P_{\chi^2}\left[> \left(\frac{u^{(i_k;j_1j_2)}_{21i_k}}{\sigma_{21}} \right)^2 \right] \tag{7.75}$$

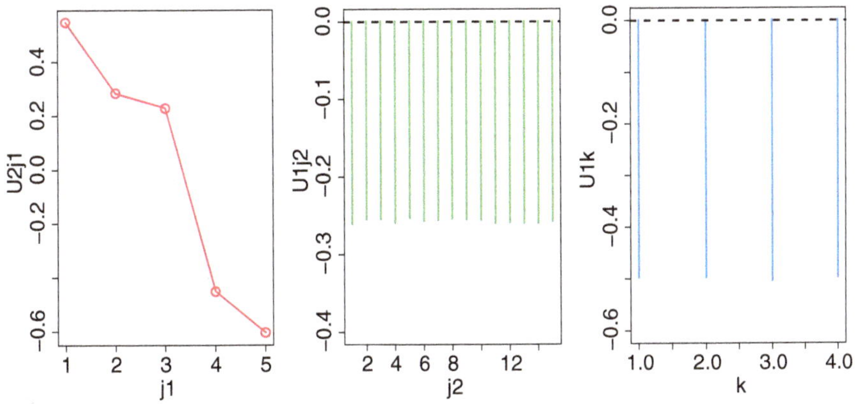

Fig. 7.22 Singular value vectors of interest. Left: $u^{(j_1)}_{2j_1}$, middle: $u^{(j_2)}_{1j_2}$, right: $u^{(k)}_{1k}$. Pearson correlation coefficient between j_1 and $u^{(j_1)}_{2j_1}$ is -0.96 ($P = 0.01$)

P values are corrected by BH criterion and i_ks associated with adjusted P values less than 0.01 ($k = 1, 2$, i.e., methylation and gene expression) and 0.05 ($k = 3, 4$, i.e., two proteomes). The number of selected i_ks is listed in Table 7.39

Although the readers of this book might not be interested in the biological point of view, the only possible validation is enrichment analysis. Gene symbols associated with selected features are uploaded to Enrichr (Tables 7.40, 7.41, 7.42, and 7.43). Table 7.40 lists the top ten enriched terms for "dbGaP" category. dbGap is the list of phenotype-related methylation. Thus, methylation naturally affects the phenotypes shown in Table 7.40. Thus, selected features, i_k, are related and biologically reasonable.

Table 7.39 The number of selected i_ks for methylation, gene expression (and those of associated gene symbols), and two proteomes

			Proteome	
	Methylation	Gene expression	Whole blood cells	Plasma
Probes	2077	11	–	–
Gene symbols	1355	8	24	22

Table 7.40 Top ten enriched terms in "dbGaP" category (methylation)

Term	Overlap	P value	Adjusted P value
Body mass index	80/437	9.56×10^{-17}	2.47×10^{-14}
Blood pressure	78/454	8.78×10^{-15}	1.14×10^{-12}
Lymphocytes	19/44	1.15×10^{-11}	9.93×10^{-10}
Cholesterol	45/268	8.92×10^{-9}	5.77×10^{-7}
Body height	57/385	1.17×10^{-8}	6.07×10^{-7}
Echocardiography	45/273	1.59×10^{-8}	6.88×10^{-7}
Stroke	46/289	3.33×10^{-8}	1.23×10^{-6}
Respiratory function tests	30/151	6.01×10^{-8}	1.95×10^{-6}
Cholesterol, HDL	51/357	2.16×10^{-7}	6.21×10^{-6}
Heart failure	33/187	2.56×10^{-7}	6.64×10^{-6}

Table 7.41 Top ten enriched terms in "GO Biological Process 2023" category (gene expression)

Term	Overlap	P value	Adjusted P value
Oxygen transport (GO:0015671)	2/10	6.29×10^{-6}	6.38×10^{-4}
Carbon dioxide transport (GO:0015670)	2/11	7.69×10^{-6}	6.38×10^{-4}
Gas transport (GO:0015669)	2/15	1.47×10^{-5}	8.11×10^{-4}
Hydrogen peroxide catabolic process (GO:0042744)	2/20	2.65×10^{-5}	1.10×10^{-3}
Positive regulation of cellular process (GO:0048522)	4/594	4.90×10^{-5}	1.63×10^{-3}
One-carbon compound transport (GO:0019755)	2/35	8.28×10^{-5}	2.29×10^{-3}
Response to hydrogen peroxide (GO:0042542)	2/46	1.44×10^{-4}	3.41×10^{-3}
Positive regulation of cell death (GO:0010942)	2/50	1.70×10^{-4}	3.52×10^{-3}
Regulation of cell death (GO:0010941)	2/59	2.37×10^{-4}	4.37×10^{-3}
Response to reactive oxygen species (GO:0000302)	2/66	2.96×10^{-4}	4.92×10^{-3}

Table 7.42 Top ten enriched terms in "GO Biological Process 2023" category (whole blood cells)

Term	Overlap	*P* value	Adjusted *P* value
Oxygen transport (GO:0015671)	2/10	6.17×10^{-5}	3.13×10^{-3}
Carbon dioxide transport (GO:0015670)	2/11	7.54×10^{-5}	3.13×10^{-3}
Gas transport (GO:0015669)	2/15	1.44×10^{-4}	3.97×10^{-3}
Hydrogen peroxide catabolic process (GO:0042744)	2/20	2.59×10^{-4}	5.37×10^{-3}
One-carbon compound transport (GO:0019755)	2/35	8.01×10^{-4}	1.11×10^{-2}
Platelet aggregation (GO:0070527)	2/35	8.01×10^{-4}	1.11×10^{-2}
Homotypic cell–cell adhesion (GO:0034109)	2/46	1.38×10^{-3}	1.43×10^{-2}
Response to hydrogen peroxide (GO:0042542)	2/46	1.38×10^{-3}	1.43×10^{-2}
Positive regulation of cell death (GO:0010942)	2/50	1.63×10^{-3}	1.51×10^{-2}
Regulation of cell death (GO:0010941)	2/59	2.26×10^{-3}	1.88×10^{-2}

Table 7.43 Top ten enriched terms in "GO Biological Process 2023" category (plasma)

Term	Overlap	*P* value	Adjusted *P* value
Regulation of heterotypic Cell–cell adhesion (GO:0034114)	3/24	1.71×10^{-6}	4.03×10^{-4}
Positive regulation of substrate adhesion-dependent cell spreading (GO:1900026)	3/35	5.48×10^{-6}	6.47×10^{-4}
Regulation of substrate adhesion-dependent cell spreading (GO:1900024)	3/51	1.73×10^{-5}	9.39×10^{-4}
Positive regulation of apoptotic cell clearance (GO:2000427)	2/7	1.99×10^{-5}	9.39×10^{-4}
Regulation of apoptotic cell clearance (GO:2000425)	2/7	1.99×10^{-5}	9.39×10^{-4}
Plasminogen activation (GO:0031639)	2/8	2.65×10^{-5}	9.51×10^{-4}
Positive regulation of phagocytosis (GO:0050766)	3/60	2.82×10^{-5}	9.51×10^{-4}
Protein polymerization (GO:0051258)	3/63	3.27×10^{-5}	9.64×10^{-4}
Positive regulation of cell–substrate adhesion (GO:0010811)	3/71	4.68×10^{-5}	1.23×10^{-3}
Fibrinolysis (GO:0042730)	2/12	6.23×10^{-5}	1.34×10^{-3}

When 24 gene symbols associated with identified proteins in whole blood cells ($k = 3$), we got the enriched terms in "GO Biological Process 2023" category of Enrichr (Table 7.42). The selected enriched terms are highly coincident with those in Table 7.41 where gene expression was considered; eight out of ten enriched terms are common between Tables 7.41 and 7.42 in spite of the distinct number of genes selected, 8 vs. 24. This can be a side proof that our strategy selected reasonable sets of genes.

When proteins identified in plasma ($k = 4$) were uploaded to Enrichr, we got enriched terms listed in Table 7.42. Since it is the measurement in plasma, although the enriched terms differ from those in Tables 7.41 and 7.42 where the measurements were in whole blood, the enriched terms are still significant enough. Thus, we can conclude that our strategy was successful for all four omics profiles.

One might wonder how the present strategy described in Sect. 5.7.1 differs from case I type II approach in Table 5.3. To see this, we reanalyzed data set treated in Sect. 7.10 (i.e., TCGA data set and GEO data set) where case I type II strategy was employed. First, we generated

$$x_{jj'1} = \sum_{i=1}^{N} x_{ij} x_{ij'} \tag{7.76}$$

$$x_{jj'2} = \sum_{k=1}^{K} x_{kj} x_{kj'} \tag{7.77}$$

and applied HOSVD to $x_{jj'p} \in \mathbb{R}^{M \times M \times 2}$, and we got

$$x_{jj'p} = \sum_{\ell_1=1}^{M} \sum_{\ell_2=1}^{M} \sum_{\ell_3=1}^{2} G(\ell_1 \ell_2 \ell_3) u_{\ell_1 j}^{(j)} u_{\ell_2 j'}^{(j')} u_{\ell_3 p}^{(k)}. \tag{7.78}$$

At first, we need to identify which $u_{\ell_1 j}^{(j)}$s are coincident with class labels (i.e., distinct between tumors and normal controls) for TCGA and GEO data sets, and we realized that $u_{2j}^{(j)}$ are distinct between tumors and healthy controls for TCGA and GEO data sets (Fig. 7.23). Singular value vectors attributed to i and k can be recovered as

$$u_{2i}^{(i;j)} = \sum_{j=1}^{M} x_{ij} u_{2j}^{(j)} \tag{7.79}$$

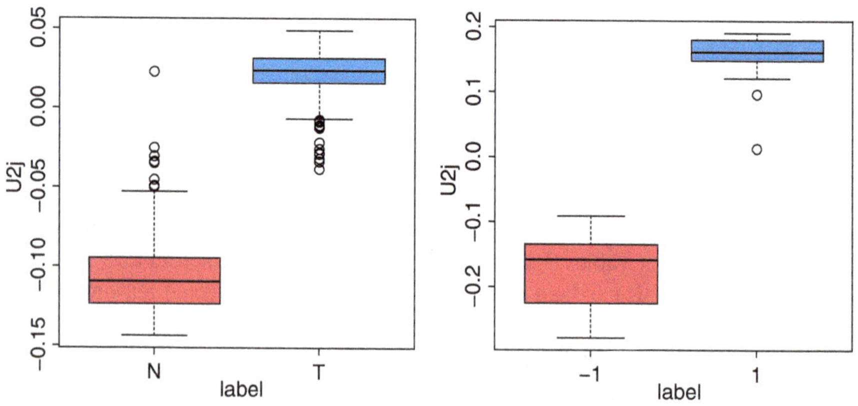

Fig. 7.23 Singular value vectors of interest, $u_{2j}^{(j)}$. Left: TCGA, right: GEO. P values computed by t-test are 8.49×10^{-47} and 4.07×10^{-17}, respectively

Table 7.44 Confusion matrix between genes selected in TCGA and GEO data sets. P value computed by Fisher's exact test is 6.67×10^{-5}. (multiomics integration)

		GEO	
		Not selected	Selected
TCGA	Not selected	17,269	101
	Selected	65	5

$$u_{2k}^{(k;j)} = \sum_{j=1}^{M} x_{kj} u_{2j}^{(j)} \tag{7.80}$$

and P values are computed as

$$P_i = P_{\chi^2}\left[> \left(\frac{u_{2i}^{(i;j)}}{\sigma_2} \right)^2 \right]$$

$$P_k = P_{\chi^2}\left[> \left(\frac{u_{2k}^{(k;j)}}{\sigma_2'} \right)^2 \right]$$

and P values are corrected by BH criterion and is and ks associated with adjusted P values less than 0.01 are selected. The confusion matrix of genes between TCGA and GEO data sets is listed in Table 7.44. This is even worth than PCA-based unsupervised FE (Table 7.36). Thus, although this strategy can reduce required memory drastically, the performance is worth than case I type II strategy (Table 7.33). Thus, if we need to integrate only two profiles, it is better to employ case I type II strategy in Table 5.3.

Before closing this section, we try to understand why this strategy works well. At first, we compare $x_{kj_1j_2j_1'j_2'}$ between ks (Table 7.45). Gene expression and two proteomes are obviously highly intercorrelated (since methylation is not correlated, it is ignored). It is possibly one of the reasons why integrated analysis of $x_{kj_1j_2j_1'j_2'}$ is successful. This is also true for the coincidence between TCGA and GEO data sets. Figure 7.24 shows the scatter plot of $x_{jj'1}$ (gene expression) and $x_{jj'2}$ (miRNA). Since it is obvious that they are significantly correlated, the commonly selected genes (Table 7.44) are also interpreted because of this correlation.

In conclusion, although the integrated analysis introduced in this section can reduce the required memory size drastically, it can be successful only when generated square matrices, $x_{kj_1j_2j_1'j_2'}$ and $x_{jj'p}$, are intercorrelated with each other.

Table 7.45 Correlation coefficients between gene expression and two proteomes ($x_{kj_1j_2j'_1j'_2}$, $2 \leq k \leq 4$). Upper half: Pearson correlation coefficient, lower half: P values

		Correlation coefficients		
		Gene expression	Whole blood cells	Plasma
	k	2	3	4
P values	Gene expression	–	0.1279405	0.2192163
	Whole blood cells	1.34×10^{-11}	–	0.4384998
	Plasma	1.52×10^{-31}	8.98×10^{-131}	–

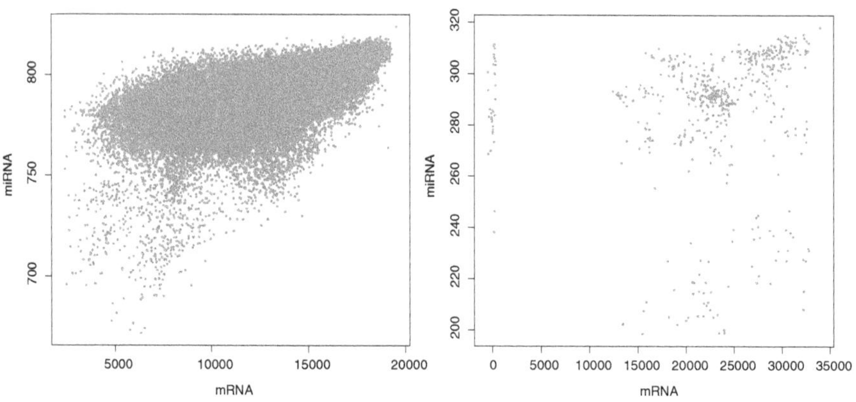

Fig. 7.24 Scatter plot of $x_{jj'1}$ (gene expression) and $x_{jj'2}$ (miRNA). Left: TCGA (Pearson correlation coefficient=0.627, $P = 0.00$), right: GEO (Pearson correlation coefficient = 0.349, $P = 2.32 \times 10^{-34}$)

7.12 Single-Cell Analyses

In this section, we introduce how we can apply TD-based unsupervised FE to single-cell analysis. Since the number of cells can be large, although one might wonder if TD-based unsupervised FE developed for *large p small n* problem can also work well for single-cell analyses, it works pretty well as you can see in the following.

7.12.1 Human and Mouse Midbrain Development

In this subsection, we apply TD-based unsupervised FE to single-cell RNA-seq data to which PCA-based unsupervised FE was applied in §6.7 [99]. The data set analyzed in this subsection is identical to those analyzed in §6.7. As in §6.7, we denote human and mouse data sets as data sets 1 and 2, respectively.

We format the data sets 1 and 2 as $x_{ij} \in \mathbb{R}^{N \times M}$ and $x_{ik} \in \mathbb{R}^{N \times K}$, respectively, where is are restricted to 13,889 (= N) common genes between human and mouse. $M(= 1,977)$ and $K(= 1,907)$ are the number of single cells of human and mouse,

Table 7.46 The number of single cells that belong to individual development stages

t	1	2	3	4	5	6	7
Human	6	7	8	9	10	11	(weeks)
Cells	287	131	331	322	509	397	
Mouse	E11.5	E12.5	F13.5	E14.5	E15.5	E18.5	Unknown
Cells	349	350	345	308	356	142	57

respectively (Table 7.46). Following the strategy case II type II in Table 5.3, we generated a tensor

$$x_{ijk} = x_{ij} x_{ik} \in \mathbb{R}^{N \times M \times K} \tag{7.81}$$

and by taking partial summation, we get a matrix

$$\tilde{x}_{jk} = \sum_{i=1}^{N} x_{ijk} \in \mathbb{R}^{M \times K} \tag{7.82}$$

to which SVD is applied and we get

$$\tilde{x}_{jk} = \sum_{\ell=1}^{L} \lambda_\ell u_{\ell j}^{(j)} u_{\ell k}^{(k)}. \tag{7.83}$$

Missing singular value vectors attributed to features can be recovered as

$$u_{\ell i}^{(i;j)} = \sum_{j=1}^{M} x_{ij} u_{\ell j}^{(j)} \tag{7.84}$$

$$u_{\ell i}^{(i;k)} = \sum_{k=1}^{K} x_{ik} u_{\ell k}^{(k)} \tag{7.85}$$

after identifying $u_{\ell j}^{(j)}$ and $u_{\ell k}^{(k)}$ of interest. P values attribute to is as

$$P_i = P_{\chi^2}\left[> \sum_{\ell \in \Omega_\ell} \left(\frac{u_{\ell i}^{(i;j)}}{\sigma_\ell} \right)^2 \right] \tag{7.86}$$

$$= P_{\chi^2}\left[> \sum_{\ell \in \Omega'_\ell} \left(\frac{u_{\ell i}^{(i;k)}}{\sigma'_\ell} \right)^2 \right] \tag{7.87}$$

Table 7.47 Confusion matrix between genes selected in mouse and human. P values computed by Fisher's exact test are 6.33×10^{-53} (PCA) and 6.94×10^{-320} (TD)

		Human			
		PCA-based unsupervised FE		TD-based unsupervised FE	
		Not selected	Selected	Not selected	Selected
Mouse	Not selected	13,730	67	12,705	330
	Selected	57	35	454	400

and is associated with adjusted P values less than 0.01 are selected for human and mouse, respectively.

Before proceeding further, we try to compare the performance achieved by TD-based unsupervised FE and that by PCA-based unsupervised FE by counting the number of commonly selected genes. With restricting genes considered to the 13,889 genes common between human and mouse, we get the confusion matrix for PCA-based unsupervised FE from Table 6.48 as shown in Table 7.47. We also selected genes by TD-based unsupervised FE with $\Omega_\ell = \Omega'_\ell = \{1 \leq \ell \leq 3\}$ since $\ell \leq 2$ for human and $\ell \leq 3$ for mouse (Table 7.47). Although the improvement between Tables 7.33 (TD) and 7.36 (PCA) is at most twice, TD-based unsupervised FE identified ten times more common genes than PCA-based unsupervised FE in Table 7.47. This definitely demonstrates the superiority of TD-based unsupervised FE over PCA-based unsupervised FE.

Since we have demonstrated the superiority of TD-based unsupervised feature extraction over PCA-based unsupervised feature extraction that has already been shown to outperform other methods, we are delving deeper into the performance of TD-based unsupervised feature extraction by focusing on how well TD-based unsupervised feature extraction detects common properties between humans and mice. To achieve this goal, we conduct a thorough investigation into which singular value vectors are associated with time dependence. We applied categorical regression analysis to $u^{(j)}_{\ell j}$ and $u^{(k)}_{\ell k}$ as

$$u^{(j)}_{\ell j} = a_\ell + \sum_{t=1}^{6} a_{\ell t}\delta_{jt} \tag{7.88}$$

$$u^{(k)}_{\ell k} = b_\ell + \sum_{t=1}^{7} b_{\ell t}\delta_{kt} \tag{7.89}$$

where $a_\ell, a_{\ell t}, b_\ell$, and $b_{\ell t}$ are regression coefficients and δ_{jt} or δ_{kt} takes 1 only when j or k is measured at time point, t (Table 7.46). P values are computed by `lm` function in R and are corrected by BH criterion. ℓs associated with adjusted P values less than 0.01 are selected for human and mice, respectively. Table 7.48 shows the confusion matrix of singular value indices, ℓ, between human and mice. It is clear that they show a high degree of coincidence with one another. This strongly

Table 7.48 Confusion matrix between singular value vector index, ℓ, selected in mouse and human. *P* value computed by Fisher's exact test is 1.24×10^{-44} and odds ratio is 2.08×10^{2}

		Human	
		Not selected	Selected
Mouse	Not selected	1840	23
	Selected	12	32

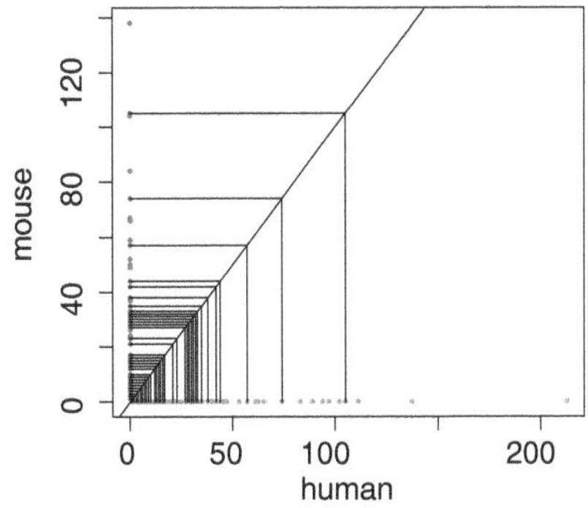

Fig. 7.25 Coincidence of ℓs between human and mouse. Blue and red dots represent the selected ℓs for human and mouse. Co-selected ℓs are connected by segments

suggests that the integrated analysis (case II type II in Table 5.3) has successfully merged the two gene expression profiles-otherwise, we would not be seeing such a level of overlap. As can be seen in Fig. 7.25, most of co-selected ℓs are smaller ones; this means that singular value vectors with larger contribution have more tendency to be co-selected between human and mice. It is reasonable, since singular value vectors with more contribution are more biologically reasonable. This also suggests the reliability of the strategy.

Next, we evaluate how many genes are selected commonly. Genes selected are 456 human genes and 507 mouse genes, and 305 genes are commonly selected among common 13,889 genes. Although the number of commonly selected genes is less than that in Table 7.47, commonly selected genes are still large enough compared with 456 human genes and 507 mouse genes selected. It is important if these commonly selected genes are biologically reasonable. Then we uploaded the commonly selected 305 genes to Enrichr (Table 7.49). It is obvious that they have strong tissue specificity with brain regions, since "GTEx Tissues V8 2023" category, which is based upon single-cell experiment, reports tissue specificity, whereas the targeted experiments are also single-cell-based experiments. Since the experiment is that of midbrain development, the proposed strategy seems to work very well. To further confirm that we have successfully identified the biologically reasonable genes, we identified 87 TFs that target the selected genes using the "ENCODE and ChEA Consensus TFs from ChIP-X" category in Enrichr. We uploaded these 87 TFs to TRRUST [33] to see if they are intercorrelated (Fig. 7.26). Since there are 27 inter-regulated genes although we are not willing to discuss the biological aspect of the regulatory network in Fig. 7.26, they are highly inter-regulated. Thus, our strategy could identify functionally reasonable genes.

Table 7.49 Top ten enriched tissues in "GTEx Tissues V8 2023" category in Enrichr

Term	Overlap	*P* value	Adjusted *P* value
Brain-hypothalamus male 20–29 up	10/100	3.03×10^{-6}	9.51×10^{-4}
Ovary female 70–79 up	9/100	2.24×10^{-5}	3.51×10^{-3}
Brain-amygdala female 50–59 up	8/100	1.47×10^{-4}	7.71×10^{-3}
Brain-frontal cortex (BA9) female 50–59 up	8/100	1.47×10^{-4}	7.71×10^{-3}
Brain-frontal cortex (BA9) male 40–49 up	8/100	1.47×10^{-4}	7.71×10^{-3}
Brain-anterior cingulate cortex (BA24) female 20–29 up	8/100	1.47×10^{-4}	7.71×10^{-3}
Ovary female 60–69 up	7/100	8.55×10^{-4}	1.92×10^{-2}
Brain-anterior cingulate cortex (BA24) female 60–69 up	7/100	8.55×10^{-4}	1.92×10^{-2}
Ovary female 50–59 up	7/100	8.55×10^{-4}	1.92×10^{-2}
Brain-anterior cingulate cortex (BA24) female 40–49 up	7/100	8.55×10^{-4}	1.92×10^{-2}

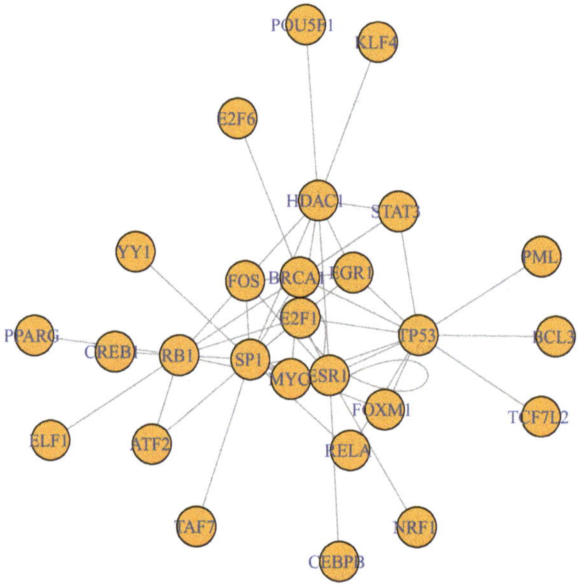

Fig. 7.26 Regulatory network between TFs identified TD identified by Enrichr

Before closing this section, we show uniform manifold approximation and projection (UMAP) [64] embedding of human and mouse cells with the selected 44 $u^{(k)}_{\ell k}$ and 55 $u^{(j)}_{\ell j}$, respectively (Fig. 7.27). Although we are not willing to discuss the comparison in detail; at a glance, the time development of single-cell distribution in embedding space is quite similar between human and mouse. For example, the distributions of both species do not change so much among the first three time points (6–8 weeks for human and e11.5–e13.5 for mouse). This also suggests the

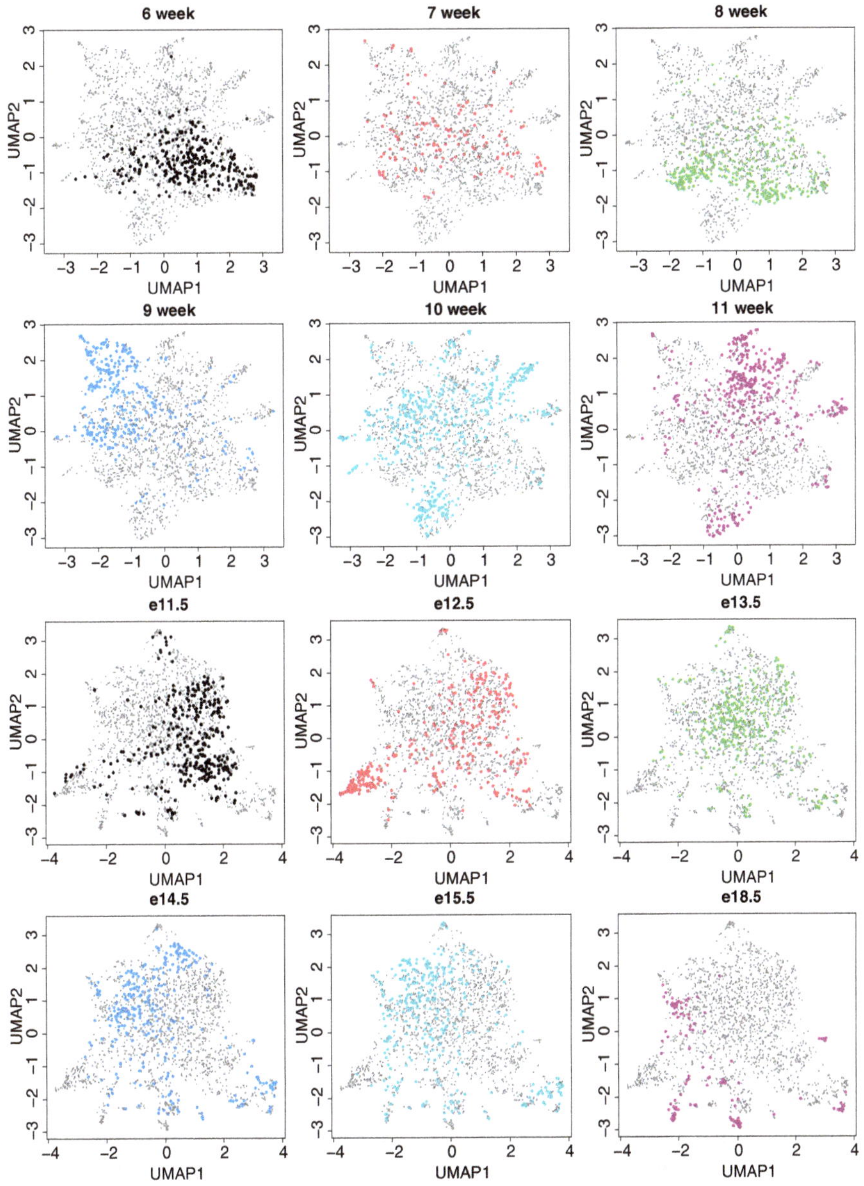

Fig. 7.27 Umap embedding of human (first two rows) and mouse (the last two rows) single cells with the selected 44 $u_{\ell k}^{(k)}$ and 55 $u_{\ell j}^{(j)}$

Table 7.50 The number of single cells that belong to individual cell types

	Cell types							
Acute stress	Astrocytes	Endotherial	Ependymal	Microglia	Neurons	Oligos	VSM	Total
Without	135	169	211	34	628	570	38	1785
With	132	71	145	14	270	431	33	1096

VSM vascular smooth muscles

successful outcomes by the integrated analysis of human and mouse single-cell experiments with TD-based unsupervised FE.

7.12.2 Mouse Hypothalamus with and Without Acute Formalin Stress

The second single-cell experiment to which case II type II tensor strategy was applied is mouse hypothalamus with and without acute formalin stress [99]. Since the procedure applied to this second data set is identical to that in the previous subsection, we do not detail it in this subsection but list some information altered from the previous subsection. $N = 24{,}341$, $M = 1{,}785$ (without acute formalin stress) and $K = 1{,}096$ (with acute formalin stress). In contrast to the data set analyzed in the previous subsection where time points are labels, the data set in this subsection has cell types as labels (Table 7.50). Thus, the categorical regression analysis should be

$$u_{\ell j}^{(j)} = a_\ell + \sum_{s=1}^{7} a_{\ell s}\delta_{js} \tag{7.90}$$

$$u_{\ell k}^{(k)} = b_\ell + \sum_{s=1}^{7} b_{\ell s}\delta_{ks} \tag{7.91}$$

where s should stand for one of seven cell types in Table 7.50.

Following the similar procedures described in the previous subsection, we identified 30 and 24 singular value vectors significantly associated with class labels (cell types) for single cells without and with acute formalin stress, respectively. To validate the coincidence between two data sets (without and with acute formalin stress), we applied Fisher's exact test to the confusion matrix (Table 7.51). Although P value is slightly more (less significant) than that in Table 7.48, the odds ratio is more than ten times larger than that in Table 7.48. Thus, the integrated analysis is at least comparatively successful, even more successful than that in the previous subsection. We also visually showed the coincidence (Fig. 7.28). As in Fig. 7.25, most of co-selected ℓs are smaller one; this also means that singular value vectors

Table 7.51 Confusion matrix between singular value vector index, ℓ, selected in single cells with and without acute formalin stress. P value computed by Fisher's exact test is 1.92×10^{-40} and odds ratio is 2.48×10^{3}

		With stress	
		Not selected	Selected
Without stress	Not selected	1065	1
	Selected	7	23

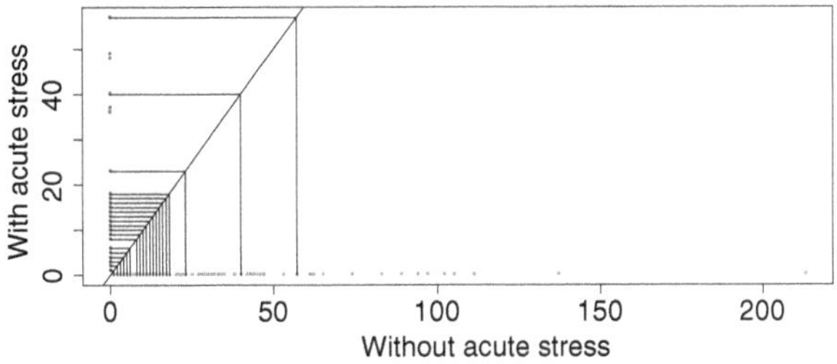

Fig. 7.28 Coincidence of ℓs between single cells without and with acute formalin stress. Blue and red dots represent the selected ℓs for single cells without and with acute formalin stress. Co-selected ℓs are connected by segments

Table 7.52 The ten hypothalamus in "GTEx Tissues V8 2023" category in Enrichr

Term	Overlap	P value	Adjusted P value
Brain-hypothalamus male 20–29 up	40/100	2.06×10^{-8}	4.39×10^{-7}
Brain-hypothalamus female 70–79 up	36/100	2.09×10^{-6}	3.02×10^{-5}
Brain-hypothalamus male 30–39 up	34/100	1.65×10^{-5}	1.80×10^{-4}
Brain-hypothalamus male 50–59 up	32/100	1.09×10^{-4}	9.75×10^{-4}
Brain-hypothalamus female 40–49 up	31/100	2.64×10^{-4}	2.23×10^{-3}
Brain-hypothalamus male 60–69 up	27/100	5.74×10^{-3}	3.36×10^{-2}
Brainhypothalamus female 50–59 up	27/100	5.74×10^{-3}	3.36×10^{-2}
Brain-hypothalamus male 40–49 up	27/100	5.74×10^{-3}	3.36×10^{-2}
Brain-hypothalamus female 60–69 up	26/100	1.11×10^{-2}	5.85×10^{-2}
Brain-hypothalamus male 70–79 up	25/100	2.04×10^{-2}	9.63×10^{-2}

with larger contribution have more tendency to be co-selected between single cells without and with acute formalin stress.

As in the previous subsection, we identified 3,319 common genes selected between single cells without and with acute formalin stress. Although there are ten hypothalamus profiles in "GTEx Tissues V8 2023" category in Enrichr, eight out of ten are significantly enriched with commonly selected 3,319 genes (Table 7.52). We also identified 77 TFs that significantly targeted 3,317 genes using "ENCODE and ChEA Consensus TFs from ChIP-X" category in Enrichr. Among 77 Tfs, there are 27 TFs that are inter-regulated based upon TRRUST (Fig. 7.29). Thus, TD-based unsupervised FE successfully integrates again two single-cell experiments.

Figure 7.30 shows umap embedding of single cells using the selected 30 $u_{\ell j}^{(j)}$ and 24 $u_{\ell k}^{(k)}$. As in Fig. 7.27, not all individual cell types are separately embedded, but

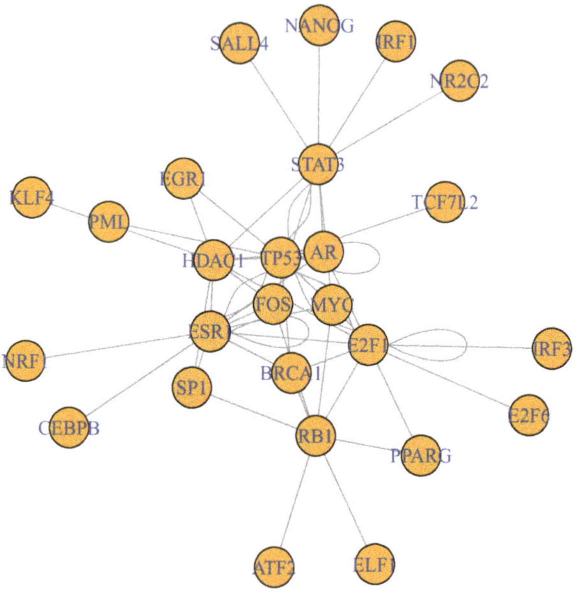

Fig. 7.29 Regulstory network between TFs identified TD identified by Enrichr

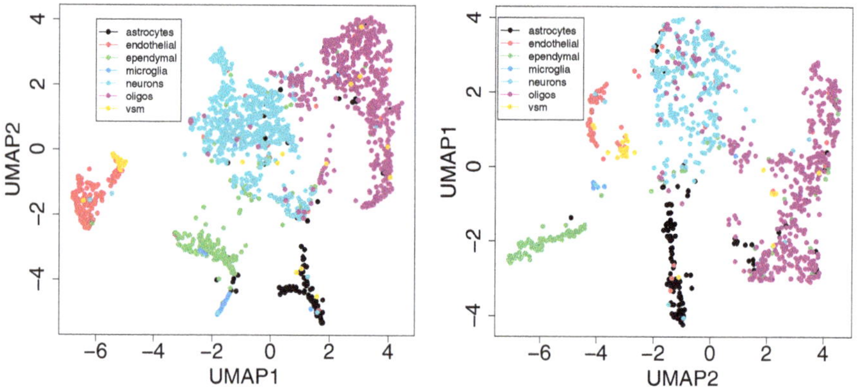

Fig. 7.30 Umap embedding of single cells without (left) and with (right) acute formalin stress using the selected 30 $u_{\ell j}^{(j)}$ and 24 $u_{\ell k}^{(k)}$, respectively

also global alignment of cell types is quite similar between two single-cell profiles. Thus, TD-based unsupervised FE successfully integrates two single-cell profiles.

7.12.3 Aging Genes in Mouse and Drug Discovery

In this subsection, we discuss aging genes related to Alzheimer's disease (AD) [98]. Mouse gene expression profile (GSE127892_microglia.kw.SeuratNorm.tsv.gz retrieved from GEO with GEO ID GSE127892) of samples composed of two genotypes (APP_NL-F-G and C57Bl/6), two tissues (Cortex and Hippocampus), four ages (3, 6, 12, and 21 weeks), two sexes (male and female), and four 96-well plates. Thus, they are maximally divergent samples. Nevertheless, the tensor is very fitted to deal with this very complicated data set since it can be formatted as a tensor $x_{j_1 j_2 j_3 j_4 j_5 j_6 i} \in \mathbb{R}^{96\times 2\times 2\times 4\times 2\times 4\times 29,341}$ that represents gene expression of ith gene of samples in j_1th well, j_2th genotype, j_3th tissue, j_4th age, j_5th sex, and j_6th plate. Applying HOSVD to $x_{j_1 j_2 j_3 j_4 j_5 j_6 i}$, we get

$$x_{j_1 j_2 j_3 j_4 j_5 j_6 i} = \sum_{\ell_1=1}^{96} \sum_{\ell_2=1}^{2} \sum_{\ell_3=1}^{2} \sum_{\ell_4=1}^{4} \sum_{\ell_5=1}^{2} \sum_{\ell_6=1}^{4} \sum_{\ell_7=1}^{29,341} G(\ell_1 \ell_2 \ell_3 \ell_4 \ell_5 \ell_6 \ell_7) \times u^{(j_1)}_{\ell_1 j_1} u^{(j_2)}_{\ell_2 j_2} u^{(j_3)}_{\ell_3 j_3} u^{(j_4)}_{\ell_4 j_4} u^{(j_5)}_{\ell_5 j_5} u^{(j_6)}_{\ell_6 j_6} u^{(i)}_{\ell_7 i} \quad (7.92)$$

where $G \in \mathbb{R}^{96\times 2\times 2\times 4\times 2\times 4\times 29,341}$ is the core tensor, and $u^{(j_1)}_{\ell_1 j_1} \in \mathbb{R}^{96\times 96}$, $u^{(j_2)}_{\ell_2 j_2} \in \mathbb{R}^{2\times 2}$, $u^{(j_3)}_{\ell_3 j_3} \in \mathbb{R}^{2\times 2}$, $u^{(j_4)}_{\ell_4 j_4} \in \mathbb{R}^{4\times 4}$, $u^{(j_5)}_{\ell_5 j_5} \in \mathbb{R}^{2\times 2}$, $u^{(j_6)}_{\ell_6 j_6} \in \mathbb{R}^{4\times 4}$, and $u^{(i)}_{\ell_7 i} \in \mathbb{R}^{29,341\times 29,341}$ are singular value matrices and orthogonal matrices.

As usual, we need to identify singular value vectors of interest attributed to samples, $u_{\ell_1 j_1}, u_{\ell_2 j_2}, u_{\ell_3 j_3}, u_{\ell_4 j_4}, u_{\ell_5 j_5}$, and $u_{\ell_6 j_6}$. The conditions required are

- $u^{(j_1)}_{\ell_1 j_1}$ is independent of j_1, i.e., wells.
- $u^{(j_2)}_{\ell_2 j_2}$ is independent of j_2, i.e., genotype.
- $u^{(j_3)}_{\ell_3 j_3}$ is independent of j_3, i.e., tissues.
- $u^{(j_4)}_{\ell_4 j_4}$ is dependent on j_4, i.e., days.
- $u^{(j_5)}_{\ell_5 j_5}$ is independent of j_5, i.e., sex.
- $u^{(j_6)}_{\ell_6 j_6}$ is independent of j_6, i.e., plate.

We aim to identify time-dependent genes that might be related to aging. Figure 7.31 shows the singular value vectors attributed to samples satisfying these requirements, i.e., $\ell_1 = \ell_2 = \ell_3 = \ell_5 = \ell_6 = 1$ and $\ell_4 = 2$. To see which $u^{(i)}_{\ell_7 i}$ is mostly associated with these singular value vectors attributed to samples, we plot $G(1, 1, 1, 2, 1, 1, \ell_7)$ in Fig. 7.32; $u^{(i)}_{\ell_7 i}$ with the largest absolute value $G(1, 1, 1, 2, 1, 1, \ell_7)$ should be used to select is which are supposed to be associated with the dependence upon j_1 to j_6 as shown in Fig. 7.31. As a result, $\ell_7 = 2$ was employed. Then, P values, P_i, are attributed to i as

$$P_i = P_{\chi^2}\left[> \left(\frac{u^{(i)}_{2i}}{\sigma_2} \right)^2 \right]. \quad (7.93)$$

The obtained P values are corrected by BH criterion and 401 is associated with adjusted P values less than 0.01 are selected.

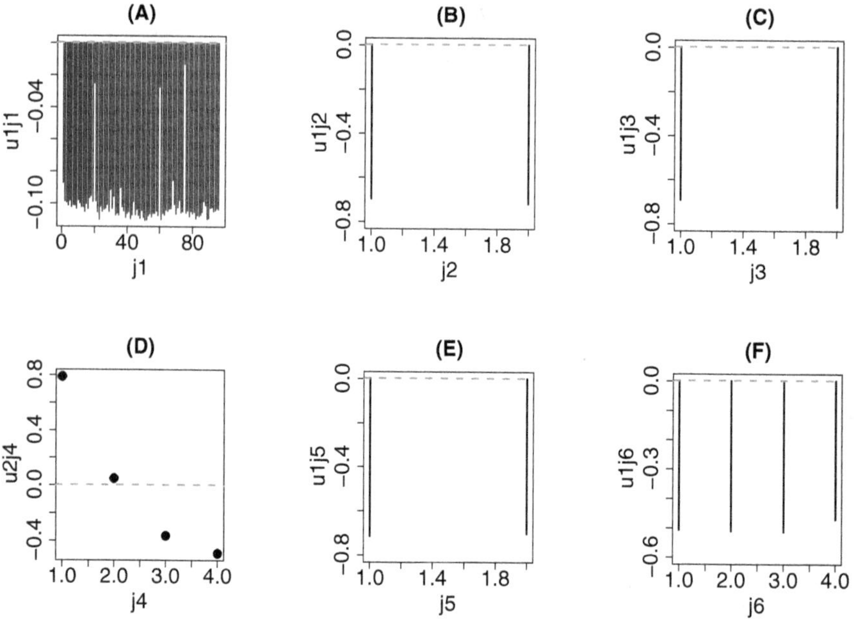

Fig. 7.31 Singular value vectors obtained by applying HOSVD to gene expression profile retrieved from GSE127892. (**a**) $u_{1j_1}^{(j_1)}$, (**b**) $u_{1j_2}^{(j_2)}$, (**c**) $u_{1j_3}^{(j_3)}$, (**d**) $u_{2j_4}^{(j_4)}$, (**e**) $u_{1j_5}^{(j_5)}$, (**f**) $u_{1j_6}^{(j_6)}$. The broken horizontal blue lines are the baselines

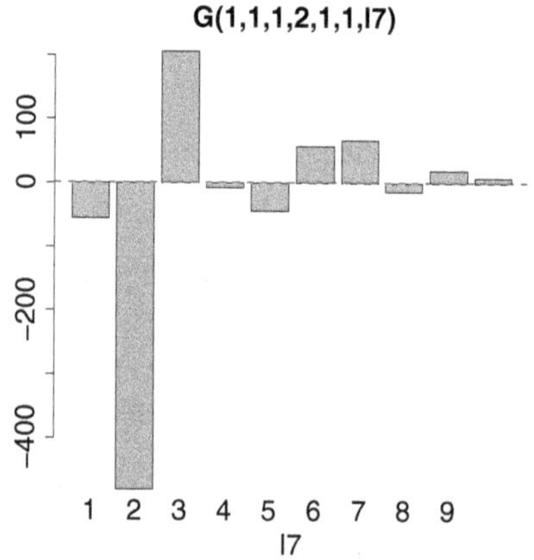

Fig. 7.32 $G(1, 1, 1, 2, 1, 1, \ell_7)$. The horizontal red broken line is the baseline

Next, we try to identify which chemical compounds can be drug candidates. To do this, we upload 401 genes to Enrichr and check various enrichment analyses related to drug treatment. Table 7.53 shows the results of "LINCS L1000 Chem Pert Consensus Sigs." Although not all of them are promising candidates, it includes something promising. For example, The fifth-ranked compound, triptolide, was recently reported to reduce Alzheimer's disease-like pathology in mice [60].

Table 7.53 The ten chemical compounds in "LINCS L1000 Chem Pert Consensus Sigs" category in Enrichr

Term	Overlap	*P* value	Adjusted *P* value
TP-0903 up	31/248	4.07×10^{-16}	4.41×10^{-12}
Bruceantin down	30/248	3.04×10^{-15}	1.65×10^{-11}
WZ-3105 up	29/243	1.28×10^{-14}	4.62×10^{-11}
Castanospermine up	29/247	1.96×10^{-14}	5.32×10^{-11}
Triptolide up	28/241	7.31×10^{-14}	1.59×10^{-10}
Tiludronate down	28/245	1.11×10^{-13}	1.90×10^{-10}
Doxorubicin up	28/246	1.23×10^{-13}	1.90×10^{-10}
Hyperforin down	27/244	6.73×10^{-13}	9.13×10^{-10}
YL-54 down	27/247	9.03×10^{-13}	1.08×10^{-9}
Emetine down	27/248	9.95×10^{-13}	1.08×10^{-9}

Table 7.54 The ten chemical compounds in "DrugMatrix" category in Enrichr

Term	Overlap	*P* value	Adjusted *P* value
Cyclosporin A-350 mg/kg in corn oil-rat-bone marrow-5d-up	51/315	2.26×10^{-31}	1.78×10^{-27}
Isoprenaline-4.2 mg/kg in saline-rat-heart-5d-up	49/304	4.54×10^{-30}	1.79×10^{-26}
Chlorambucil-0.6 mg/kg in corn oil-rat-spleen-0.25d-up	47/314	2.13×10^{-27}	5.59×10^{-24}
Hydroxyurea-400 mg/kg in saline-rat-bone marrow-5d-up	46/307	7.53×10^{-27}	1.48×10^{-23}
Tobramycin-40 mg/kg in saline-rat-kidney-28d-up	45/311	1.26×10^{-25}	1.42×10^{-22}
Gemcitabine-11 mg/kg in saline-rat-bone marrow-3d-up	47/344	1.27×10^{-25}	1.42×10^{-22}
Lead (II) acetate-600 mg/kg in saline-rat-spleen-5d-up	44/296	1.44×10^{-25}	1.42×10^{-22}
Cyclosporin A-350 mg/kg in corn oil-rat-bone marrow-3d-up	45/312	1.44×10^{-25}	1.42×10^{-22}
Cisplatin-0.5 mg/kg in saline-rat-kidney-28d-up	46/330	1.80×10^{-25}	1.50×10^{-22}
Netilmicin-40 mg/kg in saline-rat-kidney-28d-up	45/314	1.90×10^{-25}	1.50×10^{-22}

Conversely, the seventh compound, doxorubicin, was reported to cause cognitive decline [45]. It is not useful for drug repositioning but can be evidence that this list included compounds that affect cognition. The eighth-ranked one, hyperforin, was reported for its potential in Alzheimer's disease therapy [32]. The tenth-ranked one, emetine, was also investigated for its potential as an Alzheimer's disease therapy drug [3].

Next we checked "Drugmatrix" category in Enrichr (Table 7.54). The top and eighth-ranked compound is cyclosporin A, which was also previously tested for

Table 7.55 The ten chemical compounds in "Drug Perturbations from GEO up" category in Enrichr

Term	Overlap	*P* value	Adjusted *P* value
Imatinib DB00619 mouse GSE51698 sample 2522	81/288	2.24×10^{-70}	2.03×10^{-67}
Bleomycin DB00290 mouse GSE2640 sample 2851	80/329	6.03×10^{-64}	2.72×10^{-61}
Coenzyme Q10 5281915 mouse GSE15129 sample 3464	76/302	6.77×10^{-62}	2.04×10^{-59}
Calcitonin 16132288 mouse GSE60761 sample 3446	65/220	8.44×10^{-58}	1.91×10^{-55}
PRISTANE 15979 mouse GSE17297 sample 3229	71/291	1.03×10^{-56}	1.85×10^{-54}
N-METHYLFORMAMIDE 31254 rat GSE5509 sample 3570	70/283	2.37×10^{-56}	3.57×10^{-54}
Calcitonin 16132288 mouse GSE60761 sample 3447	59/177	5.83×10^{-56}	7.53×10^{-54}
Bexarotene DB00307 human GSE6914 sample 2680	55/147	2.04×10^{-55}	2.31×10^{-53}
Cyclophosphamide 2907 mouse GSE2254 sample 3626	78/413	2.45×10^{-53}	2.46×10^{-51}
Soman 7305 rat GSE13428 sample 2640	86/532	3.84×10^{-53}	3.48×10^{-51}

AD [35]. The fourth-ranked hydroxyurea was reported to decrease a characterization of neuronal membrane aging in Alzheimer's disease [125]. Thus, this list also includes some promising candidate compounds. Finally, we checked "Drug Perturbations from GEO up" category in Enrichr (Table 7.55). The top-ranked compound is imatinib, which was also previously tested for AD [23]. The second-ranked one, bleomycin, was known to cause risk of AD [66]; although it is not a piece of evidence to be used as a promising drug, it also suggests that TD-based unsupervised FE correctly identified some drugs related to AD. The third-ranked one, coenzyme, was also tested as a promising drug [21]. The fourth- and seventh-ranked one, calcitonin, was recognized as therapy target [88]. The eighth-ranked compound, bexarotene, was also once recognized as a promising candidate compound [112]. The ninth-ranked one, cyclophosphamide, was reported to cause cognitive decline [45]; it also suggests the usefulness of TD-based unsupervised FE as in the case of bleomycin above. In conclusion, regardless of the database used, the expression of the selected 401 genes is always altered by the treatment of promising AD candidate drugs. This suggests the usefulness of TD-based unsupervised FE toward drug repositioning for AD.

In spite of these successful results, one might still wonder if it is an accidental success because of the specific threshold *P* value, 0.01. To see if the results above are robust, we tested additional two threshold *P* values, 0.005 and 0.05 (Table 7.56). Although they are not completely the same, at least, not only the numbers of the selected genes are little dependent on the threshold *P* values, but also the discussed

Table 7.56 Dependence of selected genes and drugs upon the threshold P values

Threshold P value	0.005	0.01	0.05
The number of selected genes	370	401	498
LINCS L1000 chem pert consensus sigs			
Triptolide	7th	5th	8th
Doxorubicin	5th	7th	10th
Hyperforin	15th	8th	15th
Emetine	8th	10th	7th
Drugmatrix			
Cyclosporin A	2nd and 3rd,	Top and 8th,	2nd and 5th and 7th
Hydroxyurea	6th	4th	3rd
Drug perturbations from GEO up			
Imatinib	1st	1st	1st
Bleomycin	3rd	2nd	2nd
Calcitonin	4th and 5th	4th and 7th	5th and 7th
Bexarotene	13th	8th	14th
Cyclophosphamide	7th	9th	8th

drugs above are usually ranked high enough regardless of the threshold P values. Thus, the results above can be regarded to be relatively independent of the threshold P values.

Although the purpose of this subsection is not to identify genes related to AD but to identify drug candidate compounds that might be effective toward AD since we did not deal with AD directly at all, it is not a bad idea to see if the expression of the selected genes are altered by AD. Although there are seven AD profiles in "Disease Perturbations from GEO up" category in Enrichr, some of them are significantly enriched by the selected genes regardless of the threshold P values (Table 7.57). Thus, more or less, the selected genes are related to AD.

7.12.4 Single-Cell Multiomics Data Analysis

One of the most difficult fields of single-cell analysis is multiomics analysis. This is because of the heavy sparsity of obtained profiles. In general, not all epigenetic sites (e.g., DNA methylation sites) are either active or modified. When the states of individual sites are common over all samples that have distinct conditions, these sites are not informative. In addition to this, because of the small amount of DNA/RNA retrieved from single cells, most of the sites are not even investigated. Thus, only a limited number of sites among all numerous epigenetic sites are considered. This results in heavy sparsity of the profiles.

Table 7.57 AD profiles of "Disease Perturbations from GEO up" category in Enrichr

	Alzheimer's disease DOID-10652 human		
Threshold P value	0.005		
GEO ID/sample	Overlap	P value	Adj P value
GSE4757 592	49/231	1.14×10^{-37}	1.80×10^{-36}
GSE36980 522	15/301	5.22×10^{-4}	8.40×10^{-4}
GSE36980 521	17/419	2.20×10^{-3}	3.41×10^{-3}
GSE36980 519	9/220	2.17×10^{-2}	3.13×10^{-2}
GSE36980 523	7/341	4.44×10^{-1}	5.21×10^{-1}
GSE36980 524	6/308	5.07×10^{-1}	5.87×10^{-1}
GSE36980 520	5/275	5.78×10^{-1}	6.58×10^{-1}
Threshold P value	0.01		
GEO ID/sample	Overlap	P value	Adj P value
GSE4757 592	52/231	1.52×10^{-39}	2.23×10^{-38}
GSE36980 522	15/301	1.18×10^{-3}	1.84×10^{-3}
GSE36980 521	18/419	2.13×10^{-3}	3.25×10^{-3}
GSE36980 519	9/220	3.39×10^{-2}	4.80×10^{-2}
GSE36980 523	7/341	5.28×10^{-1}	6.09×10^{-1}
GSE36980 524	6/308	5.85×10^{-1}	6.65×10^{-1}
GSE36980 520	5/275	6.49×10^{-1}	7.25×10^{-1}
Threshold P value	0.05		
GEO ID/sample	Overlap	P value	Adj P value
GSE4757 592	57/231	2.19×10^{-40}	2.51×10^{-39}
GSE36980 522	18/301	5.85×10^{-4}	9.04×10^{-4}
GSE36980 521	23/419	3.68×10^{-4}	5.75×10^{-4}
GSE36980 519	12/220	9.34×10^{-3}	1.37×10^{-2}
GSE36980 523	8/341	6.17×10^{-1}	6.92×10^{-1}
GSE36980 524	10/308	2.40×10^{-1}	2.97×10^{-1}
GSE36980 520	7/275	5.30×10^{-1}	6.07×10^{-1}

Possibly because of the abovementioned difficulty, no established methods for multiomics single-cell analyses yet exist. For example, Lee et al. classified multiomics single-cell analyses into two categories:

- Correlation analysis between single-cell mono-omics data
- The analysis of one type of single-cell data followed by the integration of another single-cell data type

in recent reviews [50]. This means that there is only integration of pairwise analyses.

Nevertheless, if we employ the tensor-based method, we can directly integrate more than two omics profiles simultaneously without passing through pairwise analyses as can be seen below [106]. Suppose that we have K omics profiles each of which is formatted as a tensor

Table 7.58 The number of single cells within individual cell types included in data set 1

FGO	GO1	GO2	Granulosa	Immune	MI	MII	StromaC1	StromaC2	Total
81	40	46	93	20	155	90	189	185	899

Table 7.59 The number of single cells at four embryonic time points included in data set 2. For E7.5, the gene expression profiles of 296 single cells were measured

E4.5–5.5	E6.5	E6.75	E7.5	Total
267	98	97	390 (296)	852 (758)

$$x_{i_k jk} \in \mathbb{R}^{N_k \times M \times K} \tag{7.94}$$

which represents the amount of kth omics profile of the i_kth feature of the jth single cell. Since the number of features, N_k, is not common among multiple omics profiles at all, we cannot directly apply TD to $x_{i_k jk}$ as it is.

In order to address this problem, we employed the strategy introduced in Sect. 5.8.1 as shown in Eqs. (5.32), (5.33), and (5.34). Specifically, we have employed the following two sets of omics profiles.

- GSE154762: Hereafter denoted as data set 1 in this subsection. It is composed of 899 single cells for which gene expression, DNA methylation, and DNA accessibility were measured. These single cells represent human oocyte maturation (Table 7.58). Details about how to load omics profiles into R can be found in the original research paper [106].
- GSE121708: Hereafter denoted as data set 2 in this subsection. It is composed of 852 single cells for which gene expression profiles and 758 single cells for which gene expression were measured. These single cells represent the four time points of the mouse embryo (Table 7.59). Details about how to load omics profiles into R can be found in the original research paper [106] as well.

In these two profiles, $k = 1$ was attributed to gene expression, $k = 2$ was attributed to DNA methylation, and $k = 3$ was attributed to DNA accessibility. As a result, $K = 3$.

In these single-cell experiments, N_k can be too large to deal with them. In order to avoid this problem, by making use of the abovementioned sparsity, we employed sparse matrix format [10] to store $x_{i_k jk}$. In sparse matrix format, not all values of all elements but only those of nonzero elements are stored with the row and column numbers. Thus, the amount of required memory for the sparse matrix format of sparse matrix can be drastically reduced. In addition to this, SVD was also performed by the truncated SVD [6]. In the truncated SVD, the diagonalization was performed by the Lanczos method where only iterative multiplication of sparse matrix toward a vector is required. Thus, we do not need to alter the sparse matrix format during diagonalization.

Preprocessing of DNA methylation profiles is as follows: First, we collected genomic positions for which at least one measurement was performed for at least one single cell (i.e., union). Then, for each genomic position, three integers, -1, 0, and 1, were assigned. When the genomic position was measured in a single cell, and its state was methylated (nonmethylated), we attributed 1 (-1) to the genomic position of the single cell. Otherwise (i.e., missing observation), we attributed 0 to the genomic transition in a single cell.

Preprocessing of DNA accessibility is as follows: First, we divided the whole genome into 200 nucleotide regions, and DNA accessibility was summed up within individual regions. These values, which show the summation of DNA accessibility within individual regions, are regarded as DNA accessibility at the individual 200 nucleotide regions, each of which is supposed to approximately correspond to a single nucleosome that is composed of 140-length DNA that wraps around histones and 80-length linker DNA. In this study, these 200 nucleotide regions are called "nucleosome regions."

Before applying SVD to individual profiles, the following normalization was applied to individual profiles.

- $x_{i_k jk}$s with $k = 2, 3$ (i.e., DNA methylation and DNA accessibility) for data set 1 were normalized as

$$\sum_{i_k=1}^{N_k} \left| x_{i_k jk} \right| = N_k \tag{7.95}$$

- $x_{i_k j1}$s (i.e., gene expression) for data set 1 were normalized as

$$\sum_{i_k=1}^{N_k} x_{i_k j1} = 0 \tag{7.96}$$

$$\sum_{i_k=1}^{N_k} x_{i_k j1}^2 = N_k \tag{7.97}$$

- No normalization was applied to $x_{i_k jk}$s with $k = 2, 3$ for data set 2.

The reason why DNA methylation and DNA accessibility in data set 2 are not normalized is that some single cells are associated with very small amount of $x_{i_k jk}$. If we normalized these cells with a very small amount of $x_{i_k jk}$ as well, they have too large weight that can skew the outcomes. Thus, we could not apply normalization to DNA methylation in DNA accessibility in data set 2.

After getting $v_{\ell j}^k$ and $u_{\ell_2 j}^{(j)}$ by using Eqs. (5.32) and (5.33) with setting $L = 10$, respectively, to select $v_{\ell j}^k$s and $u_{\ell_2 j}^{(j)}$s associated classification shown in Tables 7.58 and 7.59, we applied categorical regression analysis as

$$v_{\ell j}^k = \sum_s a_{\ell s}^k \delta_{js} + b_\ell^k \tag{7.98}$$

Table 7.60 Number of singular value vectors coincident with classification shown in Tables 7.58 and 7.59

Adjusted P value	SVD $\left(v_{\ell j}^k\right)$			
	Gene expression	DNA methylation	DNA accessibility	
Table 7.58 (Data set 1)				
≤0.01	10	7	1	
>0.01	0	3	9	
Table 7.59 (Data set 2)				
≤0.01	10	7	5	
>0.01	0	3	5	
Adjusted P value	HOSVD $\left(u_{\ell_2 j}^{(j)}\right)$			
	DNA methylation and DNA accessibility	Gene expression and DNA methylation	Gene expression and DNA accessibility	All
Table 7.58 (Data set 1)				
≤0.01	10	13	11	18
>0.01	10	7	9	12
Table 7.59 (Data set 2)				
≤0.01	10	8	16	18
>0.01	10	12	4	12

$$u_{\ell_2 j}^{(j)} = \sum_s a_{\ell_2 s}\delta_{js} + b_{\ell_2} \tag{7.99}$$

where $a_{\ell s}^k$, b_ℓ^k, $a_{\ell_2 s}$, and b_{ℓ_2} are regression coefficients. δ_{is} takes 1 only when the jth sample belongs to the sth category otherwise 0 and they were performed by `lm` function in R. Obtained P values are corrected by BH criterion and ℓs and ℓ_2s associated with adjusted P values less than 0.01 were selected (Table 7.60). Since $L = 10$, total number of $v_{\ell j}^k$ is $L = 10$ whereas that of $u_{\ell_2 j}^{(j)}$ for two omics profiles and all are $2L = 20$ and $3L = 30$, respectively.

Although in most cases we could identify the singular value vectors associated with class labels, we are not sure if the integrated analysis using TD is better than that using SVD. In order to do this, we applied UMAP [64] to $v_{\ell j}^k$s and $u_{\ell_2 j}^{(j)}$s (Figs. 7.33 and 7.34). The option "custom.config$n_neighbors" was set to 100, and other options were default.

The top three UMAP embedding of the left column is for individual omics and that of the right column is for the three combinations of two of three omics. The bottom one in the left column is that of the integration of all three omics. In Fig. 7.33 for data set 1, the combination of gene expression with DNA methylation or DNA accessibility does not look much different from only gene expression. The only improvement is the better separation between MI and MII. Embedding using only

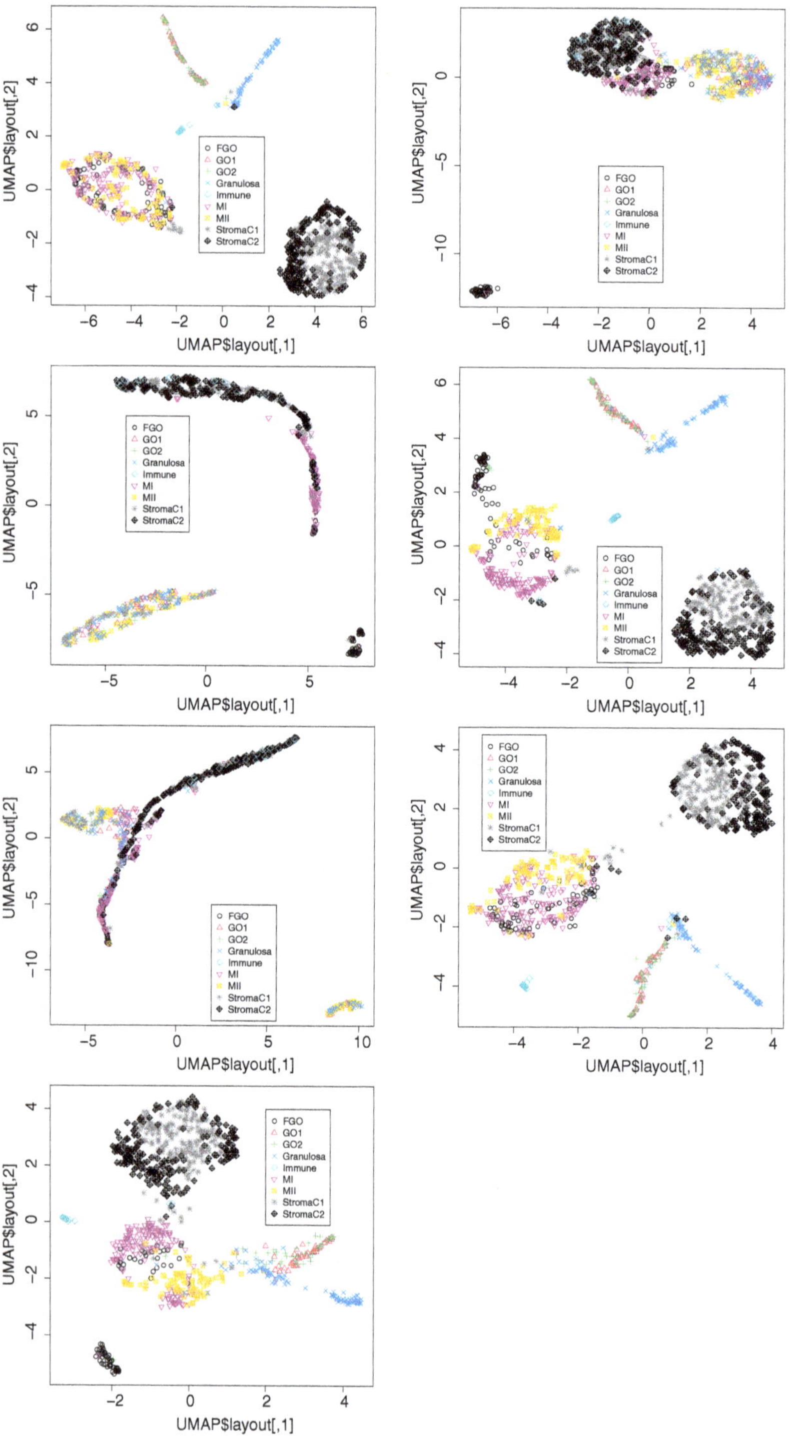

Fig. 7.33 UMAP embedding for data set 1. Left, from top to bottom, gene expression, DNA accessibility, DNA methylation, all three omics. Right, from top to bottom, DNA methylation and accessibility, gene expression and DNA methylation, gene expression and DNA accessibility

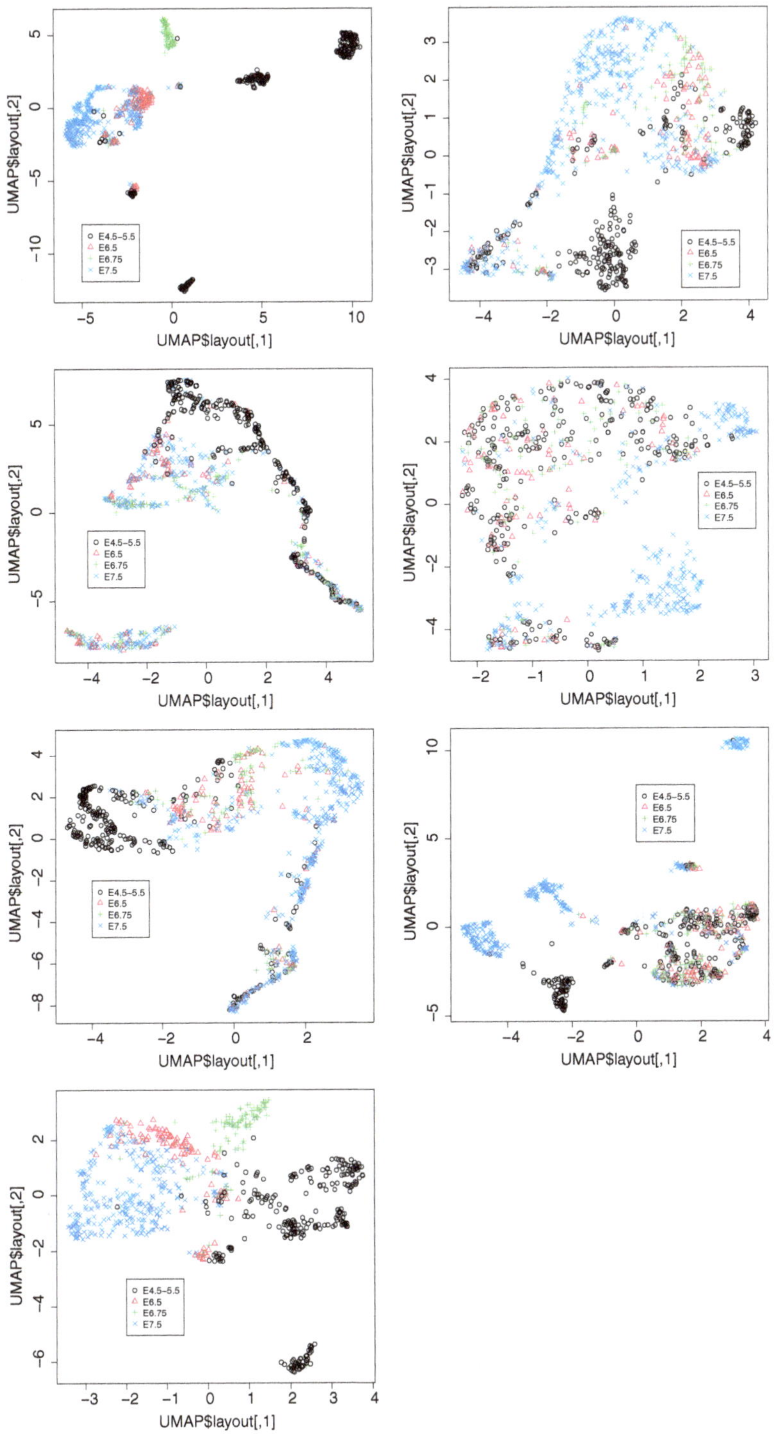

Fig. 7.34 UMAP embedding for data set 2. Left, from top to bottom, gene expression, DNA accessibility, DNA methylation, all three omics. Right, from top to bottom, DNA methylation and accessibility, gene expression, and DNA methylation, gene expression and DNA accessibility

Table 7.61 Number of single cells, features, nonzero components, and their ratios

Numbers	Expression	DNA methylation	DNA accessibility
	Data set 1		
Single cells	899	899	899
Features	26,500	26,438,807	15,478,375
Total components	2.38×10^7	2.38×10^{10}	1.39×10^{10}
Nonzero components	6.76×10^6	5.50×10^8	3.85×10^8
The ratio of nonzero components	0.28	0.02	0.03
	Data set 2		
Single cells	758	852	852
Features	22,084	20,106,507	13,627,678
Total components	1.67×10^7	1.71×10^{10}	1.16×10^{10}
Nonzero components	4.87×10^6	6.96×10^8	7.87×10^8
The ratio of nonzero components	0.29	0.04	0.07

DNA methylation and DNA accessibility and the integration of DNA methylation and DNA accessibility do not show any coincidence with classification. Integration of three omics is the best one since classification between two similar classes (MI vs. MII and GO1 vs. GO2) is better than other embedding. Thus, although the improvement is not so striking, TD-based unsupervised FE successfully integrate three omics profiles.

The improvement by the integration of three omics is more conspicuous in Fig. 7.34 for data set 2. It is rather obvious that only the integration of all three omics exhibits a clear separation between class labels. Thus, considering the results when the TD-based unsupervised FE was applied to these two data sets, the integration of three omics profiles, gene expression, DNA methylation, and DNA accessibility, in single-cell experiments is a powerful tool.

The reason why it works pretty well is related to the treatment of missing values. As denoted above, single-cell measurements are full of missing values (Table 7.61). Even in gene expression profiles, at most less than 30% are not missing values. For DNA methylation and DNA accessibility, only a few % are not missing values. The treatment of these missing values that are majority is usually problematic. Usually, it is forced to substitute some values. Nevertheless, there is no consensus about what to fill. Typically, zero or mean values are substituted, but we are not sure what was caused by this substitution. Nevertheless, TD itself is known to have the ability to treat missing values [55]. Thus, without specific procedures targeting the treatment of missing values, TD-based unsupervised FE can successfully deal with profiles with massive missing values. This is a great advantage of TD-based unsupervised FE.

It is more important to see if TD-based unsupervised FE can identify biologically reasonable genes. In order to do this, we attribute P values to gene i_1 for data set 1 as

$$u_{\ell_2 i_1}^{(i_1;j)} = \sum_{j=1}^{M} x_{i_1 j 1} u_{\ell_2 j}^{(j)} \tag{7.100}$$

$$P_{i_1} = P_{\chi^2}\left[> \sum_{\ell_2 \in \Omega_{\ell_2}} \left(\frac{u_{\ell_2 i_1}^{(i_1;j)}}{\sigma_{\ell_2}} \right)^2 \right] \tag{7.101}$$

where Ω_{ℓ_2} are 18 ℓ_2s associated with adjusted P values less than 0.01 in Table 7.60. Then we found 370 genes associated with adjusted P values less than 0.01.

There are many genes related to oocyte maturation in these 370 genes. Mutations in TUBB8 and ZP3 cause human oocyte maturation arrest and female infertility [53]. The concentration of oocyte-secreted GDF9 decreases with MII transition during human IVM [15]. HMGB1 is implicated in preimplantation embryo development in the mouse [19]. PCNA belongs to a class of proteins which are stockpiled during oogenesis in order to be utilized later for early embryogenesis [130]. SP1 governs primordial folliculogenesis by regulating pregranulosa cell development in mice [16]. NLRP2 deficiency blocks early embryogenesis in the mouse [73]. Mouse REC114 is essential for meiotic DNA double-strand break formation and forms a complex with MEI4 [48]. Deletion of Pdcd5 in mice led to the deficiency of placenta development and embryonic lethality [52]. Expression of maternally derived OOEP is associated with oocyte developmental competence in the ovine species [11]. Immunolocalization of HMGN2 is observed in early-developing parthenogenetic bovine embryos derived from oocytes of high and low developmental competence [9]. Regulation of mouse oocyte microtubule and organelle dynamics by PADI6 and the cytoplasmic lattices were reported [43].

The number of genes listed above is as small as 11 which is much smaller than the total number of the selected genes, 370. Nevertheless, if we consider the 370 genes are at most less than 2% of the total number of genes, 26,000, even this small number is unlikely accidental. In fact, if we upload 370 genes to Enrichr, "oocyte meiosis" is associated with adjusted P values less than 0.05 in some categories (7.494×10^{-4} for "BioPlanet 2019" and 2.629×10^{-3} for "KEGG 2021 Human"). This suggests that TD-based unsupervised FE successfully identified genes related to oocyte maturation.

We have repeated the above procedure for data set 2 and identified as many as 583 genes associated with Ensembl gene ID. After converting them to 584 gene symbols with Gene ID Conversion Tool in DAVID, we uploaded these gene symbols to Enrichr (Table 7.62). Table 7.63 shows the top ten enriched terms in "Mouse Gene Atlas" of Enrichr. The first and second terms in Table 7.63 are embryonic stem lines. As can be seen, various embryo-related markers are enriched. There are also at least ten experiments in "ProteomicsDB 2020" significantly associated with embryonic stem (Table 7.64). Thus, TD-based unsupervised FE could identify biologically reasonable genes for data set 2 as well.

In conclusion, TD-based unsupervised FE successfully integrated gene expression, DNA methylation, and DNA accessibility in single-cell experiments.

Table 7.62 Embryo related markers in "CellMarker Augmented 2021" category of Enrichr

Term	Overlap	*P* value	Adjusted *P* value
Neural progenitor cell:embryonic prefrontal cortex	34/166	2.05×10^{-19}	1.64×10^{-17}
Morula cell (blastomere):embryo	22/99	8.51×10^{-14}	3.73×10^{-12}
Pluripotent stem cell:embryo	10/93	3.88×10^{-4}	5.50×10^{-3}
Embryonic stem cell:tooth	9/90	1.27×10^{-3}	1.46×10^{-2}
Pluripotent very small embryonic-like cell:Ovary	9/92	1.48×10^{-3}	1.66×10^{-2}
Interneuron:embryonic prefrontal cortex	5/32	2.20×10^{-3}	2.21×10^{-2}
Embryonic stem cell:undefined	4/20	2.42×10^{-3}	2.33×10^{-2}
Embryonic stem cell:embryo	3/11	3.45×10^{-3}	3.13×10^{-2}

Table 7.63 Top ten enriched terms in "Mouse Gene Atlas" of Enrichr

Term	Overlap	*P* value	Adjusted *P* value
Embryonic stem line Bruce4 p13	73/876	5.55×10^{-16}	5.05×10^{-14}
Embryonic stem line V26 2 p16	59/728	1.44×10^{-12}	6.54×10^{-11}
Mega erythrocyte progenitor	33/562	1.28×10^{-4}	3.88×10^{-3}
B cells GL7negative Alum	20/379	8.40×10^{-3}	1.91×10^{-1}
Baf3	20/440	3.53×10^{-2}	6.23×10^{-1}
Stem cells HSC	10/186	4.73×10^{-2}	6.23×10^{-1}
B cells GL7 negative KLH	11/212	4.79×10^{-2}	6.23×10^{-1}
Placenta	18/411	5.90×10^{-2}	6.71×10^{-1}
mIMCD-3	13/300	1.04×10^{-1}	1.00×10^{0}
neuro2a	17/425	1.21×10^{-1}	1.00×10^{0}

7.13 Integration of Multiomics Profiles Without Gene Expression

Although it is very usual to integrate multiomics profiles, also as in this book, they typically include gene expression. Integration of multiomics without gene expression is not easy since gene expression is the most informative. Without gene expression, the integration of multiomics hardly can identify biologically reasonable outcomes. In this section, we introduce two studies where multiomics data sets not including gene expression were integrated with TD-based unsupervised FE.

7.13.1 Histone Modification Bookmarks in Postmitotic Transcriptional Reactivation

The first example of integration of multiomics not including gene expression is that of histone modification [107]. Before starting the analysis, we would like to briefly discuss the biological background of the analysis.

Table 7.64 Ten enriched experiments related to embryonic stem in "ProteomicsDB 2020" of Enrichr

Term	Overlap	*P* value	Adjusted *P* value
Embryonic stem HES-3 PDB:200034 IMR90 mix1 hESC	51/223	5.10×10^{-31}	1.63×10^{-29}
Embryonic stem hES H1 PDB:200035 8plex protein rep3-1	66/627	1.23×10^{-19}	1.16×10^{-18}
Embryonic stem hES H7 PDB:200036 8plex protein rep3-2	66/627	1.23×10^{-19}	1.16×10^{-18}
Embryonic stem hES H9 PDB:200038 8plex protein rep3-3	66/627	1.23×10^{-19}	1.16×10^{-18}
Embryonic stem hES H14 PDB:200037 8plex protein rep3-4	66/629	1.46×10^{-19}	1.31×10^{-18}
Embryonic stem hES H7 PDB:200036 8plex protein rep1-2	54/506	1.76×10^{-16}	1.21×10^{-15}
Embryonic stem hES H1 PDB:200035 8plex protein rep1-1	54/507	1.92×10^{-16}	1.31×10^{-15}
Embryonic stem hES H14 PDB:200037 8plex protein rep1-4	54/508	2.09×10^{-16}	1.40×10^{-15}
Embryonic stem hES H9 PDB:200038 8plex protein rep1-3	54/508	2.09×10^{-16}	1.40×10^{-15}
Embryonic stem hES H7 PDB:200036 8plex protein rep2-2	65/728	9.57×10^{-16}	6.03×10^{-15}
Embryonic stem hES H14 PDB:200037 8plex protein rep2-4	65/729	1.02×10^{-15}	6.32×10^{-15}

Histone modifications are known to regulate transcription through chromatin remodeling, which means the restructuring of chromatin structure. If DNA binding is too tight, no functional proteins can reach the DNA region from which mRNA must be transcribed. Thus, controlling chromatin structure can regulate mRNA transcription. Since there are various types in histone modification that can either suppress or activate transcription, the successful integration of histone modification can be a rich information source about how individual genes act in a context-dependent manner.

One of the most interesting features of how histone modification regulates transcription is postmitotic regulation of gene expression. During the cell division process, histone modification must be released since DNA binding to core proteins, histone, must be tight during cell division. On the other hand, once the cell division process completes, the original (those before cell division starts) structure, i.e., histone modification, must be recovered to maintain cell functions. But how? If histone modification is to be completely initialized, there will be no way to recover the original state. There must be some "memory" to store the original state of histone modification. The purpose of this subsection is to seek candidate factors that can store histone modification before cell division starts during cell division.

Table 7.65 Combinations of experimental conditions. Individual conditions are associated with two replicates ($s = 1, 2$)

	Histone modifications (k)							
	H3K27ac ($k = 1$)		H3K4me1 ($k = 2$)		H3K4me3 ($k = 3$)		Input ($k = 4$)	
	Cell lines ($j = 1$: RPE1, $j = 2$:USO2)							
Phases (m)	RPE1	U2OS	RPE1	U2OS	RPE1	U2OS	RPE1	U2OS
Interphase ($m = 1$)	○	○	○	○	○	○	○	○
Prometaphase ($m = 2$)	○	○	○	○	○	○	○	○
Anaphase/telophase ($m = 3$)	○	○	○	○	○	○	○	○

The basic procedure is the same as in Sect. 7.5, i.e., histone modification are summed over fragmented DNA regions of the length of 25,000 nucleotides. Through this preprocessing, individual histone modification profiles can have the same number of features, N, which enables us to integrate these histone modification profiles using TD.

Histone modification used is as follows: There are three time points measured: interphase, prometaphase, and anaphase/telophase. The types of histone modification are H3K27ac, H3K4me1, H3K4me3, and input which is used as a reference. There are two cell lines, RPE1 and U2OS, used. In total, 3(time points) $\times$ 4(histone modicaication) $\times$ 2(cell lines) $=$ 24 conditions are available. Since there are two biological replicates in individual conditions, there are $24 \times 2 = 48$ samples in total (Table 7.65). These 48 samples' histone modification profiles are formatted as tensor, $x_{ijkms} \in \mathbb{R}^{N \times 2 \times 4 \times 3 \times 2}$ that represents the amount of kth histone modification ($k = 1$: acetylation, H3K27ac; $k = 2$: H3K4me1; $k = 3$: H3K4me3; and $k = 4$:Input) of the jth cell line at the mth time point of the sth biological replicates at the ith region. $N = 123{,}817$ which is approximately as many as 120,000 regions that are obtained as 3,000,000,000 necleotide length, which is human genome length divided by 25,000, the length of individual (fragmented) DNA regions. x_{ijkms} was normalized as

$$\sum_{i=1}^{N} x_{ijkms} = 0 \tag{7.102}$$

$$\sum_{i=1}^{N} x_{ijkms}^2 = N. \tag{7.103}$$

Applying HOSVD to x_{ijkms}, we get

$$x_{ijkms} = \sum_{\ell_1=1}^{2} \sum_{\ell_2=1}^{4} \sum_{\ell_3=1}^{3} \sum_{\ell_4=1}^{2} \sum_{\ell_5=1}^{N} G(\ell_1 \ell_2 \ell_3 \ell_4 \ell_5) u_{\ell_1 j}^{(j)} u_{\ell_2 k}^{(k)} u_{\ell_3 m}^{(m)} u_{\ell_4 s}^{(s)} u_{\ell_5 i}^{(i)} \tag{7.104}$$

where $u^{(j)}_{\ell_1 j} \in \mathbb{R}^{2\times 2}$, $u^{(k)}_{\ell_2 k} \in \mathbb{R}^{4\times 4}$, $u^{(m)}_{\ell_3 m} \in \mathbb{R}^{3\times 3}$, $u^{(s)}_{\ell_4 s} \in \mathbb{R}^{2\times 2}$, and $u^{(i)}_{\ell_5 i} \in \mathbb{R}^{N\times N}$, are singular value matrices, which are all orthogonal matrices, and $G(\ell_1\ell_2\ell_3\ell_4\ell_5) \in \mathbb{R}^{2\times 4\times 3\times 2\times N}$ is a core tensor.

At first, we need to identify which singular value vectors attributed to samples are associated with the desired properties:

- Since we do not expect any dependence upon cell lines (j) or biological replicates (s), singular value vectors attributed to j and s, $u^{(j)}_{\ell_1 j}$ and $u^{(s)}_{\ell_4 s}$, should be constant.
- Since we are interested in the reactivation, singular value vectors attributed to m, $u^{(m)}_{\ell_3 m}$, should also represent reactivation. This means that $u^{(m)}_{\ell_3 m}$ should show up-down or down-up profiles dependent upon m.
- As for histone modification, since input ($k = 4$) should be distinct from the other three, H3K27ac, H3K4me1, and H3K4me3, $u^{(k)}_{\ell_2 4}$ should be distinct from, at least, one of $u^{(k)}_{\ell_2 k}, k \leq 3$.

First, we investigate the cell cycle dependence, m (Fig. 7.35). Since we are interested in reactivation after the cell division is completed, the singular value vectors of interest should once decrease from $m = 1$ to $m = 2$, and then go up again from $m = 2$ to $m = 3$. At a glance, ℓ_3 which is most coincident with this requirement is $\ell_3 = 3$. Thus, in the following, we consider $\ell_3 = 3$.

Next, we consider the distinction of H3K27ac, H3K4me1, and H3K4me3 from input (Fig. 7.36). Among four singular value vectors shown in Fig. 7.36, $\ell_2 = 2$ is the most distinct between H3K27ac, H3K4me1, H3K4me3, and input. Thus, in the following, we consider $\ell_2 = 2$. As for j and s independence, we found that $\ell_1 = \ell_4 = 1$ should be employed (Fig. 7.37).

Based upon the selection $\ell_1 = 1$, $\ell_2 = 2$, $\ell_3 = 3$ and $\ell_4 = 1$, we need to find which ℓ_5 is associated with the largest $|G(1, 2, 3, 1, \ell_5)|$ (Fig. 7.38). Since $\ell_5 = 2$ is

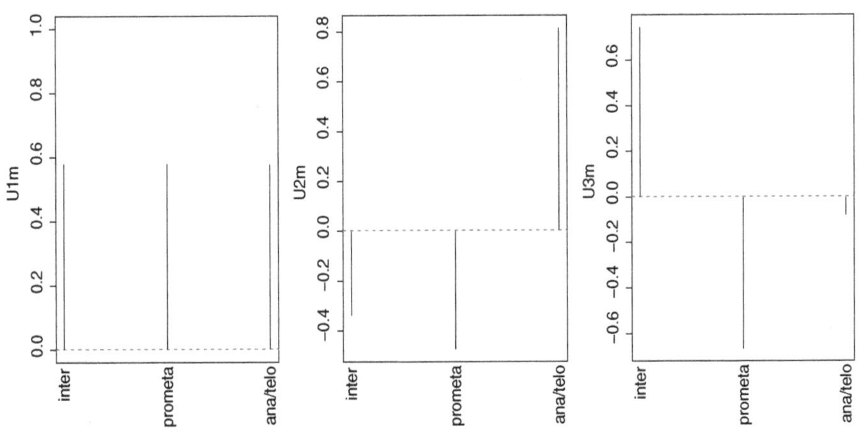

Fig. 7.35 $u^{(m)}_{\ell_3 m}$ that represents cell cycle phase dependence. From left to right, $\ell_2 = 1, 2, 3$. The horizontal red broken lines are the baselines

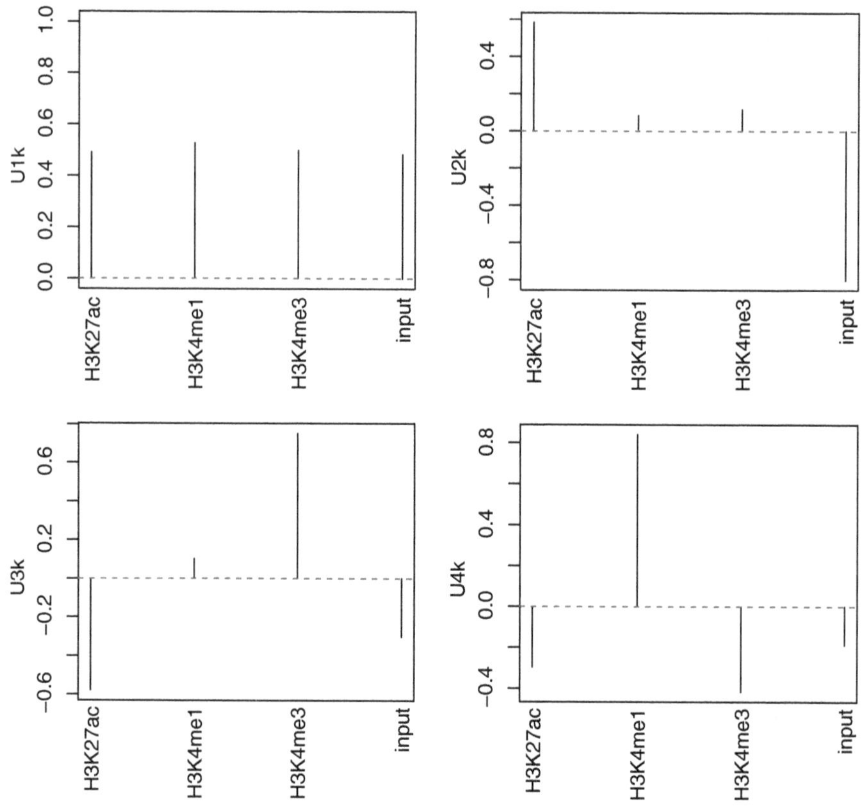

Fig. 7.36 $u_{\ell_2 k}^{(k)}$ that represents cell cycle phase dependence. Top left: $\ell_2 = 1$, top right: $\ell_2 = 2$, bottom left: $\ell_2 = 3$, and bottom right: $\ell_2 = 4$. The horizontal red broken lines are the baselines

associated with the largest absolute $G(1, 2, 3, 1, \ell_5)$, we attributed P values to the ith region as

$$P_i = P_{\chi^2}\left[> \left(\frac{u_{2i}^{(i)}}{\sigma_2} \right)^2 \right]. \tag{7.105}$$

The attributed P values are corrected by the BH criterion, and the 700 regions associated with adjusted P values less than 0.01 are selected. At first, we checked if the histone modification of the selected regions is coincident with reactivation as well as greater than the input (Table 7.66); from tests 5–7, we can see that the histone modification is larger than the input. From test 2, we can see that H3K4me1 can be bookmarked since the distinction was not detected between three phases while the other two histone modifications H3K27ac or H3K4me3 cannot, since H3K27ac and

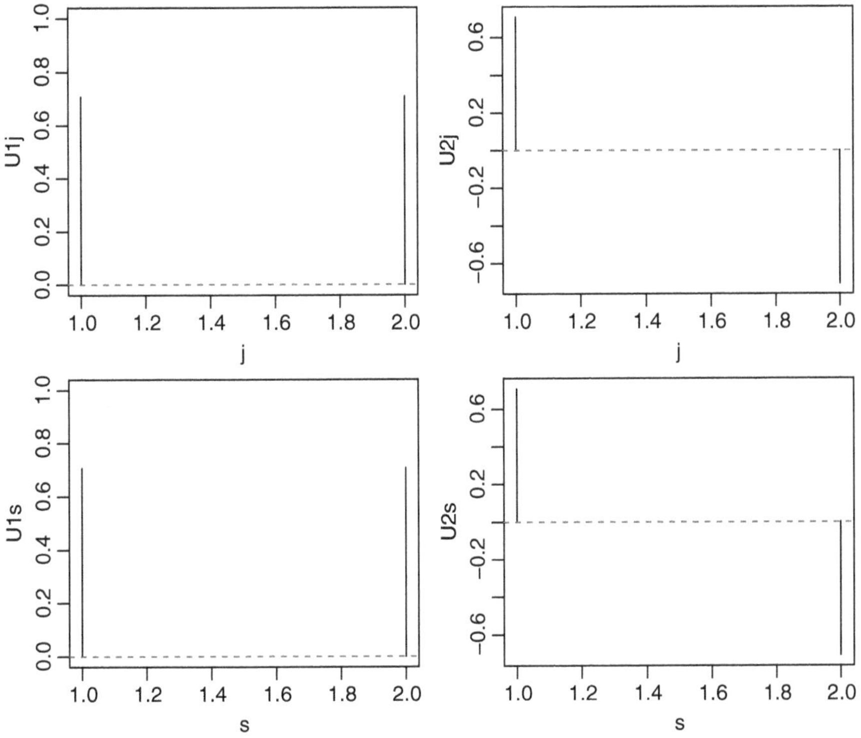

Fig. 7.37 $u^{(j)}_{\ell_1 j}$ (top row) and $u^{(s)}_{\ell_4 s}$ (bottom row) that represent dependence upon cell lines and biological replicates, respectively. Left column: $\ell_1 = \ell_4 = 1$, right column: $\ell_1 = \ell_4 = 2$. The horizontal red broken lines are the baselines

Fig. 7.38 $|G(1, 2, 3, 1, \ell_5)|$. $\ell_5 = 2$ is associated with the largest absolute $G(1, 2, 3, 1, \ell_5)$

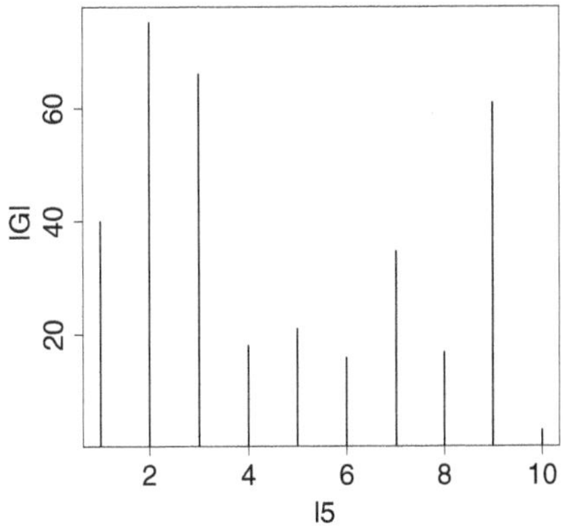

H3K4me3 are distinct between three phases (see tests 1 and 3). This suggests that TD-based unsupervised FE could successfully identify the regions with the desired properties.

Table 7.66 *t*-tests that check if histone modification of the selected 700 regions are coincident with reactivation and greater than inputs

Test	Alternative hypothesis	*P* value	Description of desired relationship
1	$\{x_{ij1ms} \| m = 1, 3\} > \{x_{ij12s}\}$	9.13×10^{-3}	H3K27ac reactivation (int & ana/tel > pro)
2	$\{x_{ij2ms} \| m = 1, 3\} \neq \{x_{ij22s}\}$	6.25×10^{-1}	H3K4me1 bookmark (int & ana/tel = pro)
3	$\{x_{ij3ms} \| m = 1, 3\} < \{x_{ij32s}\}$	1.61×10^{-3}	H3K4me3 anti-reactivation (int & ana/tel < pro)
4	$\{x_{ij4ms} \| m = 1, 3\} \neq \{x_{ij42s}\}$	9.56×10^{-1}	Input as control (int & ana/tel = pro)
5	$\{x_{ij2ms}\} > \{x_{ij4ms}\}$	6.13×10^{-3}	H3K4me1 > Input
6	$\{x_{ij3ms}\} > \{x_{ij4ms}\}$	1.85×10^{-50}	H3K4me3 > Input
7	$\{x_{ij1ms}\} > \{x_{ij4ms}\}$	1.69×10^{-78}	H3K27ac > Input

Next, we tried to biologically evaluate the selected regions. To do this, 1214 gene symbols included in the 700 regions were retrieved using biomaRt [22]; these 1214 gene symbols were uploaded to Enrichr. Table 7.67 shows the enriched histone modification in "ENCODE Histone Modifications 2015" category of Enrichr. At least one experiment is associated with all three histone modifications, H3K27ac, H3K4me1, and H3K4me3.

Next, we tried to check if the selected genes are related to mitosis which takes place during the cell division process targeted in this study. Table 7.68 summarizes the various enriched terms related to mitosis in several categories in Enrichr. It is obvious that the selected 1214 genes are widely related to mitosis.

Since we have successfully confirmed that we could identify the reasonable set of genes, next, we try to identify factors related to bookmark function, i.e., maintaining pre-mitoic state information to postmitotic state. In the study [44] that derived the data set analyzed in this study, the researchers discussed some transcription factors. In the following, we evaluate if these transcription factors significantly target these 1214 gene symbols. Prior to this investigation, we checked how many transcription factors were judged to significantly target these 1,214, gene symbols in various categories in Enrichr (Table 7.69). It is obvious that TD-based unsupervised FE could successfully identify transcription factors that significantly target these 1,214 gene symbols.

Next, we investigated some specific transcription factors discussed in the original research [44], RUNX, TEAD, JUN, FOXO, and FosL (Table 7.70). All of these four transcription factors are identified at least in one of seven transcription factor–related categories in Table 7.69. Thus, TD-based unsupervised FE could identify promising transcription factors as well.

Before closing this subsection, we tried to compare our performance with other conventional methods. Nevertheless, when the original paper was published [107], there were no comparative methods because the comparative methods must satisfy the following two conditions:

Table 7.67 H3K27ac, H3K4me1, and H3K4me3 in "ENCODE Histone Modifications 2015" of Enrichr

Term	Overlap	*P* value	Adjusted *P* value
H3K27ac MCF-7 hg19	166/2000	1.50×10^{-5}	1.10×10^{-4}
H3K27ac fibroblast of lung hg19	266/3617	2.70×10^{-4}	1.85×10^{-3}
H3K27ac HCT116 hg19	157/2000	4.01×10^{-4}	2.47×10^{-3}
H3K27ac ES-Bruce4 mm9	155/2000	7.68×10^{-4}	4.52×10^{-3}
H3K27ac CH12.LX mm9	154/2000	1.05×10^{-3}	6.02×10^{-3}
H3K27ac thymus mm9	151/2000	2.58×10^{-3}	1.42×10^{-2}
H3K27ac H1-hESC hg19	146/2000	9.91×10^{-3}	4.86×10^{-2}
H3K4me1 HCT116 hg19	163/2000	4.78×10^{-5}	3.34×10^{-4}
H3K4me3 fibroblast of foreskin hg19	136/1505	1.41×10^{-6}	1.17×10^{-5}
H3K4me3 NB4 hg19	168/2000	6.65×10^{-6}	5.08×10^{-5}
H3K4me3 fibroblast of mammary gland hg19	158/2005	3.22×10^{-4}	2.07×10^{-3}
H3K4me3 Jurkat hg19	162/2071	3.79×10^{-4}	2.40×10^{-3}
H3K4me3 BE2C hg19	129/1594	4.17×10^{-4}	2.53×10^{-3}
H3K4me3 SK-N-MC hg19	112/1414	2.08×10^{-3}	1.17×10^{-2}
H3K4me3 MCF-7 hg19	144/1890	2.31×10^{-3}	1.29×10^{-2}
H3K4me3 fibroblast of the aortic adventitia hg19	125/1648	5.16×10^{-3}	2.76×10^{-2}
H3K4me3 fibroblast of gingiva hg19	152/2061	5.98×10^{-3}	3.12×10^{-2}
H3K4me3 choroid plexus epithelial cell hg19	151/2069	8.84×10^{-3}	4.50×10^{-2}
H3K4me3 HCT116 hg19	149/2040	9.04×10^{-3}	4.54×10^{-2}
H3K4me3 HCT116 hg19	149/2040	9.04×10^{-3}	4.54×10^{-2}

- The method must compare more than two states, i.e., not only pairwise comparison, since three phases must be compared in this study.
- The method also must be able to consider pairwise comparison between treated (H3K4me1, H3K4me3, and H3K27ac) with input.

There were no methods that could perform the above two-way comparisons simultaneously for ChIP-seq data sets. This point is definitely a great advantage of TD-based method which can easily perform comparisons in multiways simultaneously as demonstrated above.

7.13.2 Prostate Cancer Multiomics Data

In the previous subsection, we demonstrated how TD-based unsupervised FE could analyze multiple histone modification not in pairwise manner but in an integrative manner. In this subsection, not only histone modification but also ChIP-seq, i.e., transcription factor binding to DNA, were analyzed together with histone modification [101], which is much more complicated than that in the previous subsection.

Table 7.68 Enriched terms related to mitosis in several categories of Enrichr

Term	Overlap	*P* value	Adjusted *P* value
Reactome 2022			
Cell cycle, mitotic R-HSA-69278	53/523	1.77×10^{-4}	5.17×10^{-3}
Mitotic G1 phase and G1/S transition R-HSA-453279	20/147	5.84×10^{-4}	1.37×10^{-2}
FBXL7 Downregulates AURKA during mitotic entry and in early mitosis R-HSA-8854050	10/54	1.35×10^{-3}	2.50×10^{-2}
Mitotic metaphase and anaphase R-HSA-2555396	26/233	2.02×10^{-3}	3.37×10^{-2}
BioPlanet 2019			
Ran role in mitotic spindle regulation	6/10	8.40×10^{-6}	6.52×10^{-4}
Mitotic G1-G1/S phases	19/135	5.19×10^{-4}	1.31×10^{-2}
BioCarta 2016			
Role of Ran in mitotic spindle regulation Homo sapiens h ranMSpathway	6/11	1.75×10^{-5}	5.78×10^{-4}
Jensen COMPARTMENTS			
Mitotic spindle microtubule	9/37	2.87×10^{-4}	1.17×10^{-2}
RNAseq automatic GEO signatures human down			
Regulation protein translation mitosis GSE67902 1	50/250	5.54×10^{-14}	3.07×10^{-12}
Hnrnp key synthesis mitosis GSE83493 1	37/250	4.50×10^{-7}	5.31×10^{-6}
Modifier mitotic orchestrating non-synchronized GSE123957 1	27/250	2.68×10^{-3}	1.11×10^{-2}
Mitotically mancr cycle triple GSE102155 1	26/250	5.24×10^{-3}	1.97×10^{-2}
Mitocpr surveillance mitochondria import GSE96726 2	25/250	9.87×10^{-3}	3.40×10^{-2}
RNAseq automatic GEO signatures human up			
Hnrnp key synthesis mitosis GSE83493 1	53/250	7.44×10^{-16}	5.12×10^{-14}
Major determinant mitotic fate GSE68219 4	42/250	1.74×10^{-9}	3.67×10^{-8}
Cohesin removal mitotic entry GSE139845 1	41/250	5.60×10^{-9}	1.04×10^{-7}
Per2 synchronizes mitotic decidual GSE62854 1	40/250	1.75×10^{-8}	2.95×10^{-7}
Major determinant mitotic fate GSE68219 3	40/250	1.75×10^{-8}	2.95×10^{-7}
Major determinant mitotic fate GSE68219 2	37/250	4.50×10^{-7}	5.14×10^{-6}
Modifier mitotic orchestrating non-synchronized GSE123957 1	35/250	3.35×10^{-6}	3.12×10^{-5}
Major determinant mitotic fate GSE68219 1	33/250	2.20×10^{-5}	1.63×10^{-4}
Mitocpr surveillance mitochondria import GSE96726 2	28/250	1.32×10^{-3}	5.99×10^{-3}
V336Y mitochondrial ribosomal Hek293 GSE101503 1	28/250	1.32×10^{-3}	5.99×10^{-3}
Effect mitochondrial uncoupling triple-negative GSE161502 1	25/250	9.87×10^{-3}	3.45×10^{-2}

The data sets analyzed in this study were retrieved from GEO using GEO ID GSE130408. Three transcription factor binding, AR, FOXA1, and HOXB13, four histone modification, H3K27AC, H3K27me3, H3K4me3, and K4me2, and

Table 7.69 Number of transcription factors (TFs) associated with adjusted P values less than 0.05 in various TF-related Enrichr categories

	Terms	Adjusted P values	
		>0.05	<0.05
(I)	ChEA 2022	398	353
(II)	ENCODE and ChEA consensus TFs from ChIP-X	25	80
(III)	ARCHS4 TFs Coexp	1384	335
(IV)	TF perturbations followed by expression	836	1117
(V)	Enrichr submissions TF-gene coocurrence	408	1313
(VI)	ENCODE TF ChIP-seq 2015	517	299
(VII)	TF-LOF expression from GEO	239	25

Table 7.70 Enriched transcription factors discussed in [44]

Transcription factor	(I)	(II)	(III)	(IV)	(V)	(VI)	(VII)
RUNX1	○	○	○	○	○		
RUNX2	○				○		
RUNX3					○		
TEAD1			○		○		
TEAD2			○		○		
TEAD3			○		○		
TEAD4			○		○	○	
JUN	○		○	○	○		
CJUN	○						
JUND	○		○	○	○	○	
JUNB			○	○			
FOXO1	○			○	○		
FOXO3					○		
FOXO4					○		
FOXO6					○		
FOSL1	○		○	○	○	○	
FOSL2		○	○		○	○	

ATAC-seq which measures the opening of DNA; in total, eight omics profiles were measured in each sample. The total number of samples varies from omics to omics. We used two AR, four FOXA1, four HOXB13, ten H3K27AC, one H3K27me3, one H3K4me3, one K4me2, and one ATAC-seq. In total, 24 multiomics measurement groups were constructed. Each group was composed of six samples, i.e., three biological replicates times two tissue subclasses (tumor and normal prostate). Table 7.71 summarizes the number of samples in groups. The amount of values in individual omics profiles are averaged over 25,000 nucleotides as in the previous subsection. The total number of 25,000 nucleotide length regions is 123,817.

These data sets were formatted as a tensor, $x_{ijkm} \in \mathbb{R}^{N \times 24 \times 2 \times 3}$, which represents the averaged value of the ith regions of the jth omics measurement group (see

Table 7.71 Summary of sample numbers in the measurement groups

Omics	Number of Multiomics	Tissues	Biological replicates	j
AR	2	2	3	$1 \le j \le 2$
FOXA1	4	2	3	$3 \le j \le 6$
HOXB13	4	2	3	$7 \le j \le 10$
H3K27AC	10	2	3	$11 \le j \le 20$
H3K27me3	1	2	3	$j = 21$
H3K4me3	1	2	3	$j = 22$
K4me2	1	2	3	$j = 23$
ATAC	1	2	3	$j = 24$
Total	24	–	–	–

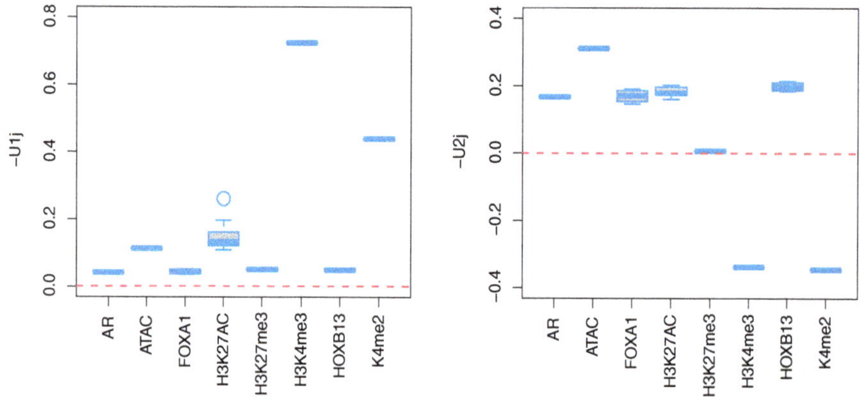

Fig. 7.39 Boxplots of $u_{1j}^{(j)}$ (left) and $u_{2j}^{(j)}$ (right). The horizontal red broken lines are the baselines

Table 7.71) of the kth tissue ($k = 1$: tumor, $k = 2$: normal prostate) of the mth biological replicates ($1 \le m \le 3$). Here, $N = 123{,}817$ is the total number of genomic regions of 25,000 nucleotide length.

After applying HOSVD to x_{ijkm}, we get

$$x_{ijkm} = \sum_{\ell_1=1}^{24} \sum_{\ell_2=1}^{2} \sum_{\ell_3=1}^{3} \sum_{\ell_4=1}^{N} G(\ell_1\ell_2\ell_3\ell_4) u_{\ell_1 j}^{(j)} u_{\ell_2 k}^{(k)} u_{\ell_3 m}^{(m)} u_{\ell_4 i}^{(i)} \tag{7.106}$$

where $G \in \mathbb{R}^{24\times 2\times 3\times N}$ is a core tensor, $u_{\ell_1 j}^{(j)} \in \mathbb{R}^{24\times 24}$, $u_{\ell_2 k}^{(k)} \in \mathbb{R}^{2\times 2}$, $u_{\ell_3 m}^{(m)} \in \mathbb{R}^{3\times 3}$, and $u_{\ell_4 i}^{(i)} \in \mathbb{R}^{N\times N}$ are singular value matrices and orthogonal matrices.

At first, we need to find $u_{\ell_1 j}^{(j)}$s of interest. Figure 7.39 shows $\ell_1 = 1, 2$. To explain why $u_{1j}^{(j)}$ and $u_{2j}^{(j)}$ are of interest, we need to discuss their biological aspects. At first, the values are highly coincident with class labels and have very small diversities

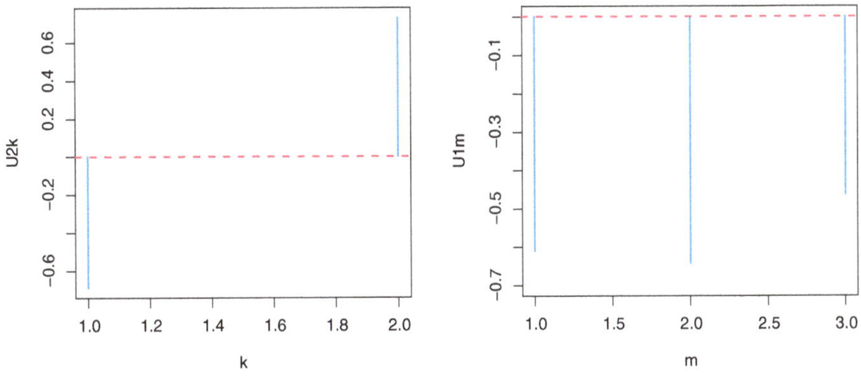

Fig. 7.40 Left: $u_{2k}^{(k)}$, right: $u_{1m}^{(m)}$

within classes. Second, histone modification and transcription factor binding have larger values together in $u_{1j}^{(j)}$ and $u_{2j}^{(j)}$, respectively. On the other hand, ATAC-seq, which means DNA opening, has larger values for both. This suggests that $u_{1j}^{(j)}$ and $u_{2j}^{(j)}$ represent activation of histone modification and transcription factor binding, respectively. Thus, we decided to employ these two singular value vectors as the ones of interest.

Next, we tried to select $u_{\ell_2 k}^{(k)}$ and $u_{\ell_3 m}^{(m)}$ of interest. Since $u_{\ell_2 k}^{(k)}$ should represent distinction between tumor ($k = 1$) and normal prostate ($k = 2$), $u_{\ell_2 1}^{(k)} = -u_{\ell_2 2}^{(k)}$ is desired. We found that $\ell_2 = 2$ satisfies this requirement (Fig. 7.40). On the other hand, since $u_{\ell_3 m}^{(m)}$ should represent the independence of biological replicates, $u_{\ell_3 m}^{(m)}$ should take constant value regardless of m. We found that $\ell_3 = 1$ satisfies this requirement (Fig. 7.40). Then in the following, we consider $\ell_1 = 1, 2$, $\ell_2 = 2$, and $\ell_3 = 1$.

To identify which $u_{\ell_4 i}^{(i)}$ should be used for the selection of regions, is, we need to see which $G(1, 2, 1, \ell_4)$ and $G(2, 2, 1, \ell_4)$ have the largest absolute values (Fig. 7.41). It is obvious that $\ell_4 = 8$ is associated with the largest $|G|$ regardless of ℓ_1. Thus, we decided to use $u_{8i}^{(i)}$ to attribute P values to the ith region

$$P_i = P_{\chi^2}\left[> \left(\frac{u_{8i}^{(i)}}{\sigma_8}\right)^2\right]. \tag{7.107}$$

P values were corrected and the 1,447 regions associated with adjusted P values less than 0.01 were selected.

Although there are multiple ways to evaluate the 1,447 selected regions, one possible evaluation is to count the number of the protein-coding genes included in these 1,447 selected regions. Since this number of regions is less than 2% of the total number of regions, the expected number of genes included is 2% of human

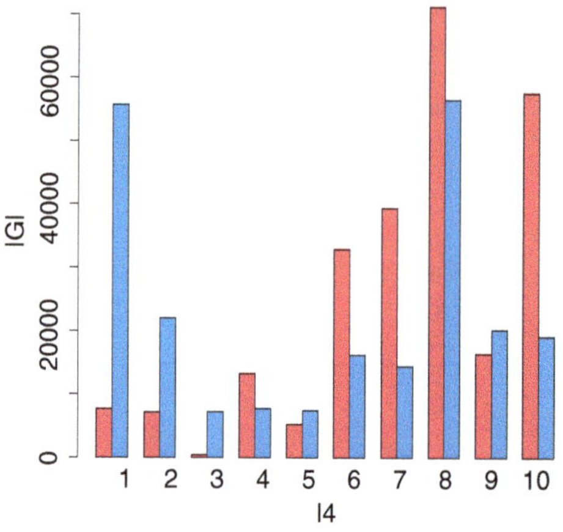

Fig. 7.41 $|G(1, 2, 1, \ell_4)|$ (red) and $|G(2, 2, 1, \ell_4)|$ (blue)

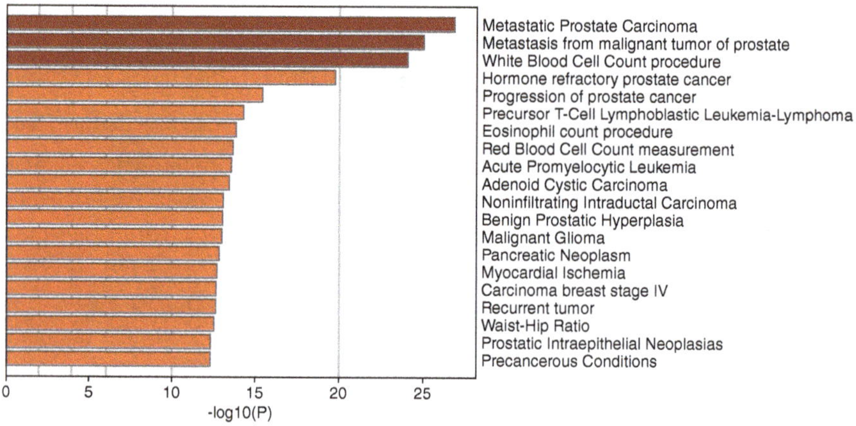

Fig. 7.42 Enrichment analysis in "DisGeNET" of Metascape

genes. Since the number of human genes is about 2×10^4, the expected number of genes included in the 1,447 regions is as small as a few hundreds. Then we evaluated the protein-coding genes included in these 1,447 regions using biomaRt as in the previous subsection. The number of genes is as many as 2,168 which is as many as ten times larger than the expected numbers. We can conclude that TD-based unsupervised FE correctly selected the regions including the protein-coding regions.

To validate further the selected 2,166 genes, we uploaded them to Metascape [128] (Fig. 7.42). Since it is obvious that top-ranked terms include many prostate cancer–related terms with significant P values, TD-based unsupervised FE could select biologically reasonable genes. In addition to this, LNCAP is the top most significant cell line in 'PaGenBase" category of Metascape (Fig. 7.43); LNCAP is the cancer cell line derived from prostate cancer. "Tissue-

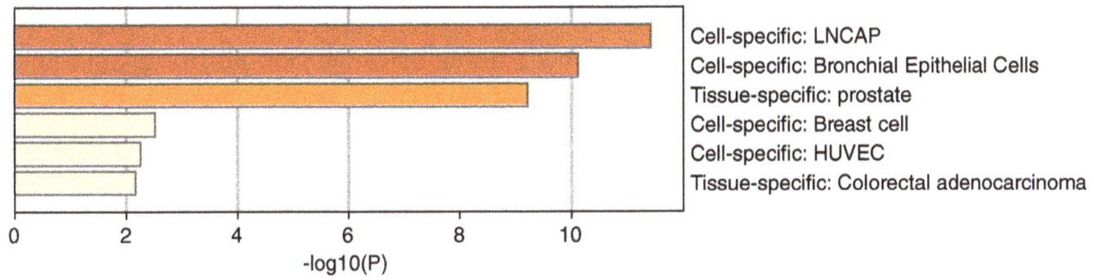

Fig. 7.43 Enrichment analysis in "PaGenBase" of Metascape

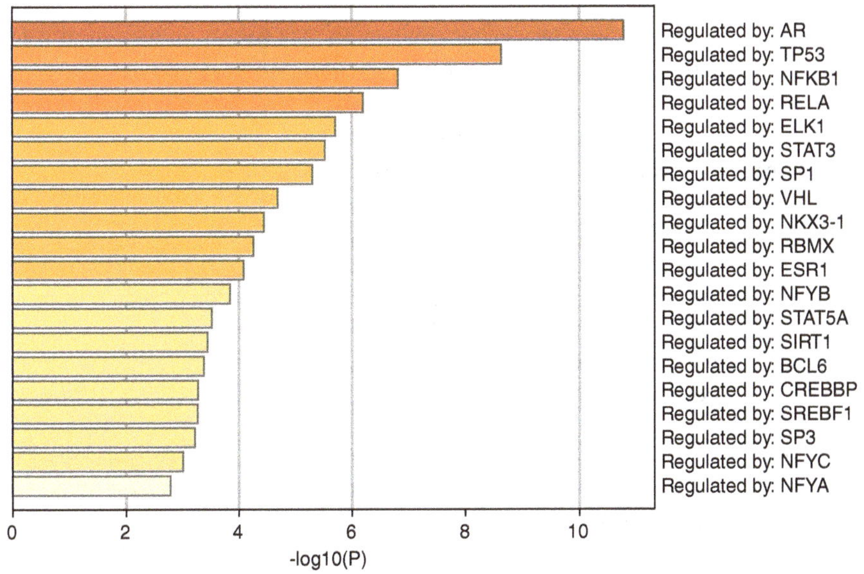

Fig. 7.44 Enrichment analysis in "TRRUST" of Metascape

specific:prostate" is also ranked as the third most significant tissue specificity (Fig. 7.43).

As for transcription factor binding, AR is the most significant transcription factor that targets the uploaded 2,168 genes in "TRRUST" category of Metascape (Fig. 7.44). These also support the success of TD-based unsupervised FE.

Next, we tried to compare the performance of TD-based unsupervised FE with other methods. Although the above performance of TD-based unsupervised FE was excellent, if other much simpler or more conventional methods can achieve similar or superior performance, TD-based unsupervised FE is useless. In contrast to the problem in the previous subsection where only histone modification was integrated, there are no *de facto* standard methods that can integrate histone modification, transcription binding, and ATAC-seq simultaneously. Thus, we were forced to employ more general methods not specific to genomic science.

The first method to be compared is the random forest [36] that is a voting-based tree collection method. It also implements the feature selection ability in it. Thus, it is suitable to be compared with TD-based unsupervised FE. When random forest was applied to the present data sets, the data sets were regarded to be composed of 144 samples which are equal to 24 × 2 × 3 and are supposed to be classified into one of 16 classes each of which is a pair of one of the eight measurement group (Table 7.70) and one of two tissues (e.g., "AR and normal prostate"). The random forest was performed using randomForest package [54] in R. Although it is a result of only one trial (random forest has some dependence upon random seed), 93 out of 144 samples are correctly predicted.

To compare discrimination performance with TD-based unsupervised FE, HOSVD was performed again only with 1,447 regions and 16 classes were tried to be predicted with linear discrimination analysis using $u_{1j}^{(j)} u_{2j}^{(k)} u_{1m}^{(m)}$ as input (leave one out cross validation eave one out cross validation was employed for cross validation). This predicts 90 out of 144 samples correctly.

Although the number of correctly predicted samples is a bit smaller for TD-based unsupervised FE, if we consider other aspects together, the impression might be reversed. For example, whereas TD-based unsupervised FE selected as small as 1,447 regions, random forest selected as many as 11,278 regions which is eight times larger than TD-based unsupervised FE. If we consider the number of selected regions, the superiority of random forest performance, i.e., only three more samples are predicted correctly, is too small to be regarded to be superior to TD-based unsupervised FE.

In addition to this, not only the number of correctly predicted samples but also the probability of being correctly predicted is considered, and the situation changes. Since there are as many as 16 classes and the prediction was performed based upon the class associated with the largest probability, even if the probability assigned to the correct class is very small (e.g., only a few percentages), the prediction can be correct. To evaluate the possibility of being predicted correctly, we define the probability of P_s where s is one of 144 samples. Then we compute two values $\sum_{s=1}^{144} \log P_s$ and $\sum_{s=1}^{144} P_s$ whose values are slightly larger for TD-based unsupervised FE than random forest (Table 7.72). In conclusion, the classification performance of TD-based unsupervised FE is better than random forest.

Although the classification performance of random forest is inferior to TD-based unsupervised FE, biological reliability of the selected genes might be better than TD-based unsupervised FE. To confirm this point, we uploaded 1,267 protein-coding genes associated with top-ranked (i.e., with more importance which random

Table 7.72 Various measures to compare classification performance between random forest and TD-based unsupervised FE

	The ratio of correct prediction	$\sum_{s=1}^{144} \log P_s$	$\sum_{s=1}^{144} P_s$
Random forest	93/144	−151.6276	63.02549
TD-based unsupervised FE	90/144	−148.7659	65.74874

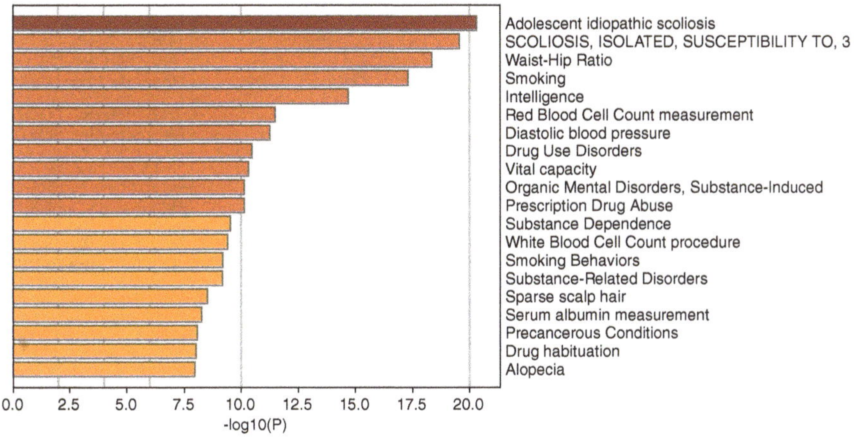

Fig. 7.45 Enrichment analysis in "DisGeNET" of Metascape when 1267 protein-coding genes included in top-ranked 1447 regions were selected by random forest

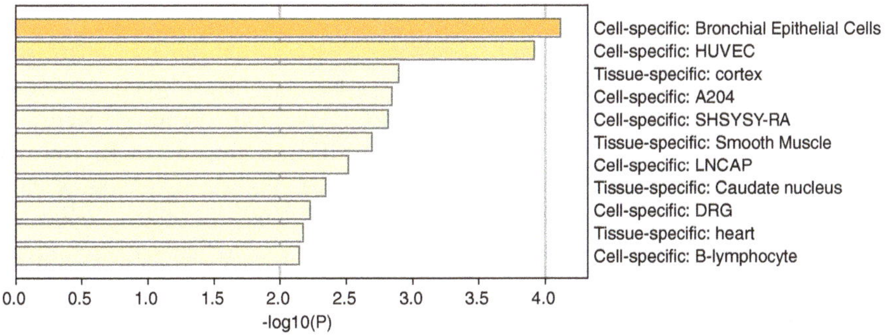

Fig. 7.46 Enrichment analysis in "PaGenBase" of Metascape when 1267 protein-coding genes included in top-ranked 1447 regions are selected by random forest

forest attributes to individual features) 1,447 regions, which is equal to the number of selected regions by TD-based unsupervised FE, to Metascape, since 11,278 regions selected by random forest are too many to upload all protein-coding genes included in such a large number of regions to Metascape. As can be seen in Fig. 7.45, in contrast to Fig. 7.42 where many prostate cancer–related terms are top ranked, no prostate cancer–related terms are included. Also in Fig. 7.46, LNCAP, which is top ranked in Fig. 7.43, was in the eighth rank with nonsignificant *P* values. AR which was top ranked in Fig. 7.44 was also not ranked at all in Fig. 7.47. In conclusion, the random forest's ability to select biologically reasonable genes is substantially inferior to that of TD-based unsupervised FE.

The second method to be compared with TD-based unsupervised FE is categorical regression analysis.

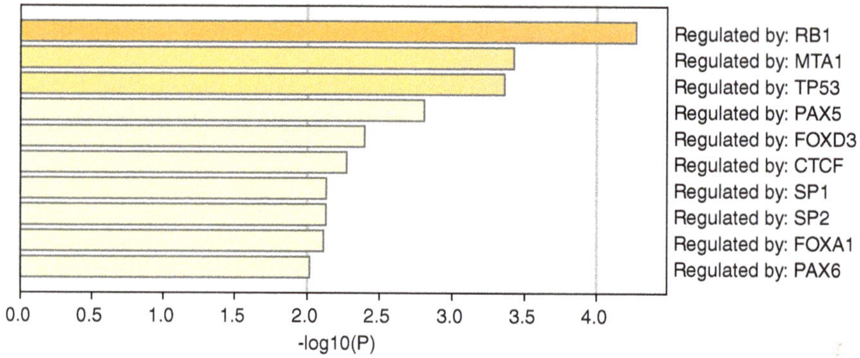

Fig. 7.47 Enrichment analysis in "TRRUST" of Metascape when 1267 protein-coding genes included in top-ranked 1447 regions are selected by random forest

$$x_{ijkm} = \sum_{J=1}^{8} a_J \delta_{Jj} + \sum_{K=1}^{2} b_K \delta_{Kk} \tag{7.108}$$

where a_J and b_K are regression coefficients, δ_{Jj} and δ_{Kk} take 1 when j belongs to the Jth measurement group and k belongs to the Kth tissue, respectively. Categorical regression analysis was performed using `lm` function in R, and the obtained P values are corrected by BH criterion; then 106,761 regions associated with adjusted P values less than 0.01 were selected. Since the total number of regions is 123,817, categorical regression analysis has selected almost all regions. Thus, it is useless. The reason why this happens is because the categorical regression analysis detects any kind of distinction within classes. For example, even if there is distinction between "AR and tumor" and "H3K27ac and normal prostate," categorical regression analysis can attribute the significant P values although it is biologically meaningless; it is impossible for the categorical regression analysis to avoid this biologically meaningless situation. Thus, it detects too many regions associated with small enough P values.

One might still wonder if top-ranked regions are selected, genes included in the selected regions by categorical regression analysis might be enriched in some biological terms. To see this, we considered only top-ranked 1,447 genomic regions and identified 962 protein-coding genes which were uploaded to Metascape (Figs. 7.48, 7.49, and 7.50). It is obvious that they are inferior to those when TD-based unsupervised FE was used, since no prostate cancer–related terms, no prostate tumor–originated cell lines, or no transcription factors targeted by ChIP-seq were included.

Finally, we compared the performance achieved by TD-based unsupervised FE with those by other unsupervised methods. At first, we employed PCA-based unsupervised FE. To apply PCA to the tensor, x_{ijkm} was unfolded as a matrix, $x_{i(jkm)} \in \mathbb{R}^{N \times 144}$. Figure 7.51 shows the boxplots of the first to third PC loading

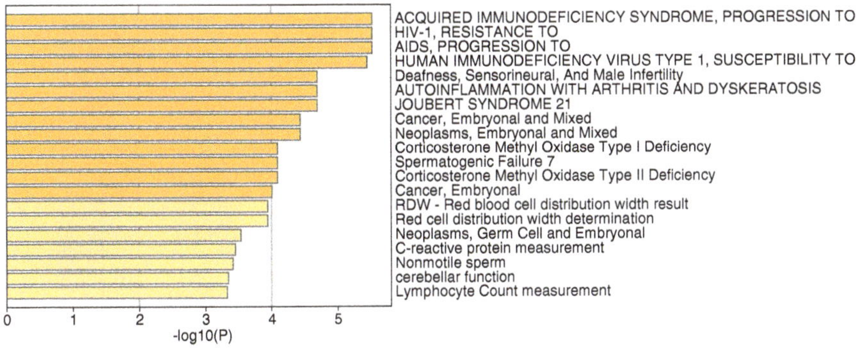

Fig. 7.48 Enrichment analysis in "DisGeNET" of Metascape when 962 protein-coding genes included in top-ranked 1447 regions are selected by categorical regression

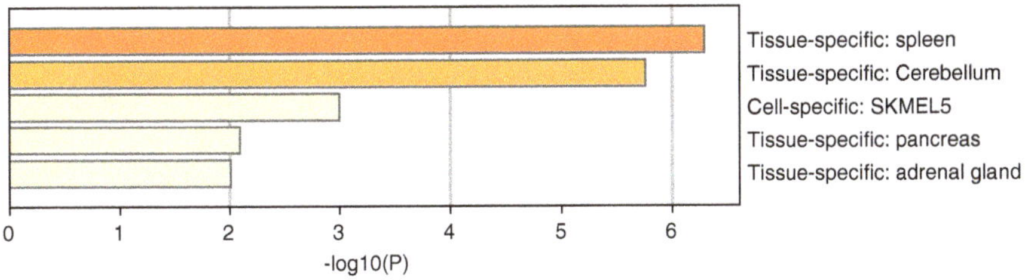

Fig. 7.49 Enrichment analysis in "PaGenBase" of Metascape when 962 protein-coding genes included in top-ranked 1447 regions are selected by categorical regression

attributed to 144 samples. The coincidence with class labels (i.e., measurement groups in Table 7.71) is clearly inferior to that by TD shown in Fig. 7.39, since divergence in individual classes is much larger than that in Fig. 7.39. Thus, PCA-based unsupervised FE is clearly inferior to TD-based unsupervised FE.

As yet another unsupervised method, we employed MNMF [7], which is a multimodal version of nonnegative matrix factorization. It is obvious that MNMF is inferior to TD-based unsupervised FE, since the coincidence with class labels is less than TD-based unsupervised FE (Fig. 7.52).

In conclusion, not only TD-based unsupervised FE could perform very well but also other methods could not compete with TD-based unsupervised FE.

7.14 Effect of Drug Treatment to Gene Expression

In this book, we have repeatedly used gene expression for drug repositioning. Nevertheless, it is unclear how drug treatment affects gene expression. In this section, we investigate the effect of drug treatment on gene expression from a more fundamental perspective, beyond its application in drug repositioning.

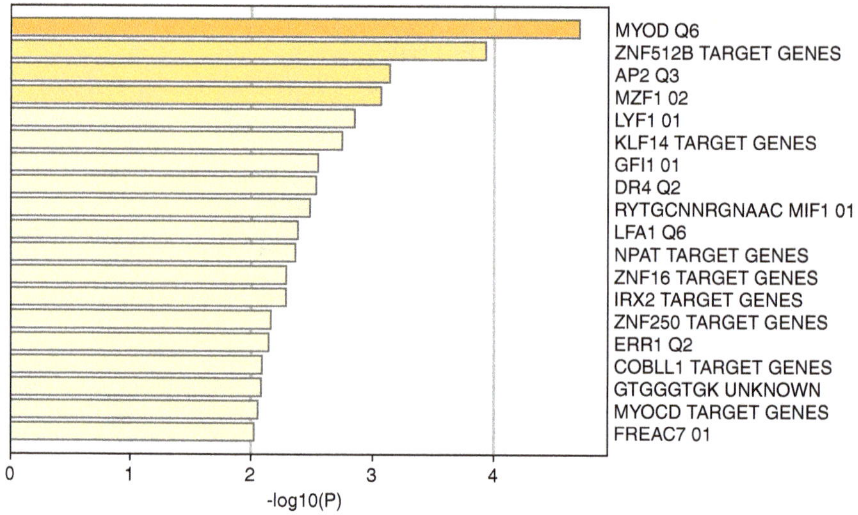

Fig. 7.50 Enrichment analysis in transcription factor targets of Metascape when 962 protein-coding genes included in top-ranked 1447 regions are selected by categorical regression

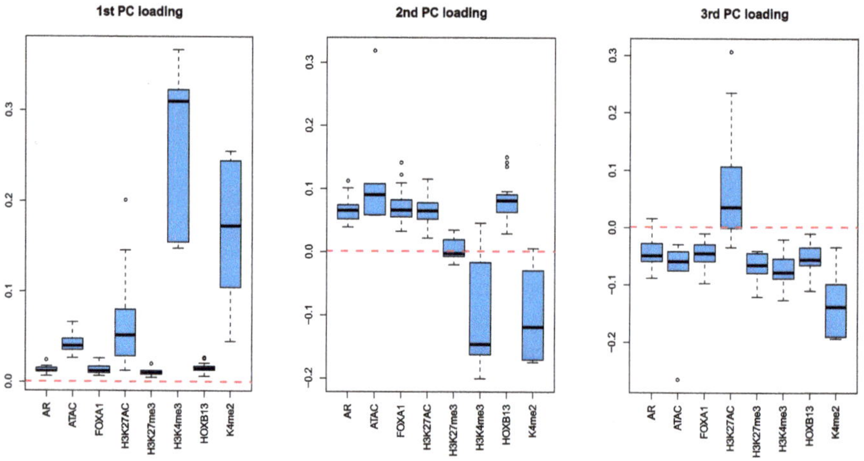

Fig. 7.51 Boxplots of the first to third PC loading attributed to 144 samples classified into eight subclasses, each of which corresponds to a multiomics measurement. The horizontal red broken lines are the baselines (zero)

7.14.1 Drug–Drug Interaction Detection Based on Gene Expression Profiles

As demonstrated above, TD was mainly used in the case where "multiple conditions" × "biological replicates" × "omics profiles" although "omics profiles" can

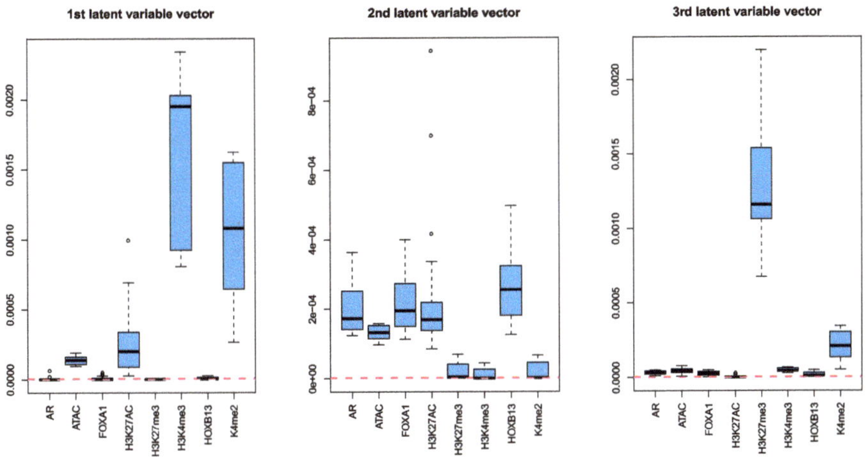

Fig. 7.52 Boxplots of the first to third latent variable vectors attributed to 144 samples classified into eight subclasses, each of which corresponds to a multiomics measurement. The horizontal red broken lines are the baselines (zero)

be composed of either single omics or multiple omics. When "multiple conditions" is composed of more than one type, e.g., human subjects and tissues, the mode of tensor can be more than three. Nevertheless, if TD is the format of "multiple conditions" × "multiple conditions" × "omics profiles," we can investigate the interaction between conditions. In this section, we employed drug treatment as such a condition [105].

The purpose of this kind of analysis, i.e., the investigation of drug–drug interaction, is to find how pairs of drug treatment can alter gene expression in a combinatorial manner. In this sense, the combinatorial effects should not be represented by the simple linear combination of profiles derived by single drug treatment. It is important to know that even if we can observe nonlinear behavior, i.e., that which cannot be represented by the linear combination of the effects of single drug treatments, in the combinatorial treatment of drugs, we are not sure that the origin of the nonlinearity is really the interaction between the drugs.

To see this, we need to compare single drug treatment with pairs of drugs treatment. To do this, we used gene expression profiles of budding yeast treated with either pairs of drugs or a single drug by multiple dose [58], which is available from GEO with GEO ID, GSE138256. The four drugs, myriocin, rapamycin, LiCl, and cycloheximide were tested. The number of combinatorial doses varies from pair to pair (Table 7.73). Since we would like to format gene expression profiles as a tensor, we employ 16 profiles (i.e., minimum number of samples in individual pairs of treatment) among the profiles available. Within each profile, the dose of drugs varies in an antiparallel way, i.e., a dose of one drug decreases as that of another drug increases. Then we get a tensor, $x_{ijk} \in \mathbb{R}^{N \times 16 \times 6}$, which represents the expression of ith gene of jth combination of dose of the kth pair of drugs (Table 7.73). $N = 6{,}717$

Table 7.73 Number of doses tested for drug combinations

	Empty cell	Myriocin	Rapamycin	LiCl
Cycloheximide	22	20	19	
LiCl	27	18		
Rapamycin	16			

is the total number of genes. Gene expression profiles were normalized as

$$\sum_{i=1}^{N} x_{ijk} = 0 \tag{7.109}$$

$$\sum_{i=1}^{N} x_{ijk}^2 = N. \tag{7.110}$$

HOSVD was applied to x_{ijk} and we got

$$x_{ijk} = \sum_{\ell_1=1}^{16} \sum_{\ell_2=1}^{6} \sum_{\ell_3=1}^{N} G(\ell_1 \ell_2 \ell_3) u_{\ell_1 j}^{(j)} u_{\ell_2 k}^{(k)} u_{\ell_3 i}^{(i)} \tag{7.111}$$

where $G \in \mathbb{R}^{16 \times 6 \times N}$ is a core tensor, $u_{\ell_1 j}^{(j)} \in \mathbb{R}^{16 \times 16}$, $u_{\ell_2 k}^{(k)} \in \mathbb{R}^{6 \times 6}$, and $u_{\ell_3 i}^{(i)} \in \mathbb{R}^{N \times N}$ are singular value matrices and are orthogonal matrices. Figure 7.53 shows the lowess smoothed [12] $u_{\ell_1 j}^{(j)}, 1 \leq \ell_1 \leq 6$ where lowess is the abbreviation of "locally weighted scatterplot smoother," which means smoothing by taking weighted local averages of plots. Although one might wonder if convex dependence shown in $\ell_1 = 3, 4$ might be because of nonlinear interaction between drugs, the convex dependence was also observed in Fig. 7.3 (green plots with $\ell_2 = 3$) where only single drug treatment was considered. Thus, the observation of convex dependence upon dose might not always be because of nonlinear interaction between drugs.

To see this point, we retrieved single drug experiments provided together with pairs of drug treatment (Table 7.74) in GSE138256. Gene expression profiles of single drug treatment are formatted as a tensor, $x_{ijk} \in \mathbb{R}^{N \times 14 \times 4}$, which represents the expression of ith gene of jth single drug dose of the kth drug. For $k = 2$, cycloheximide, there are only 11 unique doses. In order to have 14 doses as many as those of the other 3 drugs, 3 out of 11 doses are duplicated using replicates. x_{ijk} is normalized as Eqs. (7.109) and (7.110). HOSVD was applied to x_{ijk}, and we got

$$x_{ijk} = \sum_{\ell_1=1}^{14} \sum_{\ell_2=1}^{4} \sum_{\ell_3=1}^{N} G(\ell_1 \ell_2 \ell_3) u_{\ell_1 j}^{(j)} u_{\ell_2 k}^{(k)} u_{\ell_3 i}^{(i)}. \tag{7.112}$$

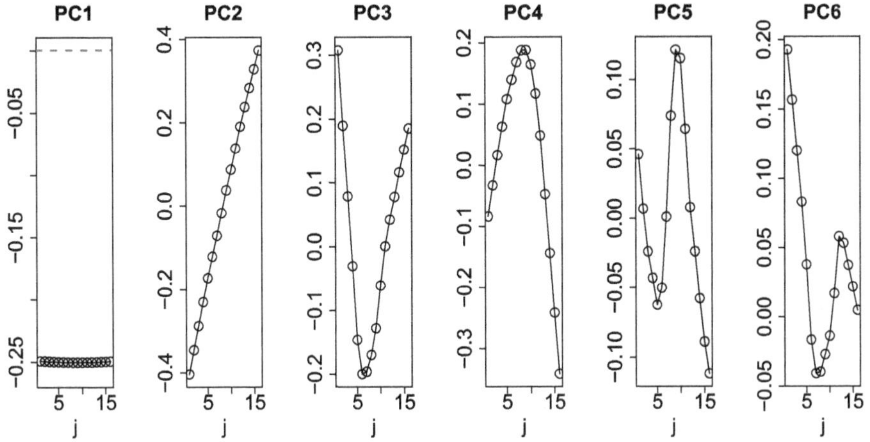

Fig. 7.53 Lowess-smoothed $u^{(j)}_{\ell_1 j}$, $1 \le \ell_1 \le 6$ for combinatorial drug treatments

Table 7.74 Number of doses tested for individual drugs. Numbers in parentheses indicate unique doses

k	Drug	Number of samples
1	Myriocin	25 (14)
2	Cycloheximide	23 (11)
3	LiCl	28 (14)
4	Rapamycin	30 (14)

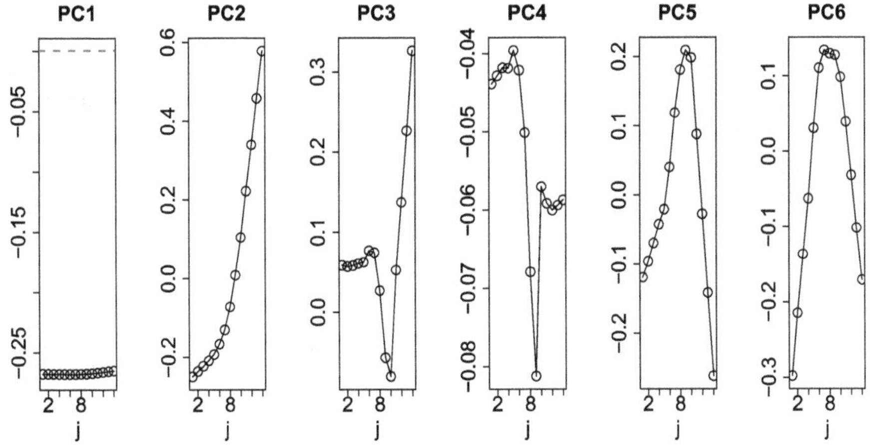

Fig. 7.54 Lowess-smoothed $u^{(j)}_{\ell_1 j}$, $1 \le \ell_1 \le 6$ for single drug treatments

Figure 7.54 shows $u^{(j)}_{\ell_1 j}$, $1 \le \ell_1 \le 6$. In contrast to the expectation that single-dose treatment should not represent nonlinearity, $u^{(j)}_{\ell_1 j}$s surely exhibit nonlinear dependence upon j, although they are the results of single drug treatments. Thus, the nonlinearity shown in Fig. 7.53 is not always due to the interaction between drugs.

One might wonder if nonlinearity seen in Figs. 7.53 and 7.54 might be an artifact caused by applying TD and cannot be seen in individual gene expression. To deny this possibility, we showed two representative gene expressions of individual genes for combinatorial drug treatment (Fig. 7.55) and single drug treatment (Fig. 7.56). It is obvious that nonlinear dose dependence can be seen in individual gene expression profiles as well. Thus, the nonlinearity in Figs. 7.53 and 7.54 are also unlikely to be an artifact.

Next, we investigated the biological function of genes whose expression is associated with convex dependence. For this purpose, we need to find which $u_{\ell_3 i}^{(i)}$s are mostly associated with the convex dependence upon j. First, we consider combinatorial drug treatment. After investigation of $u_{\ell_2 k}^{(k)}$, we realized that $\ell_2 = 1$ is associated with constant value, i.e., independence of combination of drug treatment or single drug treatment (not shown here). Then we seek which $G(\ell_1, 1, \ell_3)$ with $\ell_1 = 3, 4$ has the largest absolute values (Fig. 7.57) since $\ell_1 = 3, 4$ are associated with convex dependence (Fig. 7.53). Then we found that $4 \leq \ell_3 \leq 6$ have the larger $|G(\ell_1, 1, \ell_3)|$ with $\ell_1 = 3, 4$. P values were attributed to is with

$$P_i = P_{\chi^2}\left[> \sum_{\ell_3=4}^{6} \left(\frac{u_{\ell_3 i}^{(i)}}{\sigma_{\ell_3}} \right)^2 \right]. \tag{7.113}$$

P values were corrected by BH criterion and 157 genes associated with adjusted P values less than 0.01 are selected. The 157 genes were uploaded to Metascape, and Fig. 7.58 shows the top-level gene ontology biological processes. Metabolic and cellular processes are the top two–ranked terms. Thus, convex dependence upon dose is related to quite general cellular functions. Next, we consider single drug treatment. After the investigation of $u_{\ell_2 k}^{(k)}$, we again realized that $\ell_2 = 1$ is associated with constant value, i.e., independence upon combination of drug treatment or single drug treatment. Then we seek which $G(\ell_1, 1, \ell_3)$ with $\ell_1 = 5, 6$ has the largest absolute values (Fig. 7.59) since $\ell_1 = 5, 6$ are associated with convex dependence (Fig. 7.54). Then we found that $\ell_3 = 4$ has the largest $|G(\ell_1, 1, \ell_3)|$ with $\ell_1 = 5, 6$. P values were attributed to is with

$$P_i = P_{\chi^2}\left[> \left(\frac{u_{4i}^{(i)}}{\sigma_4} \right)^2 \right]. \tag{7.114}$$

P values were corrected by BH criterion, and 77 genes associated with adjusted P values less than 0.01 are selected. The 77 genes were uploaded to Metascape and Fig. 7.60 shows the top-level gene ontology biological processes. Metabolic and cellular processes are the top two–ranked terms as in Fig. 7.58. Thus, convex dependence upon dose is related to quite general cellular functions as in combinatorial dose. Figure 7.61 shows the Venn diagram of genes selected in combinatorial drug treatment and single drug treatment. They are highly overlapped. Since we could

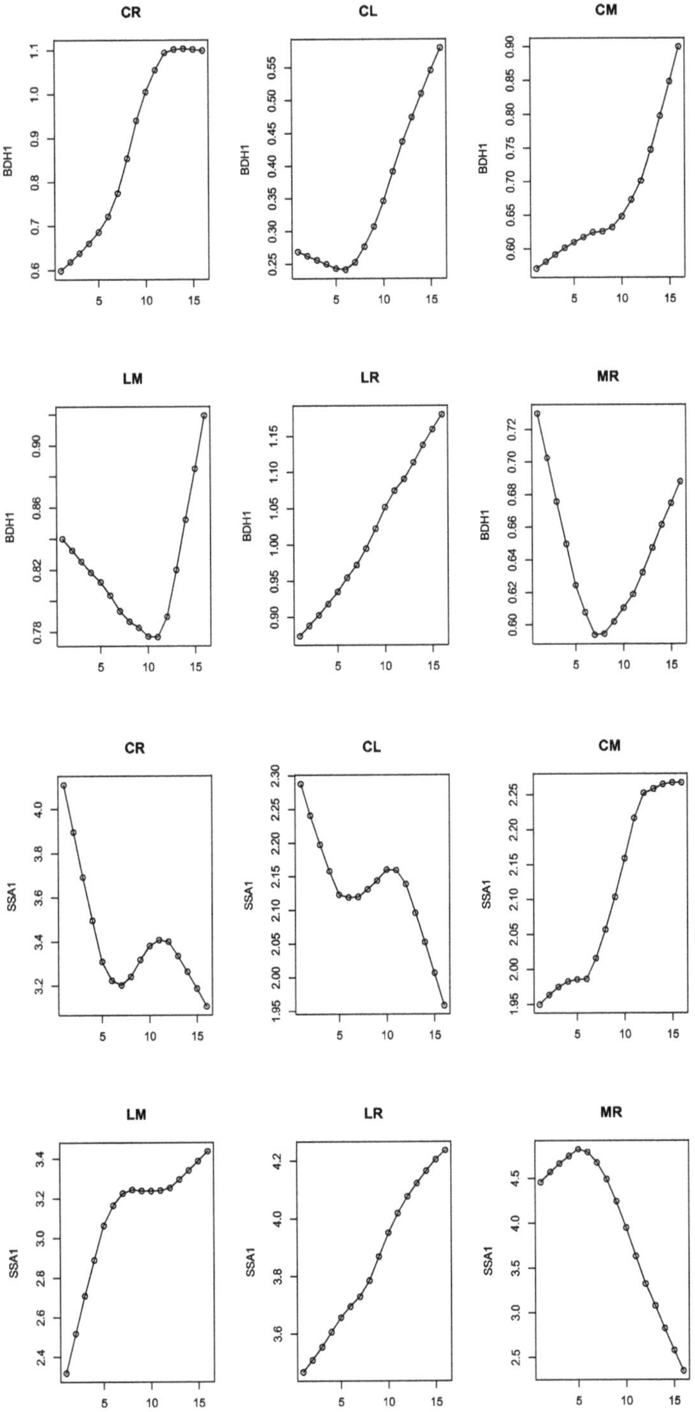

Fig. 7.55 Lowess-smoothed gene expression profiles for BDH1 (the first and second rows) and SSA1 (the third and fourth rows). Two letters above each panel show the combinations of drugs: M: myriocin C: cycloheximide, L: LiCl, R: rapamycin

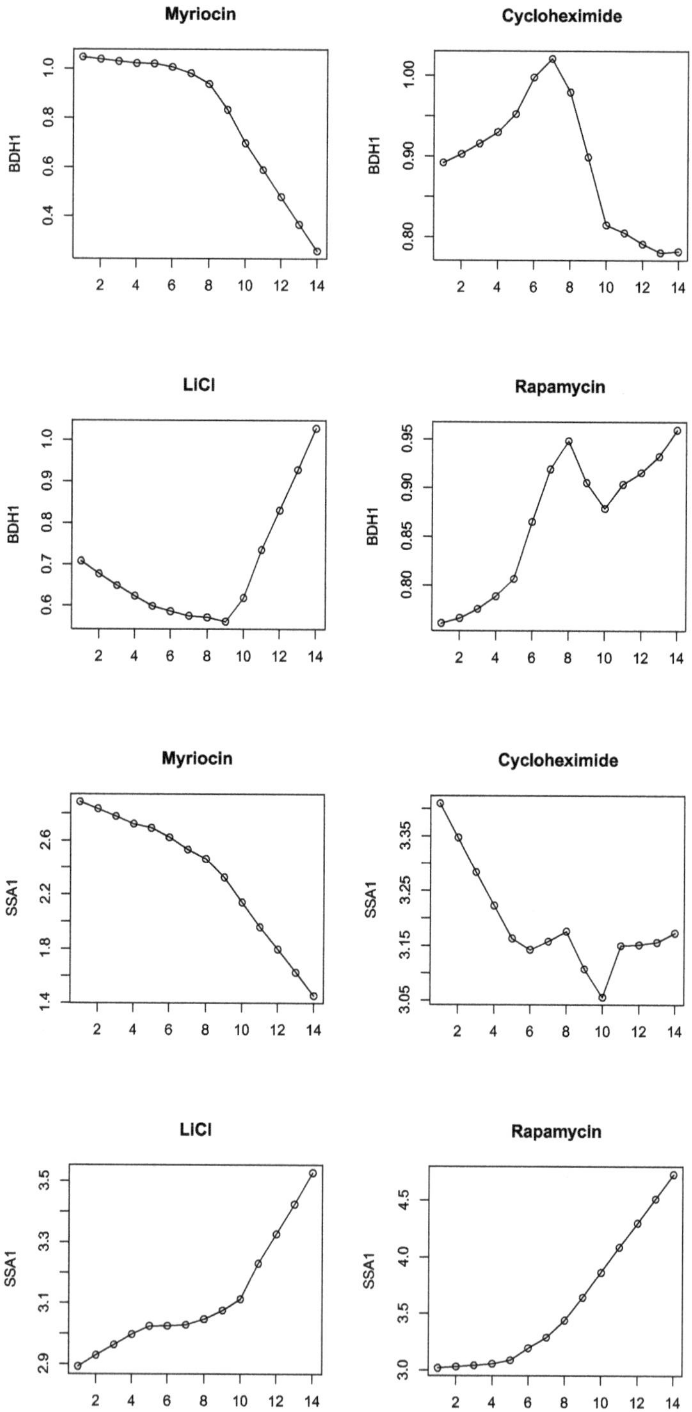

Fig. 7.56 Lowess-smoothed gene expression profiles for BDH1 (the first and second rows) and SSA1 (the third and fouth rows)

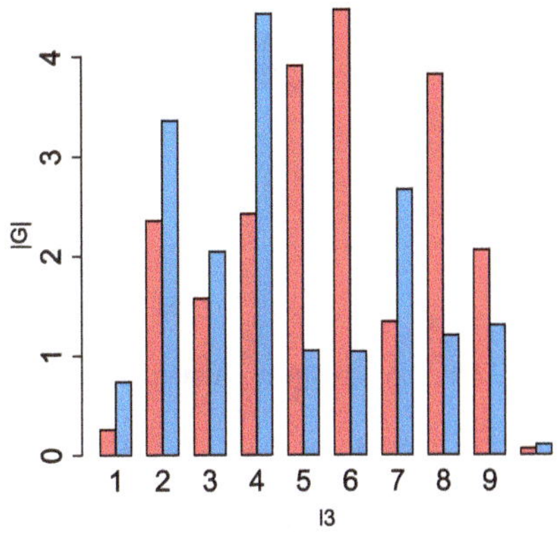

Fig. 7.57 $|G(\ell_1, 1, \ell_3)|$ where $\ell_1 = 3$ (red) or $\ell_1 = 4$ (blue) for combinatorial drug treatment

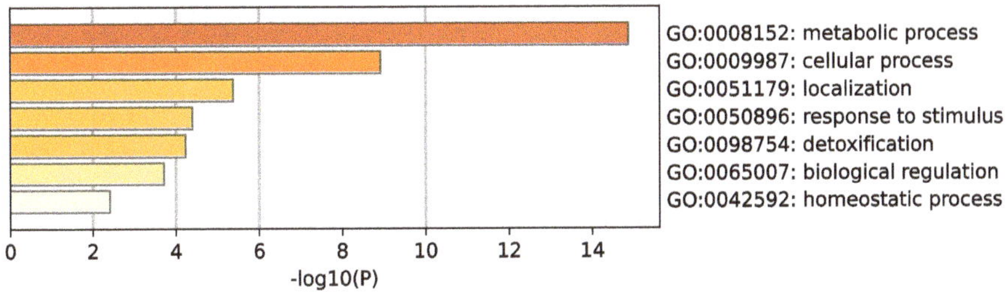

Fig. 7.58 Top-level gene ontology biological processes by Metascape

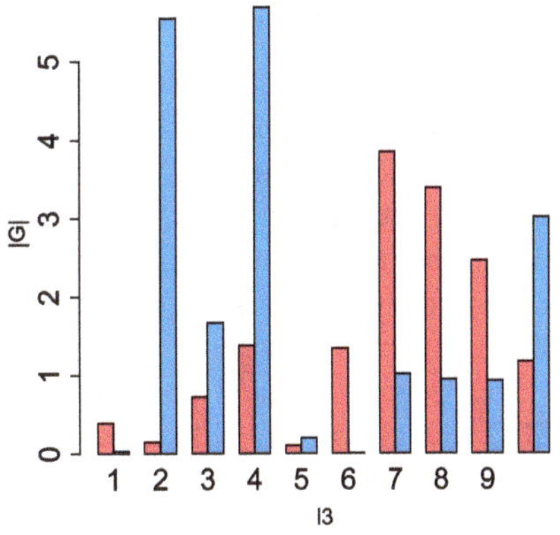

Fig. 7.59 $|G(\ell_1, 1, \ell_3)|$ where $\ell_1 = 5$ (red) or $\ell_1 = 6$ (blue) for combinatorial drug treatment

not find any critical difference of genes associated with convex dose dependence between combinatoral drug treatment and single grug treatment, nonlinearity is unlikely because of interaction between drug treatments.

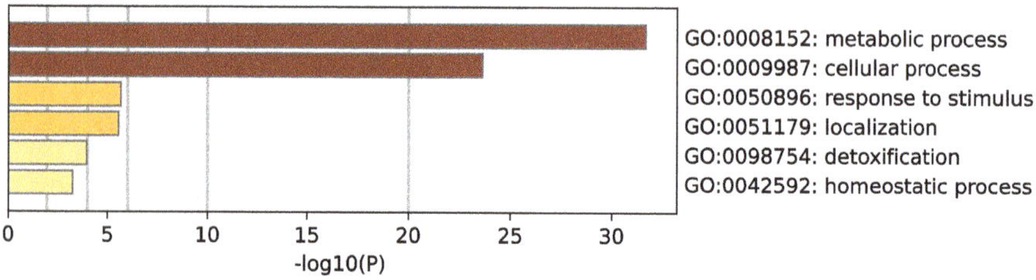

Fig. 7.60 Top-level gene ontology biological processes by Metascape

Fig. 7.61 Venn diagram of 157 genes selected for combinatorial drug treatments union 77 genes selected for single drug treatments

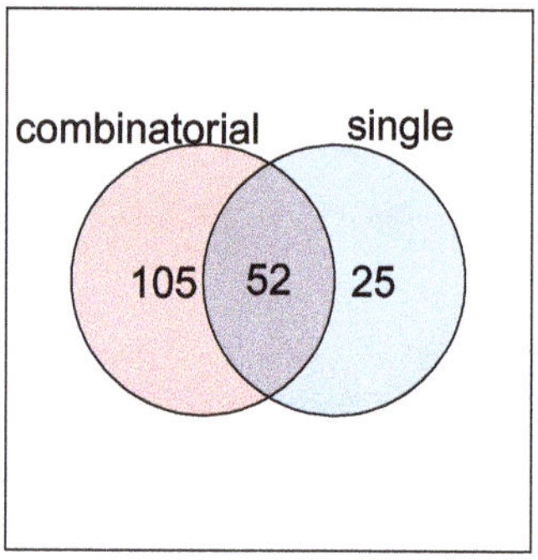

7.14.2 *Dependency of Gene Expression on Tissue and Drug Treatment*

In contrast to the previous subsection where the combinatorial effect of drug treatment was investigated, in this subsection, we focus on yet another important aspect of drug treatment: i.e., tissue specificity. Drugs are expected to affect gene expression profiles in the tissue-specific manner. But how? To see this, we applied TD-based unsupervised FE to mouse gene expression in various tissues with various drug treatments [102].

The gene expression profiles analyzed in this subsection were retrieved from GEO with GEO ID GSE142068. It is composed of mouse gene expression profiles measured in as many as 24 tissues treated with 15 drugs (with some controls). The gene expression profiles are formatted as a tensor as usual, $x_{ijkm} \in \mathbb{R}^{N \times 24 \times 18 \times 2}$ that represents the expression of ith gene at jth tissue with kth drug treatment of mth biological replicates where N is the number of genes. It is also normalized as usual

$$\sum_{i=1}^{N} x_{ijkm} = 0 \tag{7.115}$$

$$\sum_{i=1}^{N} x_{ijkm}^2 = N. \tag{7.116}$$

HOSVD was applied to x_{ijkm}, and we got

$$x_{ijkm} = \sum_{\ell_1=1}^{24} \sum_{\ell_2=1}^{18} \sum_{\ell_3=1}^{2} \sum_{\ell_4=1}^{N} G(\ell_1 \ell_2 \ell_3 \ell_4) u_{\ell_1 j}^{(j)} u_{\ell_2 k}^{(k)} u_{\ell_3 m}^{(m)} u_{\ell_4 i}^{(i)} \tag{7.117}$$

where $G \in \mathbb{R}^{24 \times 18 \times 2 \times N}$ is a core tensor, $u_{\ell_1 j}^{(j)} \in \mathbb{R}^{24 \times 24}$, $u_{\ell_2 k}^{(k)} \in \mathbb{R}^{18 \times 18}$, $u_{\ell_3 m}^{(m)} \in \mathbb{R}^{2 \times 2}$, and $u_{\ell_4 i}^{(i)} \in \mathbb{R}^{N \times N}$ are singular value matrices and orthogonal matrices.

As usual, we need to find which singular value vectors attributed to tissues or drug treatment are of interest. First, we consider those attributed to tissues. Figure 7.62 shows $u_{\ell_1 j}^{(j)}, 1 \leq \ell_1 \leq 6$. Those with $\ell_1 \geq 2$ reveal clear tissue specificity. $u_{2j}^{(j)}$ shows larger absolute values for brain, eye, pituitary gland, and testis. Three out of four, brain, eye, and pituitary gland, are nervous tissues. Although testis is not a nervous tissue, brain and testis are known to be more similar with each other than previously thought [63]. Thus, co-expression between testis and neuronal tissues is not an artifact. Hereafter we call these four tissues as neuron-specific tissues. Since $u_{3j}^{(j)}$ has the larger absolute value only for one tissue, we simply do not consider it further. $u_{4j}^{(j)}$ exhibits larger values for SkMuscle (skeletal muscle) and heart. Since these two are muscle tissues, hereafter we call these two tissues muscle-specific tissues. $u_{5j}^{(j)}$ and $u_{6j}^{(j)}$ have larger absolute values for Pancreas and Stomach, although two values have the opposite signs for $u_{6j}^{(j)}$. Hereafter we call these two gastrointestinal-specific tissues. Thus, $u_{\ell_1 j}^{(j)}$ exhibit tissue-specific expression coincident with their functions. We consider $\ell_1 = 2, 4, 5$, and 6 for the further analyses.

Next, we need to specify $u_{\ell_2 k}^{(k)}$ of interest, which is attributed to drug treatment. Figure 7.63 shows the scatter plot of $u_{2k}^{(k)}$ and $u_{3k}^{(k)}$. Since $u_{1k}^{(k)}$ did not show any k dependence, it was ignored (not shown here). In this plane spanned by $u_{2k}^{(k)}$ and $u_{3k}^{(k)}$, drug treatments are separated into two groups: the smaller one including mainly control groups and the larger one including other than controls. In the latter (the larger one), drug treatments are aligned perpendicular to the direction that points from controls to others. Thus, in the following, we consider $\ell_2 = 2, 3$ for further analyses.

The next step is to identify $u_{\ell_3 i}^{(i)}$ used for the selection of is (genes). Since $u_{\ell_1 j}^{(j)}$ exhibits the complicated tissue dependency upon ℓ_1, it is not straight forward. In the following, we selected genes associated with tissue specificity one by one. Since $\ell_3 = 1$ is always associated with a constant value of $u_{\ell_3 m}^{(m)}$, i.e, independent of biological replicates, we check $|G(\ell_1, \ell_2, 1, \ell_4)|$ for the ℓ_1s and ℓ_2s selected above (Fig. 7.64). Then we selected $\ell_4 = 2$ for $\ell_1 = 2$, $\ell_4 = 4$ for $\ell_1 = 4$, $\ell_4 = 5$ for $\ell_1 = 5$, and $\ell_4 = 6, 7$ for $\ell_1 = 6$.

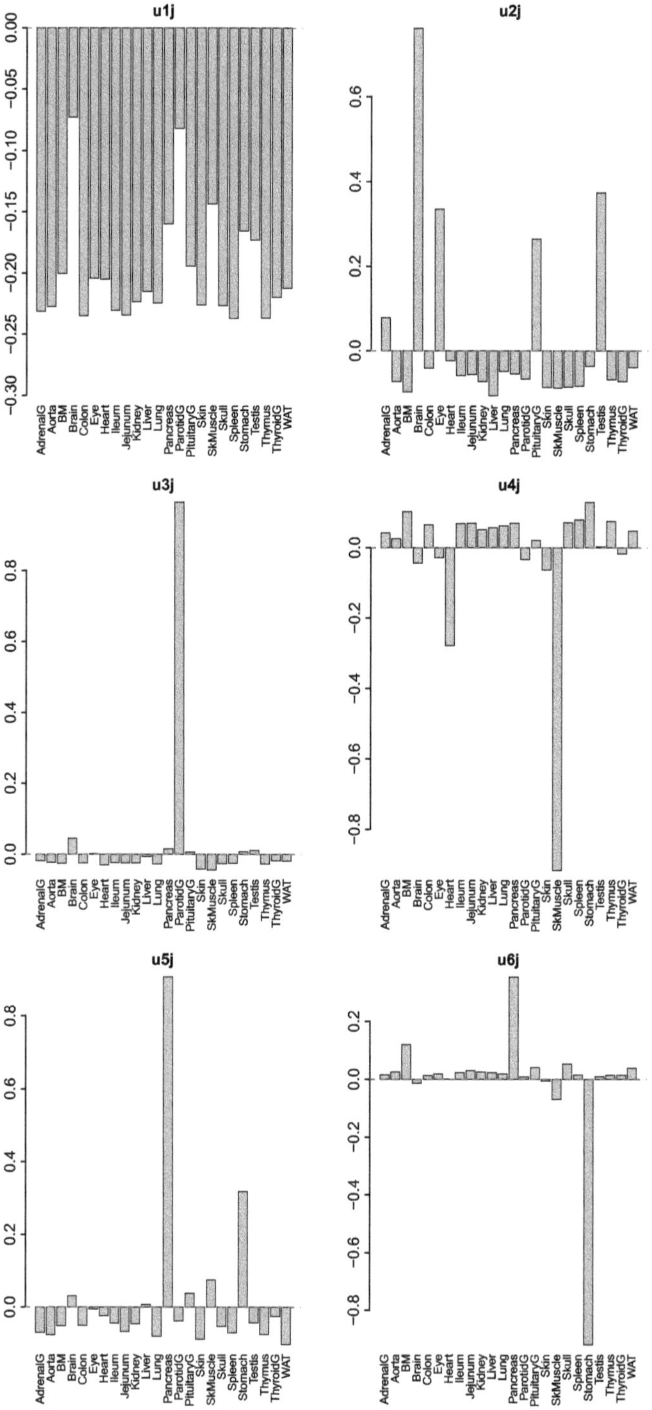

Fig. 7.62 Singular value vectors, $u^{(j)}_{\ell_1 j}$, attributed to tissues. u_{1j}: no tissue specificity. u_{2j}: Brain, eye, pituitary, and testis, thus mostly neuron-specific. u_{3j}: Parotid-specific, u_{4j}: Heart and SkMuscle, thus muscle-specific, u_{5j} and u_{6j}: stomach and pancreas, thus, gastrointestinal-specific

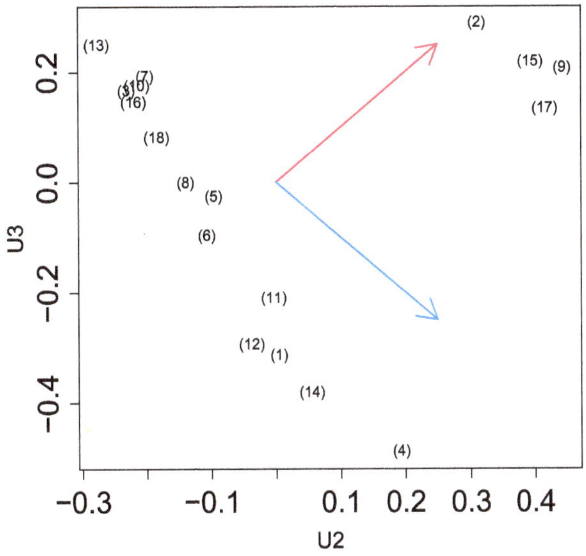

Fig. 7.63 Scatter plot between $u_{2k}^{(k)}$ (horizontal axis) and $u_{3k}^{(3)}$ (vertical axis) attributed to drug treatment. Red and blue arrows represent distinct controls and drug treatments, and diversity among drug treatments, respectively. (1) Alendronate, (2) APAP, (3) aripiprazole, (4) asenapine, (5) cisplatin, (6) clozapine, (7) Dox, (8) EMPA, (9) FivePercentSucrose, (10) lenalidomide, (11) lurasidone, (12) olanzapine, (13) repatha, (14) risedronate, (15) sofosbuvir, (16) teriparatide, (17) WT.No.treated, (18) 5%CMC0.25%Tween80

We attributed P values to is in the tissue-specific manners as

$$P_i = P_{\chi^2}\left[> \left(\frac{u_{2i}^{(i)}}{\sigma_2}\right)^2\right], \text{Neuron} \tag{7.118}$$

$$= P_{\chi^2}\left[> \left(\frac{u_{4i}^{(i)}}{\sigma_4}\right)^2\right], \text{Muscle} \tag{7.119}$$

$$= P_{\chi^2}\left[> \left(\frac{u_{5i}^{(i)}}{\sigma_5}\right)^2\right], \text{Gastric 1} \tag{7.120}$$

$$= P_{\chi^2}\left[> \sum_{\ell_4=6}^{7}\left(\frac{u_{\ell_4 i}^{(i)}}{\sigma_{\ell_4}}\right)^2\right], \text{Gastric 2} \tag{7.121}$$

P values were corrected by BH criterion, and genes associated with adjusted P values less than 0.01 are selected (Table 7.75).

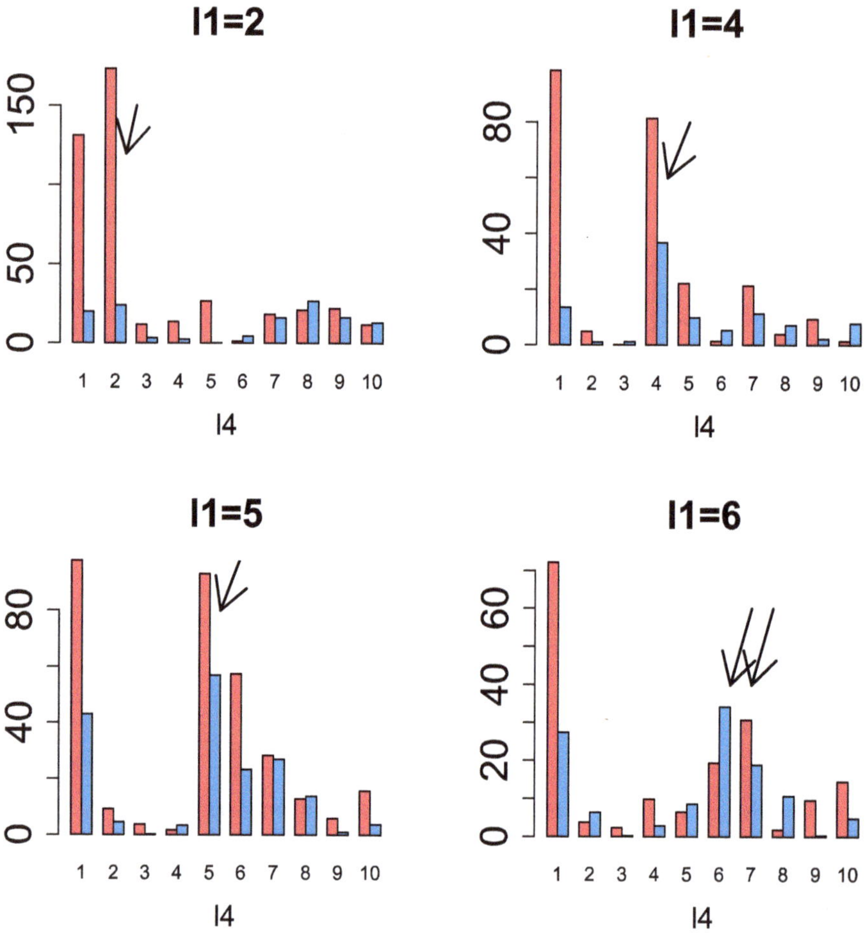

Fig. 7.64 $|G(\ell_1, \ell_2, 1, \ell_4)|$. Red: $\ell_2 = 2$, blue: $\ell_2 = 3$. Black arrows indicate the selected ℓ_4s. $\ell_1 = 2$: $\ell_4 = 2$, $\ell_1 = 4$: $\ell_4 = 4$, $\ell_1 = 5$: $\ell_4 = 5$, $\ell_1 = 6$: $\ell_4 = 6, 7$

We have also checked if the selected genes are distinct between two groups (selected tissues vs. others, or control vs. drug treatment) and confirmed that their expression is distinct between two classes (Table 7.75). Suppose that a set I is that of genes selected. For the comparisons between selected tissues (i.e., neuron-specific tissues, muscle-specific tissues, and gastrointestinal tissues), js that belong to the specified group are defined as a set J and others as $\overline{J}$. Then $\{x_{ijkm} | j \in J, i \in I\}$ and $\{x_{ijkm} | j \in \overline{J}, i \in I\}$ are compared by t-test. For the comparisons between controls and drug treatments, suppose that K is a set of drug treatments and $\overline{K}$ is a set of controls (i.e., (2), (9), (15), and (17) in Fig. 7.63). Then $\{x_{ijkm} | k \in K, i \in I\}$ and $\{x_{ijkm} | k \in \overline{K}, i \in I\}$ are compared by t-test and Wilcoxon test.

Table 7.75 The number of genes selected and the results of statistical tests applied to these sets of genes

				P values by statistical tests		
				Tissues	Drug treatment	
ℓ_1	Tissue specificity	Number of genes	Specified tissues	*t*-test	Wilcoxon test	*t*-test
2	Neuron	18	Brain, eye, pituitary, testis	2.14×10^{-24}	9.65×10^{-49}	0.22
4	Muscle	51	Heart, SkMuscle	1.99×10^{-55}	2.67×10^{-77}	0.04
5	Gastrointestinal	97	Pancreas, stomach	8.48×10^{-11}	2.74×10^{-40}	8.13×10^{-22}
6		128		6.67×10^{-8}	8.89×10^{-90}	2.83×10^{-27}

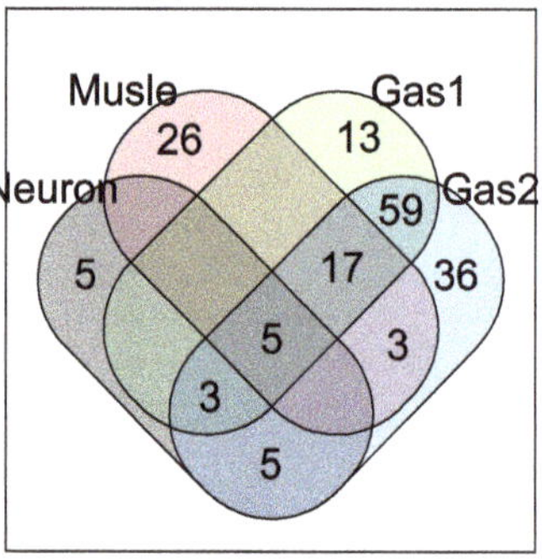

Fig. 7.65 Venn diagram of genes selected by TD-based unsupervised FE. Neuron: genes associated with $u^{(j)}_{2j}$, which is supposed to be neuron-specific. Muscle: genes associated with $u^{(j)}_{4j}$, which is supposed to be muscle-specific. Gas1 and Gas2: genes associated with $u^{(j)}_{5j}$ and $u^{(j)}_{6j}$, respectively, which are supposed to be gastrointestinal-specific

Figure 7.65 shows Venn diagram of selected genes. They are distinct from one another in tissue-specific manners. There are in total 172 genes and only five genes are selected by all four tissue-specific groups. On the other hand, five genes in neuron-specific tissue groups, 26 genes in muscle tissue-specific groups, and $13 + 59 + 36 = 108$ genes in two gastrointestinal-specific tissue groups, in total $5 + 26 + 108 = 139$ genes among 172 genes were selected from only one of three tissue-specific groups. Thus, TD-based unsupervised FE successfully identified tissue-specific gene groups apart from one another. Lastly, we delved into whether this group of genes, which are anticipated to exhibit tissue-specific characteristics, is linked to tissue specificity from a biological functional perspective. Figure 7.66 is an enrichment analysis obtained when four sets of genes were uploaded to Metascape. Their enriched terms are well coincident with tissue specificity. For example, for neuron-specific tissue group, "sperm DNA condensation" and "visual perception" were enriched. It is coincident with the fact that neuron-specific tissue group is composed of nervous tissues and testis. For the muscle tissue group, "striated muscle contraction" and "regulation of muscle system process" were reasonably detected. The gastrointestinal-specific tissue groups were associated with "pancreatic secretion." All of these are coincident with the supposed tissue specificity. Thus, TD-based unsupervised FE successfully identified tissue-specific genes.

Next, we seek if the selected genes are known to be related to diseases. To do this, we upload four sets of genes to Enrichr and check the category, "Orphanet Augmented 2021" (Table 7.76). It is obvious that there are many tissue-specific diseases listed. For neuron-specific tissues, various cataract-related terms were listed. For muscle-specific tissues, there are some terms related to muscle deficiency. For two gastrointestinal-specific tissue groups, some pancreatic diseases are included. Thus, drug treatment altered the expression of disease-related genes in these tissue-specific groups as expected.

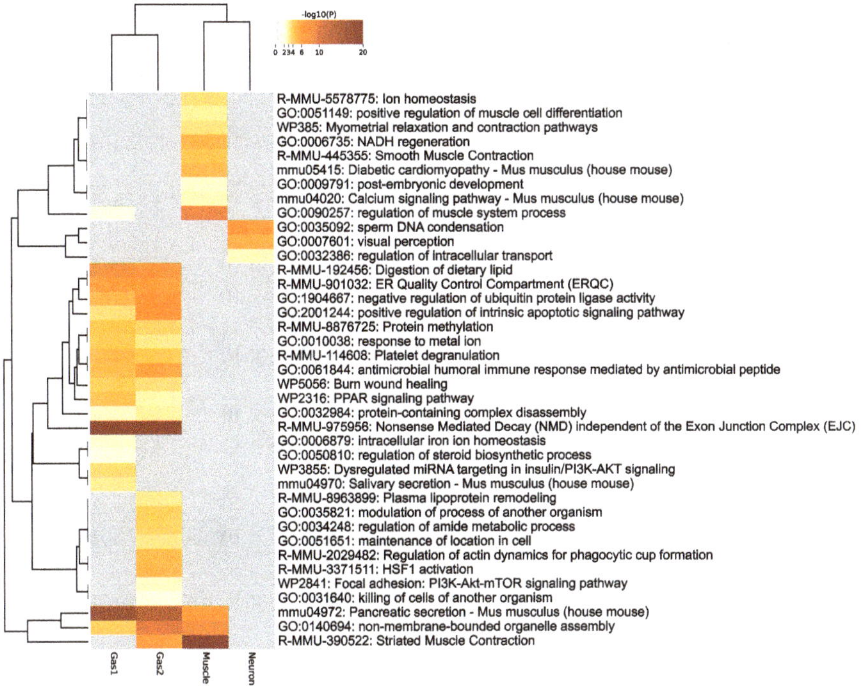

Fig. 7.66 Heatmap of enrichment analysis provided by Metascape

Next, we investigated if these drugs are related to diseases in a way coincident with these tissue specificity (Table 7.77). Then we found that these drugs are frequently related to nervous, muscle, and gastrointestinal diseases. Although most of the individual terms in Table 7.77 are associated with citations, listing all of them here might be simply annoying; if the readers are interested in actual references, they can refer to the original paper [102] of this study.

In conclusion, in this subsection, we could successfully demonstrate that TD-based unsupervised FE has the ability to integrate gene expression measures in various tissues treated by various drugs.

7.15 Drug Repositioning for SARS-CoV-2

From the methodological point of view, the content of this section might not be very novel. Nevertheless, from the application and historical point of view, attacking COVID-19 should be full of interest. Thus, we decided to introduce two studies where drug repositioning targets COVID-19.

Table 7.76 Enriched terms in "Orphanet Augmented 2021" category of Enrichr. For the tissues other than neuron-specific tissues, only seemingly related terms are listed

Term	Overlap	P value	Adjusted P value
Neuron-specific tissues			
Early-onset anterior polar cataract ORPHA:98988	3/104	1.05×10^{-4}	3.72×10^{-3}
Cataract-microcornea syndrome ORPHA:1377	3/105	1.08×10^{-4}	3.72×10^{-3}
Early-onset lamellar cataract ORPHA:441452	3/108	1.18×10^{-4}	3.72×10^{-3}
Pulverulent cataract ORPHA:98984	3/109	1.21×10^{-4}	3.72×10^{-3}
NON RARE IN EUROPE: Partial color blindness, protan type ORPHA:319691	2/101	3.66×10^{-3}	4.77×10^{-2}
Klippel-Feil anomaly-myopathy-facial dysmorphism syndrome ORPHA:447974	2/101	3.66×10^{-3}	4.77×10^{-2}
Retinitis punctata albescens ORPHA:52427	2/101	3.66×10^{-3}	4.77×10^{-2}
X-linked cleft palate and ankyloglossia ORPHA:324601	2/101	3.66×10^{-3}	4.77×10^{-2}
Abruzzo-Erickson syndrome ORPHA:921	2/101	3.66×10^{-3}	4.77×10^{-2}
Cerulean cataract ORPHA:98989	2/104	3.88×10^{-3}	4.77×10^{-2}
Muscle-specific tissues			
Congenital myopathy with reduced type 2 muscle fibers ORPHA:544602	17/101	4.16×10^{-27}	1.65×10^{-25}
Glycogen storage disease due to muscle beta-enolase deficiency ORPHA:99849	13/101	3.37×10^{-19}	4.22×10^{-18}
Rippling muscle disease ORPHA:97238	12/101	2.45×10^{-17}	2.79×10^{-16}
Glycogen storage disease due to muscle glycogen phosphorylase deficiency ORPHA:368	10/101	9.07×10^{-14}	9.14×10^{-13}
Myostatin-related muscle hypertrophy ORPHA:275534	3/101	2.31×10^{-3}	1.16×10^{-2}
Gastrointestinal-specific tissue 1			
NON RARE IN EUROPE: Recurrent acute pancreatitis ORPHA:64740	9/103	1.92×10^{-9}	2.37×10^{-7}
Hereditary chronic pancreatitis ORPHA:676	8/102	3.73×10^{-8}	3.07×10^{-6}
Partial pancreatic agenesis ORPHA:2805	4/100	1.48×10^{-3}	4.30×10^{-2}
Gastrointestinal-specific tissue 2			
NON RARE IN EUROPE: Recurrent acute pancreatitis ORPHA:64740	9/103	2.19×10^{-8}	1.68×10^{-6}
Hereditary chronic pancreatitis ORPHA:676	8/102	3.18×10^{-7}	2.03×10^{-5}

7.15.1 Using Mouse Gene Expression

The first example is not a human sample, but a model animal, mouse [103]. Generally speaking, it is not easy to investigate human diseases using model animals that do not always exhibit symptoms similar to that of humans. Especially, at the very early stage of COVID-19 pandemic, the SAR-CoV-2 virus that causes COVID-19 was a newcomer whose detailed property was not known at all. Under such

Table 7.77 Previously reported drug effects on neuron (brain and eye), muscle, and pancreas tissues. The associated citations can be seen in the original paper [102]. (*): Reported side effects

	Tissue types		
Drugs	Neuron	Muscle	Pancreas or stomach
Alendronate	Brain calcification	Muscle mass	Pancreatitis
Acetaminophen (APAP)	Brain	Skeletal muscle	Pancreatitis
Aripiprazole	Brain Activation	Muscle spasms (*)	Pancreatitis
Asenapine	Cognitive and monoamine dysfunction	Muscle rigidity (*)	–
Cisplatin	Prefrontal cortex	Muscle atrophy	Pancreas
Clozapine	Brain	Myotoxicity	Pancreatitis
Doxycycline	Brain	Smooth Muscle	Acute pancreatitis
Empagliflozin	Neurovascular unit and neuroglia	Muscle sympathetic nerve activity	Pancreatitis
Lenalidomide	Memory loss	Muscle cramp	Pancreatic cancer
Lurasidone	Acute schizophrenia	Muscle (*)	–
Olanzapine	Brain stem	Acute muscle toxicity	Pancreatitis
Repatha (Evolocumab)	–	Muscle-related statin Intolerance	–
Risedronate (actonel)	Ocular myasthenia	Muscle weakness	Gastrointestinal cancer
Sofosbuvir	Ocular surface	Myositis	Pancreatitis
Teriparatide	–	Muscle Cramp	Pancreatitis (*)

circumstances, we are not even sure if we can successfully establish model animals or not, since there is no concrete criterion, based on which we can judge whether the established model animals are suitable to be used for the investigation of human diseases. Thus, it is important to see if the outcome, e.g., gene expression profile alteration associated with infection, is coincident with human disease or not.

To see this, we investigate mouse hepatitis virus (MHV) infection that is supposed to be a good model of SARS-CoV-2 infection [47]. If genes whose expression is altered by MHV infection are common in those with SARS-CoV-2 infection in humans, a mouse infected by MHV infection is potentially a model animal of COVID-19. Gene expression of liver and spleen from female mice experimentally infected with MHV or injected with phosphate-buffered saline (PBS) as a control group for comparison are downloaded from GEO using GEO ID GSE146074. Since it is also associated with two genotypes (Ly6e-knockout (KO) and wide type (WT)) and three biological replicates, each of which is further associated with technical replicates, the overall number of samples included is as in Table 7.78.

Although it is reasonable to investigate the infection of MHV to liver apparently because of its name, "hepatitis", one might wonder if it is suitable since SARS-CoV-2 infects not the liver but the lung. To address this doubt, we also investigated

Table 7.78 Number of biological replicates. Two technical replicates are available for each biological replicate

	PBS day 5		MHV day 3		MHV day 5	
	Liver	Spleen	Liver	Spleen	liver	Spleen
WT	3	3	3	3	3	3
KO	3	3	3	3	3	3

Table 7.79 Number of biological replicates of SARS-CoV infection toward mouse lung gene expression profiles

GSE33266					GSE50000				
Days	D1	D2	D4	D7		d1	d2	d4	d7
Mock	3	3	3	3	BatSRBD	5	5	5	4
10^2 pfu	5	5	5	5	icSARS	4	5	5	5
10^3 pfu	5	5	5	5	Mock	4	4	4	4
10^4 pfu	5	5	5	5					
10^5 pfu	5	5	5	5					

two additional gene expression profiles of SARS-CoV, not SARA-CoV-2, infection to mouse lung retrieved also from GEO with GEO IDs GSE33266 and GSE50000, respectively (Table 7.79). The reason why we used SARS-CoV instead of SARS-CoV-2 is because there were no data sets of SARS-CoV-2 at the early stage of COVID-19 pandemic when the study was performed.

As usual, the gene expression profiles in Tables 7.78 and 7.79 must be formatted as a form of tensors as follows: The gene expression profiles of MHV infection in Table 7.78 are formatted as a tensor, $x_{ijkmnp} \in \mathbb{R}^{N \times 2 \times 2 \times 3 \times 3 \times 2}$, which represents the expression of the $i$the gene that belongs to the $j$the genotype ($j = 1$ KO and $j = 2$ WT) of kth tissue ($k = 1$ liver and $k = 2$ spleen) under the mth treatment ($m = 1$ PBS day 5, $m = 2$ MHV day 3, and $m = 3$ MHV day 5) of the nth biological and the pth biological replicates. $N = 25,669$.

The gene expression profiles in GSE33266 are formatted as a tensor, $x_{ijkn} \in \mathbb{R}^{N \times 5 \times 4 \times 5}$, which represents the expression of the ith gene of jth experimental conditions ($j = 1$: Mock, $j = 2$: 10^2 pfu, $j = 3$: 10^3 pfu, $j = 4$: 10^4 pfu, and $j = 5$: 10^5 pfu) at the kth time point ($k = 1$: D1, $k = 2$: D2, $k = 3$: D4, and $k = 4$: D7) of the nth boological replicates. $N = 45,018$.

Similarly, the gene expression profiles in GSE50000 are formatted as a tensor, $x_{ijkn} \in \mathbb{R}^{N \times 3 \times 4 \times 5}$, which represents the expression of the ith gene of jth experimental conditions ($j = 1$: BatSRBD , $j = 2$: icSARS, and $j = 3$: Mock) at the kth time point ($k = 1$: d1, $k = 2$: d2, $k = 3$: d4, and $k = 4$: d7) of the nth boological replicates. $N = 41,174$.

Although the numbers of biological replicates are assumed to be five in the two above tensors for GSE33266 and GSE50000, not all cases are associated with as many as five biological replicates; for cases with less than five biological replicates in GSE33266 and GSE50000, we used some replicates more than once, in order to have five biological replicates for individual cases.

Next we applied HOSVD to these tensors and got

7.15 Drug Repositioning for SARS-CoV-2

$$x_{ijkmnp} = \sum_{\ell_1=1}^{2}\sum_{\ell_2=1}^{2}\sum_{\ell_3=1}^{3}\sum_{\ell_4=1}^{3}\sum_{\ell_5=1}^{2}\sum_{\ell_6=1}^{N} G(\ell_1\ell_2\ell_3\ell_4\ell_5\ell_6)u_{\ell_1 j}^{(j)}u_{\ell_2 k}^{(k)}u_{\ell_3 m}^{(m)}u_{\ell_4 n}^{(n)}u_{\ell_5 p}^{(p)}u_{\ell_6 i}^{(i)} \quad (7.122)$$

for GSE146074 where $u_{\ell_1 j}^{(j)} \in \mathbb{R}^{2\times 2}$, $u_{\ell_2 k}^{(k)} \in \mathbb{R}^{2\times 2}$, $u_{\ell_3 m}^{(m)} \in \mathbb{R}^{3\times 3}$, $u_{\ell_4 n}^{(n)} \in \mathbb{R}^{3\times 3}$, $u_{\ell_5 p}^{(p)} \in \mathbb{R}^{2\times 2}$, and $u_{\ell_6 i}^{(i)} \in \mathbb{R}^{N\times N}$ are singular matrices and orthogonal matrices and $G \in \mathbb{R}^{2\times 2\times 3\times 3\times 2\times N}$ is a core tensor.

HOSVD is also applied to x_{ijkm} for GSE33266 and we get

$$x_{ijkn} = \sum_{\ell_1=1}^{5}\sum_{\ell_2=1}^{4}\sum_{\ell_3=1}^{5}\sum_{\ell_4=1}^{N} G(\ell_1\ell_2\ell_3\ell_4)u_{\ell_1 j}^{(j)}u_{\ell_2 k}^{(k)}u_{\ell_3 n}^{(n)}u_{\ell_4 i}^{(i)} \quad (7.123)$$

where $u_{\ell_1 j}^{(j)} \in \mathbb{R}^{5\times 5}$, $u_{\ell_2 k}^{(k)} \in \mathbb{R}^{4\times 4}$, $u_{\ell_3 n}^{(n)} \in \mathbb{R}^{5\times 5}$, and $u_{\ell_4 i}^{(i)} \in \mathbb{R}^{N\times N}$ are singular matrices and orthogonal matrices and $G \in \mathbb{R}^{5\times 4\times 5\times N}$ is a core tensor.

TD for GSE50000 also can be obtained by applying HOSVD to x_{ijkm} as

$$x_{ijkn} = \sum_{\ell_1=1}^{3}\sum_{\ell_2=1}^{4}\sum_{\ell_3=1}^{5}\sum_{\ell_4=1}^{N} G(\ell_1\ell_2\ell_3\ell_4)u_{\ell_1 j}^{(j)}u_{\ell_2 k}^{(k)}u_{\ell_3 n}^{(n)}u_{\ell_4 i}^{(i)} \quad (7.124)$$

where $u_{\ell_1 j}^{(j)} \in \mathbb{R}^{3\times 3}$, $u_{\ell_2 k}^{(k)} \in \mathbb{R}^{4\times 4}$, $u_{\ell_3 n}^{(n)} \in \mathbb{R}^{5\times 5}$, and $u_{\ell_4 i}^{(i)} \in \mathbb{R}^{N\times N}$ are singular matrices and orthogonal matrices and $G \in \mathbb{R}^{3\times 4\times 5\times N}$ is a core tensor.

As usual, we need to identify which singular value vectors attributed to samples are of interest. Figure 7.67 shows $u_{1j}^{(j)}$, $u_{1k}^{(k)}$, $u_{2m}^{(m)}$, $u_{1n}^{(n)}$, and $u_{1p}^{(p)}$ for GSE146074. Four of those other than $u_{2m}^{(m)}$ exhibit constant values which means independence of genotype, tissue, biological, and technical replicates. $u_{2m}^{(m)}$ represents dependence upon time. They require gene expression profile change as time goes but take constant regardless of the other conditions. Thus, we decided that singular value vectors of interest are $\ell_1 = \ell_2 = \ell_4 = \ell_5 = 1$ and $\ell_3 = 2$ for GSE146074.

Figure 7.68 shows $u_{2j}^{(j)}$, $u_{2k}^{(k)}$, and $u_{1n}^{(n)}$ for GSE33266. $u_{2j}^{(j)}$ and $u_{2k}^{(k)}$ represent dependence upon pfu and time, respectively whereas $u_{1n}^{(n)}$ represents independence of biological replicates. They require gene expression profile change as pfu varies and time goes but take constant regardless of biological replicates. Thus, we decided that singular value vectors of interest are $\ell_1 = \ell_2 = 2$ and $\ell_3 = 1$ for GSE33266.

Figure 7.69 shows $u_{2j}^{(j)}$, $u_{2k}^{(k)}$, and $u_{1n}^{(n)}$ for GSE50000. $u_{2j}^{(j)}$ and $u_{2k}^{(k)}$ represent distinction between infection and mock and the dependence upon time, respectively, whereas $u_{1n}^{(n)}$ represents independence of biological replicates. They require gene expression profiles to be distinct between infection and mock and vary as time goes but take constant regardless of biological replicates. Thus, we decided that singular value vectors of interest are $\ell_1 = \ell_2 = 2$ and $\ell_3 = 1$ for GSE50000.

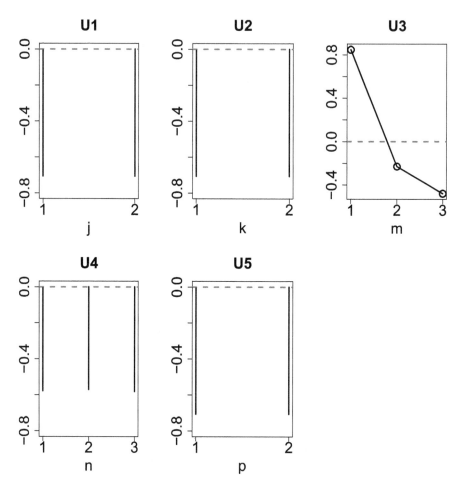

Fig. 7.67 Singular value vectors attributed to experimental conditions and of interest for GSE146074, $u_{1j}^{(j)}, u_{1k}^{(k)}, u_{2m}^{(m)}, u_{1n}^{(n)}$, and $u_{1p}^{(p)}$

Next we need to identify which $|G|$s have the larger absolute values with fixed ℓ_1, ℓ_2, etc selected above to decide which singular value vectors attributed to is are used to select is. Figure 7.70 shows $|G(1, 1, 2, 1, 1, \ell_6)|$ for GSE146074. Since it is obvious that $\ell_6 = 3, 4$ have larger $|G|$, we decided to employ $\ell_6 = 3, 4$ for selecting is for GSE146074. Figure 7.71 shows $|G(2, 2, 1, \ell_4)|$ for GSE33266. Since it is obvious that $\ell_4 = 3, 4$ have larger $|G|$, we decided to employ $\ell_4 = 3, 4$ for selecting is for GSE33266 as well. Figure 7.72 shows $|G(2, 2, 1, \ell_4)|$ for GSE33266. Since it is obvious that $\ell_4 = 3$ has larger $|G|$, we decided to employ $\ell_4 = 3$ for selecting is for GSE50000.

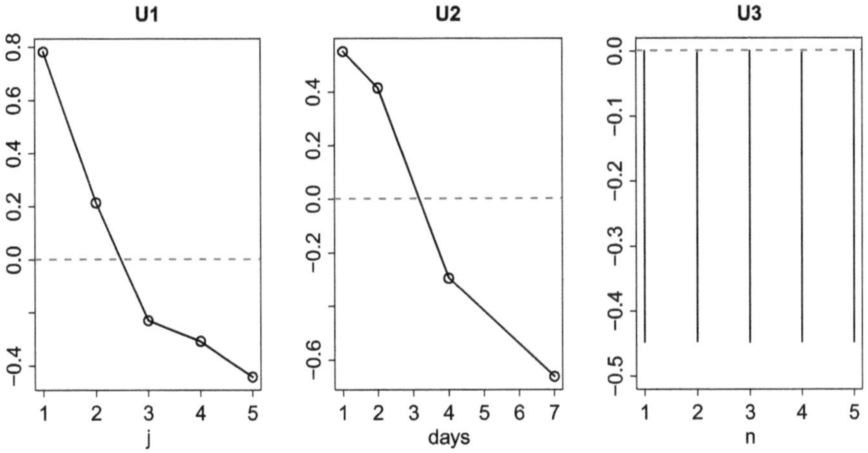

Fig. 7.68 Singular value vectors attributed to experimental conditions and of interest for GSE33266, $u_{2j}^{(j)}$, $u_{2k}^{(k)}$, and $u_{1n}^{(n)}$

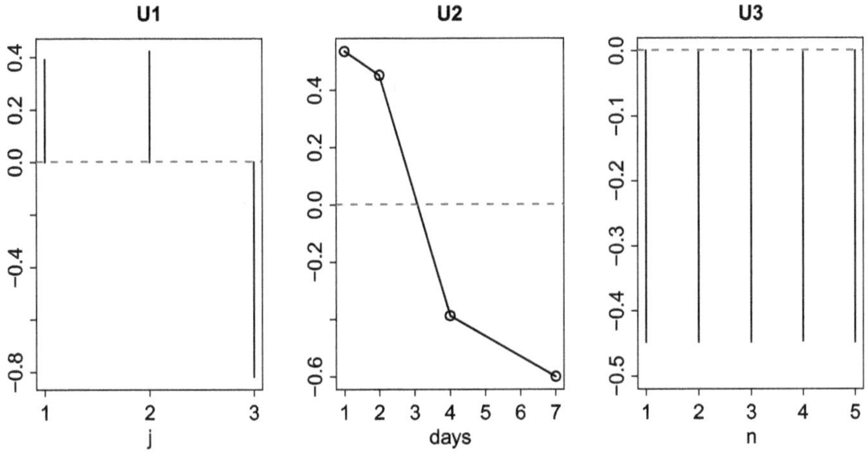

Fig. 7.69 Singular value vectors attributed to experimental conditions and of interest for GSE50000, $u_{2j}^{(j)}$, $u_{2k}^{(k)}$, and $u_{1n}^{(n)}$

Then we can attribute P values to is using the corresponding singular value vectors. For GSE146074,

$$P_i = P_{\chi^2}\left[> \sum_{\ell_6=3}^{4} \left(\frac{u_{\ell_6 i}^{(i)}}{\sigma_{\ell_6}} \right)^2 \right], \tag{7.125}$$

for GSE33266,

Fig. 7.70 $|G(1, 1, 2, 1, 1, \ell_6)|$ for GSE146074

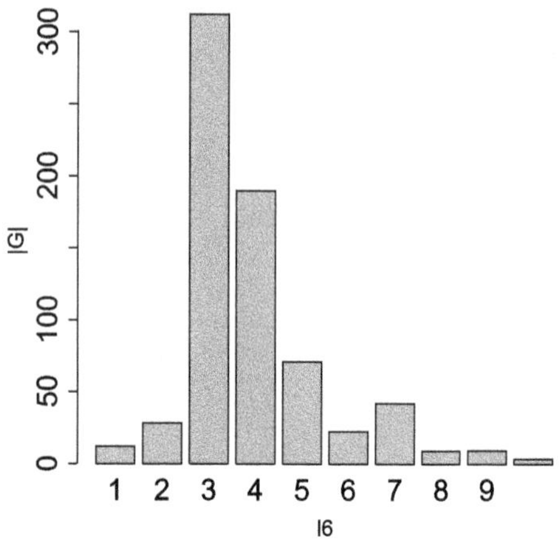

Fig. 7.71 $|G(2, 2, 1, \ell_4)|$ for GSE33266

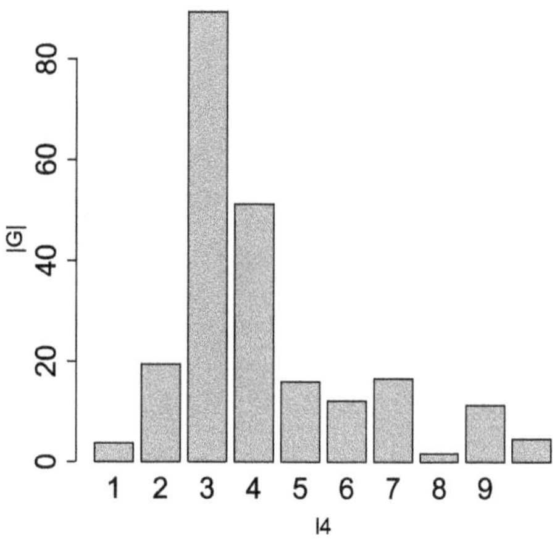

Fig. 7.72 $|G(2, 2, 1, \ell_4)|$ for GSE50000

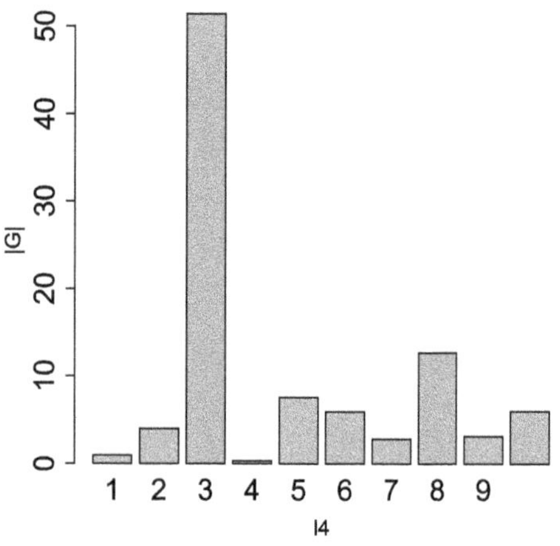

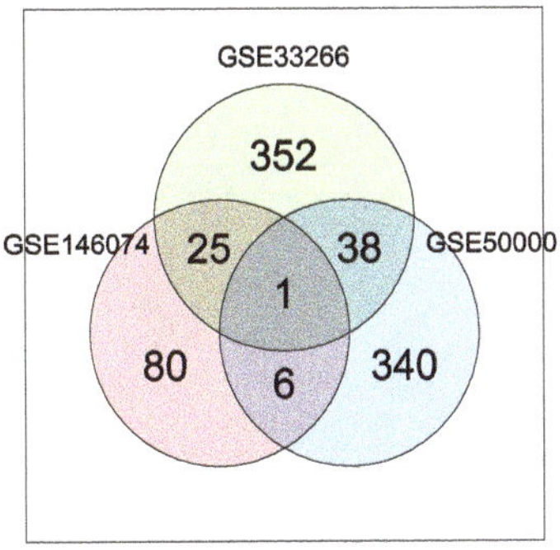

Fig. 7.73 Venn diagram of gene symbols associated selected is for GSE146074, GSE33266, and GSE50000

$$P_i = P_{\chi^2}\left[> \sum_{\ell_4=3}^{4} \left(\frac{u_{\ell_4 i}^{(i)}}{\sigma_{\ell_4}} \right)^2 \right], \tag{7.126}$$

and for GSE50000

$$P_i = P_{\chi^2}\left[> \left(\frac{u_{3i}^{(i)}}{\sigma_3} \right)^2 \right]. \tag{7.127}$$

Attributed P values are corrected by BH criterion and 112, 692, and 498 is associated with adjusted P values less than 0.01 were selected for GSE146074, GSE33266, and GSE50000. Figure 7.73 shows a Venn diagram of gene symbols associated with these is. They are substantially distinct from one another.

To validate these gene symbols biologically, they are uploaded to Enrichr. Then we found 15 experiments associated with adjusted P values less than 0.05 directly related to SARS-CoV-2 infection in "COVID-19-Related Gene Sets 2021" category when gene symnbols associated with 112 is identified for GSE146074 (Table 7.80). Thus, even if we consider neither human samples nor SARS-CoV-2 infection, we can correctly predict genes associated with SARS-CoV-2 infection when we employ TD-based unsupervised FE for GSE146074. The same can stand for the other two gene expression profiles, GSE33266 and GSE50000, although the number of experiments is much more. Although we identified as many as 142 experiments for GSE33266 because of the greater number of gene symbols uploaded than GSE146074, we listed only the top ten in Table 7.81. It also supports that we can identify genes whose expression is altered by SARS-CoV-2 infection even if we do not have direct experiment of SARS-CoV-2 infection. We also could identify as many as 85 experiments for GSE50000 and list the top ten in Table 7.82. In conclusion, in spite of quite distinct genes identified in three gene expression profiles (Fig. 7.73), they are enriched with genes whose expression is altered by SARS-COV-2 infection without dealing with direct experiments of SARS-CoV-2 infection if we employed TD-based unsupervised FE, which is quite a robust method.

Table 7.80 Enriched "SARS-CoV-2" terms in "COVID-19 Related Gene Sets 2021" category of Enrichr for GSE146074

Rank	Term	Overlap	*P* value	Adjusted *P* value
15	500 genes downregulated by SARS-CoV-2 in human Caco2 cells at 24h from GSE148729	12/490	1.96×10^{-5}	5.13×10^{-4}
19	500 genes upregulated by SARS-CoV-2 in human lung organoid cells at 24 hpi from GSE148697	11/464	5.95×10^{-5}	1.15×10^{-3}
20	Top 500 upregulated genes for SARS-CoV-2 infection in human lung organoids from GSE148697	11/464	5.95×10^{-5}	1.15×10^{-3}
21	500 genes upregulated by SARS-CoV-2 in A549-ACE2 cells from GSE154613 RS504393	11/467	6.30×10^{-5}	1.15×10^{-3}
24	Top 500 down genes for SARS-CoV-2 early infection in human female blood from GSE161731	11/498	1.11×10^{-4}	1.77×10^{-3}
27	500 genes upregulated by SARS-CoV-2 in human lung tissue from GSE147507	10/461	2.67×10^{-4}	3.62×10^{-3}
29	500 genes upregulated by SARS-CoV-2 in human lung cells from GSE147507	10/489	4.25×10^{-4}	5.56×10^{-3}
30	Top 500 down genes for SARS-CoV-2 middle-stage infection in human female blood from GSE161731	10/496	4.75×10^{-4}	6.00×10^{-3}
31	500 genes upregulated by SARS-CoV-2 in A549-ACE2 cells from GSE154613 terfenadine	9/412	5.19×10^{-4}	6.30×10^{-3}
32	Top 500 up genes for SARS-CoV-2 infection in Mesocricetus auratus hamster blood Day 14 from GSE162208	7/251	5.34×10^{-4}	6.30×10^{-3}
33	500 genes upregulated by SARS-CoV-2 in A549-ACE2 cells from GSE154613 berbamine	9/463	1.18×10^{-3}	1.33×10^{-2}
34	500 genes upregulated by SARS-CoV-2 in human Organoids cells from GSE154613	9/464	1.20×10^{-3}	1.33×10^{-2}
35	Top 500 upregulated genes in mouse liver with SARS-CoV-2 infection (Day 3) from GEO GSE162113	8/402	1.92×10^{-3}	2.04×10^{-2}
36	Top 500 down genes for SARS-CoV-2 late-stage infection in human female blood from GSE161731	9/498	1.95×10^{-3}	2.04×10^{-2}
38	Top 500 up genes for SARS-CoV-2 infection in Rhesus macaques at group 1 dose in PBMCs at 10 DPI from GSE156701	8/458	4.25×10^{-3}	4.20×10^{-2}

Table 7.81 Top ten-ranked enriched "SARS-CoV-2" terms in "COVID-19 Related Gene Sets 2021" category of Enrichr for GSE33266

Rank	Term	Overlap	*P* value	Adjusted *P* value
36	500 genes upregulated by SARS-CoV-2 in human Calu3 cells at 24h from GSE148729 s1 polyA	42/497	1.13×10^{-14}	1.46×10^{-13}
38	Top 500 upregulated genes in mouse lung with SARS-CoV-2 infection (day 3) from GEO GSE162113	35/360	3.40×10^{-14}	4.16×10^{-13}
40	500 genes upregulated by SARS-CoV-2 in A549-ACE2 cells from GSE154613 terfenadine	36/412	3.68×10^{-13}	4.20×10^{-12}
42	Top 500 up genes from control vs. Ad5-hACE2 for SARS-CoV-2 infection in mouse lung from GSE158069	34/391	1.96×10^{-12}	2.16×10^{-11}
43	Top 500 up genes for SARS-CoV-2 infection 48 hpi in human alveolar organoids for GSE152586	37/460	2.09×10^{-12}	2.25×10^{-11}
44	Top 500 up genes for SARS-CoV-2 infection in Rhesus macaques at Group 1 dose in PBMCs at 1 DPI from GSE156701	37/471	4.21×10^{-12}	4.43×10^{-11}
45	500 top downregulated genes for SARS-CoV-2 infection vs. mock from GSE161881	38/498	5.13×10^{-12}	5.27×10^{-11}
46	Top 500 upregulated genes for SARS-CoV-2 in human colon organoid from GSE148696	36/452	5.60×10^{-12}	5.63×10^{-11}
47	500 genes upregulated by SARS-CoV-2 in human lung tissue from GSE147507	36/461	9.89×10^{-12}	9.41×10^{-11}
49	Top 500 upregulated genes in mouse heart with SARS-CoV-2 infection from GEO GSE162113	33/392	1.01×10^{-11}	9.41×10^{-11}

As denoted at the beginning of the section, the purpose of this section is drug repositioning. Fortunately, Enrichr implements various categories where we can identify drugs whose treatment significantly alters the expression of the uploaded set of genes. We have checked four categories, "LINCS L1000 Chem Pert Consensus Sigs," "DrugMatrix," "Drug Perturbations from GEO down," and "Drug Perturbations from GEO up" and identified a huge number of drugs (Number of hits in Table 7.83). Thus, apparently, this trial was successful. Nevertheless, we are not sure if the identified genes are really promising ones or not. To confirm the effectiveness of identified drugs, we compared the identified drugs with the potential drugs reported recently in an independent work [85] (Table 7.83). As can be seen, drugs identified in the independent study are largely identified by us too. Thus, it is likely that our identified drug should include many promising drug compounds.

Table 7.82 Top ten-ranked enriched "SARS-CoV-2" terms in "COVID-19 Related Gene Sets 2021" category of Enrichr for GSE50000

Rank	Term	Overlap	*P* value	Adjusted *P* value
73	Top 500 up genes for SARS-CoV-2 infection in Rhesus macaques at group 2 dose in bronchoalveolar lavage at 2 DPI from GSE156701	40/468	2.57×10^{-15}	1.59×10^{-14}
74	500 genes upregulated by SARS-CoV-2 in human A549 cells from GSE147507 Series2	41/495	3.32×10^{-15}	2.02×10^{-14}
75	Top 500 up genes for SARS-CoV-2 infection in Rhesus macaques at group 3 dose in PBMCs at 2 DPI from GSE156701	40/474	3.95×10^{-15}	2.37×10^{-14}
76	Top 500 up genes for SARS-CoV-2 infection in Rhesus macaques at group 3 dose in PBMCs at 4 DPI from GSE156701	40/477	4.88×10^{-15}	2.89×10^{-14}
77	Top 500 up genes for SARS-CoV-2 infection in Mesocricetus auratus hamster lung day 2 from GSE162208	35/368	6.40×10^{-15}	3.74×10^{-14}
78	Top 500 upregulated genes in mouse heart with SARS-CoV-2 infection from GEO GSE162113	36/392	7.64×10^{-15}	4.41×10^{-14}
80	Top 500 up genes for SARS-CoV-2 infection in Mesocricetus auratus hamster blood day 2 from GSE162208	34/363	2.49×10^{-14}	1.40×10^{-13}
81	Top 500 up genes for SARS-CoV-2 infection in Rhesus macaques at group 3 dose in PBMCs at 10 DPI from GSE156701	38/463	4.85×10^{-14}	2.69×10^{-13}
82	Top 500 up genes for SARS-CoV-2 infection in Rhesus macaques at group 1 dose in PBMCs at 4 DPI from GSE156701	38/464	5.19×10^{-14}	2.85×10^{-13}
83	Top 500 upregulated genes in mouse lung with SARS-CoV-2 infection (day 3) from GEO GSE162113	33/360	1.12×10^{-13}	6.09×10^{-13}

Before closing this subsection, we compared the performance with some other conventional methods. Since the data structure of the above three gene expression profiles is complicated, there unlikely exists *de facto* standard method to be applied. For simplicity, we consider GSE146074 where we have identified the smallest number of genes which can be most likely increased by other methods. The categorical regression analysisis

$$x_{ijkmnp} = \alpha_{ijkm} + \beta_i, \tag{7.128}$$

Table 7.83 The number of drugs identified with various enrichment analyses in Enrichr and the overlap with drugs identified in [85]. Enrichr categories employed are (1) "LINCS L1000 Chem Pert Consensus Sigs," (2) "DrugMatrix," (3) "Drug Perturbations from GEO down," and (4) "Drug Perturbations from GEO up." Number of hits means drugs associated with adjusted *P* values less than 0.05. The number of overlaps is that of drugs in [85] among those associated with adjusted *P* values less than 0.05

	GSE146074				GSE33266				GSE50000			
Drug	(1)	(2)	(3)	(4)	(1)	(2)	(3)	(4)	(1)	(2)	(3)	(4)
Olaparib					○				○			
Cisplatin	○	○	○	○	○	○	○	○	○		○	○
Gemcitabine					○	○			○			
Etoposide		○			○	○			○	○		
Palbociclib					○				○			
Cinnarizine		○			○	○			○			
Fenofibrate	○	○			○	○			○			
Clofibrate		○			○	○			○			
Sertraline		○			○	○			○			
Rucaparib					○				○			
Clotrimazole		○			○	○			○			
Podofilox												
Paclitaxel			○	○	○				○		○	
Daunorubicin		○			○	○			○	○		
Daunorubicin citrate												
Mitoxantrone		○			○	○						
Doxorubicin		○	○	○	○	○	○	○	○	○	○	○
Epirubicin		○			○	○			○			
Idarubicin		○			○	○			○			
Vincristine		○		○	○	○	○		○			
Hydroquinone						○						
Digitoxin		○			○	○			○			
Teniposide					○				○			
Dexrazoxane												
Amsacrine					○				○			
Valrubicin					○				○			
Fluorouracil		○			○	○			○			
Cyclophosphamide		○	○	○	○	○	○		○			○
Capecitabine					○				○			
Gefitinib			○	○	○		○		○			
Raloxifene		○			○	○						
Acriflavine												
Prochlorperazine edisylate												
Methacycline												
Gallium nitrate												
Cytarabine		○	○	○	○	○	○	○		○		

(continued)

Table 7.83 (continued)

	GSE146074				GSE33266				GSE50000			
Drug	(1)	(2)	(3)	(4)	(1)	(2)	(3)	(4)	(1)	(2)	(3)	(4)
Fludarabine					o				o			
Clofarabine					o				o			
Cladribine					o				o			
Hydroxyurea		o			o	o						
Dofetilide					o				o			
Medroxyprogesterone acetate												
Amiodarone		o			o	o			o			
Guanidine												
Sotalol	o	o			o	o			o			
Dalfampridine	o				o				o			
Number of overlaps	4	21	6	7	36	23	6	3	32	4	3	3
Number of hits	568	517	177	187	5337	637	143	145	4214	25	101	104

this means that gene expression is independent of biological and technical replicates. P values are attributed to each i using `lm` function implemented in R, and is associated with adjusted P values less than 0.01 were selected. Unfortunately, as many as 25,609 is among 25,669 is are selected. Thus, categorical regression analysis is definitely useless. To check the usefulness of top-ranked ones, we selected top-ranked 112 is which are as many as those selected by TD-based unsupervised FE and uploaded to Enrichr. The number of terms selected in "COVID-19 Related Gene Sets 2021" category is 10 which is smaller than 15 which is the number of terms selected when TD-based unsupervised FE was employed. Thus, categorical regression analysis is not only inferior to TD-based unsupervised FE from the point of the number of selected is but also less reliably from the point of the selected genes.

At last, we employed a hybrid approach. From Fig. 7.67, we know the reliable dependence upon j, k, m, n, p. To reflect this

$$x_{ijkmnp} = \alpha_i m + \beta'_i, \tag{7.129}$$

this means that gene expression is dependent upon only days (m) and is proportional to m. We found that there are as many as 4,174 genes selected. By uploading these genes to Enrichr, we found that 182 terms are selected in "COVID-19 Related Gene Sets 2021" category. Thus, the hybrid approach is very useful but categorical regression analysis is not only very promising at all.

7.15.2 Using Human Cell Line Expression

In this subsection, we tried drug repositioning using human cell line gene expression [100]. Compared with the cases in the previous subsection, since in this subsection we deal with the infection of SARS-CoV-2 to human cells lines, it is expected to get more biologically reasonable results. On the other hand, since infection to cell lines, i.e., in vitro infection, is less realistic compared to model animals, i.e., in vivo infection, it might reduce the effectiveness of the study. Thus, it is important to validate outcome carefully.

To do this, we employed gene expression profile retrieved from GEO with GEO ID GSE147507, which is composed of gene expression of five human lung cancer cell lines, infected or not infected by SARS-CoV-2, each of which is associated with three biological replicates. Thus, in total, there are $5 \times 2 \times 3 = 30$ samples. The file GSE147507_RawReadCounts_Human.tsv.gz was downloaded from Supplementary file section in GSE147507. Since some other profiles not used in the analyses are included in this file, we have to retrieve 30 profiles from the file. The gene expression profiles are formatted as $x_{ijkm} \in \mathbb{R}^{N \times 5 \times 2 \times 3}$ where N is the number of genes whose expression is measured. x_{ijkm} represents the expression of the ith gene of jth cell ($j = 1$: Calu3, $j = 2$: NHBE, $j = 3$: A549 MOI 0.2, $j = 4$: A549 MOI 2,0, $j = 5$: A549 ACE2 expressed) line at the kth treatment ($k = 1$: Mock and $k = 2$: SARS-CoV-2 infected) of the mth biological replicates. x_{ijkm} is decomposed into TD by applying HOSVD

$$x_{ijkm} = \sum_{\ell_1=1}^{5} \sum_{\ell_2=1}^{2} \sum_{\ell_3=1}^{3} \sum_{\ell_4=1}^{N} G(\ell_1 \ell_2 \ell_3 \ell_4) u_{\ell_1 j}^{(j)} u_{\ell_2 k}^{(k)} u_{\ell_3 m}^{(m)} u_{\ell_4 i}^{(i)} \tag{7.130}$$

where $G \in \mathbb{R}^{5 \times 2 \times 3 \times N}$ is a core tensor, and $u_{\ell_1 j}^{(j)} \in \mathbb{R}^{5 \times 5}$, $u_{\ell_2 k}^{(k)} \in \mathbb{R}^{2 \times 2}$, $u_{\ell_3 m}^{(m)} \in \mathbb{R}^{3 \times 3}$, $u_{\ell_4 i}^{(i)} \in \mathbb{R}^{N \times N}$ are singular value matrices and orthogonal matrices. As usual, we need to identify which singular value vectors attributed to cell lines, infection conditions, and biological replicates are associated with the desired properties: distinction between mock and SARS-CoV-2 infection ($k = 1, 2$) as well as independence of cell lines (j) and replicates (m). As can be seen in Fig. 7.74, $\ell_1 = \ell_3 = 1$ and $\ell_2 = 2$ satisfy the required conditions since $u_{21}^{(k)} = -u_{22}^{(k)}$ whereas $u_{1j}^{(j)}$s and $u_{1m}^{(m)}$s are constant (i.e., independent of js and ms).

Next, we need which $u_{\ell_4 i}^{(i)}$ is most associated with these above selected singular value vectors: $u_{1j}^{(j)}$, $u_{2k}^{(k)}$, and $u_{1m}^{(m)}$. To do this, we need to find the largest absolute value of $G(1, 2, 1, \ell_4)$. Figure 7.75 shows the $|G(1, 2, 1, \ell_4)|$ and $\ell_4 = 5$ is associated with the largest $|G(1, 2, 1, \ell_4)|$. Thus we decided to use $u_{5i}^{(i)}$ to select genes.

P values are attributed to ith genes as

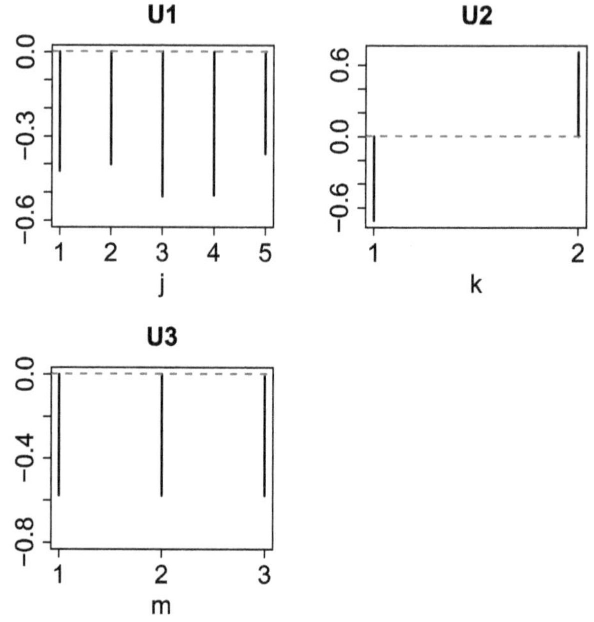

Fig. 7.74 Singular value vectors obtained by the HOSVD algorithm. U1: $u_{1j}^{(j)}$, U2: $u_{2k}^{(k)}$, U3: $u_{1m}^{(m)}$

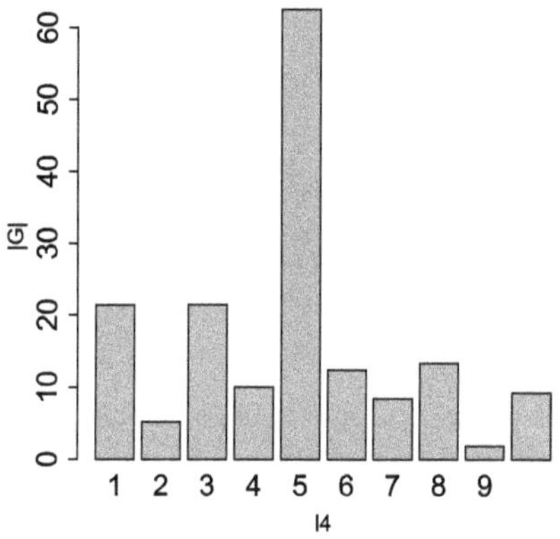

Fig. 7.75 $|G(1, 2, 1, \ell_4)|$

$$P_i = P_{\chi^2}\left[> \left(\frac{u_{5i}^{(i)}}{\sigma_5}\right)^2 \right]. \tag{7.131}$$

P values are corrected by BH criterion, and 163 is associated with adjusted P values less than 0.01 are selected. In order to validate these 163 genes, they are uploaded to Enrichr. There are as many as 172 SARS-CoV-2 infection experiments associated with adjusted P values less than 0.05 in "SARS-CoV-2" terms in "COVID-19 Related Gene Sets 2021" category. This number is ten times larger than the 15 experiments listed in Table 7.80. This number, 172, is also larger than 142 and 85, which are the numbers of experiments identified in GSE33266 and GSE50000,

Table 7.84 Top ten-ranked enriched "SARS-CoV-2" terms in "COVID-19 Related Gene Sets 2021" category of Enrichr

Term	Overlap	*P* value	Adjusted *P* value
Genes whose expression is altered by SARS-COV-2 infection	118/118	5.91×10^{-273}	2.62×10^{-270}
500 genes downregulated by SARS-CoV-2 (0.2 MOI) in human A549 cells transduced with ACE2 vector from GSE147507	43/494	4.13×10^{-32}	4.57×10^{-30}
500 genes upregulated by SARS-CoV-2 in human lung organoid cells at 24 hpi from GSE148697	35/464	5.74×10^{-24}	2.82×10^{-22}
Top 500 upregulated genes for SARS-CoV-2 infection in human lung organoids from GSE148697	35/464	5.74×10^{-24}	2.82×10^{-22}
500 genes upregulated by SARS-CoV-2 in A549-ACE2 cells from GSE154613 RS504393	35/467	7.12×10^{-24}	3.15×10^{-22}
500 genes upregulated by SARS-CoV-2 in human A549 cells from GSE147507 Series5	34/494	5.29×10^{-22}	1.46×10^{-20}
500 genes downregulated by SARS-CoV-2 (2 MOI) in human A549 cells transduced with ACE2 vector from GSE147507	33/492	5.15×10^{-21}	1.27×10^{-19}
500 genes upregulated by SARS-CoV-2 (2 MOI) in human A549 cells from GSE147507	32/470	1.39×10^{-20}	3.09×10^{-19}
500 genes upregulated by SARS-CoV-2 in human NHBE cells from GSE147507	31/486	3.94×10^{-19}	8.32×10^{-18}
Top 500 upregulated genes in human nasal epithelial cells with SARS-CoV-2 infection (mutant, 8 hpi) from GEO GSE162131	30/454	5.88×10^{-19}	1.18×10^{-17}

respectively. Thus as expected, infection to human cell lines is more likely close to human SARS-CoV-2 infection than those in the previous section. Table 7.84 shows top ten–ranked "SARS-CoV-2" terms in "COVID-19 Related Gene Sets 2021" category of Enrichr.

Although we have successfully identified SARS-CoV-2-related genes as shown above, our main purpose in this subsection is to identify known drugs that might be possible candidates toward COVID-19. To do this, we investigated various drug-related categories in Enrichr.

The first category we investigated is "LINCS L1000 Chem Pert Consensus Sigs." Table 7.85 shows the top ten–ranked drugs. These drugs include some promising candidates. The top one, fenretinide, was once regarded to be the promising drug for COVID-19 [70]. The eighth one, ercalcitriol, is the active circulating metabolite of vitamin D2; the improvement in oxygenation among hospitalized patients with COVID-19 treated with vitamin D was observed [24].

The second category we investigated is "DrugMatrix." Table 7.86 shows the top ten–ranked drugs. These drugs also include some promising candidates. The third

Table 7.85 Top ten-ranked enriched terms in "LINCS L1000 Chem Pert Consensus Sigs" category of Enrichr

Term	Overlap	P value	Adjusted P value
Fenretinide up	36/242	3.14×10^{-35}	3.40×10^{-31}
Etilefrine down	35/248	2.03×10^{-33}	1.10×10^{-29}
TUL-XX025TFA down	31/244	3.66×10^{-28}	1.13×10^{-24}
BG-1024 up	31/245	4.16×10^{-28}	1.13×10^{-24}
SSR-69071 up	31/247	5.38×10^{-28}	1.16×10^{-24}
SKI-II up	30/238	3.68×10^{-27}	6.64×10^{-24}
K784-3188 up	30/244	7.84×10^{-27}	1.06×10^{-23}
Ercalcitriol up	30/244	7.84×10^{-27}	1.06×10^{-23}
K784-3187 up	30/245	8.88×10^{-27}	1.07×10^{-23}
Macelignan up	30/249	1.45×10^{-26}	1.57×10^{-23}

Table 7.86 Top ten-ranked enriched terms in "DrugMatrix" category of Enrichr

Term	Overlap	P value	Adjusted P value
2-Amino-4-nitrophenol-625 mg/kg in CMC-rat-kidney-1d-up	26/300	2.01×10^{-19}	1.59×10^{-15}
Allyl alcohol-32 mg/kg in saline-rat-liver-1d-up	25/291	1.30×10^{-18}	5.12×10^{-15}
Meloxicam-33 mg/kg in corn oil-rat-kidney-5d-up	23/261	1.96×10^{-17}	4.64×10^{-14}
Lipopolysaccharide E. coli O55:B5-1.25 mg/kg in saline-rat-kidney-1d-up	24/295	2.36×10^{-17}	4.64×10^{-14}
44'-Methylenedianiline-81 mg/kg in corn oil-rat-liver-3d-up	25/333	3.27×10^{-17}	5.16×10^{-14}
Gentamicin-40 mg/kg in saline-rat-kidney-14d-up	24/309	6.83×10^{-17}	7.68×10^{-14}
Lead(IV) acetate-600 mg/kg in saline-rat-kidney-5d-up	24/309	6.83×10^{-17}	7.68×10^{-14}
Dibromochloromethane-325 mg/kg in CMC-rat-kidney-3d-up	24/312	8.51×10^{-17}	8.38×10^{-14}
Allopurinol-175 mg/kg in corn oil-rat-kidney-3d-up	24/329	2.84×10^{-16}	2.40×10^{-13}
Benzyl acetate-1868 mg/kg in CMC-rat-kidney-3d-up	24/330	3.05×10^{-16}	2.40×10^{-13}

one, meloxicam, was once reported to be effective to COVID-19 [78]. The fourth one, lipopolysaccharide, was reported to be related to COVID-19 [84]. The sixth one, gentamicin, was once recommended to be used in the management of COVID-19 [2]. The ninth one, allopurinol, was reported to increase the risk of COVID-19 hospitalization [127]. The tenth one, benzyl acetate, was reported to be a promising inhibitor [126].

The third category we investigated is "Drug Perturbations from GEO down." Table 7.87 shows the top ten–ranked drugs. These drugs also include some promising candidates. The first one, gatifloxacin, was evaluated to interact tightly

7.15 Drug Repositioning for SARS-CoV-2

Table 7.87 Top ten-ranked enriched terms in "Drug Perturbations from GEO down" category of Enrichr

Term	Overlap	P value	Adjusted P value
Gatifloxacin 5379 human GSE9166 sample 2626	48/266	1.61×10^{-51}	1.42×10^{-48}
Atorvastatin DB01076 human GSE2450 sample 2484	46/250	1.02×10^{-49}	4.52×10^{-47}
Bexarotene DB00307 human GSE12791 sample 2681	46/253	1.84×10^{-49}	5.42×10^{-47}
Clinafloxacin 60063 human GSE9166 sample 2625	55/470	1.30×10^{-48}	2.88×10^{-46}
Motexafin gadolinium (12 h) DB05428 human GSE2189 sample 3127	48/320	1.89×10^{-47}	3.33×10^{-45}
BPDE 41322 human GSE19510 sample 3379	47/300	2.36×10^{-47}	3.47×10^{-45}
Trovafloxacin 62959 human GSE9166 sample 2629	53/459	1.98×10^{-46}	2.50×10^{-44}
HYPOCHLOROUS ACID 24341 human GSE11630 sample 3199	40/204	2.85×10^{-44}	3.14×10^{-42}
Trovafloxacin DB00685 human GSE9166 sample 3036	51/451	4.05×10^{-44}	3.97×10^{-42}
Doxycycline DB00254 human GSE2624 sample 3077	48/391	3.82×10^{-43}	3.37×10^{-41}

with spike protein [37]. The second one, atorvastatin, was reported to reduce the severity of COVID-19 [17]. The third one, bexarotene, was reported to bind tightly with main protease and ACE2 receptors [86]. The fourth one, clinafloxacin, was reported to interact with SARS-CoV-2 protease [61]. The seventh and ninth one, trovafloxacin, was reported to be SARS-CoV-2 M-pro inhibitor [31]. The tenth one, doxycycline, was reported to reduce the need for ICU admission when added to SoC [20].

The fourth category we investigated is "Drug Perturbations from GEO up." Table 7.88 shows the top ten–ranked drugs. These drugs also include some promising candidates. The second one, fluticasone, was reported to suppress the SARS-CoV-2-induced increase in respiratory epithelial permeability in vitro [62]. The fourth and eighth one, quercetin, was once regarded to be one of the most promising [29]. The seventh one, apratoxin, was reported to inhibit SARS-CoV-2 [75]. The tenth one, rosiglitazone, was one of 27 molecules as potential candidates [42].

The fifth category we investigated is "Proteomics Drug Atlas 2023." Table 7.89 shows the top ten–ranked drugs. These drugs also include some promising candidates. The second one, rameltemon, was once tested as a promising drug [121]. The seventh one, iloprost, was also once considered to be promising [26]. The tenth one, osilodrostat, was also reported to be related to COVID-19 [14].

Since the drugs listed in all five tables include promising candidates, our proposed drugs will be promising. One might wonder why drug repositioning looks

Table 7.88 Top ten-ranked enriched terms in "Drug Perturbations from GEO up" category of Enrichr

Term	Overlap	P value	Adjusted P value
MK-886 CID 3651377 human GSE3202 sample 3193	54/368	3.90×10^{-53}	3.50×10^{-50}
Fluticasone 5311101 human GSE15823 sample 3090	53/351	8.70×10^{-53}	3.91×10^{-50}
1-Naphthyl isothiocyanate 11080 rat GSE5509 sample 3568	50/301	7.71×10^{-52}	2.31×10^{-49}
Quercetin 5280343 human GSE7259 sample 3416	50/327	6.03×10^{-50}	1.35×10^{-47}
N-METHYLFORMAMIDE 31254 rat GSE5509 sample 3570	46/283	4.37×10^{-47}	7.86×10^{-45}
NICKEL 935 human GSE6907 sample 3531	46/288	1.02×10^{-46}	1.53×10^{-44}
Apratoxin A 6326668 human GSE2742 sample 3070	43/246	2.30×10^{-45}	2.95×10^{-43}
Quercetin 5280343 human GSE7259 sample 3415	47/336	6.05×10^{-45}	6.79×10^{-43}
Sapphyrin PCI-2050 (1.25 μ M) 9855235 human GSE6400 sample 3101	48/367	1.71×10^{-44}	1.70×10^{-42}
Rosiglitazone DB00412 human GSE7035 sample 2810	43/281	9.52×10^{-43}	8.55×10^{-41}

Table 7.89 Top ten-ranked enriched terms in "Proteomics Drug Atlas 2023" category of Enrichr

Term	Overlap	P value	Adjusted P value
STF-31 up	20/141	2.10×10^{-19}	3.48×10^{-16}
Ramelteon up	19/140	4.13×10^{-18}	3.41×10^{-15}
Molindone up	17/142	2.22×10^{-15}	1.16×10^{-12}
Phenylboronic acid up	17/144	2.81×10^{-15}	1.16×10^{-12}
RN-1734 up	15/141	6.11×10^{-13}	2.02×10^{-10}
Tanomastat up	15/145	9.24×10^{-13}	2.18×10^{-10}
Iloprost up	15/145	9.24×10^{-13}	2.18×10^{-10}
STK066844 up	15/149	1.38×10^{-12}	2.85×10^{-10}
GS-9620 up	14/141	9.60×10^{-12}	1.75×10^{-9}
Osilodrostat up	14/142	1.06×10^{-11}	1.75×10^{-9}

so successful although we have simply selected genes whose expression is distinct between control cell lines and infected ones. To clarify this point, we uploaded 163 genes to Rummagene [18] that can compare uploaded gene sets with those included in various supplementary materials attached with the published papers. Although there were as many as 115,380 gene sets that have significant overlaps with the 163 genes, restricting them to those filtered by the term "SARS-CoV-2," there remained only as small as 78 gene sets. Table 7.90 lists top ten-ranked gene sets and associated papers. The top four studies list human genes that interact with SARS-CoV-2 proteins or RNAs. This suggests that 163 genes are likely to be those

7.15 Drug Repositioning for SARS-CoV-2

Table 7.90 Top ten-ranked gene sets in Rummagene when 163 genes were uploaded. GSS: gene set size, OL: overlap. PMC is paper IDs in PubMed central

Paper	Title	GSS	OL	Odds	P value	Adjusted P value
PMC9040432	Characterization and functional interrogation of the SARS-CoV-2 RNA interactome	501	50	12.6	3.30×10^{-41}	8.42×10^{-39}
PMC7951565	Discovery and functional interrogation of SARS-CoV-2 RNA–host protein interactions	1165	66	7.13	9.24×10^{-40}	2.09×10^{-37}
PMC7553159	Systematic discovery and functional interrogation of SARS-CoV-2 viral RNA–host protein interactions during infection	1165	66	7.13	9.24×10^{-40}	2.09×10^{-37}
PMC9040432	Characterization and functional interrogation of the SARS-CoV-2 RNA interactome	1001	58	7.29	5.72×10^{-35}	8.97×10^{-33}
PMC8521106	Defining the innate immune responses for SARS-CoV-2-human macrophage interactions]	1761	69	4.93	1.26×10^{-31}	1.54×10^{-29}
PMC8299217	Impaired local intrinsic immunity to SARS-CoV-2 infection in severe COVID-19	837	45	6.77	2.57×10^{-25}	2.07×10^{-23}
PMC7899452	Impaired local intrinsic immunity to SARS-CoV-2 infection in severe COVID-19	837	45	6.77	2.57×10^{-25}	2.07×10^{-23}
PMC8521106	Defining the innate immune responses for SARS-CoV-2-human macrophage interactions	2408	68	3.55	1.05×10^{-22}	7.06×10^{-21}

(continued)

Table 7.90 (continued)

Paper	Title	GSS	OL	Odds	P value	Adjusted P value
PMC7439834	Candesartan could ameliorate the COVID-19 cytokine storm	1177	47	5.03	4.89×10^{-21}	2.98×10^{-19}
PMC8605816	Identification of the susceptibility genes for COVID-19 in lung adenocarcinoma with global data and biological computation methods	285	24	10.6	5.52×10^{-18}	2.77×10^{-16}

interacting SARS-CoV-2 proteins or RNAs. Possibly, this is the reason why drugs that target 163 genes are those once identified as promising drugs for COVID-19. If drugs can affect human proteins to which SARS-CoV-2 proteins intend to bind, the infection process will be disturbed to some extent by interrupting binding to human proteins. This might suppress the infection.

Before closing this subsection, we compared our performance with other methods: t-test, SAM, limma, and DESeq2 [57]. Since these four methods are unable to treat five cell lines at once, we applied individual methods to cell lines one by one. The results for DESeq2 were retrieved from the study [13] that generated data sets analyzed in this study.

Table 7.91 lists the numbers of DEGs identified by four methods. It is obvious that none of the four methods identify reasonable numbers of DEGs for all five cell lines. t-test and SAM identified too small DEGs and limma identified too many DEGs. The number of genes that DESeq2 identified varies from cell line to cell line. Thus, it is obvious that TD-based unsupervised FE is superior to all four methods based on the number of DEGs identified.

7.16 Integrated Analysis of Epitranscriptome and mRNA Expression

Epitranscriptome [117], which is the modification to RNAs, is one of the new targets of genomic analysis, although the transcriptome is not the genome itself but those transcribed from it. Since TD-based unsupervised FE can deal with multiomics analysis, worthwhile to see whether epitranscriptome can also be used as one of the multiomics. In this section, we considered m^6A, which is one of the most popular epitranscriptome. In the following two subsections, we focus on the relation to two specific topics.

Table 7.91 DEG identification by *t*-test, SAM, limma, and DESeq2. Genes associated with adjusted *P* values less than 0.01 are regarded to be DEGs

	t-test		SAM		Limma		DESeq2	
Cell lines	$P \geq 0.01$	$P < 0.01$	$P \geq 0.01$	$P < 0.01$	$P \geq 0.01$	$P < 0.01$	$P \geq 0.01$	$P < 0.01$
Calu3	21,754	43	21,797	0	42	13,380	7278	16,432
NHBE	21,797	0	21,797	0	41	13,328	23,383	327
A549 MOI 0.2	21,797	0	21,797	0	50	13,867	7858	15,852
A549 MOI 2.0	21,472	325	21,797	0	15	13,823	16,279	7431
A549 ACE2 expressed	21,796	1	21,797	0	111	11,403	16,201	7509

7.16.1 m^{6}A I: Hypoxia

In this subsection, we introduce integrated analysis of epitranscriptome and mRNA expression [82] that integrate a matrix and a tensor that can share samples.

Gene expression profiles and m^{6}A profiles were retrieved from GEO with GEO ID GSE141941. For m^{6}A profiles, the file GSE141941_RAW.tar available as part of the Supplementary Information in GSE141941 was retrieved. m^{6}A profiles were summed up within 25,000-nucleotide intervals sequentially divided over the whole genome, as has been done in Sect. 7.5. As a result, 123,817 genomic regions of 25,000 nucleotides in length were obtained. For gene expression, four profiles included in GSE141941_normoxiaVShypoxia6h.12h.24h_RNA-seq.PROCESSED.DATA.xlsx, which is also available as a part of the Supplementary Information in GSE141941, was retrieved. Eight files for m^{6}A were composed of four time points (including the control), time input, or treated files. Four profiles for gene expression were composed of four times points as well.

In the following, we performed the integrated analysis of gene expression and m^{6}A profiles. The alternative data set with only m^{6}A profiles that lack associated gene expression profiles were retrieved from GEO with GEO ID GSE120860 to show that integrated analysis of gene expression and m^{6}A profiles is critical to get reasonable results.

Before integrated analysis of gene expression and m^{6}A profiles, we performed analysis using either gene expression or m^{6}A profiles. At first, PCA-based unsupervised FE was applied to gene expression profiles, formatted as a matrix, $x_{it} \in \mathbb{R}^{N \times 4}$ that represents the gene expressionn of $i$the gene at the tth time points (N is the number of genes). Only for PCA, x_{it} is normalized as

$$\sum_{i=1}^{N} x_{it} = 0 \tag{7.132}$$

$$\sum_{i=1}^{N} x_{it}^2 = N \tag{7.133}$$

Suppose that X is $N \times 4$ matrix, we solved eigenvalue problem to get PC score attributed to i as

$$XX^T \boldsymbol{u}_\ell = \lambda_\ell \boldsymbol{u}_\ell \tag{7.134}$$

where $\boldsymbol{u}_\ell \in \mathbb{R}^N$ is a unit vector. On the other hand, PC loading attributed to time point t, $\boldsymbol{v}_\ell \in \mathbb{R}^M$ can be computed as

$$\boldsymbol{v}_\ell = X^T \boldsymbol{u}_\ell. \tag{7.135}$$

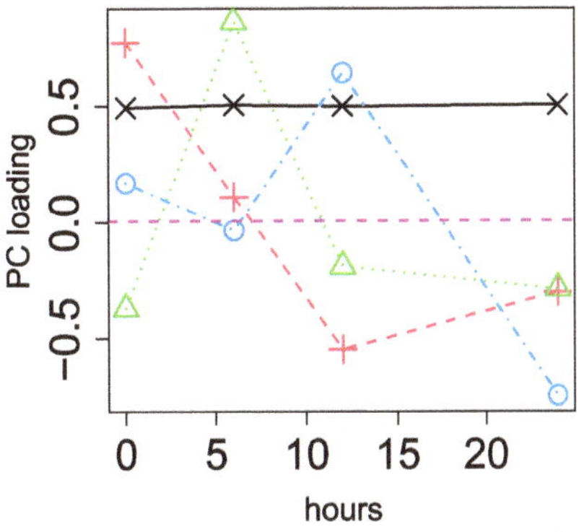

Fig. 7.76 $\boldsymbol{v}_\ell$, black solid line (black crosses): $\ell = 1$, red broken line (red crosses): $\ell = 2$, green dot line (green triangles): $\ell = 3$, and blue block dot line (blue open circles): $\ell = 4$. The magenta horizontal broken line is the baseline. Correlation coefficients between $v_{\ell t}$ and t are 0.61 ($\ell = 1$), -0.76 ($\ell = 2$), -0.24 ($\ell = 3$), and -0.60 ($\ell = 4$)

Figure 7.76 shows the dependence of $\boldsymbol{v}_\ell$ upon t (hours). Since $\boldsymbol{v}_2$ has the strongest correlation with t, we decided to select i with $\boldsymbol{u}_2$. P values were attributed to is as

$$P_i = P_{\chi^2}\left[> \left(\frac{u_{2i}}{\sigma_2}\right)^2 \right]. \tag{7.136}$$

P values were corrected by BH criterion, and and 52 is associated with adjusted P values less than 0.01 were selected. Since gene IDs attributed to is are Ensembl gene IDs, they are converted to gene symbols with DAVID gene ID conversion. The gene symbols were uploaded to Rummagene to evaluate these genes. Although there are as many as 38,135 statistically significant matches, we can have 113 statistically significant matches by screening them with "hypoxia." Table 7.92 lists the top ten–ranked gene sets. Although the ninth and tenth do not include "hypoxia" in paper titles, gene sets investigated are associated with column names that include "hypoxia" (not shown here). They are not only directly related to hypoxia but also related to cancers; it is reasonable since the cell lines used were cancer cell lines. This suggests that our analysis successfully identified reasonable genes.

Next, we applied TD-based unsupervised FE to m^{6}A profiles, which is formatted as a tensor, $x_{ktj} \in \mathbb{R}^{K\times 4\times 2}$ ($K = 123{,}817$ is the number of regions), which represents the amount of m^{6}A in the kth region at the tth time point under the jth condition ($j = 1$: control and $j = 2$: m^{6}A). We get TD as

$$x_{ktj} = \sum_{\ell_1=1}^{K}\sum_{\ell_2=1}^{4}\sum_{\ell_3=1}^{2} G(\ell_1\ell_2\ell_3)u^{(k)}_{\ell_1 k}u^{(t)}_{\ell_2 t}u^{(j)}_{\ell_3 j} \tag{7.137}$$

where $G \in \mathbb{R}^{K\times 4\times 2}$ is a core tensor, $u^{(k)}_{\ell_1 k} \in \mathbb{R}^{K\times K}$, $u^{(t)}_{\ell_2 t} \in \mathbb{R}^{4\times 4}$, and $u^{(t)}_{\ell_3 j} \in \mathbb{R}^{2\times 2}$ are singular value matrices and orthogonal matrices. To decide which $u^{(k)}_{\ell_1 k}$ is used the selection of regions, we need to find which $u^{(t)}_{\ell_2 t}$ is correlated with t and which $u^{(j)}_{\ell_3 j}$ is distinct between $j = 1$ and $j = 2$. Figure 7.77 shows $\boldsymbol{u}^{(t)}_{\ell_2}$ and the comparison

Table 7.92 Top ten-ranked gene sets in Rummagene when 52 genes were uploaded. GSS: gene set size, OL: overlap. PMC is paper IDs in PubMed central

Paper	Title	GSS	OL	Odds	*P* value	Adjusted *P* value
PMC4887009	Identification and validation of reference genes for RT-qPCR studies of hypoxia in squamous cervical cancer patients	405	27	27.2	2.40×10^{-33}	1.63×10^{-30}
PMC9232399	A multiomics deep learning model for hypoxia phenotype to predict tumor aggressiveness and prognosis in uveal melanoma for rationalized hypoxia-targeted therapy	96	11	46.8	4.25×10^{-16}	3.56×10^{-14}
PMC10287711	Hypoxia-induced responses are reflected in the stromal proteome of breast cancer	1860	27	5.93	7.11×10^{-16}	5.81×10^{-14}
PMC6405289	Mutations in an innate immunity pathway are associated with poor overall survival outcomes and hypoxic signaling in cancer	235	13	22.6	1.03×10^{-14}	7.47×10^{-13}
PMC9218734	A novel 7-hypoxia-related long noncoding RNA signature associated with prognosis and proliferation in melanoma	237	13	22.4	1.15×10^{-14}	8.30×10^{-13}
PMC10287711	Hypoxia-induced responses are reflected in the stromal proteome of breast cancer	1188	21	7.22	1.12×10^{-13}	7.32×10^{-12}
PMC5061082	The BET inhibitor JQ1 selectively impairs tumor response to hypoxia and downregulates CA9 and angiogenesis in triple negative breast cancer	49	8	66.7	2.95×10^{-13}	1.85×10^{-11}
PMC6405289	Mutations in an innate immunity pathway are associated with poor overall survival outcomes and hypoxic signaling in cancer	51	8	64.1	4.14×10^{-13}	2.57×10^{-11}
PMC6411608	Circulating miRNA profiling of women at high risk for ovarian cancer	51	8	64.1	4.14×10^{-13}	2.57×10^{-11}
PMC9515221	TP53 mutations and RNA-binding protein MUSASHI-2 drive resistance to PRMT5-targeted therapy in B cell lymphoma	72	8	45.4	7.50×10^{-12}	4.11×10^{-10}

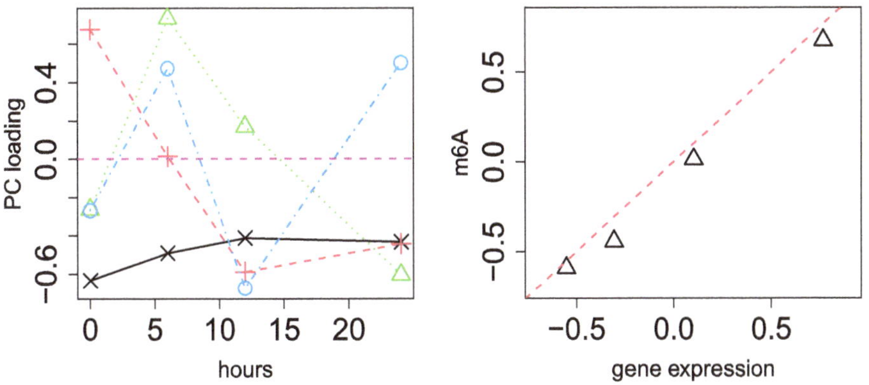

Fig. 7.77 Left: $\boldsymbol{v}_\ell$, black solid line (black crosses): $\ell = 1$, red broken line (red crosses): $\ell = 2$, green dot line (green triangles): $\ell = 3$, and blue block dot line (blue open circles): $\ell = 4$. The magenta horizontal broken line is the baseline. Correlation coefficients between $v_{\ell t}$ and t are 0.78 ($\ell = 1$), -0.80 ($\ell = 2$), -0.47 ($\ell = 3$), and 0.36 ($\ell = 4$). Right: the scatter plot of $\boldsymbol{v}_2$ for gene expression (horizontal) and $\boldsymbol{u}_2^{(t)}$ for m^{6}A (vertical)

between $\boldsymbol{u}_2^{(t)}$ and $\boldsymbol{v}_2$. Since $\boldsymbol{u}_2$ is most correlated with t, we decided to employ $\ell_2 = 2$. Interestingly, $\boldsymbol{u}_2^{(t)}$ and $\boldsymbol{v}_2$ are very close to each other. If we consider that these two are derived from very distinct variables, gene expression and m^{6}A, this coincident is a bit surprising. Since $u_{21}^{(j)}$ takes the opposite sign of $u_{22}^{(j)}$ (Fig. 7.78), we decided to employ $\boldsymbol{u}_{\ell_1}^{(k)}$ associated with the largest absolute $G(\ell_1, 2, 2)$ to select regions, k. Figure 7.79 shows $|G(\ell_1, 2, 2)|$ and since $|G(2, 2, 2)|$ is the largest, we decided to employ $\boldsymbol{u}_2^{(k)}$ to select regions. P values are attributed to ks as

$$P_k = P_{\chi^2}\left[> \left(\frac{u_{2k}^{(k)}}{\sigma_2}\right)^2\right]. \tag{7.138}$$

P values were corrected by BH criterion, and 106 ks that are associated with adjusted P values less than 0.01 and include 196 unique gene symbols are selected. The 196 genes were uploaded to Rummagene, and 43,107 statistically significant matches were identified. 43,107 matches were screened with the term "hypoxia," and there remained as many as 120 statistically significant matches. This number is a bit larger than those identified with gene expression, although the significance is a bit less (i.e., P values are larger). Table 7.93 shows the top ten–ranked statistically significant matches. Again they are cancer-related ones and are reasonable.

Next, to see if integrated analysis of gene expression and m^{6}A profiles can improve the performance, we applied TD-based unsupervised FE to the integrated analysis of gene expression and m^{6}A profiles. At first, we generate a reduced matrix from gene expression profiles formatted as a matrix, x_{it} and a reduced tensor from m^{6}A profiles formatted as a tensor, x_{Ktj} as

Fig. 7.78 $u_2^{(j)}$

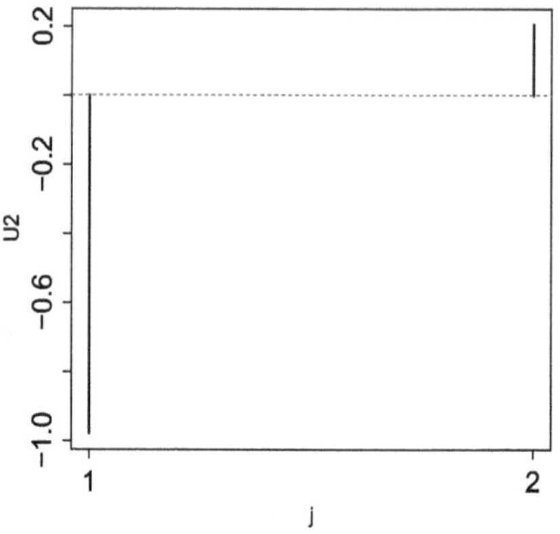

Fig. 7.79 $|G(\ell_1, 2, 2)|$

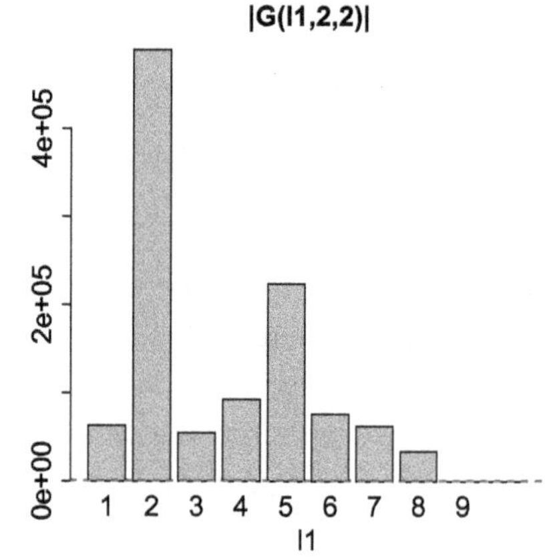

$$x_{tt'} = \sum_{i=1}^{N} x_{it} x_{it'} \in \mathbb{R}^{4\times 4}, \tag{7.139}$$

$$x_{tjt'j'} = \sum_{k=1}^{K} x_{ktj} x_{kt'j'} \in \mathbb{R}^{4\times 2\times 4\times 2}, \tag{7.140}$$

From these, we generate a tensor, $\tilde{x}_{tjt'j'}$ as

$$\tilde{x}_{tjt'j'} = \sum_{t''=1}^{4} x_{tt''} x_{t''jt'j'} \tag{7.141}$$

to which HOSVD was applied, and we get

Table 7.93 Top ten-ranked gene sets in Rummagene when 196 genes were uploaded. GSS: gene set size, OL: overlap. PMC is paper IDs in PubMed central

Paper	Title	GSS	OL	Odds	*P* value	Adjusted *P* value
PMC10287711	Hypoxia-induced responses are reflected in the stromal proteome of breast cancer	1860	46	4.23	1.99×10^{-18}	7.20×10^{-15}
PMC10287711	Hypoxia-induced responses are reflected in the stromal proteome of breast cancer	1188	37	5.33	8.66×10^{-18}	2.35×10^{-14}
PMC9081503	Development and validation of a novel hypoxia score for predicting prognosis and immune microenvironment in rectal cancer	341	19	9.53	1.12×10^{-13}	5.71×10^{-11}
PMC8505699	Identification and validation of hypoxia-related lncRNA signature as a prognostic model for hepatocellular carcinoma	76	10	22.5	2.16×10^{-11}	5.05×10^{-9}
PMC4887009	Identification and validation of reference genes for RT-qPCR studies of hypoxia in squamous cervical cancer patients	405	17	7.18	2.11×10^{-10}	3.70×10^{-8}
PMC6405289	Mutations in an innate immunity pathway are associated with poor overall survival outcomes and hypoxic signaling in cancer	235	13	9.46	1.20×10^{-9}	1.70×10^{-7}
PMC9585224	A novel hypoxia and lactate metabolism-related signature to predict prognosis and immunotherapy responses for breast cancer by integrating machine learning and bioinformatic analyses	471	17	6.17	2.10×10^{-9}	2.77×10^{-7}
PMC5495571	HIF-1α is required for disturbed flow-induced metabolic reprogramming in human and porcine vascular endothelium	166	11	11.3	3.69×10^{-9}	4.49×10^{-7}
PMC8523849	Characterizing intercellular communication of pan-cancer reveals SPP1+ tumor-associated macrophage expanded in hypoxia and promoting cancer malignancy through single-cell RNA-Seq aata	495	17	5.87	4.42×10^{-9}	5.25×10^{-7}
PMC6405289	Mutations in an innate immunity pathway are associated with poor overall survival outcomes and hypoxic signaling in cancer	44	7	27.2	6.22×10^{-9}	7.07×10^{-7}

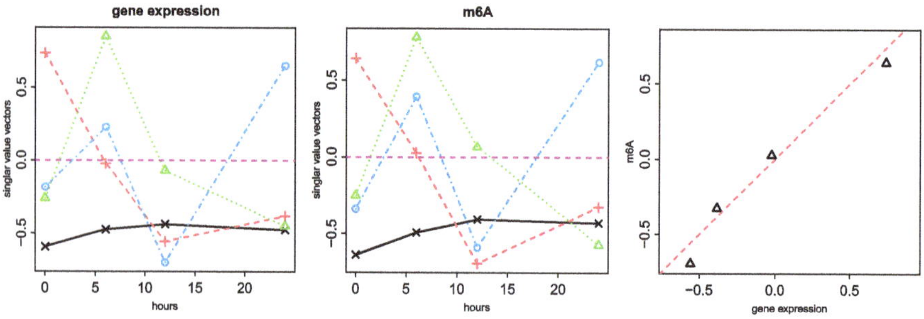

Fig. 7.80 Left: $\boldsymbol{u}_{\ell_1}^{(t)}$, middle: $\boldsymbol{u}_{\ell_3}^{(t')}$. black solid lines (black crosses): $\ell_1 = \ell_3 = 1$, red broken lines (red crosses): $\ell_1 = \ell_3 = 2$, green dot lines (green triangles): $\ell_1 = \ell_3 = 3$, and blue block dot lines (blue open circles): $\ell_1 = \ell_3 = 4$. The magenta horizontal broken lines are the baselines. Correlation coefficients between $\boldsymbol{u}_{\ell_1}^{(t)}$ and t are 0.61 ($\ell_1 = 1$), −0.77 ($\ell_1 = 2$), −0.41 ($\ell_1 = 3$), and 0.48 ($\ell_1 = 4$). Those between $\boldsymbol{u}_{\ell_3}^{(t')}$ and t' are 0.78 ($\ell_3 = 1$), −0.70 ($\ell_3 = 2$), −0.47 ($\ell_3 = 3$), and 0.52 ($\ell_3 = 4$). Right: the scatter plot of $\boldsymbol{u}_{\ell_1}^{(t)}$ for gene expression (horizontal) and $\boldsymbol{u}_{\ell_3}^{(t')}$ for m^{6}A (vertical)

$$\tilde{x}_{tjt'j'} = \sum_{\ell_1=1}^{4}\sum_{\ell_2=1}^{2}\sum_{\ell_3=1}^{4}\sum_{\ell_4=1}^{2} G(\ell_1\ell_2\ell_3\ell_4) u_{\ell_1 t}^{(t)} u_{\ell_2 j}^{(j)} u_{\ell_3 t'}^{(t')} u_{\ell_4 j'}^{(j')} \tag{7.142}$$

where $G(\ell_1\ell_2\ell_3\ell_4) \in \mathbb{R}^{4\times2\times4\times2}$ is a core tensor, $u_{\ell_1 t}^{(t)}, u_{\ell_3 t'}^{(t')} \in \mathbb{R}^{4\times4}$ and $u_{\ell_2 j}^{(j)}, u_{\ell_4 j'}^{(j')} \in \mathbb{R}^{2\times2}$ are singular value matrices and orthogonal matrices. Here one should notice that $u_{\ell_2 j}^{(j)}$ and $u_{\ell_4 j'}^{(j')}$ are identical since they come from j in m^{6}A profiles whereas $u_{\ell_1 t}^{(t)}$ and $u_{\ell_3 t'}^{(t')}$ are not since the former comes from gene expression and the latter comes from m^{6}A profiles. To see which $u_{\ell_1 t}^{(t)}$ and $u_{\ell_3 t'}^{(t')}$ have the strongest correlation with time points, t or t', we shows them in Fig. 7.80. Again, $\ell_1 = \ell_3 = 2$ are associated with the strongest correlation with t or t'. It is also impressive that $\boldsymbol{u}_2^{(t)}$ and $\boldsymbol{u}_2^{(t')}$ are identical as well. We also need to know which $\boldsymbol{u}_{\ell_1}^{(j)}$ (or $\boldsymbol{u}_{\ell_3}^{(j')}$) is distinct between js (or j's), i.e., control and m^{6}A. From Fig. 7.81, it is obvious that $\ell_2 = 2$ is associated with the distinction between $j = 1$ and $j = 2$.

Using $\boldsymbol{u}_2^{(t)}$, $\boldsymbol{u}_2^{(t')}$, and $\boldsymbol{u}_2^{(j)}$, we can recover singular value vectors attributed to is and ks as

$$u_{2i}^{(i;t)} = \sum_{t=1}^{4} x_{it} u_{2t}^{(t)} \tag{7.143}$$

$$u_{22k}^{(k;jt')} = \sum_{j=1}^{2}\sum_{t'=1}^{4} x_{kt'j} u_{2j}^{(j)} u_{2t'}^{(t')} \tag{7.144}$$

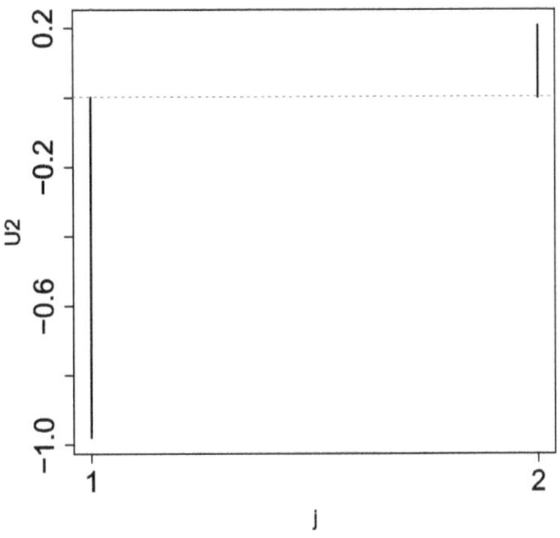

Fig. 7.81 $u_2^{(j)}$

Then P values were attributed to is and ks as

$$P_i = P_{\chi^2}\left[> \left(\frac{u_{2i}^{(i;t)}}{\sigma_2}\right)^2\right], \tag{7.145}$$

$$P_k = P_{\chi^2}\left[> \left(\frac{u_{22k}^{(k;jt')}}{\sigma_{22}}\right)^2\right]. \tag{7.146}$$

P values were corrected by BH criterion, and 53 is and 128 ks associated with adjusted P values less than 0.01 were selected; 200 gene symbols are included in the 128 regions that correspond to 128 ks. Ensembl gene IDs attributed to 128 is were converted to 53 gene symbols with the gene ID conversion tool in DAVID. Two sets of gene symbols were uploaded to Rummagene and 38,571 and 39,981 statistically significant matches for gene expression and m^{6}A profiles, respectively. They were further screened by the term "hypoxia," and 113 and 128 statistically significant matches remained. The former, 113, is as many as the one identified only by gene expression, and the latter, 128, is larger than the one identified by m^{6}A profiles. Thus, the integrated analysis definitely improved performance.

Tables 7.94 and 7.95 list top ten–ranked statistically significant matches for gene expression and m^{6}A profiles. They are also related to cancers; thus they are reasonable.

Before closing this section, we compared the other methods' performances with those of TD-based unsupervised FE that employs integrated analysis of gene expression and m^{6}A. The methods compared with TD-based unsupervsied FE are linear regression, SAM, limma, and random forest. For the linear regression, we used the following ones,

$$x_{it} = a_i + b_i T(t) \tag{7.147}$$

Table 7.94 Top ten-ranked gene sets in Rummagene when 53 genes were uploaded. GSS: gene set size, OL: overlap. PMC is paper IDs in PubMed central

Paper	Title	GSS	OL	Odds	P value	Adjusted P value
PMC4887009	Identification and validation of reference genes for RT-qPCR studies of hypoxia in squamous cervical cancer patients	405	29	28.7	1.17×10^{-36}	1.05×10^{-33}
PMC10287711	Hypoxia-induced responses are reflected in the stromal proteome of breast cancer	1860	28	6.02	1.16×10^{-16}	9.69×10^{-15}
PMC9232399	A multiomics deep learning model for hypoxia phenotype to predict tumor aggressiveness and prognosis in uveal melanoma for rationalized hypoxia-targeted therapy	96	11	45.9	5.43×10^{-16}	4.26×10^{-14}
PMC10287711	Hypoxia-induced responses are reflected in the stromal proteome of breast cancer	1188	21	7.07	1.83×10^{-13}	1.13×10^{-11}
PMC5061082	The BET inhibitor JQ1 selectively impairs tumor response to hypoxia and downregulates CA9 and angiogenesis in triple negative breast cancer	49	8	65.3	3.50×10^{-13}	2.11×10^{-11}
PMC6405289	Mutations in an innate immunity pathway are associated with poor overall survival outcomes and hypoxic signaling in cancer	235	12	20.4	4.20×10^{-13}	2.51×10^{-11}
PMC9218734	A novel 7-hypoxia-related long noncoding RNA signature associated with prognosis and proliferation in melanoma	237	12	20.3	4.64×10^{-13}	2.76×10^{-11}
PMC6405289	Mutations in an innate immunity pathway are associated with poor overall survival outcomes and hypoxic signaling in cancer	51	8	62.8	4.92×10^{-13}	2.92×10^{-11}
PMC6411608	Circulating miRNA profiling of women at high risk for ovarian cancer	51	8	62.8	4.92×10^{-13}	2.92×10^{-11}
PMC9515221	TP53 mutations and RNA-binding protein MUSASHI-2 drive resistance to PRMT5-targeted therapy in B cell lymphoma	72	8	44.5	8.90×10^{-12}	4.73×10^{-10}

Table 7.95 Top ten-ranked gene sets in Rummagene when 200 genes were uploaded. GSS: gene set size, OL: overlap. PMC is paper IDs in PubMed central

Paper	Title	GSS	OL	Odds	*P* value	Adjusted *P* value
PMC10287711	Hypoxia-induced responses are reflected in the stromal proteome of breast cancer	1860	50	4.24	6.01×10^{-20}	1.86×10^{-15}
PMC6405289	Mutations in an innate immunity pathway are associated with poor overall survival outcomes and hypoxic signaling in cancer	235	22	14.8	1.02×10^{-19}	2.64×10^{-15}
PMC9881958	Secretion encoded single-cell sequencing (SEC-seq) uncovers gene expression signatures associated with high VEGF-A secretion in mesenchymal stromal cells	179	20	17.6	1.55×10^{-19}	3.60×10^{-15}
PMC5495571	HIF-1α is required for disturbed flow-induced metabolic reprogramming in human and porcine vascular endothelium	95	16	26.5	9.26×10^{-19}	1.45×10^{-14}
PMC9081503	Development and validation of a novel hypoxia score for predicting prognosis and immune microenvironment in rectal cancer	341	24	11.1	1.52×10^{-18}	2.12×10^{-14}
PMC9218734	A novel 7-hypoxia-related long noncoding RNA signature associated with prognosis and proliferation in melanoma	237	21	14	2.33×10^{-18}	2.93×10^{-14}
PMC10287711	Hypoxia-induced responses are reflected in the stromal proteome of breast cancer	1188	39	5.17	3.53×10^{-18}	3.44×10^{-14}
PMC8548828	Identification of a hypoxia-related gene signature for predicting systemic metastasis in prostate cancer	185	19	16.2	6.69×10^{-18}	6.04×10^{-14}
PMC9165826	Identification and validation of a prognostic signature–related to hypoxic tumor microenvironment in cervical cancer	190	19	15.8	1.11×10^{-17}	8.29×10^{-14}
PMC9792869	A hypoxia risk score for prognosis prediction and tumor microenvironment in adrenocortical carcinoma	191	19	15.7	1.23×10^{-17}	8.96×10^{-14}

$$x_{ktj} = a_k + b_k T(t) j \tag{7.148}$$

where $T(t)$ is the hours at the tth time point and $(T(1), T(2), T(3), T(4)) = (0, 6, 12, 24)$ and a_i, b_i, a_k, and b_k were regression coefficients. For SAM and limma, they are assumed to be four and eight classes for gene expression and m^6A, respectively (in other words, each sample is supposed to belong to its own class without any other members). For linear regression, no is or ks were associated with adjusted P values less than 0.01. For SAM, we could not apply SAM to these data sets, since SAM cannot accept data without replicates in individual classes. For limma, there are no genes associated with adjusted P values less than 0.01 although there were as many as 722 genomic regions associated with adjusted P values less than 0.01. Only random forest among the four compared methods could identify nonzero is and ks associated with adjusted P values less than 0.01: 480 is and 722 ks. Since these numbers are much more than those identified by TD-based unsupervised FE with integrated analysis, we evaluate them by uploading to Rummagene (for regions, genes included in these regions were uploaded). 9121 and 16,789 statistically significant matches were identified for gene expression and m^6A profiles, respectively. These numbers were substantially less than any numbers obtained above regardless of the combinations of methods and data sets. After screening them by the term "hypoxia," there remained only ten and five statistically significant matches for gene expression and m^6A profiles. These numbers are ten times less than any numbers obtained above regardless of the combinations of methods and data sets. Thus, the methods tested above outperformed all four conventional methods.

Finally, we consider one alternative set retrieved from GEO with GEO ID GSE120860, which contains only m^6A profile. m^6A profiles are formatted as a tensor, $x_{ijkm} \in \mathbb{R}^{N \times 4 \times 2 \times 2}$, which represents m^6A profiles of the ith gene at the jth subject of the kth group ($k = 1$: annotated as 0204 in GEO, $k = 2$: annotated as patient in GEO) of the mth tissue ($m = 1$:tumor, $m = 2$:paratumor). HOSVD was applied to x_{ijkm} and we got

$$x_{ijkm} = \sum_{\ell_1=1}^{N} \sum_{\ell_2=1}^{4} \sum_{\ell_3=1}^{2} \sum_{\ell_4=1}^{2} G(\ell_1 \ell_2 \ell_3 \ell_4) u^{(i)}_{\ell_1 i} u^{(j)}_{\ell_2 j} u^{(k)}_{\ell_3 k} u^{(m)}_{\ell_4 m} \tag{7.149}$$

where $G \in \mathbb{R}^{N \times 4 \times 2 \times 2}$ is a core tensor, $u^{(i)}_{\ell_1 i} \in \mathbb{R}^{N \times N}$, $u^{(j)}_{\ell_2 j} \in \mathbb{R}^{4 \times 4}$, and $u^{(k)}_{\ell_3 k}, u^{(m)}_{\ell_4 m} \in \mathbb{R}^{2 \times 2}$ are singular value vectors and orthogonal vectors.

The desired properties of singular value vectors are independence of subjects, independence of groups, and distinction between tissues; Figure 7.82 lists $\boldsymbol{u}^{(j)}_1, \boldsymbol{u}^{(k)}_1$, and $\boldsymbol{u}^{(m)}_2$, which seemingly satisfy the required properties. Thus, $\ell_2 = \ell_3 = 1$ and $\ell_4 = 2$ are selected. Figure 7.83 shows $|G(\ell_1, 1, 1, 2)|$. Since $G(4, 1, 1, 2)$ has the largest absolute value, we decided to select $\boldsymbol{u}^{(i)}_4$ to select is. P values were attributed to is as

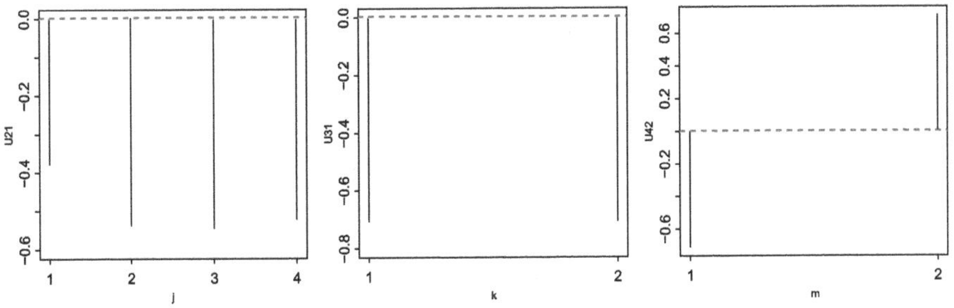

Fig. 7.82 Left: $\boldsymbol{u}_1^{(j)}$, middle: $\boldsymbol{u}_1^{(k)}$, right: $\boldsymbol{u}_2^{(m)}$

Fig. 7.83 $|G(\ell_1, 1, 1, 2)|$

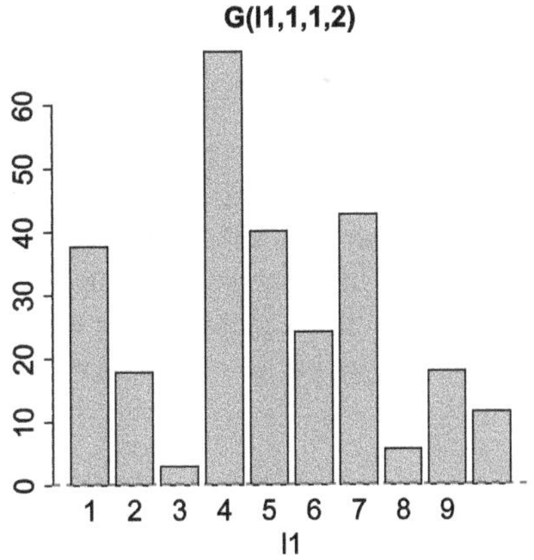

$$P_i = P_{\chi^2}\left[> \left(\frac{u_{4i}^{(i)}}{\sigma_4} \right)^2 \right]. \tag{7.150}$$

P values were corrected by BH criterion and 54 is associated with adjusted P values less than 0.01 are selected. Forty-six gene symbols associated with 54 is were uploaded to Rammagene. 12,697 statistically significant matches were identified, and 167 statistically significant matches remain after screening "cancer." Thus, our results are robust.

7.16.2 m^{6}A II : Human vs. Mouse

In the previous subsection, we integrated m^{6}A profiles with gene expression profiles. In this subsection, we integrated m^{6}A with histone modifications in addition to gene expression in single-cell-based measurements [110]. Gene expression profiles, two histone modification profiles (H3K4me3and H3K27ac), and m^{6}A profiles for human cell lines and mouse embryonic stem cells (mESC) were retrieved from

Table 7.96 Number of samples used in this study. MeRIP: m^6A, nuclearRNA: gene expression

	H3K4me3	H3K27ac	MeRIP	nuclearRNA
Human cell line (HEC-1-A)				
Control	2	2	2	2
KD	2	2	2	2
Mouse embryonic stem cells (mESC)				
Control	2	2	2	2
KO1	2	2	2	2
KO2	2	2	2	2

GEO with GEO ID, GSE140561 and GSE133600, respectively. Table 7.96 lists the number of samples used in this study. HEC-1-A and mESC data sets were formatted as tensors, $x_{ijks} \in \mathbb{R}^{N \times 4 \times K \times 2}$, which represent jth measurements ($j = 1$: H3K4me3, $j = 2$: H3K27ac, $j = 3$: MeRIP, and $j = 4$: nascent nuclear RNA expression) of ith gene (Ensembl gene ID) of sth biological replicate of kth sample (for HEC-1-A, $K = 2$, $k = 1$: control and $k = 2$:KO; for mESC, $K = 3$, $k = 1$: control, $k = 2$: the first KO, and $k = 3$: the second KO). Since m^6A profiles and histone modifications are not attributed to individual genes, to attribute them to individual genes, values within genomic regions within individual genes are summed to get values attributed to genes. Finally, x_{ijks} is normalized as

$$\sum_{i=1}^{N} x_{ijks} = 0 \tag{7.151}$$

$$\sum_{i=1}^{N} x_{ijks}^2 = N. \tag{7.152}$$

HOSVD was applied to x_{ijks} and we got

$$x_{ijks} = \sum_{\ell_1=1}^{4} \sum_{\ell_2=1}^{K} \sum_{\ell_3=1}^{2} \sum_{\ell_4=1}^{N} G(\ell_1 \ell_2 \ell_3 \ell_4) u_{\ell_1 j}^{(j)} u_{\ell_2 k}^{(k)} u_{\ell_3 s}^{(s)} u_{\ell_4 i}^{(i)} \tag{7.153}$$

where $G \in \mathbb{R}^{4 \times K \times 2 \times N}$ is a core tensor and $u_{\ell_1 j}^{(j)} \in \mathbb{R}^{4 \times 4}$, $u_{\ell_2 k}^{(k)} \in \mathbb{R}^{K \times K}$, $u_{\ell_3 s}^{(s)} \in \mathbb{R}^{2 \times 2}$, and $u_{\ell_4 i}^{(i)} \in \mathbb{R}^{N \times N}$ are singular value matrices and are orthogonal matrices.

At first, we need to find which singular value vectors, $u_{\ell_1 j}^{(j)}$, $u_{\ell_2 k}^{(k)}$, and $u_{\ell_3 s}^{(s)}$, have the required properties, i.e.,

- Independence of omics (i.e., two histone modifications, m^6A, and gene expression)
- Distinction between control and knockout
- Independence of biological replicates

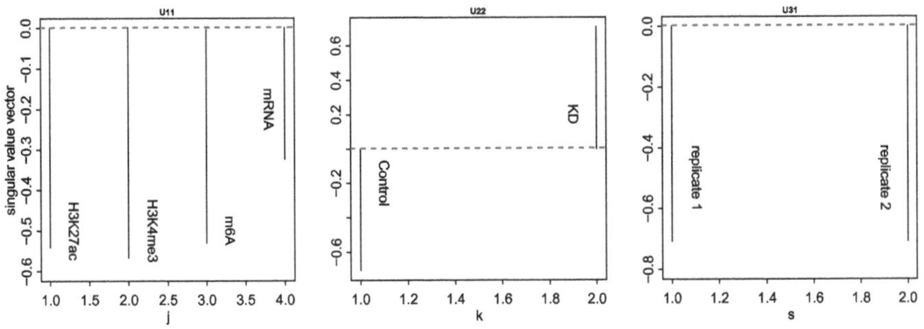

Fig. 7.84 Human cell line (HEC-1-A). Left: $\boldsymbol{u}_1^{(j)}$, middle: $\boldsymbol{u}_2^{(k)}$, right: $\boldsymbol{u}_1^{(s)}$

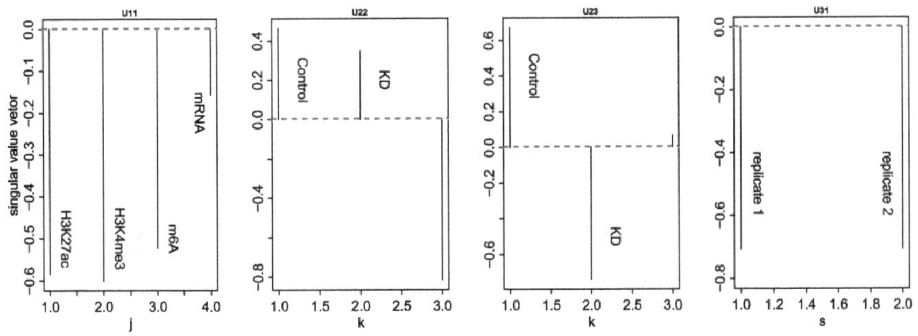

Fig. 7.85 mESC. From left to rigt: $\boldsymbol{u}_1^{(j)}$, $\boldsymbol{u}_2^{(k)}$, $\boldsymbol{u}_3^{(k)}$, $\boldsymbol{u}_1^{(s)}$

Figure 7.84 shows $\boldsymbol{u}_1^{(j)}$, $\boldsymbol{u}_2^{(k)}$, and $\boldsymbol{u}_1^{(s)}$ for human cell lines. We can see that $\ell_1 = \ell_3 = 1$ and $\ell_2 = 2$ satisfy the desired properties. Figure 7.85 shows $\boldsymbol{u}_1^{(j)}$, $\boldsymbol{u}_2^{(k)}$, $\boldsymbol{u}_3^{(k)}$, and $\boldsymbol{u}_1^{(s)}$ form ESC. We can see that $\ell_1 = \ell_3 = 1$ and $\ell_2 = 2, 3$ satisfy the desired properties. To see which ℓ_4 is associated with $\ell_1 = \ell_3 = 1$ and $\ell_2 = 2$ (and $\ell_2 = 3$ only for mouse), we show $|G(1, 2, 1, \ell_4)|$ (and $|G(1, 3, 1, \ell_4)|$ only for mouse) in Fig. 7.86. Then we found that $\ell_4 = 5$ is associated with the largest $|G(1, 2, 1, \ell_4)|$ and $\ell_4 = 7$ is associated with the largest $|G(1, 3, 1, \ell_4)|$ for mouse.

P values were attributed to ith as

$$P_i = P_{\chi^2}\left[> \left(\frac{u_{\ell_4 i}}{\sigma_{\ell_4}} \right)^2 \right] \tag{7.154}$$

using the specified ℓ_4s. 741 (for human), 668 (for mouse, $\ell_4 = 5$), and 684 (for mouse, $\ell_4 = 7$) is (Ensembl gene IDs) associated with adjusted P values less than 0.01 were selected. Figure 7.87 shows the Venn diagram of gene symbols associated with these 741, 668 and 684 Ensembl gene IDs. If we consider $N = 58{,}721$ for human and $N = 54{,}446$, this intersection is highly significant since only at most a few percentages of genes are selected for three set of genes. To see if the selected

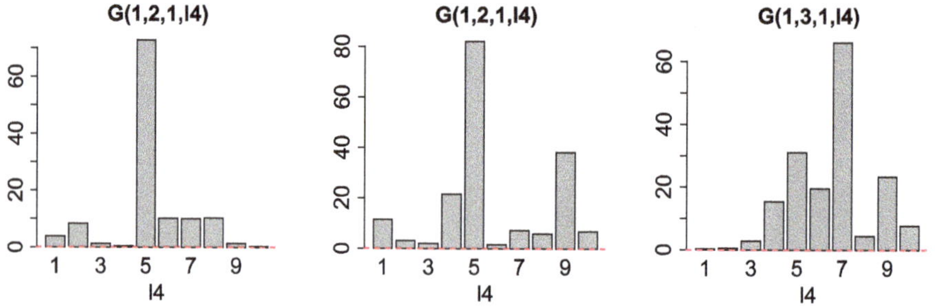

Fig. 7.86 From left to right, for human cell line $|G(1,2,1,\ell_4)|$, for mESC $|G(1,2,1,\ell_4)|$ and $|G(1,3,1,\ell_4)|$

Fig. 7.87 Venn diagram of gene symbols selected using TD-based unsupervised FE for mice (mESC, mouse1 for $\ell_3 = 5$, mouse2 fro $\ell_3 = 7$) and humans (HEC-1-A)

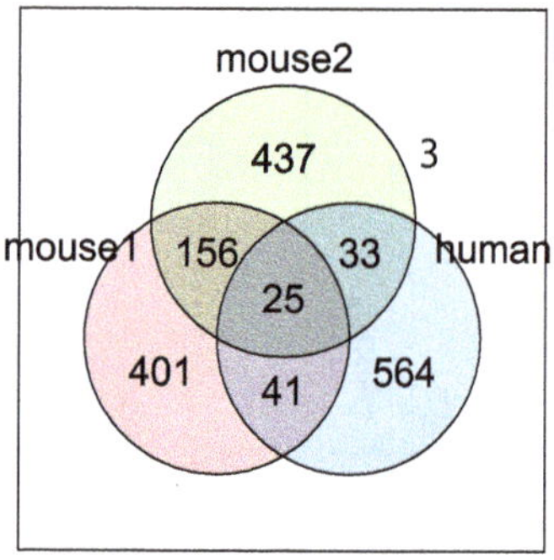

gene sets are overlapped with known m^{6}A related gene sets, we uploaded three sets of gene symbols to Rummagene. 79,652 for human, 69,955 for mouse1, and 57,582 for mouse2, statistically significant matches were identified. After screening "m^{6}A," 40 for human, 39 for mouse1, 80 for mouse2 (31 are common between two) remained. Thus, three sets of genes are significantly related with m^{6}A definitely.

Now we can see what happens with these data sets since we could successfully show that the sets of selected genes are related to m^{6}A. At first, we upload these sets of gene to Enrichr. The category that we evaluated first is "ENCODE and ChEA Consensus TFs from ChIP-X." Figure 7.88 shows the Venn diagram of TFs associated with adjusted P values less than 0.05. In contrast to Fig. 7.87 where only small parts of genes are overlapped, the majority of the selected TFs are common. Thus, more common features between human and mouse were selected than in the case of genes (Table 7.97).

Figure 7.89 shows the Venn diagram of the enriched (associated with adjusted P values less than 0.05) tissues in "Allen Brain Atlas 10x scRNA 2021." Not only there are many enriched tissues, but also there are many commonly selected tissues. Among 43 enriched tissues, as many as 13 tissues are related to oligodendrocyte whose relation to m^{6}A was reported [118, 120]. Although cell lines used are not directly related to brains, we could identify the relationship between m^{6}A and brain

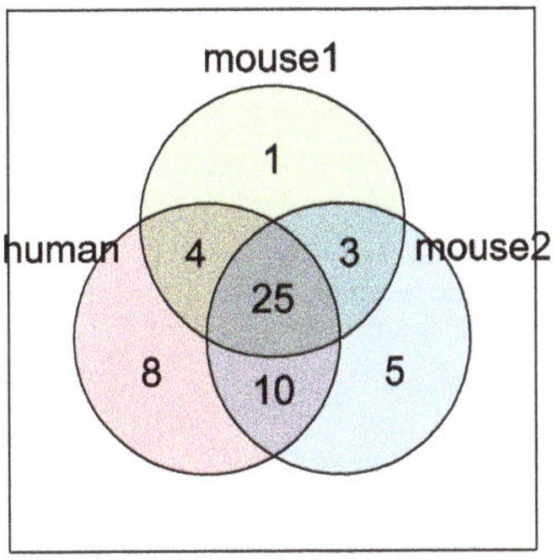

Fig. 7.88 Venn diagram of TFs in "ENCODE and ChEA Consensus TFs from ChIP-X" category, selected by uploading genes selected by TD-based unsupervised FE to Enrichr for mice (mESC, mouse1 for $\ell_3 = 5$, mouse2 fro $\ell_3 = 7$) and humans (HEC-1-A)

Table 7.97 Twenty-six commonly selected TFs from Fig. 7.88

Databases	TFs
ENCODE	CHD1, GATA2, PBX3, TAF1 TCF3, UBTF, UBTF, YY1, ZNF384, ZBTB7A,
CHEA	AR, E2F1 ESR1, GATA1, MYC, NANOG, NFE2L2, KLF4, POU5F1, SALL4, SMAD4, SOX2, STAT3, TCF3, TP53, TP63

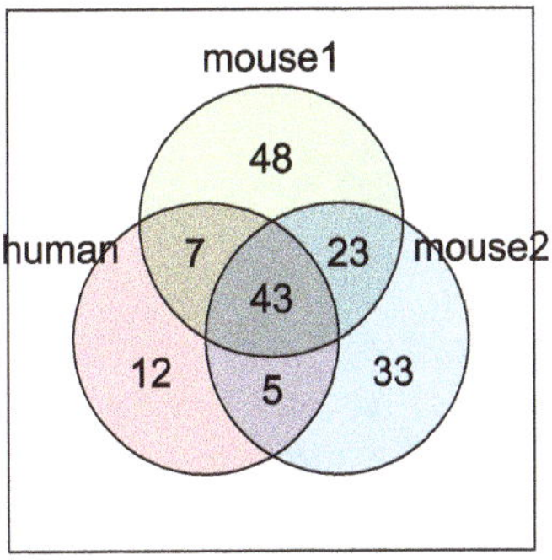

Fig. 7.89 Venn diagram of tissues in "Allen Brain Atlas 10x scRNA 2021" category, selected by uploading genes selected by TD-based unsupervised FE to Enrichr for mice (mESC, mouse1 for $\ell_3 = 5$, mouse2 fro $\ell_3 = 7$) and humans (HEC-1-A)

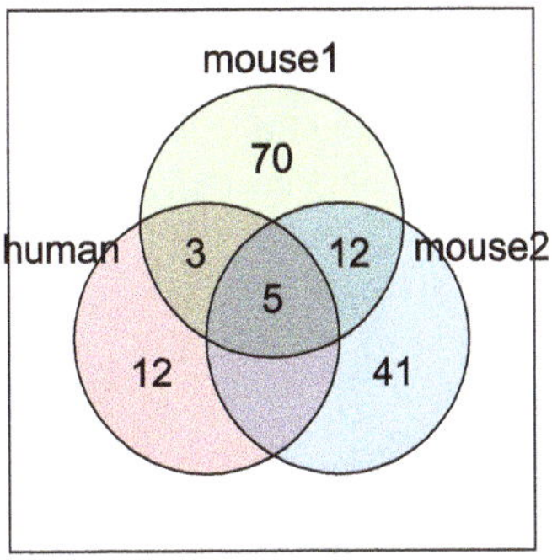

Fig. 7.90 Venn diagram of pathways in "Reactome 2022" category, selected by uploading genes selected by TD-based unsupervised FE to Enrichr for mice (mESC, mouse1 for $\ell_3 = 5$, mouse2 fro $\ell_3 = 7$) and humans (HEC-1-A)

in the level of subregions of brains. This suggests the effectiveness of integrated analysis using TD-based unsupervised FE for transcriptome.

Figure 7.90 shows the Venn diagram of the enriched (associated with adjusted P values less than 0.05) pathways in "Reactome 2022." Not only there are many

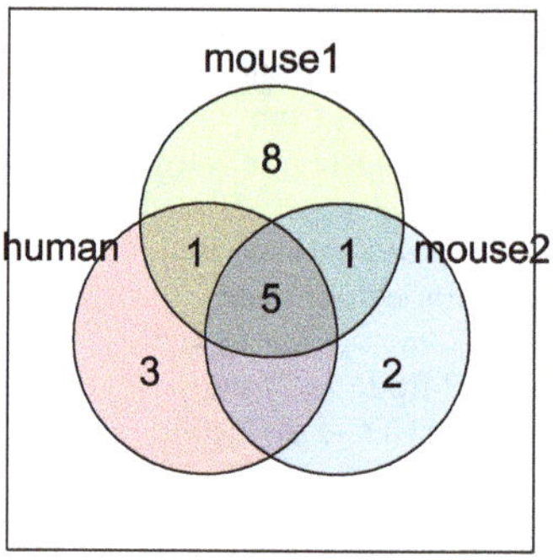

Fig. 7.91 Venn diagram of pathways in "Jensen DISEASES" category, selected by uploading genes selected by TD-based unsupervised FE to Enrichr for mice (mESC, mouse1 for $\ell_3 = 5$, mouse2 for $\ell_3 = 7$) and humans (HEC-1-A)

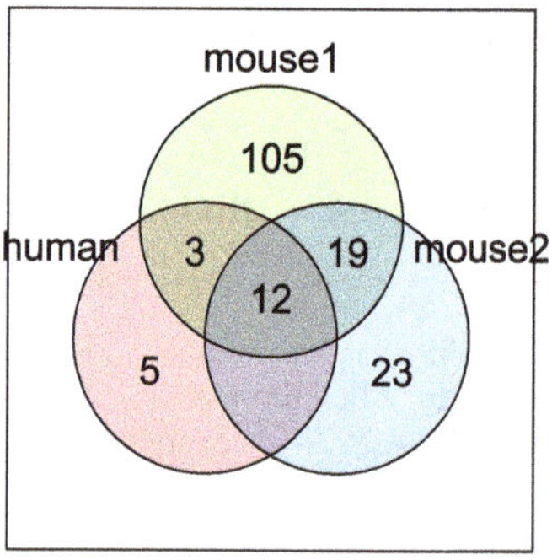

Fig. 7.92 Venn diagram of pathways in "MGI Mammalian Phenotype Level 4 2021" category, selected by uploading genes selected by TD-based unsupervised FE to Enrichr for mice (mESC, mouse1 for $\ell_3 = 5$, mouse2 fro $\ell_3 = 7$) and humans (HEC-1-A)

enriched pathways, but also there are multiple commonly selected pathways. Among five enriched pathways, as many as three pathways are related to Rho GTPases whose relation to m^6A in cancers was reported [83]; we could identify the relationship between m^6A and cancers. This also suggests the effectiveness of integrated analysis using TD-based unsupervised FE for transcriptome.

Figure 7.91 shows the Venn diagram of the enriched (associated with adjusted P values less than 0.05) diseases in "Jensen DISEASES." There were five commonly selected diseases, four of which are Kidney cancer [87], liver cancer [77], endometrial cancer [56], and breast cancer [129] and were reported to be related to m^6A. This also suggests the effectiveness of integrated analysis using TD-based unsupervised FE for transcriptome.

Figure 7.92 shows the Venn diagram of the enriched (associated with adjusted P values less than 0.05) diseases in "MGI Mammalian Phenotype Level 4 2021." There were 12 commonly selected terms (Table 7.98) which are related to "decrease," "lethal," or "incomplete." Since the lack of m^6A is known to be related to lethality [30], it is reasonable. This also suggests the effectiveness of integrated analysis using TD-based unsupervised FE for transcriptome.

In conclusion, integrated analysis of epitranscriptome can be a powerful tool to investigate functionalities of epitranscriptome.

Table 7.98 Twelve commonly selected terms from Fig. 7.92

Terms
"embryonic growth retardation MP:0003984," "premature death MP:0002083," "decreased fetal size MP:0004200," "preweaning lethality, incomplete penetrance MP:0011110," "decreased embryo size MP:0001698," "preweaning lethality, complete penetrance MP:0011100," "embryonic lethality during organogenesis, complete penetrance MP:0011098," "decreased body weight MP:0001262," "decreased body size MP:0001265," "decreased fetal weight MP:0009431," "lethality throughout fetal growth and development, complete penetrance MP:0011099," "abnormal heart development MP:0000267"

7.17 Gene Expression Analysis Without Sample Matching

In this subsection, we demonstrate the application in Sect. 5.8.2, where how to integrate multiple profiles sharing only samples was discussed, to the real examples [108].

7.17.1 Integrated Analysis of Three Gene Expression Profiles

We start the integration of three gene expression profiles that share the most of genes but lack sample correspondence. Usually, we need some sample correspondence between samples to compare multiple profiles. For example, if two profiles share the class label, e.g., healthy control and patients, it is easy to integrate two profiles. However, if we have two profiles without sample matching, it is not easy. we introduced the case II approach in Table 5.3. Nevertheless, when we have more than two profiles, it turned out that case II approach cannot be very effective although it is mathematically possible to apply the case II approach to the cases with more than two profiles.

In contrast to the case II approach, that in Sect. 5.8.2 is practically useful even when applied to the case with more than two profiles. Three gene expression profiles, all of which are related to Alzheimer's disease (AD) in some sense, were retrieved from GEO with GEO IDs, GSE160224, GSE155567, and GSE162873. Hereafter, they will be referred to as data sets 1, 2, and 3. All data sets in the following were normalized as

$$\sum_{i=1}^{N} x_{ij_kk} = 0 \tag{7.155}$$

$$\sum_{i=1}^{N} x_{ij_kk}^2 = N \tag{7.156}$$

Data set 1 is composed of three classes, nondemented controls (NDC), APP duplications, and isogenically corrected induced pluripotent stem-cell lines, each of which has three replicates. Thus, there are nine samples in total. The number of genes whose expression is measured is 58,302. In the following, NDC is regarded to be control and others are regarded to be treated samples. Thus, they are regarded to be composed of two classes. These profiles are formatted as a matrix $x_{ij_11} \in \mathbb{R}^{58,302\times 9}$.

Data set 2 is composed of four classes, which are orthogonal products of two sets of two classes; the first set of two classes is CD33 KO or not and the second set of two classes is PTPN6 KO or not. Although each of the four classes has six replicates, only 23 out of 24 samples are associated with measured gene expression profiles. The number of genes whose expression is measured is 60,617. These profiles are formatted as a matrix $x_{ij_22} \in \mathbb{R}^{60,617\times 23}$.

Data set 3 is composed of two classes, AD and normal cell lines, each of which has four replicates. Thus, there are eight samples in total. The number of genes whose expression is measured is 47,749. These profiles are formatted as a matrix $x_{ij_33} \in \mathbb{R}^{47,749\times 8}$.

After getting $v^k_{\ell j_k}$s using Eq. (5.35), instead of

$$x_{i\ell k} = u^k_{\ell i} \tag{7.157}$$

in Eq. (5.36),

$$x_{i\ell k} = \sum_{j_k=1}^{M_k} x_{ij_kk} v^k_{\ell j_k} = \lambda_\ell u^k_{\ell i}. \tag{7.158}$$

Thus, $x_{i\ell k}$ above differs from that in Eq. (5.36) by the factor of λ_ℓ.

Then $x_{i\ell k}$s were formatted as a tensor $x_{i\ell k} \in \mathbb{R}^{60,617\times 8\times 3}$. Since data sets 1 and 3 have a smaller number of genes than data set 2, missing values are filled with zero, and up to the eighth singular values are considered.

$u^{(i)}_{\ell_1 i}$ was computed by Eq. (5.36). Then $u^{(j_k;i)}_{\ell_1 j_k}$ was computed by Eq. (5.30). Figure 7.93 shows $u^{(j_k;i)}_{1 j_k}$ where P values were computed by the categorical regression analysis

$$u^{(j_k;i)}_{\ell_1 j_k} = a_{\ell_1} + \sum_s b_{\ell_1 s}\delta_{j_k s} \tag{7.159}$$

where a_{ℓ_1} and $b_{\ell_1 s}$ are regression coefficients and $\delta_{j_k s}$ takes one only when j_k belongs to sth category otherwise zero. Figure 7.93 suggests that the proposed method has the ability to relate three independent data sets without sharing labels. This means that genes whose expression is distinct between control and AD in data set 1 should express the gene expression distinct between four classes in data set 2 and between three classes in data set 3 in the way shown in Fig. 7.93. Usually,

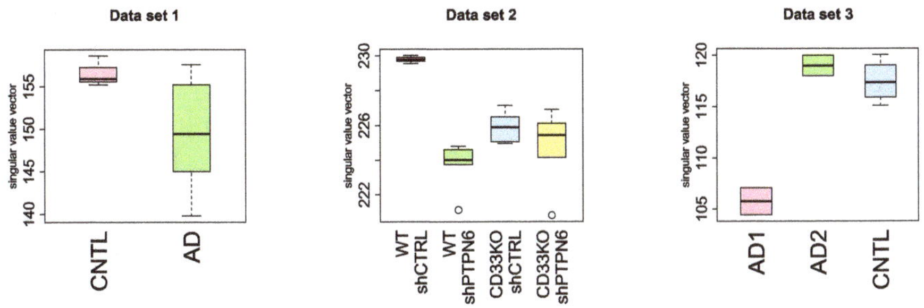

Fig. 7.93 Boxplots of $u_{1j_1}^{(j_1;i)}$ (left, $P = 4.86 \times 10^{-2}$), $u_{1j_2}^{(j_2;i)}$ (middle, $P = 1.94 \times 10^{-6}$), $u_{1j_3}^{(j_3;i)}$ (right, $P = 1.55 \times 10^{-3}$). P values were computed by the categorical regression

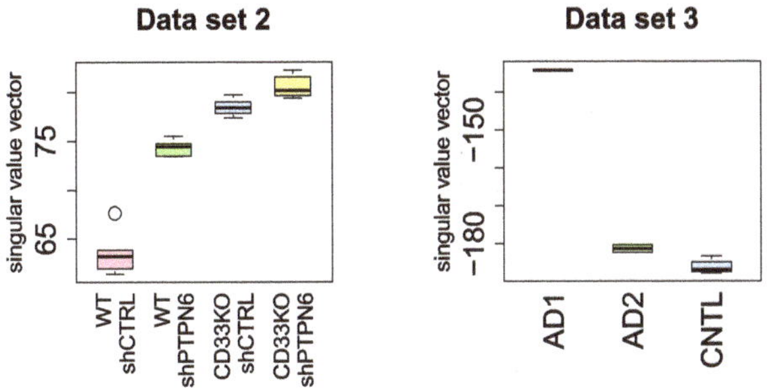

Fig. 7.94 Boxplots of $u_{2j_2}^{(j_2;i)}$ (left, $P = 7.81 \times 10^{-14}$), $u_{2j_3}^{(j_3;i)}$ (right, $P = 8.23 \times 10^{-7}$). P values were computed by the categorical regression analysis

although this kind of analysis cannot be performed when data sets do not share labels, the proposed method can do this. This is the primary motivation for the introduction of the proposed method.

Although $u_{\ell_1 i}^{(i)}$ other than $u_{1i}^{(i)}$ cannot provide us $u_{\ell_1 j_k}^{(j_k;i)}$ associated with distinction between classes in all three data sets, they can give $u_{\ell_1 j_k}^{(j_k;i)}$ which are distinct between classes in some of three data sets. Figure 7.94 shows $u_{2j_1}^{(j_1;i)}$ for data sets 2 and 3. It also shows the coincidence between data sets 2 and 3 which lack the direct correspondence between class labels. On the other hand, $u_{3j_2}^{(j_2;i)}$ exhibits the distinction between class labels only in data set 2 (Fig. 7.95). Thus, $u_{\ell_1 j_2}^{(j_2;i)}$ does not always provide us the information about the coincidence between distinct (independent) data sets.

Figures 7.96 and 7.97 show the boxplots of $u_{4j_k}^{(j_k;i)}$ and $u_{5j_k}^{(j_k;i)}$, respectively, which exhibit distinction between classes only for data sets 2 and 3 as in the case of $u_{3j_k}^{(j_k;i)}$ above. In conclusion, although in which data sets $u_{\ell_1 j_k}^{(j_k;i)}$ can exhibit distinction between class labels vary from case to case, the proposed method has potential activities to show coincidence of class labels between distinct data sets.

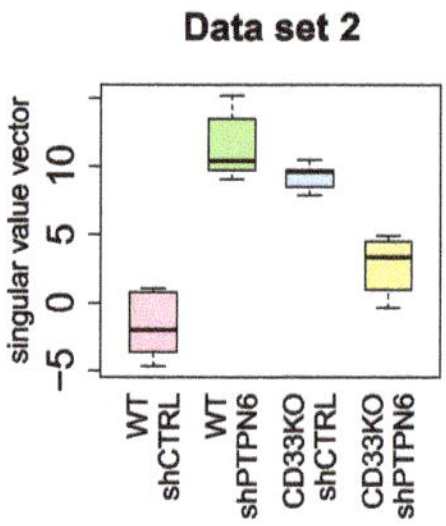

Fig. 7.95 Boxplots of $u_{3j_2}^{(j_2;i)}$ ($P = 4.15 \times 10^{-9}$). P value was computed by the categorical regression analysis

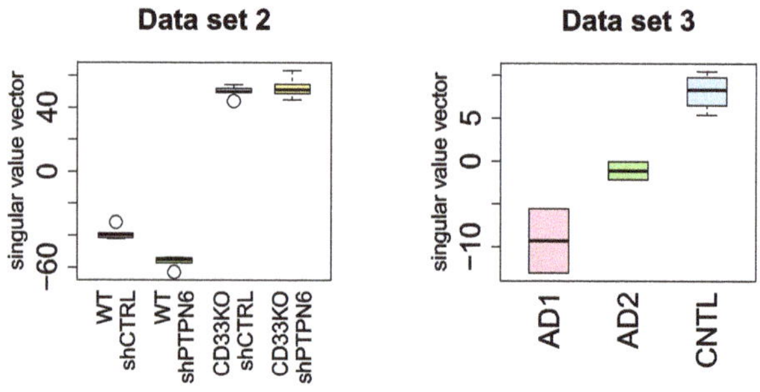

Fig. 7.96 Boxplots of $u_{4j_2}^{(j_2;i)}$ (left, $P < 2.2 \times 10^{-16}$), $u_{4j_3}^{(j_3;i)}$ (right, $P = 2.74 \times 10^{-3}$). P values were computed by the categorical regression analysis

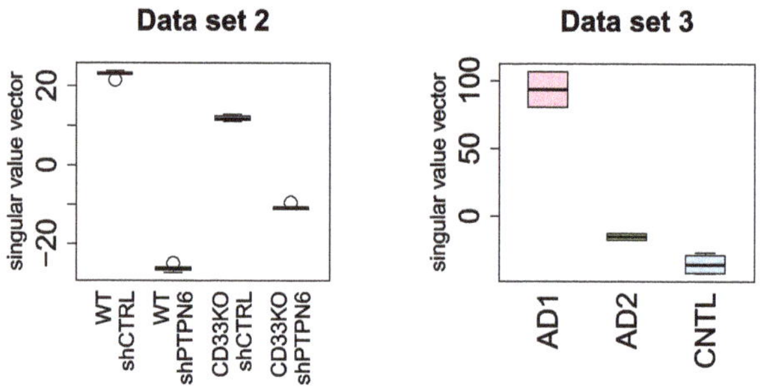

Fig. 7.97 Boxplots of $u_{5j_2}^{(j_2;i)}$ (left, $P < 2.2 \times 10^{-16}$), $u_{5j_3}^{(j_3;i)}$ (right, $P = 7.77 \times 10^{-5}$). P values were computed by the categorical regression analysis

Next, we try to select genes associated with $u_{\ell_1 i}^{(i)}$. Since all $1 \leq \ell_1 \leq 5$ are related to the distinction between class labels in any of three data sets, we decided to use all five $u_{\ell_1 i}^{(i)}$ for $1 \leq \ell_1 \leq 5$ to select genes. P values were attributed to i as

$$P_i = P_{\chi^2}\left[> \sum_{\ell_1=1}^{5} \left(\frac{u_{\ell_1 i}^{(i)}}{\sigma_{\ell_1}} \right)^2 \right] \tag{7.160}$$

Table 7.99 Top ten-ranked enriched terms in "KEGG 2021 HUMAN" category of Enrichr

Term	Overlap	*P* value	Adjusted *P* value
Ribosome	68/158	1.73×10^{-63}	4.33×10^{-61}
Coronavirus disease	71/232	5.82×10^{-54}	7.27×10^{-52}
Parkinson's disease	38/249	1.17×10^{-17}	9.77×10^{-16}
Pathways of neurodegeneration	48/475	1.23×10^{-14}	7.68×10^{-13}
Amyotrophic lateral sclerosis	41/364	3.06×10^{-14}	1.53×10^{-12}
Salmonella infection	32/249	6.11×10^{-13}	2.55×10^{-11}
Alzheimer's disease	39/369	1.01×10^{-12}	3.62×10^{-11}
Huntington's disease	32/306	1.52×10^{-10}	4.75×10^{-9}
Prion disease	30/273	1.72×10^{-10}	4.77×10^{-9}
Pathogenic Escherichia coli infection	20/197	6.97×10^{-7}	1.74×10^{-5}

and *P* values are corrected by BH criterion. Finally, 565 is associated with adjusted *P* values less than 0.01 were selected. Since gene IDs are Ensembl gene IDs, they are converted to 558 gene symbols by DAVID gene ID converter. These gene symbols were uploaded to Enrichr for evaluation. At first, we checked "KEGG 2021 HUMAN" category (Table 7.99). From third to nine terms are related to neurodegenerative diseases, including Alzheimer's disease ranked as the seventh. Thus, the proposed method not only could relate the class labels with one another among distinct data sets but also could select a reasonable set of genes.

7.17.2 Drug Repositioning Using the Tensor Obtained with Data Sets 1, 2, and 3

In Sects. 7.3 and 7.6.2, we have shown how gene expression can be used for drug repositioning. Especially we related gene expression profiles of diseases to those of drug-treated-model animals to perform drug repositioning in Sect. 7.6.2. Nevertheless, we could integrate a model animal and a disease gene expression profile. In this subsection, we demonstrate how we can integrate multiple disease gene expression profiles with those of drug-treated model animals.

To do this, we integrate three gene expression profiles analyzed in the previous subsection with a drug-treated gene expression profile retrieved from GEO with GEO ID, GSE164788, which will be referred to as data set 4. Data set 4 is composed of an in vitro differentiated mixture of neuron and glial cells derived from the ReNcell VM neural progenitor cell line treated with 80 different compounds. Among the 80 compounds, we select single drug treatments as well as their combinations (in total 94 cases), from which at least two doses are tested. For each dose, three biological replicates are provided. When more than three are used, we randomly select three of them. Since the number of genes whose expression is measured is 28,044, gene expression profiles in data set 4 is formatted as a tensor,

$x_{ij_4^{[1]}j_4^{[2]}j_4^{[3]}} \in \mathbb{R}^{28,044 \times 94 \times 4 \times 3}$ that represents the expression of the ith gene of the $j_4^{[1]}$th drug treatment at the $j_4^{[2]}$th dose of the $j_4^{[3]}$th replicate, to which HOSVD was applied and we got

$$x_{ij_4^{[1]}j_4^{[2]}j_4^{[3]}} = \sum_{\ell_1=1}^{94} \sum_{\ell_2=1}^{4} \sum_{\ell_3=1}^{3} \sum_{\ell_4=1}^{28,044} G(\ell_1\ell_2\ell_3\ell_4) u^{[4]}_{\ell_1 j_4^{[1]}} u^{[4]}_{\ell_2 j_4^{[2]}} u^{[4]}_{\ell_3 j_4^{[3]}} u^{[4]}_{\ell_4 i} \tag{7.161}$$

where $G \in \mathbb{R}^{94 \times 4 \times 3 \times 28,044}$ is a core tensor and $u^{[4]}_{\ell_1 j_4^{[1]}} \in \mathbb{R}^{94 \times 94}$, $u^{[4]}_{\ell_2 j_4^{[2]}} \in \mathbb{R}^{4 \times 4}$, $u^{[4]}_{\ell_3 j_4^{[3]}} \in \mathbb{R}^{3 \times 3}$, and $u^{[4]}_{\ell_4 i} \in \mathbb{R}^{28,044 \times 28,044}$ are singular value matrices and are orthogonal matrices. To get $x_{i\ell 4} \in \mathbb{R}^{28,044 \times L}$, we do

$$x_{i\ell 4} = \sum_{j_4^{[1]}=1}^{94} \sum_{j_4^{[2]}=1}^{4} \sum_{j_4^{[3]}=1}^{3} x_{ij_4^{[1]}j_4^{[2]}j_4^{[3]}} u^{[4]}_{\ell_1 j_4^{[1]}} u^{[4]}_{\ell_2 j_4^{[2]}} u^{[4]}_{\ell_3 j_4^{[3]}},$$

$$1 \le \ell_s \le L_4^{[s]} (\le M_4^{[s]}), 1 \le \ell \le L = \prod_{s=1}^{3} L_4^{[s]} \tag{7.162}$$

where $M_4^{[1]} = 94$, $M_4^{[2]} = 4$, $M_4^{[3]} = 3$ and $L_4^{[s]}$ is the number of used singular value vectors, ℓ_s, and $L_4^{[1]} = 4$, $L_4^{[2]} = 2$, $L_4^{[3]} = 1$, thus $L = 8$. Then HOSVD was applied to $x_{i\ell k} \in \mathbb{R}^{60,617 \times 8 \times 4}$. $u^{(i)}_{\ell_1 i}$ was computed by Eq. (5.36) and $u^{(j_k;i)}_{\ell_1 j_k}$ was computed by Eq. (5.30).

At first, we evaluate if $u^{(j_k;i)}_{\ell_1 j_k}$ is associated with the distinction between class labels in data sets 1, 2, and 3. Figure 7.98 shows $u^{(j_k;i)}_{1 j_k}$. It is obvious that $u^{(j_k;i)}_{1 j_k}$ is associated with the distinction between class labels in data sets 1, 2, and 3. As in the previous subsection, no other $u^{(j_k;i)}_{\ell_1 j_k}$ other than $u^{(j_k;i)}_{1 j_k}$ is associated with distinction between class labels in all three data sets.

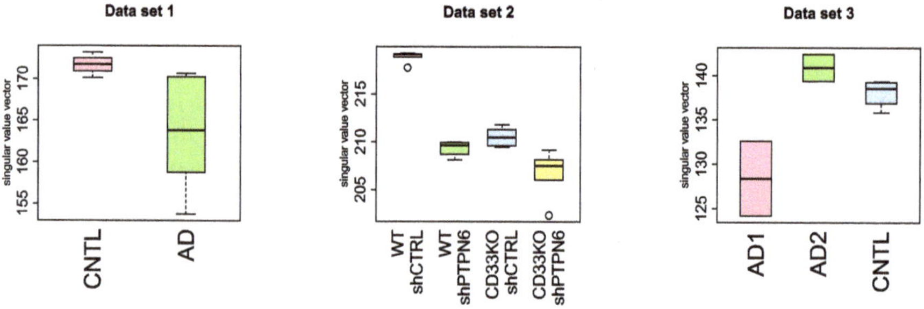

Fig. 7.98 Boxplots of $u^{(j_1;i)}_{1 j_1}$ (left, $P = 2.82 \times 10^{-2}$), $u^{(j_2;i)}_{1 j_2}$ (middle, $P = 4.04 \times 10^{-11}$), $u^{(j_3;i)}_{1 j_3}$ (right, $P = 2.02 \times 10^{-2}$). P values were computed by the categorical regression analysis

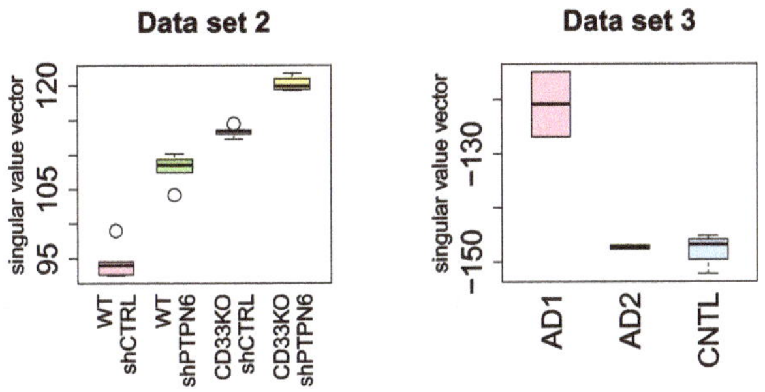

Fig. 7.99 Boxplots of $u_{2j_2}^{(j_2;i)}$ (left, $P = 3.95 \times 10^{-15}$), $u_{2j_3}^{(j_3;i)}$ (right, $P = 2.09 \times 10^{-2}$). P values were computed by the categorical regression analysis

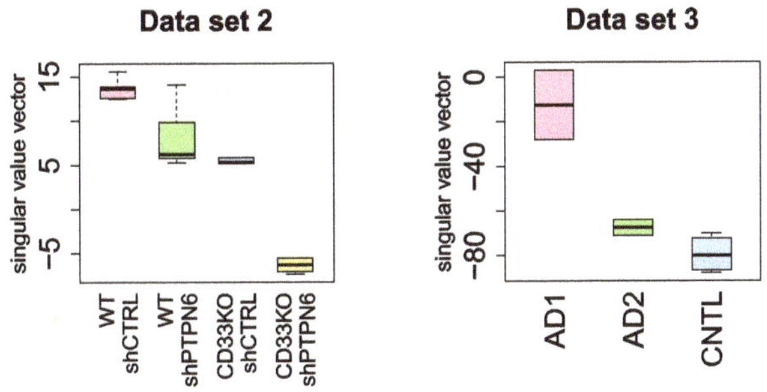

Fig. 7.100 Boxplots of $u_{3j_2}^{(j_2;i)}$ (left, $P = 2.05 \times 10^{-12}$), $u_{3j_3}^{(j_3;i)}$ (right, $P = 2.53 \times 10^{-3}$). P values were computed by the categorical regression analysis

Fig. 7.101 Boxplots of $u_{4j_2}^{(j_2;i)}$ ($P = 3.05 \times 10^{-10}$). P value was computed by the categorical regression analysis

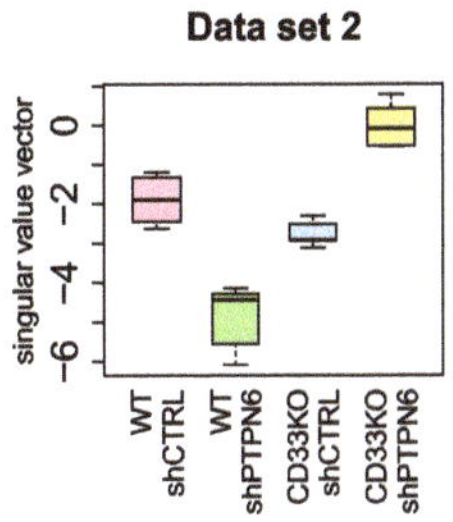

Figures 7.99 and 7.100 show $u_{2j_k}^{(j_k;i)}$ and $u_{3j_k}^{(j_k;i)}$, which are associated with the distinction between class labels in only data sets 2 and 3. On the other hand, $u_{4j_k}^{(j_k;i)}$ and $u_{5j_k}^{(j_k;i)}$ are associated with the distinction between class labeled only in data set 2 (Figs. 7.101 and 7.102). Anyway, as in the previous subsection, the proposed method has potential activities to show the coincidence of class labels between distinct data sets.

Next, we try to select the genes associated with $u_{\ell_1 i}^{(i)}$. As in the previous subsection, we employed all five $u_{\ell_1 i}^{(i)}$ for $1 \le \ell_1 \le 5$ to select genes. After

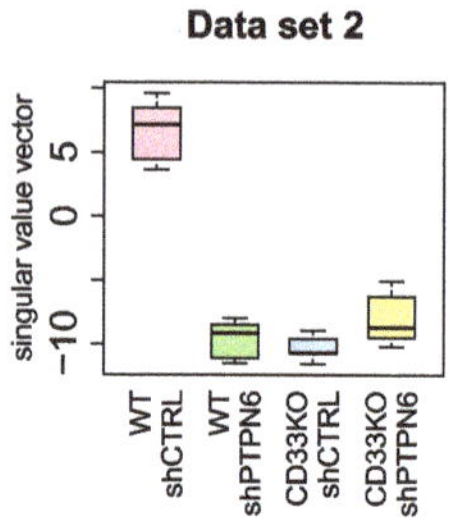

Fig. 7.102 Boxplots of $u^{(j_2;i)}_{5j_2}$ ($P = 1.62 \times 10^{-12}$). P value was computed by the categorical regression analysis

Table 7.100 Top ten-ranked enriched terms in "KEGG 2021 HUMAN" category of Enrichr

Term	Overlap	P value	Adjusted P value
Ribosome	76/158	1.38×10^{-77}	3.38×10^{-75}
Coronavirus disease	79/232	3.65×10^{-66}	4.48×10^{-64}
Parkinson's disease	41/249	4.27×10^{-21}	3.49×10^{-19}
Pathways of neurodegeneration	52/475	2.87×10^{-18}	1.76×10^{-16}
Alzheimer's disease	43/369	2.86×10^{-16}	1.40×10^{-14}
Amyotrophic lateral sclerosis	42/364	9.19×10^{-16}	3.75×10^{-14}
Prion disease	33/273	3.26×10^{-13}	1.14×10^{-11}
Huntington's disease	33/306	7.87×10^{-12}	2.41×10^{-10}
Diabetic cardiomyopathy	25/203	1.69×10^{-10}	4.59×10^{-9}
Salmonella infection	27/249	5.74×10^{-10}	1.41×10^{-8}

following the same procedure, we got 544 is associated with adjusted P values less than 0.01. Then 527 gene symbols associated with these is are uploaded to Enrichr for evaluation. Again, we checked "KEGG 2021 HUMAN" category (Table 7.100). Similar to Table 7.99, From third to nine terms are related to neurodegenerative diseases, including Alzheimer's disease ranked as the fifth, although individual rank order differs. Thus, again, the proposed method not only could relate the class labels with one another among distinct data sets but also could select a reasonable set of genes.

Finally, we perform drug discovery. $x_{\ell_1 j_4^{[1]} j_4^{[2]} j_4^{[3]}}$ was computed by

$$x_{\ell_1 j_4^{[1]} j_4^{[2]} j_4^{[3]}} = \sum_{i=1}^{60,617} u^{(i)}_{\ell_1 i} x_{i j_4^{[1]} j_4^{[2]} j_4^{[3]}} \tag{7.163}$$

to which HOSVD is applied and we got

$$x_{\ell_1 j_4^{[1]} j_4^{[2]} j_4^{[3]}} = \sum_{\tilde{\ell}_1=1}^{94} \sum_{\tilde{\ell}_2=1}^{4} \sum_{\tilde{\ell}_3=1}^{3} \tilde{G}^{[\ell_1]}(\tilde{\ell}_1 \tilde{\ell}_2 \tilde{\ell}_3) \tilde{u}^{[\ell_1]}_{\tilde{\ell}_1 j_4^{[1]}} \tilde{u}^{[\ell_1]}_{\tilde{\ell}_2 j_4^{[2]}} \tilde{u}^{[\ell_1]}_{\tilde{\ell}_3 j_4^{[3]}} \tag{7.164}$$

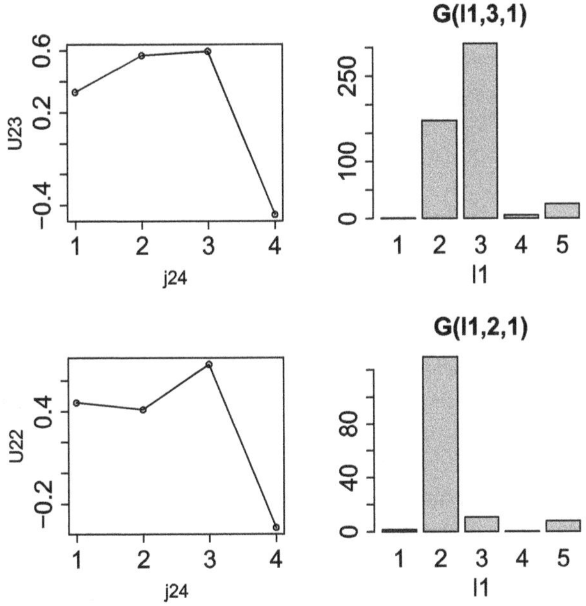

Fig. 7.103 Top left: $\tilde{u}^{[4]}_{3j_4^{[2]}}$, top right: $|\tilde{G}^{[4]}(\tilde{\ell}_1 31)|$, bottom left: $\tilde{u}^{[5]}_{2j_4^{[2]}}$, bottom right: $|\tilde{G}^{[5]}(\tilde{\ell}_1 21)|$

where $\tilde{u}^{[\ell_1]}_{\tilde{\ell}_1 j_4^{[1]}} \in \mathbb{R}^{94\times 94}$ is a singular value vector attributed to drugs, $\tilde{u}^{[\ell_1]}_{\tilde{\ell}_2 j_4^{[2]}}$ is a singular value vector attributed to dose density, $\tilde{u}^{[\ell_1]}_{\tilde{\ell}_3 j_4^{[3]}}$ is a singular value vector attributed to biological replicates.

To find which drugs are effective, we investigated $\tilde{u}^{[\ell_1]}_{\tilde{\ell}_2 j_4^{[2]}}$ that have reasonable dose dependence and we found that $\tilde{u}^{[4]}_{3j_4^{[2]}}$ and $\tilde{u}^{[5]}_{2j_4^{[2]}}$ are associated with reasonable dose dependence (Fig. 7.103). Since $\tilde{\ell}_3 = 1$ is always associated with the independence of biological replicates (not shown here), to see which $\tilde{u}^{[\ell_1]}_{\tilde{\ell}_1 j_4^{[1]}}$ is most associated with selected $\tilde{u}^{[\ell_1]}_{\tilde{\ell}_2 j_4^{[2]}}$ and $\tilde{u}^{[\ell_1]}_{\tilde{\ell}_3 j_4^{[3]}}$, we investigated $|\tilde{G}^{[\ell_1]}(\tilde{\ell}_1\tilde{\ell}_2\tilde{\ell}_3)|$. Then we find that $\tilde{u}^{[4]}_{3j_4^{[1]}}$ and $\tilde{u}^{[5]}_{2j_4^{[1]}}$ are most associated with $\tilde{u}^{[4]}_{3j_4^{[2]}}$ and $\tilde{u}^{[5]}_{2j_4^{[2]}}$ that have reasonable dose dependence, respectively. Table 7.101 lists the top three–ranked drugs with the largest absolute values of $\tilde{u}^{[4]}_{3j_4^{[1]}}$ and $\tilde{u}^{[5]}_{2j_4^{[1]}}$ and most of them are evaluated as an effective drug in the original study [80]. Thus, the proposed method could perform drug repositioning with integrating mulitple gene expression profiles other than those of drug-treated cell lines.

Table 7.101 Top three–ranked drugs with the largest absolute values of $\tilde{u}^{[4]}_{3j_4^{[1]}}$ and $\tilde{u}^{[5]}_{2j_4^{[1]}}$. The drugs asterisked were evaluated as effecive in the original study

Singular value vectors	Drugs
$\tilde{u}^{[4]}_{3j_4^{[1]}}$	"a443654*," "baricitinib* + dsRNA,", "dovitinib*"
$\tilde{u}^{[5]}_{2j_4^{[1]}}$	"a443654*," "torin1*," "bortezomib"

7.17.3 Transfer Learning

Another advantage of this kind of integrated analysis is a transfer learning [116]. Transfer learning is a short time or light additional learning after massive learning using a massive data set; in spite of the small amount of learning, transfer learning can achieve excellent performance that cannot be achieved by starting from scratch. Transfer learning is usually used for deep learning. It is not believed that transfer learning can be performed for classical linear algebra-based machine learning, to which TD belongs to.

In this subsection, we demonstrate how we can perform transfer learning using integrated analysis of data sets 1, 2, and 3. To do this, we consider additional data set 5 retrieved from GEO with GEO ID GSE164642. This is composed of two classes corresponding to ABCC1-activated cells by distinct RNA or control cells. They include three biological replicates (18 total samples). The purpose of the original study [39] is to identify how overexpression of ABCC1 progresses AD.

It is formatted as a tensor, $x_{ij_5^{[1]}j_5^{[2]}j_5^{[3]}} \in \mathbb{R}^{58,033\times 3\times 2\times 3}$, which represents expression of the ith gene of $j_5^{[1]}$th replicates, of $j_5^{[2]}$th treatment ($j_5^{[2]} = 1$: control, $j_5^{[2]} = 2$: treated) at the $j_5^{[3]}$th RNA-seq method ($j_5^{[3]} = 1$ by TrueSeq, $j_5^{[3]} = 2, 3$ by SMARTer), to which HOSVD was applied and we got

$$x_{ij_5^{[1]}j_5^{[2]}j_5^{[3]}} = \sum_{\ell_1=1}^{3}\sum_{\ell_2=1}^{3}\sum_{\ell_3=1}^{3}\sum_{\ell_4=1}^{58,033} G(\ell_1\ell_2\ell_3\ell_4)u^{[5]}_{\ell_1 j_5^{[1]}}u^{[5]}_{\ell_2 j_5^{[2]}}u^{[5]}_{\ell_3 j_5^{[3]}}u^{[5]}_{\ell_4 i} \tag{7.165}$$

where $G \in \mathbb{R}^{3\times 2\times 3\times 58,033}$ is a core tensor and $u^{[5]}_{\ell_1 j_4^{[1]}} \in \mathbb{R}^{3\times 3}$, $u^{[5]}_{\ell_2 j_4^{[2]}} \in \mathbb{R}^{2\times 2}$, $u^{[5]}_{\ell_3 j_4^{[3]}} \in \mathbb{R}^{3\times 3}$, and $u^{[5]}_{\ell_4 i} \in \mathbb{R}^{58,033\times 58,033}$ are singular value matrices and are orthogonal matrices. To get $x_{i\ell 5} \in \mathbb{R}^{58,033\times L}$, we do

$$x_{i\ell 5} = \sum_{j_5^{[1]}=1}^{3}\sum_{j_5^{[2]}=1}^{2}\sum_{j_5^{[3]}=1}^{3} x_{ij_5^{[1]}j_5^{[2]}j_5^{[3]}}u^{[5]}_{\ell_1 j_5^{[1]}}u^{[5]}_{\ell_2 j_5^{[2]}}u^{[5]}_{\ell_3 j_5^{[3]}},$$

$$1 \le \ell_s \le L_5^{[s]}(\le M_5^{[s]}), 1 \le \ell \le L = \prod_{s=1}^{3} L_5^{[s]} \tag{7.166}$$

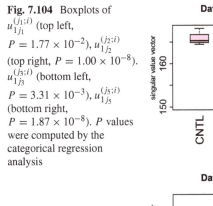

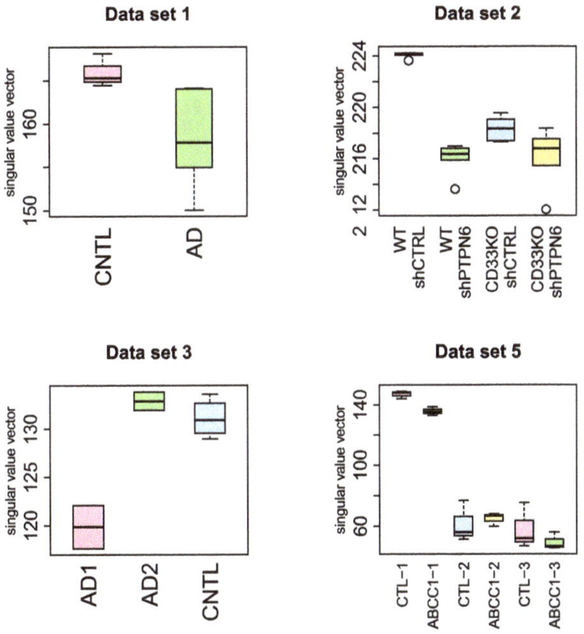

Fig. 7.104 Boxplots of $u^{(j_1;i)}_{1j_1}$ (top left, $P = 1.77 \times 10^{-2}$), $u^{(j_2;i)}_{1j_2}$ (top right, $P = 1.00 \times 10^{-8}$). $u^{(j_3;i)}_{1j_3}$ (bottom left, $P = 3.31 \times 10^{-3}$), $u^{(j_5;i)}_{1j_5}$ (bottom right, $P = 1.87 \times 10^{-8}$). P values were computed by the categorical regression analysis

where $M_5^{[1]} = 3$, $M_5^{[2]} = 2$, $M_5^{[3]} = 3$ and $L_5^{[s]}$ is the number of used singular value vectors, ℓ_s and $L_5^{[1]} = L_5^{[2]} = L_5^{[3]} = 2$, thus $L = 8$. Then HOSVD was applied to $x_{i\ell k} \in \mathbb{R}^{60,617 \times 8 \times 4}$ where $k = 5$ was replaced with $k = 4$ in the previous subsection. $u^{(i)}_{\ell_1 i}$ was computed by Eq. (5.36) and $u^{(j_k;i)}_{\ell_1 j_k}$ was computed by Eq. (5.30).

Figure 7.104 shows $u^{(j_k;i)}_{1j_k}$. In contrast to the expectation, the primary distinction in data set 5 is not that between treated (ABCC1 overexpression) and control samples, but that between sequencing methods: TruSeq and SMARTer. Although many studies report the difference between the two methods [71, 89, 114], this study is the first one that reports the biological significance of this difference. Without transfer learning, this kind of finding is impossible, since genes whose expression is distinct between two sequencing methods are usually regarded to be a batch effect and are ignored.

Figures 7.105, 7.106, 7.107 and 7.108 show $u^{(j_k;i)}_{\ell_1 j_k}$, $2 \leq \ell_1 \leq 5$. Those other than $\ell_1 = 4$ which shows the distinction between class labels only in data sets 2 and 5 show the distinction between class labels in data sets 2, 3, and 5.

Next, we try to select genes associated with $u^{(i)}_{\ell_1 i}$. As in the previous subsection, we employed all five $u^{(i)}_{\ell_1 i}$ for $1 \leq \ell_1 \leq 5$ to select genes. After following the same procedure, we got 660 is associated with adjusted P values less than 0.01. Then 647 gene symbols associated with these is are uploaded to Enrichr for the evaluation. Again, we checked "KEGG 2021 HUMAN" category (Table 7.102). Similar to Tables 7.99 and 7.100, seven out of ten terms are related to neurodegerative diseases, including Alzheimer's disease ranked as the fifth, although individual rank order

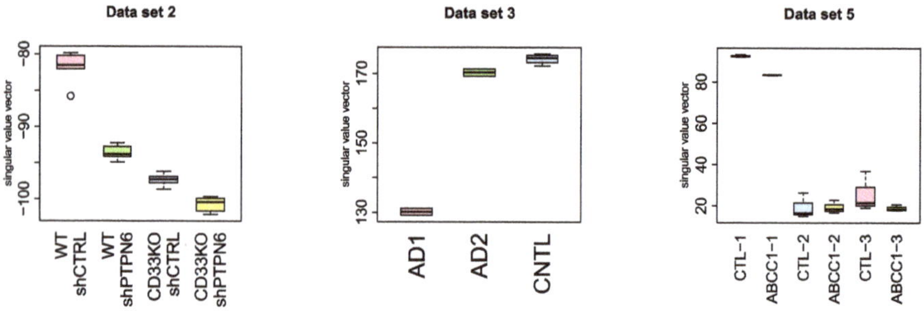

Fig. 7.105 Boxplots of $u^{(j_2;i)}_{2j_2}$ (left, $P = 9.07 \times 10^{-15}$). $u^{(j_3;i)}_{2j_3}$ (middle, $P = 9.14 \times 10^{-7}$), $u^{(j_5;i)}_{2j_5}$ (right, $P = 2.05 \times 10^{-10}$). P values were computed by the categorical regression analysis

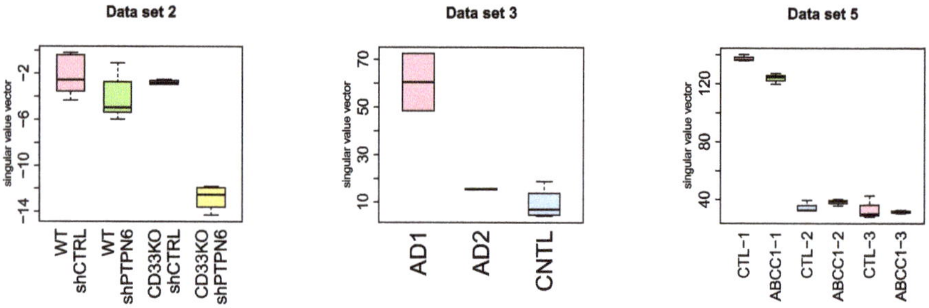

Fig. 7.106 Boxplots of $u^{(j_2;i)}_{3j_2}$ (left, $P = 1.15 \times 10^{-10}$). $u^{(j_3;i)}_{3j_3}$ (middle, $P = 3.35 \times 10^{-3}$), $u^{(j_5;i)}_{3j_5}$ (right, $P = 4.24 \times 10^{-13}$). P values were computed by the categorical regression analysis

Fig. 7.107 Boxplots of $u^{(j_2;i)}_{4j_2}$ (left, $P = 1.05 \times 10^{-13}$), $u^{(j_5;i)}_{4j_5}$ (right, $P = 4.69 \times 10^{-12}$). P values were computed by the categorical regression analysis

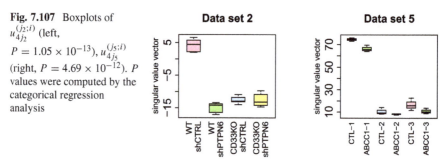

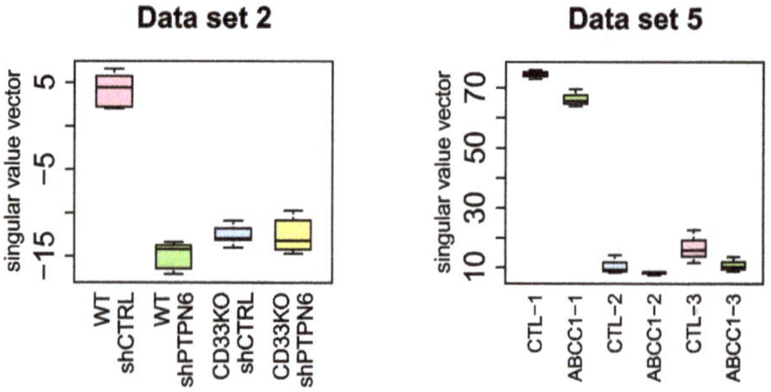

differs. Thus, again, the proposed method not only could relate the class labels with one another among distinct data sets but also could select a reasonable set of genes.

7.17.4 Single-Cell Analysis

One possible application of the present approach is a single cell that usually lacks class labels. To see if it works, we consider single-cell gene expression profiles

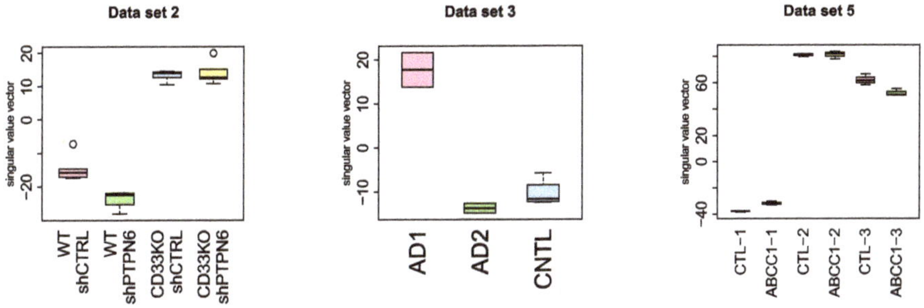

Fig. 7.108 Boxplots of $u_{5j_2}^{(j_2;i)}$ (left, $P = 2.33 \times 10^{-15}$). $u_{5j_3}^{(j_3;i)}$ (middle, $P = 4.57 \times 10^{-4}$), $u_{5j_5}^{(j_5;i)}$ (right, $P = 3.17 \times 10^{-16}$). P values were computed by the categorical regression analysis

Table 7.102 Top ten-ranked enriched terms in "KEGG 2021 HUMAN" category of Enrichr

Term	Overlap	P value	Adjusted P value
Ribosome	68/158	4.97×10^{-59}	1.29×10^{-56}
Coronavirus disease	71/232	2.00×10^{-49}	2.60×10^{-47}
Parkinson's disease	40/249	4.29×10^{-17}	3.72×10^{-15}
Pathways of neurodegeneration	50/475	2.13×10^{-13}	1.39×10^{-11}
Amyotrophic lateral sclerosis	42/364	9.31×10^{-13}	4.84×10^{-11}
Alzheimer's disease	41/369	5.96×10^{-12}	2.58×10^{-10}
Salmonella infection	31/249	1.39×10^{-10}	5.16×10^{-9}
Prion disease	31/273	1.39×10^{-9}	4.51×10^{-8}
Huntington's disease	30/306	7.51×10^{-8}	2.17×10^{-6}
Pathogenic Escherichia coli infection	22/197	4.56×10^{-7}	1.19×10^{-5}

retrieved from GEO with GEO ID GSE163577, which will be referred to as data set 6. It includes both AD and healthy controls from 25 hippocampus and superior frontal cortex samples across 17 control and eight AD patients. A total of twenty-five individual sets of data have been formatted as a matrix, $x_{ij_6^{[c]}} \in \mathbb{R}^{33,538 \times M_6^{[c]}}, 1 \leq c \leq 25$, representing the expression of the ith gene at the $j_6^{[c]}$th cell within the cth single-cell profile. The number of cells, $M_6^{[c]}$, is roughly 10^4. $x_{i\ell c} \in \mathbb{R}^{33,538 \times 10 \times 25}$ was computed by Eq. (7.158) to which HOSVD was applied and we get $u_{\ell_3 c}^{(c)} \in \mathbb{R}^{25 \times 25}$ computed by Eq. (5.36). To see which $u_{\ell_3 c}^{(c)}$ is differed between four classes, including either AD or healthy controls in either hippocampus or cortex brain regions, we applied categorical regression analysis to $u_{\ell_3 c}^{(c)}$

$$u_{\ell_3 c}^{(c)} = a_{\ell_3} + \sum_{s=1}^{4} b_{s\ell_3} \delta_{cs} \tag{7.167}$$

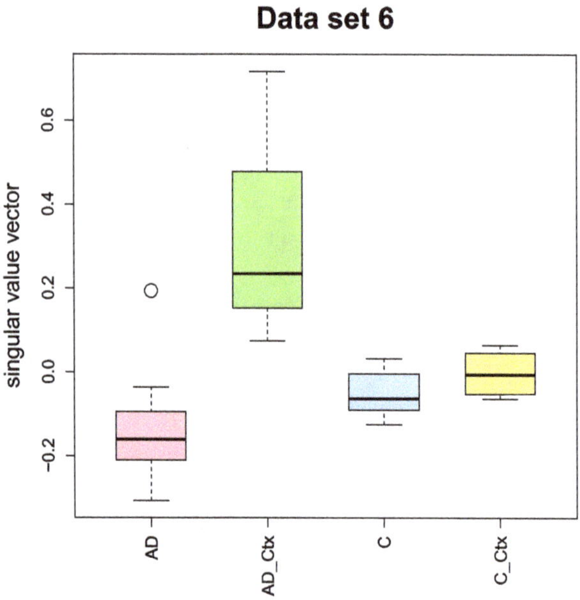

Fig. 7.109 Boxplots of $u_{6c}^{(c)}$ ($P = 5.04 \times 10^{-4}$). P value was computed by the categorical regression analysis. AD: AD in hippocampus, AD_Ctx: AD in cortex, C: control in hippocampus, C_Ctx: control in cortex

where a_{ℓ_3} and $b_{s\ell_3}$ are regression coefficients and δ_{cs} takes 1 only when cth single-cell profile is associated with the sth class label ($1 \leq s \leq 4$). Then we found that only $u_{6c}^{(c)}$ is associated with significant distinction between four classes (Fig. 7.109).

Next we need to select is and evaluate them biologically to see if the analysis is successful or not. To see which $u_{\ell_1 i}^{(i)}$ is most associated with $u_{6c}^{(c)}$, we compute $\sum_{\ell_2=1}^{10} G(\ell_1 \ell_2 6)^2$ (Fig. 7.110) since there are no reasons to use any specific ℓ_2. Since $\sum_{\ell_2=1}^{10} G(6\ell_2 6)^2$ is the largest, we decided to use $u_{6i}^{(i)}$ to select genes. P values are attributed to ith gene as

$$P_i = P_{\chi^2}\left[> \left(\frac{u_{6i}^{(i)}}{\sigma_6} \right)^2 \right] \tag{7.168}$$

and P values were corrected by BH criterion. One hundred seventy-seven is associated with adjusted P values less than 0.01 were selected. Gene IDs (Ensembl gene IDs) were converted to gene symbols using DAVID gene ID converter and 171 gene symbols associated with 177 is were uploaded to Enrichr for evaluation. Table 7.103 shows "Azimuth 2023" category in Enrichr, which is a web application that uses an annotated reference data set to automate the processing, analysis, and interpretation of a new single-cell RNA-seq or ATAC-seq experiment [34]. It is

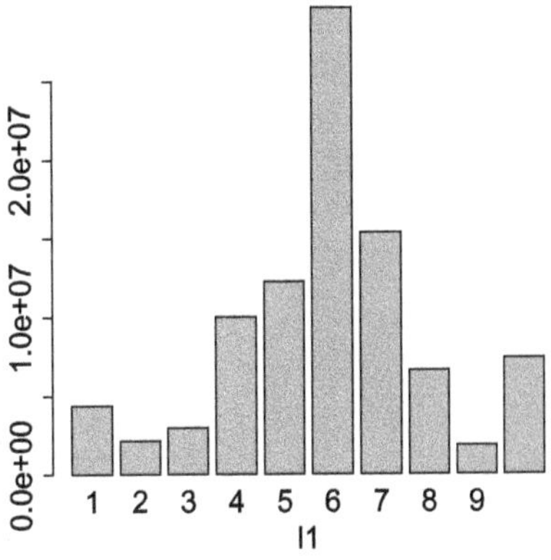

Fig. 7.110 $\sum_{\ell_2=1}^{10} G(\ell_1 \ell_2 6)^2$

Table 7.103 Top ten-ranked enriched terms in "Azimuth 2023" category of Enrichr

Term	Overlap	*P* value	Adjusted *P* value
Motor cortex-subclass-oligodendrocyte	9/9	1.77×10^{-19}	8.18×10^{-17}
Motor cortex-class-nonneuronal cell	7/9	9.63×10^{-14}	8.90×10^{-12}
Motor cortex-cluster-OPALIN+ FTH1P3+ layer 2-6 oligodendrocyte	7/9	9.63×10^{-14}	8.90×10^{-12}
Motor cortex-cluster-OPALIN+ LDLRAP1+ layer 5-6 oligodendrocyte	7/9	9.63×10^{-14}	8.90×10^{-12}
Mouse motor cortex-cross-species cluster-oligodendrocyte 2	7/9	9.63×10^{-14}	8.90×10^{-12}
Fetal development-L1-inhibitory interneurons	6/9	2.74×10^{-11}	1.81×10^{-9}
Fetal development-L2-cerebellum-inhibitory interneurons	6/9	2.74×10^{-11}	1.81×10^{-9}
Motor cortex-cross-species cluster-oligodendrocyte 2	5/9	4.98×10^{-9}	1.21×10^{-7}
Heart-L1-neuronal	5/9	4.98×10^{-9}	1.21×10^{-7}
Heart-L2-neuronal	5/9	4.98×10^{-9}	1.21×10^{-7}

obvious that all top ten–ranked terms are related to brain/neuron including some malfunction. Thus the proposed methods could identify a set of genes that not only its expression is distinct between class labels but also its function is biologically reasonable.

7.17.5 Comparison with Other Methods

Before closing this section, we compare the performances with those of other methods. Although we tried to find genome-oriented methods that can be applicable to gene expression profiles without shared labels, we could not find any of such ones. Thus we were forced to seek the methods to be compared with outside genomic science. To be compared, we employed three methods, collective matrix factorization (CMF) [46], group factor analysis (GFA) [51], and a simple concatenation.

For CMF, $x_{ij_1 1} \in \mathbb{R}^{60,617\times 9}$, $x_{ij_2 2} \in \mathbb{R}^{60,617\times 23}$, and $x_{ij_3 3} \in \mathbb{R}^{60,617\times 8}$ were employed such that they shared the same number of genes. Missing values were filled with zero and were normalized to have zero mean and a standard deviation of one, as denoted at the beginning of this section. The structure of assumed modeling is given as

$$x_{ij_k k} = \sum_{\ell=1}^{L} u_{\ell i} u_{\ell j_k}^{[k]} + b_i^{[k]} + b_{j_k}^{[k]} + \epsilon_{ij_k}^{[k]}. \tag{7.169}$$

Thus, these three profiles were analyzed separately other than $u_{\ell i}$ shared among three profiles. $u_{\ell i} \in \mathbb{R}^{L\times 60,617}$, $u_{\ell j_k}^{[k]} \in \mathbb{R}^{L\times M_k}$, M_k is the number of sample in data set k, $M_1 = 9$, $M_2 = 23$, $M_3 = 8$. $b_i^{[k]} \in \mathbb{R}^{60,617}$,$b_{j_k}^{[k]} \in \mathbb{R}^{M_k}$, $\epsilon_{ij_k}^{[k]} \in \mathbb{R}^{60,617\times M_k}$. To use CMF, we need to specify the distribution that $x_{ij_k k}$ is supposed to obey. We assumed a Poisson distribution, since the negative signed binary distribution that the RNA-seq data set is supposed to obey was not in the set of options. $L = 4$ is assumed. The coincidence between $u_{\ell j_k}^{[k]}$ and class labels is evaluated with categorical regression analysis.

On the other hand, GFA is supposed to obey a much simpler model,

$$x_{ij_k k} = \sum_{\ell=1}^{L} u_{\ell i} u_{\ell j_k}^{[k]} + \epsilon_{ij_k}^{[k]} \tag{7.170}$$

where no distribution that $x_{ij_k k}$ obeys is assumed. Again, the three profiles, $x_{ij_1 k}, 1 \le k \le 3$, were analyzed separately other than $u_{\ell i}$ shared among three profiles. $L = 5$ was used. The coincidence between $u_{\ell j_k}^{[k]}$ and class labels is evaluated with categorical regression analysis.

For simple concatenation, we constructed a matrix, $x_{ij} \in \mathbb{R}^{60,617\times 40}$ where

$$x_{ij} = \begin{cases} x_{ij_1 1}, \ 1 \le j = j_1 \le M_1 = 9 \\ x_{ij_2 2}, \ M_1 < j = j_2 + M_1 \le M_1 + M_2, (M_2 = 23) \\ x_{ij_3 3}, \ M_1 + M_2 < j = j_3 + M_1 + M_2 \le M_1 + M_2 + M_3 = 40. \end{cases} \tag{7.171}$$

SVD was applied to x_{ij} and we got

$$x_{ij} = \sum_{\ell=1}^{L} \lambda_\ell u_{\ell i} v_{\ell j} \tag{7.172}$$

as usual. $u_{\ell i}$s are used to select genes (see below) and the coincidence between $v_{\ell j}$s and class labels is evaluated with categorical regression analysis.

Table 7.104 Coincidence between latent variables and classification

TD-based unsupervised FE				
Latent variables	1st	2nd	3rd	4th
Data set 1	4.86×10^{-2}	–	–	–
Data set 2	1.94×10^{-6}	7.81×10^{-14}	4.15×10^{-9}	$<2.2 \times 10^{-16}$
Data set 3	1.55×10^{-3}	8.23×10^{-7}	–	2.74×10^{-3}
CMF				
Latent variables	1st	2nd	3rd	4th
Data set 1	–	–	–	–
Data set 2	–	–	–	–
Data set 3	–	–	–	–
GFA				
Latent variables	1st	2nd	3rd	4th
Data set 1	–	–	–	–
Data set 2	1.94×10^{-6}	7.81×10^{-14}	4.15×10^{-9}	$<2.2 \times 10^{-16}$
Data set 3	–	3.40×10^{-4}	1.26×10^{-2}	3.41×10^{-4}
Simple concatenation				
Latent variables	1st	2nd	3rd	4th
Data set 1	4.86×10^{-2}	–	–	–
Data set 2	6.88×10^{-5}	$<2.2 \times 10^{-16}$	$<2.2 \times 10^{-16}$	$<2.2 \times 10^{-16}$
Data set 3	1.55×10^{-3}	8.23×10^{-7}	–	2.74×10^{-3}

We summarize the performance of the other three methods, CMF, GFA, and a simple concatenation, in Table 7.104 where that of TD-based unsupervised FE was also listed for the comparison. Neither the two advanced methods, CFT and GFA, could compete with TD-based unsupervised FE whereas simply simple concatenation ironically could. However, enrichment analysis of selected 147 gene symbols by simple concatenation (Table 7.105) was clearly inferior to that of TD-based unsupervised FE (Table 7.100) since associated P values were clearly larger (less significant) and less number of neurodegenerative diseases even not including AD were ranked. Here is associated with adjusted P values less than 0.01 by BH criterion were converted to gene symbols with DAVID gene ID converter after P values were addressed to is using

$$P_i = P_{\chi^2}\left[> \sum_{\ell=1}^{5} \left(\frac{u_{\ell i}}{\sigma_\ell}\right)^2 \right]. \tag{7.173}$$

Thus, we could not find any competitive methods toward TD-based unsupervised FE.

Table 7.105 Top ten-ranked enriched terms in "KEGG 2021 HUMAN" category of Enrichr

Term	Overlap	*P* value	Adjusted *P* value
Ribosome	68/158	3.23×10^{-63}	8.13×10^{-61}
Coronavirus disease	72/232	6.26×10^{-55}	7.89×10^{-53}
Salmonella infection	27/249	2.31×10^{-9}	1.94×10^{-7}
Parkinson's disease	23/249	6.55×10^{-7}	4.13×10^{-5}
Fluid shear stress and atherosclerosis	16/139	1.95×10^{-6}	9.81×10^{-5}
Hippo signaling pathway	17/163	3.69×10^{-6}	1.36×10^{-4}
Legionellosis	10/57	3.79×10^{-6}	1.36×10^{-4}
Pathways of neurodegeneration	32/475	5.36×10^{-6}	1.69×10^{-4}
Shigellosis	21/246	6.94×10^{-6}	1.78×10^{-4}
Protein processing in endoplasmic reticulum	17/171	7.06×10^{-6}	1.78×10^{-4}

7.18 KTD Applied to Real Data Set

In Sect. 5.9, we introduced KTD. In this section, we introduce some of real examples to which KTD was applied [104]. As can be seen below, KTD can be an effective method even when it is applied to real data sets.

7.18.1 *COVID-19*

At first, we re-analyze the data set dealt with in Sect. 7.15.2 using RBF kernel. Since there is a freedom to select σ, we specifically selected $\sigma = 10$. Based upon Eq. (7.130), we define RBF kernel and its KTD as

$$K(x_{ijkm}, x_{ij'k'm'}) = \exp\left\{-\frac{\sum_{i=1}^{N}\left(x_{ijkm} - x_{ij'k'm'}\right)^2}{\sigma^2}\right\}$$

$$= \sum_{\ell_1=1}^{5}\sum_{\ell_2=1}^{2}\sum_{\ell_3=1}^{3}\sum_{\ell'_1=1}^{5}\sum_{\ell'_2=1}^{2}\sum_{\ell'_3=1}^{3} G(\ell_1\ell_2\ell_3\ell'_1\ell'_2\ell'_3)$$

$$\times u^{(j)}_{\ell_1 j} u^{(k)}_{\ell_2 k} u^{(m)}_{\ell_3 m} u^{(j')}_{\ell'_1 j'} u^{(k')}_{\ell'_2 k'} u^{(m')}_{\ell'_3 m'}. \qquad (7.174)$$

Figure 7.111 shows singular value vectors. In contrast to Fig. 7.74 no $u^{(j)}\ell_1 j$ exhibit independence of j, cell lines.

To perform feature selection, we need some evaluation of the coincidence between singular value vectors by removing one i and class label. For this evaluation, we compute the correlation coefficient of $u^{(j)}_{1j} u^{(k)}_{2k} u^{(m)}_{3m}$ between TD in Sect. 7.15.2 and KTD with removing one i. Then we select top-ranked is which

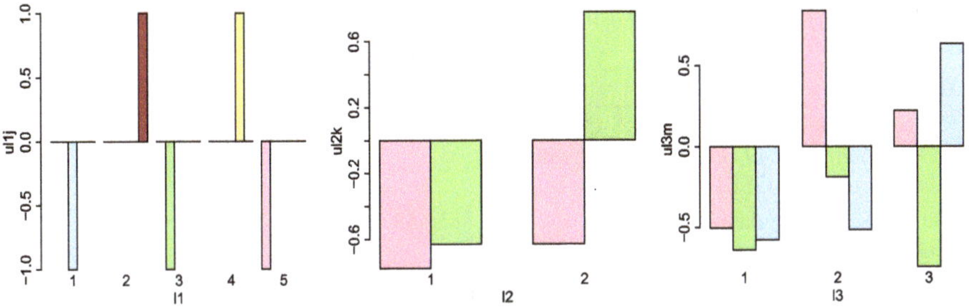

Fig. 7.111 Left: $u^{(j)}_{\ell_1 j} \in \mathbb{R}^{5\times 5}$, middle: $u^{(k)}_{\ell_2 k} \in \mathbb{R}^{2\times 2}$, right: $u^{(m)}_{\ell_3 m} \in \mathbb{R}^{3\times 3}$. ℓ_1, ℓ_2, and ℓ_3 are aligned from left to right in individual plots and distinct colors correspond to distinct j, k, m

result in a larger decrease in the correlation. Since KTD does not give us any P values, we select top-ranked 500 is associated with a larger decrease of correlation.

To evaluate the selected 500 is (genes), these genes were uploaded to Enrichr and "COVID-19 Related Gene Sets 2021" category check as usual (Table 7.106). If compared with Table 7.84, these "SARS-CoV-2" terms are more significant (smaller P values) than those in Table 7.84. Thus KTD could successfully more biologically reasonable genes than TD.

On problem of KTD is that we need to tune the parameter σ as well as the number of the selected genes. Generally, in the present case smaller σ and the selected genes reduce the significance substantially. Thus, if we can improve the performance by the employment of KTD is highly dependence upon the cases.

7.18.2 Kidney Cancer

Next, we consider the data set used in Sect. 7.10. At first, we introduced two RBF kernels from mRNA expression and miRNA expression profiles, respectively, as

$$x^{\text{mRNA}}_{jj'} = K(x_{ij}, x_{ij'}) = \exp\left\{-\frac{\sum_{i=1}^N \left(x_{ij} - x_{ij'}\right)^2}{\sigma^2}\right\} \tag{7.175}$$

$$x^{\text{miRNA}}_{jj'} = K(x_{kj}, x_{kj'}) = \exp\left\{-\frac{\sum_{k=1}^K \left(x_{kj} - x_{kj'}\right)^2}{\sigma'^2}\right\}. \tag{7.176}$$

Then $x_{jj'} \in \mathbb{R}^{M\times M}$ was computed as

$$x_{jj'} = \sum_{j''=1}^{M} x^{\text{mRNA}}_{jj''} x^{\text{miRNA}}_{j''j'} \tag{7.177}$$

Table 7.106 Top ten-ranked enriched "SARS-CoV-2" terms in "COVID-19 Related Gene Sets 2021" category of Enrichr

Term	Overlap	*P* value	Adjusted *P* value
500 genes downregulated by SARS-CoV-2 (0.2 MOI) in human A549 cells transduced with ACE2 vector from GSE147507	96/494	6.64×10^{-58}	1.01×10^{-55}
500 genes downregulated by SARS-CoV-2 in A549-ACE2 cells from GSE154613 trifluoperazine	85/458	2.07×10^{-49}	2.36×10^{-47}
500 genes downregulated by SARS-CoV-2 (2 MOI) in human A549 cells transduced with ACE2 vector from GSE147507	79/492	4.91×10^{-41}	3.73×10^{-39}
Genes whose expression is altered by SARS-COV-2 infection	44/118	4.80×10^{-40}	3.12×10^{-38}
500 top downregulated genes for SARS-CoV-2 infection vs. mock from GSE161881	75/498	5.82×10^{-37}	3.32×10^{-35}
500 genes downregulated by SARS-CoV-2 in human Calu3 cells at 4h from GSE148729 mock s1 polyA	67/497	4.87×10^{-30}	1.85×10^{-28}
500 genes upregulated by SARS-CoV-2 in A549-ACE2 cells from GSE154613 amlodipine	63/454	5.51×10^{-29}	1.67×10^{-27}
500 genes up-re500 genes downregulated by SARS-CoV-2 in human Calu3 cells at 24h from GSE148729 total RNA	65/495	1.77×10^{-28}	4.48×10^{-27}
500 genes downregulated by SARS-CoV-2 in A549-ACE2 cells from GSE154613 berbamine	61/455	3.02×10^{-27}	7.24×10^{-26}
500 genes downregulated by SARS-CoV-2 in A549-ACE2 cells from GSE154613 RS504393	61/462	7.00×10^{-27}	1.60×10^{-25}

to which SVD was applied and we got

$$x_{jj'} = \sum_{\ell=1}^{M} \lambda_\ell u_{\ell j} v_{\ell j'} \tag{7.178}$$

where $u_{\ell j} \in \mathbb{R}^{M \times M}$ is attributed to mRNA samples and $v_{\ell j} \in \mathbb{R}^{M \times M}$ is attributed to miRNA samples. The performance of the above RBF kernel is compared with the approach in Sect. 7.10 and the following linear kernel

$$x_{jj'}^{\text{mRNA}} = K(x_{ij}, x_{ij'}) = \sum_{i=1}^{N} x_{ij} x_{ij'} \tag{7.179}$$

$$x_{jj'}^{\text{miRNA}} = K(x_{kj}, x_{kj'}) = \sum_{k=1}^{K} x_{kj} x_{kj'}. \tag{7.180}$$

Table 7.107 Performances of the linear and RBF kernel TD-based unsupervised FE and that achieved in Sect. 7.10. The P values for tumor vs. normal kidney were computed by t-test and those marked by an asterisk correspond to the most significant values

	Tumor vs. normal kidney		Correlation coefficient
	u_{2j} (mRNA)	v_{2j} (miRNA)	u_{2j} vs. v_{2j}
1st data set (TCGA)			
Linear kernel	1.27×10^{-38}	1.76×10^{-62}	$0.912\ (1.28 \times 10^{-126})$
RBF kernel, $\sigma = 10^5, \sigma' = 1$	1.99×10^{-14}	1.99×10^{-14}	1.00 (0.00*)
RBF kernel, $\sigma = 10^6, \sigma' = 10$	1.06×10^{-44}*	4.24×10^{-38}	$0.974\ (6.93 \times 10^{-212})$
RBF kernel, $\sigma = 10^7, \sigma' = 10^2$	4.70×10^{-43}	2.89×10^{-66}	$0.914\ (2.00 \times 10^{-128})$
Section 7.10	7.10×10^{-39}	2.13×10^{-71}*	$0.905\ (1.63 \times 10^{-121})$
2nd data set (GEO)			
Linear kernel	5.36×10^{-23}*	2.76×10^{-18}	$0.937\ (3.91 \times 10^{-16})$
RBF kernel, $\sigma = 10^2, \sigma' = 1$	5.61×10^{-7}	5.61×10^{-7}	1.00 (0.00*)
RBF kernel, $\sigma = 10^3, \sigma' = 10$	5.41×10^{-15}	1.17×10^{-13}	$0.966\ (2.06 \times 10^{-20})$
RBF kernel, $\sigma = 10^4, \sigma' = 10^2$	6.93×10^{-18}	2.893×10^{-15}	$0.921\ (1.16 \times 10^{-14})$
Section 7.10	6.74×10^{-22}	2.54×10^{-18}*	$0.931(1.58 \times 10^{-15})$

as well. The evaluation is to measure how much $u_{\ell j}$ and $v_{\ell j'}$ are coincident with class labels, i.e., distinction between tumors and normal kidney and are correlated with each other (Table 7.107).

It is obvious that the most significant P values (asterisked) are not always given by RBF kernel. This again suggests that we need to seek the best hyper parameters although RBF-kernel-based KTD potentially has the ability to outperform linear methods.

7.19 Summary

Because TD-based unsupervised FE was more recently proposed than PCA-based unsupervised FE, the examples of applications of TD-based unsupervised FE introduced in this chapter is very limited. In spite of that, it still covers a wide range of applications tried in the previous chapter using PCA-based unsupervised FE: analysis of time course data set, integrated analysis of multiomics data set, and identification of disease-causing genes. In addition to this, it has a new application target, e.g., application to in silico drug discovery.

The general procedure of application of TD-based unsupervised FE is as follows: If there are no tensors available, generate case I or case II tensor of type I. Occasionally, it might be required to generate type II tensor in order to reduce the required computational memory. If generated type II tensor is matrix, apply PCA. If not, apply HOSVD. If the type II tensor is employed, generate missing singular value vectors by multiplying the original tensor to obtain singular value vectors. Seek singular value vectors attributed to samples coincident with the desired property, e.g., the distinction between controls and treated samples. Then, in order to select singular value vectors attributed to features used for FE, the core tensor is investigated. Singular value vectors that share core tensor with larger absolute values with singular value vectors attributed to samples associated with desired properties are selected. P values are attributed to features using selected singular value vectors attributed to features with assuming χ^2 distributions. P values are corrected by BH criterion and features associated with adjusted P values less than 0.01.

This general procedure can be applied to a wide range of bioinformatics topics depending upon what kind of singular value vectors attributed to samples are selected. In this sense, TD-based unsupervised FE is expected to be applicable to a wider range of biological problems other than those treated in this chapter.

Appendix

Universality of miRNA Transfection

Study 1

This data set includes transfection of three miRNAs, miR-200a, 200b, and 200c. The number of probes in the microarray is as many as 43,376. For each of the three, two-paired experiments of treated and control samples, treated and control sample measurement is performed by one microarray. Thus, these two must be retrieved from it (columns annotated as gProcessedSignal and rProcessedSignal). Then, it is possible to make a tensor, $x_{ij_1j_2j_3} \in \mathbb{R}^{43,376\times 3\times 2\times 2}$ where i stands for probes, j_1 stands for miRNAs, j_2 stands for two replicates, and j_3 stands for control vs. treated samples. Nevertheless, it is not suitable for this specific case. If the number of components is two, automatically the two components of singular value vectors are $u_j = u_{j'}$ and $u_j = -u_{j'}$ where j and j' are each of two categories. The present purpose is to see if the components independent of category exist. This means the setup that always results in the components independent of category is not good. Therefore, in this specific case, we format mRNA expression profiles as $x_{ij} \in \mathbb{R}^{43,376\times 12}$ where $1 \leq j \leq 6$ and $7 \leq j \leq 12$ are control and treated samples, respectively. PCA is applied to x_{ij} such that PC score, $\boldsymbol{u}_\ell \in \mathbb{R}^{43,376}$, and PC loading, $\boldsymbol{v}_\ell \in \mathbb{R}^{12}$, are attributed to probes and samples, respectively. As a result, we find that $\boldsymbol{v}_2$ represents distinct expression between control and treated samples, but independent of miRNAs transfected (Fig. 7.112). This suggests that there are nonnegligible number of mRNAs affected by sequence-nonspecific off-target regulation. P values are attributed to probes using the second PC score $\boldsymbol{u}_2$ with assuming χ^2 distribution as

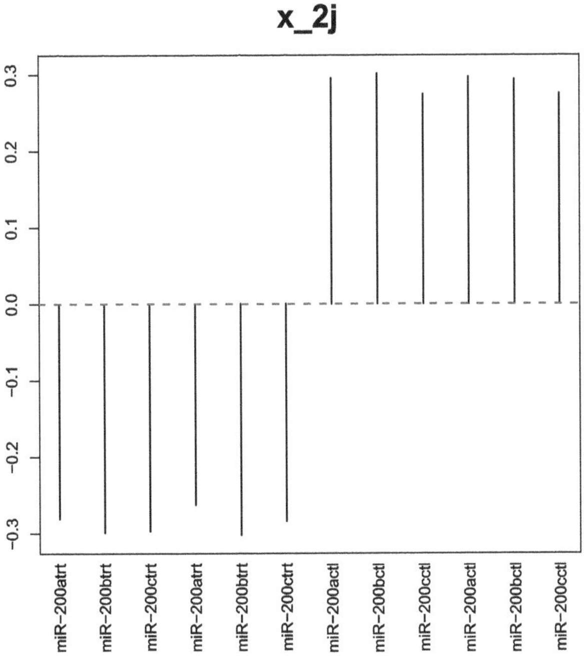

Fig. 7.112 The second PC loading, $\boldsymbol{v}_2$, obtained by PCA applied to x_{ij} made out of study 1

$$P_i = P_{\chi^2}\left[> \left(\frac{u_{2i}}{\sigma_2}\right)^2 \right]. \tag{7.181}$$

P values are corrected by BH criterion and probes associated with adjusted P values less than 0.01 are selected.

Study 2

This data set includes two miR-7 transfection experiments, two miR-128 transfection experiments, and three control experiments, normalized by mas5 procedure [74]. As mentioned at the beginning of the previous chapter, microarray technology measure photoemission of hybridized probes. Thus, various normalization procedures are applied. mas5 is one of such popular procedures, although I do not intend to explain mas5 in more details, because it is beyond the scope of this book. Because number of experiments of treated and control samples, they are difficult to be formatted in tensor. Thus it is instead as matrix, $x_{ij} \in \mathbb{R}^{54,675\times 7}$, where $j = 1, 2$ corresponds to miR-7 transfection $j = 3, 4$ corresponds to miR-128 transfection and $5 \le j \le 7$ correspond to control samples. PCA is applied to x_{ij} such that PC score, $\boldsymbol{u}_\ell \in \mathbb{R}^{54,675}$, and PC loading, $\boldsymbol{v}_\ell \in \mathbb{R}^7$, are

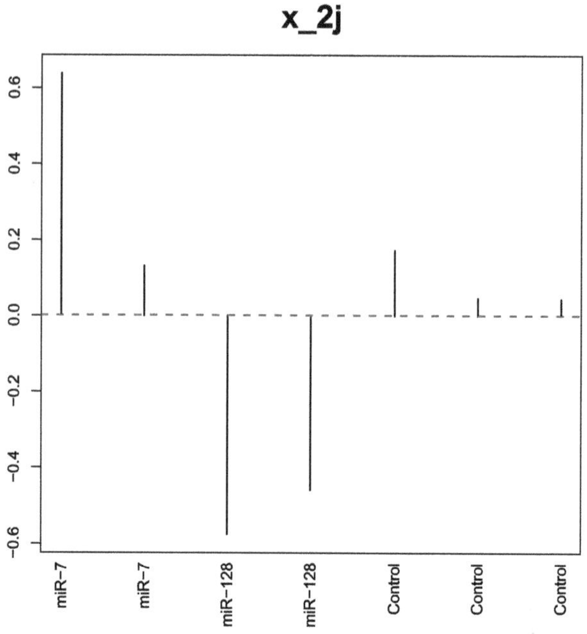

Fig. 7.113 The second PC loading, $\boldsymbol{v}_2$, obtained by PCA applied to x_{ij} made out of study 2

attributed to probes and samples, respectively. The result is a bit disappointing. In contrast to Fig. 7.112, we cannot find any PC loading that is constant independent of miRNAs transfected. Figure 7.113 shows the second PC loading, $\boldsymbol{v}_2$, which exhibits opposite signs between miR-7 transfection and miR-128 transfection. In spite of that, Fig. 7.113 still suggests the possibility of sequence-nonspecific off-target regulation. As mentioned previously, the only canonical function of miRNA is to downregulate target mRNAs. With only this function, it is impossible to assign opposite signs toward controls between miR-7 and miR-128 transfection as shown in Fig. 7.113. Downregulation can result in only the same signs toward controls. At least, either of miR-7 or miR-128 transfetion must be associated with sequence-nonspecific off-target regulation that can cause upregulation. Thus, we keep the selection of the second PC loading and assign P values to probes as Eq. (7.181). P values are corrected by BH criterion and probes associated with adjusted P values less than 0.01 are selected.

Study 3

This data set includes two miR-7 transfection experiments, two miR-128 transfection experiments, and six control experiments, normalized by plier procedure [76]. Plier is yet another procedure that nomalizes microarray, although I do not intend to

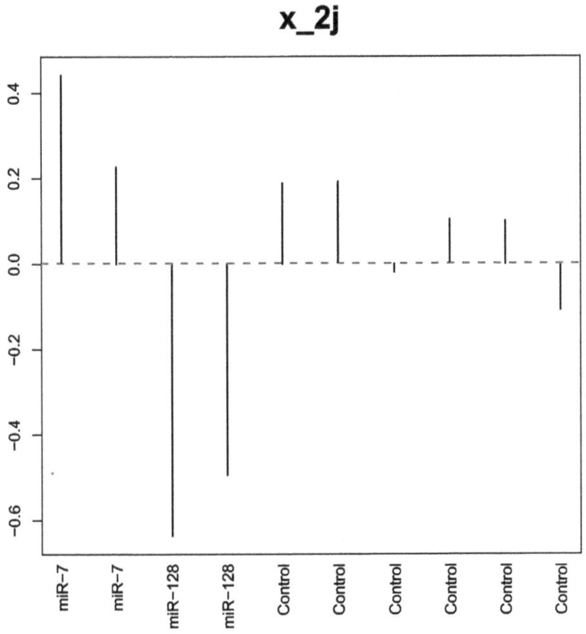

Fig. 7.114 The second PC loading, $\boldsymbol{v}_2$, obtained by PCA applied to x_{ij} made out of study 3

explain plier in more details, because it is beyond the scope of this book. Because number of experiments of treated and control samples, they are difficult to be formatted in tensor. Thus, it is instead as matrix, $x_{ij} \in \mathbb{R}^{54,675\times 10}$, where $j = 1, 2$ corresponds to miR-7 transfection $j = 3, 4$ corresponds to miR-128 transfection and $5 \leq j \leq 10$ correspond to control samples. PCA is applied to x_{ij} such that PC score, $\boldsymbol{u}_\ell \in \mathbb{R}^{54,675}$, and PC loading, $\boldsymbol{v}_\ell \in \mathbb{R}^{10}$, are attributed to probes and samples, respectively. The result is similar to study 2. In contrast to Fig. 7.112, we cannot find any PC loading that is constant independent of miRNAs transfected. Figure 7.114 shows the second PC loading, $\boldsymbol{v}_2$, which exhibits opposite signs between miR-7 transfection and miR-128 transfection. As in study 2, we keep the selection of the second PC loading and assign P values to probes as Eq. (7.181). P values are corrected by BH criterion and probes associated with adjusted P values less than 0.01 are selected.

Study 4

This data set includes two replicates of nine transfected miRNAs (miR-7/ 9/ 122a/ 128a/ 132/ 133a/ 142/ 148b/ 181a) and corresponding 18 control samples. Thus, the total number of samples is 36. This is successfully formatted as tensor, $x_{ijk} \in \mathbb{R}^{23,651\times 18\times 2}$ where i stands for probes, j stands for nine miRNAs transfection times

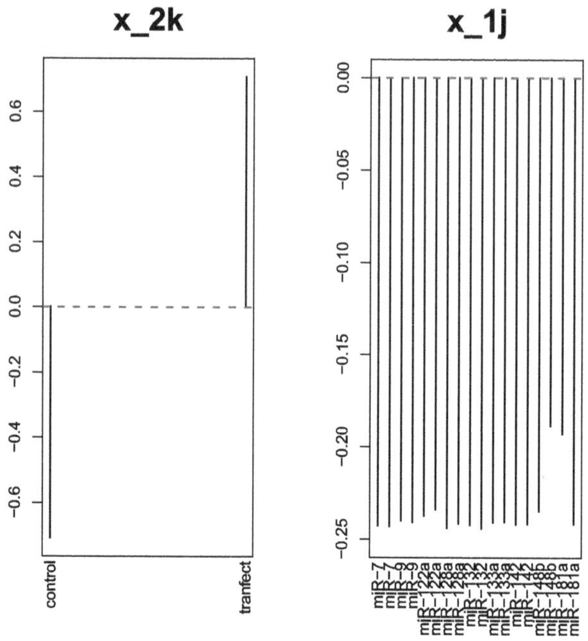

Fig. 7.115 The second singular value vector, $\boldsymbol{u}_2^{(k)}$, attributed to control and treated samples, and the first singular value vector, $\boldsymbol{u}_1^{(j)}$, attributed to miRNAs and replicates, obtained by HOSVD applied to x_{ijk} made out of study 4

two biological replicates and k is control and treated samples. We apply HOSVD algorithm, Fig. 3.8, to x_{ijk} as

$$x_{ijk} = \sum_{\ell_1=1}^{23,651} \sum_{\ell_2=1}^{18} \sum_{\ell_3=1}^{2} G(\ell_1, \ell_2, \ell_3) u_{\ell_1}^{(i)} u_{\ell_2}^{(j)} u_{\ell_3}^{(k)} \tag{7.182}$$

where $\boldsymbol{u}_{\ell_1}^{(i)} \in \mathbb{R}^{23,651}, \boldsymbol{u}_{\ell_2}^{(j)} \in \mathbb{R}^{18}, \boldsymbol{u}_{\ell_3}^{(k)} \in \mathbb{R}^2$ are singular value vectors and $G(\ell_1, \ell_2, \ell_3) \in \mathbb{R}^{23,651\times 18\times 2}$ is a core tensor. Now we need to find $\boldsymbol{u}_{\ell_3}^{(k)}$ satisfying $u_{\ell_3 1}^{(k)} = -u_{\ell_3 2}^{(k)}$; $\ell_3 = 2$ turns out to satisfy this requirement. On the other hand, we need to find $\boldsymbol{u}_{\ell_2}^{(j)}$ satisfying $u_{\ell_2 j}^{(j)} = \text{constant}$; $\ell_2 = 1$ turns out to satisfy this requirement (Fig. 7.115). After investigating which $G(\ell_1, 1, 2)$ has the largest absolute value, we find that $\ell_1 = 6$. P values are attributed to probes using the sixth PC score $\boldsymbol{u}_6^{(i)}$ with assuming χ^2 distribution as

$$P_i = P_{\chi^2}\left[> \left(\frac{u_{6i}^{(i)}}{\sigma_6} \right)^2 \right]. \tag{7.183}$$

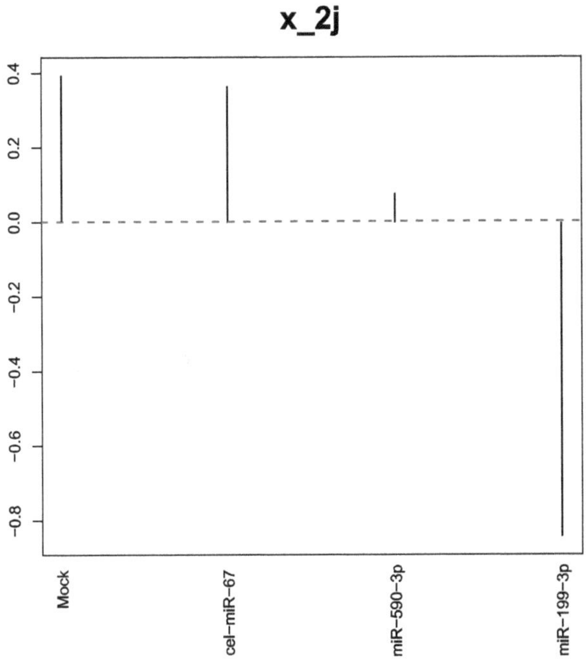

Fig. 7.116 The second PC loading, $\boldsymbol{v}_2$, obtained by PCA applied to x_{ij} made out of study 5

P values are corrected by BH criterion and probes associated with adjusted P values less than 0.01 are selected.

Study 5

This data set includes four profiles to which mock and cel-miR-67 miR-509/199a-3p are transfected. We format it to matrix $x_{ij} \in \mathbb{R}^{41,539 \times 4}$. PCA is applied to x_{ij} and the second PC loading, $\boldsymbol{v}_2$, is selected as that exhibits distinction between mock + cel-miR-67 and miR-509/199a-3p (Fig. 7.116). Although outcome cannot be said very promising, because $\boldsymbol{v}_2$ is best fitted with the requirement, P values are attributed to probes using Eq. (7.181). P values are corrected by BH criterion and probes associated with adjusted P values less than 0.01 are selected.

Study 6

This data set includes transfection of eight miRNAs, miR-10a-5p, 150-3p/5p, 148a-3p/5p, 499a-5p, and 455-3p. Number of probes in microarray is as many as 62,976. The number of samples is 16 composed of a combination of miRNAs and cell lines. Not all miRNAs are used equally. For each of the 16, two paired experiments of

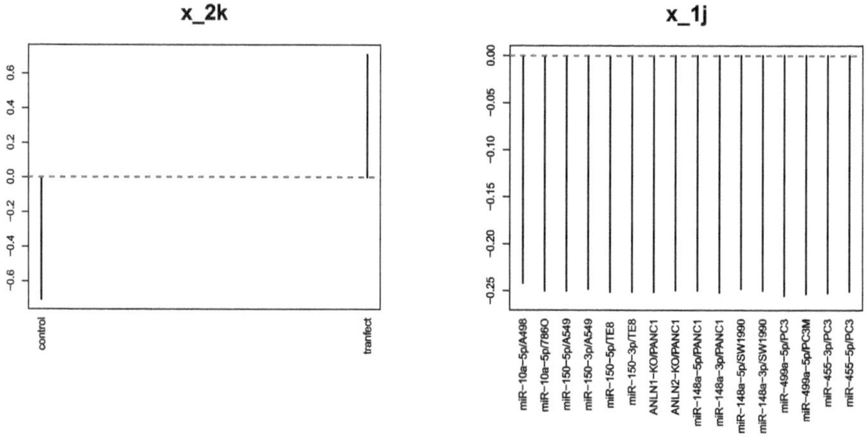

Fig. 7.117 The second singular value vector, $\boldsymbol{u}_2^{(k)}$, attributed to the various combinations of control and cell lines, and the first singular value vector, $\boldsymbol{u}_1^{(j)}$, attributed to miRNAs and replicates, obtained by HOSVD applied to x_{ijk} made out of study 6

treated and control samples. Treated and control sample measurement is performed by one microarray. Thus these two must be retrieved from it (columns annotated as gProcessedSignal and rProcessedSignal). This is successfully formatted as a tensor, $x_{ijk} \in \mathbb{R}^{62,976 \times 16 \times 2}$ where i stands for probes, j stands for combinations of eight miRNAs transfection and cell lines and k is control and treated samples. We apply HOSVD algorithm, Fig. 3.8, to x_{ijk} as

$$x_{ijk} = \sum_{\ell_1=1}^{62,976} \sum_{\ell_2=1}^{16} \sum_{\ell_3=1}^{2} G(\ell_1, \ell_2, \ell_3) u_{\ell_1}^{(i)} u_{\ell_2}^{(j)} u_{\ell_3}^{(k)} \tag{7.184}$$

where $\boldsymbol{u}_{\ell_1}^{(i)} \in \mathbb{R}^{62,976}, \boldsymbol{u}_{\ell_2}^{(j)} \in \mathbb{R}^{16}, \boldsymbol{u}_{\ell_3}^{(k)} \in \mathbb{R}^2$ are singular value vectors and $G(\ell_1, \ell_2, \ell_3) \in \mathbb{R}^{62,976 \times 16 \times 2}$ is a core tensor. Now we need to find $\boldsymbol{u}_{\ell_3}^{(k)}$ satisfying $u_{\ell_3 1}^{(k)} = -u_{\ell_3 2}^{(k)}$; $\ell_3 = 2$ turns out to satisfy this requirement. On the other hand, we need to find $\boldsymbol{u}_{\ell_2}^{(j)}$ satisfying $u_{\ell_2 j}^{(j)} =$ constant; $\ell_2 = 1$ turns out to satisfy this requirement (Fig. 7.117). After investigating which $G(\ell_1, 1, 2)$ has the largest absolute value, we find that $\ell_1 = 7$. P values are attributed to probes using the seventh PC score $\boldsymbol{u}_7^{(i)}$ with assuming χ^2 distribution as

$$P_i = P_{\chi^2}\left[> \left(\frac{u_{7i}^{(i)}}{\sigma_7} \right)^2 \right]. \tag{7.185}$$

P values are corrected by BH criterion and probes associated with adjusted P values less than 0.01 are selected.

Study 7

This data set includes transfection of nine miR-205/29a/144-3p/5p, 210, 23b, 221/ 222/ 223. The number of probes in microarray is as many as 62,976. The number of samples is 19 composed of a combination of miRNAs and cell lines. Not all miRNAs are used equally. For each of 19, two paired experiments of treated and control samples. Treated and control sample measurement is performed by one microarray. Thus these two must be retrieved from it (columns annotated as gProcessedSignal and rProcessedSignal). This is successfully formatted as tensor, $x_{ijk} \in \mathbb{R}^{62,976 \times 19 \times 2}$ where i stands for probes, j stands for combinations of eight miRNAs transfection and cell lines and k is control and treated samples. We apply HOSVD algorithm, Fig. 3.8, to x_{ijk} as

$$x_{ijk} = \sum_{\ell_1=1}^{62,976} \sum_{\ell_2=1}^{19} \sum_{\ell_3=1}^{2} G(\ell_1, \ell_2, \ell_3) u_{\ell_1}^{(i)} u_{\ell_2}^{(j)} u_{\ell_3}^{(k)} \tag{7.186}$$

where $\boldsymbol{u}_{\ell_1}^{(i)} \in \mathbb{R}^{62,976}, \boldsymbol{u}_{\ell_2}^{(j)} \in \mathbb{R}^{19}, \boldsymbol{u}_{\ell_3}^{(k)} \in \mathbb{R}^2$ are singular value vectors and $G(\ell_1, \ell_2, \ell_3) \in \mathbb{R}^{62,976 \times 19 \times 2}$ is a core tensor. Now we need to find $\boldsymbol{u}_{\ell_3}^{(k)}$ satisfying $u_{\ell_3 1}^{(k)} = -u_{\ell_3 2}^{(k)}$; $\ell_3 = 2$ turns out to satisfy this requirement. On the other hand, we need to find $\boldsymbol{u}_{\ell_2}^{(j)}$ satisfying $u_{\ell_2 j}^{(j)} =$ constant; $\ell_2 = 1$ turns out to satisfy this requirement (Fig. 7.118). After investigating which $G(\ell_1, 1, 2)$ has the larger absolute values, we find that $\ell_1 = 2, 3$. P values are attributed to probes using the second and third PC scores $\boldsymbol{u}_{\ell_1}^{(i)}, \ell_1 = 2, 3$ with assuming χ^2 distribution as

$$P_i = P_{\chi^2}\left[> \sum_{\ell_1=2}^{3} \left(\frac{u_{\ell_1 i}^{(i)}}{\sigma_{\ell_1}} \right)^2 \right]. \tag{7.187}$$

P values are corrected by BH criterion and probes associated with adjusted P values less than 0.01 are selected.

Study 8

This data set includes the transfection of two miRNAs, miR-146a/b. The number of probes in the microarray is as many as 43,379. The number of samples is 18 composed of six miR-146a OE, four miR-146b OE, and eight miR-146a KO. For each of 19, two paired experiments of treated and control samples. Treated and

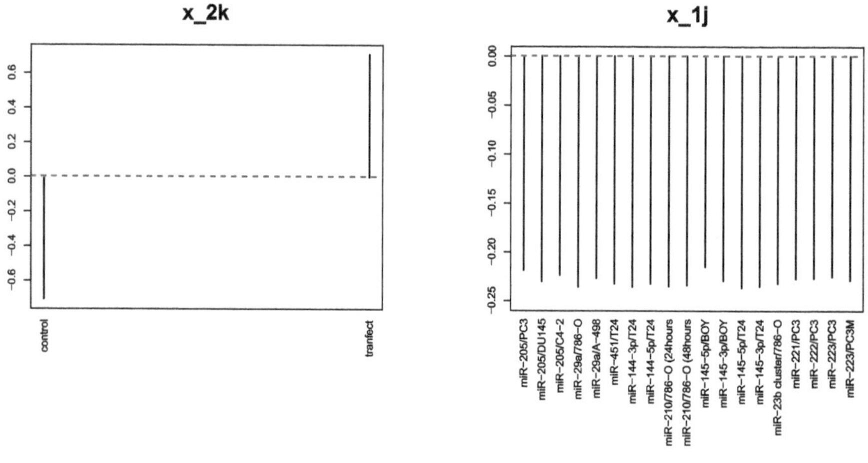

Fig. 7.118 The second singular value vector, $\boldsymbol{u}_2^{(k)}$, attributed to control and treated samples, and the first singular value vector, $\boldsymbol{u}_1^{(j)}$, attributed to the combinations of miRNAs and cell lines, obtained by HOSVD applied to x_{ijk} made out of study 7

control sample measurement is performed by one microarray. Thus, these two must be retrieved from it (columns annotated as gProcessedSignal and rProcessedSignal). This is successfully formatted as tensor, $x_{ijk} \in \mathbb{R}^{43,379 \times 18 \times 2}$ where i stands for probes, j stands for combinations of eight miRNAs transfection and cell lines and k is control and treated samples. We apply HOSVD algorithm, Fig. 3.8, to x_{ijk} as

$$x_{ijk} = \sum_{\ell_1=1}^{43,379} \sum_{\ell_2=1}^{18} \sum_{\ell_3=1}^{2} G(\ell_1, \ell_2, \ell_3) u_{\ell_1}^{(i)} u_{\ell_2}^{(j)} u_{\ell_3}^{(k)} \tag{7.188}$$

where $\boldsymbol{u}_{\ell_1}^{(i)} \in \mathbb{R}^{43,379}, \boldsymbol{u}_{\ell_2}^{(j)} \in \mathbb{R}^{18}, \boldsymbol{u}_{\ell_3}^{(k)} \in \mathbb{R}^2$ are singular value vectors and $G(\ell_1, \ell_2, \ell_3) \in \mathbb{R}^{43,379 \times 18 \times 2}$ is a core tensor. Now we need to find $\boldsymbol{u}_{\ell_3}^{(k)}$ satisfying $u_{\ell_3 1}^{(k)} = -u_{\ell_3 2}^{(k)}$; $\ell_3 = 2$ turns out to satisfy this requirement. On the other hand, we need to find $\boldsymbol{u}_{\ell_2}^{(j)}$ satisfying $u_{\ell_2 j}^{(j)} =$ constant; $\ell_2 = 1$ turns out to satisfy this requirement (Fig. 7.119). After investigating which $G(\ell_1, 1, 2)$ has the largest absolute value, we find that $\ell_1 = 5$. P values are attributed to probes using the fifth PC score $\boldsymbol{u}_5^{(i)}$ with assuming χ^2 distribution as

$$P_i = P_{\chi^2}\left[> \left(\frac{u_{5i}^{(i)}}{\sigma_5} \right)^2 \right]. \tag{7.189}$$

P values are corrected by BH criterion and probes associated with adjusted P values less than 0.01 are selected.

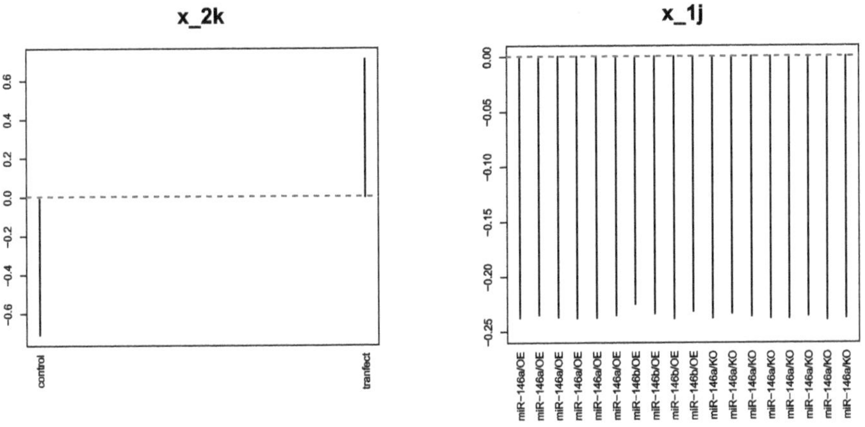

Fig. 7.119 The second singular value vector, $\boldsymbol{u}_2^{(k)}$, attributed to control and treated samples, and the first singular value vector, $\boldsymbol{u}_1^{(j)}$, attributed to miRNAs and replicates, obtained by HOSVD applied to x_{ijk} made out of study 8

Study 9

This data set includes transfection of two miRNAs, miR-107/181b. transfected to HeLa cell lines. Number of probes in microarray is as many as 9987. The number of samples is 18 composed of six controls, two anti-miR-107, four miR-107, two anti-miR-181b, and four miR-181b transfected samples. This is successfully formatted as a tensor, $x_{ijk} \in \mathbb{R}^{9987 \times 16 \times 3}$ where i stands for probes, j stands for repliactes and k is control, miR-107 and miR-181b. We apply HOSVD algorithm, Fig. 3.8, to x_{ijk} as

$$x_{ijk} = \sum_{\ell_1=1}^{9987} \sum_{\ell_2=1}^{6} \sum_{\ell_3=1}^{3} G(\ell_1, \ell_2, \ell_3) u_{\ell_1}^{(i)} u_{\ell_2}^{(j)} u_{\ell_3}^{(k)} \tag{7.190}$$

where $\boldsymbol{u}_{\ell_1}^{(i)} \in \mathbb{R}^{9987}, \boldsymbol{u}_{\ell_2}^{(j)} \in \mathbb{R}^6, \boldsymbol{u}_{\ell_3}^{(k)} \in \mathbb{R}^3$ are singular value vectors and $G(\ell_1, \ell_2, \ell_3) \in \mathbb{R}^{9987 \times 6 \times 3}$ is a core tensor. Now we need to find $\boldsymbol{u}_{\ell_3}^{(k)}$ satisfying $u_{\ell_3 1}^{(k)} = -u_{\ell_3 2}^{(k)} = -u_{\ell_3 3}^{(k)}$; $\ell_3 = 2$ turns out to satisfy this requirement. On the other hand, we need to find $\boldsymbol{u}_{\ell_2}^{(j)}$ satisfying $u_{\ell_2 j}^{(j)}$ = constant; $\ell_2 = 1$ turns out to satisfy this requirement (Fig. 7.120). After investigating which $G(\ell_1, 1, 2)$ has the largest absolute value, we find that $\ell_1 = 2$. P values are attributed to probes using the second PC score $\boldsymbol{u}_2^{(i)}$ with assuming χ^2 distribution as Eq. (7.181). P values are corrected by BH criterion and probes associated with adjusted P values less than 0.01 are selected.

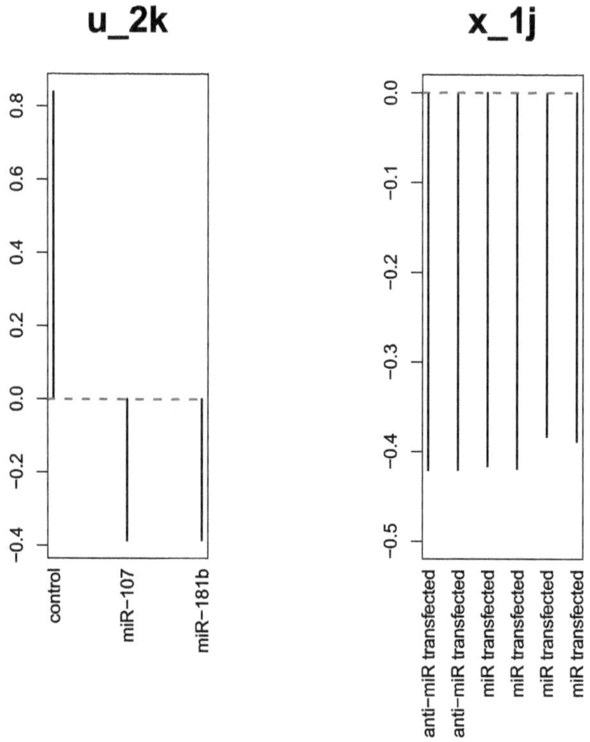

Fig. 7.120 The second singular value vector, $u_2^{(k)}$, attributed to control, miR-107 and miR-181b transfection, and the first singular value vector, $u_1^{(j)}$, attributed to miRNAs and replicates, obtained by HOSVD applied to x_{ijk} made out of study 9

Study 10

Everything is the same as study 9 other than that transfected cell line is HEK 293 cell line (see Fig. 7.121 for singular value vectors selected).

Study 11

This data set includes transfection of a miRNA, miR-181b. transfected to the SH-SY5Y cell line. Number of probes in microarray is as many as 9987. The number of samples is 8 composed of four controls, two anti-miR-181b, and two miR-181b transfected samples. This is successfully formatted as tensor, $x_{ijk} \in \mathbb{R}^{9987 \times 4 \times 2}$ where i stands for probes, j stands for replicactes and k is control and miR-181b. We apply HOSVD algorithm, Fig. 3.8, to x_{ijk} as

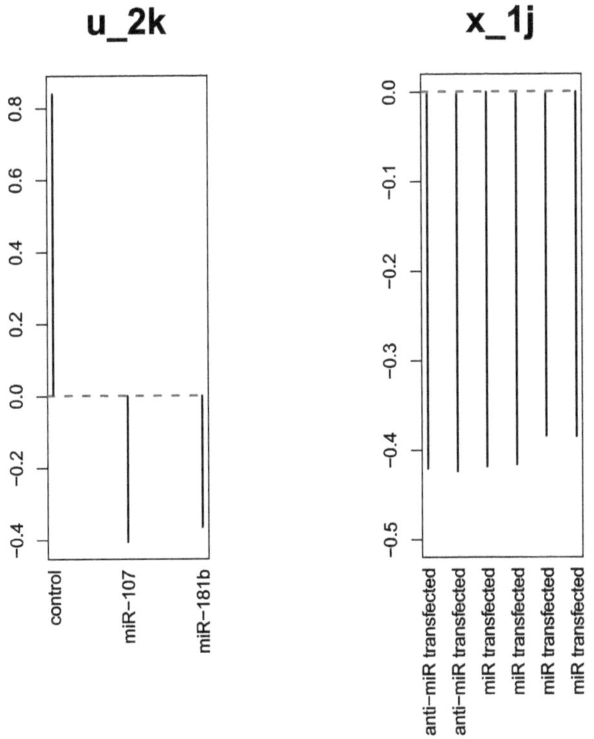

Fig. 7.121 The second singular value vector, $\boldsymbol{u}_2^{(k)}$, attributed to control, miR-107 and miR-181b transfection, and the first singular value vector, $\boldsymbol{u}_1^{(j)}$, attributed to miRNAs and replicates, obtained by HOSVD applied to x_{ijk} made out ofstudy 10

$$x_{ijk} = \sum_{\ell_1=1}^{9987} \sum_{\ell_2=1}^{4} \sum_{\ell_3=1}^{2} G(\ell_1, \ell_2, \ell_3) u_{\ell_1}^{(i)} u_{\ell_2}^{(j)} u_{\ell_3}^{(k)} \tag{7.191}$$

where $\boldsymbol{u}_{\ell_1}^{(i)} \in \mathbb{R}^{9987}, \boldsymbol{u}_{\ell_2}^{(j)} \in \mathbb{R}^4, \boldsymbol{u}_{\ell_3}^{(k)} \in \mathbb{R}^2$ are singular value vectors and $G(\ell_1, \ell_2, \ell_3) \in \mathbb{R}^{9987\times4\times2}$ is a core tensor. Now we need to find $\boldsymbol{u}_{\ell_3}^{(k)}$ satisfying $u_{\ell_3 1}^{(k)} = -u_{\ell_3 2}^{(k)}$; $\ell_3 = 2$ turns out to satisfy this requirement. On the other hand, we need to find $\boldsymbol{u}_{\ell_2}^{(j)}$ satisfying $u_{\ell_2 j}^{(j)}$ = constant; $\ell_2 = 1$ turns out to satisfy this requirement (Fig. 7.122). After investigating which $G(\ell_1, 1, 2)$ has the largest absolute value, we find that $\ell_1 = 2$. P values are attributed to probes using the second PC score $\boldsymbol{u}_2^{(i)}$ with assuming χ^2 distribution as Eq. (7.181). P values are corrected by BH criterion and probes associated with adjusted P values less than 0.01 are selected.

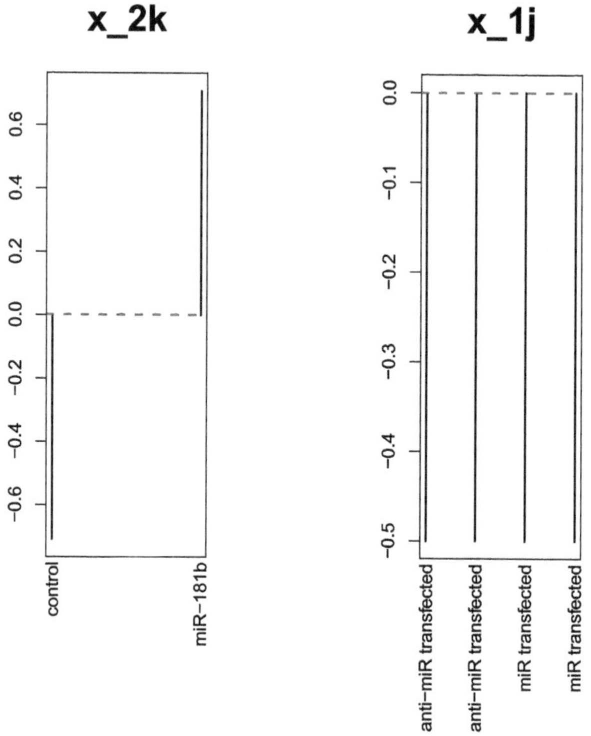

Fig. 7.122 The second singular value vector, $u_2^{(k)}$, attributed to control and miR-181b transfection, and the first singular value vector, $u_1^{(j)}$, attributed to miRNAs and replicates, obtained by HOSVD applied to x_{ijk} made out of study 11

Drug Discovery from Gene Expression: II

Heart Failure

Human gene expression profiles are downloaded from GEO with GEO ID 57345. The file used is GSE57345-GPL11532_series_matrix.txt.gz. Rat heart gene expression profiles are downloaded from GEO with GEO ID GSE59905. Files used are GSE59905-GPL5426_series_matrix.txt.gz, and GSE59905-GPL5425_series_matrix.txt.gz. 3937 genes are shared between human and rat. Case II tensor, $x_{ij_1j_2j_3}$, is generated as

$$x_{ij_1j_2j_3} = x_{ij_1j_2}x_{ij_3}. \tag{7.192}$$

HOSVD algorithm, Fig. 3.8, is applied to $x_{ij_1j_2j_3}$.

At first, we try to find $u_{\ell_3}^{(j_3)}$ associated with significant distinction between three classes, healthy control, idiopathic dilated cardiomyopathy, ischemic stroke, by applying categorical regression

$$u^{(j_3)}_{\ell_3 j_3} = a_{\ell_3} + \sum_{s=1}^{3} b_{\ell_3 s}\delta_{j_3 s} \tag{7.193}$$

P values computed by categorical regression analysis are corrected by BH criterion. Then we found that $\ell_3 = 2, 3, 5, 17, 313$ are associated with adjusted P values less than 0.01, raw P values of which are 1.65×10^{-17}, 1.00×10^{-39}, 1.29×10^{-4}, 4.97×10^{-6} and 1.554×10^{-4}. Among them we select $\ell_3 = 2, 3$ because they have more contribution than others. Figure 7.123a shows the $\boldsymbol{u}^{(j_3)}_{\ell_3}$, $1 \leq \ell_3 \leq 3$.

Next we try to identify $\boldsymbol{u}^{(j_2)}_{\ell_2}$ associated with significant time dependence. Figure 7.123b shows the $\boldsymbol{u}^{(j_2)}_{\ell_2}$, $1 \leq \ell_2 \leq 4$. The correlation coefficients between $\boldsymbol{u}^{(j_2)}_{\ell_2}$ and (1/4,1,3,5) are −0.72, −0.82, 0.51, and −0.09. Then $\ell_2 = 2$ with largest absolute value is selected. Then we need to find $\boldsymbol{u}^{(j_1)}_{\ell_1}$ and $\boldsymbol{u}^{(i)}_{\ell_4}$ associated with larger absolute $G(\ell_1, 2, 2, \ell_4)$ or $G(\ell_1, 2, 3, \ell_4)$ in order to select compounds j_1 and genes i associated with time dependence and distinction between patients and healthy controls simultaneously. Thus, we list top 20 $G(\ell_1, 2, 2, \ell_4)$ or $G(\ell_1, 2, 3, \ell_4)$ (Table 7.108). Because G gradually decreases, we cannot select specific cutoff. Thus, tentatively, we select ℓ_1 and ℓ_4 associated with top ten Gs; $\ell_1 = 2$ and $\ell_4 = 21, 25, 27, 28, 33, 36, 37, 38, 41, 42$. Figure 7.123c shows $\boldsymbol{u}^{(j_1)}_2$. Forty three outlier drugs, $\left|u^{(j_1)}_{2 j_1}\right| > 0.1$, blue parts, are selected, by visual inspection, because P values computed from $\boldsymbol{u}^{(j_1)}_2$ and corrected by BH criterion cannot be less than 0.01. On the other hand P values are attributed to ith gene as

$$P_i = P_{\chi^2}\left[> \sum_{\ell_4=21,25,27,28,33,36,37,38,41,42} \left(\frac{u_{\ell_4 i}}{\sigma_{\ell_4}}\right)^2 \right] \tag{7.194}$$

P values are corrected by BH criterion and 274 genes associated with adjusted P values less than 0.01 are selected.

PTSD

PTSD model rat amygdala and hippocampus gene expression are downloaded from GEO with GEO ID GSE60304. A file GSE60304_series_matrix.txt.gz is used. Gene expression profiles of the brain for drug treatments of rats are downloaded from GEO with GEO ID GSE59895. Files used are GSE59895-GPL5425_series_matrix.txt.gz and GSE59895-GPL5426_series_matrix.txt.gz. Case II tensor, $x_{ij_1j_2j_3j_4j_5}$, is generated as

$$x_{ij_1j_2j_3j_4j_5} = x_{ij_1j_2} x_{ij_3j_4} x_{ij_3j_5}. \tag{7.195}$$

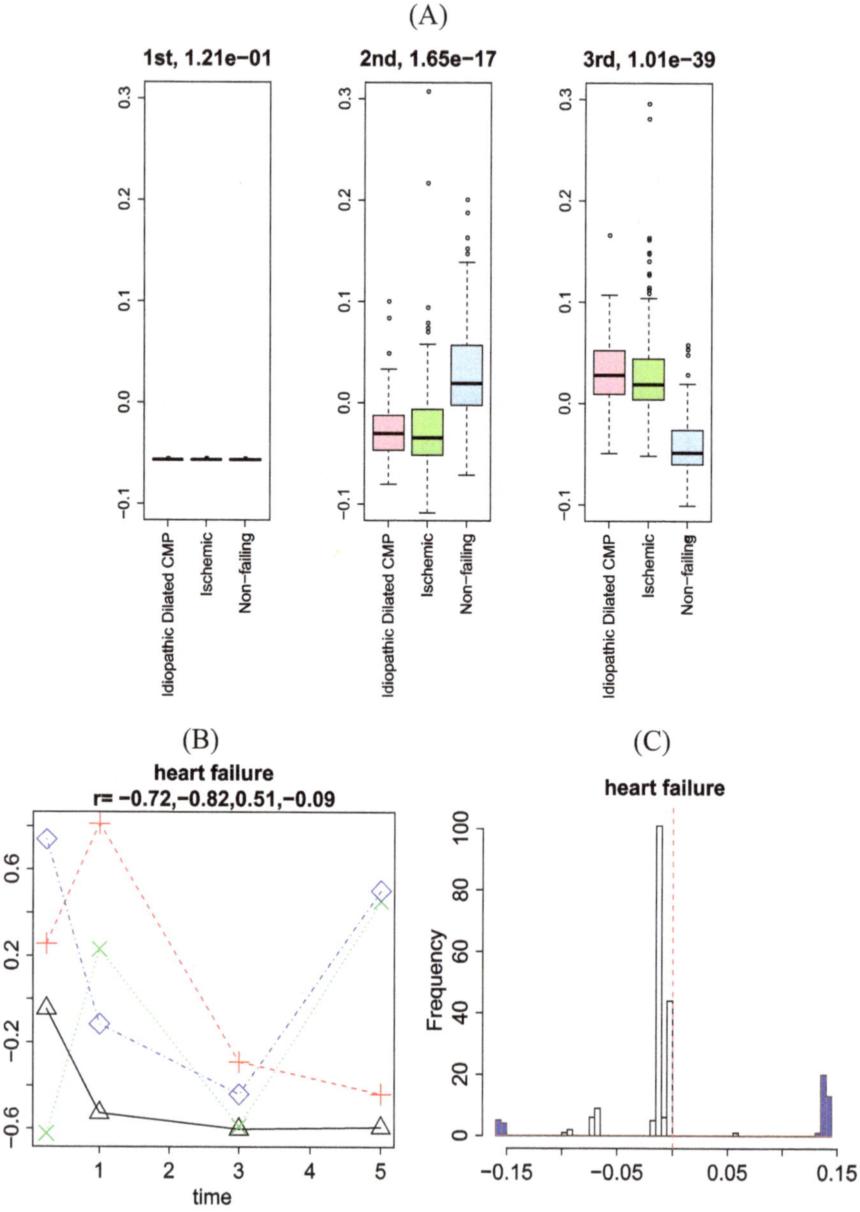

Fig. 7.123 (**a**) $u_{\ell_3}^{(j_3)}, 1 \leq \ell_3 \leq 3$, P values are computed by categorical regression analysis, Eq. (7.193). (**b**) $u_{\ell_2}^{(j_2)}, 1 \leq \ell_2 \leq 4$, $\Delta : \ell_2 = 1$, $+ : \ell_2 = 2$, $\times : \ell_2 = 3$, $\diamond : \ell_2 = 4$. r: correlation coefficient. (**c**) Histogram of $u_2^{(j_1)}$. Blue parts are selected ones. Vertical red broken line is 0

Table 7.108 Top 20 $G(\ell_1, 2, 2, \ell_4)$ or $G(\ell_1, 2, 3, \ell_4)$

Rank	1	2	3	4	5	6	7	8	9	10
ℓ_1	2	2	2	2	2	2	2	2	2	2
ℓ_3	2	3	2	2	3	3	2	3	3	2
ℓ_4	27	38	33	28	41	37	21	36	42	25
$G(\ell_1, 2, \ell_3, \ell_4)$	66.2	−43.7	40.7	−40.2	38.2	−31.6	28.5	−26.8	−26.2	−26.2
Rank	11	12	13	14	15	16	17	18	19	20
ℓ_1	2	2	2	2	2	2	2	2	2	2
ℓ_3	2	3	2	2	3	3	2	2	2	3
ℓ_4	40	29	31	39	32	33	26	11	18	31
$G(\ell_1, 2, \ell_3, \ell_4)$	−25.5	25.2	−22.6	21.8	20.7	−19.7	−19.5	−18.2	−17.3	15.4

HOSVD algorithm, Fig. 3.8, is applied to $x_{ij_1j_2j_3j_4j_5}$.

In order to identify $\boldsymbol{u}_{\ell_4}^{(j_4)}$ and $\boldsymbol{u}_{\ell_5}^{(j_5)}$ associated with three classes, control samples, minimal behavioral response samples, and extreme behavioral response samples, by applying categorical regression analysis,

$$u_{\ell j_4}^{(j_4)} = a_\ell + \sum_{s=1}^{3} b_{\ell s}\delta_{j_4 s} \tag{7.196}$$

$$u_{\ell j_5}^{(j_5)} = a_\ell + \sum_{s=1}^{3} b_{\ell s}\delta_{j_5 s} \tag{7.197}$$

where regression coefficients are shared between $\ell_4 = \ell_5 = \ell$. P values computed by categorical regression analysis are corrected by BH criterion. Then, only $\ell = 3$ is associated with adjusted P values less than 0.05 (Fig. 7.124a).

Next we try to identify $\boldsymbol{u}_{\ell_2}^{(j_2)}$ associated with significant time dependence. Figure 7.124b shows the $\boldsymbol{u}_{\ell_2}^{(j_2)}, 1 \leq \ell_2 \leq 4$. The correlation coefficients between $\boldsymbol{u}_{\ell_2}^{(j_2)}$ and (1/4,1,3,5) are −0.75, −0.81, −0.30, and 0.50. Then $\ell_2 = 2$ with largest absolute value is selected. Then we need to find $\boldsymbol{u}_{\ell_1}^{(j_1)}$ and $\boldsymbol{u}_{\ell_6}^{(i)}$ associated with larger absolute $G(\ell_1, 2, \ell_3, 3, 3, \ell_6)$ in order to select compounds j_1 and genes i associated with time dependence and distinction between patients and healthy controls simultaneously. Thus, we list top 20 $G(\ell_1, 2, \ell_3, 3, 3, \ell_6)$ (Table 7.109). Because G gradually decreases, we cannot select specific cutoff. Thus, tentatively, we select ℓ_1 and ℓ_4 associated with top ten Gs; $\ell_1 = 2$ and $\ell_6 = 75, 77, 81, 83, 84, 85, 88, 89, 90, 102$. Figure 7.124c shows $\boldsymbol{u}_2^{(j_1)}$. Six outlier drugs, $u_{2j_1}^{(j_1)} < -0.2$ and $u_{1j_1}^{(j_1)} < -0.15$, blue parts, are selected, by visual inspection, because P values computed from $\boldsymbol{u}_2^{(j_1)}$ and corrected by BH criterion cannot be less than 0.01. On the other hand P values are attributed to ith gene as

$$P_i = P_{\chi^2}\left[> \sum_{\ell_6=75,77,81,83,84,85,88,89,90,102} \left(\frac{u_{\ell_6 i}}{\sigma_{\ell_6}}\right)^2\right] \tag{7.198}$$

P values are corrected by BH criterion and 374 genes associated with adjusted P values less than 0.01 are selected.

All

Bone marrow gene expression profiles of drug-treated rats are downloaded from GEO with GEO ID GSE59894, and ALL human bone marrow gene expression is from GEO with GEO ID GSE67684. Used files are GSE67684-GPL570_series_matrix.txt.gz, GSE67684-GPL96_series_matrix.txt.gz, GSE59894-

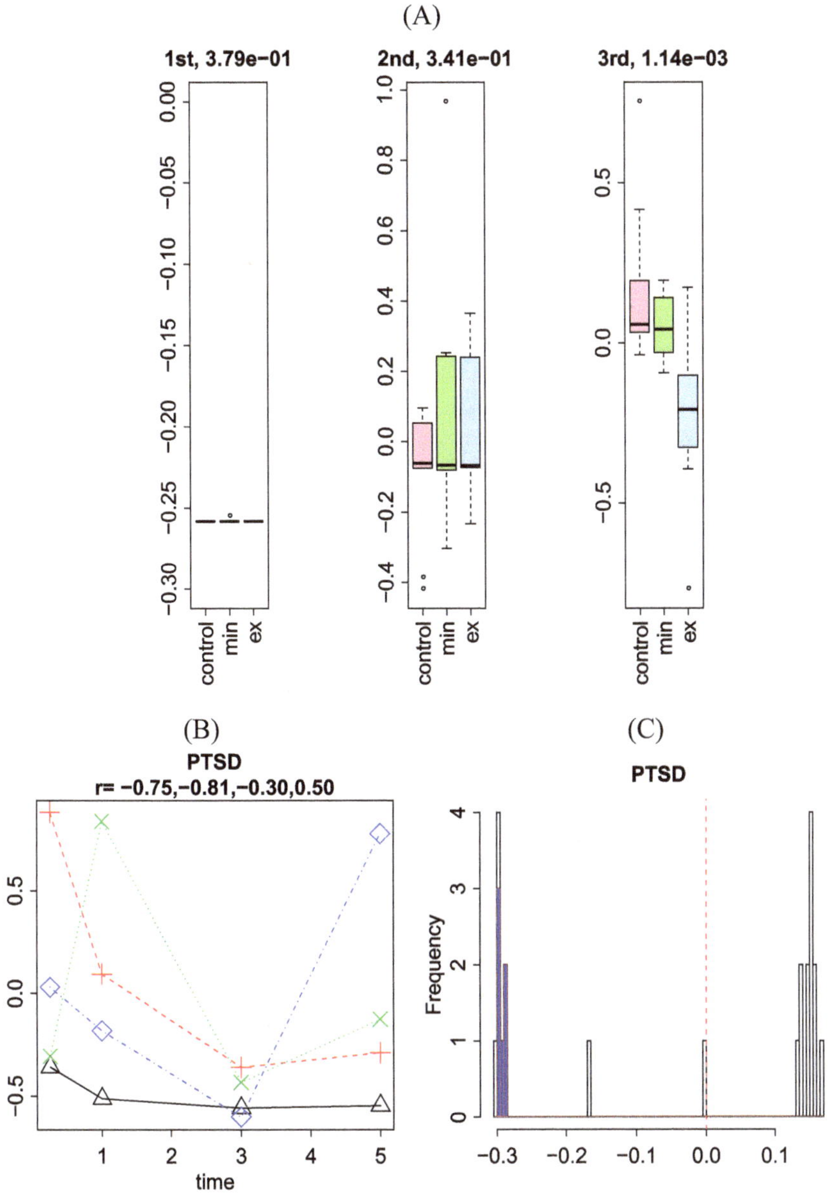

Fig. 7.124 (**a**) $u^{(j_3)}_{\ell_3}$, $1 \leq \ell_3 \leq 3$, P values are computed by categorical regression analysis, Eqs. (7.196) and (7.197). (**b**) $u^{(j_2)}_{\ell_2}$, $1 \leq \ell_2 \leq 4$, $\Delta : \ell_2 = 1$, $+ : \ell_2 = 2$, $\times : \ell_2 = 3$, $\diamond : \ell_2 = 4$. r: correlation coefficient. (**c**) Histogram of $u^{(j_1)}_2$. Blue parts are selected ones. Vertical red broken line is 0

Table 7.109 Top 20 $G(\ell_1, 2, \ell_3, 3, 3, \ell_6)$

Rank	1	2	3	4	5	6	7	8	9	10
ℓ_1	2	2	2	2	2	2	2	2	2	2
ℓ_3	1	1	1	1	1	1	1	1	1	1
ℓ_6	81	84	88	77	85	75	83	90	90	102
$G(\ell_1, 2, \ell_3, 3, 3, \ell_6)$	−0.133	0.112	0.110	−0.078	0.075	−0.075	0.074	0.069	0.069	−0.063
Rank	11	12	13	14	15	16	17	18	19	20
ℓ_1	2	2	2	2	2	2	2	2	2	2
ℓ_3	1	1	2	2	2	1	2	2	1	2
ℓ_6	76	80	94	76	128	285	86	286	92	282
$G(\ell_1, 2, \ell_3, 3, 3, \ell_6)$	−0.063	0.062	0.054	−0.054	−0.053	−0.052	0.048	0.047	0.045	0.045

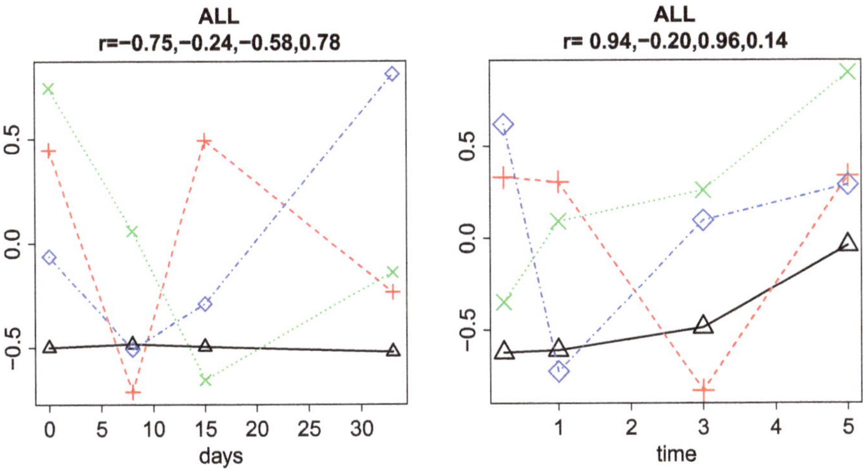

Fig. 7.125 (**a**) $\boldsymbol{u}^{(j_3)}_{\ell_3}$, $1 \le \ell_2 \le 4$, (**b**) $\boldsymbol{u}^{(j_2)}_{\ell_2}$, $1 \le \ell_2 \le 4$, $\triangle : \ell_2 = 1$, $+ : \ell_2 = 2$, $\times : \ell_2 = 3$, $\diamond : \ell_2 = 4$. r: correlation coefficient

GPL5425_series_matrix.txt.gz, and GSE59894-GPL5426_series_matrix.txt.gz. In this case, both gene expression profiles are time dependent. ALL human bone marrow gene expression profiles are measured at four times points, 0, 8, 15, and 33 days after a remission induction therapy. Case II tensor, $x_{ij_1j_2j_3j_4}$ is obtained as

$$x_{ij_1j_2j_3j_4} = x_{ij_1j_2}x_{ij_3j_4} \tag{7.199}$$

HOSVD algorithm, Fig. 3.8, is applied to $x_{ij_1j_2j_3j_4}$.

We compute correlation coefficients between $\boldsymbol{u}^{(j_3)}_{\ell_3}$ and days after a remission induction therapy, we decide to select $\ell_3 = 4$ because it has the largest absolute value of correlation coefficient (Fig. 7.125a).

Next we try to identify $\boldsymbol{u}^{(j_2)}_{\ell_2}$ associated with significant time dependence. Figure 7.125b shows the $\boldsymbol{u}^{(j_2)}_{\ell_2}$, $1 \le \ell_2 \le 4$. The correlation coefficients between $\boldsymbol{u}^{(j_2)}_{\ell_2}$ and (1/4,1,3,5) are 0.94, −0.20, 0.96, and 0.14. Then $\ell_2 = 3$ with largest absolute value is selected. Then we need to find $\boldsymbol{u}^{(j_1)}_{\ell_1}$ and $\boldsymbol{u}^{(i)}_{\ell_5}$ associated with larger absolute $G(\ell_1, 3, 4, \ell_4, \ell_5)$ in order to select compounds j_1 and genes i associated with time dependence and distinction between patients and healthy controls simultaneously. Thus, we list top 20 $G(\ell_1, 3, 4, \ell_4, \ell_5)$ (Table 7.110). For ℓ_1 and ℓ_5, we decide to select those associated with top ten Gs. As a result, $\ell_1 = 2, 3, 5, 6, 9, 10$ and $\ell_5 = 1, 2, 3, 5$ are selected. P values are attributed to j_1 and i as

Table 7.110 Top 20 $G(\ell_1, 3, 4, \ell_4, \ell_5)$

Rank	1	2	3	4	5	6	7	8	9	10
ℓ_1	3	5	2	3	10	9	6	3	2	9
ℓ_4	4	4	4	7	4	4	4	5	4	4
ℓ_5	1	1	1	5	3	3	1	5	2	2
$G(\ell_1, 2, \ell_3, 3, 3, \ell_6)$	260.6	−40.2	40.6	−20.9	20.7	20.4	−19.9	−18.0	16.8	−15.0
Rank	11	12	13	14	15	16	17	18	19	20
ℓ_1	8	6	14	3	3	13	12	2	1	3
ℓ_4	4	4	4	8	2	4	4	4	4	2
ℓ_5	6	4	2	5	5	4	2	3	4	1
$G(\ell_1, 2, \ell_3, 3, 3, \ell_6)$	−13.9	13.3	13.2	−12.8	12.3	11.6	11.4	11.3	10.5	−10.5

$$P_{j_1} = P_{\chi^2}\left[> \sum_{\ell_1=2,3,5,6,9,10} \left(\frac{u_{\ell_1 i}}{\sigma_{\ell_1}}\right)^2 \right], \tag{7.200}$$

$$P_i = P_{\chi^2}\left[> \sum_{\ell_5=1,2,3,5} \left(\frac{u_{\ell_5 i}}{\sigma_{\ell_5}}\right)^2 \right]. \tag{7.201}$$

P values are corrected by BH criterion and two compounds and 24 genes associated with adjusted P values less than 0.01 are selected.

Diabetes

Drug-treated rat kidney gene expression profiles are downloaded from GEO with GEO ID GSE59913. Human diabetic kidney gene expression profiles are downloaded from GEO with GEO ID GSE30122. Files used are GSE59913-GPL5425_series_matrix.txt.gz, GSE59913-GPL5426_series_matrix.txt.gz, and GSE30122_series_matrix.txt.gz. Case II tensor, $x_{ij_1j_2j_3}$, is generated as

$$x_{ij_1j_2j_3} = x_{ij_1j_2}x_{ij_3}. \tag{7.202}$$

HOSVD algorithm, Fig. 3.8, is applied to $x_{ij_1j_2j_3}$.

At first, we try to find $\boldsymbol{u}_{\ell_3}^{(j_3)}$ associated with significant distinction between four classes, normal human glomeruli, normal human kidney, normal Human Tubuli, and diabetic human Kidney, by applying categorical regression

$$u_{\ell_3 j_3}^{(j_3)} = a_{\ell_3} + \sum_{s=1}^{4} b_{\ell_3 s}\delta_{j_3 s} \tag{7.203}$$

P values computed by categorical regression analysis are corrected by BH criterion. Then we found that $\ell_3 = 1, 4$ are associated with adjusted P values less than 0.01, raw P values of which are 2.69×10^{-9} and 1.66×10^{-9} and are selected. Figure 7.126a shows the $\boldsymbol{u}_{\ell_3}^{(j_3)}, 1 \le \ell_3 \le 4$.

Next, we try to identify $\boldsymbol{u}_{\ell_2}^{(j_2)}$ associated with significant time dependence. Figure 7.126b shows the $\boldsymbol{u}_{\ell_2}^{(j_2)}, 1 \le \ell_2 \le 4$. The correlation coefficients between $\boldsymbol{u}_{\ell_2}^{(j_2)}$ and (1/4,1,3,5) are −0.60, −0.85, 0.53, and 0.20. Then $\ell_2 = 2$ with largest absolute value is selected. Then we need to find $\boldsymbol{u}_{\ell_1}^{(j_1)}$ and $\boldsymbol{u}_{\ell_4}^{(i)}$ associated with larger absolute $G(\ell_1, 2, 1, \ell_4)$ or $G(\ell_1, 2, 4, \ell_4)$ in order to select compounds j_1 and genes i associated with time dependence and the distinction between patients and healthy controls simultaneously. Thus, we list top 20 $G(\ell_1, 2, 1, \ell_4)$ or $G(\ell_1, 2, 4, \ell_4)$ (Table 7.111). Because top two Gs are outstandingly large, we select $\ell_1 = 2$ and $\ell_4 = 1, 4$ associated with top two Gs.

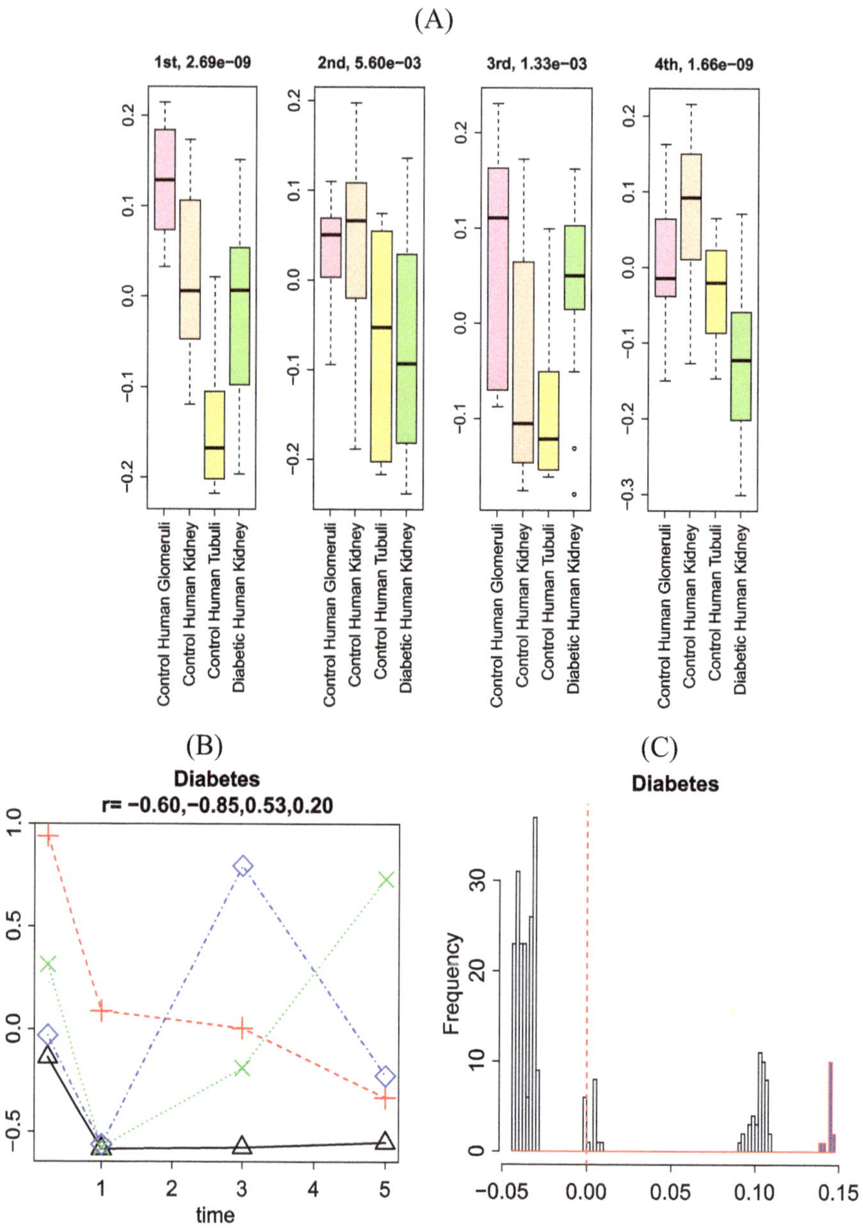

Fig. 7.126 (**a**) $\boldsymbol{u}^{(j_3)}_{\ell_3}$, $1 \leq \ell_3 \leq 4$, P values are computed by categorical regression analysis, Eq. (7.203). (**b**) $\boldsymbol{u}^{(j_2)}_{\ell_2}$, $1 \leq \ell_2 \leq 4$, $\triangle : \ell_2 = 1$, $+ : \ell_2 = 2$, $\times : \ell_2 = 3$, $\diamond : \ell_2 = 4$. r: correlation coefficient. (**c**) Histogram of $\boldsymbol{u}^{(j_1)}_2$. Blue parts are selected ones. Vertical red broken line is 0

Table 7.111 Top 20 $G(\ell_1, 2, 1, \ell_4)$ or $G(\ell_1, 2, 4, \ell_4)$

Rank	1	2	3	4	5	6	7	8	9	10
ℓ_1	2	2	3	4	4	3	9	11	2	4
ℓ_3	1	4	1	1	4	4	1	1	1	1
ℓ_4	1	4	1	1	4	4	48	59	42	42
$G(\ell_1, 2, \ell_3, \ell_4)$	−1410	955	−75	74	−53	51	38	34	−34	34
Rank	11	12	13	14	15	16	17	18	19	20
ℓ_1	2	9	2	9	11	9	6	11	9	4
ℓ_3	4	1	1	1	1	4	1	1	4	1
ℓ_4	40	29	31	39	32	33	26	11	18	31
$G(\ell_1, 2, \ell_3, \ell_4)$	−33	33	−32	31	31	−30	−30	−29	−29	28

Figure 7.126c shows $\boldsymbol{u}_2^{(j_1)}$. Fourteen outlier drugs, $u_{2j_1}^{(j_1)} > 0.13$, blue parts, are selected, by visual inspection, because P values computed from $\boldsymbol{u}_2^{(j_1)}$ and corrected by BH criterion cannot be less than 0.01. On the other hand P values are attributed to ith gene as

$$P_i = P_{\chi^2}\left[> \sum_{\ell_4=1,4} \left(\frac{u_{\ell_4 i}}{\sigma_{\ell_4}}\right)^2 \right] \tag{7.204}$$

P values are corrected by BH criterion and 65 genes associated with adjusted P values less than 0.01 are selected.

Renal Carcinoma

Drug-treated rat kidney gene expression profiles are downloaded from GEO with GEO ID GSE59913. Human renal cancer gene expression profiles are downloaded from GEO with GEO ID GSE40435. Files used are GSE59913-GPL5425_series_matrix.txt.gz, GSE59913-GPL5426_series_matrix.txt.gz, and GSE40435_series_matrix.txt.gz. Case II tensor, $x_{ij_1j_2j_3}$, is generated as

$$x_{ij_1j_2j_3} = x_{ij_1j_2} x_{ij_3}. \tag{7.205}$$

HOSVD algorithm, Fig. 3.8, is applied to $x_{ij_1j_2j_3}$.

At first, we try to find $\boldsymbol{u}_{\ell_3}^{(j_3)}$ associated with significant distinction between two classes, normal and cancer kidney, by applying categorical regression

$$u_{\ell_3 j_3}^{(j_3)} = a_{\ell_3} + \sum_{s=1}^{2} b_{\ell_3 s} \delta_{j_3 s} \tag{7.206}$$

P values computed by categorical regression analysis are corrected by BH criterion. Then we found that $\ell_3 = 13, 15, 30, 33, 35$ are associated with adjusted P values less than 0.05, raw P values of which are 3.4×10^{-4}, 1.1×10^{-3}, 2.7×10^{-4}, 1.1×10^{-4}, and 2.4×10^{-4} and are selected. Figure 7.127a shows the $\boldsymbol{u}_{\ell_3}^{(j_3)}, \ell_3 = 13, 15, 30, 33, 35$.

Next we try to identify $\boldsymbol{u}_{\ell_2}^{(j_2)}$ associated with significant time dependence. Figure 7.127b shows the $\boldsymbol{u}_{\ell_2}^{(j_2)}, 1 \leq \ell_2 \leq 4$. The correlation coefficients between $\boldsymbol{u}_{\ell_2}^{(j_2)}$ and (1/4,1,3,5) are −0.60, −0.84, 0.54, and 0.21. Then $\ell_2 = 2$ with largest absolute value is selected. Then we need to find $\boldsymbol{u}_{\ell_1}^{(j_1)}$ and $\boldsymbol{u}_{\ell_4}^{(i)}$ associated with larger absolute $G(\ell_1, 2, \ell_3, \ell_4), \ell_3 = 13, 15, 30, 33, 35$ in order to select compounds j_1 and genes i associated with time dependence and distinction between patients and healthy controls simultaneously. Thus, we list top 20 $G(\ell_1, 2, \ell_3, \ell_4), \ell_3 = 13, 15, 30, 33, 35$ (Table 7.112). For top 20 Gs, it is always that $\ell_1 = 2$. On the other hand, because G gradually changes, we cannot decide threshold values. Thus,

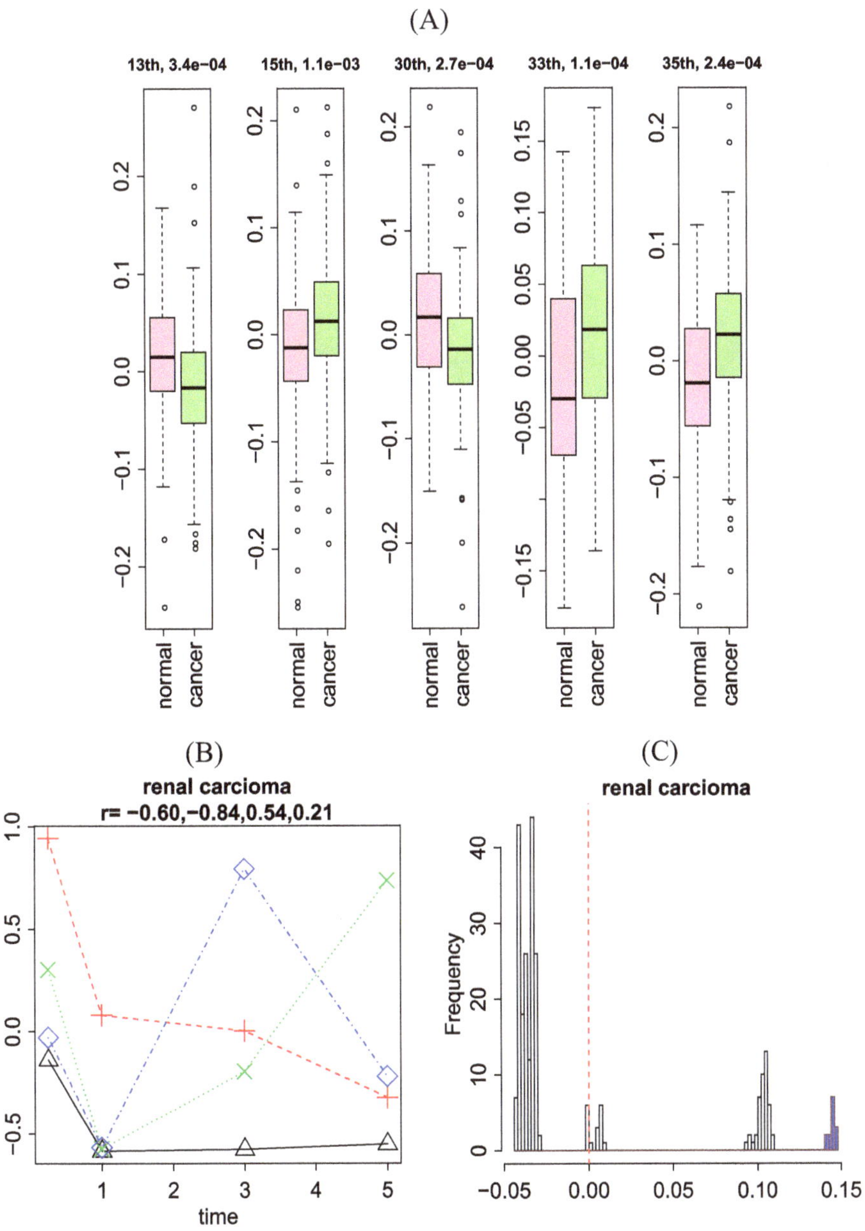

Fig. 7.127 (**a**) $\boldsymbol{u}_{\ell_3}^{(j_3)}, \ell_3 = 13, 15, 30, 33, 35$, P values are computed by categorical regression analysis, Eq. (7.206). (**b**) $\boldsymbol{u}_{\ell_2}^{(j_2)}, 1 \leq \ell_2 \leq 4$, $\Delta : \ell_2 = 1$, $+ : \ell_2 = 2$, $\times : \ell_2 = 3$, $\diamond : \ell_2 = 4$. r: correlation coefficient. (**c**) Histogram of $\boldsymbol{u}_2^{(j_1)}$. Blue parts are selected ones. The vertical red broken line is 0

Table 7.112 Top 20 $G(\ell_1, 2, \ell_3, \ell_4)$, $\ell_3 = 13, 15, 30, 33, 35$

Rank	1	2	3	4	5	6	7	8	9	10
ℓ_1	2	2	2	2	2	2	2	2	2	2
ℓ_3	13	13	13	13	15	15	13	13	13	15
ℓ_4	215	269	233	186	309	312	251	244	274	318
$G(\ell_1, 2, \ell_3, \ell_4)$	5.63	−5.30	5.08	−5.06	−4.84	4.78	4.66	4.61	4.57	−4.56
Rank	11	12	13	14	15	16	17	18	19	20
ℓ_1	2	2	2	2	2	2	2	2	2	2
ℓ_3	13	15	15	15	15	13	15	13	13	15
ℓ_4	289	399	336	206	363	255	375	219	342	297
$G(\ell_1, 2, \ell_3, \ell_4)$	−4.53	4.43	4.37	4.24	−4.19	−4.05	4.04	−3.97	−3.88	3.86

we tentatively decide that $\ell_4 = 186, 215, 233, 244, 251, 269, 274, 309, 312, 318$ associated with top ten Gs.

Figure 7.127c shows $\boldsymbol{u}_2^{(j_1)}$. Fourteen outlier drugs, $u_{2j_1}^{(j_1)} > 0.13$, blue parts, are selected, by visual inspection, because P values computed from $\boldsymbol{u}_2^{(j_1)}$ and corrected by BH criterion cannot be less than 0.01. On the other hand P values are attributed to ith gene as

$$P_i = P_{\chi^2}\left[> \sum_{\ell_4=186,215,233,244,251,269,274,309,312,318} \left(\frac{u_{\ell_4 i}}{\sigma_{\ell_4}}\right)^2\right] \tag{7.207}$$

P values are corrected by BH criterion and 225 genes associated with adjusted P values less than 0.01 are selected.

Cirrhosis

Drug-treated rat liver gene expression profiles are downloaded from GEO with GEO ID GSE59923. Cirrhosis patient human liver gene expression profile is downloaded from GEO with GEO ID GSE15654. File used are GSE15654_series_matrix.txt.gz, GSE59923-GPL5424_series_matrix.txt.gz, GSE59923-GPL5425_series_matrix.txt.gz, and GSE59923-GPL5426_series_matrix.txt.gz. Case II tensor, $x_{ij_1j_2j_3}$, is generated as

$$x_{ij_1j_2j_3} = x_{ij_1j_2} x_{ij_3}. \tag{7.208}$$

HOSVD algorithm, Fig. 3.8, is applied to $x_{ij_1j_2j_3}$.

At first, we try to find $\boldsymbol{u}_{\ell_3}^{(j_3)}$ associated with significant distinction between three classes, good, intermediate, and poor prognosis, by applying categorical regression

$$u_{\ell_3 j_3}^{(j_3)} = a_{\ell_3} + \sum_{s=1}^{3} b_{\ell_3 s}\delta_{j_3 s} \tag{7.209}$$

P values computed by categorical regression analhysis are corrected by BH criterion. Then we found that $\ell_3 = 2, 6$ are associated with adjusted P values less than 0.01, raw P values of which are 2.3×10^{-14} and 1.0×10^{-9} and are selected. Figure 7.128a shows the $\boldsymbol{u}_{\ell_3}^{(j_3)}, \ell_3 = 2, 6$.

Next we try to identify $\boldsymbol{u}_{\ell_2}^{(j_2)}$ associated with significant time dependence. Figure 7.128b shows the $\boldsymbol{u}_{\ell_2}^{(j_2)}, 1 \leq \ell_2 \leq 4$. The correlation coefficients between $\boldsymbol{u}_{\ell_2}^{(j_2)}$ and (1/4,1,3,5) are −0.56, −0.78, 0.52 and 0.36. Then $\ell_2 = 2$ with largest absolute value is selected. Then we need to find $\boldsymbol{u}_{\ell_1}^{(j_1)}$ and $\boldsymbol{u}_{\ell_4}^{(i)}$ associated with larger absolute $G(\ell_1, 2, \ell_3, \ell_4), \ell_3 = 2, 6$ in order to select compounds j_1 and genes i associated with time dependence and distinction between patients and healthy controls simultaneously. Thus, we list top 20 $G(\ell_1, 2, \ell_3, \ell_4), \ell_3 = 2, 6$ (Table 7.113). For top 20 Gs, it is always that $\ell_1 = 2$. On the other hand, because G

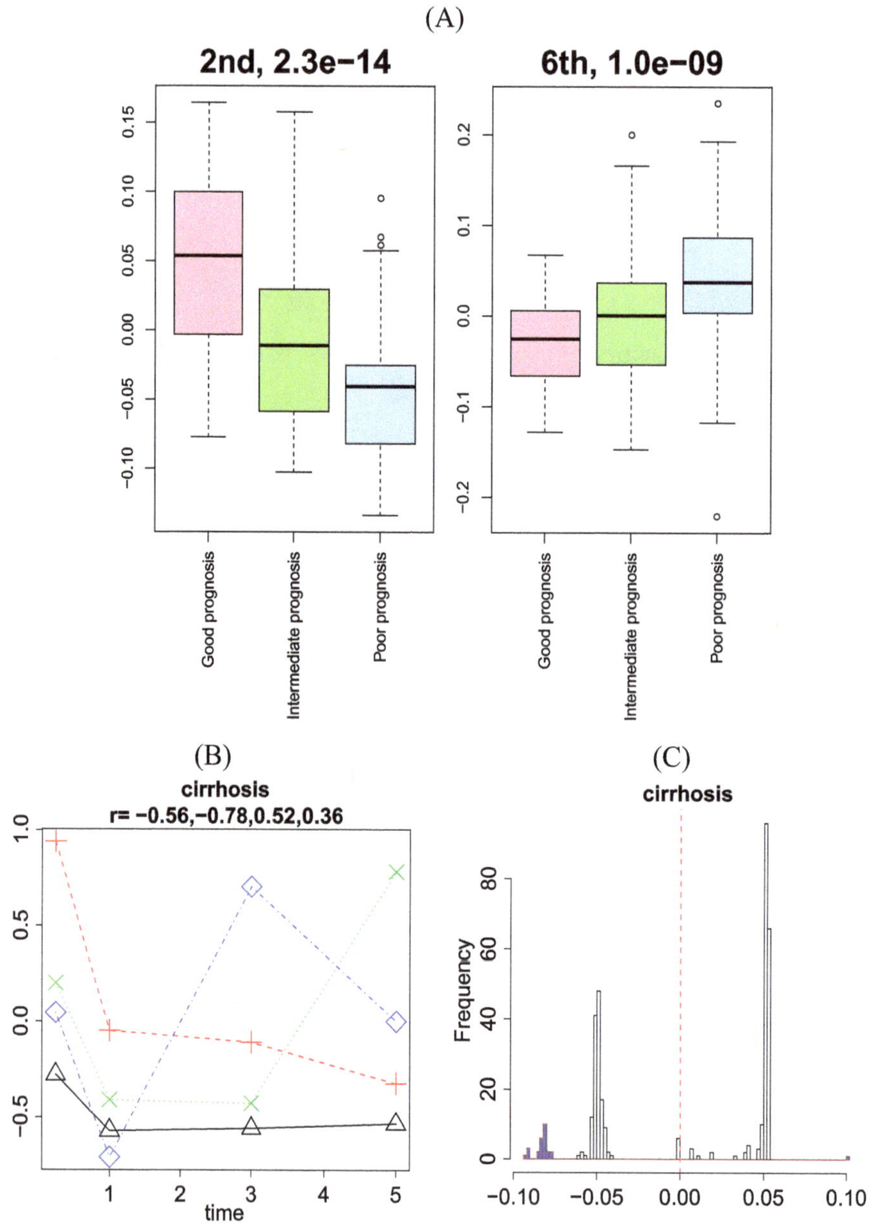

Fig. 7.128 (**a**) $u^{(j_3)}_{\ell_3}, \ell_3 = 2, 6$, P values are computed by categorical regression analysis, Eq. (7.209). (**b**) $u^{(j_2)}_{\ell_2}, 1 \le \ell_2 \le 4$, $\triangle : \ell_2 = 1$, $+ : \ell_2 = 2$, $\times : \ell_2 = 3$, $\diamond : \ell_2 = 4$. r: correlation coefficient. (**c**) Histogram of $u^{(j_1)}_2$. Blue parts are selected ones. Vertical red broken line is 0

Table 7.113 Top 20 $G(\ell_1, 2, \ell_3, \ell_4)$, $\ell_3 = 2, 6$

Rank	1	2	3	4	5	6	7	8	9	10
ℓ_1	2	2	2	2	2	2	2	2	2	2
ℓ_3	2	6	6	6	6	6	2	6	2	2
ℓ_4	2	8	7	6	9	10	6	5	4	3
$G(\ell_1, 2, \ell_3, \ell_4)$	−945	310	278	194	−123	93	77	−76	−73	−67
Rank	11	12	13	14	15	16	17	18	19	20
ℓ_1	2	2	2	2	2	2	2	2	2	2
ℓ_3	6	6	6	6	6	6	6	2	6	2
ℓ_4	4	11	12	17	13	3	16	7	23	5
$G(\ell_1, 2, \ell_3, \ell_4)$	−59	49	43	40	33	−32	−31	27	25	−23

gradually changes, we cannot decide threshold values. Thus, we tentatively decide to select $2 \leq \ell_4 \leq 10$ associated with top ten Gs.

Figure 7.128c shows $\boldsymbol{u}_2^{(j_1)}$. Twenty-seven outlier drugs, $\left|u_{2j_1}^{(j_1)}\right| > 0.075$, blue parts, are selected, by visual inspection, because P values computed from $\boldsymbol{u}_2^{(j_1)}$ and corrected by BH criterion cannot be less than 0.01. On the other hand P values are attributed to ith gene as

$$P_i = P_{\chi^2}\left[> \sum_{2 \leq \ell_4 \leq 10} \left(\frac{u_{\ell_4 i}}{\sigma_{\ell_4}}\right)^2 \right] \tag{7.210}$$

P values are corrected by BH criterion and 132 genes associated with adjusted P values less than 0.01 are selected.

References

1. Acharya, C., Coop, A., Polli, J.E., MacKerell, A.D.: Recent advances in ligand-based drug design: relevance and utility of the conformationally sampled pharmacophore approach. Current Comput. Aided-Drug Design **7**(1), 10–22 (2011). https://doi.org/10.2174/157340911793743547
2. Adebisi, Y.A., Jimoh, N.D., Ogunkola, I.O., Uwizeyimana, T., Olayemi, A.H., Ukor, N.A., Lucero-Prisno, D.E.: The use of antibiotics in COVID-19 management: a rapid review of national treatment guidelines in 10 African countries. Trop. Med. Health **49**(1), 51 (2021). https://doi.org/10.1186/s41182-021-00344-w
3. Ahmad, S.S., Khan, H., Khalid, M., Almalki, A.S.: Emetine and indirubin- 3- monoxime interaction with human brain acetylcholinesterase: a computational and statistical analysis. Cell. Mol. Biol. **67**(4), 106–114 (2022). https://doi.org/10.14715/cmb/2021.67.4.12. https://cellmolbiol.org/index.php/CMB/article/view/4042
4. Albrecht, M., Stichel, D., Müller, B., Merkle, R., Sticht, C., Gretz, N., Klingmüller, U., Breuhahn, K., Matthäus, F.: TTCA: an R package for the identification of differentially expressed genes in time course microarray data. BMC Bioinfor. **18**(1), 33 (2017). https://doi.org/10.1186/s12859-016-1440-8
5. Anderson, A.C.: The process of structure-based drug design. Chem. Biol. **10**(9), 787–797 (2003). https://doi.org/10.1016/j.chembiol.2003.09.002. http://www.sciencedirect.com/science/article/pii/S1074552103001947
6. Baglama, J., Reichel, L., Lewis, B.W.: Irlba: Fast Truncated Singular Value Decomposition and Principal Components Analysis for Large Dense and Sparse Matrices (2022). https://CRAN.R-project.org/package=irlba. R package version 2.3.5.1
7. Baldwin, E., Han, J., Luo, W., Zhou, J., An, L., Liu, J., Zhang, H.H., Li, H.: On fusion methods for knowledge discovery from multiomics data sets. Comput. Struct. Biotechnol. J. **18**, 509–517 (2020). https://doi.org/10.1016/j.csbj.2020.02.011
8. Bandola-Simon, J., Roche, P.A.: Dysfunction of antigen processing and presentation by dendritic cells in cancer. Mol. Immunol. (2018). https://doi.org/10.1016/j.molimm.2018.03.025. http://www.sciencedirect.com/science/article/pii/S0161589018301044
9. Bastos, G.M., Gonçalves, P.B.D., Bordignon, V.: Immunolocalization of the high-mobility group N2 protein and acetylated histone H3K14 in early developing parthenogenetic bovine embryos derived from oocytes of high and low developmental competence. Mol. Reprod. Develop. **75**(2), 282–290 (2008). https://doi.org/10.1002/mrd.20798. https://onlinelibrary.wiley.com/doi/abs/10.1002/mrd.20798

10. Bates, D., Maechler, M., Jagan, M.: Matrix: Sparse and Dense Matrix Classes and Methods (2022). https://CRAN.R-project.org/package=Matrix. R package version 1.5-3
11. Bebbere, D., Ariu, F., Bogliolo, L., Masala, L., Murrone, O., Fattorini, M., Falchi, L., Ledda, S.: Expression of maternally derived KHDC3, NLRP5, OOEP and TLE6 is associated with oocyte developmental competence in the ovine species. BMC Developm. Biol. **14**(1), 40 (2014). https://doi.org/10.1186/s12861-014-0040-y
12. Berger, J.A., Hautaniemi, S., Järvinen, A.K., Edgren, H., Mitra, S.K., Astola, J.: Optimized LOWESS normalization parameter selection for DNA microarray data. BMC Bioinfor. **5**(1), 194 (2004). https://doi.org/10.1186/1471-2105-5-194
13. Blanco-Melo, D., Nilsson-Payant, B.E., Liu, W.C., Uhl, S., Hoagland, D., Møller, R., Jordan, T.X., Oishi, K., Panis, M., Sachs, D., Wang, T.T., Schwartz, R.E., Lim, J.K., Albrecht, R.A., tenOever, B.R.: Imbalanced host response to SARS-CoV-2 drives development of COVID-19. Cell **181**(5), 1036–1045.e9 (2020). https://doi.org/10.1016/j.cell.2020.04.026. https://www.sciencedirect.com/science/article/pii/S009286742030489X
14. Bogus lawska, A., Minasyan, M., Hubalewska-Dydejczyk, A., Gilis-Januszewska, A.: COVID-19 infection in a patient with Cushing's disease on osilodrostat treatment. Endokrynologia Polska **74**(3), 342–343 (2023). https://journals.viamedica.pl/endokrynologia_polska/article/view/EP.a2023.0041
15. Cadenas, J., Pors, S.E., Kumar, A., Kalra, B., Kristensen, S.G., Andersen, C.Y., Mamsen, L.S.: Concentrations of oocyte secreted GDF9 and BMP15 decrease with MII transition during human IVM. Reprod. Biol. Endocrinol. **20**(1), 126 (2022). https://doi.org/10.1186/s12958-022-01000-6
16. Cai, H., Liu, B., Wang, H., Sun, G., Feng, L., Chen, Z., Zhou, J., Zhang, J., Zhang, T., He, M., Yang, T., Guo, Q., Teng, Z., Xin, Q., Zhou, B., Zhang, H., Xia, G., Wang, C.: SP1 governs primordial folliculogenesis by regulating pregranulosa cell development in mice. J. Mol. Cell Biol. **12**(3), 230–244 (2019). https://doi.org/10.1093/jmcb/mjz059
17. Cho, D.H., Choi, J., Gwon, J.G.: Atorvastatin reduces the severity of COVID-19: a nationwide, total population-based, case-control study. COVID **2**(3), 398–406 (2022). https://doi.org/10.3390/covid2030028. https://www.mdpi.com/2673-8112/2/3/28
18. Clarke, D.J.B., Marino, G.B., Deng, E.Z., Xie, Z., Evangelista, J.E., Ma'ayan, A.: Rummagene: Mining gene sets from supporting materials of PMC publications. bioRxiv (2023). https://doi.org/10.1101/2023.10.03.560783. https://www.biorxiv.org/content/early/2023/10/05/2023.10.03.560783
19. Cui, X.S., Shen, X.H., Kim, N.H.: High mobility group box 1 (HMGB1) is implicated in preimplantation embryo development in the mouse. Mol. Reprod. Develop. **75**(8), 1290–1299 (2008). https://doi.org/10.1002/mrd.20694. https://onlinelibrary.wiley.com/doi/abs/10.1002/mrd.20694
20. Dhar, R., Kirkpatrick, J., Gilbert, L., Khanna, A., Modi, M.M., Chawla, R.K., Dalal, S., Maturu, V.N., Stern, M., Keppler, O.T., Djukanovic, R., Gadola, S.D.: Doxycycline for the prevention of progression of COVID-19 to severe disease requiring intensive care unit (ICU) admission: A randomized, controlled, open-label, parallel group trial (doxprevent.icu). PLOS ONE **18**(1), 1–16 (2023). https://doi.org/10.1371/journal.pone.0280745
21. Dumont, M., Kipiani, K., Yu, F., Wille, E., Katz, M., Calingasan, N.Y., Gouras, G.K., Lin, M.T., Beal, M.F.: Coenzyme Q10 decreases amyloid pathology and improves behavior in a transgenic mouse model of Alzheimer's disease. J. Alzheimer's Disease **27**(1), 211–223 (2011). https://doi.org/10.3233/jad-2011-110209
22. Durinck, S., Moreau, Y., Kasprzyk, A., Davis, S., De Moor, B., Brazma, A., Huber, W.: BioMart and Bioconductor: a powerful link between biological databases and microarray data analysis. Bioinformatics **21**, 3439–3440 (2005)
23. Eisele, Y.S., Baumann, M., Klebl, B., Nordhammer, C., Jucker, M., Kilger, E.: Gleevec increases levels of the amyloid precursor protein intracellular domain and of the amyloid-β-degrading enzyme neprilysin. Mol. Biol. Cell **18**(9), 3591–3600 (2007). https://doi.org/10.1091/mbc.e07-01-0035. PMID: 17626163

24. Elamir, Y.M., Amir, H., Lim, S., Rana, Y.P., Lopez, C.G., Feliciano, N.V., Omar, A., Grist, W.P., Via, M.A.: A randomized pilot study using calcitriol in hospitalized COVID-19 patients. Bone **154**, 116175 (2022). https://doi.org/10.1016/j.bone.2021.116175. https://www.sciencedirect.com/science/article/pii/S8756328221003410
25. Evans, W.E., Guy, R.K.: Gene expression as a drug discovery tool. Nat. Genet. **36**(3), 214–215 (2004). https://doi.org/10.1038/ng0304-214
26. Faggioli, P.M., Mumoli, N., Mazzone, A.: Iloprost in COVID-19: the rationale of therapeutic benefit. Front. Cardiovasc. Med. **8** (2021). https://doi.org/10.3389/fcvm.2021.649499. https://www.frontiersin.org/articles/10.3389/fcvm.2021.649499
27. Farazi, T.A., Horlings, H.M., ten Hoeve, J.J., Mihailovic, A., Halfwerk, H., Morozov, P., Brown, M., Hafner, M., Reyal, F., van Kouwenhove, M., Kreike, B., Sie, D., Hovestadt, V., Wessels, L.F., van de Vijver, M.J., Tuschl, T.: Microrna sequence and expression analysis in breast tumors by deep sequencing. Cancer Res. **71**(13), 4443–4453 (2011). https://doi.org/10.1158/0008-5472.CAN-11-0608. http://cancerres.aacrjournals.org/content/71/13/4443
28. Farhadi, T.: Advances in protein tertiary structure prediction. Biomed. Biotechnol. Res. J. **2**(1), 20 (2018). https://doi.org/10.4103/bbrj.bbrj_94_17
29. Gasmi, A., Mujawdiya, P.K., Lysiuk, R., Shanaida, M., Peana, M., Gasmi Benahmed, A., Beley, N., Kovalska, N., Bjørklund, G.: Quercetin in the prevention and treatment of coronavirus infections: A focus on SARS-CoV-2. Pharmaceuticals **15**(9) (2022). https://doi.org/10.3390/ph15091049. https://www.mdpi.com/1424-8247/15/9/1049
30. Geula, S., Moshitch-Moshkovitz, S., Dominissini, D., Mansour, A.A., Kol, N., Salmon-Divon, M., Hershkovitz, V., Peer, E., Mor, N., Manor, Y.S., Ben-Haim, M.S., Eyal, E., Yunger, S., Pinto, Y., Jaitin, D.A., Viukov, S., Rais, Y., Krupalnik, V., Chomsky, E., Zerbib, M., Maza, I., Rechavi, Y., Massarwa, R., Hanna, S., Amit, I., Levanon, E.Y., Amariglio, N., Stern-Ginossar, N., Novershtern, N., Rechavi, G., Hanna, J.H.: m^6A mRNA methylation facilitates resolution of naïve pluripotency toward differentiation. Science **347**(6225), 1002–1006 (2015). https://doi.org/10.1126/science.1261417. https://www.science.org/doi/abs/10.1126/science.1261417
31. Gimeno, A., Mestres-Truyol, J., Ojeda-Montes, M.J., Macip, G., Saldivar-Espinoza, B., Cereto-Massagué, A., Pujadas, G., Garcia-Vallvé, S.: Prediction of novel inhibitors of the main protease (M-pro) of SARS-CoV-2 through consensus docking and drug reposition. Int. J. Mol. Sci. **21**(11) (2020). https://doi.org/10.3390/ijms21113793. https://www.mdpi.com/1422-0067/21/11/3793
32. Griffith, N.T., Varela-Nallar, L., Dinamarca, C.M., Inestrosa, C.N.: Neurobiological effects of hyperforin and its potential in Alzheimers disease therapy. Current Med. Chem. **17**(5), 391–406 (2010). https://doi.org/10.2174/092986710790226156. http://www.eurekaselect.com/article/15778
33. Han, H., Cho, J.W., Lee, S., Yun, A., Kim, H., Bae, D., Yang, S., Kim, C.Y., Lee, M., Kim, E., Lee, S., Kang, B., Jeong, D., Kim, Y., Jeon, H.N., Jung, H., Nam, S., Chung, M., Kim, J.H., Lee, I.: TRRUST v2: an expanded reference database of human and mouse transcriptional regulatory interactions. Nucl. Acids Res. **46**(D1), D380–D386 (2017). https://doi.org/10.1093/nar/gkx1013
34. Hao, Y., Hao, S., Andersen-Nissen, E., Mauck, W.M., Zheng, S., Butler, A., Lee, M.J., Wilk, A.J., Darby, C., Zager, M., Hoffman, P., Stoeckius, M., Papalexi, E., Mimitou, E.P., Jain, J., Srivastava, A., Stuart, T., Fleming, L.M., Yeung, B., Rogers, A.J., McElrath, J.M., Blish, C.A., Gottardo, R., Smibert, P., Satija, R.: Integrated analysis of multimodal single-cell data. Cell **184**(13), 3573–3587.e29 (2021). https://doi.org/https://doi.org/10.1016/j.cell.2021.04.048. https://www.sciencedirect.com/science/article/pii/S0092867421005833
35. Heuvel, C.V.D., Donkin, J.J., Finnie, J.W., Blumbergs, P.C., Kuchel, T., Koszyca, B., Manavis, J., Jones, N.R., Reilly, P.L., Vink, R.: Downregulation of amyloid precursor protein (APP) expression following post-traumatic cyclosporin-a administration. J. Neurotrauma **21**(11), 1562–1572 (2004). https://doi.org/10.1089/neu.2004.21.1562. PMID: 15684649
36. Ho, T.K.: Random decision forests. In: Proceedings of 3rd International Conference on Document Analysis and Recognition, vol. 1, pp. 278–282. IEEE (1995)

37. Jamal, Q.M.S., Ahmad, V., Alharbi, A.H., Ansari, M.A., Alzohairy, M.A., Almatroudi, A., Alghamdi, S., Alomary, M.N., AlYahya, S., Shesha, N.T., Rehman, S.: Therapeutic development by repurposing drugs targeting SARS-CoV-2 spike protein interactions by simulation studies. Saudi J. Biol. Sci. **28**(8), 4560–4568 (2021). https://doi.org/10.1016/j.sjbs.2021.04.057. https://www.sciencedirect.com/science/article/pii/S1319562X21003181
38. Jareborg, N., Birney, E., Durbin, R.: Comparative analysis of noncoding regions of 77 orthologous mouse and human gene pairs. Genome Res. **9**(9), 815–824 (1999). https://doi.org/10.1101/gr.9.9.815. http://genome.cshlp.org/content/9/9/815.abstract
39. Jepsen, W.M., De Both, M., Siniard, A.L., Ramsey, K., Piras, I.S., Naymik, M., Henderson, A., Huentelman, M.J.: Adenosine triphosphate binding cassette subfamily C member 1 (ABCC1) overexpression reduces APP processing and increases alpha- versus beta-secretase activity, in vitro. Biol. Open **10**(1), bio054627 (2021). https://doi.org/10.1242/bio.054627
40. Jin, H.Y., Gonzalez-Martin, A., Miletic, A.V., Lai, M., Knight, S., Sabouri-Ghomi, M., Head, S.R., Macauley, M.S., Rickert, R.C., Xiao, C.: Transfection of microrna mimics should be used with caution. Front. Genet. **6**, 340 (2015). https://doi.org/10.3389/fgene.2015.00340. https://www.frontiersin.org/article/10.3389/fgene.2015.00340
41. Jonic, S., Vénien-Bryan, C.: Protein structure determination by electron cryo-microscopy. Curr. Opin. Pharmacol. **9**(5), 636–642 (2009). https://doi.org/10.1016/j.coph.2009.04.006
42. Kabir, E.R., Mustafa, N., Nausheen, N., Sharif Siam, M.K., Syed, E.U.: Exploring existing drugs: proposing potential compounds in the treatment of COVID-19. Heliyon **7**(2), e06284 (2021). https://doi.org/10.1016/j.heliyon.2021.e06284. https://www.sciencedirect.com/science/article/pii/S2405844021003893
43. Kan, R., Yurttas, P., Kim, B., Jin, M., Wo, L., Lee, B., Gosden, R., Coonrod, S.A.: Regulation of mouse oocyte microtubule and organelle dynamics by PADI6 and the cytoplasmic lattices. Develop. Biol. **350**(2), 311–322 (2011). https://doi.org/10.1016/j.ydbio.2010.11.033. https://www.sciencedirect.com/science/article/pii/S0012160610012388
44. Kang, H., Shokhirev, M.N., Xu, Z., Chandran, S., Dixon, J.R., Hetzer, M.W.: Dynamic regulation of histone modifications and long-range chromosomal interactions during postmitotic transcriptional reactivation. Genes Develop. **34**(13-14), 913–930 (2020). https://doi.org/10.1101/gad.335794.119. http://genesdev.cshlp.org/content/34/13-14/913.abstract
45. Keeney, J.T., Ren, X., Warrier, G., Noel, T., Powell, D.K., Brelsfoard, J.M., Sultana, R., Saatman, K.E., St. Clair, D.K., Butterfield, D.A.: Doxorubicin-induced elevated oxidative stress and neurochemical alterations in brain and cognitive decline: protection by MESNA and insights into mechanisms of chemotherapy-induced cognitive impairment ("chemobrain"). Oncotarget **9**(54), 30324–30339 (2018). https://doi.org/10.18632/oncotarget.25718. https://www.oncotarget.com/article/25718/
46. Klami, A., Bouchard, G., Tripathi, A.: Group-sparse Embeddings in Collective Matrix Factorization. In Proceedings of International Conference on Learning Representations (ICLR) 2014. International Conference on Learning Representations, Banff, Canada, 14/04/2014 (2014). http://arxiv.org/pdf/1312.5921v2
47. Körner, R.W., Majjouti, M., Alcazar, M.A.A., Mahabir, E.: Of mice and men: the coronavirus MHV and mouse models as a translational approach to understand SARS-CoV-2. Viruses **12**(8) (2020). https://doi.org/10.3390/v12080880. https://www.mdpi.com/1999-4915/12/8/880
48. Kumar, R., Oliver, C., Brun, C., Juarez-Martinez, A.B., Tarabay, Y., Kadlec, J., de Massy, B.: Mouse REC114 is essential for meiotic DNA double-strand break formation and forms a complex with mei4. Life Sci. Alliance **1**(6) (2018). https://doi.org/10.26508/lsa.201800259. https://www.life-science-alliance.org/content/1/6/e201800259
49. Lachmann, A., Rouillard, A.D., Monteiro, C.D., Gundersen, G.W., Jagodnik, K.M., Jones, M.R., Kuleshov, M.V., McDermott, M.G., Fernandez, N.F., Duan, Q., Jenkins, S.L., Koplev, S., Wang, Z., Ma'ayan, A.: Enrichr: a comprehensive gene set enrichment analysis web server 2016 update. Nucl. Acids Res. **44**(W1), W90–W97 (2016). https://doi.org/10.1093/nar/gkw377

50. Lee, J., Hyeon, D.Y., Hwang, D.: Single-cell multiomics: technologies and data analysis methods. Experim. Molecul. Med. **52**(9), 1428–1442 (2020). https://doi.org/10.1038/s12276-020-0420-2
51. Leppäaho, E., ud din, M.A., Kaski, S.: GFA: Exploratory analysis of multiple data sources with group factor analysis. J. Mach. Learn. Res. **18**(39), 1–5 (2017). http://jmlr.org/papers/v18/16-509.html
52. Li, G., Xu, C., Lin, X., Qu, L., Xia, D., Hongdu, B., Xia, Y., Wang, X., Lou, Y., He, Q., Ma, D., Chen, Y.: Deletion of Pdcd5 in mice led to the deficiency of placenta development and embryonic lethality. Cell Death Disease **8**(5), e2811–e2811 (2017). https://doi.org/10.1038/cddis.2017.124
53. Li, W., Li, Q., Xu, X., Wang, C., Hu, K., Xu, J.: Novel mutations in TUBB8 and ZP3 cause human oocyte maturation arrest and female infertility. Eur. J. Obstet. Gynecol. Reprod. Biol. **279**, 132–139 (2022). https://doi.org/10.1016/j.ejogrb.2022.10.017
54. Liaw, A., Wiener, M.: Classification and regression by randomforest. R News **2**(3), 18–22 (2002). https://CRAN.R-project.org/doc/Rnews/
55. Liu, J., Musialski, P., Wonka, P., Ye, J.: Tensor completion for estimating missing values in visual data. IEEE Trans. Pattern Analy. Mach. Intell. **35**(1), 208–220 (2013). https://doi.org/10.1109/TPAMI.2012.39
56. Liu, J., Eckert, M.A., Harada, B.T., Liu, S.M., Lu, Z., Yu, K., Tienda, S.M., Chryplewicz, A., Zhu, A.C., Yang, Y., Huang, J.T., Chen, S.M., Xu, Z.G., Leng, X.H., Yu, X.C., Cao, J., Zhang, Z., Liu, J., Lengyel, E., He, C.: m^6A mRNA methylation regulates AKT activity to promote the proliferation and tumorigenicity of endometrial cancer. Nature Cell Biol. **20**(9), 1074–1083 (2018). https://doi.org/10.1038/s41556-018-0174-4
57. Love, M.I., Huber, W., Anders, S.: Moderated estimation of fold change and dispersion for RNA-seq data with DESeq2. Genome Biol. **15**(12), 550 (2014). https://doi.org/7:10.1186/s13059-014-0550-8
58. Lukačišin, M., Bollenbach, T.: Emergent gene expression responses to drug combinations predict higher-order drug interactions. Cell Syst. **9**(5), 423–433.e3 (2019). https://doi.org/10.1016/j.cels.2019.10.004
59. Maglott, D., Ostell, J., Pruitt, K.D., Tatusova, T.: Entrez gene: gene-centered information at NCBI. Nucl. Acids Res. **39**(suppl_1), D52–D57 (2011). https://doi.org/10.1093/nar/gkq1237
60. Mao, R., Xu, S., Sun, G., Yu, Y., Zuo, Z., Wang, Y., Yang, K., Zhang, Z., Yang, W.: Triptolide injection reduces Alzheimer's disease-like pathology in mice. Synapse **77**(3), e22261 (2023). https://doi.org/10.1002/syn.22261. https://onlinelibrary.wiley.com/doi/abs/10.1002/syn.22261
61. Marciniec, K., Beberok, A., Pęcak, P., Boryczka, S., Wrześniok, D.: Ciprofloxacin and moxifloxacin could interact with SARS-CoV-2 protease: preliminary in silico analysis. Pharmacol. Rep. **72**(6), 1553–1561 (2020). https://doi.org/10.1007/s43440-020-00169-0
62. Martens, K., Vanhulle, E., Viskens, A.S., Hellings, P., Vermeire, K.: Fluticasone propionate suppresses the SARS-CoV-2 induced increase in respiratory epithelial permeability in vitro. Rhinol. J. **0**(0), 0–0 (2022). https://doi.org/10.4193/rhin22.223
63. Matos, B., Publicover, S.J., Castro, L.F.C., Esteves, P.J., Fardilha, M.: Brain and testis: more alike than previously thought? Open Biol. **11**(6), 200322 (2021). https://doi.org/10.1098/rsob.200322. https://royalsocietypublishing.org/doi/abs/10.1098/rsob.200322
64. McInnes, L., Healy, J., Saul, N., Großberger, L.: UMAP: Uniform manifold approximation and projection. J. Open Source Softw. **3**(29), 861 (2018). https://doi.org/10.21105/joss.00861
65. Merritt, M.A., Cramer, D.W.: Molecular pathogenesis of endometrial and ovarian cancer. Cancer Biomarkers **9**(1-6), 287–305 (2011). https://doi.org/10.3233/cbm-2011-0167
66. Montoya, S.E., Aston, C.E., DeKosky, S.T., Kamboh, M.I., Lazo, J.S., Ferrell, R.E.: Bleomycin hydrolase is associated with risk of sporadic Alzheimer's disease. Nat. Genetics **18**(3), 211–212 (1998). https://doi.org/10.1038/ng0398-211
67. Moustafa, A.A., Gilbertson, M.W., Orr, S.P., Herzallah, M.M., Servatius, R.J., Myers, C.E.: A model of amygdala-hippocampal-prefrontal interaction in fear conditioning and extinction in animals. Brain Cognit. **81**(1), 29 – 43 (2013). https://doi.org/10.1016/j.bandc.2012.10.005. http://www.sciencedirect.com/science/article/pii/S0278262612001418

68. National Toxicology Program: DrugMatrix (2010). https://ntp.niehs.nih.gov/drugmatrix/index.html
69. Ng, K.L., Taguchi, Y.H.: Identification of miRNA signatures for kidney renal clear cell carcinoma using the tensor-decomposition method. Sci. Rep. **10**(1), 15149 (2020). https://doi.org/10.1038/s41598-020-71997-6
70. Orienti, I., Gentilomi, G.A., Farruggia, G.: Pulmonary delivery of fenretinide: a possible adjuvant treatment in COVID-19. Int. J. Mol. Sci. **21**(11) (2020). https://doi.org/10.3390/ijms21113812. https://www.mdpi.com/1422-0067/21/11/3812
71. Palomares, M.A., Dalmasso, C., Bonnet, E., Derbois, C., Brohard-Julien, S., Ambroise, C., Battail, C., Deleuze, J.F., Olaso, R.: Systematic analysis of TruSeq, SMARTer and SMARTer Ultra-Low RNA-seq kits for standard, low and ultra-low quantity samples. Sci. Rep. **9**(1), 7550 (2019). https://doi.org/10.1038/s41598-019-43983-0
72. Patalano, S., Vlasova, A., Wyatt, C., Ewels, P., Camara, F., Ferreira, P.G., Asher, C.L., Jurkowski, T.P., Segonds-Pichon, A., Bachman, M., González-Navarrete, I., Minoche, A.E., Krueger, F., Lowy, E., Marcet-Houben, M., Rodriguez-Ales, J.L., Nascimento, F.S., Balasubramanian, S., Gabaldon, T., Tarver, J.E., Andrews, S., Himmelbauer, H., Hughes, W.O.H., Guigó, R., Reik, W., Sumner, S.: Molecular signatures of plastic phenotypes in two eusocial insect species with simple societies. Proc. Natl. Acad. Sci. **112**(45), 13970–13975 (2015). https://doi.org/10.1073/pnas.1515937112. https://www.pnas.org/content/112/45/13970
73. Peng, H., Liu, H., Liu, F., Gao, Y., Chen, J., Huo, J., Han, J., Xiao, T., Zhang, W.: NLRP2 and FAF1 deficiency blocks early embryogenesis in the mouse. Reproduction **154**(3), 245 – 251 (2017). https://doi.org/10.1530/REP-16-0629. https://rep.bioscientifica.com/view/journals/rep/154/3/REP-16-0629.xml
74. Pepper, S.D., Saunders, E.K., Edwards, L.E., Wilson, C.L., Miller, C.J.: The utility of mas5 expression summary and detection call algorithms. BMC Bioinf. **8**(1), 273 (2007). https://doi.org/10.1186/1471-2105-8-273
75. Pohl, M.O., Martin-Sancho, L., Ratnayake, R., White, K.M., Riva, L., Chen, Q.Y., Lieber, G., Busnadiego, I., Yin, X., Lin, S., Pu, Y., Pache, L., Rosales, R., Déjosez, M., Qin, Y., Jesus, P.D.D., Beall, A., Yoh, S., Hale, B.G., Zwaka, T.P., Matsunaga, N., García-Sastre, A., Stertz, S., Chanda, S.K., Luesch, H.: Sec61 inhibitor apratoxin s4 potently inhibits SARS-CoV-2 and exhibits broad-spectrum antiviral activity. ACS Infect. Diseases **8**(7), 1265–1279 (2022). https://doi.org/10.1021/acsinfecdis.2c00008
76. Qu, Y., He, F., Chen, Y.: Different effects of the probe summarization algorithms plier and rma on high-level analysis of affymetrix exon arrays. BMC Bioinfor. **11**(1), 211 (2010). https://doi.org/10.1186/1471-2105-11-211
77. Qu, N., Bo, X., Li, B., Ma, L., Wang, F., Zheng, Q., Xiao, X., Huang, F., Shi, Y., Zhang, X.: Role of N6-methyladenosine (m6A) methylation regulators in hepatocellular carcinoma. Front. Oncol. **11** (2021). https://doi.org/10.3389/fonc.2021.755206. https://www.frontiersin.org/articles/10.3389/fonc.2021.755206
78. Reese, J.T., Coleman, B., Chan, L., Blau, H., Callahan, T.J., Cappelletti, L., Fontana, T., Bradwell, K.R., Harris, N.L., Casiraghi, E., Valentini, G., Karlebach, G., Deer, R., McMurry, J.A., Haendel, M.A., Chute, C.G., Pfaff, E., Moffitt, R., Spratt, H., Singh, J.A., Mungall, C.J., Williams, A.E., Robinson, P.N.: NSAID use and clinical outcomes in COVID-19 patients: a 38-center retrospective cohort study. Virol. J. **19**(1), 84 (2022). https://doi.org/10.1186/s12985-022-01813-2
79. Ritchie, M.E., Phipson, B., Wu, D., Hu, Y., Law, C.W., Shi, W., Smyth, G.K.: limma powers differential expression analyses for RNA-sequencing and microarray studies. Nucl. Acids Res. **43**(7), e47 (2015). https://doi.org/10.1093/nar/gkv007
80. Rodriguez, S., Hug, C., Todorov, P., Moret, N., Boswell, S.A., Evans, K., Zhou, G., Johnson, N.T., Hyman, B.T., Sorger, P.K., Albers, M.W., Sokolov, A.: Machine learning identifies candidates for drug repurposing in Alzheimer's disease. Nat. Commun. **12**(1), 1033 (2021). https://doi.org/10.1038/s41467-021-21330-0
81. Roider, H.G., Pavlova, N., Kirov, I., Slavov, S., Slavov, T., Uzunov, Z., Weiss, B.: Drug2gene: an exhaustive resource to explore effectively the drug-target relation network. BMC Bioinf. **15**(1), 68 (2014). https://doi.org/10.1186/1471-2105-15-68

82. Roy, S.S., Taguchi, Y.H.: Identification of genes associated with altered gene expression and m6A profiles during hypoxia using tensor decomposition based unsupervised feature extraction. Sci. Rep. **11**(1), 8909 (2021). https://doi.org/10.1038/s41598-021-87779-7
83. Saliani, M., Mirzaiebadizi, A., Mosaddeghzadeh, N., Ahmadian, M.R.: RHO GTPase-related long noncoding RNAs in human cancers. Cancers **13**(21) (2021). https://doi.org/10.3390/cancers13215386. https://www.mdpi.com/2072-6694/13/21/5386
84. Samsudin, F., Raghuvamsi, P., Petruk, G., Puthia, M., Petrlova, J., MacAry, P., Anand, G.S., Bond, P.J., Schmidtchen, A.: SARS-CoV-2 spike protein as a bacterial lipopolysaccharide delivery system in an overzealous inflammatory cascade. J. Mol. Cell Biol. **14**(9), mjac058 (2022). https://doi.org/10.1093/jmcb/mjac058
85. Samy, A., Maher, M.A., Abdelsalam, N.A., Badr, E.: SARS-CoV-2 potential drugs, drug targets, and biomarkers: a viral-host interaction network-based analysis. Sci. Rep. **12**(1), 11934 (2022). https://doi.org/10.1038/s41598-022-15898-w
86. Shahabadi, N., Zendehcheshm, S., Mahdavi, M., Khademi, F.: Inhibitory activity of FDA-approved drugs cetilistat, abiraterone, diiodohydroxyquinoline, bexarotene, remdesivir, and hydroxychloroquine on COVID-19 main protease and human ACE2 receptor: a comparative in silico approach. Inf. Med. Unlocked **26**, 100745 (2021). https://doi.org/10.1016/j.imu.2021.100745. https://www.sciencedirect.com/science/article/pii/S2352914821002215
87. Shi, Y., Dou, Y., Zhang, J., Qi, J., Xin, Z., Zhang, M., Xiao, Y., Ci, W.: The RNA N6-methyladenosine methyltransferase METTL3 promotes the progression of kidney cancer via N6-methyladenosine-dependent translational enhancement of ABCD1. Front. Cell Develop. Biol. **9** (2021). https://doi.org/10.3389/fcell.2021.737498. https://www.frontiersin.org/articles/10.3389/fcell.2021.737498
88. Singh, Y., Gupta, G., Shrivastava, B., Dahiya, R., Tiwari, J., Ashwathanarayana, M., Sharma, R.K., Agrawal, M., Mishra, A., Dua, K.: Calcitonin gene-related peptide (CGRP): A novel target for Alzheimer's disease. CNS Neurosci. Therapeut. **23**(6), 457–461 (2017). https://doi.org/10.1111/cns.12696. https://onlinelibrary.wiley.com/doi/abs/10.1111/cns.12696
89. Song, Y., Milon, B., Ott, S., Zhao, X., Sadzewicz, L., Shetty, A., Boger, E.T., Tallon, L.J., Morell, R.J., Mahurkar, A., Hertzano, R.: A comparative analysis of library prep approaches for sequencing low input translatome samples. BMC Genomics **19**(1), 696 (2018). https://doi.org/10.1186/s12864-018-5066-2
90. Subramanian, A., Narayan, R., Corsello, S.M., Peck, D.D., Natoli, T.E., Lu, X., Gould, J., Davis, J.F., Tubelli, A.A., Asiedu, J.K., Lahr, D.L., Hirschman, J.E., Liu, Z., Donahue, M., Julian, B., Khan, M., Wadden, D., Smith, I.C., Lam, D., Liberzon, A., Toder, C., Bagul, M., Orzechowski, M., Enache, O.M., Piccioni, F., Johnson, S.A., Lyons, N.J., Berger, A.H., Shamji, A.F., Brooks, A.N., Vrcic, A., Flynn, C., Rosains, J., Takeda, D.Y., Hu, R., Davison, D., Lamb, J., Ardlie, K., Hogstrom, L., Greenside, P., Gray, N.S., Clemons, P.A., Silver, S., Wu, X., Zhao, W.N., Read-Button, W., Wu, X., Haggarty, S.J., Ronco, L.V., Boehm, J.S., Schreiber, S.L., Doench, J.G., Bittker, J.A., Root, D.E., Wong, B., Golub, T.R.: A next generation connectivity map: L1000 platform and the first 1,000,000 profiles. Cell **171**(6), 1437–1452.e17 (2017). https://doi.org/10.1016/j.cell.2017.10.049. http://www.sciencedirect.com/science/article/pii/S0092867417313090
91. Suzuki, A., Kawano, S., Mitsuyama, T., Suyama, M., Kanai, Y., Shirahige, K., Sasaki, H., Tokunaga, K., Tsuchihara, K., Sugano, S., Nakai, K., Suzuki, Y.: DBTSS/DBKERO for integrated analysis of transcriptional regulation. Nucl. Acids Res. **46**(D1), D229–D238 (2018). https://doi.org/10.1093/nar/gkx1001
92. Taguchi, Y.H.: One-class differential expression analysis using tensor decomposition-based unsupervised feature extraction applied to integrated analysis of multiple omics data from 26 lung adenocarcinoma cell lines. In: 2017 IEEE 17th International Conference on Bioinformatics and Bioengineering (BIBE), pp. 131–138 (2017). https://doi.org/10.1109/BIBE.2017.00-66
93. Taguchi, Y.H.: Tensor decomposition-based unsupervised feature extraction applied to matrix products for multi-view data processing. PLOS ONE **12**(8), 1–36 (2017). https://doi.org/10.1371/journal.pone.0183933

94. Taguchi, Y.H.: Tensor decomposition-based unsupervised feature extraction identifies candidate genes that induce post-traumatic stress disorder-mediated heart diseases. BMC Med. Genomics **10**(4), 67 (2017). https://doi.org/10.1186/s12920-017-0302-1
95. Taguchi, Y.H.: Tensor decomposition-based and principal-component-analysis-based unsupervised feature extraction applied to the gene expression and methylation profiles in the brains of social insects with multiple castes. BMC Bioinf. **19**(4), 99 (2018). https://doi.org/10.1186/s12859-018-2068-7
96. Taguchi, Y.H.: Tensor decomposition-based unsupervised feature extraction can identify the universal nature of sequence-nonspecific off-target regulation of mrna mediated by microrna transfection. Cells **7**(6) (2018). https://doi.org/10.3390/cells7060054. http://www.mdpi.com/2073-4409/7/6/54
97. Taguchi, Y.H.: Drug candidate identification based on gene expression of treated cells using tensor decomposition-based unsupervised feature extraction for large-scale data. BMC Bioinf. **19**(13), 388 (2019). https://doi.org/10.1186/s12859-018-2395-8
98. Taguchi, Y.h., Turki, T.: Neurological disorder drug discovery from gene expression with tensor decomposition. Curr. Pharm. Des. **25**(43), 4589–4599 (2019). https://doi.org/10.2174/1381612825666191210160906. http://www.eurekaselect.com/article/102901
99. Taguchi, Y.h., Turki, T.: Tensor decomposition-based unsupervised feature extraction applied to single-cell gene expression analysis. Front. Genet. **10** (2019). https://doi.org/10.3389/fgene.2019.00864. https://www.frontiersin.org/articles/10.3389/fgene.2019.00864
100. Taguchi, Y.h., Turki, T.: A new advanced in silico drug discovery method for novel coronavirus (SARS-CoV-2) with tensor decomposition-based unsupervised feature extraction. PLOS ONE **15**(9), 1–16 (2020). https://doi.org/10.1371/journal.pone.0238907
101. Taguchi, Y.h., Turki, T.: Tensor-decomposition-based unsupervised feature extraction applied to prostate cancer multiomics data. Genes **11**(12) (2020). https://doi.org/10.3390/genes11121493. https://www.mdpi.com/2073-4425/11/12/1493
102. Taguchi, Y.H., Turki, T.: Universal nature of drug treatment responses in drug-tissue-wide model-animal experiments using tensor decomposition-based unsupervised feature extraction. Front. Genet. **11** (2020). https://doi.org/10.3389/fgene.2020.00695. https://www.frontiersin.org/articles/10.3389/fgene.2020.00695
103. Taguchi, Y.H., Turki, T.: Application of tensor decomposition to gene expression of infection of mouse hepatitis virus can identify critical human genes and efffective drugs for SARS-CoV-2 infection. IEEE J. Sel. Top. Signal Process. **15**, 1–1 (2021). https://doi.org/10.1109/jstsp.2021.3061251
104. Taguchi, Y.H., Turki, T.: Mathematical formulation and application of kernel tensor decomposition based unsupervised feature extraction. Knowl.-Based Syst. **217**, 106834 (2021). https://doi.org/10.1016/j.knosys.2021.106834. https://www.sciencedirect.com/science/article/pii/S0950705121000976
105. Taguchi, Y.H., Turki, T.: Novel method for the prediction of drug-drug interaction based on gene expression profiles. Eur. J. Pharmaceut. Sci. **160**, 105742 (2021). https://doi.org/10.1016/j.ejps.2021.105742. https://www.sciencedirect.com/science/article/pii/S0928098721000440
106. Taguchi, Y.h., Turki, T.: Tensor-decomposition-based unsupervised feature extraction in single-cell multiomics data analysis. Genes **12**(9) (2021). https://doi.org/10.3390/genes12091442. https://www.mdpi.com/2073-4425/12/9/1442
107. Taguchi, Y.h., Turki, T.: Unsupervised tensor decomposition-based method to extract candidate transcription factors as histone modification bookmarks in post-mitotic transcriptional reactivation. PLOS ONE **16**(5), 1–20 (2021). https://doi.org/10.1371/journal.pone.0251032
108. Taguchi, Y.h., Turki, T.: A tensor decomposition-based integrated analysis applicable to multiple gene expression profiles without sample matching. Sci. Rep. **12**(1), 21242 (2022). https://doi.org/10.1038/s41598-022-25524-4
109. Taguchi, Y.h., Turki, T.: Novel feature selection method via kernel tensor decomposition for improved multiomics data analysis. BMC Med. Genomics **15**(1), 37 (2022). https://doi.org/10.1186/s12920-022-01181-4

110. Taguchi, Y.h., Dharshini, S.A.P., Gromiha, M.M.: Identification of transcription factors, biological pathways, and diseases as mediated by N6-methyladenosine using tensor decomposition-based unsupervised feature extraction. Appl. Sci. **11**(1) (2021). https://doi.org/10.3390/app11010213. https://www.mdpi.com/2076-3417/11/1/213
111. Tomczak, K., Czerwińska, P., Wiznerowicz, M.: The cancer genome atlas (TCGA): an immeasurable source of knowledge. Contemporary Oncology/Współczesna Onkologia **2015**, 68–77 (2015). https://doi.org/10.5114/wo.2014.47136
112. Tousi, B.: The emerging role of bexarotene in the treatment of Alzheimer's disease: current evidence. Neuropsychiatric Dis. Treat. **2015**, 311 (2015). https://doi.org/10.2147/ndt.s61309
113. Tusher, V.G., Tibshirani, R., Chu, G.: Significance analysis of microarrays applied to the ionizing radiation response. Proc. Natl. Acad. Sci. **98**(9), 5116–5121 (2001). https://doi.org/10.1073/pnas.091062498
114. Ura, H., Togi, S., Niida, Y.: A comparison of mRNA sequencing (RNA-Seq) library preparation methods for transcriptome analysis. BMC Genomics **23**(1), 303 (2022). https://doi.org/10.1186/s12864-022-08543-3
115. Weiner, S.A., Toth, A.L.: Epigenetics in social insects: a new direction for understanding the evolution of castes. Genet. Res. Int. **2012**, 1–11 (2012). https://doi.org/10.1155/2012/609810
116. Weiss, K., Khoshgoftaar, T.M., Wang, D.: A survey of transfer learning. J. Big Data **3**(1), 9 (2016). https://doi.org/10.1186/s40537-016-0043-6
117. Wiener, D., Schwartz, S.: The epitranscriptome beyond m6A. Nat. Rev. Genet. **22**(2), 119–131 (2020). https://doi.org/10.1038/s41576-020-00295-8
118. Wu, R., Li, A., Sun, B., Sun, J.G., Zhang, J., Zhang, T., Chen, Y., Xiao, Y., Gao, Y., Zhang, Q., Ma, J., Yang, X., Liao, Y., Lai, W.Y., Qi, X., Wang, S., Shu, Y., Wang, H.L., Wang, F., Yang, Y.G., Yuan, Z.: A novel m^6A reader Prrc2a controls oligodendroglial specification and myelination. Cell Res. **29**(1), 23–41 (2019). https://doi.org/10.1038/s41422-018-0113-8
119. Xie, X., Luo, X., Xie, M., Liu, Y., Wu, T.: Risk of lung cancer in parkinson's disease. Oncotarget **7**(47) (2016). https://doi.org/10.18632/oncotarget.12964. https://doi.org/10.18632/oncotarget.12964
120. Xu, H., Dzhashiashvili, Y., Shah, A., Kunjamma, R.B., lan Weng, Y., Elbaz, B., Fei, Q., Jones, J.S., Li, Y.I., Zhuang, X., li Ming, G., He, C., Popko, B.: m6A mRNA methylation is essential for oligodendrocyte maturation and CNS myelination. Neuron **105**(2), 293–309.e5 (2020). https://doi.org/10.1016/j.neuron.2019.12.013
121. Yadalam, P.K., Balaji, T.M., Varadarajan, S., Alzahrani, K.J., Al-Ghamdi, M.S., Baeshen, H.A., Alfarhan, M.F.A., Khurshid, Z., Bhandi, S., Jagannathan, R., Patil, V.R., Raj, A.T., Ratnayake, J., Patil, S.: Assessing the therapeutic potential of agomelatine, ramelteon, and melatonin against SARS-CoV-2. Saudi J. Biol. Sci. **29**(5), 3140–3150 (2022). https://doi.org/10.1016/j.sjbs.2022.01.049. https://www.sciencedirect.com/science/article/pii/S1319562X22000493
122. Yamanishi, Y., Kotera, M., Moriya, Y., Sawada, R., Kanehisa, M., Goto, S.: DINIES: drug-target interaction network inference engine based on supervised analysis. Nucl. Acids Res. **42**(W1), W39–W45 (2014). https://doi.org/10.1093/nar/gku337
123. Yan, H., Bonasio, R., Simola, D.F., Liebig, J., Berger, S.L., Reinberg, D.: DNA methylation in social insects: how epigenetics can control behavior and longevity. Ann. Rev. Entomol. **60**(1), 435–452 (2015). https://doi.org/10.1146/annurev-ento-010814-020803. PMID: 25341091
124. Yoo, M., Shin, J., Kim, J., Ryall, K.A., Lee, K., Lee, S., Jeon, M., Kang, J., Tan, A.C.: DSigDB: drug signatures database for gene set analysis. Bioinformatics **31**(18), 3069–3071 (2015). https://doi.org/10.1093/bioinformatics/btv313
125. Yu, Q., Cheng, X.: Hydroxyurea-induced membrane fluidity decreasing as a characterization of neuronal membrane aging in Alzheimer's disease. Aging **13**(9), 12817–12832. https://doi.org/10.18632/aging.202949
126. Zhang, G.N., Zhao, J., Li, Q., Wang, M., Zhu, M., Wang, J., Cen, S., Wang, Y.: Discovery and optimization of 2-((1H-indol-3-yl)thio)-N-benzyl-acetamides as novel SARS-CoV-2 RdRp inhibitors. Eur. J. Med. Chem. **223**, 113622 (2021). https://doi.org/10.1016/j.ejmech.2021.113622. https://www.sciencedirect.com/science/article/pii/S0223523421004712

127. Zhang, K., Chen, R., Jiang, Q.: Allopurinol increased the risk of COVID-19 hospitalization mediated by e-selectin downregulation. J. Infect. **86**(6), 620–621 (2023). https://doi.org/10.1016/j.jinf.2023.02.030
128. Zhou, Y., Zhou, B., Pache, L., Chang, M., Khodabakhshi, A.H., Tanaseichuk, O., Benner, C., Chanda, S.K.: Metascape provides a biologist-oriented resource for the analysis of systems-level data sets. Nat. Commun. **10**(1), 1523 (2019). https://doi.org/10.1038/s41467-019-09234-6
129. Zhou, M., Dong, M., Yang, X., Gong, J., Liao, X., Zhang, Q., Liu, Z.: The emerging roles and mechanism of m6a in breast cancer progression. Front. Genet. **13** (2022). https://doi.org/10.3389/fgene.2022.983564. https://www.frontiersin.org/articles/10.3389/fgene.2022.983564
130. Zuber, M., Yasui, W., Tan, E.M., Ryoji, M.: Quantitation and subcellular localization of proliferating cell nuclear antigen (PCNA/cyclin) in oocytes and eggs of Xenopus laevis. Experim. Cell Res. **182**(2), 384–393 (1989). https://doi.org/10.1016/0014-4827(89)90243-7. https://www.sciencedirect.com/science/article/pii/0014482789902437

Chapter 8
Theoretical Investigation of TD- and PCA-Based Unsupervised FE

Did she make that promise to you or to the person you were pretending to be?
Yang Xiao Long, RWBY, Volume 6, Episode 78

8.1 Introduction

In the preceding two chapters, we applied PCA- and TD-based unsupervised FE to a wide range of genomic problems. When I first started applying PCA to these problems, the microarray was the main method. Since then, although the primary method shifted from microarray to high throughput sequencing, now further to single-cell sequencing, PCA- and TD-based unsupervised FE remained effective. What kind of hidden mechanism allowed these methods to be even applicable to rapidly growing measurement methods? There should be some reasons behind this robustness. In this chapter, we try to figure out what the principal mechanism behind these technologies is as much as possible with developing new methodologies.

8.2 Projection in Genomic Analysis

8.2.1 Projection Pursuit

Projection pursuit [6] (PP) is an old fashioned technology proposed several years ago that aims to tackle the so-called large p small n problem, i.e., the main topics aimed to be solved in this book. In contrast to other statistical methods supposed to be applied to the "large p small n" problem where what to be aimed is clearly presented, e.g., minimization of residues, PP is a quite subjective method since it tries to find "interesting" projections of high dimensional data sets; what is interesting is, of course, quite person- or study-dependent. In this subsection, we aim to interpret unsupervised FE based on PCA in the context of PP [26].

Y-h. Taguchi, *Unsupervised Feature Extraction Applied to Bioinformatics*, Unsupervised and Semi-Supervised Learning,
https://doi.org/10.1007/978-3-031-60982-4_8

Regardless to say, PCA and SVD, and also HOSVD since HOSVD is essentially SVD of unfolded tensor, are in some sense, projection. In fact, PCA can be interpreted as a branch of PP [6]. PCA as PP is to look for the projection with the maximum standard deviation (or minimum of residues) as "interesting" projection direction. PP usually seeks projection with non-Gaussian distribution; PCA- and TD-based unsupervised FE seek the projection with Gaussian distribution, but also associated with some outliers supposed to be selected features in the context of PP. "Interesting" projection in PCA- and TD-based FE is that coincides with class labels (e.g., distinction between patients and healthy control). Thus, PCA- and TD-based unsupervised FE can be regarded as a variant of PP.

8.2.2 Kidney Cancer

To demonstrate in which sense TD-based unsupervised FE can be regarded as PP, we employ data sets used in Sect. 7.10. To see this, we apply PP to these data sets. In PP, for the data set formatted as matrix $x_{ij} \in \mathbb{R}^{N \times M}$ where N is the number of dimensions and M is the number of samples, the projection vector $\boldsymbol{y} \in \mathbb{R}^M$ is defined and the projection $\boldsymbol{b} = X\boldsymbol{y} \in \mathbb{R}^N$ is computed where X is a matrix $N \times M$ whose element is x_{ij}; $\boldsymbol{y}$ is selected to represent "interesting" projection. In the study in Sect. 7.10, the "interesting" projection is supposed to represent the distinction between healthy controls and patients. In TD-based unsupervised FE, $\boldsymbol{y}$ is sought within singular value vectors attributed to samples, js. To emphasize TD-based unsupervised FE as PP, instead of seeking $\boldsymbol{y}$ within singular value vectors attributed to samples, j, we manually define $\boldsymbol{y}$ as

$$y_j = \begin{cases} -\frac{M}{M_N} & \text{when the } j\text{th sample is normal kidney} \\ \frac{M}{M_T} & \text{when the } j\text{th sample is tumor,} \end{cases} \tag{8.1}$$

where M_N and M_T are the number of normal kidney samples and tumor samples, respectively. Then $\boldsymbol{b}$ was computed as

$$b_i = \sum_{j=1}^{M} y_j x_{ij} \text{ (for mRNA)} \tag{8.2}$$

$$b_k = \sum_{j=1}^{M} y_j x_{kj} \text{ (for miRNA).} \tag{8.3}$$

P values are attributed to is and ks as

$$P_i = P_{\chi^2}\left[> \left(\frac{b_i}{\sigma_b}\right)^2 \right] \tag{8.4}$$

$$P_k = P_{\chi^2}\left[> \left(\frac{b_k}{\sigma_b}\right)^2 \right]. \tag{8.5}$$

P values were corrected as usual, and is (mRNAs) and ks (miRNAs) associated with adjusted P values less than 0.01 were selected. As a result, 78 mRNAs and 13 miRNAs for the first data set (TCGA) and 194 mRNAs and 3 miRNAs for the second data set (GEO), associated with adjusted P values less than 0.01, respectively. Table 8.1 shows the confusion matrix of selected mRNAs and miRNAs between those by PP and those in Sect. 7.10. Since it is obvious that they are highly coincident with each other, TD-based unsupervised FE can be regarded as a PP conceptually.

Before closing this subsection, we investigated if the overall distribution of P values is coincident with each other between TD-based unsupervised FE and PP. To do this, we draw QQplot [34] that visualizes the coincidence between two distributions for P values (Fig. 8.1). When two distributions are coincident with each other, QQplot exhibits straight lines. Since QQplots are straight lines, PP can produce the almost identical distribution of P values to those of TD-based unsupervised FE.

8.2.3 COVID-19

To avoid the possibility that the demonstration in the previous subsection is accidentally successful, we deal with yet another data set, i.e., that in Sect. 7.15.2. Since the labels used in Sect. 7.15.2 are not vectors but tensors, we need projection tensor, y_{jkm}, instead of projection vector, y_j, to get projection $\boldsymbol{b}$. $y_{jkm} \in \mathbb{R}^{5\times2\times3}$ is defined as

$$y_{jkm} = \alpha_j \beta_k \gamma_m \tag{8.6}$$

$$\alpha_j = 1 \in \mathbb{R}^5 \tag{8.7}$$

$$\beta_k = (-1)^k \in \mathbb{R}^2 \tag{8.8}$$

$$\gamma_m = 1 \in \mathbb{R}^3, \tag{8.9}$$

which means the independence of j (cell lines) and m (replicates) and distinction between infection and not-infection (k). The projection $\boldsymbol{b}$ was computed as

$$b_i = \sum_{j=1}^{5}\sum_{k=1}^{2}\sum_{m=1}^{3} y_{jkm} x_{ijkm}, \tag{8.10}$$

and P values were attributed to is with Eq. (8.4) and were corrected by BH criterion. is associated with adjusted P values less than 0.01 were selected. Table 8.2 shows

Table 8.1 Confusion matrix of selected mRNAs and miRNAs between TD-based unsupervised FE and PP in the first and second data sets. P values were computed by Fisher's exact test

1st data set (TCGA)			
mRNA	$P = 1.90 \times 10^{-149}$		
TD-based unsupervised FE		PP	
		Adjusted P values > 0.01	Adjusted P values < 0.01
	Adjusted P values > 0.01	19447	17
	Adjusted P values < 0.01	11	61
miRNA	$P = 2.76 \times 10^{-23}$		
TD-based unsupervised FE		Adjusted P values > 0.01	Adjusted P values < 0.01
	Adjusted P values > 0.01	812	2
	Adjusted P values < 0.01	0	11
2nd data set (GEO)			
mRNA	$P = 0$		
TD-based unsupervised FE		Adjusted P values > 0.01	Adjusted P values < 0.01
	Adjusted P values > 0.01	33781	8
	Adjusted P values < 0.01	23	186
miRNA	$P = 1.87 \times 10^{-7}$		
TD-based unsupervised FE		Adjusted P values > 0.01	Adjusted P values < 0.01
	Adjusted P values > 0.01	316	0
	Adjusted P values < 0.01	0	3

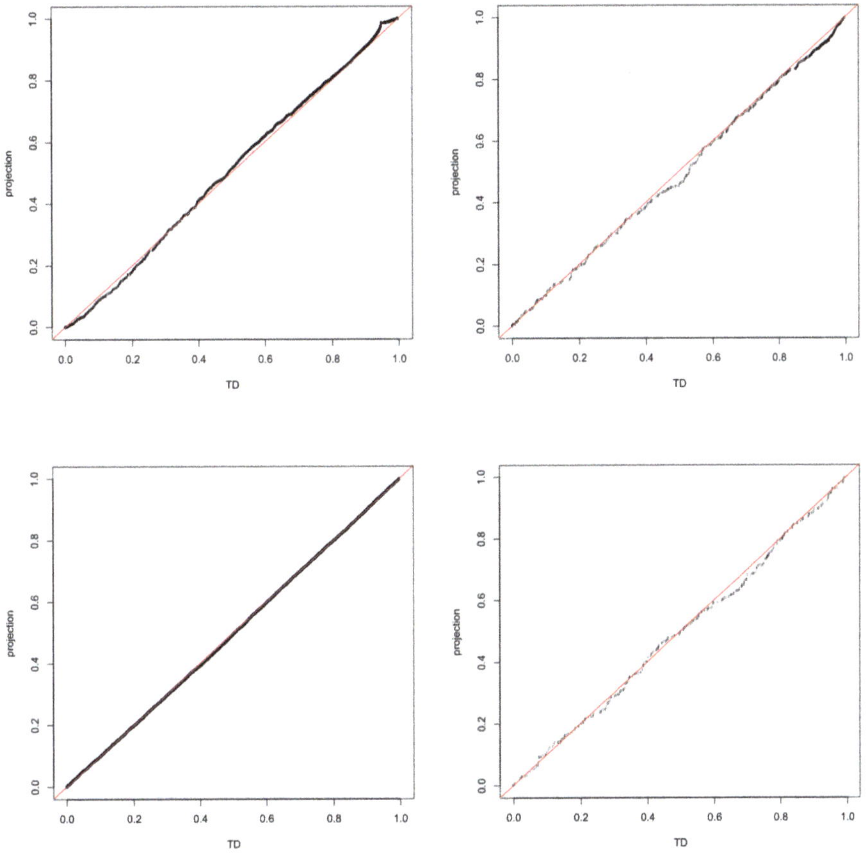

Fig. 8.1 QQplot between P values computed by TD-based unsupervised FE and PP. Top row: the first data set (TCGA), bottom row :the second data set (GEO), left column: mRNA, right column: miRNA. The red straight lines are diagonal lines, i.e., two distributions are identical

Table 8.2 Confusion matrix of selected mRNAs between TD-based unsupervised FE and PP in corona infection data set. P value was computed by Fisher's exact test

mRNA	$P = 1.40 \times 10^{-241}$		
		PP	
TD-based unsupervised FE		Adjusted P values > 0.01	Adjusted P values < 0.01
	Adjusted P values > 0.01	21582	52
	Adjusted P values < 0.01	60	103

the confusion matrix of selected mRNA between PP and TD-based unsupervised FE (Sect. 7.15.2). Again, it is obvious that PP is highly coincident with TD-based unsupervised FE. Thus, the agreement between PP and TD-based unsupervised FE is unlikely accidental.

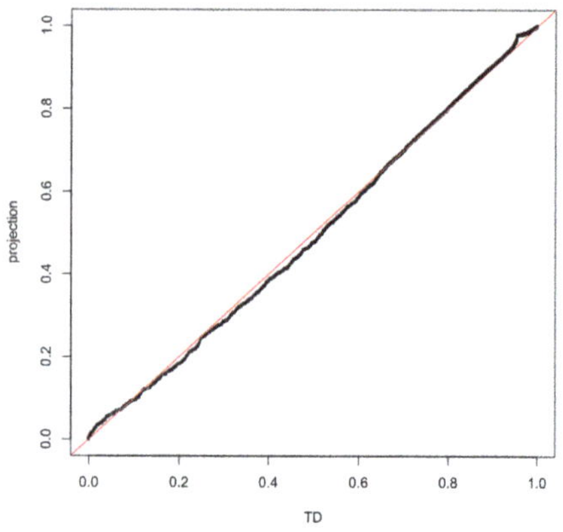

Fig. 8.2 QQplot between P values computed by TD-based unsupervised FE and PP. The red straight line is a diagonal line, i.e., two distributions are identical

Before closing this subsection, we investigated if the overall distribution of P values is coincident with each other between TD-based unsupervised FE and PP. To do this, we again draw QQplot (Fig. 8.2). Since QQplot is a straight line, PP can produce the almost identical distribution of P values to that of TD-based unsupervised FE as well.

8.2.4 Rationalization of Gaussian Distribution

Although we have successfully demonstrated that TD- and PCA-based unsupervised FE can be regarded as an advanced version of old fashioned technology, projection pursuit, we have yet another issue to be clarified. In the feature selection, we assumed the individual components of singular value vectors obey Gaussian and attributed P values to features using cumulative χ^2 distribution. Nevertheless, it is not guaranteed that the individual components of singular value vectors obey Gaussian. In fact, the individual components of singular value vectors do not always obey Gaussian. If so, why this criterion works pretty well as demonstrated in the preceding two chapters although it seems to be based upon the wrong assumption?

To understand this point, we tried to perform feature selection without explicitly assuming a Gaussian distribution. To do this, we generate the null distribution by shuffling the order of is in individual samples and recompute TD with shuffled tensor. We repeated this procedure 100 times. Then we rank singular value vectors obtained without shuffling within these sets of null distribution. Since there are $100N$ values generated by shuffling, if there are N_0 values larger than the ith component of singular value vectors, P_i attributed to i with null hypothesis is $\frac{N_0}{100N}$. The attributed P_is are corrected for by the BH criterion as usual. Unfortunately, if the threshold P values are assumed to be as small as 0.01 as usual, no features can

Table 8.3 Confusion matrix of selected mRNAs and miRNAs between TD-based unsupervised FE and selection using the null distribution generated by the shuffling in kidney cancer data set and COVID-19 data set. P values were computed by Fisher's exact test

1st data set (TCGA)			
mRNA	$P = 6.68 \times 10^{-194}$		
		The null distribution generated by shuffling	
		Adjusted P values > 0.1	Adjusted P values < 0.1
TD-based unsupervised FE	Adjusted P values > 0.01	19574	0
	Adjusted P values < 0.01	3	69
miRNA	$P = 4.25 \times 10^{-24}$		
		Adjusted P values > 0.1	Adjusted P values < 0.1
TD-based unsupervised FE	Adjusted P values > 0.01	813	1
	Adjusted P values < 0.01	0	11
2nd data set (GEO)			
mRNA	$P = 0$		
		Adjusted P values > 0.1	Adjusted P values < 0.1
TD-based unsupervised FE	Adjusted P values > 0.01	33736	53
	Adjusted P values < 0.01	0	209
COVID-19			
mRNA	$P = 1.37 \times 10^{-103}$		
		Adjusted P values > 0.1	Adjusted P values < 0.1
TD-based unsupervised FE	Adjusted P values > 0.01	19153	0
	Adjusted P values < 0.01	115	48

pass the screening. With the threshold P value of 0.1, the null distribution generated by shuffling can give us the reasonable results. Table 8.3 shows the confusion matrices of the selected genes between TD-based unsupervised FE and PP. It is rather obvious that they are highly coincident with each other. Although we do not know why the feature selection assuming Gaussian distribution can match with that employing more rigorous criterion, i.e., the null hypothesis generated from random shuffling, the threshold value of adjusted P values of 0.01 under the assumption of Gaussian distribution empirically well matches with that of 0.1 under the null distribution. This might be the reason why the usage of an unjustified Gaussian distribution could give us good performance, although the true distribution is far from Gaussian.

8.3 Optimization of the Standard Deviations

8.3.1 How to Optimize the Standard Deviations

In the preceding two chapters, we demonstrated that PCA- and TD-based unsupervised FE work pretty well for genomic science. Nevertheless, the number of features not selected was often too high, resulting in a significant number of false negatives. Too small selected features might mean that P values attributed to features were overestimated; overestimated P values result in less significance, which causes too small number of selected features. The problem is what causes the overestimation.

Since P values were attributed to features using cumulated χ^2 distribution, one possible cause of overestimation of P values is that of standard deviation; larger standard deviation reduces an argument for cumulated χ^2 distribution, i.e., the amount of individual features divided by the standard deviation.

The next problem to be discussed is what causes the overestimation of the standard deviation. One possible cause is the contribution from the selected features which should be more apart from mean than the others. If the standard deviation is estimated including the selected features that do not belong to Gaussian distribution, the standard deviation of Gaussian distribution which is supposed to be used as null hypothesis will be overestimated. In this sense, the correct standard deviation of Gaussian distribution used as the null hypothesis must be estimated by excluding selected features.

This is a bit controversial. Without estimating the standard deviation, we cannot compute P values to be used for the feature selection, whereas the correct standard deviation must be computed with excluding the selected feature. Thus the standard deviation must be estimated such that the distribution of not selected features obeys Gaussian distribution as much as possible.

For this purpose, we have done as follows. At first, we attribute P_i to the ith feature, u_i as

$$P_i = P_{\chi^2}\left[> \left(\frac{u_i}{\sigma}\right)^2\right] \tag{8.11}$$

with the specified standard deviation, σ. Suppose that $h_s(1 - P_i)$ is the sth $(1 \leq s \leq S)$ histogram of $1-P_i$ where *the selected features*, i.e., those associated with the adjusted P value less than threshold, *are excluded.* Then we compute the standard deviation of h_s, σ_h as

$$\langle h_s \rangle = \frac{1}{S}\sum_s h_s \tag{8.12}$$

$$\sigma_h = \sqrt{\frac{1}{S}\sum_s (h_s - \langle h_s \rangle)^2}. \tag{8.13}$$

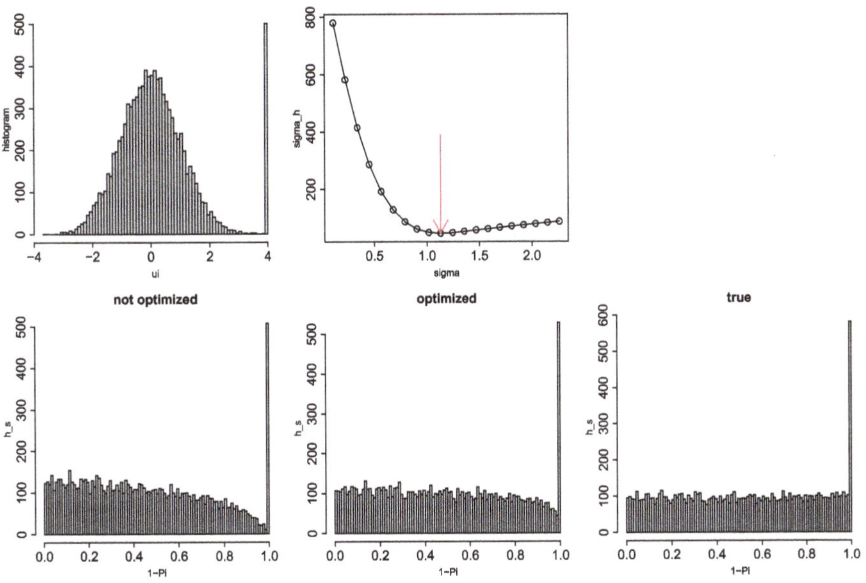

Fig. 8.3 Top left: the histogram of u_i given by Eq. (8.14), top middle: scatter plot of σ_h (vertical axis) and σ (horizontal axis), red arrow indicates the selected σ, bottom left: h_s when $\sigma = 1.31$ (computed using all u_is), bottom middle: h_s when $\sigma = 1.13$ (computed using u_i excluding the selected ones, i.e., optimized), h_s when $\sigma = 1$ (i.e., true)

Then we tried to minimize σ_h with changing σ. This allows us to find the optimal σ that generates the distribution of not selected features, u_i, obeying Gaussian distribution as much as possible, since h_s should be flat when the distribution of not selected features obeys Gaussian distribution (minimal σ_h should give us the most flat h_s).

To confirm that the proposed strategy can work well, we have done followings. At first, we prepare $u_i \in \mathbb{R}^{10^4}$ as

$$u_i \sim \begin{cases} \mathcal{N}(0, 1), & i \leq 9500 \\ 4, & i > 9500 \end{cases}, \tag{8.14}$$

where $\mathcal{N}(\mu, \sigma)$ is the Gaussian distribution that has a mean of μ and the standard deviation of σ. The purpose of the following analysis is to select $i > 9500$ as those that do not obey the Gaussian distribution.

To do this, the above procedure was applied to the u_i defined by Eq. (8.14); Fig. 8.3 shows the results when $S = 100$ and the threshold P value is set to be 0.01. By definition, u_is include outliers ($u_i = 4$, top left panel of Fig. 8.3). σ_h takes minimum at some value of σ as expected (the vertical solid red arrow in the top middle panel of Fig. 8.3). When the standard deviation of u_i, σ, was computed using all u_i, h_s cannot exhibit flat one (bottom left panel of Fig. 8.3) which is expected

Table 8.4 Confusion matrix between u_is not obeying Gaussian distribution ($i > 9500$) and those identified as not obeying the Gaussian distribution (associated with adjusted P values less than 0.01)

			$i \leq 9500$	$i > 9500$
True	($\sigma = 1$)	Adjusted P values > 0.01	9495	0
		Adjusted P values < 0.01	5	500
Optimized	($\sigma = 1.13$)	Adjusted P values > 0.01	9500	0
		Adjusted P values < 0.01	0	500
Not optimized	($\sigma = 1.31$)	Adjusted P values > 0.01	9500	500
		Adjusted P values < 0.01	0	0

when $\sigma = 1$ (true one of Gaussian distribution, bottom right panel of Fig. 8.3). On the other hand, when σ was computed with excluding the u_is supposed to be selected as those not obeying Gaussian distribution (i.e., optimized), h_s exhibits more flat one (bottom middle panel of Fig. 8.3).

More important point is that the usage of the optimized σ enables us to correctly estimate u_is not obeying Gaussian distribution (i.e., $i > 9500$, Table 8.4). Although with not optimized σ(=1.31), no u_is are selected as those not obeying Gaussian distribution but with optimized σ(=1.13), all u_is not obeying the Gaussian ($i > 9500$) were correctly identified as in the case of the usage of true σ(=1). Thus, we can successfully identify the features not obeying Gaussian distribution. This might reduce the false negatives. In the following, we tried to apply this strategy to various problems.

8.3.2 *Gene Expression*

The first example to which the method proposed in the previous subsection, which is hereafter often referred to as "TD (or PCA) based unsupervised FE with optimized standard deviation (SD)," is applied is gene expression [25], since PCA- and TD-based unsupervised FE were most frequently applied to gene expression profiles as shown in this book. To demonstrate the effectiveness of the proposed method, we employed the highly curated gene expression profiles, the MicroArray Quality Control (MAQC) [22]. MAQC is composed of seven human brain expression profiles and seven universal human reference (UHR) expression profiles. As the name of MAQC indicates, it is high-quality data set. They were formatted to a matrix $x_{ij} \in \mathbb{R}^{40,933 \times 14}$, which represents the expression of the ith gene at the jth sample. How to convert fastq files into x_{ij} is detailedly described in the original paper [25]. x_{ij} was normalized as

$$\sum_{i=1}^{40,933} x_{ij} = 0 \tag{8.15}$$

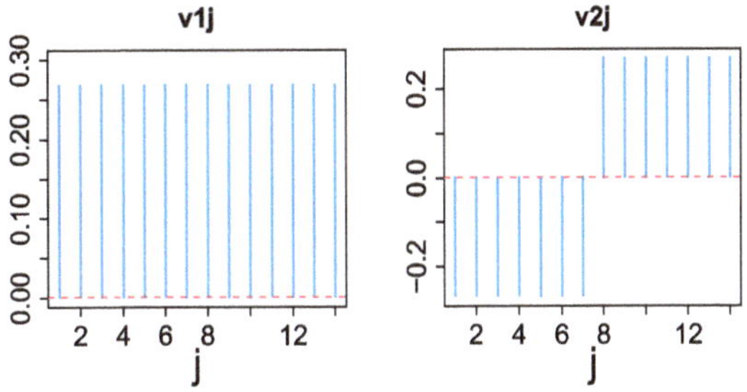

Fig. 8.4 Left: $\boldsymbol{v}_1$, right: $\boldsymbol{v}_2$. $1 \le j \le 7$: UHR, $8 \le j \le 14$: brain

$$\sum_{i=1}^{40933} x_{ij}^2 = 40,933. \tag{8.16}$$

Then PC score, $\boldsymbol{u}_\ell \in \mathbb{R}^{40,933}$, and PC loading, $\boldsymbol{v}_\ell \in \mathbb{R}^{14}$, can be obtained by applying SVD to x_{ij} as

$$x_{ij} = \sum_{\ell=1}^{14} \lambda_\ell u_{\ell i} v_{\ell j}. \tag{8.17}$$

At first, we need to identify which $\boldsymbol{v}_\ell$ is associated with the distinction between UHR and brain. Figure 8.4 shows $\boldsymbol{v}_1$ and $\boldsymbol{v}_2$, respectively. It is obvious that $\boldsymbol{v}_2$ exhibits the distinction between UHR and brain, whereas $\boldsymbol{v}_1$ takes constant values regardless of j. Thus we decided to employ $\ell = 2$ to select genes expressed distinctly between UHR and brain. When P values are simply attributed to is

$$P_i = P_{\chi^2}\left[> \left(\frac{u_{2i}}{\sigma_2}\right)^2\right] \tag{8.18}$$

using σ_2 computed with all u_{2i}, the histogram of $1 - P_i$ is far from that expected for the Gaussian distribution (Fig. 8.5, left). Then we tried to optimize SD, σ_2, by minimizing σ_h (Fig. 8.6). In Fig. 8.6, we decided to employ $\sigma_2 \sim 0.19$ as the optimal value of SD used to compute P_i with assuming the null hypothesis that u_{2i} obeys the Gaussian. Figure 8.5 right shows h_s, the histogram of $1 - P_i$, with optimized σ_2. It is more likely to coincide with the null hypothesis that u_{2i} obeys Gaussian than without SD optimization, σ_2.

Next, we investigate how optimized SD affects gene selection. At first, we compare the number of genes selected with and without SD optimization. Without optimization, as little as 268 genes were associated with adjusted P values less than 0.01, whereas as many as 5,308 genes were associated with adjusted P values less than 0.01. One of the reasons why we needed to improve PCA- and TD-based unsupervised FE was too small genes selected; this problem was clearly resolved by the optimization of SD.

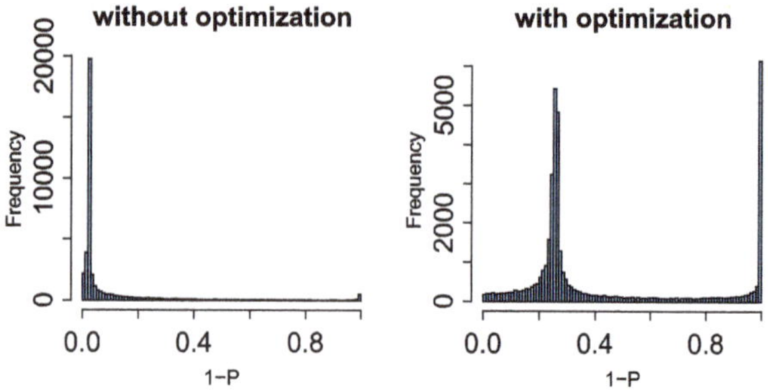

Fig. 8.5 Histogram of $1 - P_i$. Left: $\sigma_2 = 2.07$ (without optimization), right: $\sigma_2 \sim 0.19$ (with optimization)

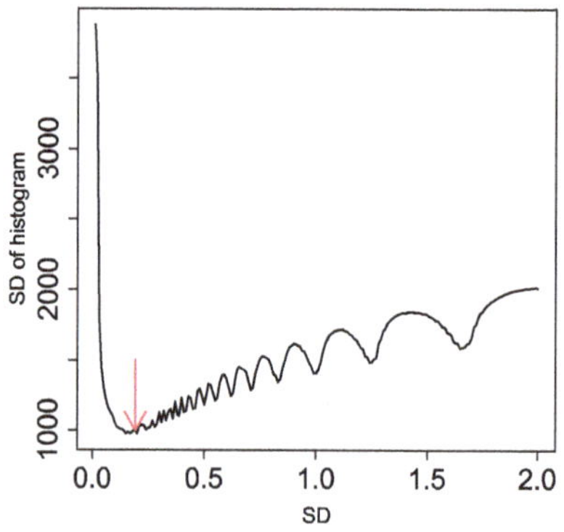

Fig. 8.6 σ_h as a function of σ_2. The vertical red arrow suggests the optimized value of σ_2

One might wonder if 5,308 genes are too many to be expressed distinctly between UHR and brain. To check this point, we applied DESeq2, a *de facto* standard method, to the present data set (detailed setup on how to apply DESeq2 to the present data set can be found in the original paper [25]). Then we found that as many as 17,244 genes were detected as being distinctly expressed between UHR and brain. Thus, 5,308 genes are not too many to be expressed differently between UHR and brain. In actual, Costa-Silva et al. [3] carefully evaluated DEGs in MAQC and found that about half of investigated genes are differently expressed between UHR and brain. Thus, 5,308 might not be too large.

Although we have successfully increased the number of DEGs identified by PCA-based unsupervised FE with optimized SD, could we do it in biologically reasonable way? In order to validate this point, we need to evaluate the selected genes biologically. To do this, we uploaded sets of selected genes to Enrichr as usual. At first, we considered "Jensen TISSUES" category of Enrichr (Fig. 8.7). Although two top, eye and intestine, are not related to brains, the remaining is related to brain. It is also obvious that genes selected by PCA-based unsupervised FE can give much more significance than those selected by DESeq2. Thus, not only PCA-based unsupervised FE with optimized SD could identify the reduced number of DEGs than DESeq2, but also PCA-based unsupervised FE with optimized SD could identify biologically more reasonable DEGs than DESeq2.

One may still wonder if the other state-of-the-art methods might be better than PCA-based unsupervised FE with optimized SD. To deny this possibility, we bio-

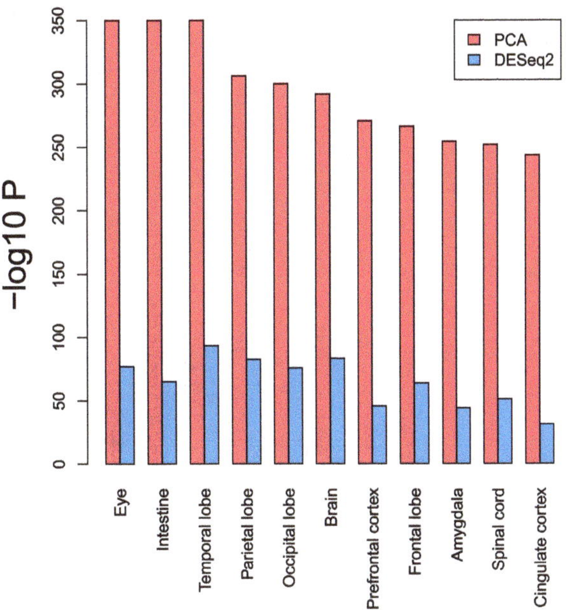

Fig. 8.7 Enrichment analysis of the selected genes, by uploading gene symbols associated with ensemble gene IDs selected by either PCA-based unsupervised FE with optimized SD (denoted as PCA) or DEseq2 (denoted as DESeq2), respectively, based upon "Jensen Tissues" category in Enrichr. Three terms associated with $-\log_{10} P = 350$ are linked with ∞ since $P = 0$

logically evaluated the genes selected for MAQC using edgeR [20], voom [13], and NOISeq [30]. We have evaluated the performance of these methods including PCA-based unsupervised FE with optimized SD and DESeq2 for "KEGG 2021 Human" category of Erichr with focusing neuron/synapse/axon related terms (Fig. 8.8); it is obvious that these three methods are even inferior to DESeq2 biologically. Next, we consider "GO Biological Process 2023" with focusing neuron/synapse/axon related terms (Fig. 8.9). Although PCA-based unsupervised FE with optimized SD is comparative or inferior to other methods in some axon related terms (2nd, 9th, 24th, and 25th), it still outperformed other methods in most of terms. Finally, we consider "Human Gene Atlas" with focusing brain related tissues (Fig. 8.10). This time, PCA-based unsupervised FE with optimized SD definitely outperformed other methods.

In the above, PCA-based unsupervised FE with optimized SD could identify DEGs that are biologically more reliable than other conventional methods. In the following, it will be shown that PCA-based unsupervised FE with optimized SD naturally satisfy the frequently required requirement, i.e., more expressed genes should be more likely to be DEGs. In the conventional methods, the statistical significance and fold change are two major criteria to identify DEGs. Nevertheless, these two measures have some limitation, i.e., they cannot take the amount of expression into account. Suppose that there are two genes whose expression under the distinct two conditions is 10 vs. 100 and 100 vs. 1000, respectively. Either t-test or fold change cannot identify the difference of two genes. P values computed by t-test do not vary when expressions of all genes considered are doubled and fold change does not change either when expressions of all genes considered are doubled.

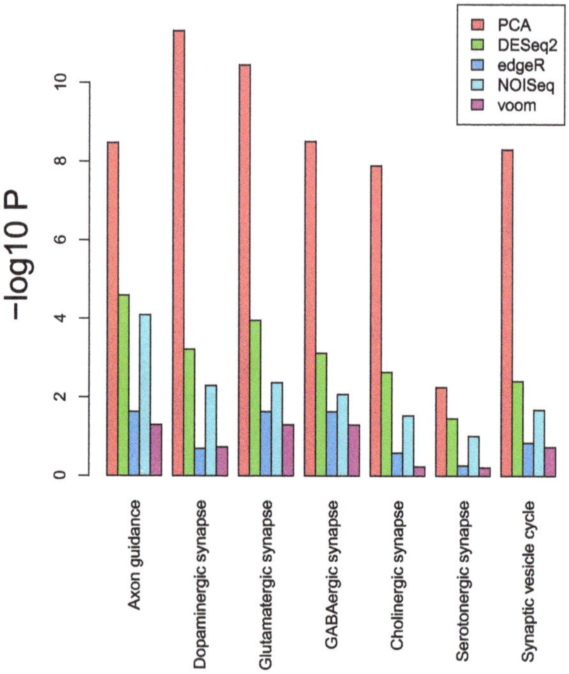

Fig. 8.8 Enrichment analysis for MAQC with other methods in "KEGG 2021 Human" category of Enrichr with focusing neuron/synapse/axon related terms

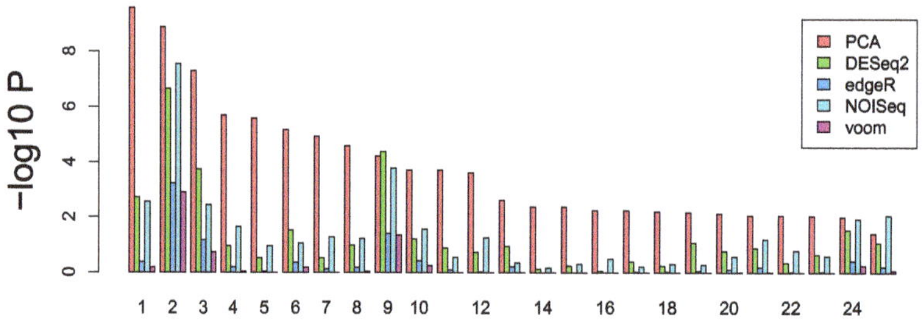

Fig. 8.9 Enrichment analysis for MAQC with other methods in "GO Biological Process 2023" category of Enrichr with focusing neuron/synapse/axon related terms. Numbers correspond to [1] "AxonDevelopment," [2] "Axonogenesis," [3] "SynapseOrganization," [4] "RegulationOfAxonogenesis," [5] "PositiveRegulationOfAxonogenesis," [6] "RegulationOfAxonExtension," [7] "SignalReleaseFromSynapse," [8] "PositiveRegulationOfExcitatoryPostsynapticPotential," [9] "AxonGuidance," [10] "ModulationOfExcitatoryPostsynapticPotential," [11] "PresynapticEndocytosis," [12] "RegulationOfSynapseAssembly," [13] "RetrogradeAxonalTransport," [14] "RegulationOfSynapseOrganization," [15] "Vesicle-MediatedTransportInSynapse," [16] "SemaphorinPlexinSignalingPathwayInvolvedInAxonGuidance," [17] "PositiveRegulationOfSynapseMaturation," [18] "RegulationOfSynapseMaturation," [19] "AnterogradeAxonalTransport," [20] "PositiveRegulationOfAxonExtension," [21] "SynapseAssembly," [22] "PostsynapticMembraneOrganization," [23] "PositiveRegulationOfSynapseAssembly," [24] "NegativeRegulationOfAxonogenesis," [25] "AxonExtension"

Fig. 8.10 Enrichment analysis for MAQC with other methods in "Human Gene Atlas" category of Enrichr with focusing brain related tissues

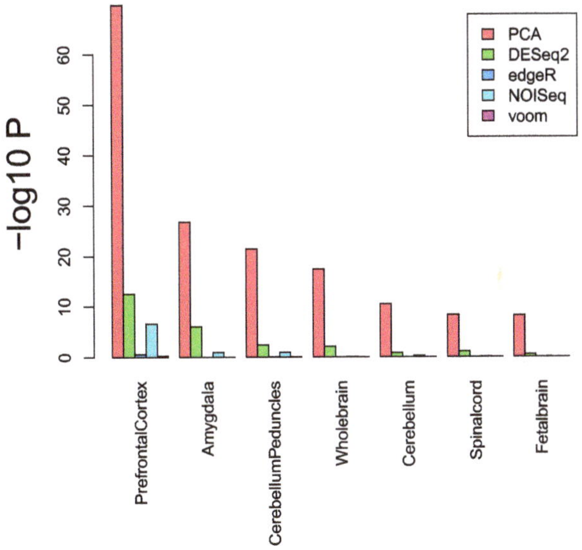

In spite of that, biologists often feel that DEGs with more expression should be biologically important, since low expressed genes are unlikely to play important roles.

To address this point, some empirical procedure is often taken. For example, the so-called dispersion relation

$$\frac{\alpha(\mu)}{\mu^2} = \alpha_0 + \frac{\alpha_1}{\mu} \tag{8.19}$$

is often assumed [15, 20]. Here $\alpha(\mu)$ is the variance supposed to be a function of μ, that is, mean. Due to the second term on the right hand side of the above equation, genes whose mean values of expression are low are forced to have a larger ratio of the variance to the squared mean. This inevitably makes genes whose expression is low difficult to be identified as DEGs since DEG must have larger difference between two classes, which generally increases as μ increases, than standard deviation which is a root squared variance, $\alpha(\mu)$. The problem is that the above equation is purely empirical and has no biological justification other than the nontrivial assumption that the gene associated with more expression should be more likely to be DEGs.

In PCA-based unsupervised FE, DEGs were identified by selecting genes whose projection of expression onto $v_{2j}^{(j)}$ is larger. As can be seen in Fig. 8.4, $v_{2j}^{(j)}$ represents the difference between two classes. If we denote the mean values of two classes x_A and x_B, respectively,

$$v_{2j}^{(j)} \sim x_A - x_B. \tag{8.20}$$

On the other hand, fold change is assumed to be $\frac{x_A}{x_B}$. Then

$$\frac{x_A}{x_B} = \frac{v_{2j}^{(j)} + x_B}{x_B} = \frac{v_{2j}^{(j)}}{x_B} + 1. \tag{8.21}$$

This means if we select genes associated with $u_{2i}^{(i)}$ that corresponds to $v_{2j}^{(j)}$ and is larger than some threshold values, these genes become more likely selected in spite of that fold change decreases as x_B increases. This tendency is qualitatively equivalent to that in Eq. (8.19). Thus, without empirical assumption like dispersion relation, Eq. (8.19), PCA-based unsupervised FE satisfies the requirement that genes with more expression should be more likely to be DEGs. The above discussion suggests that PCA-based unsupervised FE is more likely biologically reasonable than the other conventional methods that require some additional unjustified constraint.

To visualize qualitative equivalence between Eqs. (8.19) and (8.21), we show MAPlot in Fig. 8.11. The vertical axis of MAPlot represents a logarithmic ratio of mean values in each of two classes, FC, i.e.,

$$\log \mathrm{FC} = \log \left(\frac{\sum_{j=1}^{7} x_{ij}}{\sum_{j=8}^{14} x_{ij}} \right), \tag{8.22}$$

where the numerator and the denominator are the means of x_{ij}s that belong to brain and UTR, respectively. The horizontal axis of MAPlot represents logarithmic mean,

$$\log \left(\frac{1}{14} \sum_{j=1}^{14} x_{ij} \right). \tag{8.23}$$

Two plots in Fig. 8.11 represent the tendency that genes with smaller expression (horizontally left hand side) must be vertically apart from horizontal axis to be selected even if they have the same amount of fold changes, although PCA-based unsupervised FE with optimized SD that employs Eq. (8.21) is associated with a more enhanced tendency. This means that PCA-based unsupervised FE with optimized SD not only can identify more biologically meaningful genes but also can deny the necessity of unjustified requirement like dispersion relation, Eq. (8.19).

Although PCA-based unsupervised FE with optimized SD seems to work pretty well, since it was tested toward one highly qualified gene expression profile, the good performance might be accidental. To deny this possibility, we applied PCA-based unsupervised FE with optimized SD to more number of profiles. For this purpose, we employed SEQC data set [21] including as many as 13 highly qualified gene expression profiles. Although all of 13 profiles are composed of two classes, the numbers of genes and samples vary from profiles to profiles (Table 8.5).

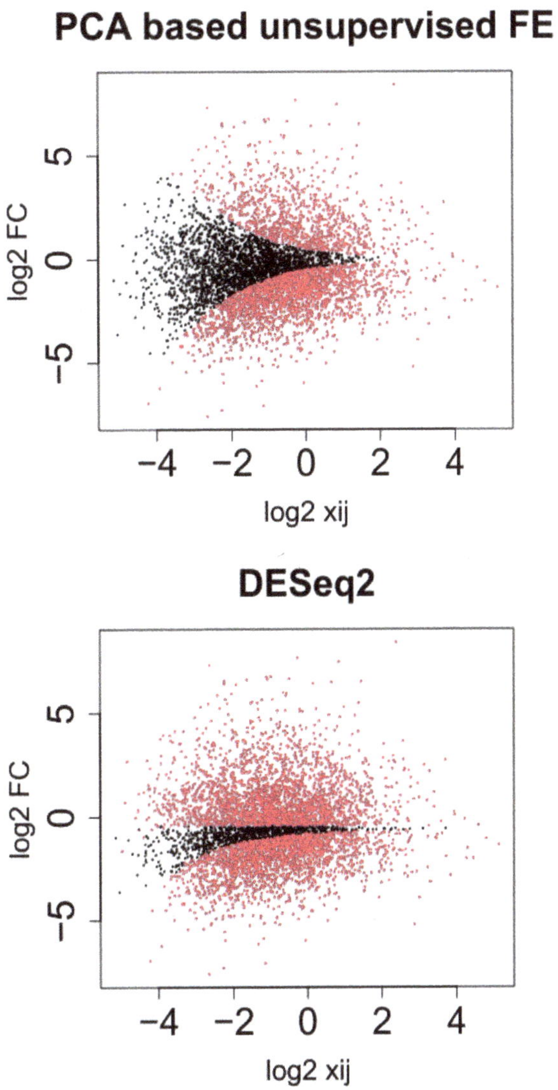

Fig. 8.11 MAPlot with selected genes colored in red. Upper: PCA-based unsupervised FE with optimized SD, lower: DESeq2

Figure 8.12 shows the results when PCA-based unsupervised FE with optimized SD and DESeq2 were applied to 13 profiles. The number of genes selected highly varied from profiles to profiles when DESeq2 was used, whereas PCA-based unsupervised FE with optimized SD identified a relatively stable number of genes. This suggests that the good performances obtained by PCA-based unsupervised FE with optimized SD when applied to MAQC is not accidental but robust.

Although PCA-based unsupervised FE with optimized SD achieved the excellent performance, it was applied to only highly qualified gene expression profiles. To confirm that PCA-based unsupervised FE with optimized SD is also effective toward

Table 8.5 The number of genes and samples in 13 SEQC data sets

Profiles	The number of genes	The number of samples	
		Class 1	Class 2
BGI	14261	80	80
AGR	15601	64	64
CNL	14028	75	75
COH	16224	32	32
MAY	13753	80	80
NVS	15600	64	64
LIV	16135	10	10
NWU	11842	60	60
PSU	10745	60	60
LIF_SQW	11659	60	60
MGP	11055	2	2
NYU	11672	2	2
ROC_SQW	10613	2	2

more realistic gene expression profiles, we applied TD-based unsupervised FE with optimized SD to the data set used in Sect. 7.15.2. Then we got 2792 genes associated with adjusted P values less than 0.01 instead of 163 genes identified by TD-based unsupervised FE without optimized SD applied to the data set (Fig. 8.13).

We have uploaded these 2792 genes to Enrichr. Table 8.6 lists top ten-ranked enriched "SARS-CoV-2" terms in "COVID-19 Related Gene Sets 2021" category of Enrichr. If compared with Tables 7.84 and 7.106, the significance is increased since adjusted P values are smaller. The most remarkable point is that all top ten-ranked profiles are those retrieved from GSE147507 that is nothing but GEO data set analyzed in Sect. 7.15.2. Although one might think that it is not remarkable, neither of Tables 7.106, nor 7.84, could do this. It is not easy to find single gene set which simultaneously overlaps with all five cell lines in GSE147507. Thus TD-based unsupervised FE with optimized SD could outperform TD-based unsupervised FE without optimized SD as well as kernel TD-based unsupervised FE even when applied to more realistic data sets.

We have also evaluated the repositioning of the drug toward SARS-CoV-2 based on the 2792 genes identified, as has been done in Sect. 7.15.2. The first category we investigated is "LINCS L1000 Chem Pert Consensus Sigs." Table 8.7 shows the top 10 drugs ranked. Although Table 7.85 includes only two promising candidates, Fenretinide and ercalcitriol, in the top ten-ranked drugs, Table 8.7 includes as many as four promising candidates, Birinapant [14], tyrphostin-A9 [8], Fenretinide [17], and Niclosamide [1, 33] in the top ten-ranked drugs. Thus, TD-based unsupervised FE with optimized SD can provide us more promising candidates for drug repositioning toward SARS-CoV-2.

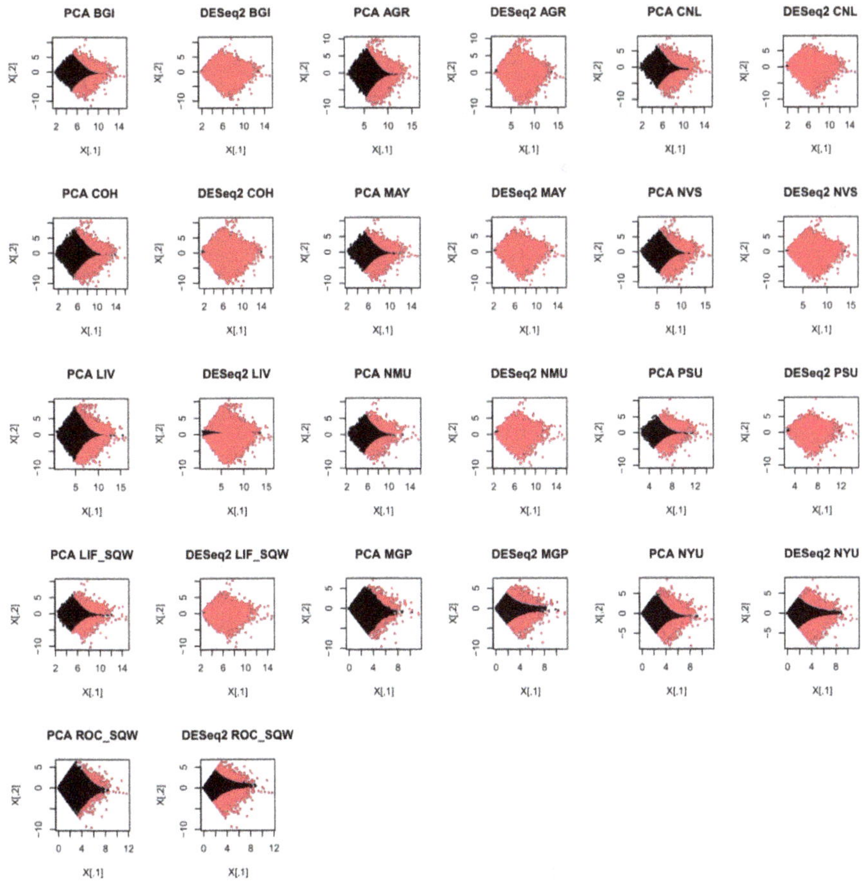

Fig. 8.12 MAPlot with selected genes colored in red for 13 profiles in SEQC data sets

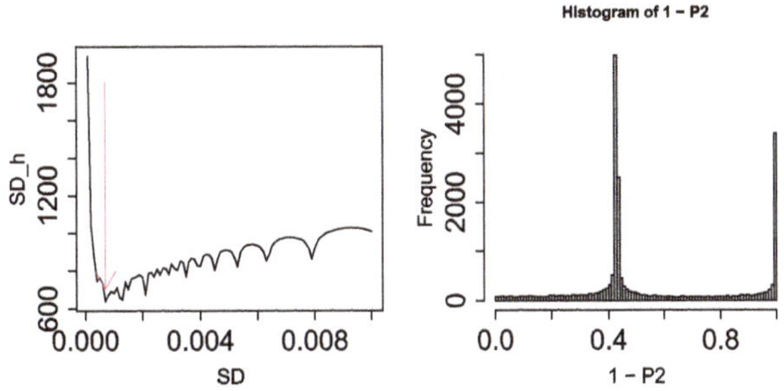

Fig. 8.13 Optimization of σ_ℓ. Left column: σ_h as a function of σ_ℓ. The vertical red arrows show the selected σ_ℓ, 7.00×10^{-4}. Right column: the histogram $1 - P_i$ with optimized σ_ℓ

Table 8.6 Top ten-ranked enriched "SARS-CoV-2" terms in "COVID-19 Related Gene Sets 2021" category of Enrichr

Term	Overlap	P value	Adjusted P value
500 genes downregulated by SARS-CoV-2 in human A549 cells from GSE147507 Series5	323/494	3.02×10^{-157}	1.43×10^{-154}
500 genes downregulated by SARS-CoV-2 (2 MOI) in human A549 cells from GSE147507	319/489	8.54×10^{-155}	2.02×10^{-152}
500 genes downregulated by SARS-CoV-2 (2 MOI) in human A549 cells transduced with ACE2 vector from GSE147507	297/492	2.77×10^{-130}	2.62×10^{-128}
500 genes downregulated by SARS-CoV-2 in human Calu3 cells from GSE147507	290/491	1.37×10^{-123}	1.08×10^{-121}
500 genes upregulated by SARS-CoV-2 in human A549 cells from GSE147507 Series5	290/494	1.30×10^{-122}	8.82×10^{-121}
500 genes downregulated by SARS-CoV-2 (0.2 MOI) in human A549 cells transduced with ACE2 vector from GSE147507	287/494	1.13×10^{-119}	6.70×10^{-118}
500 genes downregulated by SARS-COV-2 infection of Calu3 cells	281/479	1.30×10^{-118}	5.62×10^{-117}
500 genes downregulated by SARS-CoV-2 (2 MOI) in human Calu-3 cells from GSE147507	281/479	1.30×10^{-118}	5.62×10^{-117}
500 genes upregulated by SARS-CoV-2 in human Calu3 cells from GSE147507	283/494	8.30×10^{-116}	3.28×10^{-114}
500 genes upregulated by SARS-COV-2 infection of Calu3 cells	277/476	1.17×10^{-115}	3.95×10^{-114}

Table 8.7 Top ten-ranked enriched terms in "LINCS L1000 Chem Pert Consensus Sigs" category of Enrichr

Term	Overlap	P value	Adjusted P value
Salermide Up	144/239	1.12×10^{-62}	1.22×10^{-58}
Niclosamide Up	141/235	4.70×10^{-61}	2.55×10^{-57}
I-606051 Up	142/241	4.66×10^{-60}	1.69×10^{-56}
Fenretinide Up	142/242	9.79×10^{-60}	2.65×10^{-56}
Tyrphostin-A9 Up	138/234	1.88×10^{-58}	4.08×10^{-55}
Vinorelbine Up	141/248	6.36×10^{-57}	1.15×10^{-53}
Ingenol-Mebutate Up	141/249	1.27×10^{-56}	1.97×10^{-53}
Heliomycin Up	136/237	1.33×10^{-55}	1.61×10^{-52}
O-Anisidinehcl Down	136/237	1.33×10^{-55}	1.61×10^{-52}
Birinapant Up	138/247	1.86×10^{-54}	2.02×10^{-51}

8.3.3 DNA Methylation

In this subsection, we applied TD- and PCA-based unsupervised FE with optimized SD to the analysis of DNA methylation [28]. The remarkable point of this subsection is that TD- and PCA-based unsupervised FE with optimized SD developed for gene

Table 8.8 DNA methylation profiles used in this subsection

ID	Sites	Samples	Platform
GSE77965	485,577	Six colorectal cancer vs. six adjacent normal colon tissue	Illumina HumanMethylation450 BeadChip.
GSE42308		Three HCT116 cell lines vs. three HCT116 mutants	Illumina HumanMethylation450 BeadChip.
GSE34864	Not specified	Two oocyte vs. two zygote profiles	BSseq
EH1072		Three Treg cell profiles taken from fat, liver, skin, and lymph nodes and T-cell control taken from lymph nodes	BSseq

expression analysis is applicable to DNA methylation without any modification. In the standard analyses applied to gene expression and DNA methylation, we need to assume some distribution of gene expression and DNA methylation. The distribution of gene expression is often assumed to obey a negative binomial distribution, whereas that of DNA methylation is often assumed to obey beta distribution. Thus, it is impossible to develop one method applicable to both gene expression and DNA methylation. Nevertheless, TD- and PCA-based unsupervised FE with optimized SD is applicable to both gene expression and DNA methylation since it is unsupervised method that does not assume any specific distribution of gene expression and DNA methylation in advance. TD and PCA-based unsupervised FE with optimized SD deal with singular value vectors generated by SVD and TD, to which the assumption of Gaussian distribution seems to work pretty well.

To demonstrate the ability of TD- and PCA-based unsupervised FE with optimized SD applied to DNA methylation, we employed various data sets of DNA methylation. The purpose of the analysis is to identify differentially methylated cytosines (DMC). Another advantage of TD- and PCA-based unsupervised FE with optimized SD is that it is seamlessly applicable to profiles retrieved by sequencing and microarray. The data sets analyzed are listed in Table 8.8. Since DNA methylation is represented as relative ratio, it should be formatted as $x_{ij} \in [0, 1]^{N \times M}$ that represents the DNA methylation at the ith site of jth sample. Missing values were filled with zero. x_{ij} is normalized as

$$\sum_{i=1}^{N} x_{ij} = 0 \tag{8.24}$$

$$\sum_{i=1}^{N} x_{ij}^2 = N. \tag{8.25}$$

At first, PCA was applied to four x_{ij}s, and we got

$$x_{ij} = \sum_{\ell=1}^{L} \lambda_\ell u_{\ell i}^{(i)} v_{\ell j}^{(j)}, \tag{8.26}$$

where $\boldsymbol{u}_\ell^{(i)} \in \mathbb{R}^N$ and $\boldsymbol{v}_\ell^{(j)} \in \mathbb{R}^M$ are PC score and PC loading, respectively. To decide which $u_{\ell i}^{(i)}$ is used to select is, we need to find which $v_{\ell j}^{(j)}$ is different between the class labels (Fig. 8.14). For GSE77965, GSE42308, and EH1072, $v_{2j}^{(j)}$ exhibits significant association with class labels. Although there are no $v_{\ell j}^{(j)}$s associated with significant P values for GSE34864, we decided to employ $\ell = 3$ since $v_{3j}^{(j)}$ is most associated with the distinction between two classes.

We have attributed P values to is using

$$P_i = P_{\chi^2}\left[> \left(\frac{u_{\ell i}^{(i)}}{\sigma_\ell} \right)^2 \right] \tag{8.27}$$

with $\ell = 2$ for GSE77965, GSE42308, and EH1072 and $\ell = 3$ for GSE34864. At first, σ_ℓ was optimized such that $u_{\ell i}^{(i)}$ obeys Gaussian as much as possible (Fig. 8.15). The optimized σ_ℓs are 0.4045275 (for GSE77965), 0.2770134 (for GSE42308), 0.8568365 (for GSE34864), and 0.6019531 (for EH1072). The obtained P values are corrected by BH criterion as usual. The numbers of is associated with adjusted P values less than 0.01 are 1739 (for GSE77965), 34,626 (for GSE42308), and 89,485 (for EH1072). For GSE34864, we could not find any is associated with adjusted P values less than 0.01. Then we use $P = 0.3$ to select is. Then we selected 40,952.

Next, we evaluated selected is from the biological point of view. For microarray (GSE77965 and GSD42308), we could make use of annotations of probes. We have tested how much the selected is are overlapped with the probes that have annotations (Table 8.9). Although enhancer is not always significantly coincident with selected is, other annotations of probes are coincident with selected is for both profiles.

For HTS (GSE34864 and EH1072), no probe annotations are available. Thus, we only compared selected is with DHS sites whose information is available elsewhere (for more details, see the original paper [28]). We have compared the attributed P values between DHS sites and non-DHS sites for an individual chromosome using the t-test. We found that all P values obtained by the t-test attributed to chromosome are zero (Fig. 8.16). The P values associated with DHS sites are significantly lower than those for non-DHS sites. Consequently, it is clear that sites with lower P values, which are more likely to be selected, have a high degree of overlap with DHS sites.

Finally, we tested if PCA-based unsupervised FE with optimized SD is applicable to more realistic cases, i.e., those with clinical data. To this, we retrieve breast and colon cancer data from TCGA. For more details on how we could retrieve the data sets, see the original paper [28]. As class labels to select features, we employed "ajcc staging system edition" [4]. The American Joint Committee on Cancer (AJCC) releases the staging labels for cancer development. Thus, the staging

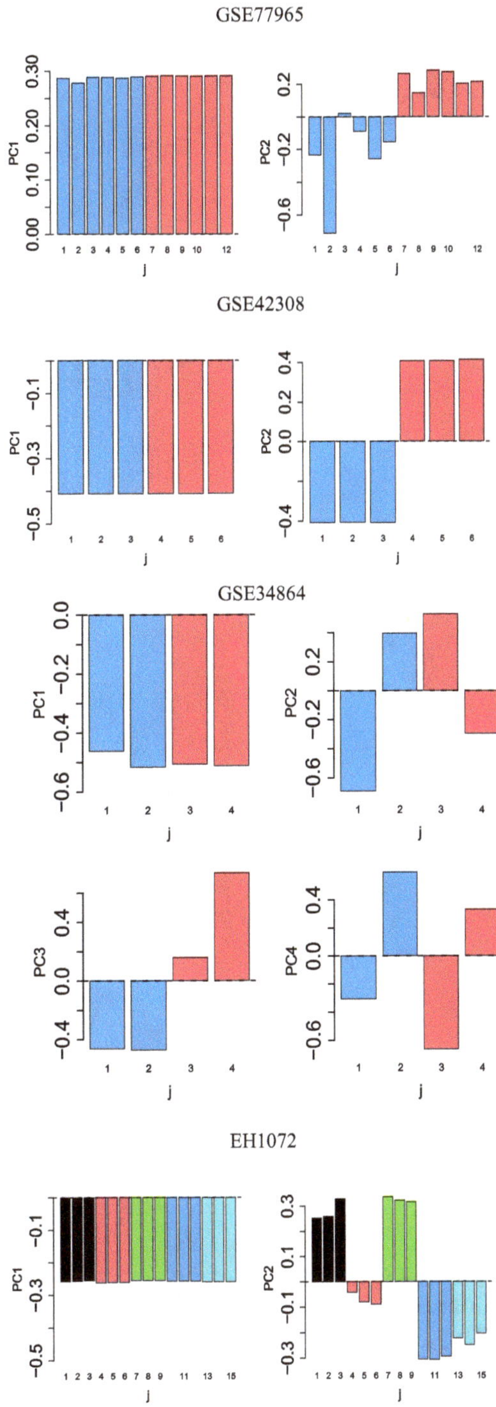

Fig. 8.14 $v_{\ell j}^{(j)}$ for GSE77965 (the 1st row, $\ell = 1, 2$, $P = 5.68 \times 10^{-3}$ for $\ell = 2$), GSE42308 (the 2nd row, $\ell = 1, 2$, $P = 9.21 \times 10^{-7}$ for $\ell = 2$), GSE34864 (the 3rd and 4th rows, $1 \leq \ell \leq 4$), and EH1072 (bottom row, $\ell = 1, 2$, $P = 5.22 \times 10^{-3}$ for $\ell = 2$). P values are computed by t-test for GSE77965, GSE42308, and GSE34864 and by categorical regression analysis for EH1072. Colors indicate distinct class labels

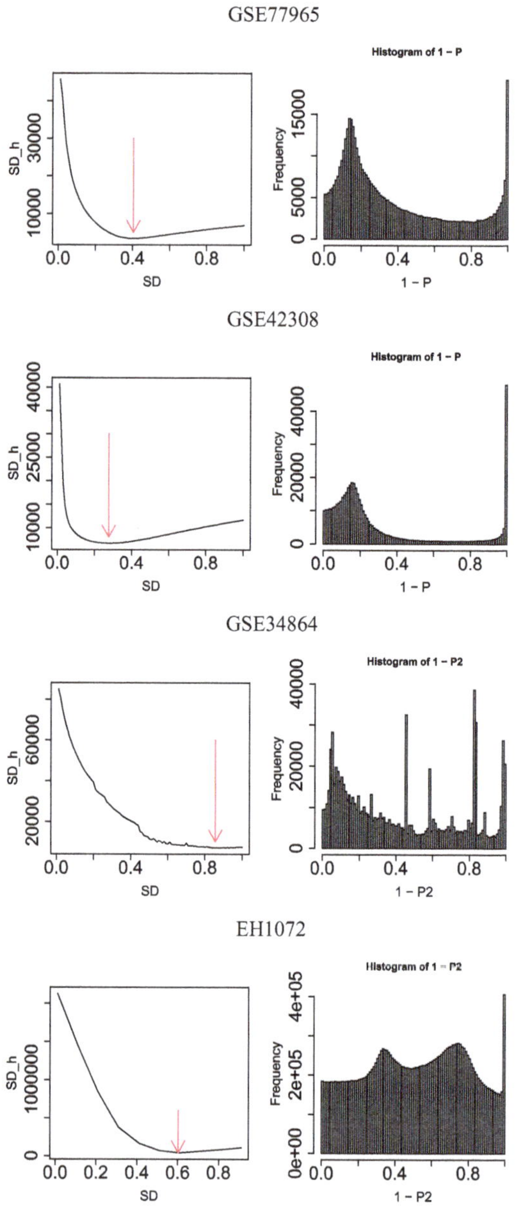

Fig. 8.15 Optimization of σ_ℓ. Left column: σ_h as a function of σ_ℓ. The vertical red arrows show the selected σ_ℓ. Right column: the histogram $1 - P_i$ with optimized σ_ℓ

system is a good indicator of the grades of individual samples. For our results, for both breast and colon cancer data sets, the second PC attributed to the samples, $v_{2j}^{(j)}$, was significantly associated with "theajcc staging system edition." Then, the PC attributed to the sites, $u_{2i}^{(i)}$, was used and SD was optimized (Fig. 8.17). After attributing P values to is and correcting the obtained P values with BH criterion,

Table 8.9 Fisher's exact tests between the selected is and probes with annotations. DMR: known differentially methylated regions, DHS: DNase I hypersensitive site

	Adjusted P value			Adjusted P value			Adjusted P value	
GSE77965								
DMR	> 0.01	< 0.01	Enhancer	> 0.01	< 0.01	DHS	> 0.01	< 0.01
NO	447734	506	No	381647	1371	No	424603	1058
YES	36104	1233	YES	102191	368	Yes	59235	681
Odds ratio	30.19			1.002			4.61	
P values	0			0.97			5.63×10^{-177}	
GSE42308								
DMR	> 0.01	< 0.01	Enhancer	> 0.01	< 0.01	DHS	> 0.01	< 0.01
NO	417585	30655	No	358362	24656	No	358362	24656
YES	33366	3971	YES	92589	9970	Yes	92589	9970
Odds ratio	1.62			1.56			1.54	
P values	5.88×10^{-147}			7.03×10^{-269}			1.16×10^{-174}	

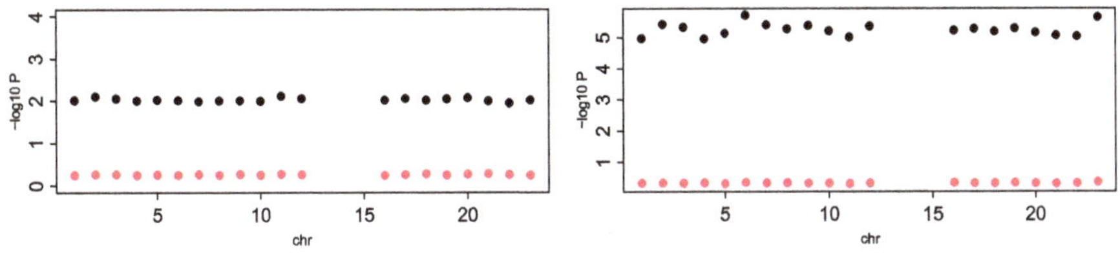

Fig. 8.16 Mean negative logarithm of mean P values of DHS (black circles) and non-DHS sites (red circles) within each chromosome (horizontal axis) when PCA-based unsupervised FE with optimized SD was applied to GSE34864 (left) and EH1072 (right). P values computed using t-test between these two were zero for all chromosomes (not shown here). Some chromosomes did not have an associated DHS

we identified 9544 and 11,447 probes that were associated with distinctions between multiple classes in "ajcc staging system edition" for breast and colon cancers, respectively (i.e., those associated with adjusted P values less than 0.01).

Now we come to the stage to evaluate the selected is biologically. Since it is a clinical data, the coincidence with various functional sites is not suitable for evaluation. Instead of that, we have evaluated them by uploading to Enrichr (Figs. 8.18, 8.19, 8.20, and 8.21).

It is obvious that PCA-based unsupervised FE with optimized SD could identify many biologically significant terms and could outperform COHCAP [32], which is one of de facto standard methods (Table 8.10). In conclusion, PCA-based unsupervised FE with optimized SD could not only identify functionally regulated sites, but also identify clinically informative sites as well.

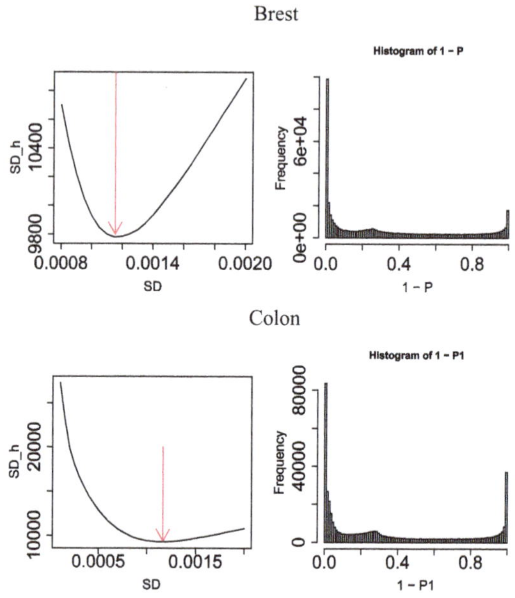

Fig. 8.17 Optimization of σ_ℓ. Left column: σ_h as a function of σ_ℓ. The vertical red arrows show the selected σ_ℓ. Right column: the histogram $1 - P_i$ with optimized σ_ℓ

8.3.4 Histone Modification

The next topics to which PCA- and TD-based unsupervised FE with optimized SD was applied is histone modification [31]. As described in Sect. 6.2.5, histone modification is one of the major factors that regulate gene expression. Thus, if PCA- and TD-based unsupervised FE with optimized SD are applicable to the analysis of histone modification, it is very useful. In contrast to DNA methylation whose state is either methylated or unmethylated, histone modification has more varieties. In addition to this, DNA methylation is the modification of DNA, whereas histone modification is not a modification of DNA itself but that of protein, histone, that binds to DNA. Thus the relationship between histone modification and DNA is indirect. This makes more difficult to detect histone modification than DNA methylation. Thus, even if PCA- and TD-based unsupervised FE with optimized SD proposed for gene expression could be applicable to DNA methylation, using PCA- and TD-based unsupervised FE with optimized SD toward histone modification is still challenging.

To demonstrate the usability of PCA- and TD-based unsupervised FE with optimized SD toward histone modification, we have retrieved various data sets from GEO (Table 8.11) that include various kinds of histone modification. These are generally taken from five "core histone marks" proposed by the Roadmap Epigenomics Consortium [12] (H3K4me1/H3K27ac, H3K4me3, H3K36me3, H3K27me3, and H3K9me3). Since all of them were retrieved by HTS, we need to preprocess them before TD or PCA was applied. To do this, we introduced genomic regions within which histone modifications were averaged or summed as in the previous section

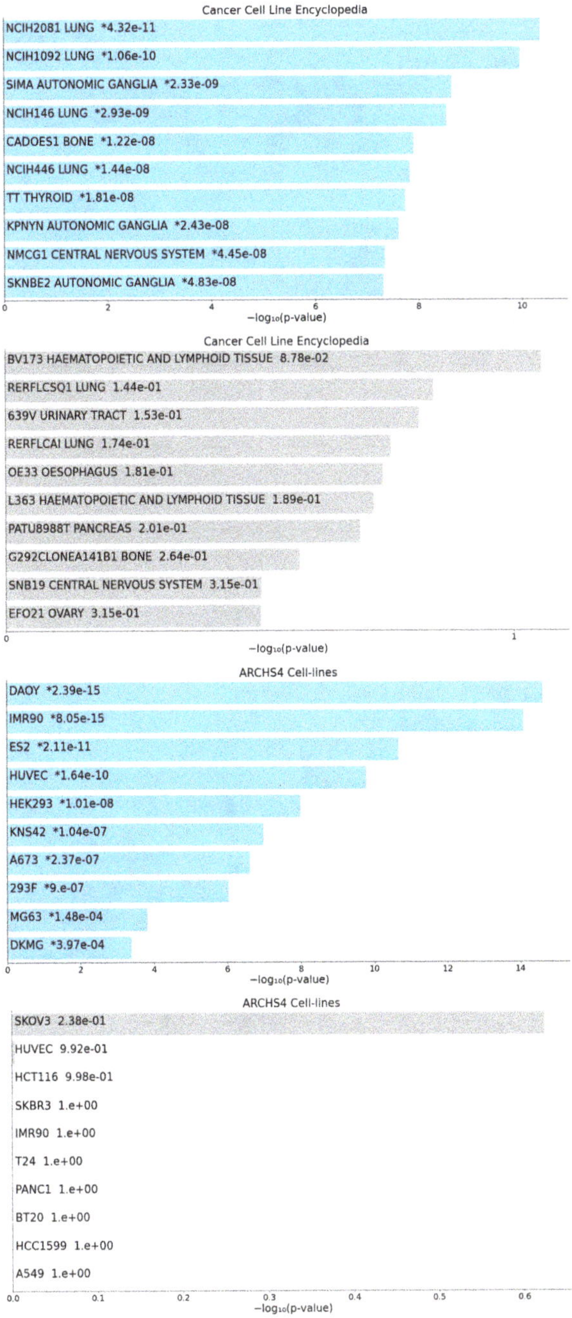

Fig. 8.18 Top ten-ranked terms of "Cancer cell line encyclpedia" and "ARCHS4 Cell-lines" categories in Enricher for breast cancer. PCA-based unsupervised FE (the top and the third row) and COHCAP (the second and the fourth row). Blue colored: raw $P < 0.05$ and asterisked: adjusted $P < 0.05$, respectively

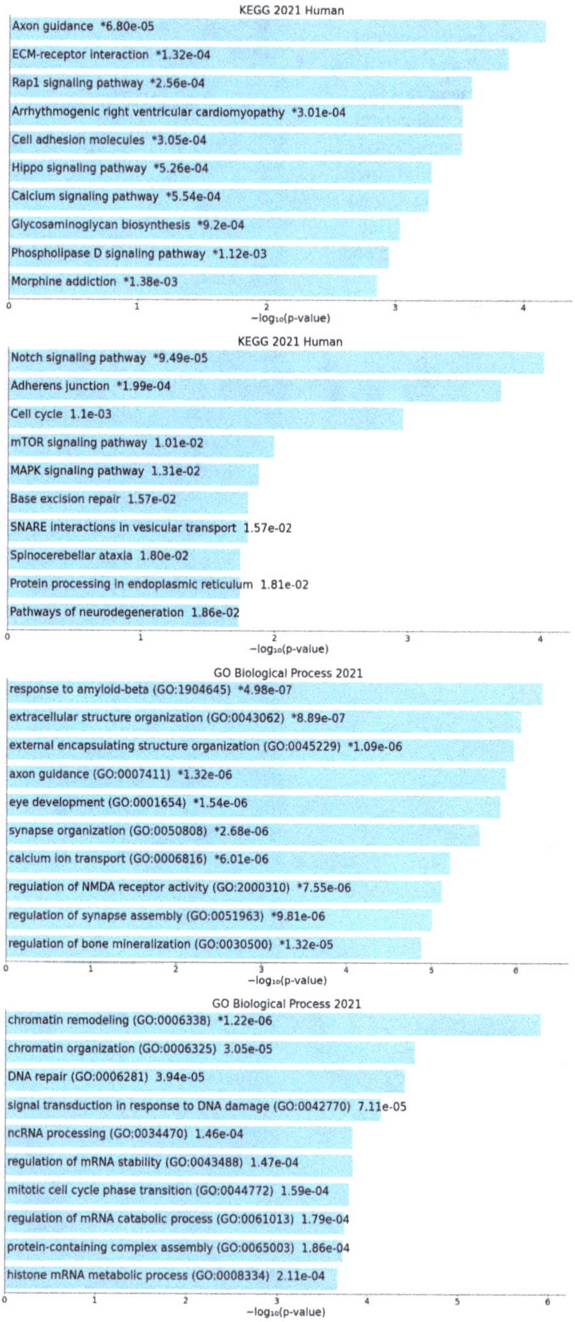

Fig. 8.19 Top ten-ranked terms of "KEGG 2021 HUMAN" and "GO Biological process 2021" categories in Enricher for breast cancer. PCA-based unsupervised FE (the top and the third row) and COHCAP (the second and the fourth row). Blue: raw $P < 0.05$ and asterisked: adjusted $P < 0.05$, respectively

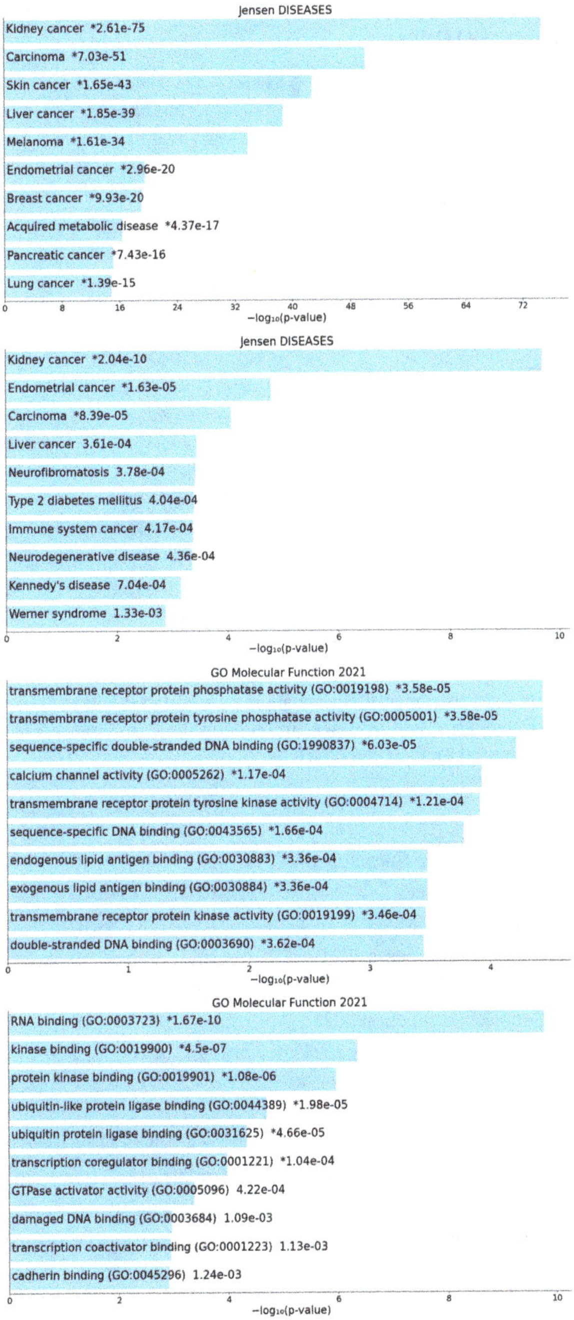

Fig. 8.20 Top ten-ranked terms of "Jensen Diseases" and "GO Molecular function 2021" categories in Enricher for breast cancer. PCA-based unsupervised FE (the top and the third row) and COHCAP (the second and the fourth row). Blue: raw $P < 0.05$ and asterisked: adjusted $P < 0.05$, respectively

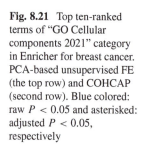

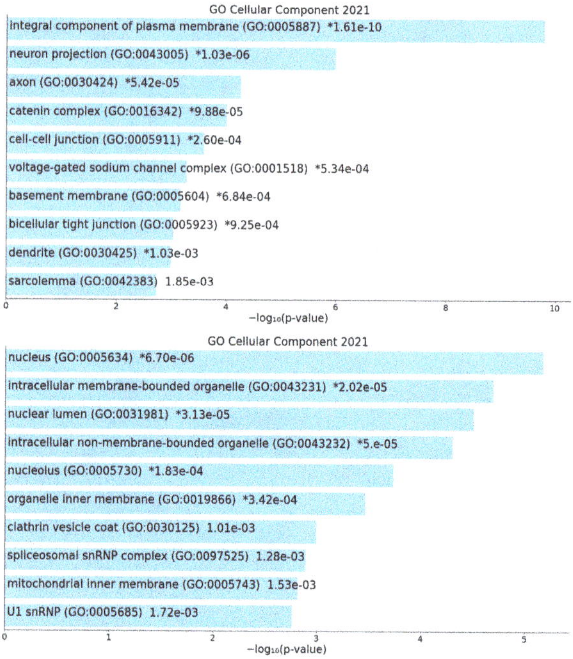

Fig. 8.21 Top ten-ranked terms of "GO Cellular components 2021" category in Enricher for breast cancer. PCA-based unsupervised FE (the top row) and COHCAP (second row). Blue colored: raw $P < 0.05$ and asterisked: adjusted $P < 0.05$, respectively

Table 8.10 Number of significant (i.e., adjusted P values are less than 0.05) variables (e.g., cell lines, pathways, and diseases etc.) among the top 10 ranked factors in individual categories for the proposed method and COHCAP

	Proposed method		COHCAP	
Category	Breast	Colon	Breast	Colon
Cancer Cell Line Encyclopedia	10	10	0	0
ARCHS4 cell lines	10	10	0	0
KEGG 2021 HUMAN	10	10	2	0
GO biological process 2021	10	10	1	9
Jensen diseases	10	10	3	7
GO molecular function 2021	10	10	6	7
GO cellular components 2021	9	10	6	0

Sect. 7.13, although length of regions, 25,000 is used only for H3K9me3the as in Sect. 7.13, whereas 1,000 is used for others. There are two kinds of measurements for histone modification. One is coarse-grained histone modification, for which histone modification is averaged within individual intervals; the other is the genomic loci, for which histone modification occurs. Coarse-grained values are further averaged with the individual regions of length L for the former, whereas the regions overlapping the regions of length L are counted for the latter.

We have applied TD-based unsupervised FE with optimized SD to these histone modification profiles and obtained the results (Table 8.12). Before applying TD-

Table 8.11 The list of histone modification profiles analyzed in this chapter

GEO ID	Histone modification	Description
GSE24850	H3K9me3	This study contained 11 H3K9me3 ChIP-seq mouse nucleus accumbens experiments comprising six controls, three saline-treated samples, and two cocaine samples. Among them, five controls and five treated samples were employed.
GSE159075	H3K4me3, H3K27me3, H3K27ac	This study contained various histone modification ChIP-seq experiments using human umbilical vein endothelial cell lines. Among them, three H3K4me3 ChIP-seq, three H3K27me3 ChIP-seq, and three H3K27ac ChIP-seq, as well as one input ChIP-seq profile were employed.
GSE74055	H3K4me1	This study contained various histone modification ChIP-seq experiments using mouse E14 or DKO cell lines. Among them, 16 H3K4me1 ChIP-seq profiles and the corresponding 16 input profiles (32 profiles in total) were downloaded.
GSE124690	H3K4me1, H3K4me3, H3K27ac	This study comprised various histone modification CUT&Tag experiments using human H1 or K562 cell lines. Among them, six bulk H3K4me1 profiles, four bulk H3K3me3 profiles, and four bulk H3K27ac profiles were downloaded.
GSE188173	H3K27ac	This study contained nine ChIP-seq H3K27ac profiles (with one control and one treated with SPT) using patient-derived xenografts of human castration-resistant prostate cancer.
GSE159022	H3K27me3	This study comprised four H3K4me3 ChIP-seq profiles, four H3K27me3 ChIP-seq profiles, and four H4K16ac ChIP-seq profiles using mouse progenitor cells (two wild type (WT) and two neurofibromin knockouts). Among them, four H3K27me3 profiles were used.
GSE168971	H3K9ac	This study contained H3K27ac and H3K9ac ChIP-seq profiles taken from various experimental conditions, six H3K9ac profiles using C3H-WT mouse liver, and two corresponding inputs were used.
GSE159411	H3K36me3	This study comprised various ChIP-seq profiles. Among them, four H3K36me3 ChIP-seq profiles (two hiPSC cardiomyocytes and two WT hiPSCs) were used.
GSE181596	H3K27me3	This study consisted of four H3K27me3 ChIP-seq profiles (two controls and two treatments) and four H3K4me3 ChIP-seq profiles (two controls and two treatments) in addition to two input profiles that used cells as odontoblasts (treatment was siRNA: si-IKBz)

based unsupervised FE with optimized SD to these histone modification profiles, $x_{ij} \in \mathbb{R}^{N \times M}$, which represents the histone modification of the ith region of the jth sample, x_{ij} is normalized as

$$\sum_{i=1}^{N} x_{ij} = 0 \tag{8.28}$$

Table 8.12 The number of regions/peaks, Entrez gene IDs, and gene symbols selected by various methods and that of associated enriched histone modification (all and targeted) in the "Epigenomics Roadmap HM ChIP-seq" category of Enrichr with adjusted P values less than 0.05 for profiles shown in Table 8.11. *: "ENCODE Histone Modifications 2015" category is used

GEO ID	Histone Modification	Regions /Peaks	Entrez Genes	Gene Symbols	Histone Modification All	Histone Modification Targeted	Species
GSE24850	H3K9me3	1,302	894	641	10	10	Mouse
GSE159075	H3K4me3	34,538	13,692	13,671	198	54	Human
	H3K27me3	62,141	5,217	5,208	83	56	
	H3K27ac	61,306	11,604	11,590	175	24	
GSE74055	H3K4me1	61,329	11,890	11,858	58*	6*	Mouse
(when PCA is replaced with TD)		70,187	14,220	14,187	102 *	10 *	
GSE124690	H3K4me1	164,466	14,893	14,866	3*	0*	Human
(CUT&Tag)	H3K4me3	37,534	14,972	14,946	200	54	
	H3K27ac	81,249	13,086	13,061	139	24	
GSE188173	H3K27ac	105,438	15,579	15,548	155	24	Human
GSE159022	H3K27me3	55,923	5,022	4,996	70	56	Mouse
GSE168971	H3K9ac	58,490	15,460	15,452	81*	6*	Mouse
GSE159411	H3K36me3	253,326	12,282	12,270	201	32	Human
GSE181596	H3K27me3	36,972	3,543	3,545	72	56	Human

$$\sum_{i=1}^{N} x_{ij}^2 = N. \tag{8.29}$$

After applying SVD to x_{ij}s as

$$x_{ij} = \sum_{\ell=1}^{L} \lambda_\ell u_{\ell i}^{(i)} v_{\ell j}^{(j)} \tag{8.30}$$

and $v_{\ell j}^{(j)}$s of interest were selected (Figs. 8.22 and 8.23). For GSE74055, SVD was applied to 16 H3K4me1 modifications divided by the corresponding input (regions, is, for which input samples were zero, were discarded in advance). Thus the number of samples is not 32 but 16.

Then using the corresponding $u_{\ell i}^{(i)}$, σ_ℓs were optimized (Figs. 8.24, 8.25, and 8.26).

They are somewhat different from those (Figs. 8.14, 8.16, and 8.21) gotten when TD-based unsupervised FE with optimized SD was applied to gene expression profile and DNA methylation in the preceding two subsections. They are not combination of flat and peak but are seemingly composed of two Gaussian.

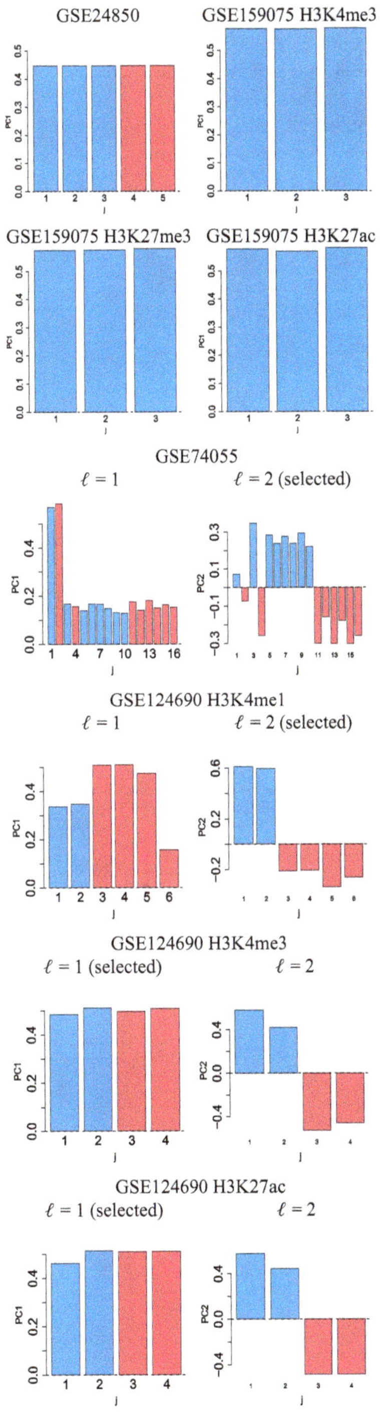

Fig. 8.22 $v_{\ell j}^{(j)}$s for histone modification profiles. For GSE24850 and GSE159075, only $\ell = 1$, which is selected, is shown. For others, $\ell = 1, 2$ are shown. Those selected is labeled as "(selected)." Blue and red correspond to three saline-treated samples versus two cocaine samples (GSE24850), E14 vs. DKO (GSE74055), and H1 vs. K562 (GSE124690)

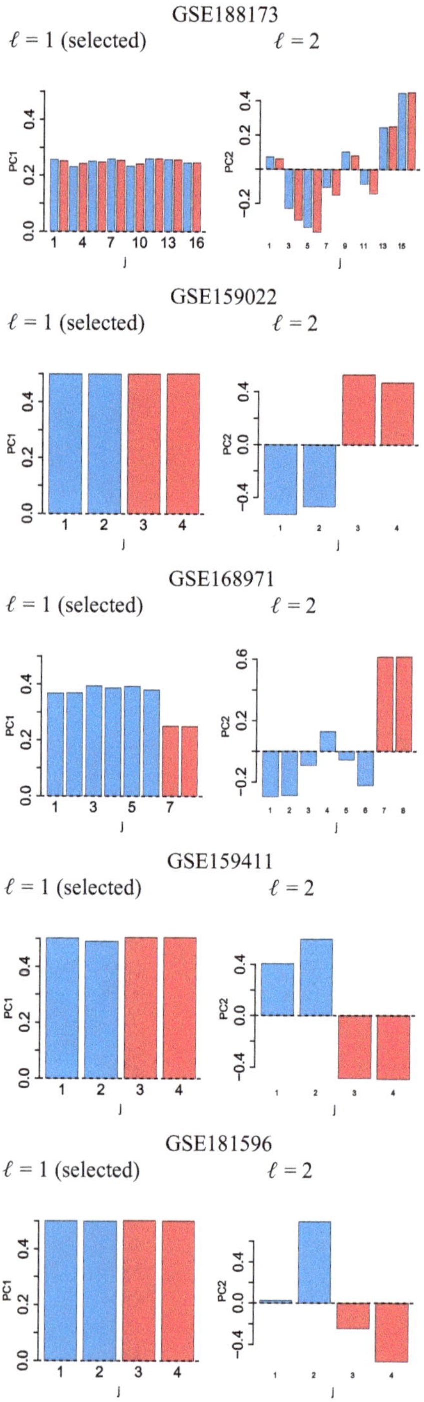

Fig. 8.23 $v_{\ell j}^{(j)}$s for histone modification profiles. $\ell = 1, 2$ are shown. Those selected are labeled as "(selected)." Blue and red colors correspond to: control vs. treated with SPT (GSE18173), Het vs. KO (GSE159022), protein binding vs. input (GSE168971), cpc vs. iPSC (GSE159411), and control vs. IKBz (GSE181596)

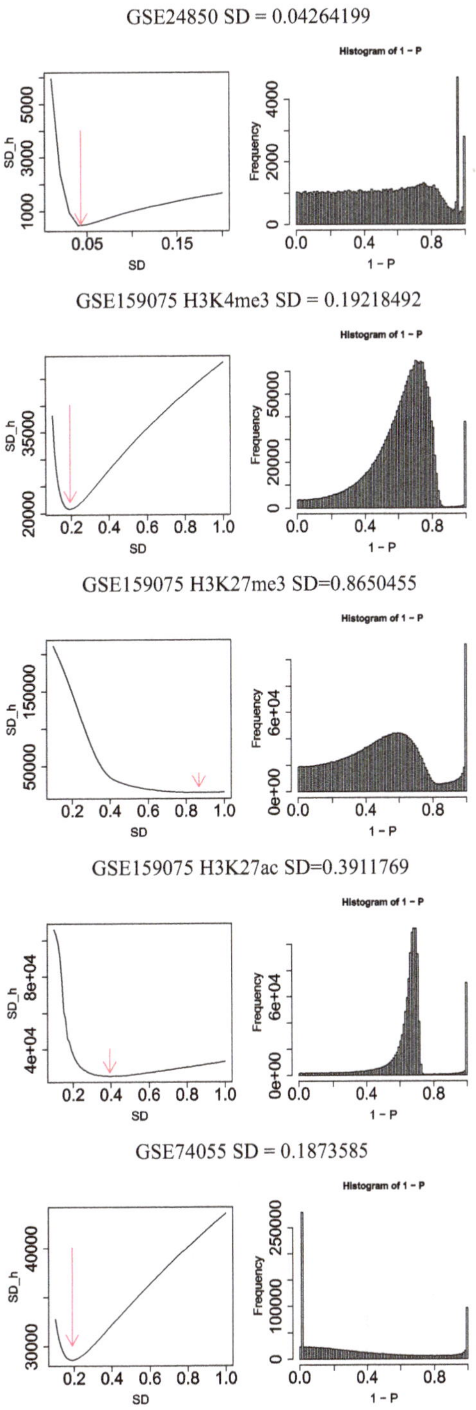

Fig. 8.24 Optimization of σ_ℓ. Left column: σ_h as a function of σ_ℓ. The vertical red arrows show the selected σ_ℓ. Right column: the histogram $1 - P_i$ with optimized σ_ℓ

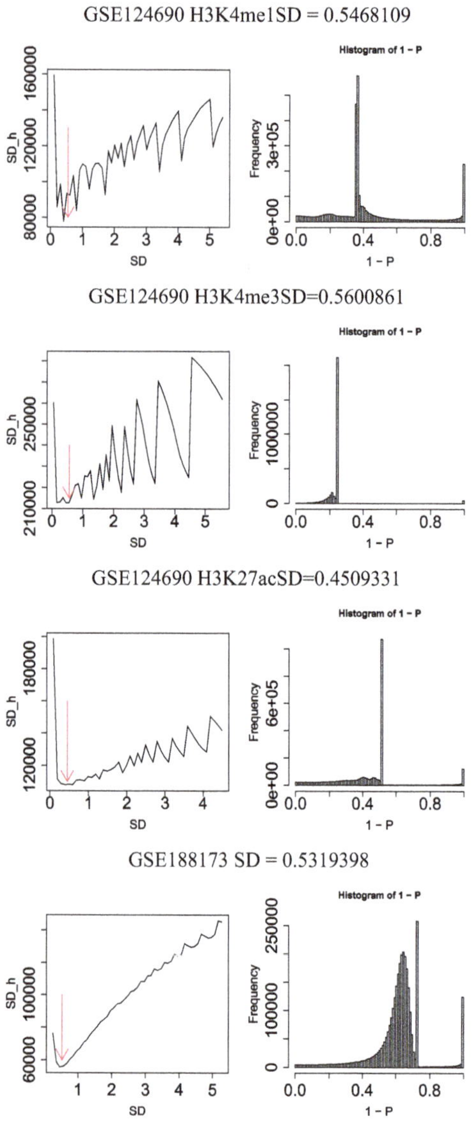

Fig. 8.25 Optimization of σ_ℓ. Left column: σ_h as a function of σ_ℓ. The vertical red arrows show the selected σ_ℓ. Right column: the histogram $1 - P_i$ with optimized σ_ℓ

Nevertheless the selected genes associated with regions with adjusted P values less than 0.01 are reasonable. To validate the selected genes, we uploaded them to Enrichr and investigated "Epigenomics Roadmap HM ChIP-seq" category (for some histone modifications we employed "ENCODE Histone Modifications 2015"). All of them are associated with some histone modifications by adjusted P values less than 0.05 (The numbers are denoted as "All" in Table 8.6). Among them, targeted histone modifications are associated excluding H3K4me1 in GSE124690. Thus, basically, TD-based unsupervised FE with optimized SD can be applicable

Fig. 8.26 Optimization of σ_ℓ. Left column: σ_h as a function of σ_ℓ. The vertical red arrows show the selected σ_ℓ. Right column: the histogram $1 - P_i$ with optimized σ_ℓ

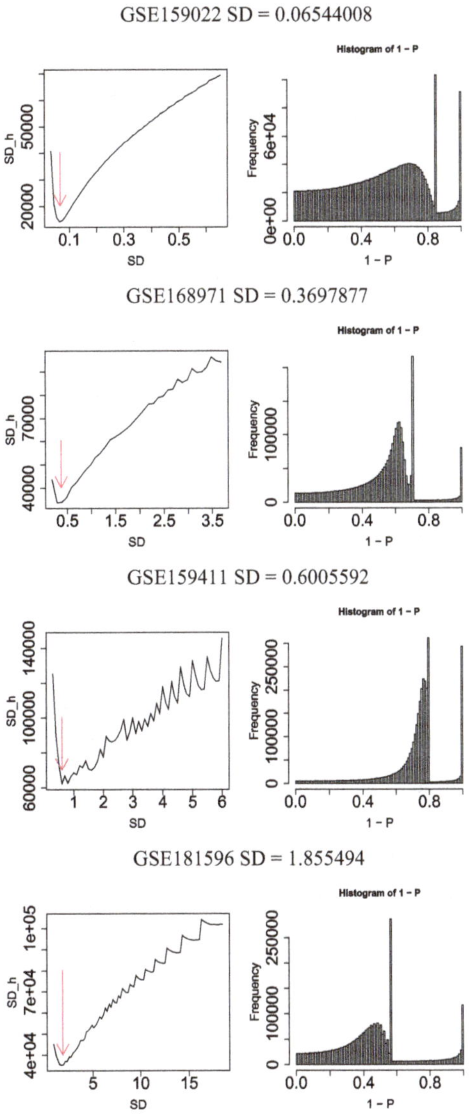

to histone modification without any modifications from those applied to gene expression and DNA methylation. It is remarkable since histone modification is usually regarded to be distinct from DNA methylation and gene expression, and the so-called peak identification algorithm was employed.

To demonstrate the effectiveness of the proposed method, we compared it with various methods (Table 8.13) with the data set GSE24850 (Table 8.14).

Since MOSAiCS and DFilter accept only pairwise comparisons, five pairs were individually tested by the methods. It is obvious the coincidence between

Table 8.13 Names of methods and descriptions

MOSAiCS
MOSAiCS [24] was implemented as a bioconductor package [7]. We installed version 2.32.0 of MOSAiCS in R [19] and applied it to GSE24850. MOSAiCS provides statistical models that are biologically motivated for reads that arise under both non-enrichment (background) and enrichment (signal). Additionally, MOSAiCS constructs a parametric background model that accounts for biases such as GC content and mappability that are inherent to ChIP-seq data. The MOSAiCS model does not assume punctuated or broad peak structures, but instead quantifies whether the ChIP reads show enrichment compared to the background reads for every genomic interval (e.g., bin) of user-defined size in the genome.
DFilter
Kumar et al. implemented DFilter [11] as a Linux command-line program. We downloaded version 1.6 from https://reggenlab.github.io/DFilter/ (accessed on Jan. 7, 2024). DFilter requires a set of sequence tags mapped to a reference genome as input. Based on the genomic distribution of tags, the algorithm classifies individual n-base-pair bins as positive (signal) or negative (noise) regions. DFilter implements linear finite-impulse-response detection, which is a windowed linear filter h of user-specified width, followed by the standard thresholding step.
F-Seq2
F-Seq2 [35] is a Linux command-line program that uses the Gaussian kernel density function to quantify the amount of protein binding. It was downloaded from https://github.com/Boyle-Lab/F-Seq2 (accessed on Jan. 7, 2024). To account for the fluctuation of total reads between different chromosomes, the total control read count was linearly scaled to be equal to the total treatment read count at the individual chromosome level.
HOMER
HOMER [5] is a Linux command-line program that was downloaded from http://homer.ucsd.edu/homer/ (accessed on Jan. 7, 2024). The latest version, HOMER 4.11, was released on October 24, 2019. For each ChIP-seq experiment, ChIP-enriched regions (peaks) were identified by first detecting significant clusters of ChIP-seq tags and then filtering these clusters for those that were significantly enriched relative to background sequencing and local ChIP-seq signal.
RSEG
RSEG [23] is a Linux command-line program that was downloaded from http://smithlabresearch.org/software/rseg/ (accessed on Jan. 7, 2024). The latest version, 0.4.9, is used to quantify thc amount of protein binding between control and treated samples using the negative binomial distribution. The NBDiff distribution, which is the discrete distribution of the difference between two independent negative binomial random variables, is employed to achieve this.

selected genes and those associated with known histone modification is poorer than TD-based unsupervised FE with optimized SD. Although F-Seq2, HOMER, and RSEQ accept the comparison between multiple pairs, F-Seq2 failed because of too much cpu memory required and RSEG could not be even installed because of outdated implementation. HOMER could be tested, but again its performance was poorer than TD-based unsupervised FE with optimized SD. In conclusion, TD-based unsupervised FE with optimized SD could outperform various state-of-the-art methods even when applied to a histone modification profile.

Since all of these above are PCA-based unsupervised FE with optimized SD, to demonstrate that TD-based unsupervised FE with optimized SD also works for

Table 8.14 The number of regions/peaks, Entrez gene IDs, and gene symbols selected by various methods and that of the H3K9me3-associated enriched histone modification in the "Epigenomics Roadmap HM ChIP-seq" category of Enrichr with adjusted P values less than 0.05 for the GSE24850 data set (H3K9me3)

Methods	Pair No.	Regions /Peaks	Entrez Genes	Gene Symbols	Enriched Histone Modification	H3K9me3
Proposed method	–	1,302	894	641	10	10
MOSAiCS	1	4,367	1,833	994	3	1
	2	3,648	1,599	851	0	0
	3	2,096	1,136	567	0	0
	4	1,985	1,018	532	2	2
	5	5,556	2,223	1,184	2	0
DFilter	1	25,080	6,286	2,621	1	0
	2	22,863	5,721	2,524	1	0
	3	21,371	5,470	2,499	1	0
	4	23,811	5,987	2,631	1	0
	5	23,369	5,902	2,544	1	0
F-Seq2	–	–	–	–	–	–
HOMER	–	114,727	6,771	6,747	1	0
RSEG	–	–	–	–	–	–

histone modification, we employed GSE74005 that was formatted as a tensor, $x_{ijk} \in \mathbb{R}^{N \times 16 \times 2}$, which represents histone modification of the ith region of jth sample at the kth treatment ($k = 1$: treated, $k = 2$: input). HOSVD was applied to x_{ijk}, and we got

$$x_{ijk} = \sum_{\ell_1=1}^{N} \sum_{\ell_2=1}^{16} \sum_{\ell_3=1}^{2} G(\ell_1 \ell_2 \ell_3) u_{\ell_1 i}^{(i)} u_{\ell_2 j}^{(j)} u_{\ell_3 k}^{(k)}, \tag{8.31}$$

where $G(\ell_1 \ell_2 \ell_3) \in \mathbb{R}^{N \times 16 \times 2}$ is a core tensor and $u_{\ell_1 i}^{(i)} \in \mathbb{R}^{N \times N}$, $u_{\ell_2 j}^{(j)} \in \mathbb{R}^{16 \times 16}$, and $u_{\ell_3 k}^{(k)} \in \mathbb{R}^{2 \times 2}$ are singular value matrices and orthogonal matrices. At first, we need to find $u_{\ell_2 j}^{(j)}$ and $u_{\ell_3 k}^{(k)}$ of interest. Figure 8.27 shows $u_{1j}^{(j)}$ and $u_{2k}^{(k)}$ that represent the independence of replicates and the distinction between treatment, respectively. Then we decided to employ these two. Next, we need to find which $u_{\ell_1 i}^{(i)}$ is most associated with these two. To do this, we investigated $G(\ell_1 12)$ and found that $|G(212)|$ is maximum (not shown here). Then $u_{2i}^{(i)}$ was used to attributed P values to i as

$$P_i = P_{\chi^2}\left[> \left(\frac{u_{2i}^{(i)}}{\sigma_2} \right)^2 \right], \tag{8.32}$$

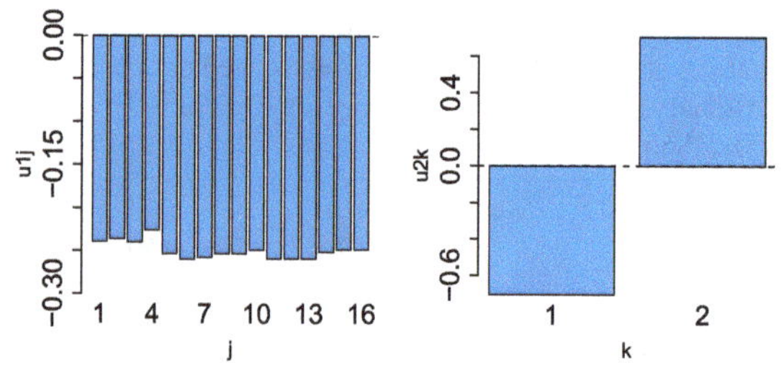

Fig. 8.27 $u_{1j}^{(j)}$ (left) and $u_{2k}^{(k)}$ (right) for GSE74005

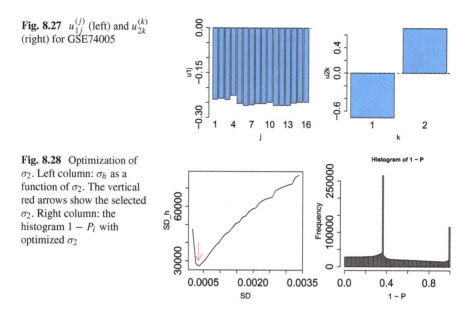

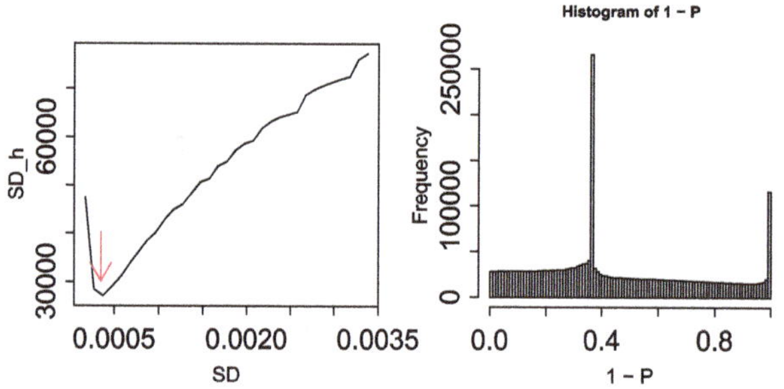

Fig. 8.28 Optimization of σ_2. Left column: σ_h as a function of σ_2. The vertical red arrows show the selected σ_2. Right column: the histogram $1 - P_i$ with optimized σ_2

where σ_2 must be optimized (Fig. 8.28). Then we found that the optimized value of σ_2 is 3.451553×10^{-4}. The number of regions selected based upon adjusted P values, those of gene IDs and gene symbols included in these regions, and the associations between genes and those associated with known histone modification were listed in Table 8.12 and are clearly improved from those by PCA-based unsupervised FE with optimized SD. This means that TD-based unsupervised FE with optimized SD can potentially improve the results obtained by PCA-based unsupervised FE with optimized SD even when applied to histone modification.

8.3.5 *scATAC-seq*

Next we consider the application of TD-based unsupervised FE with optimized SD to scATAC-seq [29]. As demonstrated in Sects. 7.12.4 and 7.13.2 (in the former, although it was not explicitly denoted, DNA accessibility was essentially measured by scATAC-seq), ATAC-seq was usually analyzed with some other omics profiles. In this subsection, we try to analyze scATAC-seq, which is single-cell version of ATAC-seq, without any other associated omics data set. To demonstrate capability of TD-based unsupervised FE with optimized SD applied to scATAC-seq, we employed two data sets retrieved from GEO with GEO IDs GSE167050 and GSE139950, respectively. The former was composed of four mice tissues (CTX, MGE, CGE, and LGE from mouse embryonic forebrain) with two replicates each, whereas the latter includes mice kidney of unilateral ureter obstruction at Days 0, 2, and 10 also with two replicates (Table 8.15). These ATAC-seq profiles were formatted as tensors, $x_{ij_kkm} \in \mathbb{R}^{N \times M_{km} \times K \times 2}$, that represent ATAC values in the ith

Table 8.15 Number of single cells included in the files analyzed in this chapter. CTX: Cortex, MGE: Medial ganglionic eminence, CGE: Caudal ganglionic eminence, LGE: Lateral ganglionic eminence

GSE167050									
Tissues	CTX1	MGE1	CGE1	LGE1	CTX2	MGE2	CGE2	LGE2	Total
Number of single cells (M_{km})	4108	6845	4013	6577	4946	3465	4530	4769	39,253

GSE139950							
	Day 0		Day 2		Day 10		
Biological replicates	1	2	1	2	1	2	Total
Number of single cells (M_{km})	10,521	7499	5828	9792	8103	5636	47,379

region of the j_{km}th sample at the kth profile ($K = 4$ for GSE167050 and $K = 3$ for GSE139950, respectively) in the mth replicate. scATAC-seq values are integrated within the 200 amino acid length regions as in Sect. 7.12.4. The number of regions is as many as 13,627,618.

To apply TD-based unsupervised FE with optimized SD to $x_{ij_{km}km}$, we need to process distinct M_{km} that prevents us from applying TD to $x_{ij_{km}km}$ as it is. To do this we apply SVD to individual $x_{ij_{km}km}$ as

$$x_{ij_{km}km} = \sum_{\ell=1}^{L} \lambda_\ell u_{\ell i}^{km} v_{\ell j_{km}}^{km} \tag{8.33}$$

after the standardization

$$\sum_i x_{ij_{km}km} = 1, \tag{8.34}$$

which means $x_{ij_{km}km}$ is replaced with $\frac{x_{ij_{km}km}}{\sum_i x_{ij_{km}km}}$. Then we collected $u_{\ell i}^{mk}$s to construct

$$U_{\ell i} = \begin{cases} u_{\ell i}^{11}, & 1 \le \ell \le L \\ u_{(\ell-L)i}^{21}, & L+1 \le \ell \le 2L \\ u_{(\ell-2L)i}^{31}, & 2L+1 \le \ell \le 3L \\ u_{(\ell-3L)i}^{41}, & 3L+1 \le \ell \le 4L \\ u_{(\ell-4L)i}^{12}, & 4L+1 \le \ell \le 5L \\ u_{(\ell-5L)i}^{22}, & 5L+1 \le \ell \le 6L \\ u_{(\ell-6L)i}^{32}, & 6L+1 \le \ell \le 7L \\ u_{(\ell-7L)i}^{42}, & 7L+1 \le \ell \le 8L \end{cases} \in \mathbb{R}^{2KL \times N} \tag{8.35}$$

for GSE167050 ($K = 4$) and

$$U_{\ell i} = \begin{cases} u^{11}_{\ell i}, & 1 \le \ell \le L \\ u^{12}_{(\ell-L)i}, & L+1 \le \ell \le 2L \\ u^{21}_{(\ell-2L)i}, & 2L+1 \le \ell \le 3L \\ u^{22}_{(\ell-3L)i}, & 3L+1 \le \ell \le 4L \\ u^{31}_{(\ell-4L)i}, & 4L+1 \le \ell \le 5L \\ u^{32}_{(\ell-5L)i}, & 5L+1 \le \ell \le 6L \end{cases} \in \mathbb{R}^{2KL \times N} \tag{8.36}$$

for GSE139950 ($K = 3$), respectively. Then SVD was applied to $U_{\ell i}$ to get

$$U_{\ell i} = \sum_{\ell'=1}^{2KL} \lambda_{\ell'} u'_{\ell' i} v'_{\ell' \ell}. \tag{8.37}$$

In the following, we used $L = 10$. Since we would like to evaluate single-cell embedding to low dimensional space, we recover singular value vectors attributed to single cells, j_k, in the kth profile as

$$V^{km}_{\ell' j_{km}} = \sum_{i=1}^{N} x_{i j_{km} km} u'_{\ell' i} \in \mathbb{R}^{2KL \times M_{km}}. \tag{8.38}$$

Then we get concatenated matrix $V_{\ell' j} \in \mathbb{R}^{2KL \times \sum_k \sum_m M_{km}}$. UMAP [16] was applied to $V_{\ell' j}$ with `umap.defaults$n_neighbors <-30` option, and we got the low dimensional embedding (Figs. 8.29 and 8.30). They are quite common between

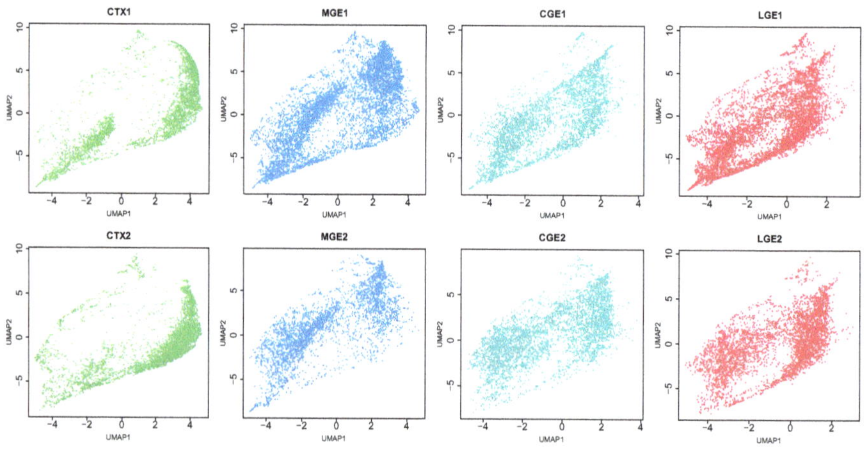

Fig. 8.29 UMAP embedding of eight samples analyzed in this chapter for GSE167050 (see Table 8.15)

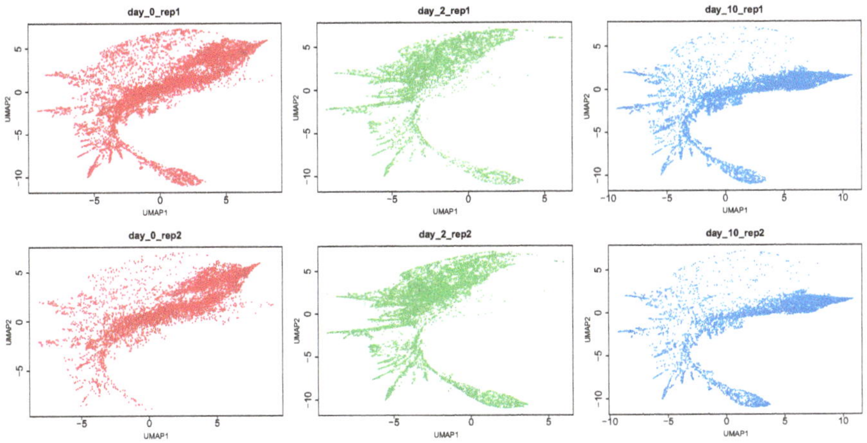

Fig. 8.30 UMAP embedding of eight samples analyzed in this chapter for GSE139950 (see Table 8.15)

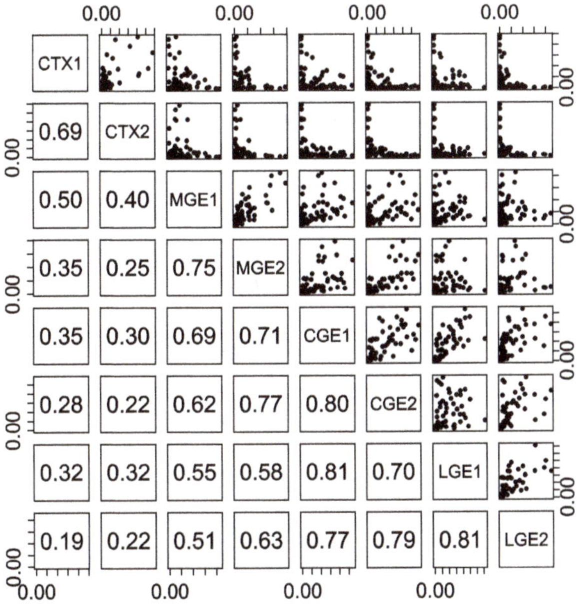

Fig. 8.31 Upper: scatter plot of the number of single cells in Fig. 8.29, $P(I, J)$, $1 \leq I, J \leq 10$, in one of 10×10 regions for GSE167050. Lower: Kendall's correlation coefficients of $P(I, J)$ between eight samples

replicates, i.e., tissue specific for GSE167050 and day specific for GSE139950, respectively.

To see whether low dimensional embedding is really common between replicates, we need to define distance measure between embedding. For this purpose, we computed coarse-grained distribution within segmented grid in low dimensional space. We divided whole plain into 10×10 grid space and counted the number of single cells included in individual grid space. Then the numbers are divided by the

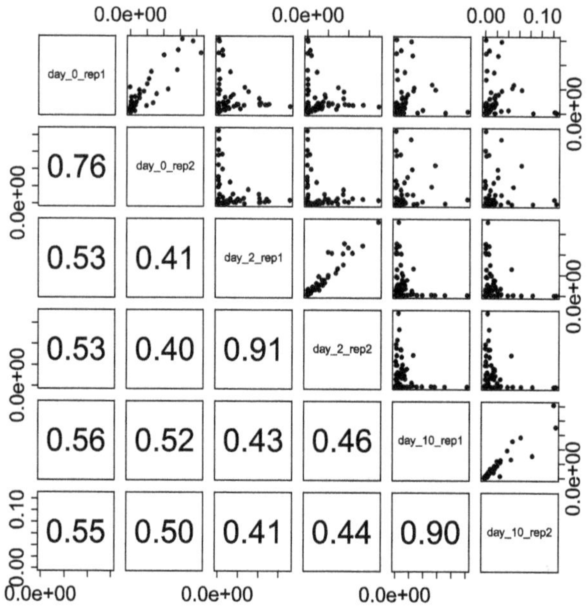

Fig. 8.32 Upper: scatter plot of the number of single cells in Fig. 8.30, $P(I, J)$, $1 \leq I, J \leq 10$, in one of 10×10 regions for GSE139950. Lower: Kendall's correlation coefficients of $P(I, J)$ between six samples

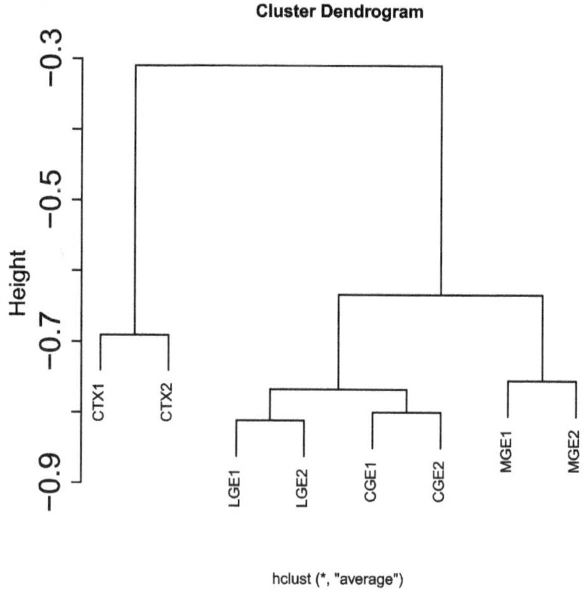

Fig. 8.33 Hierarchical clustering by UPGMA of eight profiles in GSE167050 (Table 8.15) using coordinates obtained by UMAP. Distance is represented by negatively signed correlation coefficients in Fig. 8.31

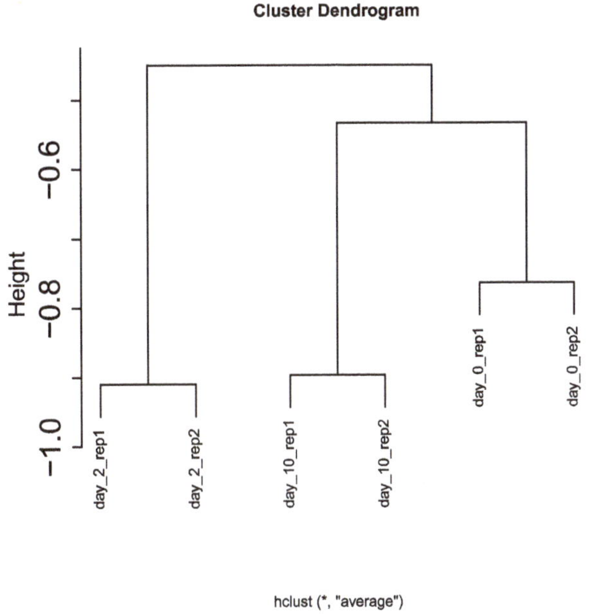

Fig. 8.34 Hierarchical clustering by UPGMA of six profiles in GSE139950 (Table 8.15) using coordinates obtained by UMAP. Distance is represented by negatively signed correlation coefficients in Fig. 8.32

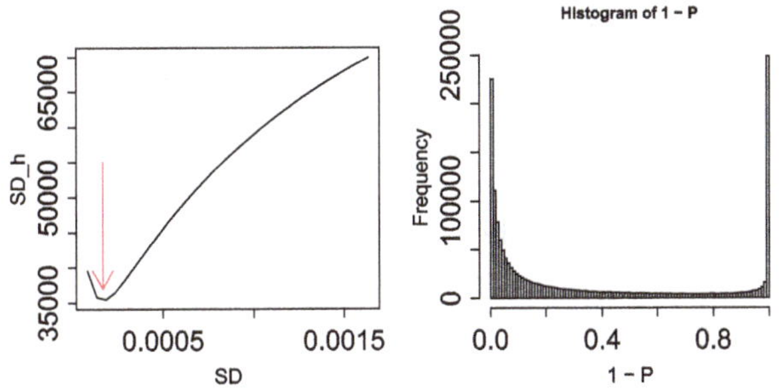

Fig. 8.35 Optimization of σ_2. Left column: σ_h as a function of σ_2. The vertical red arrows show the selected σ_2. Right column: the histogram $1 - P_i$ with optimized σ_2

total number of single cells. Then we get the probability $P(I, J), 1 \leq I, J \leq 10$ that represents the probability that the individual cells are placed in the (I, J)th grid space. Figures 8.31 and 8.32 show scatter plots of $P(I, J)$ between individual profiles. To see if $P(I, J)$ can discriminate profiles, after computing Kendall's coefficients of $P(I, J)$ between distinct profiles, UPGMA was applied with using the negative signed correlation coefficients as distance. It was performed by `hclust` function in R with `method="average"` option (Figs. 8.33 and 8.34). It is obvious that replicates are clustered together. Thus, it can discriminate distinct profiles well.

Since we have successfully confirmed that TD can generate low dimensional embedding by which UMAP can generate tissue specific low dimensional embedding, next we tried to see if TD can help us to select biologically reasonable genes. To do this, we attributed P values to the regions with the second singular value vector, u'_{2i}, for GSE167050,

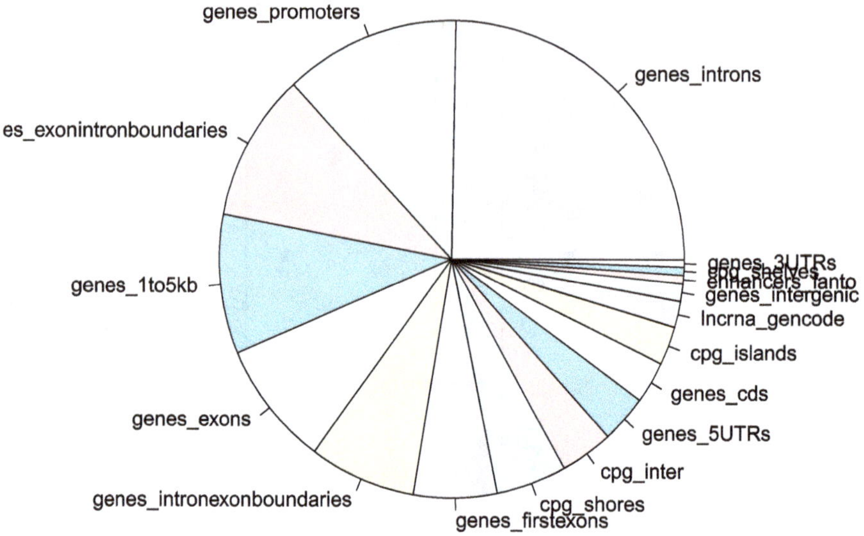

Fig. 8.36 Pie chart of annotations by annotatr about 216,875 regions selected by the proposed method

$$P_i = P_{\chi^2}\left[> \left(\frac{u'_{2i}}{\sigma_2}\right)^2 \right] \tag{8.39}$$

since the second singular value vectors are most often associated with distinction between biologically reasonable classification, where σ_2 was optimized to be 1.676208×10^{-4} (Fig. 8.35). Then we found that as many as 216,875 regions are associated with adjusted P values less than 0.01 within 1,575,170 regions selected from 13,627,618 by excluding regions without any associated scATAC-seq values. To see if these 216,875 regions are biologically reasonable, we used annotatr [2] to validate selected regions (Fig. 8.36). It is obvious that it is enriched with more functional regions than expected. The amount of the ratio of exon is as high as 8 %, which is four times larger than expectation, 2%.

To further validate the selected regions, we uploaded as many as 15,379 gene symbols included in these 216,875 regions to Enrichr. Table 8.16 lists the enriched terms in "MGI Mammalian Phenotype Level 4 2021" category in Enrichr. It is obvious that these terms are highly enriched with embryonic terms that is coincident with the fact that these samples were retrieved from mouse embryonic forebrain. To deny the possibility that it is accidental agreement, we further investigated yet another category, "KOMP2 Mouse Phenotypes 2022" (Table 8.17), where there are also many embryonic related terms included. In conclusion, TD-based unsupervised FE with optimized SD could not only generate low dimensional embedding well but also select biologically reasonable genes.

Table 8.16 Top ten-ranked enriched terms in "MGI Mammalian Phenotype 2022" category of Enrichr

Term	Overlap	P value	Adjusted P value
Preweaning lethality, complete penetrance MP:0011100	1268/1441	1.07×10^{-28}	4.93×10^{-25}
Embryonic growth retardation MP:0003984	502/559	1.09×10^{-15}	2.51×10^{-12}
Decreased embryo size MP:0001698	483/539	1.00×10^{-14}	1.47×10^{-11}
Embryonic lethality prior to organogenesis MP:0013292	335/364	1.28×10^{-14}	1.47×10^{-11}
Embryonic lethality during organogenesis, complete penetrance MP:0011098	526/600	8.39×10^{-12}	7.72×10^{-9}
Embryonic lethality prior to tooth bud stage MP:0013293	224/243	2.16×10^{-10}	1.65×10^{-7}
Embryonic lethality, complete penetrance MP:0011092	345/388	7.13×10^{-10}	4.46×10^{-7}
Open neural tube MP:0000929	134/140	7.76×10^{-10}	4.46×10^{-7}
Exencephaly MP:0000914	188/204	6.59×10^{-9}	3.37×10^{-6}
Embryonic lethality between implantation and somite formation, complete penetrance MP:0011096	238/266	8.68×10^{-8}	3.99×10^{-5}

Table 8.17 Top ten-ranked enriched terms in "KOMP2 Mouse Phenotypes 2022" category of Enrichr

Term	Overlap	P value	Adjusted P value
Preweaning lethality, complete penetrance (MP:0011100)	1631/1785	9.06×10^{-63}	4.79×10^{-60}
Embryonic lethality prior to organogenesis (MP:0013292)	482/510	5.95×10^{-28}	1.57×10^{-25}
Embryonic lethality prior to tooth bud stage (MP:0013293)	294/311	1.49×10^{-17}	2.63×10^{-15}
Abnormal embryo size (MP:0001697)	264/283	1.00×10^{-13}	1.32×10^{-11}
Embryonic growth retardation (MP:0003984)	183/198	5.59×10^{-9}	5.91×10^{-7}
Prenatal lethality prior to heart atrial septation (MP:0013294)	105/109	1.77×10^{-8}	1.56×10^{-6}
Abnormal embryo turning (MP:0001700)	69/70	2.20×10^{-7}	1.66×10^{-5}
Preweaning lethality, incomplete penetrance (MP:0011110)	668/801	2.50×10^{-6}	1.65×10^{-4}
Abnormal neural tube closure (MP:0003720)	69/73	4.95×10^{-5}	2.69×10^{-3}
Increased Startle Reflex (MP:0001488)	240/278	5.08×10^{-5}	2.69×10^{-3}

One might wonder if the other methods can achieve the similar performance. As described in the original study [29] the other conventional state-of-the-art method could not deal with the present data because of a huge number of variables, 13,627,618. Our method is the only method to deal with this size of the data set, since it can make use of sparse matrix format as described in Sect. 7.12.4.

8.3.6 Integrated Analysis of PPI and Gene Expression

In this subsection, we would like to discuss how we can make use of TD-based unsupervised FE with optimized SD to integrate gene expression and protein–protein interaction (PPI) [27]. In the previous subsections, various omics profiles were integrated. Nevertheless, all omics profiles have the shape of sample vs. features. On the other hand, a matrix representation of PPI is sample vs. sample. It is not clear how we can integrate sample vs. sample with sample vs. features.

The answer is as follows. Suppose that $x_{ij} \in \mathbb{R}^{N \times M}$ represents the expression of the ith gene of the jth sample and $n_{ii'} \in [0, 1]^{N \times N}$ represents the PPI. Then we apply SVD to x_{ij} and $n_{ii'}$ as

$$n_{ii'} = \sum_{\ell=1}^{L} \lambda_\ell u_{\ell i} u_{\ell i'}, \tag{8.40}$$

$$x_{ij} = \sum_{\ell=1}^{L} \lambda'_\ell u'_{\ell i} v_{\ell j}. \tag{8.41}$$

Then a tensor $x_{i\ell k} \in \mathbb{R}^{N \times L \times 2}$ is defined as

$$x_{i\ell k} = \begin{cases} u_{\ell i} & k = 1 \\ u'_{\ell i} & k = 2 \end{cases}, \tag{8.42}$$

and TD was applied to $x_{i\ell k}$ and we get

$$x_{i\ell k} = \sum_{\ell_1=1}^{N} \sum_{\ell_2=1}^{L} \sum_{\ell_3=1}^{2} G(\ell_1 \ell_2 \ell_3) \tilde{u}^{(i)}_{\ell_1 i} \tilde{u}^{(\ell)}_{\ell_2 \ell} \tilde{u}^{(k)}_{\ell_3 k}, \tag{8.43}$$

where $G \in \mathbb{R}^{N \times L \times 2}$, $\tilde{u}^{(i)}_{\ell_1 i} \in \mathbb{R}^{N \times N}$, $\tilde{u}^{(\ell)}_{\ell_2 \ell} \in \mathbb{R}^{L \times L}$, and $\tilde{u}^{(k)}_{\ell_3 k} \in \mathbb{R}^{2 \times 2}$ are singular value matrices and orthogonal matrices. Then the missing singular value vectors attributed to the samples, js, can be recovered as

$$\tilde{v}_{\ell_1 j} = \sum_{i=1}^{N} x_{ij} \tilde{u}^{(i)}_{\ell_1 i} \in \mathbb{R}^{L \times M}. \tag{8.44}$$

Possible evaluation of integrated analysis of gene expression with PPI is to see how much the coincidence between singular value vectors attributed to samples, $\tilde{v}_{\ell_1 j}$, and classification (class labels) is improved from those without integration, $v_{\ell j}$.

To do this, we selected the following data sets. For PPI, we used two data sets. One is the Stanford data set[1] and another is BIOGRID PPI Data Set [18]. In the former, $n_{ii'} \in [0, 1]^{N \times N}$, whereas $n_{ii'} \in \mathbb{N}^{N \times N}$ in the latter. In other words, although Stanford data set only considers if there is interaction between proteins, BIOGRID PPI data set can consider the strength of interaction, too. The number of proteins considered is also different. Stanford data set has $N = 21{,}557$ proteins, and BIOGRID data set has $N = 27{,}978$ proteins. Regarding gene expression, we used the TCGA data set included in the RTCGA package [9] which has the RTCGA.rnaseq [10] data set having RNA-seq data of up to 27 cancer cell lines (Table 8.17). They are also associated with more than or equal to one of five clinical labels, which are hereafter referred as short names, "vital_status," "pathologic_stage," "pathologic_m," "pathologic_t," and "pathologic_n." To integrate gene expression and PPI, we need to select common genes (proteins) between two. As a result, there are 16,774 genes (proteins) in common between gene expression and Stanford data set that includes 342,353 interactions, whereas there are 11,294 genes (proteins) in common between gene expression and BIOGDID data set that includes 437,679 interactions.

To validate coincidence between $v_{\ell j}$ or $\tilde{v}_{\ell_1 j}$ and class labels shown in Table 8.18, we employed categorical regression as

$$v_{\ell j} = a_\ell + \sum_s b_{\ell s} \delta_{js} \tag{8.45}$$

$$\tilde{v}_{\ell_1 j} = a'_{\ell_1} + \sum_s b'_{\ell_1 s} \delta_{js}, \tag{8.46}$$

where $a_\ell, b_{\ell s}, a'_{\ell_1}, b'_{\ell_1 s}$ are regression coefficients and δ_{js} takes one only when the jth sample is associated with the sth class in individual clinical labels otherwise zero.

To see whether integrated analysis can improve the coincidence between singular value vectors attributed to samples, $v_{\ell j}$, and class labels shown in Table 8.18, we attributed P values to ℓs or ℓ_1s using `lm` function in R. When there are L $v_{\ell j}$s or $\tilde{v}_{\ell j}$s considered in M_0 cancer cell lines (Table 8.18), there are as many as LM_0 P_ℓs or $P_{\ell 1}$s. To compare P_ℓs with P_{ℓ_1}s, we reorder each of them by ascending order as $P_h > P_{h'}$ when $h > h'$. Then, reordered P_hs were compared between P_ℓs and P_{ℓ_1}s using the one-sided Wilcoxon signed-rank sum test with the alternative hypothesis that P_hs made from P_{ℓ_1}s are lower (i.e., more coincident with class labels) than those made from P_ℓs. As a result, we can have five P values each of which is attributed to one of five clinical labels (Fig. 8.37). Although by Stanford data set, integrated analysis failed to improve the coincidence for all five clinical labels that by BIOGRID data set successfully improve coincidence. Thus, the integrated

[1] Human Protein-Protein Interaction Network. 2018. Available online: https://snap.stanford.edu/biodata/datasets/10000/10000-PP-Pathways.html (accessed on 10 Jan 2024).

Table 8.18 Availability of class labels for 27 cancer cell line data sets. (1) "patient.vital_status," (2) "pathologic_stage," (3) "pathologic_m," (4) "pathologic_t," and (5) "pathologic_n"

	ACC	BLCA	BRCA	CESC	COAD	ESCA	GBM	HNSC	KICH	KIRC	KIRP	LGG	LIHC	LUAD
(1)	○	○	○	○	○	○	○	○	○	○	○	○	○	○
(2)	○	○	○		○	○		○	○	○	○		○	○
(3)		○	○	○	○	○		○	○	○	○		○	○
(4)	○	○	○	○	○	○		○	○	○	○		○	○
(5)	○	○	○	○	○	○		○	○	○	○		○	○
	LUSC	OV	PAAD	PCPG	PRAD	READ	SARC	SKCM	STAD	TGCT	THCA	UCEC	UCS	Total ($= M_0$)
(1)	○	○	○	○	○	○	○	○	○	○	○	○	○	27
(2)	○		○			○		○	○	○	○			18
(3)	○		○			○		○	○	○	○			18
(4)	○		○		○	○		○	○	○	○			20
(5)	○		○		○	○		○	○	○	○			20

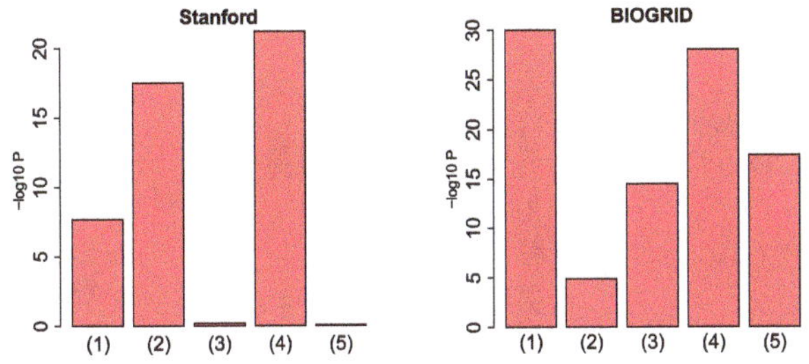

Fig. 8.37 Barplot of P values computed by the one-sided Wilcoxon signed-rank sum test to evaluate the difference in ascending ordered P_h between $v_{\ell j}$ and $\tilde{v}_{\ell j}$ when Stanford PPI (left) or BIOGRID PPI (right) was used. (1) "vital_status," (2) "pathologic_stage," (3) "pathologic_m," (4) "pathologic_t," (5) "pathologic_n"

analysis is effective to improve the coincidence. For readers interested in individual scatter plots of P_h between P_ℓs and P_{ℓ_1}s, see the original article [27].

Although we successfully demonstrated that the integrated analysis could improve the coincidence, it is much better if we can also show that integrated analysis enables us to select more biologically reasonable genes. To do this, we first find which ℓ or ℓ_1 is associated with the smallest P_ℓ or P_{ℓ_1} with each of the five clinical labels. Then P values are attributed to is with $u^{\ell i}$ or $u^{(i)}_{\ell_1 i}$ as

$$P_i^{(m)} = P_{\chi^2}\left[> \left(\frac{u_{\ell i}}{\sigma_\ell}\right)^2 \right] \tag{8.47}$$

$$\tilde{P}_i^{(m)} = P_{\chi^2}\left[> \left(\frac{u^{(i)}_{\ell_1 i}}{\sigma_{\ell_1}}\right)^2 \right] \tag{8.48}$$

with optimized σ_ℓ or σ_{ℓ_1}. Because of too many cases, we showed neither individual scatter plot of σ_h and σ_ℓ or σ_{ℓ_1} nor individual $h(1 - P_i)$ here. Then is associated with adjusted P values less than 0.01 are selected and are uploaded to Enrichr for the enrichment analysis (Fig. 8.38).

For almost all cases, the integrated analyses result in more number of enriched biological terms. Thus, we can conclude that integrated analysis can improve biological reliability (readers who are interested in more detailed results, e.g., how integrated analysis can increase the number of biological terms in clinical label specific or cell line specific ways; see the original paper [27]).

One might wonder why PPI can improve the coincidence with clinical labels since PPI unlikely has the information of clinical labels. Nevertheless, clinical labels cannot be independent of biological mechanisms behind diseases and PPI can have information about biological mechanisms. For example, proteins sharing PPI are likely up- or downregulated together in diseases. This might be able to improve the coincidence with the clinical labels.

Fig. 8.38 Barplot of the number of enriched biological terms summed over 27 cancers when Stanford PPI (upper) or BIOGRID PPI (lower) was used. (1) "vital_status," (2) "pathologic_stage," (3) "pathologic_m," (4) "pathologic_t," and (5) "pathologic_n." Red: without integration of PPI, blue: with integration of PPI

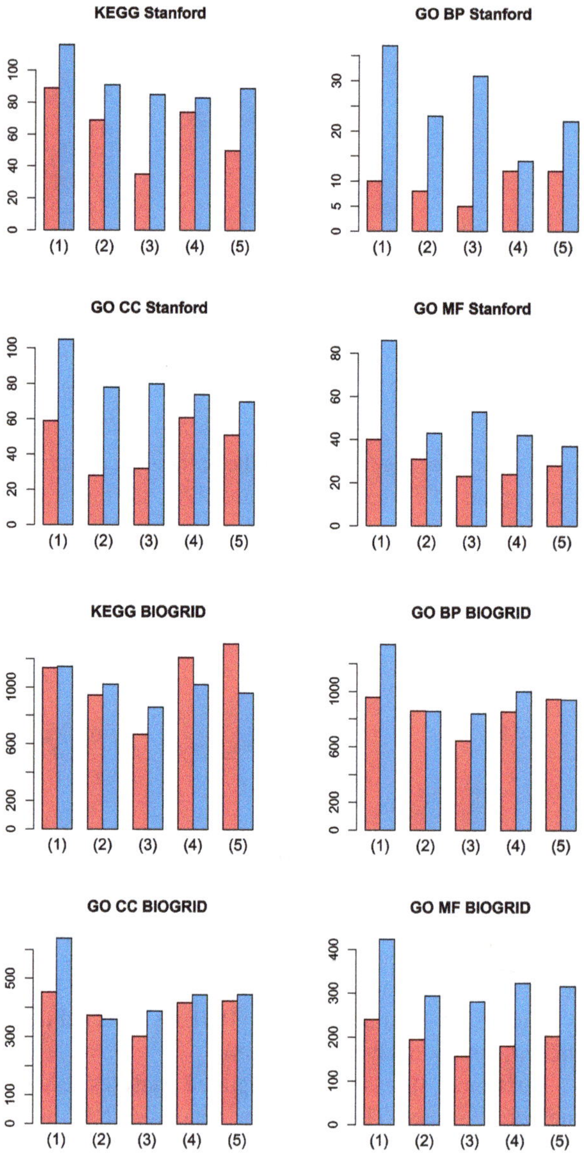

References

1. Cairns, D.M., Dulko, D., Griffiths, J.K., Golan, Y., Cohen, T., Trinquart, L., Price, L.L., Beaulac, K.R., Selker, H.P.: Efficacy of niclosamide vs placebo in SARS-CoV-2 respiratory viral clearance, viral shedding, and duration of symptoms among patients with mild to moderate COVID-19: a phase 2 randomized clinical trial. JAMA Netw. Open **5**(2), e2144942–e2144942 (2022). https://doi.org/10.1001/jamanetworkopen.2021.44942

2. Cavalcante, R.G., Sartor, M.A.: annotatr: genomic regions in context. Bioinformatics **33**(15), 2381–2383 (2017). https://doi.org/10.1093/bioinformatics/btx183
3. Costa-Silva, J., Domingues, D., Lopes, F.M.: RNA-seq differential expression analysis: an extended review and a software tool. PLoS One **12**(12), 1–18 (2017). https://doi.org/10.1371/journal.pone.0190152
4. Edge, S.B., Compton, C.C.: The American Joint Committee on Cancer: the 7th Edition of the AJCC cancer staging manual and the future of TNM. Ann. Surg. Oncol. **17**(6), 1471–1474 (2010). https://doi.org/10.1245/s10434-010-0985-4
5. Heinz, S., Benner, C., Spann, N., Bertolino, E., Lin, Y.C., Laslo, P., Cheng, J.X., Murre, C., Singh, H., Glass, C.K.: Simple combinations of lineage-determining transcription factors prime cis-regulatory elements required for macrophage and B cell identities. Mol. Cell **38**(4), 576–589 (2010). http://dx.doi.org/10.1016/j.molcel.2010.05.004
6. Huber, P.J.: Projection Pursuit. Ann. Stat. **13**(2), 435–475 (1985). https://doi.org/10.1214/aos/1176349519
7. Huber, W., Carey, V.J., Gentleman, R., Anders, S., Carlson, M., Carvalho, B.S., Bravo, H.C., Davis, S., Gatto, L., Girke, T., Gottardo, R., Hahne, F., Hansen, K.D., Irizarry, R.A., Lawrence, M., Love, M.I., MacDonald, J., Obenchain, V., Oleś, A.K., Pagès, H., Reyes, A., Shannon, P., Smyth, G.K., Tenenbaum, D., Waldron, L., Morgan, M.: Orchestrating high-throughput genomic analysis with bioconductor. Nat. Methods **12**(2), 115–121 (2015). http://dx.doi.org/10.1038/nmeth.3252
8. Karthikeyan, S., Sundaramoorthy, A., Kandasamy, S., Bharanidharan, G., Aruna, P., Suganya, R., Mangaiyarkarasi, R., Ganesan, S., Pandian, G.N., Ramamoorthi, A., Chinnathambi, S.: A biophysical approach of tyrphostin AG879 binding information in: bovine serum albumin, human ErbB2, c-RAF1 kinase, SARS-CoV-2 main protease and angiotensin-converting enzyme 2. J. Biomol. Struct. Dyn. **0**(0), 1–14 (2023). https://doi.org/10.1080/07391102.2023.2204368. PMID: 37114656
9. Kosinski, M.: RTCGA: The Cancer Genome Atlas Data Integration (2023). https://doi.org/10.18129/B9.bioc.RTCGA. https://bioconductor.org/packages/RTCGA. R package version 1.30.0
10. Kosinski, M.: RTCGA.rnaseq: Rna-seq datasets from The Cancer Genome Atlas Project (2023). https://doi.org/10.18129/B9.bioc.RTCGA.rnaseq. https://bioconductor.org/packages/RTCGA.rnaseq. R package version 20151101.30.0
11. Kumar, V., Muratani, M., Rayan, N.A., Kraus, P., Lufkin, T., Ng, H.H., Prabhakar, S.: Uniform, optimal signal processing of mapped deep-sequencing data. Nat. Biotechnol. **31**(7), 615–622 (2013). http://dx.doi.org/10.1038/nbt.2596
12. Kundaje, A., Meuleman, W., Ernst, J., Bilenky, M., Yen, A., Heravi-Moussavi, A., Kheradpour, P., Zhang, Z., Wang, J., Ziller, M.J., Amin, V., Whitaker, J.W., Schultz, M.D., Ward, L.D., Sarkar, A., Quon, G., Sandstrom, R.S., Eaton, M.L., Wu, Y.C., Pfenning, A.R., Wang, X., Claussnitzer, M., Liu, Y., Coarfa, C., Harris, R.A., Shoresh, N., Epstein, C.B., Gjoneska, E., Leung, D., Xie, W., Hawkins, R.D., Lister, R., Hong, C., Gascard, P., Mungall, A.J., Moore, R., Chuah, E., Tam, A., Canfield, T.K., Hansen, R.S., Kaul, R., Sabo, P.J., Bansal, M.S., Carles, A., Dixon, J.R., Farh, K.H., Feizi, S., Karlic, R., Kim, A.R., Kulkarni, A., Li, D., Lowdon, R., Elliott, G., Mercer, T.R., Neph, S.J., Onuchic, V., Polak, P., Rajagopal, N., Ray, P., Sallari, R.C., Siebenthall, K.T., Sinnott-Armstrong, N.A., Stevens, M., Thurman, R.E., Wu, J., Zhang, B., Zhou, X., Beaudet, A.E., Boyer, L.A., De Jager, P.L., Farnham, P.J., Fisher, S.J., Haussler, D., Jones, S.J.M., Li, W., Marra, M.A., McManus, M.T., Sunyaev, S., Thomson, J.A., Tlsty, T.D., Tsai, L.H., Wang, W., Waterland, R.A., Zhang, M.Q., Chadwick, L.H., Bernstein, B.E., Costello, J.F., Ecker, J.R., Hirst, M., Meissner, A., Milosavljevic, A., Ren, B., Stamatoyannopoulos, J.A., Wang, T., Kellis, M.: Integrative analysis of 111 reference human epigenomes. Nature **518**(7539), 317–330 (2015). http://dx.doi.org/10.1038/nature14248
13. Law, C.W., Chen, Y., Shi, W., Smyth, G.K.: voom: precision weights unlock linear model analysis tools for RNA-seq read counts. Genome Biol. **15**(2), R29 (2014). https://doi.org/10.1186/gb-2014-15-2-r29

14. Lokhande, K.B., Doiphode, S., Vyas, R., Swamy, K.V.: Molecular docking and simulation studies on SARS-CoV-2 Mpro reveals mitoxantrone, leucovorin, birinapant, and dynasore as potent drugs against COVID-19. J. Biomol. Struct. Dyn. **39**(18), 7294–7305 (2021). https://doi.org/10.1080/07391102.2020.1805019. PMID: 32815481
15. Love, M.I., Huber, W., Anders, S.: Moderated estimation of fold change and dispersion for RNA-seq data with DESeq2. Genome Biol. **15**(12), 550 (2014). https://doi.org/8:10.1186/s13059-014-0550-8
16. McInnes, L., Healy, J., Saul, N., Großberger, L.: UMAP: Uniform manifold approximation and projection. J. Open Source Softw. **3**(29), 861 (2018). https://doi.org/10.21105/joss.00861
17. Orienti, I., Gentilomi, G.A., Farruggia, G.: Pulmonary delivery of fenretinide: a possible adjuvant treatment in COVID-19. Int. J. Mol. Sci. **21**(11), 3812 (2020). https://doi.org/10.3390/ijms21113812. https://www.mdpi.com/1422-0067/21/11/3812
18. Oughtred, R., Rust, J., Chang, C., Breitkreutz, B.J., Stark, C., Willems, A., Boucher, L., Leung, G., Kolas, N., Zhang, F., Dolma, S., Coulombe-Huntington, J., Chatr-aryamontri, A., Dolinski, K., Tyers, M.: The BioGRID database: a comprehensive biomedical resource of curated protein, genetic, and chemical interactions. Protein Sci. **30**(1), 187–200 (2021). https://doi.org/10.1002/pro.3978. https://onlinelibrary.wiley.com/doi/abs/10.1002/pro.3978
19. R Core Team: R: A Language and Environment for Statistical Computing. R Foundation for Statistical Computing, Vienna (2023). https://www.R-project.org/
20. Robinson, M.D., McCarthy, D.J., Smyth, G.K.: edgeR: a Bioconductor package for differential expression analysis of digital gene expression data. Bioinformatics **26**(1), 139–140 (2009). https://doi.org/10.1093/bioinformatics/btp616.
21. SEQC/MAQC-III_Consortium: A comprehensive assessment of RNA-seq accuracy, reproducibility and information content by the sequencing quality control consortium. Nat. Biotechnol. **32**(9), 903–914 (2014). http://dx.doi.org/10.1038/nbt.2957
22. Shi, L., Shi, L., Reid, L.H., Jones, W.D., Shippy, R., Warrington, J.A., Baker, S.C., Collins, P.J., de Longueville, F., Kawasaki, E.S., Lee, K.Y., Luo, Y., Sun, Y.A., Willey, J.C., Setterquist, R.A., Fischer, G.M., Tong, W., Dragan, Y.P., Dix, D.J., Frueh, F.W., Goodsaid, F.M., Herman, D., Jensen, R.V., Johnson, C.D., Lobenhofer, E.K., Puri, R.K., Scherf, U., Thierry-Mieg, J., Wang, C., Wilson, M., Wolber, P.K., Zhang, L., Amur, S., Bao, W., Barbacioru, C.C., Lucas, A.B., Bertholet, V., Boysen, C., Bromley, B., Brown, D., Brunner, A., Canales, R., Cao, X.M., Cebula, T.A., Chen, J.J., Cheng, J., Chu, T.M., Chudin, E., Corson, J., Corton, J.C., Croner, L.J., Davies, C., Davison, T.S., Delenstarr, G., Deng, X., Dorris, D., Eklund, A.C., Fan, X.h., Fang, H., Fulmer-Smentek, S., Fuscoe, J.C., Gallagher, K., Ge, W., Guo, L., Guo, X., Hager, J., Haje, P.K., Han, J., Han, T., Harbottle, H.C., Harris, S.C., Hatchwell, E., Hauser, C.A., Hester, S., Hong, H., Hurban, P., Jackson, S.A., Ji, H., Knight, C.R., Kuo, W.P., LeClerc, J.E., Levy, S., Li, Q.Z., Liu, C., Liu, Y., Lombardi, M.J., Ma, Y., Magnuson, S.R., Maqsodi, B., McDaniel, T., Mei, N., Myklebost, O., Ning, B., Novoradovskaya, N., Orr, M.S., Osborn, T.W., Papallo, A., Patterson, T.A., Perkins, R.G., Peters, E.H., Peterson, R., Philips, K.L., Pine, P.S., Pusztai, L., Qian, F., Ren, H., Rosen, M., Rosenzweig, B.A., Samaha, R.R., Schena, M., Schroth, G.P., Shchegrova, S., Smith, D.D., Staedtler, F., Su, Z., Sun, H., Szallasi, Z., Tezak, Z., Thierry-Mieg, D., Thompson, K.L., Tikhonova, I., Turpaz, Y., Vallanat, B., Van, C., Walker, S.J., Wang, S.J., Wang, Y., Wolfinger, R., Wong, A., Wu, J., Xiao, C., Xie, Q., Xu, J., Yang, W., Zhang, L., Zhong, S., Zong, Y., Slikker, W.: The MicroArray Quality Control (MAQC) project shows inter- and intraplatform reproducibility of gene expression measurements. Nat. Biotechnol. **24**(9), 1151–1161 (2006). http://dx.doi.org/10.1038/nbt1239
23. Song, Q., Smith, A.D.: Identifying dispersed epigenomic domains from ChIP-Seq data. Bioinformatics **27**(6), 870–871 (2011). https://doi.org/10.1093/bioinformatics/btr030
24. Sun, G., Chung, D., Liang, K., Keleş, S.: Statistical analysis of ChIP-seq data with MOSAiCS. In: Shomron, N. (ed.) Deep Sequencing Data Analysis, pp. 193–212. Humana Press, Totowa (2013). https://doi.org/10.1007/978-1-62703-514-9_12
25. Taguchi, Y.h., Turki, T.: Adapted tensor decomposition and PCA based unsupervised feature extraction select more biologically reasonable differentially expressed genes than conventional methods. Sci. Rep. **12**(1), 17438 (2022). https://doi.org/10.1038/s41598-022-21474-z

26. Taguchi, Y.h., Turki, T.: Projection in genomic analysis: a theoretical basis to rationalize tensor decomposition and principal component analysis as feature selection tools. PLoS One **17**(9), 1–20 (2022). https://doi.org/10.1371/journal.pone.0275472
27. Taguchi, Y.H., Turki, T.: Integrated analysis of gene expression and protein-protein interaction with tensor decomposition. Mathematics **11**(17), 3655 (2023). https://doi.org/10.3390/math11173655. https://www.mdpi.com/2227-7390/11/17/3655
28. Taguchi, Y.H., Turki, T.: Principal component analysis- and tensor decomposition-based unsupervised feature extraction to select more suitable differentially methylated cytosines: optimization of standard deviation versus state-of-the-art methods. Genomics **115**(2), 110577 (2023). https://doi.org/10.1016/j.ygeno.2023.110577. https://www.sciencedirect.com/science/article/pii/S0888754323000216
29. Taguchi, Y.H., Turki, T.: Tensor decomposition discriminates tissues using scATAC-seq. Biochim. Biophys. Acta (BBA) Gen. Subj. **1867**(6), 130360 (2023). https://doi.org/10.1016/j.bbagen.2023.130360. https://www.sciencedirect.com/science/article/pii/S0304416523000582
30. Tarazona, S., García, F., Ferrer, A., Dopazo, J., Conesa, A.: NOIseq: a RNA-seq differential expression method robust for sequencing depth biases. EMBnet.journal **17**(B), 18–19 (2012). https://doi.org/10.14806/ej.17.B.265. http://journal.embnet.org/index.php/embnetjournal/article/view/265
31. Turki, T., Roy, S.S., Taguchi, Y.H.: Optimized tensor decomposition and principal component analysis outperforming state-of-the-art methods when analyzing histone modification chromatin immunoprecipitation profiles. Algorithms **16**(9), 401 (2023). https://doi.org/10.3390/a16090401. https://www.mdpi.com/1999-4893/16/9/401
32. Warden, C.D., Lee, H., Tompkins, J.D., Li, X., Wang, C., Riggs, A.D., Yu, H., Jove, R., Yuan, Y.C.: COHCAP: an integrative genomic pipeline for single-nucleotide resolution DNA methylation analysis. Nucleic Acids Res. **41**(11), e117–e117 (2013). https://doi.org/10.1093/nar/gkt242
33. Weiss, A., Touret, F., Baronti, C., Gilles, M., Hoen, B., Nougairède, A., de Lamballerie, X., Sommer, M.O.A.: Niclosamide shows strong antiviral activity in a human airway model of SARS-CoV-2 infection and a conserved potency against the Alpha (B.1.1.7), Beta (B.1.351) and Delta variant (B.1.617.2). PLoS One **16**(12), 1–8 (2021). https://doi.org/10.1371/journal.pone.0260958
34. Wilk, M.B., Gnanadesikan, R.: Probability plotting methods for the analysis for the analysis of data. Biometrika **55**(1), 1–17 (1968). https://doi.org/10.1093/biomet/55.1.1
35. Zhao, N., Boyle, A.P.: F-Seq2: improving the feature density based peak caller with dynamic statistics. NAR Genomics Bioinf. **3**(1), lqab012 (2021). https://doi.org/10.1093/nargab/lqab012

Appendix A
Various Implementations of TD

A.1 Introduction

Because TD is not a major technology, it might not be easy to find implementation of TDs. Thus, we list a few of implementations in various platforms. Nevertheless, since the publication of the first edition of this book, it has become much easier to find packages to perform tensor decomposition.

A.2 R

R is a major language used for data science. It has various implementation of TD.

A.2.1 rTensor

It is a part of CRAN. `rTensor`[1] can be installed via standard install command, `install.packages`. It includes the followings.

- `hosvd` that executes Tucker decomposition using HOSVD algorithm.
- `cp` that executes CP decomposition.
- `tucker` that executes Tucker decomposition using HOOI algorithm.

---

[1] https://cran.r-project.org/web/packages/rTensor/index.html.

Y-h. Taguchi, *Unsupervised Feature Extraction Applied to Bioinformatics*,
Unsupervised and Semi-Supervised Learning,
https://doi.org/10.1007/978-3-031-60982-4

A.2.2 *ttTensor*

It is a part of CRAN. `ttTensor`[2]

- `TTSVD` that executes tensor train decomposition.

A.2.3 *nnTensor*

nnTensor[3] in CRAN has some functions for performing non-negative matrix factorization, non-negative CANDECOMP/PARAFAC (CP) decomposition, non-negative Tucker decomposition, and generating toy model data.

A.2.4 *scTensor*

scTensor[4] aims detection of cell-cell interaction from single-cell RNA-seq data set by tensor decomposition in Bioconductor.

A.2.5 *DelayedTensor*

DelayedTensor[5] is a R package for sparse and out-of-core arithmetic and decomposition of Tensor in Bioconductor.

A.3 python

python is a script language, which is recently adopted to machine learning.

[2] https://cran.r-project.org/web/packages/ttTensor/index.html.

[3] https://cran.r-project.org/web/packages/nnTensor/index.html.

[4] https://doi.org/doi:10.18129/B9.bioc.scTensor.

[5] https://doi.org/doi:10.18129/B9.bioc.DelayedTensor.

A.3.1 TensorLy

TensorLy[6] is a tensor-oriented package that can perform various tensor decomposition described in this book, including Tucker, CP and tensor train.

A.3.2 HOTTBOX

HOTTBOX: Higher Order Tensors ToolBOX[7]

- `HOSVD` that executes Tucker decomposition using HOSVD algorithm.
- `CPD` that executes CP decomposition.
- `HOOI` that executes Tucker decomposition using HOOI algorithm.
- `TTSVD` that executes tensor train decomposition.

A.3.3 TensorTools

TensorTools[8] is a bare bones Python package for fitting and visualizing canonical polyadic (CP) tensor decompositions of higher-order data arrays. These pages contain documentation for its basic functionality.

A.4 MATLAB

MATLAB is a software that aims matrix manipulations.

A.4.1 Tensor Toolbox

Tensor Toolbox[9]

- `hosvd` that executes Tucker decomposition using HOSVD algorithm.
- `cp_als` that executes CP decomposition.
- `tucker_als` that executes Tucker decomposition using HOOI algorithm.

---

[6] https://tensorly.org/stable/index.htm.

[7] https://hottbox.github.io/stable/index.html.

[8] https://tensortools-docs.readthedocs.io/en/latest/.

[9] http://www.tensortoolbox.org.

A.5 julia

julia is a script language that mainly aims statistical analysis.

A.5.1 TensorDecompositions.jl

TensorDecompositions.jl[10] is a package that aims tensor decompositions.

- `hosvd` that executes Tucker decomposition using HOSVD algorithm.
- `candecomp` that executes CP decomposition.

A.6 TensorFlow

TensorFlow is a library for deep learning.

A.6.1 t3f

t3f[11] is a package that aims tensor train decomposition.

A.6.2 TensorD

TensorD[12] is a tensor decomposition library in TensorFlow.

[10] https://github.com/yunjhongwu/TensorDecompositions.jl.
[11] https://github.com/Bihaqo/t3f.
[12] https://tensord-v02.readthedocs.io/en/latest/index.html.

Appendix B
List of Published Papers Related to the Methods

Here is a comprehensive list of my papers where I applied PCA and TD-based unsupervised FE to various topics in genomic science. Some of them were also cited in preceding individual chapters.

References

1. Amakura, Y., Taguchi, Y.H.: Estimation of metabolic effects upon cadmium exposure during pregnancy using tensor decomposition. Genes **13**(10), 1698 (2022). https://doi.org/10.3390/genes13101698. https://www.mdpi.com/2073-4425/13/10/1698
2. Fujisawa, K., Shimo, M., Taguchi, Y.H., Ikematsu, S., Miyata, R.: PCA-based unsupervised feature extraction for gene expression analysis of COVID-19 patients. Sci. Rep. **11**(1), 17351 (2021). https://doi.org/10.1038/s41598-021-95698-w
3. Fujita, S., Karasawa, Y., Hironaka, K.i., Taguchi, Y.H., Kuroda, S.: Features extracted using tensor decomposition reflect the biological features of the temporal patterns of human blood multimodal metabolome. PLoS One **18**(2), 1–26 (2023). https://doi.org/10.1371/journal.pone.0281594
4. Hashimoto, S., Zhao, H., Hayakawa, M., Nakajima, K., Taguchi, Y.H., Murakami, Y.: Developing a diagnostic method for latent tuberculosis infection using circulating miRNA. Transl. Med. Commun. **5**(1), 25 (2020). https://doi.org/10.1186/s41231-020-00078-7
5. Hayakawa, M., Umeyama, H., Iwadate, M., Taguchi, Y.H., Yano, Y., Honda, T., Itami-Matsumoto, S., Kozuka, R., Enomoto, M., Tamori, A., Kawada, N., Murakami, Y.: Development of a novel anti-hepatitis B virus agent via Sp1. Sci. Rep. **10**(1), 47 (2020). https://doi.org/10.1038/s41598-019-56842-9
6. Ishida, S., Umeyama, H., Iwadate, M., Taguchi, Y.H.: Bioinformatic screening of autoimmune disease genes and protein structure prediction with FAMS for drug discovery. Protein Pept. Lett. **21**(8), 828–39 (2014). https://doi.org/10.2174/09298665113209990052

Y-h. Taguchi, *Unsupervised Feature Extraction Applied to Bioinformatics*,
Unsupervised and Semi-Supervised Learning,
https://doi.org/10.1007/978-3-031-60982-4

7. Kawamura, E., Matsubara, T., Daikoku, A., Deguchi, S., Kinoshita, M., Yuasa, H., Urushima, H., Odagiri, N., Motoyama, H., Kotani, K., Kozuka, R., Hagihara, A., Fujii, H., Uchida-Kobayashi, S., Tanaka, S., Takemura, S., Iwaisako, K., Enomoto, M., Taguchi, Y.H., Tamori, A., Kubo, S., Ikeda, K., Kawada, N.: Suppression of intrahepatic cholangiocarcinoma cell growth by SKI via upregulation of the CDK inhibitor p21. FEBS Open Bio **12**(12), 2122–2135 (2022). https://doi.org/10.1002/2211-5463.13489. https://febs.onlinelibrary.wiley.com/doi/abs/10.1002/2211-5463.13489
8. Kinoshita, R., Iwadate, M., Umeyama, H., Taguchi, Y.H.: Genes associated with genotype-specific DNA methylation in squamous cell carcinoma as candidate drug targets. BMC Syst. Biol. **8**(Suppl 1), S4 (2014). https://doi.org/10.1186/1752-0509-8-S1-S4
9. Murakami, Y., Kubo, S., Tamori, A., Itami, S., Kawamura, E., Iwaisako, K., Ikeda, K., Kawada, N., Ochiya, T., Taguchi, Y.H.: Comprehensive analysis of transcriptome and metabolome analysis in Intrahepatic Cholangiocarcinoma and Hepatocellular Carcinoma. Sci. Rep. **5**, 16294 (2015). https://doi.org/10.1038/srep16294
10. Murakami, Y., Tanahashi, T., Okada, R., Toyoda, H., Kumada, T., Enomoto, M., Tamori, A., Kawada, N., Taguchi, Y.H., Azuma, T.: Comparison of hepatocellular carcinoma miRNA expression profiling as evaluated by next generation sequencing and microarray. PloS One **9**(9), e106314 (2014). https://doi.org/10.1371/journal.pone.0106314
11. Murakami, Y., Toyoda, H., Tanahashi, T., Tanaka, J., Kumada, T., Yoshioka, Y., Kosaka, N., Ochiya, T., Taguchi, Y.H.: Comprehensive miRNA expression analysis in peripheral blood can diagnose liver disease. PloS One **7**(10), e48366 (2012). https://doi.org/10.1371/journal.pone.004836
12. Ng, K.L., Taguchi, Y.H.: Identification of miRNA signatures for kidney renal clear cell carcinoma using the tensor-decomposition method. Sci. Rep. **10**(1), 15149 (2020). https://doi.org/10.1038/s41598-020-71997-6
13. Roy, S.S., Taguchi, Y.H.: Identification of genes associated with altered gene expression and m6A profiles during hypoxia using tensor decomposition based unsupervised feature extraction. Sci. Rep. **11**(1), 8909 (2021). https://doi.org/10.1038/s41598-021-87779-7
14. Taguchi, Y.H.: Integrative analysis of gene expression and promoter methylation during reprogramming of a non-small-cell lung cancer cell line using principal component analysis-based unsupervised feature extraction. In: Intelligent Computing in Bioinformatics, pp. 445–455. Springer International Publishing, Berlin (2014). https://doi.org//10.1007/978-3-319-09330-7_52
15. Taguchi, Y.H.: Identification of aberrant gene expression associated with aberrant promoter methylation in primordial germ cells between E13 and E16 rat F3 generation vinclozolin lineage. BMC Bioinf. **16**(Suppl 18), S16 (2015). https://doi.org/10.1186/1471-2105-16-S18-S16
16. Taguchi, Y.H.: Identification of More Feasible MicroRNA-mRNA Interactions within multiple cancers using principal component analysis based unsupervised feature extraction. Int. J. Mol. Sci. **17**(5), 696 (2016). https://doi.org/10.3390/ijms17050696
17. Taguchi, Y.H.: microRNA-mRNA interaction identification in Wilms tumor using principalcomponent analysis based unsupervised feature extraction. In: The 16th annual IEEE International Conference on Bioinformatics and Bioengineering (2016). https://doi.org/10.1109/BIBE.2016.14
18. Taguchi, Y.H.: Principal component analysis based unsupervised feature extraction applied to budding yeast temporally periodic gene expression. BioData Min. **9**, 22 (2016). https://doi.org/10.1186/s13040-016-0101-9
19. Taguchi, Y.H.: Principal component analysis based unsupervised feature extraction applied to publicly available gene expression profiles provides new insights into the mechanisms of action of histone deacetylase inhibitors. Neuroepigenetics **8**, 1–18 (2016). http://dx.doi.org/10.1016/j.nepig.2016.10.001
20. Taguchi, Y.H.: Identification of candidate drugs for heart failure using tensor decomposition-based unsupervised feature extraction applied to integrated analysis of gene expression between heart failure and DrugMatrix data sets. In: Intelligent Computing Theories and

Application, pp. 517–528. Springer International Publishing, Berlin (2017). https://doi.org/10.1007/978-3-319-63312-1_45
21. Taguchi, Y.H.: Identification of candidate drugs using tensor-decomposition-based unsupervised feature extraction in integrated analysis of gene expression between diseases and drugmatrix datasets. Sci. Rep. **7**(1), 13733 (2017). https://doi.org/10.1038/s41598-017-13003-0
22. Taguchi, Y.H.: One-class differential expression analysis using tensor decomposition-based unsupervised feature extraction applied to integrated analysis of multiple omics data from 26 lung adenocarcinoma cell lines. In: 2017 IEEE 17th International Conference on Bioinformatics and Bioengineering (BIBE), pp. 131–138 (2017). https://doi.org/10.1109/BIBE.2017.00-66
23. Taguchi, Y.H.: Principal components analysis based unsupervised feature extraction applied to gene expression analysis of blood from dengue haemorrhagic fever patients. Sci. Rep. **7**, 44016 (2017). https://doi.org/10.1038/srep44016
24. Taguchi, Y.H.: Tensor decomposition-based unsupervised feature extraction applied to matrix products for multi-view data processing. PLoS ONE **12**(8), e0183933 (2017). https://doi.org/10.1371/journal.pone.0183933
25. Taguchi, Y.H.: Tensor decomposition-based unsupervised feature extraction identifies candidate genes that induce post-traumatic stress disorder-mediated heart diseases. BMC Med. Genomics **10**(Suppl 4), 67 (2017). https://doi.org/10.1186/s12920-017-0302-1
26. Taguchi, Y.H.: Principal component analysis-based unsupervised feature extraction applied to single-cell gene expression analysis. In: Huang, D.S., Jo, K.H., Zhang, X.L. (eds.) Intelligent Computing Theories and Application, pp. 816–826. Springer International Publishing, Cham (2018). https://doi.org/10.1007/978-3-319-95933-7_90
27. Taguchi, Y.H.: Tensor decomposition-based unsupervised feature extraction can identify the universal nature of sequence-nonspecific off-target regulation of mRNA mediated by microRNA transfection. Cells **7**(6), 54 (2018). https://doi.org/10.3390/cells7060054. http://www.mdpi.com/2073-4409/7/6/54
28. Taguchi, Y.H.: Tensor decomposition/principal component analysis based unsupervised feature extraction applied to brain gene expression and methylation profiles of social insects with multiple castes. BMC Bioinf. **19**(Suppl 4), 99 (2018). https://doi.org/10.1186/s12859-018-2068-7
29. Taguchi, Y.H.: Drug candidate identification based on gene expression of treated cells using tensor decomposition-based unsupervised feature extraction for large-scale data. BMC Bioinf. **19**(13), 388 (2019). https://doi.org/10.1186/s12859-018-2395-8
30. Taguchi, Y.H.: Multiomics data analysis using tensor decomposition based unsupervised feature extraction. In: Huang, D.S., Bevilacqua, V., Premaratne, P. (eds.) Intelligent Computing Theories and Application, pp. 565–574. Springer International Publishing, Cham (2019)
31. Taguchi, Y.H.: In silico drug discovery using tensor decomposition based unsupervised feature extraction. In: Roy, S.S., Taguchi, Y.H. (eds.) Handbook of Machine Learning Applications for Genomics, pp. 101–120. Springer Nature Singapore, Singapore (2022). https://doi.org/10.1007/978-981-16-9158-4_7
32. Taguchi, Y.H.: Multiomics data analysis of cancers using tensor decomposition and principal component analysis based unsupervised feature extraction. In: Roy, S.S., Taguchi, Y.H. (eds.) Handbook of Machine Learning Applications for Genomics, pp. 1–17. Springer Nature Singapore, Singapore (2022). https://doi.org/10.1007/978-981-16-9158-4_1
33. Taguchi, Y.H.: Single cell RNA-seq analysis using tensor decomposition and principal component analysis based unsupervised feature extraction. In: Roy, S.S., Taguchi, Y.H. (eds.) Handbook of Machine Learning Applications for Genomics, pp. 155–176. Springer Nature Singapore, Singapore (2022). https://doi.org/10.1007/978-981-16-9158-4_11
34. Taguchi, Y.H., Dharshini, S.A.P., Gromiha, M.M.: Identification of transcription factors, biological pathways, and diseases as mediated by N6-methyladenosine using tensor decomposition-based unsupervised feature extraction. Appl. Sci. **11**(1) (2021). https://doi.org/10.3390/app11010213. https://www.mdpi.com/2076-3417/11/1/213

35. Taguchi, Y.H., Iwadate, M., Umeyama, H.: Heuristic principal component analysis-based unsupervised feature extraction and its application to gene expression analysis of amyotrophic lateral sclerosis data sets. In: 2015 IEEE Conference on Computational Intelligence in Bioinformatics and Computational Biology (2015). https://doi.org/10.1109/CIBCB.2015.7300274
36. Taguchi, Y.H., Iwadate, M., Umeyama, H.: Principal component analysis-based unsupervised feature extraction applied to in silico drug discovery for posttraumatic stress disorder-mediated heart disease. BMC Bioinf. **16**, 139 (2015). https://doi.org/10.1186/s12859-015-0574-4
37. Taguchi, Y.H., Iwadate, M., Umeyama, H.: SFRP1 is a possible candidate for epigenetic therapy in non-small cell lung cancer. BMC Med. Genomics **9**(S1) (2016). https://doi.org/10.1186/s12920-016-0196-3
38. Taguchi, Y.H., Iwadate, M., Umeyama, H., Murakami, Y.: Principal component analysis based unsupervised feature extraction applied to bioinformatics analysis. In: Computational Methods with Applications in Bioinformatics Analysis, pp. 153–182. World Scientific, Singapore (2017). https://doi.org/10.1142/9789813207981_0008
39. Taguchi, Y.H., Iwadate, M., Umeyama, H., Murakami, Y., Okamoto, A.: Heuristic principal component analysis-based unsupervised feature extraction and its application to bioinformatics. In: Big Data Analytics in Bioinformatics and Healthcare, pp. 138–162. IGI Global, New York (2014). https://doi.org/10.4018/978-1-4666-6611-5.ch007
40. Taguchi, Y.H., Komaki, S., Sutoh, Y., Ohmomo, H., Otsuka-Yamasaki, Y., Shimizu, A.: Integrated analysis of human dna methylation, gene expression, and genomic variation in imethyl database using kernel tensor decomposition-based unsupervised feature extraction. PLoS One **18**(8), 1–24 (2023). https://doi.org/10.1371/journal.pone.0289029
41. Taguchi, Y.H., Murakami, Y.: Principal component analysis based feature extraction approach to identify circulating microRNA biomarkers. PloS One **8**(6), e66714 (2013). https://doi.org/10.1371/journal.pone.0066714
42. Taguchi, Y.H., Murakami, Y.: Universal disease biomarker: can a fixed set of blood microRNAs diagnose multiple diseases? BMC Res. Notes **7**(1), 581 (2014). https://doi.org/10.1186/1756-0500-7-581
43. Taguchi, Y.H., Ng, K.: Tensor decomposition–based unsupervised feature extraction for integrated analysis of TCGA data on microrna expression and promoter methylation of genes in ovarian cancer. In: 2018 IEEE 18th International Conference on Bioinformatics and Bioengineering (BIBE), pp. 195–200 (2018). https://doi.org/10.1109/BIBE.2018.00045
44. Taguchi, Y.H., Turki, T.: Neurological disorder drug discovery from gene expression with tensor decomposition. Curr. Pharm. Des. **25**(43), 4589–4599 (2019). https://doi.org/10.2174/1381612825666191210160906. https://www.ingentaconnect.com/content/ben/cpd/2019/00000025/00000043/art00006
45. Taguchi, Y.H., Turki, T.: Tensor decomposition-based unsupervised feature extraction applied to single-cell gene expression analysis. Front. Genet. **10**, 481094 (2019). https://doi.org/10.3389/fgene.2019.00864. https://www.frontiersin.org/articles/10.3389/fgene.2019.00864
46. Taguchi, Y.H., Turki, T.: A new advanced in silico drug discovery method for novel coronavirus (SARS-CoV-2) with tensor decomposition-based unsupervised feature extraction. PLOS One **15**(9), 1–16 (2020). https://doi.org/10.1371/journal.pone.0238907
47. Taguchi, Y.H., Turki, T.: Tensor-decomposition-based unsupervised feature extraction applied to prostate cancer multiomics data. Genes **11**(12), 1493 (2020). https://doi.org/10.3390/genes11121493. https://www.mdpi.com/2073-4425/11/12/1493
48. Taguchi, Y.H., Turki, T.: Universal nature of drug treatment responses in drug-tissue-wide model-animal experiments using tensor decomposition-based unsupervised feature extraction. Front. Genet. **11**, 544287 (2020). https://doi.org/10.3389/fgene.2020.00695. https://www.frontiersin.org/articles/10.3389/fgene.2020.00695
49. Taguchi, Y.H., Turki, T.: Application of tensor decomposition to gene expression of infection of mouse hepatitis virus can identify critical human genes and efffective drugs for SARS-CoV-2 infection. IEEE J. Sel. Top. Sign. Proces. **15**(3), 746–758 (2021). http://dx.doi.org/10.1109/JSTSP.2021.3061251

50. Taguchi, Y.H., Turki, T.: Effects of collagen-glycosaminoglycan mesh on gene expression as determined by using principal component analysis-based unsupervised feature extraction. Polymers **13**(23) (2021). https://doi.org/10.3390/polym13234117. https://www.mdpi.com/2073-4360/13/23/4117
51. Taguchi, Y.H., Turki, T.: Mathematical formulation and application of kernel tensor decomposition based unsupervised feature extraction. Knowl.-Based Syst. **217**, 106834 (2021). https://doi.org/10.1016/j.knosys.2021.106834. https://www.sciencedirect.com/science/article/pii/S0950705121000976
52. Taguchi, Y.H., Turki, T.: Novel method for the prediction of drug-drug interaction based on gene expression profiles. Eur. J. Pharm. Sci. **160**, 105742 (2021). https://doi.org/10.1016/j.ejps.2021.105742. https://www.sciencedirect.com/science/article/pii/S0928098721000440
53. Taguchi, Y.H., Turki, T.: Tensor-decomposition-based unsupervised feature extraction in single-cell multiomics data analysis. Genes **12**(9), 1442 (2021). https://doi.org/10.3390/genes12091442. https://www.mdpi.com/2073-4425/12/9/1442
54. Taguchi, Y.H., Turki, T.: Unsupervised tensor decomposition-based method to extract candidate transcription factors as histone modification bookmarks in post-mitotic transcriptional reactivation. PLoS One **16**(5), 1–20 (2021). https://doi.org/10.1371/journal.pone.0251032
55. Taguchi, Y.H., Turki, T.: A tensor decomposition-based integrated analysis applicable to multiple gene expression profiles without sample matching. Sci. Rep. **12**(1), 21242 (2022). https://doi.org/10.1038/s41598-022-25524-4
56. Taguchi, Y.H., Turki, T.: Adapted tensor decomposition and PCA based unsupervised feature extraction select more biologically reasonable differentially expressed genes than conventional methods. Sci. Rep. **12**(1), 17438 (2022). https://doi.org/10.1038/s41598-022-21474-z
57. Taguchi, Y.H., Turki, T.: Integrated analysis of tissue-specific gene expression in diabetes by tensor decomposition can identify possible associated diseases. Genes **13**(6) (2022). https://doi.org/10.3390/genes13061097. https://www.mdpi.com/2073-4425/13/6/1097
58. Taguchi, Y.H., Turki, T.: Novel feature selection method via kernel tensor decomposition for improved multi-omics data analysis. BMC Med. Genet. **15**(1), 37 (2022). https://doi.org/10.1186/s12920-022-01181-4
59. Taguchi, Y.H., Turki, T.: Projection in genomic analysis: A theoretical basis to rationalize tensor decomposition and principal component analysis as feature selection tools. PLoS One **17**(9), 1–20 (2022). https://doi.org/10.1371/journal.pone.0275472
60. Taguchi, Y.H., Turki, T.: Advanced tensor decomposition-based integrated analysis of protein-protein interaction with cancer gene expression can improve coincidence with clinical labels. In: 2023 11th International Conference on Bioinformatics and Computational Biology (ICBCB), pp. 19–25 (2023). https://doi.org/10.1109/icbcb57893.2023.10246633
61. Taguchi, Y.H., Turki, T.: Application note: TDbasedUFE and TDbasedUFEadv: bioconductor packages to perform tensor decomposition based unsupervised feature extraction. Front. Artif. Intell. **6**, 1237542 (2023). https://doi.org/10.3389/frai.2023.1237542. https://www.frontiersin.org/articles/10.3389/frai.2023.1237542
62. Taguchi, Y.H., Turki, T.: Integrated analysis of gene expression and protein-protein interaction with tensor decomposition. Mathematics **11**(17), 3655 (2023). https://doi.org/10.3390/math11173655. https://www.mdpi.com/2227-7390/11/17/3655
63. Taguchi, Y.H., Turki, T.: Principal component analysis- and tensor decomposition-based unsupervised feature extraction to select more suitable differentially methylated cytosines: Optimization of standard deviation versus state-of-the-art methods. Genomics **115**(2), 110577 (2023). https://doi.org/10.1016/j.ygeno.2023.110577. https://www.sciencedirect.com/science/article/pii/S0888754323000216
64. Taguchi, Y.H., Turki, T.: Tensor decomposition discriminates tissues using scatac-seq. Biochim. Biophys. Acta (BBA) Gen. Subj. **1867**(6), 130360 (2023). https://doi.org/10.1016/j.bbagen.2023.130360. https://www.sciencedirect.com/science/article/pii/S0304416523000582
65. Taguchi, Y.H., Wang, H.: Genetic association between Amyotrophic Lateral Sclerosis and cancer. Genes **8**(10), 243 (2017). https://doi.org/10.3390/genes8100243. http://www.mdpi.com/2073-4425/8/10/243

66. Taguchi, Y.H., Wang, H.: Exploring microRNA biomarker for Amyotrophic Lateral Sclerosis. Int. J. Mol. Sci. **19**(5), 1318 (2018). https://doi.org/10.3390/ijms19051318. http://www.mdpi.com/1422-0067/19/5/1318
67. Taguchi, Y.H., Wang, H.: Exploring microRNA biomarkers for Parkinson's disease from mRNA expression profiles. Cells **7**(12), 245 (2018). https://doi.org/10.3390/cells7120245. http://www.mdpi.com/2073-4409/7/12/245
68. Taguchi, Y.H., Wang, H.: Application of PCA based unsupervised FE to neurodegenerative diseases. In: Lee, K.C., Roy, S.S., Samui, P., Kumar, V. (eds.) Data Analytics in Biomedical Engineering and Healthcare, pp. 131–144. Academic Press, New York (2021). https://doi.org/10.1016/B978-0-12-819314-3.00008-2. https://www.sciencedirect.com/science/article/pii/B9780128193143000082
69. Turki, T., Roy, S.S., Taguchi, Y.H.: Optimized tensor decomposition and principal component analysis outperforming state-of-the-art methods when analyzing histone modification chromatin immunoprecipitation profiles. Algorithms **16**(9), 401 (2023). https://doi.org/10.3390/a16090401. https://www.mdpi.com/1999-4893/16/9/401
70. Umeyama, H., Iwadate, M., Taguchi, Y.H.: TINAGL1 and B3GALNT1 are potential therapy target genes to suppress metastasis in non-small cell lung cancer. BMC Genet. **15**(Suppl 9, S2), 1–12 (2014). https://doi.org/10.1186/1471-2164-15-S9-S2
71. Umezu, T., Tanaka, S., Kubo, S., Enomoto, M., Tamori, A., Ochiya, T., Taguchi, Y.H., Kuroda, M., Murakami, Y.: Characterization of circulating miRNAs in the treatment of primary liver tumors. Cancer Reports **7**(2), e1964 (2024). https://doi.org/10.1002/cnr2.1964. https://onlinelibrary.wiley.com/doi/abs/10.1002/cnr2.1964
72. Umezu, T., Tsuneyama, K., Kanekura, K., Hayakawa, M., Tanahashi, T., Kawano, M., Taguchi, Y.H., Toyoda, H., Tamori, A., Kuroda, M., Murakami, Y.: Comprehensive analysis of liver and blood miRNA in precancerous conditions. Sci. Rep. **10**(1), 21766 (2020). https://doi.org/10.1038/s41598-020-78500-1

Appendix C
Bioconductor Packages: TDbasedUFE and TDbasedUFEadv

To perform some of methods introduced in this book, I uploaded two Bioconductor packages, TDbasedUFE [2] and TDbasedUFEadv [3]. Although how to use them is described in the vignettes included in these packages and it will be frequently updated, I simply introduce them briefly here. One can also read an application note [4] where I demonstrated the usage of TDbasedUFE to the real example.

C.1 Bioconductor

Bioconductor [1] is a major R packages repository that aims to provide those specific to Bioinformatics. It has more than 20 years history and includes more than two thousand packages. It is not easy to submit R packages to Bioconductor because of rigorous code reviews. Bioconductor also requires contributors to write detailed vignettes by which users can use submitted packages relatively easily. Once the packages are accepted, they are tested periodically in various platforms, including Windows, Macintosh, and Linux. Packages whose compile and execution errors are not fixed by the contributor will be removed in the future release. Bioconductor release new packages sets semi-annually in accordance with new R package release. Instead of these relatively strong requirements, the packages will be maintained over a long period by Bioconductor.

C.2 TDbasedUFE

TDbasedUFE is a basic package that allows user to perform only two, but popular analysis: gene expression and multiomics. Since the former is a simple one described in Sect. 5.1, it can be applicable to many kinds of single omics data set.

Y-h. Taguchi, *Unsupervised Feature Extraction Applied to Bioinformatics*, Unsupervised and Semi-Supervised Learning,
https://doi.org/10.1007/978-3-031-60982-4

For the latter, we implemented the method described in Sect. 5.7. These two are implemented with the optimization of standard deviations described in Sect. 8.3.

C.3 TDbasedUFEadv

TDbasedUFEadv implemented more advanced features; those in Table 5.3, but restricted to only two omics profiles, and those in Sect. 5.8. These two are also implemented with the optimization of standard deviations described in Sect. 8.3.

References

1. Huber, W., Carey, V.J., Gentleman, R., Anders, S., Carlson, M., Carvalho, B.S., Bravo, H.C., Davis, S., Gatto, L., Girke, T., Gottardo, R., Hahne, F., Hansen, K.D., Irizarry, R.A., Lawrence, M., Love, M.I., MacDonald, J., Obenchain, V., Ole's, A.K., Pag'es, H., Reyes, A., Shannon, P., Smyth, G.K., Tenenbaum, D., Waldron, L., Morgan, M.: Orchestrating high-throughput genomic analysis with Bioconductor. Nat. Methods **12**(2), 115–121 (2015). http://www.nature.com/nmeth/journal/v12/n2/full/nmeth.3252.html
2. Taguchi, Y.H.: TDbasedUFE: Tensor Decomposition Based Unsupervised Feature Extraction (2023). https://doi.org/10.18129/B9.bioc.TDbasedUFE. https://bioconductor.org/packages/TDbasedUFE. R package version 1.0.0
3. Taguchi, Y.H.: TDbasedUFEadv: Advanced Package of Tensor Decomposition Based Unsupervised Feature Extraction (2023). https://doi.org/10.18129/B9.bioc.TDbasedUFEadv. https://bioconductor.org/packages/TDbasedUFEadv. R package version 1.0.0
4. Taguchi, Y.H., Turki, T.: Application note: TDbasedUFE and TDbasedUFEadv: bioconductor packages to perform tensor decomposition based unsupervised feature extraction. Front. Artif. Intell. **6**, 1237542 (2023). https://doi.org/10.3389/frai.2023.1237542. https://www.frontiersin.org/articles/10.3389/frai.2023.1237542

Glossary

linear discriminant analysis The linear method that infers class labels from the given feature variables, which is also applicable to multiple classes.

categorical regression (analysis) Linear regression analysis that predict independent variables from class labels represented as dummy variables as dependent variables.

BH criterion One of the methods that collect P-values obtained by some statistical tests with considering multiple comparisons.

adjusted or corrected P-values P-values collected with considering multiple comparisons.

epigenetics The factor that can affect the amount of transcripts without modifying genomic (DNA) sequence. Typical examples are DNA methylation, histone modification, and non-coding RNAs.

Multiomics The integration of distinct omics data, e.g., gene expression, promoter methylation, metabolome, proteome, and SNP.

χ^2 distribution The distribution that obeys sum of squared variables drawn from $\mathcal{N}(0, 1)$.

cell division cycle The biological process that duplicates a cell into two. All living organism must perform cell division, because it is the only way for them to increase the numbers.

sinusoidal regression Liner regression analysis assuming that a function obeys sinusoidal shapes.

Y-h. Taguchi, *Unsupervised Feature Extraction Applied to Bioinformatics*, Unsupervised and Semi-Supervised Learning,
https://doi.org/10.1007/978-3-031-60982-4

Solutions

Problems of Chap. 1

1.1

- book and pages
- water and temperature
- movie and running Time
- stone and pieces
- human and height
- human and weight
- book and weight
- bottle and volume
- paper and thickness
- card and width

1.2

colors (red, blue, yellow, $\cdots$), nations (Japan, USA, $\cdots$), cities (Tokyo, Beijing, Paris, $\cdots$), towns (Atherton, Corte Madera, $\cdots$), foods (apple, fish, $\cdots$), names (Ben, Taro, $\cdots$), animals (lion, tiger, $\cdots$), plants (cherry, sunflower, $\cdots$), sports (baseball, football, $\cdots$), books (novel, fiction, $\cdots$)

1.3

$x + y + z, x - y - z, 2x + 3y - 4z, x + y - z, x + 2y + z, x - y + z, 3z + 2y + 4z, 2x + 2y + 2z, x + 2y + z, x - y$

Y-h. Taguchi, *Unsupervised Feature Extraction Applied to Bioinformatics*, Unsupervised and Semi-Supervised Learning,
https://doi.org/10.1007/978-3-031-60982-4

1.4

Persons	weight
Ben	34 kg
Tom	45 kg
Mac	70 kg
Naomi	64 kg

1.5

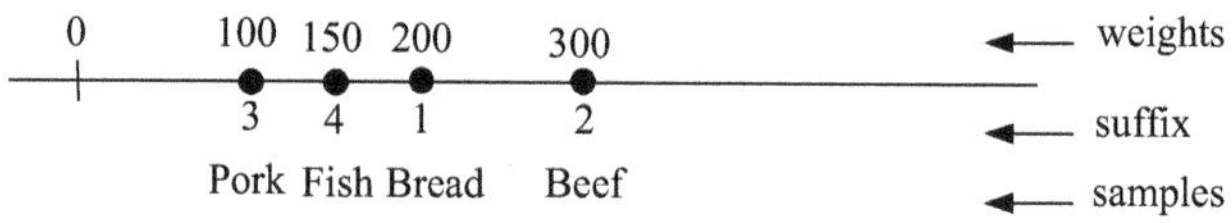

1.6
Euclidean distance between beef and bread is

$$\sqrt{(1000 - 100)^2 + (300 - 200)^2} \simeq 906 \quad \text{(C.1)}$$

1.7

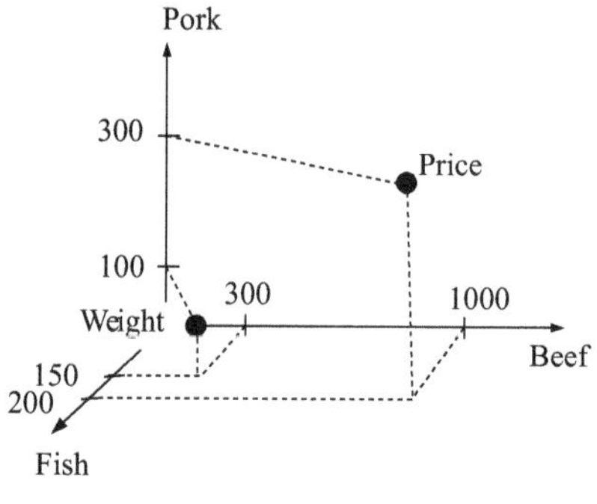

1.8
Here is a new feature, 2 × weight + 3 × price.

	weight	price	2 × weight + 3 × price
Bread	200	100	700
Beef	300	1000	3600
Pork	100	300	1100
Fish	150	200	900

1.9

Suppose, we would like to generate dummy vectors describing comic novels or movies as

Title	Hero	Heroine	villain
Superman	Clark Kent	Lois Lane	Lex Luthor
Batman	Bruce Wayne	Vicki Vale	Joker
Spiderman	Peter Parker	Mary Jane Watson	Green Goblin

1.10

$$X = \begin{pmatrix} 200 & 300 & 100 & 150 \\ 1 & 10 & 3 & 2 \end{pmatrix} \tag{C.2}$$

1.11

$$\boldsymbol{x}_1 = (200, 1) \tag{C.3}$$

$$\boldsymbol{x}_2 = (300, 10) \tag{C.4}$$

$$\boldsymbol{x}_3 = (100, 3) \tag{C.5}$$

$$\boldsymbol{x}_4 = (150, 2) \tag{C.6}$$

or

$$\boldsymbol{x}_1 = (200, 300, 100, 150) \tag{C.7}$$

$$\boldsymbol{x}_2 = (1, 10, 3, 2) \tag{C.8}$$

1.12

$$X = \begin{pmatrix} 100 & 1000 & 300 & 200 \\ 200 & 300 & 100 & 150 \end{pmatrix} \tag{C.9}$$

and

$$A = \begin{pmatrix} 1 & \frac{1}{2} \\ \frac{1}{2} & 1 \end{pmatrix} \tag{C.10}$$

then

$$X' = AX = \begin{pmatrix} 200 & 1150 & 350 & 275 \\ 250 & 800 & 250 & 250 \end{pmatrix} \tag{C.11}$$

1.13

$$x_{1jk} = \begin{pmatrix} 1 & 2 & 3 \\ 4 & 5 & 6 \\ 7 & 8 & 9 \end{pmatrix} \tag{C.12}$$

$$x_{2jk} = \begin{pmatrix} 10 & 11 & 12 \\ 13 & 14 & 15 \\ 16 & 17 & 18 \end{pmatrix} \tag{C.13}$$

$$x_{3jk} = \begin{pmatrix} 19 & 20 & 21 \\ 22 & 23 & 24 \\ 25 & 26 & 27 \end{pmatrix} \tag{C.14}$$

1.14

$$X = \begin{pmatrix} 1 & 2 & 3 & 4 & 5 & 6 & 7 & 8 & 9 \\ 10 & 11 & 12 & 13 & 14 & 15 & 16 & 17 & 18 \\ 19 & 20 & 21 & 22 & 23 & 24 & 25 & 26 & 27 \end{pmatrix} \tag{C.15}$$

1.15
The 3 mode tensor defined in exercise 1–13 is used as $\mathcal{X}$. A is supposed to be

$$\begin{pmatrix} 1 & 1 & 1 \\ 1 & 1 & 1 \\ 1 & 1 & 1 \end{pmatrix} \tag{C.16}$$

then

$$(A \times_i \mathcal{X})_{1jk} = (A \times_i \mathcal{X})_{2jk} = (A \times_i \mathcal{X})_{3jk} = \begin{pmatrix} 30 & 39 & 48 \\ 33 & 42 & 51 \\ 36 & 45 & 54 \end{pmatrix} \tag{C.17}$$

1.16

$$\boldsymbol{a} = (1, 2, 3) \tag{C.18}$$

and

$$\boldsymbol{b} = (4, 5, 6) \tag{C.19}$$

then

$$\boldsymbol{a} \times^0 \boldsymbol{b} = \begin{pmatrix} 4 & 5 & 6 \\ 8 & 10 & 12 \\ 12 & 15 & 18 \end{pmatrix} \tag{C.20}$$

Problems of Chap. 2

2.1

$$A = \begin{pmatrix} 0 & 1 & 0 \\ 0 & 0 & 1 \\ 1 & 0 & 0 \end{pmatrix} \tag{C.21}$$

$$B = \begin{pmatrix} 0 & 0 & 1 \\ 1 & 0 & 0 \\ 0 & 1 & 0 \end{pmatrix} \tag{C.22}$$

2.2

$$F = \begin{pmatrix} 350 & 87.5 & 400 & 400 \\ -100 & 37.5 & -250 & 425 \\ 250 & 137.5 & 400 & 450 \\ 300 & 112.5 & 450 & -225 \end{pmatrix} \tag{C.23}$$

2.3

$$U = \begin{pmatrix} -0.5 & -0.5 \\ -0.5 & -0.5 \\ 0.5 & -0.5 \\ -0.5 & 0.5 \end{pmatrix} \tag{C.24}$$

$$\Sigma = \begin{pmatrix} 2 & 0 \\ 0 & 2 \end{pmatrix} \tag{C.25}$$

$$V = \begin{pmatrix} -1 & 0 \\ 0 & -1 \end{pmatrix} \tag{C.26}$$

2.4

$$S = \begin{pmatrix} 0 & 0 & 0 & 0 \\ 0 & 0 & 0 & 0 \\ 0 & 0 & 2 & -2 \\ 0 & 0 & -2 & 2 \end{pmatrix} \tag{C.27}$$

then

$$U = \begin{pmatrix} 0 & 0 & 0 & 1 \\ 0 & 0 & 1 & 0 \\ -\frac{1}{\sqrt{2}} & \frac{1}{\sqrt{2}} & 0 & 0 \\ \frac{1}{\sqrt{2}} & \frac{1}{\sqrt{2}} & 0 & 0 \end{pmatrix} \tag{C.28}$$

and

$$SU = \begin{pmatrix} 0 & 0\,0\,0 \\ 0 & 0\,0\,0 \\ -2\sqrt{2} & 0\,0\,0 \\ 2\sqrt{2} & 0\,0\,0 \end{pmatrix} \tag{C.29}$$

2.5 They differ from each other because of mean extraction.

2.6

$$X^T U = \begin{pmatrix} \sqrt{2} & 0\,0\,0 \\ -\sqrt{2} & 0\,0\,0 \end{pmatrix} \tag{C.30}$$

Thus variance along the first direction is 4. Others are zero.

2.7 Residuals are zero.

2.8

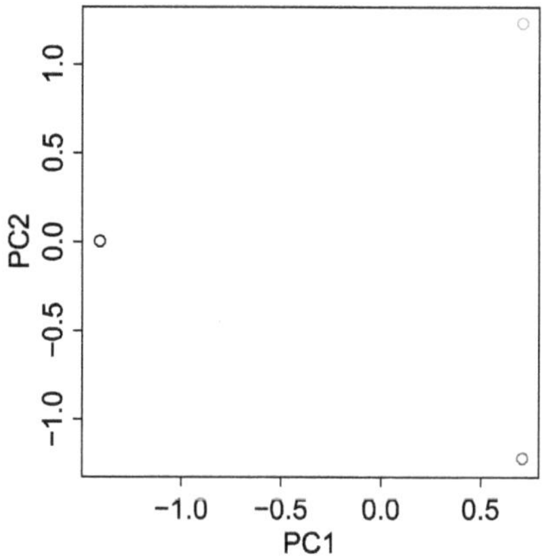

The first vs the second PC scores. Black open circles: $1 \leq i \leq 3$, red open circles: $4 \leq i \leq 6$, green open circles: $7 \leq i \leq 9$.

Problems of Chap. 3

3.1
Suppose $\boldsymbol{u} = (1, 1, 1)$ and $\mathcal{X}$ is a tensor whose component is $x_{ijk} = 1$. Then,

$$\mathcal{X} = \boldsymbol{u} \times^0 \boldsymbol{u} \times^0 \boldsymbol{u} \tag{C.31}$$

CP decomposition: In Eq. (3.1), $L = 1$, $\lambda_1 = 1$, $\boldsymbol{u}_1^{(i)} = \boldsymbol{u}_1^{(j)} = \boldsymbol{u}_1^{(k)} = \boldsymbol{u}$.

Tucker decomposition: In Eq. (3.2), $G(1, 1, 1) = 1$ and other Gs are zero. $\boldsymbol{u}_1^{(i)} = \boldsymbol{u}_1^{(j)} = \boldsymbol{u}_1^{(k)} = \boldsymbol{u}$. Other $\boldsymbol{u}_{\ell_1}^{(i)}, \boldsymbol{u}_{\ell_2}^{(j)}, \boldsymbol{u}_{\ell_3}^{(k)}$ are zero.

Tensor train decomposition: In Eq. (3.3), $R_1 = R_2 = 1$. $G^{(i)}(i, 1) = G^{(j)}(j, 1, 1) = G^{(k)}(k, 1) = 1$.

3.2

When we add the term with $\ell_1 = 1, \ell_2 = \ell_3 = 2$,

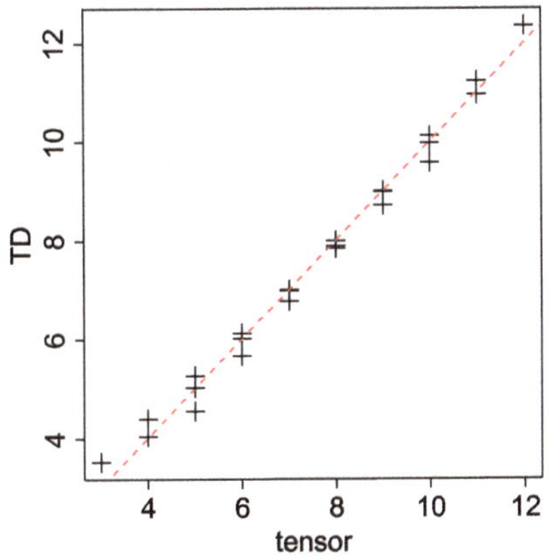

3.3

When $L = 2$,

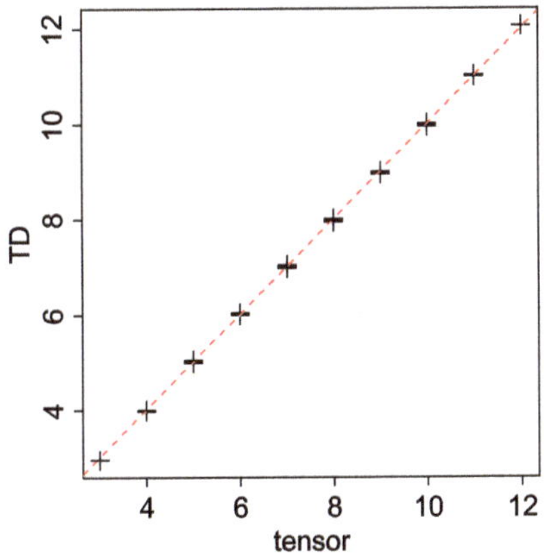

3.4

Because there are no ways to take partial summation of tensor train decomposition, we cannot draw something that corresponds to these figures.

3.5

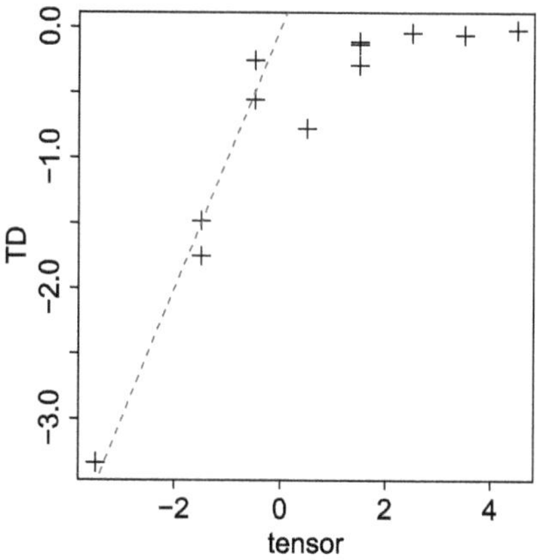

3.6
Assume

$$\mathcal{X} = \boldsymbol{a} \times^0 \boldsymbol{b} \times^0 \boldsymbol{c} \tag{C.32}$$

Then we get

$$a_1 b_1 c_1 = 1 \tag{C.33}$$
$$a_1 b_2 c_1 = 2 \tag{C.34}$$
$$a_2 b_1 c_1 = 3 \tag{C.35}$$
$$a_2 b_2 c_1 = 4 \tag{C.36}$$
$$a_1 b_1 c_2 = 5 \tag{C.37}$$
$$a_1 b_2 c_2 = 6 \tag{C.38}$$
$$a_2 b_1 c_2 = 7 \tag{C.39}$$
$$a_2 b_2 c_2 = 8 \tag{C.40}$$

From these, we get

$$a_1 = \frac{14}{(b_1 + b_2)(c_1 + c_2)} \tag{C.41}$$
$$a_2 = \frac{22}{(b_1 + b_2)(c_1 + c_2)} \tag{C.42}$$
$$b_1 = \frac{16}{(c_1 + c_2)(a_1 + a_2)} \tag{C.43}$$

$$b_2 = \frac{20}{(c_1 + c_2)(a_1 + a_2)} \tag{C.44}$$

$$c_1 = \frac{10}{(a_1 + a_2)(b_1 + b_2)} \tag{C.45}$$

$$c_2 = \frac{26}{(a_1 + a_2)(b_1 + b_2)} \tag{C.46}$$

Starting

$$\boldsymbol{a} = \boldsymbol{b} = \boldsymbol{c} = \begin{pmatrix} 1 \\ 1 \end{pmatrix} \tag{C.47}$$

In each iteration,

$$\boldsymbol{a} = \begin{pmatrix} \frac{7}{2} \\ \frac{11}{2} \end{pmatrix}, \boldsymbol{b} = \begin{pmatrix} 1 \\ 1 \end{pmatrix}, \boldsymbol{c} = \begin{pmatrix} 1 \\ 1 \end{pmatrix} \tag{C.48}$$

$$\boldsymbol{a} = \begin{pmatrix} \frac{7}{2} \\ \frac{11}{2} \end{pmatrix}, \boldsymbol{b} = \begin{pmatrix} \frac{8}{9} \\ \frac{10}{9} \end{pmatrix}, \boldsymbol{c} = \begin{pmatrix} 1 \\ 1 \end{pmatrix} \tag{C.49}$$

$$\boldsymbol{a} = \begin{pmatrix} \frac{7}{2} \\ \frac{11}{2} \end{pmatrix}, \boldsymbol{b} = \begin{pmatrix} \frac{8}{9} \\ \frac{10}{9} \end{pmatrix}, \boldsymbol{c} = \begin{pmatrix} \frac{5}{9} \\ \frac{13}{9} \end{pmatrix} \tag{C.50}$$

This is the converged solution.

3.7

$$X^{i\times(jk)} = \begin{pmatrix} 1\,3\,5\,7 \\ 2\,4\,6\,8 \end{pmatrix} \tag{C.51}$$

then

$$X^{i\times(jk)} \left(X^{i\times(jk)}\right)^T = \begin{pmatrix} 84 & 100 \\ 100 & 120 \end{pmatrix} = 4 \begin{pmatrix} 21 & 25 \\ 25 & 30 \end{pmatrix} \tag{C.52}$$

We would like to find eigen values and eigen vectors of $\begin{pmatrix} 21 & 25 \\ 25 & 30 \end{pmatrix}$. In order that, we need to solve eigen equation,

$$\begin{vmatrix} 21 - \lambda & 25 \\ 25 & 30 - \lambda \end{vmatrix} = 0 \tag{C.53}$$

$$(21 - \lambda)(30 - \lambda) - 25^2 = 0 \tag{C.54}$$

$$\lambda^2 - 51\lambda + 5 = 0 \tag{C.55}$$

$$\lambda = \frac{51 \pm \sqrt{2581}}{2} \tag{C.56}$$

On the other hand, if $\boldsymbol{u}$ is an eigen vector,

$$\begin{pmatrix} 21-\lambda & 25 \\ 25 & 30-\lambda \end{pmatrix} \begin{pmatrix} u_1 \\ u_2 \end{pmatrix} = 0 \tag{C.57}$$

then

$$u_1 = \frac{\lambda - 30}{25} u_2 \tag{C.58}$$

Since

$$\frac{\lambda - 30}{25} = \frac{-9 \pm \sqrt{2581}}{50} \tag{C.59}$$

thus

$$u_1 = \frac{-9 \pm \sqrt{2581}}{50} u_2 \tag{C.60}$$

In order that $|\boldsymbol{u}| = 1$,

$$u_1^2 + u_2^2 = 1 \tag{C.61}$$

$$\left\{ 1 + \left(\frac{-9 \pm \sqrt{2581}}{50} \right)^2 \right\} u_2^2 = 1 \tag{C.62}$$

$$u_1 = \frac{\pm 1}{\sqrt{1 + \left(\frac{-9 \pm \sqrt{2581}}{50} \right)^2}} \tag{C.63}$$

$$u_2 = \frac{\pm \frac{-9 \pm \sqrt{2581}}{50}}{\sqrt{1 + \left(\frac{-9 \pm \sqrt{2581}}{50} \right)^2}} \tag{C.64}$$

Then if we define

$$u_1^+ = \frac{1}{\sqrt{1 + \left(\frac{-9 + \sqrt{2581}}{50} \right)^2}} \tag{C.65}$$

$$u_2^+ = \frac{\frac{-9+\sqrt{2581}}{50}}{\sqrt{1+\left(\frac{-9+\sqrt{2581}}{50}\right)^2}} \tag{C.66}$$

$$u_1^- = \frac{1}{\sqrt{1+\left(\frac{-9-\sqrt{2581}}{50}\right)^2}} \tag{C.67}$$

$$u_2^- = \frac{\frac{-9-\sqrt{2581}}{50}}{\sqrt{1+\left(\frac{-9-\sqrt{2581}}{50}\right)^2}} \tag{C.68}$$

we can have

$$U^{(i)} = \begin{pmatrix} u_1^+ & u_1^- \\ u_2^+ & u_2^- \end{pmatrix} \tag{C.69}$$

using the representation of Eq. (3.55). With applying similar computation to

$$X^{j\times(ik)} = \begin{pmatrix} 1 & 2 & 5 & 6 \\ 3 & 4 & 7 & 8 \end{pmatrix} \tag{C.70}$$

and

$$X^{k\times(ij)} = \begin{pmatrix} 1 & 2 & 3 & 4 \\ 5 & 6 & 7 & 8 \end{pmatrix} \tag{C.71}$$

we can get $U^{(j)}$ and $U^{(k)}$ as well. Then G can be computed by

$$G = \mathcal{X} \times_{\ell_1} \left(U^{(i)}\right)^T \times_{\ell_2} \left(U^{(j)}\right)^T \times_{\ell_3} \left(U^{(k)}\right)^T \tag{C.72}$$

3.8

Equations (3.26) to (3.28) are solutions.

Index

Y-h. Taguchi, *Unsupervised Feature Extraction Applied to Bioinformatics*, Unsupervised and Semi-Supervised Learning,
https://doi.org/10.1007/978-3-031-60982-4